Springer Series in Operations Research

Springer Science+Business Media, LLC

Springer Series in Operations Research

Altiok: Performance Analysis of Manufacturing Systems
Birge and Louveaux: Introduction to Stochastic Programming
Bonnans and Shapiro: Perturbation Analysis of Optimization Problems
Bramel and Simchi-Levi: The Logic of Logistics: Theory, Algorithms, and Applications for Logistics Management
Dantzig and Thapa: Linear Programming 1: Introduction
Drezner (Editor): Facility Location: A Survey of Applications and Methods
Facchinei and Pang: Finite-Dimensional Variational Inequalities and Complementarity Problems, Volume I
Facchinei and Pang: Finite-Dimensional Variational Inequalities and Complementarity Problems, Volume II
Fishman: Discrete-Event Simulation: Modeling, Programming, and Analysis
Fishman: Monte Carlo: Concepts, Algorithms, and Applications
Haas: Stochastic Petri Nets: Modeling, Stability, Simulation
Klamroth: Single-Facility Location Problems with Barriers
Nocedal and Wright: Numerical Optimization
Olson: Decision Aids for Selection Problems
Whitt: Stochastic-Process Limits: An Introduction to Stochastic-Process Limits and Their Application to Queues
Yao (Editor): Stochastic Modeling and Analysis of Manufacturing Systems
Yao and Zheng: Dynamic Control of Quality in Production-Inventory Systems: Coordination and Optimization

Francisco Facchinei   Jong-Shi Pang

Finite-Dimensional Variational Inequalities and Complementarity Problems

Volume I

With 18 Figures

Springer

Francisco Facchinei
Dipartimento di Informatica e Sistemistica
Università di Roma "La Sapienza"
Rome I-00185
Italy
soler@dis.uniroma1.it

Jong-Shi Pang
Department of Mathematical Sciences
The Johns Hopkins University
Baltimore, MD 21218-2682
USA
pang@mts.jhu.edu

Mathematics Subject Classification (2000): 90-01, 90C33, 65K05, 47J20

Library of Congress Cataloging-in-Publication Data
Facchinei, Francisco
Finite-dimensional variational inequalities and complementarity problems / Francisco Facchinei, Jong-Shi Pang.
p. cm.—(Springer series in operations research)
Includes bibliographical references and indexes.

DOI 10.1007/978-0-387-21814-4

1. Variational inequalities (Mathematics) 2. Linear complementarity problem. I. Facchinei, Francisco. II. Title. III. Series.
QA316 .P36 2003
515′.64—dc21 2002042739

Printed on acid-free paper.

9 8 7 6 5 4 3 2 1 SPIN 10892611

Typesetting: Pages created by the authors in LaTeX2e.

www.springer-ny.com
www.springer.com/mycopy

Preface

The finite-dimensional nonlinear complementarity problem (NCP) is a system of finitely many nonlinear inequalities in finitely many nonnegative variables along with a special equation that expresses the complementary relationship between the variables and corresponding inequalities. This complementarity condition is the key feature distinguishing the NCP from a general inequality system, lies at the heart of all constrained optimization problems in finite dimensions, provides a powerful framework for the modeling of equilibria of many kinds, and exhibits a natural link between smooth and nonsmooth mathematics. The finite-dimensional variational inequality (VI), which is a generalization of the NCP, provides a broad unifying setting for the study of optimization and equilibrium problems and serves as the main computational framework for the practical solution of a host of continuum problems in the mathematical sciences.

The systematic study of the finite-dimensional NCP and VI began in the mid-1960s; in a span of four decades, the subject has developed into a very fruitful discipline in the field of mathematical programming. The developments include a rich mathematical theory, a host of effective solution algorithms, a multitude of interesting connections to numerous disciplines, and a wide range of important applications in engineering and economics. As a result of their broad associations, the literature of the VI/CP has benefited from contributions made by mathematicians (pure, applied, and computational), computer scientists, engineers of many kinds (civil, chemical, electrical, mechanical, and systems), and economists of diverse expertise (agricultural, computational, energy, financial, and spatial). There are many surveys and special volumes, [67, 240, 243, 244, 275, 332, 668, 687], to name a few.

Written for novice and expert researchers and advanced graduate students in a wide range of disciplines, this two-volume monograph presents a comprehensive, state-of-the-art treatment of the finite-dimensional variational inequality and complementarity problem, covering the basic theory, iterative algorithms, and important applications. The materials presented herein represent the work of many researchers worldwide. In undertaking this ambitious project, we have attempted to include every major aspect of the VI/CP, beginning with the fundamental question of existence and uniqueness of solutions, presenting the latest algorithms and results, extending into selected neighboring topics, summarizing many classical source problems, and including novel application domains. Despite our efforts, there are omissions of topics, due partly to our biases and partly to the scope of the presentation. Some omitted topics are mentioned in the notes and comments.

A Bird's-Eye View of the Subject

The subject of variational inequalities has its origin in the calculus of variations associated with the minimization of infinite-dimensional functionals. The systematic study of the subject began in the early 1960s with the seminal work of the Italian mathematician Guido Stampacchia and his collaborators, who used the variational inequality as an analytic tool for studying free boundary problems defined by nonlinear partial differential operators arising from unilateral problems in elasticity and plasticity theory and in mechanics. Some of the earliest papers on variational inequalities are [333, 512, 561, 804, 805]. In particular, the first theorem of existence and uniqueness of the solution of VIs was proved in [804]. The books by Baiocchi and Capelo [35] and Kinderlehrer and Stampacchia [410] provide a thorough introduction to the application of variational inequalities in infinite-dimensional function spaces; see also [39]. The lecture notes [362] treat complementarity problems in abstract spaces. The book by Glowinski, Lions, and Trémolière [291] is among the earliest references to give a detailed numerical treatment of such VIs. There is a huge literature on the subject of infinite-dimensional variational inequalities and related problems. Since a VI in an abstract space is in many respects quite distinct from the finite-dimensional VI and since the former problem is not the main concern of this book, in this section we focus our introduction on the latter problem only.

The development of the finite-dimensional variational inequality and nonlinear complementarity problem also began in the early 1960s but followed a different path. Indeed, the NCP was first identified in the 1964 Ph.D. thesis of Richard W. Cottle [135], who studied under the supervision of the eminent George B. Dantzig, "father of linear programming." Thus, unlike its infinite-dimensional counterpart, which was conceived in the area of partial differential systems, the finite-dimensional VI/CP was born in the domain of mathematical programming. This origin has had a heavy influence on the subsequent evolution of the field; a brief account of the history prior to 1990 can be found in the introduction of the survey paper [332]; see also Section 1.2 in [331]. In what follows, we give a more detailed account of the evolutionary process of the field, covering four decades of major events and notable highlights.

In the 1960s, largely as a result of the celebrated almost complementary pivoting algorithm of Lemke and Howson for solving a bimatrix game formulated as a linear complementarity problem (LCP) [491] and the subsequent extension by Lemke to a general LCP [490], much focus was devoted to the study of the latter problem. Cottle, Pang, and Stone presented a comprehensive treatment of the LCP in the 1992 monograph [142]. Among other things, this monograph contains an extensive bibliography of the LCP up to 1990 and also detailed notes, comments, and historical accounts about this fundamental problem. Today, research on the LCP remains active and new applications continue to be uncovered. Since much of the pre-1990 details about the LCP are already documented in the cited monograph, we rely on the latter for most of the background results for the LCP and will touch on the more contemporary developments of this problem where appropriate.

In 1967, Scarf [759] developed the first constructive iterative method for approximating a fixed point of a continuous mapping. Scarf's seminal work led to the development of the entire family of fixed-point methods and of the piecewise homotopy approach to the computation of economic equilibria. The field of *equilibrium programming* was thus born. In essence, the term "equilibrium programming" broadly refers to the modeling, analysis, and computation of equilibria of various kinds via the methodology of mathematical programming. Since the infant days of linear programming, it was clear that complementarity problems have much to do with equilibrium programs. For instance, the primal-dual relation of a linear program provides clear evidence of the interplay between complementarity and equilibrium. Indeed, all the equilibrium problems that were amenable to solution by the fixed-point methods, including the renowned Walrasian problem in general equilibrium theory and variations of this problem [760, 842, 866], were in fact VIs/CPs.

The early research in equilibrium programming was to a large extent a consequence of the landmark discoveries of Lemke and Scarf. In particular, the subject of fixed-point computations via piecewise homotopies dominated much of the research agenda of equilibrium programming in the 1970s. A major theoretical advantage of the family of fixed-point homotopy methods is their global convergence. Attracted by this advantage and the novelty of the methods, many well-known researchers including Eaves, Garcia, Gould, Kojima, Megiddo, Saigal, Todd, and Zangwill all made fundamental contributions to the subject. The flurry of research activities in this area continued for more than a decade, until the occurrence of several significant events that provided clear evidence of the practical inadequacy of this family of methods for solving realistic equilibrium problems. These events, to be mentioned momentarily, marked a turning point whereby the fixed-point/homotopy approach to the computation of equilibria gave way to an alternative set of methods that constitute what one may call a contemporary variational inequality approach to equilibrium programming. For completeness, we mention several prominent publications that contain important works on the subject of fixed-point computation via the homotopy approach and its applications [10, 11, 34, 203, 205, 206, 211, 251, 252, 285, 403, 440, 729, 760, 841, 879]. For a recent paper on this approach, see [883].

In the same period and in contrast to the aforementioned algorithmic research, Karamardian, in a series of papers [398, 399, 400, 401, 402], developed an extensive existence theory for the NCP and its cone generalization. In particular, the basic connection between the CP and the VI, Proposition **1.1.3**, appeared in [400]. The 1970s were a period when many fundamental articles on the VI/CP first appeared. These include the paper by Eaves [202] where the natural map $\mathbf{F}_K^{\mathrm{nat}}$ was used to prove a basic theorem of complementarity, important studies by Moré [623, 624] and Moré and Rheinboldt [625], which studied several distinguished classes of nonlinear functions and their roles in complementarity problems, and the individual and joint work of Kojima and Megiddo [441, 599, 600, 601], which investigated the existence and uniqueness of solutions to the NCP.

Although the initial developments of infinite-dimensional variational inequalities and finite-dimensional complementarity problems had followed different paths, there were attempts to bring the two fields more closely together, with the International School of Mathematics held in Summer 1978 in Erice, Italy, being the most prominent one. The proceedings of this conference were published in [141]. The paper [138] is among the earliest that describes some physical applications of VIs in infinite dimensions solvable by LCP methods.

One could argue that the final years of the 1970s marked the beginning of the contemporary chapter on the finite-dimensional VI/CP. During that time, the U.S. Department of Energy was employing a market equilibrium system known as the Project Independent Evaluation System (PIES) [350,

351] for energy policy studies. This system is a large-scale variational inequality that was solved on a routine basis by a special iterative algorithm known as the PIES algorithm, yielding remarkably good computational experience. For a detailed account of the PIES model, see the monograph by Ahn [5], who showed that the PIES algorithm was a generalization of the classical Jacobi iterative method for solving system of nonlinear equations [652]. For the convergence analysis of the PIES algorithm, see Ahn and Hogan [6]; for a recent update of the PIES model, which has become the National Energy Modeling System (NEMS), see [278].

The original PIES model provided a real-life economic model for which the fixed-point methods mentioned earlier were proved to be ineffective. This experience along with several related events inspired a new wave of research into iterative methods for solving VIs/CPs arising from various applied equilibrium contexts. One of these events is an important algorithmic advance, namely, the introduction of Newton's method for solving generalized equations (see below).

At about the same time as the PIES model appeared, Smith [793] and Dafermos [151] formulated the traffic equilibrium problem as a variational inequality. Parallel to the VI formulation, Aashitiani and Magnanti [1] introduced a complementarity formulation for Wardrop's user equilibrium principle [868] and established existence and uniqueness results of traffic equilibria using fixed-point theorems; see also [20, 253]. Computationally, the PIES algorithm had served as a model approach for the design of iterative methods for solving the traffic equilibrium problem [2, 254, 259]. More broadly, the variational inequality approach has had a significant impact on the contemporary point of view of this problem and the closely related spatial price equilibrium problem.

In two important papers [594, 595], Mathiesen reported computational results on the application of a sequential linear complementarity (SLCP) approach to the solution of economic equilibrium problems. These results firmly established the potential of this approach and generated substantial interest among many computational economists, including Manne and his (then Ph.D.) students, most notably, Preckel, Rutherford, and Stone. The volume edited by Manne [581] contains the papers [697, 814], which give further evidence of the computational efficiency of the SLCP approach for solving economic equilibrium problems; see also [596].

The SLCP method, as it was called in the aforementioned papers, turned out to be Newton's method developed and studied several years earlier by Josephy [389, 390, 391]; see also the later papers by Eaves [209, 210]. While the results obtained by the computational economists clearly established the practical effectiveness of Newton's method through sheer numerical experience, Josephy's work provided a sound theoretical foundation for the fast convergence of the method. In turn, Josephy's results were based on the seminal research of Robinson, who in several landmark papers [728, 730, 732, 734] introduced the *generalized equations* as a unifying mathematical framework for optimization problems, complementarity problems, variational inequalities, and related problems. As we explain below, in addition to providing the foundation for the convergence theory of Newton's method, Robinson's work greatly influenced the modern development of sensitivity analysis of mathematical programs.

While Josephy's contributions marked a breakthrough in algorithmic advances of the field, they left many questions unanswered. From a computational perspective, Rutherford [754] recognized early on the lack of robustness in Newton's method applied to some of the most challenging economic equilibrium problems. Although ad hoc remedies and specialized treatments had lessened the numerical difficulty in solving these problems, the heuristic aids employed were far from satisfactory in resolving the practical deficiency of the method, which was caused by the lack of a suitable stabilizing strategy for global convergence. Motivated by the need for a computationally robust Newton method with guaranteed global convergence, Pang [663] developed the B-differentiable Newton method with a line search and established that the method is globally convergent and locally superlinearly convergent. While this is arguably the first work on global Newton methods for solving nonsmooth equations, Pang's method suffers from a theoretical drawback in that its convergence requires a Fréchet differentiability assumption at a limit point of the produced sequence.

Newton's method for solving nondifferentiable equations had been investigated before Pang's work. Kojima and Shindo [454] discussed such a method for PC^1 functions. Kummer [466] studied this method for general nondifferentiable functions. Both papers dealt with the local convergence but did not address the globalization of the method. Generalizing the class of semismooth functions of one variable defined by Mifflin [607], Qi and Sun [701] introduced the class of vector semismooth functions and established the local convergence of Newton's method for this class of functions. The latter result of Qi and Sun is actually a special case of the general theory of Kummer. Since its introduction, the class of vector semismooth functions has played a central role throughout the subsequent algorithmic developments of the field. Although focused mainly on the smooth case, the two recent papers [282, 878] present an enlightening summary of the historical developments of the convergence theory of Newton's method.

As an alternative to Pang's line search globalization strategy, Ralph [710] presented a path search algorithm that was implemented by Dirkse and Ferris in their highly successful PATH solver [187], which was awarded the 1997 Beale-Orchard-Hays Prize for excellence in computational mathematical programming; the accompanying paper [186] contains an extensive collection of MiCP test problems. In an important paper that dealt with an optimization problem [247], Fischer proposed the use of what is now called the Fischer-Burmeister function to reformulate the Karush-Kuhn-Tucker conditions arising from an inequality constrained optimization problem as a system of nonsmooth equations. Collectively, these works paved the way for an outburst of activities that started with De Luca, Facchinei, and Kanzow [162]. The latter paper discussed the application of a globally convergent semismooth Newton method to the Fischer-Burmeister reformulation of the nonlinear complementarity problem; the algorithm described therein provided a model approach for many algorithms that followed. The semismooth Newton approach led to algorithms that are conceptually and practically simpler than the B-differentiable Newton method and the path Newton method, and have, at the same time, better convergence properties.

The attractive theoretical properties of the semismooth methods and their good performance in practice spurred much research to investigate further this class of methods and inspired much of the subsequent studies. In the second half of the 1990s, a large number of papers was devoted to the improvement, extension, and numerical testing of semismooth algorithms, bringing these algorithms to a high level of sophistication. Among other things, these developments made it clear that the B-differentiable Newton method is intimately related to the semismooth Newton method applied to the min reformulation of the complementarity problem, thus confirming the breadth of the new approach.

The above overview gives a general perspective on the evolution of the VI/CP and documents several major events that have propelled this subject to its modern status as a fruitful and exciting discipline within mathematical programming. There are many other interesting developments, such as sensitivity and stability analysis, piecewise smooth functions, error bounds, interior point methods, smoothing methods, methods of the projection family, and regularization, as well as the connections with new applications and other mathematical disciplines, all of which add to the richness and vitality of the field and form the main topics in our work. The notes and comments of these developments are contained at the end of each chapter.

A Synopsis of the Book

Divided into two volumes, the book contains twelve main chapters, followed by an extensive bibliography, a summary of main results and key algorithms, and a subject index. The first volume consists of the first six chapters, which present the basic theory of VIs and CPs. The second volume consists of the remaining six chapters, which present algorithms of various kinds for solving VIs and CPs. Besides the main text, each chapter contains (a) an extensive set of exercises, many of which are drawn from published papers that supplement the materials in the text, and (b) a set of notes and comments that document historical accounts, give the sources for the results in the main text, and provide discussions and references on related topics and extensions. The bibliography contains more than 1,300 publications in the literature up to June 2002. This bibliography serves two purposes: one purpose is to give the source of the results in the chapters, wherever applicable; the other purpose is to give a documentation of papers written on the VI/CP and related topics.

Due to its comprehensiveness, each chapter of the book is by itself quite lengthy. Among the first six sections in Chapter **1**, Sections **1.1**, **1.2**, **1.3**, and **1.5** make up the basic introduction to the VI/CP. The source problems in Section **1.4** are of very diverse nature; they fall into several general categories: mathematical programming, economics, engineering, and finance. Depending on an individual's background, a reader can safely skip those subsections that are outside his/her interests; for instance, an economist can omit the subsection on frictional contact problems, a contact mechanician can omit the subsection on Nash-Cournot production models. Section **1.6** mainly gives the definition of several extended problems; except for (1.6.1), which is re-introduced and employed in Chapter **11**, this section can be omitted at first reading.

Chapters **2** and **3** contain the basic theory of existence and multiplicity of solutions. Several sections contain review materials of well-known topics; these are included for the benefit of those readers who are unfamiliar with the background for the theory. Section **2.1** contains the review of degree theory, which is a basic mathematical tool that we employ throughout the book; due to its powerful implications, we recommend this to a reader who is interested in the theoretical part of the subject. Sections **2.2**, **2.3** (except Subsection **2.3.2**), **2.4**, and **2.5** (except Subsection **2.5.3**) contain fundamental results. While Sections **2.6** and **2.8** can be skipped at first reading, Section **2.7** contains very specialized results for the discrete frictional contact problem and is included herein only to illustrate the application of the theory developed in the chapter to an important class of mechanical problems.

Section **3.1** in Chapter **3** introduces the class of B-differentiable functions that plays a fundamental role throughout the book. With the exception of the nonstandard SBCQ, Section **3.2** is a review of various well-known CQs in NLP. Except for the last two subsections in Section **3.3** and Subsection **3.5.1**, which may be omitted at first reading, the remainder of this chapter contains important properties of solutions to the VI/CP.

Chapter **4** serves two purposes: One, it is a technical precursor to the next chapter; and two, it introduces the important classes of PA functions (Section **4.2**) and PC^1 functions (Section **4.6**). Readers who are not interested in the sensitivity and stability theory of the VI/CP can skip most of this and the next chapter. Nevertheless, in order to appreciate the class of semismooth functions, which lies at the heart of the contemporary algorithms for solving VIs/CPs, and the regularity conditions, which are key to the convergence of these algorithms, the reader is advised to become familiar with certain developments in this chapter, such as the basic notion of coherent orientation of PA maps (Definition **4.2.3**) and its matrix-theoretic characterizations for the special maps $\mathbf{M}_K^{\rm nor}$ and $\mathbf{M}_K^{\rm nat}$ (Proposition **4.2.7**) as well as the fundamental role of this notion in the globally unique solvability of AVIs (Theorem **4.3.2**). The inverse function Theorem **4.6.5** for PC^1 functions is of fundamental importance in nonsmooth analysis. Subsections **4.1.1** and **4.3.1** are interesting in their own right; but they are not needed in the remainder of the book.

Chapter **5** focuses on the single topic of sensitivity and stability of the VI/CP. While stability is the cornerstone to the fast convergence of Newton's method, readers who are not interested in this specialized topic or in the advanced convergence theory of the mentioned method can skip this entire chapter. Notwithstanding this suggestion, Section **5.3** is of classical importance and contains the most basic results concerning the local analysis of an isolated solution.

Chapter **6** contains another significant yet specialized topic that can be omitted at first reading. From a computational point of view, an important goal of this chapter is to establish a sound basis for understanding the connection between the exact solutions to a given problem and the computed solutions of iterative methods under prescribed termination criteria used in practical implementation. As evidenced throughout the chapter and also in Section **12.6**, the theory of error bounds has far-reaching consequences that extend beyond this goal. For instance, since the publication of the paper [222], which is the subject of discussion in Section **6.7**, there has been an increasing use of error bounds in designing algorithms that can identify active constraints accurately, resulting in enhanced theoretical properties of algorithms and holding promise for superior computational efficiency.

Of independent interest, Chapters **7** and **8** contain the preparatory materials for the two subsequent chapters. While Sections **7.1** and **7.4** are both concerned with the fundamentals of nonsmooth functions, the former pertains to general properties of nonsmooth functions, whereas the latter focuses on the semismooth functions. As far as specific algorithms go, Algorithms **7.3.1** and **7.5.1** in Sections **7.3** and **7.5**, respectively, are the most basic and strongly recommended for anyone interested in the subsequent developments. The convergence of the former algorithm depends on the (strong) stability theory in Chapter **5**, whereas that of the latter is rather simple, provided that one has a good understanding of semismoothness. In contrast to the previous two algorithms, Algorithm **7.2.17** is closest to a straightforward application of the classical Newton method for smooth systems of equations to the NCP.

The path search Newton method **8.1.9** is the earliest algorithm to be coded in the highly successful PATH computer software [187]. Readers who are already familiar with the line search and/or trust region methods in standard nonlinear programming may wish to peruse Subsection **8.3.3** and skip the rest of Chapter **8** in order to proceed directly to the next chapter. When specialized to C^1 optimization problems, as is the focus in Chapters **9** and **10**, much of the material in Sections **8.3** and **8.4** is classical; these two sections basically offer a systematic treatment of known techniques and results and present them in a way that accommodates nonsmooth objective functions.

The last four chapters are the core of the algorithmic part of this book. While Chapter **9** focuses on the NCP, Chapter **10** is devoted to the VI. The first section of the former chapter presents a detailed exposition of algorithms based on the FB merit function and their convergence theory. The most basic algorithm, **9.1.10**, is described in Subsection **9.1.1** and is accompanied by a comprehensive analysis. Algorithm **9.2.3**, which combines the min function and the FB merit function in a line search method, is representative of a mixture of algorithms in one overall scheme. Example **9.3.3** contains several C-functions that can be used in place of the FB C-function. The box-constrained VI in Subsection **9.4.3** unifies the generalized problems in Section **9.4**.

The development in Section **10.1** is very similar to that in Section **9.1.1**; the only difference is that the analysis of the first section in Chapter **10** is tailored to the KKT system of a finitely representable VI. The other major development in this chapter is the D-gap function in Section **10.3**, which is preceded by the preparatory discussion of the regularized gap function in Subsection **10.2.1**. The implicit Lagrangian function presented in Subsection **10.3.1** is an important merit function for the NCP.

Chapter **11** presents interior and smoothing methods for solving CPs of different kinds, including KKT systems. Developed in the abstract setting of constrained equations, the basic interior point method for the implicit MiCP, Algorithm **11.5.1**, is presented in Section **11.5**. An extensive theoretical study of the latter problem is the subject of the previous Section **11.4**; in which the important mixed P_0 property is introduced (see Definition **11.4.1**). A Newton smoothing method is outlined in Subsection **11.8.1**; this method is applicable to smoothed reformulations of CPs using the smoothing functions discussed in Subsection **11.8.2**, particularly those in Example **11.8.11**.

The twelveth and last chapter discusses various specialized methods that are applicable principally to (pseudo) monotone VIs and NCPs of the P_0 type. The first four sections of the chapter contain the basic methods and their convergence theories. The theory of maximal monotone operators in Subsection **12.3.1** plays a central role in the proximal point method that is the subject of Subsection **12.3.2**. Bregman-based methods in Subsection **12.7.2** are well researched in the literature, whereas the interior/barrier methods in Subsection **12.7.4** are recent entrants to the field.

Acknowledgments

Writing a book on this subject has been the goal of the second author since he and Harker published their survey paper [332] in 1990. This goal was not accomplished and ended with Harker giving a lecture series at the Université Catholique de Louvain in 1992 that was followed by the lecture notes [331]. The second author gratefully acknowledges Harker for the fruitful collaboration and for his keen interest during the formative stage of this book project.

The first author was introduced to optimization by Gianni Di Pillo and Luigi Grippo, who did much to shape his understanding of the discipline and to inspire his interest in research. The second author has been very lucky to have several pioneers of the field as his mentors during his early career. They are Richard Cottle, Olvi Mangasarian, and Stephen Robinson. To all these individuals we owe our deepest gratitude. Both authors have benefitted from the fruitful collaboration with their doctoral students and many colleagues on various parts of the book. We thank them sincerely. Michael Ferris and Stefan Scholtes have provided useful comments on a preliminary version of the book that help to shape its final form. We wish to thank our Series Editor, Achi Dosanjh, the Production Editor, Louise Farkas, and the staff members at Springer-New York, for their skillful editorial assistance.

Facchinei's research has been supported by grants from the Italian Research Ministry, the National Research Council, the European Commission, and the NATO Science Committee. The U.S. National Science Foundation has provided continuous research grants through several institutions to support Pang's work for the last twenty-five years. The joint research with Monteiro was supported by the Office of Naval Research as well. Pang's students have also benefited from the financial support of these two funding agencies, to whom we are very grateful.

Finally, the text of this monograph was typeset by the authors using LaTeX, a document preparation system based on Knuth's TeX program. We have used the document style files of the book [142] that were prepared by Richard Cottle and Richard Stone, and based on the LaTeX book style.

Rome, Italy — Francisco Facchinei
Baltimore, Maryland, U.S.A. — Jong-Shi Pang
December 11, 2002

Contents

Contents of Volume II

Subsections are omitted; for details, see Volume II.

Acronyms

The numbers refer to the pages where the acronyms first appear.

AVI, 7	Affine Variational Inequality
B-function, 869	Box-function
CC, 1017	Coerciveness in the Complementary variables
CE, 989	Constrained Equation
C-function, 72	Complementarity function
CP, 4	Complementarity Problem
CQ, 17	Constraint Qualification
C^r, 13	Continuously differentiable of order r
$C^{1,1}$, 529	$= LC^1$
CRCQ, 262	Constant Rank Constraint Qualification
ESSC, 896	Extended Strong Stability condition
GUS, 122	Globally Uniquely Solvable
FOA, 443	First-Order Approximation
IP, 989	Interior Point
KKT, 9	Karush-Kuhn-Tucker
LC^1, 719	C^1 functions with Lipschitz continuous gradients
LCP, 8	Linear Complementarity Problem
LICQ, 253	Linear Independence Constraint Qualification
LP, 6	Linear Program
MFCQ, 252	Mangasarian-Fromovitz Constraint Qualification
MiCP, 7	Mixed Complementarity Problem
MLCP, 7	Mixed Linear Complementarity Problem
MPEC, 65	Mathematical Program with Equilibrium Constraints
MPS, 581	Minimum Principle Sufficiency
NCP, 6	Nonlinear Complementarity Problem
NLP, 13	Nonlinear Program
PA, 344	Piecewise Affine
PC^r, 384	Piecewise smooth of order r (mainly $r = 1$)

PL, 344	Piecewise Linear
QP, 15	Quadratic Program
QVI, 16	Quasi-Variational Inequality
SBCQ, 262	Sequentially Bounded Constraint Qualification
SC^1, 686	C^1 functions with Semismooth gradients
SCOC, 490	Strong Coherent Orientation Condition
SLCP, v	Sequential Linear Complementarity Problem
SMFCQ 253	Strict Mangasarian-Fromovitz Constraint Qualification
SPSD, 67	Symmetric Positive Semidefinite
SQP, 718	Sequential Quadratic Programming
VI, 2	Variational Inequality
WMPS, 1202	Weak Minimum Principle Sufficiency

Glossary of Notation

Spaces

$\mathcal{M}^n$	the subspace of symmetric matrices in $\mathbb{R}^{n \times n}$
$\mathcal{M}^n_+$	the cone of SPSD matrices of order n
$\mathcal{M}^n_{++}$	the cone of positive definite matrices in $\mathcal{M}^n$
$\mathbb{R}^n$	the real n-dimensional space
$\mathbb{R}^n_+$	the nonnegative orthant of $\mathbb{R}^n$
$\mathbb{R}^n_{++}$	the positive orthant of $\mathbb{R}^n$
$\mathbb{R}^{n \times m}$	the space of $n \times m$ real matrices

Matrices

A	$\equiv (a_{ij})$; a matrix with entries a_{ij}
$\det A$	the determinant of a matrix A
$\operatorname{tr} A$	the trace of a matrix A
A^T	the transpose of a matrix A
A^{-1}	the inverse of a matrix A
M/A	the Schur complement of A in M
A_s	$\equiv \frac{1}{2}(A + A^T)$; the symmetric part of a matrix A
$\lambda_{\max}(A)$	the largest eigenvalue of a matrix $A \in \mathcal{M}^n$
$\lambda_{\min}(A)$	the smallest eigenvalue of a matrix $A \in \mathcal{M}^n$
$\|A\|$	$\equiv \sqrt{\lambda_{\max}(A^T A)}$; the Euclidean norm of $A \in \mathbb{R}^{n \times n}$
$A \bullet B$	the Frobenius product of two matrices A and B in $\mathbb{R}^{n \times n}$
$\|A\|_F$	$\equiv \sqrt{A \bullet A}$; the Frobenius norm of $A \in \mathbb{R}^{n \times n}$
$A_{\cdot \alpha}$	the columns of A indexed by α
$A_{\alpha \cdot}$	the rows of A indexed by α
$A_{\alpha\beta}$	submatrix of A with rows and columns indexed by α and β, respectively
$\mathbf{I}_k$	identity matrix of order k (subscript often omitted)
$\operatorname{diag}(a)$	the diagonal matrix with diagonal elements equal to the components of the vector a

Scalars

IR	the real line
$\operatorname{sgn} t$	the sign, $1, -1, 0$, of a positive, negative, or zero scalar t
t_+	$\equiv \max(0, t)$; the nonnegative part of a scalar
t_-	$\equiv \max(0, -t)$; the nonpositive part of a scalar

Vectors

x^T	$\equiv (x_1, \ldots, x_n)$; the transpose of a vector x with components x_i
x^{-1}	$\equiv (1/x_i)_{i=1}^n$ for $x > 0$
x^+	$\equiv \max(0, x)$; the nonnegative part of a vector x
x^-	$\equiv \max(0, -x)$; the nonpositive part of a vector x
x_α	subvector of x with components indexed by α
$\{x^k\}$	a sequence of vectors x^1, x^2, x^3, ...
$x^T y$	the standard inner product of vectors in IR^n
$\|x\|_p$	$\equiv \left(\sum_{i=1}^n \|x_i\|^p \right)^{1/p}$; the ℓ_p-norm of a vector $x \in \mathrm{IR}^n$
$\|x\|$	the ℓ_2-norm of $x \in \mathrm{IR}^n$, unless otherwise specified
$\|x\|_\infty$	$\equiv \max_{1 \leq i \leq n} \|x_i\|$; the ℓ_∞-norm of $x \in \mathrm{IR}^n$
$\|x\|_A$	$\equiv \sqrt{x^T A x}$; the A-norm of $x \in \mathrm{IR}^n$ for $A \in \mathcal{M}_{++}^n$
$x \geq y$	the (usual) partial ordering: $x_i \geq y_i$, $i = 1, \ldots n$
$x \gneqq y$	$x \geq y$ and $x \neq y$
$x > y$	the strict ordering: $x_i > y_i$, $i = 1, \ldots, n$
$\min(x, y)$	the vector whose i-th component is $\min(x_i, y_i)$
$\max(x, y)$	the vector whose i-th component is $\max(x_i, y_i)$
$x \circ y$	$\equiv (x_i y_i)_{i=1}^n$; the Hadamard product of x and y
$x \perp y$	x and y are perpendicular
$\mathbf{1}_k$	k-vector of all ones (subscript often omitted)

Functions

$F : \mathcal{D} \to \mathcal{R}$	a mapping with domain $\mathcal{D}$ and range $\mathcal{R}$
$F\|_\Omega$	the restriction of the mapping F to the set Ω
$F \circ G$	composition of two functions F and G
F^{-1}	the inverse of a mapping F
$F'(\cdot\,; \cdot)$	directional derivative of the mapping F
JF	$\equiv \left(\dfrac{\partial F_i}{\partial x_j} \right)$; the $m \times n$ Jacobian of a mapping $F : \mathrm{IR}^n \to \mathrm{IR}^m$ $(m \geq 2)$
$J_\beta F_\alpha$	$\equiv (JF)_{\alpha\beta}$; a submatrix of JF
$J_y F(x, y)$	the partial Jacobian matrix of F with respect to y

Functions (continued)

$\nabla\theta$	$\equiv \left(\dfrac{\partial\theta}{\partial x_j}\right)$; the gradient of a function $\theta : \mathbb{R}^n \to \mathbb{R}$
$\nabla^2\theta$	$\equiv \left(\dfrac{\partial^2\theta}{\partial x_i x_j}\right)$; the Hessian matrix of $\theta : \mathbb{R}^n \to \mathbb{R}$
$\theta^D(\cdot;\cdot)$	Dini directional derivative of $\theta : \mathbb{R}^n \to \mathbb{R}$
$\operatorname{Jac} F = \partial_B F$	the limiting Jacobian or B-subdifferential of $F : \mathbb{R}^n \to \mathbb{R}^m$
∂F	$\equiv \operatorname{conv} \operatorname{Jac} F$; the Clarke Jacobian of $F : \mathbb{R}^n \to \mathbb{R}^m$
$\partial_C F$	$\equiv (\, \partial F_1(x) \times \partial F_2(x) \times \partial F_n(x)\,)^T$
$\partial^2\theta$	$\equiv \partial\nabla\theta$; the generalized Hessian of an LC^1 function $\theta : \mathbb{R}^n \to \mathbb{R}$
$\varphi^*(y)$	the conjugate of a convex function $\varphi(x)$
$\varphi_\infty(d)$	the recession function of a convex function $\varphi(x)$
$o(t)$	any function such that $\lim\limits_{t\downarrow 0} \dfrac{o(t)}{t} = 0$
$O(t)$	any function such that $\limsup\limits_{t\downarrow 0} \dfrac{\lvert O(t)\rvert}{t} < \infty$
$\deg(\Phi, \Omega, p)$	the degree of Φ at p relative to Ω
$\deg(\Phi, \Omega)$	$\equiv \deg(\Phi, \Omega, 0)$
$\operatorname{ind}(\Phi, x)$	the index of Φ at x
$\Pi_K(x)$	the Euclidean projection of x on the set K
$\Pi_{K,A}(x)$	skewed projection of x on K under the A-norm
Π_K^A	$\equiv \Pi_{K,A} \circ A^{-1}$
$\operatorname{mid}(a, b; x)$	$\equiv \Pi_{[a,b]}(x)$; the mid function for given $a, b \in \mathbb{R}^n$
$\inf_S \theta(x)$	the infimum of the function θ on S
$\sup_S \theta(x)$	the supremum of the function θ on S
$\operatorname{dist}(x, W)$	Euclidean distance function from vector x to set W
$\operatorname{dist}_\infty(x, W)$	ℓ_∞-distance function from vector x to set W
J_Φ	the resolvent of the multifunction Φ
$D_f(x, y)$	$\equiv f(x) - f(y) - \nabla f(y)^T(x - y)$; Bregman distance induced by the strictly convex function f

Sets

$\in, \notin$	element membership, non-membership in a set
$\emptyset, \subseteq, \subset$	the empty set, set inclusion, proper set inclusion
$\cup, \cap, \times$	union, intersection, Cartesian product
$\prod S_i$	Cartesian product of sets S_i
$S_1 \setminus S_2$	the difference of two sets S_1 and S_2
$S_1 + S_2$	the vector sum of two sets S_1 and S_2

Sets (continued)

$\|S\|$	the cardinality of a finite set S
$\text{aff}\, S$, $\text{lin}\, S$	the affine, linear hull of a set S, respectively
$\text{bd}\, S = \partial S$	the topological boundary of a set S
$\text{cl}\, S$, $\text{int}\, S$	the topological closure, interior of a set S, respectively
$\text{conv}\, S$	the convex hull of a set S
$\text{pos}\, A$	the conical hull of the columns of $A \in \mathbb{R}^{m \times n}$
$\text{ri}\, S$	the relative interior of a set S
S^*, S_∞	the dual cone of a set S, the recession cone of S
$S^\perp$	the orthogonal complement of a set S
$\text{dom}\, \Phi$	the domain of a (multi)function Φ
$\text{gph}\, \Phi$	the graph of a (multi)function Φ
$\text{ran}\, \Phi$	the range of a (multi)function Φ
$\mathbb{B}(x, \delta)$	the open ball with center at x and radius δ (a neighborhood $\mathcal{N}$ of x)
$\mathbb{B}(H; \varepsilon, S)$	ε-neighborhood of the function H restricted to the set S, comprising all continuous functions G such that $\\|G - H\\|_S \equiv \sup_{y \in S} \\|G(y) - H(y)\\| < \varepsilon$
$K_{\mathcal{N}}$	$\equiv K \cap \text{cl}\, \mathcal{N}$
$\text{argmax}_S \theta(x)$	the set of constrained maximizers of θ on S
$\text{argmin}_S \theta(x)$	the set of constrained minimizers of θ on S
$\text{supp}(x)$	the support of a vector x
$\mathcal{L}(x; S)$	the linearization cone of the set S at a point $x \in S$
$\mathcal{N}(x; S)$	the normal cone of the set S at a point $x \in S$
$\mathcal{T}(x; S)$	the tangent cone of the set S at a point $x \in S$
$\mathcal{C}(x; K, F)$	critical cone of the pair (K, F) at $x \in \text{SOL}(K, F)$
$\mathcal{C}_\pi(x; K)$	$\equiv \mathcal{C}(\Pi_K(x); K, \mathbf{I} - x)$; critical cone of K at $x \in \mathbb{R}^n$
$\mathcal{I}(x)$	the index set of active constraints at x
$\mathcal{M}(x)$	the set of KKT multipliers at $x \in \text{SOL}(K, F)$
$\mathcal{M}_\pi(x)$	the set of KKT multipliers at $\Pi_K(x)$
$\mathcal{M}^e(x)$	the (finite) set of extreme KKT multipliers in $\mathcal{M}(x)$
$P(A, b)$	$\{y \in \mathbb{R}^n : Ay \leq b\}$; a polyhedron
$\mathcal{I}(A, b)$	family of index sets identifying the faces of $P(A, b)$
$\mathcal{B}_{\text{bas}}(A, b)$	normal family of basis matrix of $P(A, b)$
$[x, y]$	the closed line segment joining x and y in $\mathbb{R}^n$
(x, y)	the open line segment joining x and y in $\mathbb{R}^n$
$x^\perp$	the orthogonal complement of the vector x
$\text{epi}\, \varphi$	the epigragph of a convex function φ
H_{++}	$\equiv H(\mathbb{R}^{2n}_{++} \times \mathbb{R}^m)$, used in IP theory
H_+	$\equiv H(\mathbb{R}^{2n}_{+} \times \mathbb{R}^m)$, used in IP theory

Problem Classes and Fundamental Objects

AVI (K, q, M)	AVI defined by the polyhedron K, vector q, and matrix M
CE (G, X)	constrained equation defined by the function G and the set X
CP (F, G)	vertical CP defined by two functions F and G
CP (K, F)	CP defined by the cone K and the mapping F
CP (K, q, M)	$\equiv$ CP (K, F) with $F(x) \equiv q + Mx$
$\mathcal{D}(K, M)$	VI domain of the pair (K, M)
$\mathrm{FEA}(K, F)$	the feasible region of the CP (K, F)
$\mathcal{K}(K, M)$	VI kernel of the pair (K, M)
LCP (q, M)	LCP defined by the vector q and matrix M
NCP (F)	NCP defined by the function $F : \mathbb{R}^n \to \mathbb{R}^n$
$\mathcal{R}(K, M)$	VI range of the pair (K, M)
$\mathrm{SOL}(K, F)$	solution set of the VI (K, F)
$\mathrm{SOL}(K, G, A, b)$	solution set of the VI (K, G, A, b)
$\mathrm{SOL}(K, q, M)$	solution set of the VI (K, q, M)
$\mathrm{SOL}(q, M)$	solution set of the LCP (q, M)
VI (K, F)	VI defined by the set K and the mapping F
VI (K, G, A, b)	$\equiv$ VI (K, F) with $F(x) \equiv A^T G(Ax) + b$
VI (K, q, M)	$\equiv$ VI (K, F) with $F(x) \equiv q + Mx$

Matrix Classes

column sufficient	matrices M for which $x \circ Mx \leq 0 \Rightarrow x \circ Mx = 0$
copositive	matrices M such that $x^T Mx \geq 0$ for all $x \geq 0$
nondegenerate	matrices with nonzero principal minors
positive definite	matrices M such that $x^T Mx > 0$ for all $x \neq 0$
positive semidefinite	matrices M such that $x^T Mx \geq 0$ for all x
positive semi-definite plus	positive semidefinite + $[x^T Mx = 0 \Rightarrow Mx = 0]$
row sufficient	matrices whose transpose are column sufficient
semicopositive	matrices M for which $\forall\, x \gneqq 0\ \exists\, i$ such that $x_i(Mx)_i \geq 0$ and $x_i \neq 0$
strictly copositive	matrices M such that $x^T Mx > 0$ for all $x \gneqq 0$
strictly semi-copositive	matrices M for which $\forall\, x \gneqq 0\ \exists\, i$ such that $x_i(Mx)_i > 0$
P_0	matrices M for which $\forall\, x \neq 0\ \exists\, i$ such that $x_i(Mx)_i \geq 0$ and $x_i \neq 0$
P	matrices M for which $\forall\, x \neq 0\ \exists\, i$ such that $x_i(Mx)_i > 0$

Matrix Classes (continued)

R_0	matrices M such that $\mathrm{SOL}(0, M) = \{0\}$
S_0	matrices M such that $Mx \geq 0$ for some $x \gneq 0$
S	matrices M such that $Mx > 0$ for some $x \geq 0$

CP Functions

$\psi_{\mathrm{CCK}}(a,b)$	$\equiv \psi_{\mathrm{FB}}(a,b) - \tau \max(0,a) \max(0,b)$; the Chen-Chen-Kanzow C-function
$\psi_{\mathrm{FB}}(a,b)$	$\equiv \sqrt{a^2+b^2} - a - b$; the Fischer-Burmeister C-function
$\psi_{\mathrm{FB}_\mu}(a,b)$	$\equiv \sqrt{a^2+b^2+2\mu} - a - b$; the smoothed Fischer-Burmeister function
$\psi_{\mathrm{CHKS}_\varepsilon}(a,b)$	$\equiv \sqrt{(a-b)^2+4\varepsilon} - (a+b)$; the smoothed Chen-Harker-Kanzow-Smale function
$\psi_{\mathrm{KK}}(a,b)$	$\equiv \dfrac{\sqrt{(a-b)^2+2qab} - (a+b)}{2-q}$, $q \in [0,2)$; the Kanzow-Kleinmichel C-function
$\psi_{\mathrm{LT}}(a,b)$	$\equiv \\|(a,b)\\|_q - a - b$, $q > 1$; the Luo-Tseng C-function
$\psi_{\mathrm{LTKYF}}(a,b)$	$\equiv \phi_1(ab) + \phi_2(-a,-b)$, the Luo-Tseng-Kanzow-Yamashita-Fukushima family of C-functions
$\psi_{\mathrm{Man}}(a,b)$	$\equiv \zeta(\|a-b\|) - \zeta(b) - \zeta(a)$; Mangasarian's family of C-functions, includes the min function
$\psi_{\mathrm{U}}(a,b)$	Ulbrich's C-function; see Exercise **1.8.21**
$\psi_{\mathrm{YYF}}(a,b)$	$\equiv \frac{\eta}{2}\left((ab)_+\right)^2 + \frac{1}{2}\psi_{\mathrm{FB}}(a,b)^2$, $\eta > 0$; the Yamada-Yamashita-Fukushima C-function
$\mathbf{F}_\psi(x)$	$\equiv (\psi(x_i, F_i(x)))_{i=1}^n$; the reformulation function of the NCP (F) for a given C-function ψ
$\theta_\psi(x)$	$\equiv \frac{1}{2}\sum_{i=1}^{n} \psi^2(x_i, F_i(x))$; merit function induced by the C-function ψ
$\psi_{ab}(u,v)$	$\equiv \left(\dfrac{1}{2a} - \dfrac{1}{2b}\right) v^2 + \dfrac{1}{2b}\max(0, v - bu)^2 - \dfrac{1}{2a}\max(0, v - au)^2$; for $b > a > 0$
$\theta_{ab}^{\mathrm{ncp}}(x)$	$\equiv \sum_{i=1}^{n} \psi_{ab}(x_i, F_i(x))$; the implicit Lagrangian function for the NCP (F)
$\phi_{\mathrm{Q}}(\tau, \tau'; r, s)$	Qi's B-function; see (9.4.7)
$\mathbf{H}_{\mathrm{IP}}(x,y,z)$	the IP function for implicit MiCP; see (11.1.4)
$\mathbf{H}_{\mathrm{CHKS}}(u,x,y,z)$	the IP function for the CHKS smoothing of the min function; see Exercise **11.9.8**

KKT Functions

$\mathbf{L}(x, \mu, \lambda)$ $\equiv F(x) + \sum_{j=1}^{\ell} \mu_j \nabla h_j(x) + \sum_{i=1}^{m} \lambda_i \nabla g_i(x)$; the vector Lagrangian function of the VI (K, F) with a finitely represented K

$\Phi_{\text{FB}}(x, \mu, \lambda)$ $\equiv \begin{pmatrix} \mathbf{L}(x, \mu, \lambda) \\ -h(x) \\ \psi_{\text{FB}}(-g_i(x), \lambda_i) : 1 \le i \le m \end{pmatrix}$; the FB reformulation of the KKT system of a VI

$\theta_{\text{FB}}(x, \mu, \lambda)$ $\equiv \frac{1}{2} \Phi_{\text{FB}}(x, \mu, \lambda)^T \Phi_{\text{FB}}(x, \mu, \lambda)$; the FB merit function of the KKT system of a VI

$\Phi_{\min}(x, \mu, \lambda)$ $\equiv \begin{pmatrix} \mathbf{L}(x, \mu, \lambda) \\ -h(x) \\ \min(-g(x), \lambda) \end{pmatrix}$; the min reformulation of the KKT system of a VI

$\theta_{\min}(x, \mu, \lambda)$ $\equiv \frac{1}{2} \Phi_{\min}(x, \mu, \lambda)^T \Phi_{\min}(x, \mu, \lambda)$; the min merit function of the KKT system of a VI

VI Functions

$\mathbf{F}_K^{\text{nat}}$	the natural map associated with the pair (K, F)
$\mathbf{F}_K^{\text{nor}}$	the normal map associated with the pair (K, F)
$\mathbf{F}_{K,D}^{\text{nat}}(x)$	$\equiv x - \Pi_{K,D}(x - D^{-1}F(x))$; the skewed natural map associated with the pair (K, F) using $\Pi_{K,D}$
$\mathbf{F}_{K,\tau}^{\text{nat}}$	the natural map associated with the VI $(K, \tau F)$
$\mathbf{M}_K^{\text{nat}}$	$\equiv \mathbf{F}_K^{\text{nat}}$ for $F(x) \equiv Mx$
$\mathbf{M}_K^{\text{nor}}$	$\equiv \mathbf{F}_K^{\text{nor}}$ for $F(x) \equiv Mx$
θ_{gap}	the gap function of a VI
θ_{dual}	the dual gap function of a VI
θ_c	the regularized gap function with parameter $c > 0$
θ_c^{lin}	the linearized gap function with parameter $c > 0$
θ_{ab}	$\equiv \theta_a - \theta_b$; the D-gap function for $b > a > 0$
$y_c(x)$	unique maximizer in $\theta_c(x)$
$y_c^{\text{lin}}(x)$	unique maximizer in $\theta_c^{\text{lin}}(x)$
$T_c(x; K, F)$	$\equiv \mathcal{T}(x; K) \cap (-\mathcal{T}(y_c(x); K)) \cap (-F(x))^*$
$T_{ab}(x; K, F)$	$\equiv \mathcal{T}(y_b(x); K) \cap (-\mathcal{T}(y_a(x); K)) \cap (-F(x))^*$

Selected Function Classes and Properties

co-coercive on K	functions F for which $\exists\ \eta > 0$ such that $(x-y)^T(F(x)-F(y)) \geq \eta \|F(x)-F(y)\|^2$ for all x, y in K
monotone composite	functions $F(x) \equiv A^T G(Ax) + b$, where G is monotone
monotone plus	monotone and $(F(x)-F(y))^T(x-y) = 0 \Rightarrow F(x) = F(y)$
nonexpansive	functions F for which $\|F(x)-F(y)\| \leq \|x-y\|$ $\forall\ x$ and y
norm-coercive on X	functions F for which $\lim_{\substack{x \in X \\ \|x\| \to \infty}} \|F(x)\| = \infty$
P_0, P, $\mathrm{P}_*(\sigma)$	see Definition **3.5.8**
(pseudo) monotone	see Definition **2.3.1**
(pseudo) monotone plus	see Definition **2.3.9**
strictly, strongly monotone	see Definition **2.3.1**
S	see Exercise **2.9.5**
strongly S	functions F such that $\forall\ q \in \mathbb{R}^n$, $\exists\ x \geq 0$ satisfying $F(x) > q$
symmetric	= gradient map; differentiable $F : \mathbb{R}^n \to \mathbb{R}^n$ with symmetric JF
uniformly P	see Definition **3.5.8**
univalent	= continuous plus injective
weakly univalent	uniform limit of univalent functions

Numbering System

The chapters of the book are numbered from 1 to 12; the sections are denoted by decimal numbers of the type **2.3** (meaning Section 3 of Chapter 2). Many sections are further divided into subsections; most subsections are numbered, some are not. The numbered subsections are by decimal numbers following the section numbers; e.g., Subsection **1.3.1** means Chapter 1, Section 3, Subsection 1.

All definitions, results, and miscellaneous items are numbered consecutively within each section in the form **1.3.5, 1.3.6**, meaning Items 5 and 6 in Section 3 of Chapter 1. All items are also identified by their types, for example, **1.4.1 Proposition.**, **1.4.2 Remark.** When an item is referred to in the text, it is called out as Algorithm **5.2.1**, Theorem **4.1.7**, and so forth. Equations are numbered consecutively and identified by chapter, section, and equation. Thus (3.1.4) means Equation (4) in Section 1 of Chapter 3.

Chapter 1

Introduction

In this chapter, we formally define the problems that form the main topic of this book: the variational inequality (VI) and the complementarity problem (CP). We identify many major themes that will be discussed in detail throughout the book. The principal body of the present chapter consists of three parts. In the first part, which covers Sections **1.1** through **1.3**, we introduce some basic classifications and associated terminology for various special cases of these problems. We explain the interconnection between the VI and the CP as well as their relation to a standard nonlinear program. In the second part, Section **1.4**, we present an extensive set of source problems from engineering, economics, and finance that can be modeled as VIs and/or CPs. These applied contexts provide solid evidence of the wide applicability of the VI/CP methodology in modeling diverse equilibrium phenomena. In the third part, which covers Sections **1.5** and **1.6**, we describe various equivalent formulations of VIs and CPs as systems of smooth and nonsmooth equations and also as constrained and unconstrained optimization problems. These formulations provide the basis for the development of the theory and algorithms for the VI/CP that are the main topics in the subsequent chapters.

Except for some source problems in Section **1.4**, we strive to present the materials throughout this chapter in an elementary fashion, using only concepts and results that are well known in linear and nonlinear programming. We refer the reader unfamiliar with these concepts and background results to the commentary where basic references are suggested.

1.1 Problem Description

The simplest example of a variational inequality is the classical problem of solving a system of nonlinear equations. Indeed, as we see shortly, this problem can be thought of as a VI without constraints. In its general form, a variational inequality is formally defined below.

1.1.1 Definition. Given a subset K of the Euclidean n-dimensional space $\mathbb{R}^n$ and a mapping $F : K \to \mathbb{R}^n$, the *variational inequality*, denoted VI (K, F), is to find a vector $x \in K$ such that

$$(y - x)^T F(x) \geq 0, \quad \forall y \in K. \tag{1.1.1}$$

The set of solutions to this problem is denoted SOL(K, F). □

Throughout this book, we are interested only in the situation where the set K is closed and the function F is continuous. The latter continuity of F is understood to mean the continuity of F on an open set containing K; a similar consideration applies to the differentiability (if appropriate) of F. In most realizations of the VI (K, F) discussed in the book, the set K is convex. Mathematically, some results do not require the convexity of K, however. Thus, we do not make the convexity of K a blanket assumption. See the VI in part (c) of Exercise **1.8.31** and also the one in Exercise **2.9.28**.

Since K is closed and F is continuous, it follows that SOL(K, F) is always a closed set (albeit it could be empty). Understanding further properties of the solution set of a VI is an important theme that has both theoretical and practical significance. Many results in this book address questions pertaining to this theme.

A first geometric interpretation of a VI, and more specifically of the defining inequality (1.1.1), is that a point x in the set K is a solution of the VI (K, F) if and only if $F(x)$ forms a non-obtuse angle with every vector of the form $y - x$ for all y in K. We may formalize this observation using the concept of normal cone. Specifically, associated with the set K and any vector x' belonging to K, we may define the *normal cone* to K at x' to be the following set:

$$\mathcal{N}(x'; K) \equiv \{ d \in \mathbb{R}^n : d^T(y - x') \leq 0, \forall y \in K \}. \tag{1.1.2}$$

Vectors in this set are called *normal vectors* to the set K at x'. The inequality (1.1.1) clearly says that a vector $x \in K$ solves the VI (K, F) if and only if $-F(x)$ is a normal vector to K at x; or equivalently,

$$0 \in F(x) + \mathcal{N}(x; K). \tag{1.1.3}$$

Figure 1.1 illustrates this point of view. The normal cone is known to play an important role in convex analysis and nonlinear programming. This role persists in the study of the VI. The inclusion (1.1.3) is an instance of a *generalized equation*.

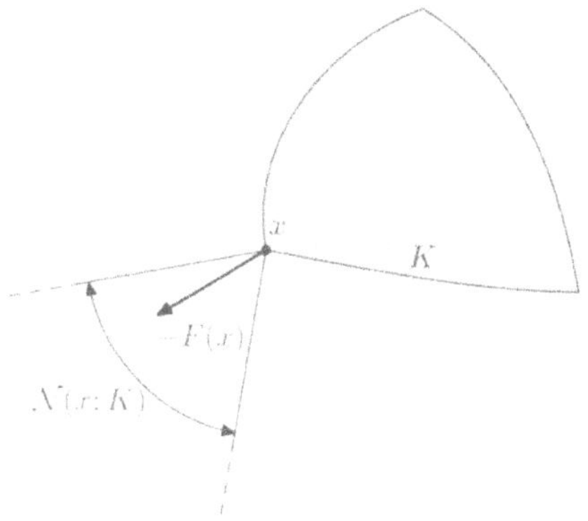

Figure 1.1: Solution and the normal cone.

In addition to providing a unified mathematical model for a variety of applied equilibrium problems, the VI includes many special cases that are important in their own right. As alluded to in the opening of this section, simplest among these cases is the problem of solving systems of nonlinear equations, which corresponds to the case where the set K is equal to the entire space $\mathbb{R}^n$. It is not difficult to show that when $K = \mathbb{R}^n$, a vector x belongs to SOL(K, F) if and only if x is a zero of the mapping F (i.e., $F(x) = 0$); in other words, SOL$(\mathbb{R}^n, F) = F^{-1}(0)$. To see this, we note that for any set K, if $x \in K$ and $F(x) = 0$, then clearly $x \in$ SOL(K, F). Thus $F^{-1}(0) \cap K$ is always a subset of SOL(K, F). To establish the reverse inclusion when $K = \mathbb{R}^n$, we note that

$$x \in \mathrm{SOL}(\mathbb{R}^n, F) \Rightarrow F(x)^T d \geq 0 \quad \forall d \in \mathbb{R}^n.$$

In particular with d taken to be $-F(x)$, we deduce that $F(x) = 0$. Thus SOL$(\mathbb{R}^n, F) \subseteq F^{-1}(0)$; hence equality holds.

The above argument applies more generally to a solution of the VI (K, F) that belongs to the topological interior of $K \subset \mathbb{R}^n$. Specifically, if x is a solution of this VI and x belongs to int K, then $F(x) = 0$. In fact, since $x \in$ int K, there exists a scalar $\tau > 0$ sufficiently small such that the vector $y \equiv x - \tau F(x)$ belongs to K. Substituting this vector into (1.1.1), we deduce that $-F(x)^T F(x) \geq 0$, which implies $F(x) = 0$.

In all interesting realizations of the VI, it is invariably the case that none of the zeros of F, if any, are solutions of the VI (K, F). In other words, the VI is a genuinely nontrivial generalization of the classical problem of solving equations. Nonetheless, as we see throughout the book, the theory and methods for solving equations are instrumental for the analysis and solution of the VI.

When K is a cone (i.e., $x \in K \Rightarrow \tau x \in K$ for all scalars $\tau \geq 0$), the VI admits an equivalent form known as a *complementarity problem*. (For an explanation of the term "complementarity", see the discussion below.)

1.1.2 Definition. Given a cone K and a mapping $F : K \to \mathbb{R}^n$, the *complementarity problem*, denoted CP (K, F), is to find a vector $x \in \mathbb{R}^n$ satisfying the following conditions:

$$K \ni x \perp F(x) \in K^*,$$

where the notation $\perp$ means "perpendicular" and K^* is the *dual cone* of K defined as:

$$K^* \equiv \{ d \in \mathbb{R}^n : v^T d \geq 0 \; \forall v \in K \};$$

that is, K^* consists of all vectors that make a non-obtuse angle with every vector in K. □

Figure 1.2 illustrates the dual cone. Writing out the $\perp$ notation explicitly, we obtain the CP (K, F) in the following form:

$$x \in K, \quad F(x) \in K^*, \quad \text{and} \quad x^T F(x) = 0.$$

The precise connection between the VI (K, F) and the CP (K, F) when K is a cone is described in the following elementary result.

1.1.3 Proposition. Let K be a cone in $\mathbb{R}^n$. A vector x solves the VI (K, F) if and only if x solves the CP (K, F).

Proof. Suppose that x solves the VI (K, F). Clearly x belongs to K. Since a cone must contain the origin, by taking $y = 0$ in (1.1.1), we obtain

$$x^T F(x) \leq 0.$$

Furthermore, since $x \in K$ and K is a cone, it follows that $2x \in K$. Thus

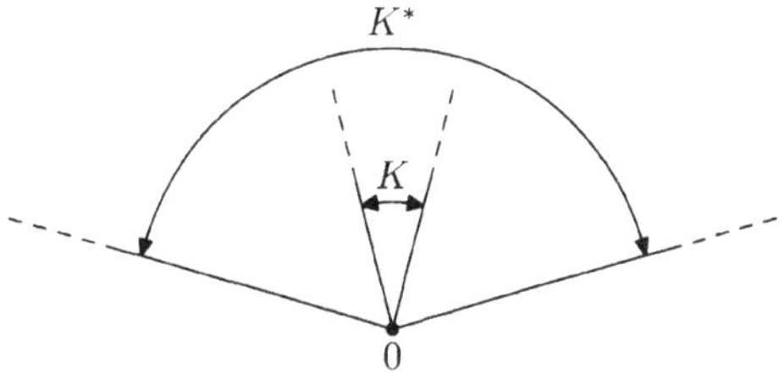

Figure 1.2: A cone and its dual.

by taking $y = 2x$ in (1.1.1), we obtain

$$x^T F(x) \geq 0.$$

Combining the above two inequalities, we deduce $x^T F(x) = 0$. In turn, this yields $y^T F(x) \geq 0$ for all $y \in K$; thus $F(x) \in K^*$. Therefore x solves the CP (K, F). Conversely, if x solves the CP (K, F), then it is trivial to show that x solves the VI (K, F). □

1.1.4 Remark. A word about the notation CP (K, F): when the acronym CP is attached to the pair (K, F), it is understood that K is a cone. □

The CP (K, F) is defined by three conditions: (i) $x \in K$, (ii) $F(x) \in K^*$, and (iii) $x^T F(x) = 0$. We introduce some concepts associated with vectors satisfying the first two conditions. Specifically, we say that a vector $x \in \mathbb{R}^n$ is *feasible* to the CP (K, F) if

$$x \in K \quad \text{and} \quad F(x) \in K^*. \tag{1.1.4}$$

We say that a vector $x \in \mathbb{R}^n$ is *strictly feasible* to the same problem if

$$x \in K \quad \text{and} \quad F(x) \in \text{int } K^*.$$

This definition implicitly assumes that $\text{int } K^*$ is nonempty, but the definition does not require $\text{int } K$ to be nonempty. If $\text{int } K \neq \emptyset$ and F is continuous, then CP (K, F) has a strictly feasible vector if and only if there exists a vector x' such that

$$x' \in \text{int } K \quad \text{and} \quad F(x') \in \text{int } K^*.$$

We say that the CP (K, F) is (strictly) feasible if it has a (strictly) feasible vector. The *feasible region* of the CP (K, F) is the set of all its feasible vectors and is denoted $\text{FEA}(K, F)$. Clearly $\text{SOL}(K, F) \subseteq \text{FEA}(K, F)$; in set notation, we can write

$$\text{FEA}(K, F) = K \cap F^{-1}(K^*).$$

For a nonlinear function F, finding a feasible point of the CP (K, F) or determining if no such point exists is not necessarily easier computationally than solving the complementarity problem itself. Nevertheless, understanding the feasibility of the CP is important because after all, feasibility is a necessary condition for solvability. When F is an affine function, there are at least two cases where determining the feasibility of the CP (K, F) is presumably easier than solving the problem itself. The first case is when K is a polyhedral cone. In this case, the feasibility condition (1.1.4) is essentially a system of linear inequalities; therefore its feasibility can be determined by solving a linear program (LP). See Subsection **1.1.1** for further discussion. The other case is when K is a special cone, known as the cone of semidefinite matrices. See Subsection **1.4.11** for discussion of such a CP.

Many special cases of the complementarity problem are very important in modeling. We now introduce two of the most important ones. When K is the nonnegative orthant of $\mathbb{R}^n$, the CP (K, F) is known as the *nonlinear complementarity problem* and denoted NCP (F). Recognizing that the dual cone of the nonnegative orthant is the nonnegative orthant itself, we have the following.

1.1.5 Definition. Given a mapping $F : \mathbb{R}^n_+ \to \mathbb{R}^n$, the NCP (F) is to find a vector $x \in \mathbb{R}^n$ satisfying

$$0 \leq x \perp F(x) \geq 0. \tag{1.1.5}$$

By expressing the orthogonality condition $x^T F(x) = 0$ in terms of the componentwise products (which is justified because x and $F(x)$ are both nonnegative vectors), we obtain the following equivalent formulation of the NCP (F):

$$\begin{aligned} 0 \leq x, \quad & F(x) \geq 0 \\ x_i F_i(x) = 0, \quad & \forall\, i = 1, \ldots, n. \end{aligned}$$

The latter formulation (or more precisely, the zero condition of the componentwise products) provides an explanation for the term "complementarity"; namely, x_i and $F_i(x)$ are complementary in the sense that if one of them is positive, then the other must be zero.

A generalization of the NCP is the *mixed complementarity problem*, abbreviated as MiCP. This corresponds to the case of the CP (K, F) where K is the special cone $\mathbb{R}^{n_1} \times \mathbb{R}^{n_2}_+$, with $n_1 + n_2 = n$. Partitioning the vectors x and $F(x)$ accordingly, we arrive at the following definition.

1.1.6 Definition. Let G and H be two mappings from $\mathbb{R}^{n_1} \times \mathbb{R}^{n_2}_+$ into $\mathbb{R}^{n_1}$ and $\mathbb{R}^{n_2}$, respectively. The MiCP (G, H) is to find a pair of vectors (u, v) belonging to $\mathbb{R}^{n_1} \times \mathbb{R}^{n_2}$ such that

$$\begin{aligned} G(u,v) &= 0, \quad u \text{ free} \\ 0 \leq v \perp &H(u,v) \geq 0. \end{aligned}$$

1.1.1 Affine problems

The CP is a special case of the VI (K, F) where the set K is a cone. In what follows, we introduce several other special cases of the VI (K, F) where either K or F has some other interesting structures. To begin, let F be the affine function given by:

$$F(x) \equiv q + Mx, \quad \forall x \in \mathbb{R}^n, \tag{1.1.6}$$

for some vector $q \in \mathbb{R}^n$ and matrix $M \in \mathbb{R}^{n \times n}$; in this case, we write VI (K, q, M) to mean VI (K, F). The solution set of the VI (K, q, M) is denoted SOL(K, q, M). If in addition K is a polyhedral set, we attach the adjective "affine" and use the notation AVI (K, q, M) to describe the all affine VI. Finally, if K is a polyhedral set but F is not necessarily affine, we use the adjective "linearly constrained" to describe the VI (K, F). Unlike the AVI where both the defining set and function are affinely structured, the VI (K, q, M) and the linearly constrained VI have only one affine member in the pair (K, F)..

An important linearly constrained VI is the *box constrained* VI (K, F), where the set K is a closed rectangle given by

$$K \equiv \{ x \in \mathbb{R}^n : a_i \leq x_i \leq b_i, \, i = 1, \ldots n \}; \tag{1.1.7}$$

here a_i and b_i are possibly infinite scalars satisfying

$$-\infty \leq a_i < b_i \leq \infty, \quad \forall i. \tag{1.1.8}$$

In particular, we allow some lower bounds a_i to be equal to $-\infty$ and some upper bounds b_i equal to ∞. This extended framework includes as a particular case the NCP (F), which corresponds to a being the zero vector and b being the vector all of whose components are equal to ∞. Similarly, the MiCP is a special case of the box constrained VI.

An MiCP (G, H) with G and H both being affine functions is called a *mixed linear complementarity problem*, abbreviated as MLCP. An NCP with an affine defining function is called a *linear complementarity problem*, abbreviated as LCP. Notationally, if F is the affine function given by (1.1.6), the NCP (F) is written as LCP (q, M):

$$0 \leq x \perp q + Mx \geq 0.$$

The solution set of this LCP is denoted $\mathrm{SOL}(q, M)$. The LCP and the MLCP are fundamental to the study of CPs and VIs. In addition to their own importance, the LCP and the MLCP often arise as "linearization" of their nonlinear counterparts; this idea of linearization lies at the heart of some of the most efficient algorithms for solving CPs and VIs. The LCP has been studied extensively; many results for this special problem provide the motivation for extensions to NCPs and VIs.

1.2 Relations Between Problem Classes

We summarize the various classes of problems and their connections in the diagram below. The notation "P $\Rightarrow$ Q" means that we can derive problem Q by specializing problem P; the notation "P $\Leftrightarrow$ Q" means that the two problems P and Q are "equivalent".

$$\begin{array}{ccccc}
\mathrm{VI}\ (K, q, M) & & & & \mathrm{VI}\ (K, q, M) \\
\Uparrow & & & & \Downarrow \\
\mathrm{VI}\ (K, F) & \Rightarrow & \text{linearly constrained VI} & \Rightarrow & \mathrm{AVI}\ (K, q, M) \\
\Downarrow & & \Updownarrow & & \Updownarrow \\
\mathrm{CP}\ (K, F) & \Rightarrow & \mathrm{MiCP}\ (F) & \Rightarrow & \mathrm{MLCP} \\
 & & \Downarrow & & \Downarrow \\
 & & \mathrm{NCP}\ (F) & \Rightarrow & \mathrm{LCP}\ (q, M).
\end{array}$$

All except two relations in the above diagram follow easily from the definitions of the respective problems. The exceptions are the equivalence between a linearly constrained VI and an MiCP and the equivalence between an AVI and an MLCP. Actually, the latter equivalence is an immediate consequence of the former, by considering an affine F. In what follows, we use a simple linear programming duality to establish an important fact that immediately yields and formally clarifies the equivalences on hand. Namely, every solution of a linearly constrained VI (K, F), where K is polyhedral, must necessarily be, along with certain auxiliary variables, a solution to a special augmented MiCP $(\tilde{F})$, where the function $\tilde{F}$ is derived from F and the linear inequalities that define K. Subsequently, we extend this fact to a non-polyhedral set K satisfying certain constraint qualifications; see Subsection **1.3.2**.

Before stating the aforementioned fact, we make a further elementary observation about the VI (K, F). The defining inequality (1.1.1) is clearly equivalent to

$$y^T F(x) \geq x^T F(x), \quad \forall y \in K.$$

Thus a vector x is a solution to the VI (K, F) if and only if x is a solution of the optimization problem in the variable y (with x considered fixed):

$$\begin{array}{ll} \text{minimize} & y^T F(x) \\ \text{subject to} & y \in K. \end{array} \tag{1.2.1}$$

Although the problem (1.2.1) is the key to the proof of Proposition **1.2.1**, this problem can not be directly used in practice for computing a solution to the VI (K, F); the reason is that the objective function of (1.2.1) is defined by a solution x of the VI (K, F) that is presumed to be known. In the following result, we write

$$K \equiv \{ x \in \mathbb{R}^n : Ax \leq b, \ Cx = d \}, \tag{1.2.2}$$

for some given matrices $A \in \mathbb{R}^{m \times n}$ and $C \in \mathbb{R}^{\ell \times n}$ and vectors $b \in \mathbb{R}^m$ and $d \in \mathbb{R}^\ell$. The resulting equivalent MiCP of the VI (K, F), which is called the *Karush-Kuhn-Tucker (KKT) system* of the VI, is dependent on the representation (1.2.2); see (1.2.3).

1.2.1 Proposition. Let K be given by (1.2.2). A vector x solves the VI (K, F) if and only if there exist vectors $\lambda \in \mathbb{R}^m$ and $\mu \in \mathbb{R}^\ell$ such that

$$\begin{aligned} 0 &= F(x) + C^T \mu + A^T \lambda \\ 0 &= d - Cx \\ 0 &\leq \lambda \perp b - Ax \geq 0. \end{aligned} \tag{1.2.3}$$

Proof. With x fixed and K being a polyhedron, (1.2.1) is a linear program in the variable y. Consequently, if x is a solution of the VI (K, F), then x is an optimal solution of (1.2.1); by linear programming duality, it follows that the desired pair (μ, λ) satisfies (1.2.3). The converse follows easily by reversing this argument. □

1.2.2 Example. To illustrate the above proposition, consider the box constrained VI (K, F) where K is the rectangle (1.1.7). Since K is clearly a polyhedral set, the VI (K, F) is therefore equivalent to its KKT conditions. In the case where all the bounds a_i and b_i are finite, these conditions can be written as:

$$\begin{aligned} 0 \leq x - a \quad &\perp \quad F(x) + v \geq 0 \\ 0 \leq b - x \quad &\perp \quad v \geq 0. \end{aligned}$$

When some of the bounds are infinite, we can write down a similar set of equivalent mixed complementarity conditions. In general, it is easy to verify that a point x in K is a solution of a box constrained VI (K, F) if and only if for every i,

$$\begin{aligned} x_i = a_i \quad &\Rightarrow \quad F_i(x) \geq 0 \\ a_i < x_i < b_i \quad &\Rightarrow \quad F_i(x) = 0 \\ x_i = b_i \quad &\Rightarrow \quad F_i(x) \leq 0, \end{aligned} \tag{1.2.4}$$

where it is understood that if $a_i = -\infty$ ($b_i = \infty$), the first (third) condition is void. □

Associated with a linearly constrained VI (K, F), define the augmented (nonlinear) function $\tilde{F}$ as follows. For all x in the domain of F and (μ, λ) in $\mathbb{R}^{\ell+m}$,

$$\tilde{F}(x, \mu, \lambda) \equiv \begin{pmatrix} F(x) + C^T\mu + A^T\lambda \\ d - Cx \\ b - Ax \end{pmatrix}. \tag{1.2.5}$$

From Proposition **1.2.1**, we conclude that $x \in \text{SOL}(K, F)$ if and only if there exists a pair of vectors (μ, λ), which are called *multipliers*, such that (x, μ, λ) solves the MiCP $(\tilde{F})$. This is the precise statement of the equivalence between a linearly constrained VI and an MiCP.

Consider the affine case where F is given by (1.1.6). The above nonlinear function $\tilde{F}$ becomes the following affine function:

$$\tilde{F} : \begin{pmatrix} x \\ \mu \\ \lambda \end{pmatrix} \in \mathbb{R}^{n+\ell+m} \mapsto \begin{pmatrix} q \\ d \\ b \end{pmatrix} + \mathbf{Q} \begin{pmatrix} x \\ \mu \\ \lambda \end{pmatrix} \in \mathbb{R}^{n+\ell+m}, \tag{1.2.6}$$

where

$$\mathbf{Q} \equiv \begin{bmatrix} M & C^T & A^T \\ -C & 0 & 0 \\ -A & 0 & 0 \end{bmatrix}. \tag{1.2.7}$$

It then follows that the AVI (K, q, M) is equivalent (in the sense made precise above) to the MLCP defined by the pair $(\tilde{F}(0), \mathbf{Q})$. We have therefore established the equivalence between an AVI and a certain augmented MLCP. Incidentally, checking the feasibility of the latter MLCP amounts to checking the consistency of the following linear inequality system:

$$\begin{aligned} 0 &= q + Mx + C^T\mu + A^T\lambda \\ 0 &= d - Cx \\ 0 &\leq b - Ax \\ 0 &\leq \lambda; \end{aligned}$$

as such, this is surely easier than solving the MLCP $(\tilde{F}(0), \mathbf{Q})$, which contains the additional complementarity condition between λ and $b - Ax$.

The matrix $\mathbf{Q}$ is the sum of two special matrices:

$$\mathbf{Q}_1 \equiv \begin{bmatrix} M & 0 & 0 \\ 0 & 0 & 0 \\ 0 & 0 & 0 \end{bmatrix} \quad \text{and} \quad \mathbf{Q}_2 \equiv \begin{bmatrix} 0 & C^T & A^T \\ -C & 0 & 0 \\ -A & 0 & 0 \end{bmatrix}$$

where $\mathbf{Q}_2$ is skew-symmetric. The matrix $\mathbf{Q}_1$ easily preserves many special properties that the matrix M might have. For instance, if M is symmetric, then so is $\mathbf{Q}_1$ although $\mathbf{Q}$ is not. Moreover if M is positive semidefinite, then so are $\mathbf{Q}_1$ and $\mathbf{Q}$. Throughout this book, a positive semidefinite matrix is not necessarily symmetric; in general an arbitrary matrix is positive semidefinite if its symmetric part is positive semidefinite in the sense of standard matrix theory.

In some cases, it is possible to convert a problem class into another. In the remaining part of this section, we consider the conversion of the KKT condition of the AVI (K, q, M) into an LCP. This conversion has both historical reasons and algorithmic implications. Specifically, consider (1.2.3) with F given by (1.1.6):

$$\begin{aligned} 0 &= q + Mx + C^T\mu + A^T\lambda \\ 0 &= d - Cx \\ 0 &\leq \lambda \perp b - Ax \geq 0. \end{aligned} \tag{1.2.8}$$

Assume that the matrix

$$\begin{bmatrix} M & C^T \\ -C & 0 \end{bmatrix} \tag{1.2.9}$$

is nonsingular. We can then use the first two equations in (1.2.8) to solve for the variables (x, μ) in terms of λ, thereby eliminating (x, μ) from the system (1.2.8) and obtaining the LCP $(\mathbf{q}, \mathbf{M})$ in the variable λ only, where

$$\mathbf{q} \equiv b + \begin{bmatrix} A & 0 \end{bmatrix} \begin{bmatrix} M & C^T \\ -C & 0 \end{bmatrix}^{-1} \begin{bmatrix} q \\ d \end{bmatrix},$$

and

$$\mathbf{M} \equiv \begin{bmatrix} A & 0 \end{bmatrix} \begin{bmatrix} M & C^T \\ -C & 0 \end{bmatrix}^{-1} \begin{bmatrix} A^T \\ 0 \end{bmatrix}.$$

Obviously, a necessary condition for the matrix (1.2.9) to be invertible is that C have full row rank. In Exercise **1.8.9**, the reader is asked to verify a necessary and sufficient condition for (1.2.9) to be nonsingular. If (1.2.9) is singular, then it is not so straightforward to transform the system (1.2.8), which is a special MLCP, into an LCP. Exercise **1.8.10** gives an extended treatment of this transformation for a general MLCP.

Traditionally, the goal to transform an MLCP into an LCP is to facilitate the application of the LCP pivotal methods for solving the former problem, most notably, the well-known Lemke almost complementary pivotal algorithm. Besides requiring some nonsingularity condition as we have seen above, such a transformation carries with it a computational burden when it comes to solving problems of large size. Namely, one has to be particularly careful not to destroy any sparsity that the original formulation might have. Although today's linear algebraic solvers are very advanced, the preservation of sparsity is always an important consideration when large problems are being solved. In Exercise **1.8.10**, the reader can see how the resulting matrices in the transformed LCP are formed and realize that it is not a trivial matter to preserve sparsity in this kind of transformation; see in particular the matrix $\tilde{M}$ is this exercise. With the advances in the AVI methodology, the need of transforming an MLCP into an LCP for computational purposes has lessened. In fact, all the algorithms presented in the later chapters of this book solve an MLCP as is, without the *a priori* conversion into an LCP. Nevertheless, such a conversion is sometimes beneficial because a pivotal algorithm such as that of Lemke can process MLCPs, which are not otherwise solvable by iterative algorithms.

1.3 Integrability and the KKT System

Variational inequalities and complementarity problems arise from a variety of interesting sources. Foremost among these sources are differentiable constrained optimization problems. Nevertheless, not all VIs or CPs are naturally derived from optimization problems. Part of the objective of this section is to show that a VI is a nontrivial extension of a standard nonlinear program (NLP). It is well known that the Karush-Kuhn-Tucker (KKT) conditions have played a key role in all aspects of nonlinear programming. It turns out that these conditions can be easily extended to the VI. This extension is the other main topic of discussion in this section.

1.3.1 Constrained optimization problems

Consider the constrained optimization problem:

$$\begin{array}{ll} \text{minimize} & \theta(x) \\ \text{subject to} & x \in K, \end{array} \tag{1.3.1}$$

where the objective function θ is defined and continuously differentiable (C^1) on an open superset of the closed set $K \subseteq \mathbb{R}^n$. By the well-known minimum principle in nonlinear programming , if the set K is convex, then any local minimizer x of (1.3.1) must satisfy:

$$(y - x)^T \nabla\theta(x) \geq 0, \quad \forall y \in K.$$

The latter is easily seen to be the VI $(K, \nabla\theta)$, which is called the *stationary point problem* associated with the optimization problem (1.3.1). A solution of the VI $(K, \nabla\theta)$ is called a *stationary point* of (1.3.1). It is further known that if θ is a convex function, then every stationary point of (1.3.1) is a global minimum of this optimization problem. Consequently, for a *convex program* , i.e., with θ a convex function and K a convex set, the VI $(K, \nabla\theta)$ is equivalent to the optimization problem (1.3.1).

The above discussion raises a question that is a main concern of this section. Given a VI (K, F) with an arbitrary vector function F and a convex set K, is this VI always the stationary point problem of some optimization problem (1.3.1) with K as the feasible set? As expected, the answer to this question is negative (otherwise, there would be no need for a comprehensive study of the VI and this book would not have been written.) A related question can be asked in terms of the function F; namely, when is a vector function F a gradient map? In general, a vector function F defined on an open subset of $\mathbb{R}^n$ and having values in $\mathbb{R}^n$ is called a *gradient map* if there exists a scalar function θ such that $F(x) = \nabla\theta(x)$ for all x in the domain of F. It turns out that a complete answer to the latter question can be obtained. In particular, this gradient condition is equivalent to the "integrability" of F on the domain in question. Specifically, the function

F is said to be *integrable* on a domain $U \subseteq \mathbb{R}^n$ if for any two vectors x and y, the line integral of F from x to y is independent of any piecewise smooth path in U that connects x to y.

The following theorem is classical. Essentially, it shows that the three concepts: gradient map, integrability, and symmetry, are all equivalent.

1.3.1 Theorem. Let $F : U \to \mathbb{R}^n$ be continuously differentiable on the open convex set $U \subseteq \mathbb{R}^n$. The following three statements are equivalent:

(a) there exists a real-valued function θ such that $F(x) = \nabla\theta(x)$ for all $x \in U$;

(b) the Jacobian matrix $JF(x)$ is symmetric for all $x \in U$;

(c) F is integrable on U.

If any one of these statements holds, then the desired scalar function θ is given by

$$\theta(x) \equiv \int_0^1 F(x^0 + t(x - x^0))^T (x - x^0)\, dt$$

for an arbitrary vector x^0 in U. □

Condition (b) in the above theorem is known as the *symmetry condition* and (c) is known as the *integrability condition*. Roughly speaking, Theorem **1.3.1** asserts that a vector function is a gradient map if and only if it is integrable and this is further equivalent to the Jacobian matrix of the given function being symmetric at all points in the domain in question. An important class of integrable functions consists of the *separable functions*. These are continuous functions $F(x)$ such that each component function $F_i(x)$ depends only on the single variable x_i; that is, $F(x) = (F_i(x_i) : i = 1, \ldots, n)$. If $F(x)$ is separable and differentiable, then the Jacobian matrix $JF(x)$ is a diagonal matrix for all x in the domain of differentiability of F.

The upshot of Theorem **1.3.1** is that if F is a gradient map defined on an open convex superset of the convex set K, then the VI (K, F) is the stationary point problem of the optimization problem (1.3.1), where $F(x) = \nabla\theta(x)$. To illustrate this conclusion, consider the affine case where F is given by (1.1.6). It follows that F is a gradient map on $\mathbb{R}^n$ if and only if M is a symmetric matrix; in this case, the VI (K, q, M) is the stationary point problem of the following optimization problem:

$$\begin{array}{ll} \text{minimize} & q^T x + \tfrac{1}{2} x^T M x \\ \text{subject to} & x \in K, \end{array} \tag{1.3.2}$$

which is a quadratic program (QP) when K is a polyhedron. This connection breaks down if M is not symmetric. Indeed, if M is not symmetric, then the stationary point problem of (1.3.2) is the VI (K, q, M_s) where M_s is the symmetric part of M; that is

$$M_s \equiv \tfrac{1}{2}(M + M^T).$$

Returning to the optimization problem (1.3.1), we consider the case where K is not convex. In this nonconvex case, one can still obtain a first-order necessary condition for a local minimizer of (1.3.1), but this no longer corresponds to a VI; see (1.3.3). However, under some suitable assumptions, a primal-dual necessary condition for optimality can be obtained that corresponds to an MiCP. The analysis that follows deals with these issues. The results we present are standard, nevertheless the discussion will give us the opportunity to introduce some important concepts that are used widely in this book.

Let us define the *tangent cone* of K at a vector $x \in K$; this cone, denoted $\mathcal{T}(x; K)$, consists of all vectors $d \in \mathbb{R}^n$, called *tangent vectors* to K at x, for which there exist a sequence of vectors $\{y^\nu\} \subset K$ and a sequence of positive scalars $\{\tau_\nu\}$ such that

$$\lim_{\nu \to \infty} y^\nu = x, \quad \lim_{\nu \to \infty} \tau_\nu = 0, \quad \text{and} \quad \lim_{\nu \to \infty} \frac{y^\nu - x}{\tau_\nu} = d.$$

Figure 1.3 illustrates the tangent cone at a point of a nonconvex set.

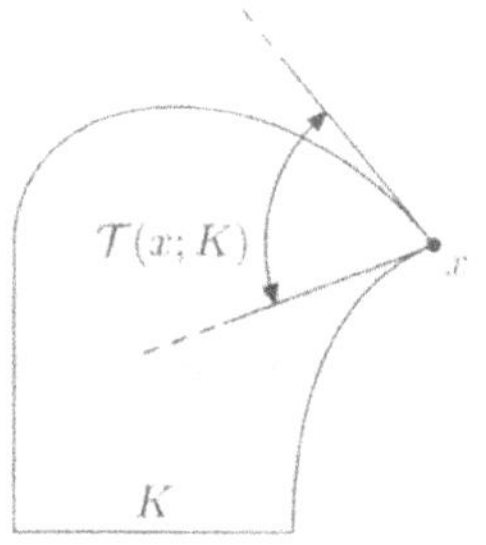

Figure 1.3: The tangent cone.

It is not hard to show that if x is a local minimizer of (1.3.1), then

$$(y - x)^T \nabla\theta(x) \geq 0, \quad \forall\, y \in x + \mathcal{T}(x; K). \tag{1.3.3}$$

In general, a vector $x \in K$ satisfying this inequality is called a *stationary point* of the (nonconvex) program (1.3.1). This is a *primal* description of a stationary point, as opposed to a primal-dual description that requires a special representation of the set K; see the next subsection.

The inequality (1.3.3) is an instance of a *quasi-variational inequality*, abbreviated as QVI. The QVI is an extension of a VI in which the defining set of the problem varies with the variable. Formally, the QVI is defined as follows. Let K be a *point-to-set* mapping from $\mathbb{R}^n$ into subsets of $\mathbb{R}^n$; that is, for every $x \in \mathbb{R}^n$, $K(x)$ is a (possibly empty) subset of $\mathbb{R}^n$. Let F be a (point-to-point) mapping from $\mathbb{R}^n$ into itself. The QVI defined by the pair (K, F) is to find a vector $x \in K(x)$ such that

$$(y - x)^T F(x) \geq 0, \quad \forall y \in K(x).$$

Notice that if $F(x)$ is identically equal to zero, the QVI reduces to the problem of finding a vector x satisfying $x \in K(x)$. Such a vector is called a *fixed point* of the point-to-set map K. See Subsection **2.1.4** for a review of point-to-set maps, including a renowned theorem that provides sufficient conditions for such a map to have a fixed point.

Thus, in addition to being an extension of the VI, the QVI also includes the classical problem of finding a fixed point of a point-to-set map as a special case. More discussion of this extended problem can be found at the end of Subsection **1.4.2**; see also Section **1.6**.

Like the normal cone, the tangent cone is another object that plays an important role in convex analysis and nonlinear programming. In the case of a convex set K, these two cones are "polar" of each other as stated in the following result.

1.3.2 Proposition. Let K be a convex subset of $\mathbb{R}^n$ and let $x \in K$ be arbitrary. It holds that

$$\mathcal{T}(x; K)^* = -\mathcal{N}(x; K).$$

Proof. Let d be an arbitrary vector belonging to the dual of the tangent cone. Let $y \in K$ be arbitrary. Since K is convex, the vector $y - x$ is a tangent vector to K at x. Therefore, $0 \leq d^T(y - x)$, which shows that $-d$ is a normal vector to K at x. The converse can be proved easily, and in fact, without even requiring the convexity of K. □

1.3.3 Remark. As we see from the above proof, the convexity of K is needed in proving one inclusion in Proposition **1.3.2**. Whereas the above definition of the tangent cone $\mathcal{T}(x; K)$ does not require K to be convex, the definition of the normal cone $\mathcal{N}(x; K)$ is tailored to a convex K; see (1.1.2). There exist various definitions of the normal cone for a nonconvex set K; these generalized definitions are beyond the scope of this book. □

Without additional description of a nonconvex set K, it is in general not easy to deal with the cone $\mathcal{T}(x; K)$. In the important case of a standard NLP , that is (1.3.1) where K is represented by finitely many differentiable inequalities and equations:

$$K \equiv \{ x \in \mathbb{R}^n : h(x) = 0, \; g(x) \leq 0 \}, \tag{1.3.4}$$

with $h : \mathbb{R}^n \to \mathbb{R}^\ell$ and $g : \mathbb{R}^n \to \mathbb{R}^m$ being vector-valued continuously differentiable functions, there are many well-known conditions on the constraint functions g and h under which $\mathcal{T}(x; K)$ becomes a polyhedral cone. More importantly, the stationarity condition (1.3.3) then has an equivalent "primal-dual" description that is very useful for both analytical and practical purposes. These conditions on the constraints are known as *constraint qualifications* (CQs) ; one of the most general of these CQs is Abadie's CQ, which simply postulates that $\mathcal{T}(x; K)$ is equal to the *linearization cone* of K at x defined as:

$$\begin{aligned} \mathcal{L}(x; K) \equiv \{ v \in \mathbb{R}^n : \quad & v^T \nabla h_j(x) = 0, \quad \forall j = 1, \ldots, \ell \\ & v^T \nabla g_i(x) \leq 0, \quad \forall i \in \mathcal{I}(x) \}, \end{aligned}$$

where $\mathcal{I}(x)$ is the *active index set* at x; i.e.,

$$\mathcal{I}(x) \equiv \{ i : g_i(x) = 0 \}.$$

It is easy to show that in general for any finitely representable set K of the form (1.3.4), the tangent cone $\mathcal{T}(x; K)$ is always contained in $\mathcal{L}(x; K)$ for any $x \in K$; see Exercise **1.8.31**. Thus Abadie's CQ is equivalent to the postulate that every vector in the linearization cone is a tangent vector. A further discussion of the Abadie CQ is given in the cited exercise. See Section **3.2** for a detailed review of various CQs.

The equality between $\mathcal{T}(x; K)$ and $\mathcal{L}(x; K)$ implies in particular that the former cone is polyhedral; it further implies, by linear programming duality, that

$$\begin{aligned} \mathcal{T}(x; K)^* &= \mathcal{L}(x; K)^* \\ &= \Big\{ v \in \mathbb{R}^n : \exists \, (\mu, \lambda_{\mathcal{I}(x)}) \text{ with } \lambda_{\mathcal{I}(x)} \geq 0 \text{ such that} \\ & \qquad 0 = v + \sum_{j=1}^{\ell} \mu_j \nabla h_j(x) + \sum_{i \in \mathcal{I}(x)} \lambda_i \nabla g_i(x) \Big\}. \end{aligned}$$

By Proposition **1.3.2**, this gives a polyhedral representation of $\mathcal{N}(x; K)$ if K is also convex. Assuming that Abadie's CQ holds at the vector $x \in K$, we can proceed to derive the equivalent primal-dual formulation of (1.3.3). Under this CQ, the condition (1.3.3) says that $\nabla\theta(x)$ belongs to the dual cone of $\mathcal{L}(x; K)$. Thus, (1.3.3) holds if and only if there exist constraint multipliers (or dual variables) $\mu \in \mathbb{R}^\ell$ and $\lambda \in \mathbb{R}^m$ such that the *Karush-Kuhn-Tucker (KKT) system* below is valid:

$$\begin{aligned} 0 &= \nabla\theta(x) + \sum_{j=1}^{\ell} \mu_j \nabla h_j(x) + \sum_{i=1}^{m} \lambda_i \nabla g_i(x) \\ 0 &= h(x) \\ 0 \leq \lambda &\perp g(x) \leq 0. \end{aligned}$$

We easily recognize the latter system as a MiCP in the variables (x, μ, λ). This KKT system is the primal-dual description of a stationary point of (1.3.1), primal-dual in the sense that the system contains both the primal variable x and the "dual" variables (or multipliers) (μ, λ) .

The above derivation of the KKT system of the nonlinear program (1.3.1) under Abadie's CQ generalizes the proof of Proposition **1.2.1**. In the latter proposition, the polyhedrality of the set K itself provides a sufficient condition for Abadie's CQ to hold.

1.3.2 The Karush-Kuhn-Tucker system

In this subsection, we extend the KKT conditions to the VI. Similar to the linearly constrained VI, we can easily establish the following proposition, part of whose proof simply consists of replacing $\nabla\theta(x)$ in the above derivation by an arbitrary vector function $F(x)$.

1.3.4 Proposition. Let K be given by (1.3.4) where the functions h_j and g_i are continuously differentiable. Let F be a mapping from K into $\mathbb{R}^n$. The following two statements are valid.

(a) Let $x \in \text{SOL}(K, F)$. If Abadie's CQ holds at x, then there exist vectors $\mu \in \mathbb{R}^\ell$ and $\lambda \in \mathbb{R}^m$ such that

$$\begin{aligned} 0 &= F(x) + \sum_{j=1}^{\ell} \mu_j \nabla h_j(x) + \sum_{i=1}^{m} \lambda_i \nabla g_i(x) \\ 0 &= h(x) \\ 0 \leq \lambda &\perp g(x) \leq 0. \end{aligned} \tag{1.3.5}$$

(b) Conversely, if each function h_j is affine and each function g_i is convex, and if (x, μ, λ) satisfies (1.3.5), then x solves the VI (K, F).

Proof. To prove (a) we note that if $x \in \mathrm{SOL}(K, F)$, then x solves the following nonlinear program in the variable y:

$$\begin{array}{ll} \text{minimize} & y^T F(x) \\ \text{subject to} & y \in K. \end{array} \tag{1.3.6}$$

Since (1.3.5) is precisely the KKT system of this nonlinear program, (a) follows readily. Conversely, if the functions h_j and g_i are affine and convex respectively, then (1.3.6) is a convex program in the variable y, so every KKT point of this program is a global solution (that is not necessarily unique). Thus x must solve the VI (K, F). □

The MiCP displayed in the above proposition is called the *KKT system* of the VI (K, F) with a "finitely representable" set K, i.e., with K as described by (1.3.4). The KKT system represents a conversion of the VI into an MiCP; this conversion gives a further example of transformations between problem classes besides those considered in the last section. Figure 1.4 illustrates the KKT conditions of a VI.

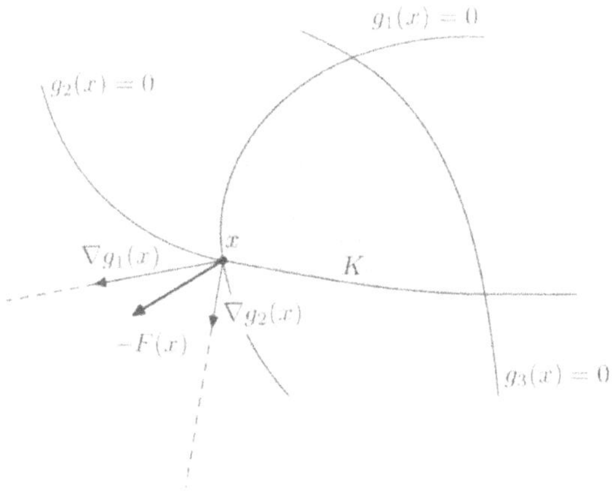

Figure 1.4: The KKT conditions of a VI.

It is useful to introduce the vector function:

$$\mathbf{L}(x, \mu, \lambda) \equiv F(x) + \sum_{j=1}^{\ell} \mu_j \nabla h_j(x) + \sum_{i=1}^{m} \lambda_i \nabla g_i(x), \quad \forall (x, \mu, \lambda) \in \mathbb{R}^{n+\ell+m},$$

which we call the (vector) *Lagrangian function* of the VI (K, F). This terminology is borrowed from the case where F is a gradient map, say $F(x) = \nabla\theta(x)$; in the latter case, we have

$$\mathbf{L}(x, \mu, \lambda) = \nabla_x L(x, \mu, \lambda),$$

where

$$L(x,\mu,\lambda) \equiv \theta(x) + \mu^T h(x) + \lambda^T g(x) \tag{1.3.7}$$

is the familiar *scalar Lagrangian function* of the differentiable NLP:

$$\begin{array}{ll} \text{minimize} & \theta(x) \\ \text{subject to} & h(x) = 0 \\ \text{and} & g(x) \leq 0. \end{array}$$

The important thing to distinguish is that for a VI (K,F), the Lagrangian function is vector valued (just like the defining function F), whereas for an NLP, the Lagrangian function is scalar valued (just like the objective function θ).

The function defining the MiCP (1.3.5) is a generalization of the function given by (1.2.5) for a non-polyhedral K; in terms of the Lagrangian function $\mathbf{L}$ for the VI (K,F), the defining function for the KKT system (1.3.5) is:

$$\tilde{F}(x,\mu,\lambda) \equiv \begin{pmatrix} \mathbf{L}(x,\mu,\lambda) \\ h(x) \\ -g(x) \end{pmatrix}, \quad \forall\,(x,\mu,\lambda) \in \mathbb{R}^{n+\ell+m}. \tag{1.3.8}$$

As an MiCP, the system (1.3.5) is well defined for all differentiable functions h and g. This system provides a set of necessary conditions that must be satisfied by all solutions to the VI that obey the Abadie CQ; conversely these conditions yield solutions to the VI if the functions h_j are affine and g_i are convex (the set K must be convex in this case). Occasionally, we work with the system (1.3.5) itself as an instance of an MiCP, without regard to whether its solutions yield solutions to the VI.

We call a triple (x,μ,λ) satisfying the system (1.3.5) a *KKT triple* of the VI (K,F), and x a *KKT point*. For any such point x, let $\mathcal{M}(x)$ denote the set of pairs $(\mu,\lambda) \in \mathbb{R}^{\ell+m}$ such that (x,μ,λ) is a KKT triple. If $x \in \mathrm{SOL}(K,F)$, the set $\mathcal{M}(x)$ obviously has an important role to play in various properties of x; for instance, in the sensitivity analysis of x as the "data" (i.e., the pair (K,F)) is slightly perturbed. In general, $\mathcal{M}(x)$ is a polyhedral convex set in $\mathbb{R}^n$ parameterized by x. Several fundamental properties of $\mathcal{M}(x)$ turn out to be equivalent to various important CQs.

1.4 Source Problems

In addition to optimization problems, many equilibrium problems from economics and important applied problems from diverse engineering fields can be profitably formulated as VIs and CPs. Several of these source problems have provided the initial impetus for the comprehensive study of the VI/CP. Starting in the next subsection, we present a fairly extensive documentation of numerous applied problems and their VI/CP formulations.

1.4.1 Saddle problems

As an extension of an optimization problem, a saddle problem is defined by a scalar function of two arguments (called a saddle function in this context) and two subsets of two possibly different Euclidean spaces. The prime example of a saddle function is the scalar Lagrangian function $L(x, \mu, \lambda)$ with the primal variable x as one argument and the dual pair (μ, λ) as the other argument; see (1.3.7). Indeed, the initial study of saddle problems was largely inspired by NLP duality theory. However, saddle problems have their own significance in modeling certain extended optimization problems.

Let $L : \mathbb{R}^{n+m} \to \mathbb{R}$ denote an arbitrary saddle function; let $X \subseteq \mathbb{R}^n$ and $Y \subseteq \mathbb{R}^m$ be two given closed sets. The saddle problem associated with this triple (L, X, Y) is to find a pair of vectors $(x, y) \in X \times Y$, called a *saddle point*, such that

$$L(x, v) \le L(x, y) \le L(u, y), \quad \forall (u, v) \in X \times Y. \tag{1.4.1}$$

When L is continuously differentiable and "convex-concave", and X and Y are convex sets, the saddle problem can be formulated as a VI. Specifically, we say that $L(x, y)$ is *convex-concave* if $L(\cdot, y)$ is convex for each fixed but arbitrary $y \in Y$ and $L(x, \cdot)$ is concave for each fixed but arbitrary $x \in X$. The second inequality in (1.4.1) says that x is a global minimum of the function $L(\cdot, y)$ on the set X; similarly, the first inequality says that y is a global maximum of $L(x, \cdot)$ on Y. Invoking the respective minimum and maximum principle for these problems, we deduce that if L is convex-concave and X and Y are closed convex sets, then (x, y) is a saddle point if and only if (x, y) solves the VI $(X \times Y, F)$ where

$$F(u, v) \equiv \begin{pmatrix} \nabla_u L(u, v) \\ \\ -\nabla_v L(u, v) \end{pmatrix}, \quad (u, v) \in \mathbb{R}^{n+m}. \tag{1.4.2}$$

A particularly important case of the saddle problem is when X and Y are polyhedral sets and L is a quadratic function:

$$L(x, y) = p^T x + q^T y + \tfrac{1}{2} x^T P x + x^T R y - \tfrac{1}{2} y^T Q y, \quad (x, y) \in \mathbb{R}^{n+m},$$

for some vectors $p \in \mathbb{R}^n$ and $q \in \mathbb{R}^m$ and matrices $P \in \mathbb{R}^{n \times n}$, $R \in \mathbb{R}^{n \times m}$ and $Q \in \mathbb{R}^{m \times m}$, with P and Q being symmetric positive semidefinite. The resulting saddle problem is called a *linear-quadratic program*.

In general, if the saddle function $L(u,v)$ is twice continuously differentiable (so that $\partial^2 L(u,v)/\partial u \partial v = \partial^2 L(u,v)/\partial v \partial u$), then the function $F(u,v)$ defined by (1.4.2) is continuously differentiable; more interestingly, its Jacobian matrix is given by

$$JF(u,v) = \begin{bmatrix} \nabla^2_{uu} L(u,v) & \nabla^2_{uv} L(u,v) \\ -\nabla^2_{uv} L(u,v)^T & -\nabla^2_{vv} L(u,v) \end{bmatrix}.$$

Since the two diagonal submatrices $\nabla^2_{uu} L(u,v)$ and $\nabla^2_{vv} L(u,v)$ are symmetric and the two off-diagonal submatrices $\nabla^2_{uv} L(u,v)$ and $-\nabla^2_{uv} L(u,v)^T$ are negative transpose of each other, it follows that $JF(u,v)$ is *bisymmetric*; this terminology originates from LCP theory and is used to describe a partitioned matrix with the kind of symmetry structure as $JF(u,v)$. This bisymmetry property of $JF(u,v)$ makes it clear that the saddle problem is not the stationary point problem of an optimization problem with feasible set $X \times Y$. If L is in addition convex-concave, then $JF(u,v)$ is a positive semidefinite, albeit asymmetric, matrix.

Associated with every saddle problem (L, X, Y) is a pair of optimization problems:

$$\left.\begin{array}{ll} \text{minimize} & \varphi(x) \\ \text{subject to} & x \in X \end{array}\right\} \quad \text{and} \quad \left\{\begin{array}{ll} \text{maximize} & \psi(y) \\ \text{subject to} & y \in Y, \end{array}\right. \tag{1.4.3}$$

where

$$\varphi(x) \equiv \sup\{\, L(x,v) \,:\, v \in Y \,\} \quad \text{and} \quad \psi(y) \equiv \inf\{\, L(u,y) \,:\, u \in X \,\}$$

are possibly extended-valued functions; that is, it is possible for $\varphi(x)$ to equal to ∞ for some $x \in X$ and $\psi(y)$ to equal to $-\infty$ for some $y \in Y$. Substituting the definitions of $\varphi(x)$ and $\psi(y)$ into (1.4.3), we can write this pair of problems as a *minimax* and a *maximin* problem, respectively,

$$\inf_{x \in X} \sup_{y \in Y} L(x,y) \quad \text{and} \quad \sup_{y \in Y} \inf_{x \in X} L(x,y).$$

The relationship between these optimization problems and the saddle problem is as follows.

1.4.1 Theorem. Let $L : X \times Y \subseteq \mathbb{R}^{n+m} \to \mathbb{R}$ be a given function. It holds that

$$\inf_{x \in X} \sup_{y \in Y} L(x,y) \geq \sup_{y \in Y} \inf_{x \in X} L(x,y). \tag{1.4.4}$$

Moreover, for a given pair $(\bar{x}, \bar{y}) \in X \times Y$, the following three statements are equivalent.

(a) $(\bar{x}, \bar{y})$ is a saddle point of L on $X \times Y$.

(b) $\bar{x}$ is a minimizer of $\varphi(x)$ on X, $\bar{y}$ is a maximizer of $\psi(y)$ on Y, and equality holds in (1.4.4).

(c) $\varphi(\bar{x}) = \psi(\bar{y}) = L(\bar{x}, \bar{y})$.

Proof. The inequality (1.4.4) is trivial to prove. We show the equivalence of (a), (b), and (c).

(a) $\Rightarrow$ (b). Suppose $(\bar{x}, \bar{y})$ is a saddle point of L on $X \times Y$. For all $(x, y) \in X \times Y$, we have

$$L(\bar{x}, y) \leq L(\bar{x}, \bar{y}) \leq L(x, \bar{y}), \tag{1.4.5}$$

From this inequality, it is not difficult to establish

$$\inf_{x \in X} \sup_{y \in Y} L(x, y) \leq \varphi(\bar{x}) \leq L(\bar{x}, \bar{y}) \leq \psi(\bar{y}) \leq \sup_{y \in Y} \inf_{x \in X} L(x, y). \tag{1.4.6}$$

Indeed, the first and last inequality are obvious, whereas the middle two inequalities are the consequence of the saddle condition (1.4.5). Thus equalities hold throughout (1.4.6). This also establishes that $\bar{x}$ minimizes $\varphi(x)$ on X and $\bar{y}$ maximizes $\psi(y)$ on Y.

(b) $\Rightarrow$ (c). If (b) holds, then

$$\varphi(\bar{x}) = \inf_{x \in X} \sup_{y \in Y} L(x, y) = \sup_{y \in Y} \inf_{x \in X} L(x, y) = \psi(\bar{y}).$$

Moreover, we have

$$\varphi(\bar{x}) \geq L(\bar{x}, \bar{y}) \geq \psi(\bar{y}).$$

Thus (c) follows.

(c) $\Rightarrow$ (a). If (c) holds, then we have

$$L(\bar{x}, \bar{y}) = \varphi(\bar{x}) \equiv \sup_{y \in Y} L(\bar{x}, y),$$

which establishes the left-hand inequality in (1.4.5). The right-hand inequality follows similarly. $\square$

Theorem **1.4.1** shows that the saddle function L is equal to a constant on the set of saddle points of the triple (L, X, Y). Unlike the optimal objective value of an optimization problem, this conclusion is not immediately obvious but is nevertheless demonstrated with the help of the pair of auxiliary optimization problems (1.4.3). The above theorem does not address the issue of the existence of a saddle point and of the solvability of the associated minimax and maximin optimization problems. These existential results are the subject of the next chapter; see Corollary **2.2.10**.

1.4.2 Nash equilibrium problems

A fundamental concept of equilibrium in noncooperative game theory was introduced by J. Nash who was awarded the Nobel Prize in Economics Sciences in 1994 for this contribution. It turns out that the computation of a "Nash equilibrium" can be accomplished by solving a variational inequality. In what follows, we explain the VI approach to the computation of Nash equilibria.

In a general noncooperative game, there are N players each of whom has a certain cost function and strategy set that may depend on the other players' actions. For the initial discussion, we assume that player i's strategy set is K_i, which is a subset of $\mathbb{R}^{n_i}$ and is independent of the other players' strategies. Player i's cost function $\theta_i(\mathbf{x})$ depends on all players' strategies, which are described by the vector $\mathbf{x}$ that consists of the subvectors $x^i \in \mathbb{R}^{n_i}$ for $i = 1, \ldots, N$. Player i's problem is to determine, for each fixed but arbitrary tuple $\tilde{\mathbf{x}}^i \equiv (x^j : j \neq i)$ of other players' strategies, an optimal strategy x^i that solves the cost minimization problem in the variable y^i:

$$\begin{array}{ll} \text{minimize} & \theta_i(y^i, \tilde{\mathbf{x}}^i) \\ \text{subject to} & y^i \in K_i. \end{array}$$

We denote the solution set of this optimization problem by $S_i(\tilde{\mathbf{x}}^i)$; note the dependence of this set on the tuple $\tilde{\mathbf{x}}^i$. A slight abuse of notation occurs in the objective function $\theta_i(y^i, \tilde{\mathbf{x}}^i)$; it is understood that this means the function θ_i evaluated at the vector whose j-th subvector is x^j for $j \neq i$ and whose i-th subvector is y^i.

A *Nash equilibrium* is a tuple of strategies $\mathbf{x} = (x^i : i = 1, \ldots, N)$ with the property that for each i, $x^i \in S(\tilde{\mathbf{x}}^i)$. In words, a Nash equilibrium is a tuple of strategies, one for each player, such that no player can lower the cost by unilaterally deviating his action from his designated strategy. The following result gives a set of sufficient conditions under which a Nash equilibrium can be obtained by solving a VI.

1.4.2 Proposition. Let each K_i be a closed convex subset of $\mathbb{R}^{n_i}$. Suppose that for each fixed tuple $\tilde{\mathbf{x}}^i$, the function $\theta_i(y^i, \tilde{\mathbf{x}}^i)$ is convex and continuously differentiable in y^i. Then a tuple $\mathbf{x} \equiv (x^i : i = 1, \ldots, N)$ is a Nash equilibrium if and only if $\mathbf{x} \in \text{SOL}(\mathbf{K}, \mathbf{F})$, where

$$\mathbf{K} \equiv \prod_{i=1}^{N} K_i \quad \text{and} \quad \mathbf{F}(\mathbf{x}) \equiv \left(\nabla_{x^i} \theta_i(\mathbf{x}) \right)_{i=1}^{N}.$$

Proof. By convexity and the minimum principle, we know that $\mathbf{x}$ is a

Nash equilibrium if and only if for each $i = 1, \ldots, N$,

$$(y^i - x^i)^T \nabla_{x^i} \theta_i(\mathbf{x}) \geq 0, \quad \forall y^i \in K_i. \tag{1.4.7}$$

Thus, if $\mathbf{x}$ is a Nash equilibrium, then by concatenating these individual VIs, it follows easily that $\mathbf{x}$ must solve the prescribed VI.

Conversely, if $\mathbf{x} \equiv (x^i : i = 1, \ldots, N)$ solves the VI $(\mathbf{K}, \mathbf{F})$, then

$$(\mathbf{y} - \mathbf{x})^T \mathbf{F}(\mathbf{x}) \geq 0, \quad \forall \mathbf{y} \in \mathbf{K}.$$

In particular, for each $i = 1, \ldots, N$, let $\mathbf{y}$ be the tuple whose j-th subvector is equal to x^j for $j \neq i$ and i-th subvector is equal to y^i, where y^i is an arbitrary element of the set K^i. The above inequality then becomes (1.4.7). □

It is worth noting that the defining set $\mathbf{K}$ of the VI in Proposition **1.4.2** is the Cartesian product of sets of lower dimensions. The Cartesian product structure of the defining set of a VI is also present in the saddle problem (see the previous subsection). Mathematically, the saddle problem is a special case of the Nash problem with $N = 2$ and $\theta_1(\mathbf{x}) = -\theta_2(\mathbf{x})$. In game theoretic terminology, the saddle problem is a "two-person zero-sum" game, meaning that the sum of the two players' cost functions is equal to zero; that is to say, one player's gain is equal to the other player's loss. More generally, the Nash equilibrium problem is sometimes called an "N-person nonzero-sum game".

In an extension of the above Nash game, each player i's strategy set $K_i(\tilde{\mathbf{x}}^i)$ can depend on other players' strategies. In this generalized context, the resulting VI is of the QVI type. More precisely, if F is defined as in Proposition **1.4.2**, and

$$\mathbf{K}(\mathbf{x}) \equiv \prod_{i=1}^{N} K_i(\tilde{\mathbf{x}}^i),$$

then the generalized Nash game can be formulated as the QVI $(\mathbf{K}, \mathbf{F})$. When each $K^i(\tilde{\mathbf{x}}^i)$ is defined by differentiable inequalities and equalities satisfying appropriate CQs, it is possible to formulate a MiCP that provides a set of necessary conditions which must be satisfied by all solutions of the QVI $(\mathbf{K}, \mathbf{F})$, as in the KKT system of a VI. Specifically, suppose that for each i:

$$K_i(\tilde{\mathbf{x}}^i) \equiv \{ x^i \in \mathbb{R}^{n_i} : g^i(\mathbf{x}) \leq 0, \; h^i(\mathbf{x}) = 0 \}$$

where each g^i and each h^j are given vector-valued functions. Then under a CQ at a solution of player i's problem, we can write down the KKT system for this player's optimization problem:

$$\begin{array}{ll} \text{minimize} & \theta_i(y^i, \tilde{\mathbf{x}}^i) \\ \text{subject to} & y^i \in K_i(\tilde{\mathbf{x}}^i), \end{array}$$

obtaining for $i = 1, \ldots, N$,

$$\begin{aligned} \nabla_{x^i}\theta(\mathbf{x}) + J_{x^i}h^i(\mathbf{x})^T\mu^i + J_{x^i}g^i(\mathbf{x})^T\lambda^i &= 0 \\ 0 &= h^i(\mathbf{x}) \\ 0 \leq \lambda^i \perp g^i(\mathbf{x}) &\leq 0, \end{aligned}$$

where the subscript x^i denotes the partial differentiation with respect to this variable. Concatenating the N KKT systems of all the players' problems results in the MiCP formulation for the generalized Nash game; see the proof of Proposition **1.4.2** for details of how this last step can be accomplished.

As generalized Nash games, QVIs have applications in various gaming contexts; they also arise from the discretization of the so-called impulse control problems introduced by Bensoussan and Lions. The latter control problems have found an increasing use in various modeling contexts. In Subsection **1.4.3**, we present an oligopolistic electricity model that has a natural QVI formulation but can be reformulated as a VI, due to the special structure of the players' individual strategy sets.

1.4.3 Nash-Cournot production/distribution

Among many practical applications of the Nash equilibrium concept, the Nash-Cournot production/distribution problem is worth mentioning. In the latter problem, a common homogeneous commodity is being produced by several producers (firms) who are the Nash players. Each firm attempts to maximize its profit by solving an optimization problem to determine the production and distribution quantities of the commodity, under the presumption that the production and distribution of the other firms are parametric inputs. In this context, a Nash equilibrium is a production/distribution pattern in which no firm can increase its profit by unilaterally changing its controlled variables; thus under this equilibrium concept, each firm determines its best response given other firms' actions.

In what follows, we present a Nash-Cournot production/distribution problem in a spatially separated market modeled by a network with node set $\mathcal{N}$ and arc set $\mathcal{A} \subseteq \mathcal{N} \times \mathcal{N}$. There are M producers of the commodity; producer f owns production plants at the subset of nodes $\mathcal{N}_f \subset \mathcal{N}$. Let

x_{fa} denote the commodity flow controlled by firm f from node $i \in \mathcal{N}$ to an adjacent node $j \in \mathcal{N}$ via the link $a = (i,j) \in \mathcal{A}$. Let s_{fi} be the total amount produced at firm f's plant $i \in \mathcal{N}_f$; the production cost there is given by the function $C_{fi}(s_{fi})$ and the production capacity is a given constant CAP_{fi}. Similarly, let d_{fj} be the total amount produced by firm f delivered to site $j \in \mathcal{N}$. The total amount of the commodity delivered to node j by all firms is thus

$$Q_j \equiv \sum_{f=1}^{M} d_{fj}.$$

The unit purchase price of the commodity at site $j \in \mathcal{N}$ is given by the inverse demand function $p_j(Q_j)$, which depends on the total delivery by all firms. The unit shipment cost incurred by firm f on arc $a = (i,j) \in \mathcal{A}$ depends on the quantity shipped and is denoted $c_{fa}(x_{fa})$. For each node $i \in \mathcal{N}$, let $\mathcal{A}_i^+$ and $\mathcal{A}_i^-$ denote, respectively, the set of arcs $a \in \mathcal{A}$ with i as the beginning and ending node.

Each firm f wishes to determine the following variables:

$$\{ d_{fj} : j \in \mathcal{N} \}, \quad \{ s_{fi} : i \in \mathcal{N}_f \}, \quad \text{and} \quad \{ x_{fa} : a \equiv (i,j) \in \mathcal{A} \}.$$

We denote the above variables controlled by firm f collectively as x^f. The constraints that must be satisfied by x^f define the firm f's feasible set of production/distribution patterns:

$$\begin{aligned} K_f \equiv \Big\{ x^f \geq 0 : \quad & s_{fi} \leq \mathrm{CAP}_{fi}, \quad \forall i \in \mathcal{N}_f \\ & d_{fi} + \sum_{a \in \mathcal{A}_i^+} x_{fa} = s_{fi} + \sum_{a \in \mathcal{A}_i^-} x_{fa}, \quad \forall i \in \mathcal{N}_f \\ & d_{fi} + \sum_{a \in \mathcal{A}_i^+} x_{fa} = \sum_{a \in \mathcal{A}_i^-} x_{fa}, \quad \forall i \in \mathcal{N} \setminus \mathcal{N}_f \Big\}. \end{aligned}$$

Let $\mathbf{x} \equiv (x^f : f = 1, \ldots, M)$ be the vector of all firms' decision variables. Firm f's profit function is equal to its revenue less costs:

$$\theta_f(\mathbf{x}) \equiv \sum_{j \in \mathcal{N}} d_{fj}\, p_j \left(\sum_{g=1}^{M} d_{gj} \right) - \sum_{i \in \mathcal{N}_f} C_{fi}(s_{fi}) - \sum_{a \in \mathcal{A}} x_{fa}\, c_a(x_{fa});$$

note that this function depends on other firms' strategies only through the demand variables d_{gj} for $g \neq f$. Firm f's profit maximization problem is: with x^g for $g \neq f$ fixed, compute x^f in order to

$$\begin{aligned} \text{maximize} \quad & \theta_f(\mathbf{x}) \\ \text{subject to} \quad & x^f \in K_f. \end{aligned} \tag{1.4.8}$$

As a Nash model, the overall equilibrium problem is to seek a partitioned vector $\mathbf{x} \equiv (x^f)$ such that each x^f solves (1.4.8), which is firm f's profit maximization problem.

As discussed previously, in order for this Nash problem to be formulated as a VI, we need each function $\theta_f(\mathbf{x})$ to be concave in the variable x^f. It is not difficult to show that this holds under some fairly standard assumptions in economics, such as (a) each inverse demand function p_j is decreasing, (b) the "industry revenue function" $Q_j p_j(Q_j)$ is concave in Q_j, (c) each cost function C_{fi} is convex, and (d) each transportation cost function $x_{fa} c_a(x_{fa})$ is convex in x_{fa}. Under these assumptions, we therefore obtain a VI formulation for the Nash-Cournot production/distribution problem.

It is interesting to examine the resulting function $\mathbf{F}(\mathbf{x})$ that defines the VI of the above Nash-Cournot problem, particularly the Jacobian matrix of $\mathbf{F}(\mathbf{x})$. In what follows, we assume that the functions p_j, C_{fi} and c_a are all twice continuously differentiable. Note that $\mathbf{F}(\mathbf{x})$ is the concatenation of $-\nabla_{x^f} \theta_f(\mathbf{x})$ for $f = 1, \ldots, M$. Since the last two summations in $\theta_f(\mathbf{x})$ are fairly simple, we focus on the first summation in $\theta_f(\mathbf{x})$ that pertains to firm f's revenue. It is not difficult to derive the following expression:

$$\frac{\partial^2 \theta_f(\mathbf{x})}{\partial d_{f'j'} \partial d_{fj}} = \begin{cases} 0 & \text{if } j' \neq j \\ p_j'(Q_j) + d_{fj}\, p_j''(Q_j) & \text{if } j' = j \text{ and } f' \neq f \\ 2p_j'(Q_j) + d_{fj}\, p_j''(Q_j) & \text{if } j' = j \text{ and } f' = f. \end{cases}$$

Based on this expression, we see that $J\mathbf{F}(\mathbf{x})$ is a block partitioned matrix $(J_{f,g}\mathbf{F}(\mathbf{x}) : f, g = 1, \ldots, M)$, where each diagonal block $J_{f,f}\mathbf{F}(\mathbf{x})$ is a diagonal matrix, which we can write in the form:

$$J_{f,f}\mathbf{F}(\mathbf{x}) = \begin{bmatrix} J_{fd}\mathbf{F}(\mathbf{x}) & 0 & 0 \\ 0 & J_{fs}\mathbf{F}(\mathbf{x}) & 0 \\ 0 & 0 & J_{fx}\mathbf{F}(\mathbf{x}) \end{bmatrix}.$$

The first diagonal matrix $J_{fd}\mathbf{F}(\mathbf{x})$ is of order $|\mathcal{N}|$ with diagonal entries equal to $-2p_j'(Q_j) - d_{fj} p_j''(Q_j)$ for all $j \in \mathcal{N}$; the second diagonal matrix $J_{fs}\mathbf{F}(\mathbf{x})$ is of order $|\mathcal{N}_f|$, with diagonal entries equal to $C_{fi}''(s_{fi})$ for all $i \in \mathcal{N}_f$; the third diagonal matrix $J_{fx}\mathbf{F}(\mathbf{x})$ is of order $|\mathcal{A}|$ with diagonal entries equal to $x_{fa} c_{fa}''(x_{fa}) + 2c_{fa}'(x_{fa})$ for all $a \in \mathcal{A}$. Partitioned in the same way, each off-diagonal block $J_{f,g}\mathbf{F}(\mathbf{x})$ for $f \neq g$ is the zero matrix of order $|\mathcal{N}| + |\mathcal{N}_f| + |\mathcal{A}|$, except for the upper left block:

$$J_{f,g}\mathbf{F}(\mathbf{x}) = \begin{bmatrix} J_{fgd}\mathbf{F}(\mathbf{x}) & 0 & 0 \\ 0 & 0 & 0 \\ 0 & 0 & 0 \end{bmatrix},$$

where $J_{fgd}\mathbf{F}(\mathbf{x})$ is the diagonal matrix of order $|\mathcal{N}|$ with diagonal entries $-p_j'(Q_j) - d_{fj}p_j''(Q_j)$ for all $j \in \mathcal{N}$. Hence the entire matrix $J\mathbf{F}(\mathbf{x})$ is highly sparse.

In the special case where each demand function $p_j(Q_j)$ is linear so that its second derivative vanishes identically, we have $J_{f,g}\mathbf{F}(\mathbf{x}) = J_{g,f}\mathbf{F}(\mathbf{x})$ for all $f \neq g$. Thus $J\mathbf{F}(\mathbf{x})$ is a symmetric matrix. Hence in this case, the Nash-Cournot problem is equivalent to a concave maximization problem on the feasible set $\prod_{f=1}^{M} K_f$. In the general case where the demand functions $p_j(Q_j)$ are nonlinear, there is no natural optimization problem defined on the same feasible set whose stationary point problem is equivalent to the Nash-Cournot problem.

Oligopolistic electricity models

Oligopolistic pricing models have wide applicability in electricity power markets. Many such models are amenable to treatment by the VI methodology via certain Nash-Cournot equilibrium formulations. Roughly, the aim of such a model is to determine the amount of electricity produced by each competing firm in a spatially separated market, the flow of power from each producing plant to the demand regions, and the transmission prices through the links of the electricity network. In general, the production and consumption of electricity requires three basic steps: generation, transmission, and distribution. The resulting models differ in how these steps are being formulated and in such additional details as market conditions, regulatory rules on transmission, power loss constraints due to network resistance, presence of arbitragers and investment strategies for plant capacities. In what follows, we describe a single-period, spatially separated, oligopolistic electricity pricing model with Cournot generators and regulated transmission as a refinement/variation of the Nash-Cournot production/distribution model discussed in the previous subsection. The refinement occurs with the introduction of multiple plants owned by each firm f at each of its sites $i \in \mathcal{N}_f$ and also of capacities of the transmission links. The variation occurs in two ways: (a) the transmission prices are some exogenous functions of all flows that reflect certain regulatory rules, and (b) each firm's flow balancing equation contains the transmission flows of its competitors, due to the link capacities. In the model, the real power losses due to the network resistance and investment strategies for generator capacities are ignored.

We introduce some additional notation used in the generation and transmission steps. Let $\mathcal{F} \equiv \{1, \ldots, M\}$ be the set of generation firms. For all $f \in \mathcal{F}$, $i \in \mathcal{N}$, and $a \in \mathcal{A}$, let

$$\begin{aligned}
G_{fi} &= \text{set of generation plants owned by firm } f \text{ at node } i \in \mathcal{N}_f; \\
\text{CAP}_{fih} &= \text{generation capacity at plant } h \in G_{fi}; \\
\text{CAP}_a &= \text{transmission capacity on link } a; \\
y_{fih} &= \text{amount produced at plant } h \in G_{fi}; \\
\rho_a &= \text{transmission price through link } a.
\end{aligned}$$

Each firm f wishes to determine the following variables:

$$\{ d_{fj} : j \in \mathcal{N} \}, \quad \{ y_{fih} : i \in \mathcal{N}_f, \, h \in G_{fi} \},$$

and

$$z^f \equiv \{ x_{fa} : a \equiv (i,j) \in \mathcal{A} \}.$$

As before, we denote the above variables controlled by firm f collectively as x^f; we let $\mathbf{x} \equiv (x^f : f \in \mathcal{F})$ be the vector of all firms' decision variables. We also let $\tilde{\mathbf{x}}^f$ be the collection all firms' except firm f's decision variables. To model the regulatory rules, we let the transmission price functions $\rho_a(\mathbf{z})$ be given, where $\mathbf{z} \equiv (z^f : f \in \mathcal{F})$. The constraints that must be satisfied by x^f define the firm f's generation/transmission/distribution patterns:

$$\begin{aligned}
K_f(\tilde{\mathbf{x}}^f) \equiv \Big\{ \quad & x^f \geq 0 : y_{fih} \leq \text{CAP}_{fih}, \quad \forall\, h \in G_{fi}, \, \forall\, i \in \mathcal{N}_f \\
& d_{fi} + \sum_{a \in \mathcal{A}_i^+} x_{fa} = \sum_{h \in G_{fi}} y_{fih} + \sum_{a \in \mathcal{A}_i^-} x_{fa}, \quad \forall\, i \in \mathcal{N}_f \\
& d_{fi} + \sum_{a \in \mathcal{A}_i^+} x_{fa} = \sum_{a \in \mathcal{A}_i^-} x_{fa}, \quad \forall\, i \in \mathcal{N} \setminus \mathcal{N}_f \\
& \sum_{f' \in \mathcal{F}} x_{f'a} \leq \text{CAP}_a, \quad a \in \mathcal{A} \Big\}.
\end{aligned}$$

Notice that firm f's feasible set $K_f(\tilde{x}^f)$, which is a polyhedron, depends on other firm's decision variables through the link capacity constraint:

$$\sum_{f' \in \mathcal{F}} x_{f'a} \leq \text{CAP}_a, \quad a \in \mathcal{A}, \tag{1.4.9}$$

which is common to all the firms. Firm f's profit function is given by:

$$\theta_f(\mathbf{x}) \equiv \sum_{j\in\mathcal{N}} d_{fj}\, p_j \left(\sum_{g=1}^{M} d_{gj} \right) - \sum_{i\in\mathcal{N}_f} \sum_{h\in G_{fi}} C_{fih}(y_{fih}) - \sum_{a\in\mathcal{A}} x_{fa}\rho_a,$$

where $C_{fih}(y_{fih})$ is the generation cost function of firm f at its plant $h \in G_{fi}$. Firm f's profit maximization problem is:

$$\begin{array}{ll} \text{maximize} & \theta_f(\mathbf{x}^f) \\ \text{subject to} & x^f \in K_f(\tilde{\mathbf{x}}^f), \end{array}$$

in which x^g for $g \neq f$ and ρ_a for all $a \in \mathcal{A}$ are taken as inputs. Thus, firm f anticipates the other firms' actions and the regulated transmission prices and treat them as endogenous when it solves its profit maximization problem. The overall equilibrium problem is to seek a vector $\mathbf{x} \equiv (x^f)$ such that x^f solves firm f's problem for each f and $\rho_a = \rho_a(\mathbf{z})$ for all a.

At first glance, this equilibrium problem is a QVI because of the dependence of the set $K_f(\tilde{\mathbf{x}}^f)$ on $\mathbf{x}$; furthermore the regulatory rule $\rho_a = \rho_a(\mathbf{z})$ for all a is seemingly imposing an extra condition on the problem. Nevertheless, it is fairly easy to derive a VI whose solution must yield a desired equilibrium point of the model. To state this VI, let $\tilde{K}_f$ denote the set $K_f(\tilde{\mathbf{x}}^f)$ without the capacity constraint (1.4.9); that is,

$$\tilde{K}_f \equiv \left\{ \begin{array}{lll} x^f \geq 0 : y_{fih} \leq \text{CAP}_{fih}, & \forall\, h \in G_{fi},\ \forall\, i \in \mathcal{N}_f \\ d_{fi} + \sum_{a\in\mathcal{A}_i^+} x_{fa} = \sum_{h\in G_{fi}} y_{fih} + \sum_{a\in\mathcal{A}_i^-} x_{fa}, & \forall\, i \in \mathcal{N}_f \\ d_{fi} + \sum_{a\in\mathcal{A}_i^+} x_{fa} = \sum_{a\in\mathcal{A}_i^-} x_{fa}, & \forall\, i \in \mathcal{N} \setminus \mathcal{N}_f \end{array} \right\}.$$

Note that $\tilde{K}_f$ is independent of $\mathbf{x}$. Further, let

$$\Omega \equiv \{\, \mathbf{x} : \mathbf{x} \text{ satisfies } (1.4.9) \,\}.$$

Let $\mathbf{d} \equiv (d_{fi} : f \in \mathcal{F}; i \in \mathcal{N})$ be the vector of electricity delivered by all the firms at the nodes; let $\mathbf{y} \equiv (y_{fih} : f \in \mathcal{F}; i \in \mathcal{N}; h \in G_{fi})$ be the vector of electricity productions at all the plants. Define the firms' marginal returns and marginal costs, i.e., the derivatives:

$$MR_{fi}(\mathbf{d}) \equiv \frac{\partial \theta_f(\mathbf{x})}{\partial d_{fi}} \quad \text{and} \quad MC_{fih}(y_{fih}) \equiv \frac{dC_{fih}(y_{fih})}{dy_{fih}},$$

and the vector function

$$\mathbf{F}(\mathbf{d},\mathbf{y},\mathbf{z}) \equiv \begin{pmatrix} -MR_{fi}(\mathbf{d}) & : & \forall\, f, i \\ MC_{fih}(y_{fih}) & : & \forall\, f, i, h \\ \rho_a(\mathbf{z}) & : & \forall\, a \end{pmatrix},$$

where the rows corresponding to the regulatory functions $\rho_a(\mathbf{z})$ are each repeated M times, one for each firm f. Finally, let

$$K \equiv \left(\prod_{f=1}^{M} \tilde{K}_f \right) \cap \Omega.$$

The following proposition gives an alternative way of computing an electricity equilibrium, which is different from the direct MiCP formulation presented at the end of Subsection **1.4.2** for a QVI.

1.4.3 Proposition. Suppose that each function $\theta_f(\mathbf{x})$ is concave in x^f for each fixed but arbitrary $\tilde{\mathbf{x}}^f$. If $\mathbf{x} \equiv (\mathbf{d},\mathbf{y},\mathbf{z})$ solves the VI $(K, \mathbf{F})$, then $\mathbf{x}$ is an equilibrium point of the oligopolistic electricity model.

Proof. Let $\mathbf{x}$ solve the VI $(K, \mathbf{F})$, we need to show that for each f, x^f solves the firm f's maximization problem with x^g for $g \neq f$ fixed and for $\rho_a \equiv \rho_a(\mathbf{z})$. Let $\hat{x}^f$ be an arbitrary vector in $K_f(\tilde{\mathbf{x}}^f)$. The vector

$$\hat{\mathbf{x}} \equiv \begin{pmatrix} \hat{x}^f \\ \tilde{\mathbf{x}}^f \end{pmatrix}$$

is easily seen to belong to the set K. Thus we have

$$(\hat{\mathbf{x}} - \mathbf{x})^T \mathbf{F}(\mathbf{x}) \geq 0,$$

which reduces to

$$(\hat{x}^f - x^f)^T \nabla_{x^f} \theta_f(\mathbf{x}) \leq 0.$$

By the concavity of the function $\theta_f(\cdot, \tilde{\mathbf{x}}^f)$, the desired claim follows readily. □

There are two interesting variations of the above model. In one variation, the generators do not control the transmission of electricity; instead, they sell their generated power to an Independent Service Operator (ISO) and pay the latter a wheeling fee for such transmission. In turn, the ISO distributes the electricity to the nodes in order to maximize its own profit, taking into account the transmission link capacities and the total amount of generated power to be distributed. Mathematically, along with some market clearing conditions (such as the regulatory prices discussed above), the ISO becomes an extra player in a Nash game and an equilibrium can be similarly defined and dealt with. This variation of the electricity power model can have a further ingredient in which there are arbitragers in the network who will eliminate price differences over space. Depending on how the firms deal with these arbitragers, one obtains either a VI formulation of the resulting model or a rather complicated Nash game in which each player of the game solves a nonconvex constrained optimization problem known as an "MPEC", whose general formulation is presented in Subsection **1.4.10**.

In summary, there exist many VI models for the study of oligopolistic market equilibria in the electricity power industry. It would not be possible for us to discuss the details of each of them. Suffices it to say that the VI/CP methodology has benefited many studies of this kind. Much more remains to be done in this vast applied area.

Markov perfect equilibria

The Markov perfect equilibrium problem is a game-theoretic oligopolistic market model in which two or more firms attempt to set the price of a homogeneous product over an infinite time horizon. For simplicity, we consider a duopoly model in which each firm takes turn to set the price and is committed to it during the period in which the action was taken and the following period when the other firm responds. The model assumes that the action of one firm at the current period depends only the opponent's action in the last period. Each firm seeks to maximize the present discounted value of its profit by choosing prices in discrete time intervals, with the instantaneous payoff of the firm being a function of the current prices and not of time. Assume that the price can take any one of n distinct values, $p(i)$ for $i = 1, \ldots, n$. The given payoff matrix of firm $k = 1, 2$ is

$$\Pi^k \equiv \begin{bmatrix} \pi^k_{11} & \cdots & \pi^k_{1n} \\ \vdots & \ddots & \vdots \\ \pi^k_{n1} & \cdots & \pi^k_{nn} \end{bmatrix},$$

where π^k_{ij} is the profit of firm k if it chooses price $p(i)$ and the other firm chooses $p(j)$. Due to the time independence assumption of the profit, the definition of π^k_{ij} does not depend on when firm k chooses the price.

Firm k's strategy is characterized by a probability transition matrix, called the reaction matrix:

$$X^k \equiv \begin{bmatrix} x_{11}^k & \cdots & x_{1n}^k \\ \vdots & \ddots & \vdots \\ x_{n1}^k & \cdots & x_{nn}^k \end{bmatrix},$$

where x_{ij}^k is the probability of firm k choosing price $p(j)$ (i.e., its j-strategy) given that the opponent firm has chosen $p(i)$ in the previous period. By definition, we have $x_{ij}^k \geq 0$ for all i, j, and

$$\sum_{j=1}^{n} x_{ij}^k = 1, \quad \forall\, i = 1, \ldots, n. \tag{1.4.10}$$

Let v_i^k be the discounted reward to firm k if the current price of the other firm is $p(i)$ and it is firm k's turn to set the price; similarly, let w_i^k be the discounted reward to firm k if it set the price to $p(i)$ last period and it is the other firm's turn to set the price. Let $\delta > 0$ be the discount factor. By a dynamic programming argument, it is possible to show that

$$\begin{aligned} v_i^k &= \max_{1 \leq j \leq n} \left(\pi_{ji}^k + \delta\, w_j^k \right), \quad \forall\, i = 1, \ldots, n \\ w_i^k &= \max_{1 \leq j \leq n} \left(\pi_{ij}^k + \delta\, v_j^k \right), \quad \forall\, i = 1, \ldots, n. \end{aligned}$$

We restrict the model by postulating the symmetry assumption; that is, the reaction matrix, the payoff matrix, and the discounted rewards for the two firms are the same, so that

$$X^1 = X^2, \quad \Pi^1 = \Pi^2,$$

and

$$v^1 = v^2, \quad w^1 = w^2.$$

With a given payoff matrix Π, a symmetric Markov perfect equilibrium is a probability transition matrix X whose entries are all nonnegative and satisfy (1.4.10), and which, along with a pair of discounted reward vectors v and w, satisfies the following conditions:

$$\begin{aligned} v_i &= \max_{1 \leq j \leq n} \left(\pi_{ji} + \delta\, w_j \right), \quad \forall\, i = 1, \ldots, n \\ w_i &= \max_{1 \leq j \leq n} \left(\pi_{ij} + \delta\, v_j \right), \quad \forall\, i = 1, \ldots, n \end{aligned} \tag{1.4.11}$$

and

$$\begin{aligned} v_i &= \sum_{j=1}^{n} x_{ij} \pi_{ji} + \delta \sum_{j=1}^{n} x_{ij} w_j, \quad \forall\, i = 1, \ldots, n \\ w_i &= \sum_{j=1}^{n} x_{ij} \pi_{ij} + \delta \sum_{j=1}^{n} x_{ij} v_j, \quad \forall\, i = 1, \ldots, n. \end{aligned} \tag{1.4.12}$$

Since the variables of this problem are x_{ij}, v_i and w_i, (1.4.12) are nonlinear equations in these unknowns.

Clearly, a symmetric Markov perfect equilibrium will not be affected if we add the same constant to all entries of the payoff matrix Π, even though such a change will cause the discounted rewards v_i and w_i to individually be modified by the same constant. Thus for the computation of such an equilibrium, we may assume without loss of generality that the payoff matrix Π is positive. Under this assumption, we have the following result.

1.4.4 Proposition. Let Π be a positive $n \times n$ matrix. An $n \times n$ matrix X is a symmetric Markov perfect equilibrium if and only if there exist two n-vectors v and w such that the triple (X, v, w) is a solution of the NCP (F), where $F : \mathbb{R}^{n^2+2n} \to \mathbb{R}^{n^2+2n}$ is defined by

$$F(X, v, w) \equiv \begin{pmatrix} v_i - \delta\, w_j - \pi_{ji} & : \quad i, j = 1, \ldots, n \\ \displaystyle\sum_{j=1}^{n} x_{ij} - 1 & : \quad i = 1, \ldots, n \\ w_i - \displaystyle\sum_{j=1}^{n} \pi_{ij} x_{ij} - \delta \sum_{j=1}^{n} v_j x_{ij} & : \quad i = 1, \ldots, n \end{pmatrix}.$$

Proof. Suppose that X is a symmetric Markov perfect equilibrium. Let v and w be two associated vectors of equilibrium discounted rewards. To show that (X, v, w) solves the NCP (F), it suffices to verify that

$$x_{ij}\,(\, v_i - \delta\, w_j - \pi_{ji}\,) = 0, \quad \forall\, i, j.$$

In turn, since x_{ij} and $v_i - \delta w_j - \pi_{ji}$ are both nonnegative, it suffices to show that

$$\sum_{j=1}^{n} x_{ij}\,(\, v_i - \delta\, w_j - \pi_{ji}\,) = 0.$$

The left-hand side is equal to

$$v_i - \delta \sum_{j=1}^{n} x_{ij} w_j - \sum_{j=1}^{n} x_{ij} \pi_{ji},$$

which is indeed equal to zero by (1.4.12).

To show the converse, let (X, v, w) solve the NCP (F). We first establish that both v and w are positive vectors. Since $\pi_{ij} > 0$, we have

$$v_i \geq \delta\, w_j + \pi_{ji} > 0.$$

It follows by complementarity that

$$\sum_{j=1}^{n} x_{ij} = 1.$$

Hence $x_{ij} > 0$ for at least one j. We have

$$w_i \geq \sum_{j=1}^{n} \pi_{ij} x_{ij} + \delta \sum_{j=1}^{n} v_j x_{ij} > 0.$$

In turn, by complementarity, we have

$$w_i = \sum_{j=1}^{n} \pi_{ij} x_{ij} + \delta \sum_{j=1}^{n} v_j x_{ij},$$

which is the second equation in (1.4.12). The first equation in (1.4.12) can be proved easily be reversing the argument in the proof of the "only if " part. Finally, the proof of the remaining two conditions in (1.4.11) is not difficult. □

By interchanging the roles of v and w, it is easy to see that an alternative NCP $(\tilde{F})$, where

$$\tilde{F}(X, w, v) \equiv \begin{pmatrix} w_i - \delta\, v_j - \pi_{ij} & : \quad i, j = 1, \ldots, n \\ \displaystyle\sum_{j=1}^{n} x_{ij} - 1 & : \quad i = 1, \ldots, n \\ \displaystyle v_i - \sum_{j=1}^{n} \pi_{ji} x_{ij} - \delta \sum_{j=1}^{n} w_j x_{ij} & : \quad i = 1, \ldots, n \end{pmatrix},$$

also provides an equivalent way of computing a symmetric Markov perfect equilibrium. Both CPs are nonlinear because they contain products of the variables.

1.4.4 Economic equilibrium problems

Many economic equilibrium models can be formulated as variational inequalities and/or complementarity problems. These include the general Walrasian equilibrium model of Arrow-Debreu to find commodity prices, sector activities, and consumer consumptions in a perfectly competitive economy, a market equilibrium model that is the basis of the "PIES" and "NEMS" energy models developed at the US Department of Energy to compute equilibrium fuel prices and quantities for the US energy sector (PIES is the acronym for Project Independence Energy System, and NEMS is for

National Energy Modeling System), and a spatial price equilibrium model for the computation of commodity prices, supplies, and demands in a network of spatially separated markets.

The Walrasian equilibrium problem

The purpose of this general equilibrium model is to predict economic activities in a closed economy; that is, to compute the equilibrium activities and prices in an economy when all interactions between the commodities comprising this economy have been incorporated. This well-known problem is the basis of much of mathematical economics and general equilibrium theory and has proven very useful in macroeconomics, particularly in analyzing tax policy, international trade, issues in energy economics, to name a few applications.

There are many mathematical formulations of the general equilibrium problem. We adopt a simplified model that easily leads to a nonlinear complementarity problem. Let m and n be, respectively, the number of economic activities and goods. The unit cost (assumed constant) of operating the i-th activity is c_i and the initial endowment of the j-th good is b_j. The unknown level of the i-th activity is denoted y_i and the price of the j-th good is denoted p_j. The demand function for the j-th good is $d_j(p)$, where $p \equiv (p_j) \in \mathbb{R}^n$ is the price vector of all goods. This function $d_j(p)$ is typically determined from an utility maximization problem; in some cases, it is positively homogeneous of degree 0; that is, $d_j(\lambda p) = d_j(p)$ for all scalars $\lambda > 0$. Moreover, certain utility functions lead to demand functions $d_j(p)$ that are defined only for nonnegative prices; an example of such a function that is derived from the class of utility functions having a "constant elasticity of substitution" is:

$$d_j(p) \equiv \frac{(p_j/\mu_j)^{r-1}\, p^T b}{\sum_{k=1}^{n} p_k^r/\mu_k^{r-1}}, \tag{1.4.13}$$

where r is a given scalar and the μ_k are positive constants. Note the positive homogeneity of this function $d_j(p)$.

The technology input-output matrix of the economy is given by the $m \times n$ matrix $A(p) \equiv (a_{ij}(p))$. The transpose of this matrix converts levels of activities into netput vectors of goods. Specifically, for a vector $y \equiv (y_i)$ of activities, $A(p)^T y$ is the vector of goods resulting from these activities; for a vector p of prices, $A(p)p$ is the vector of per unit activity returns.

A pair of activity-price patterns (y, p) is a *general equilibrium* if the following conditions are satisfied:

$$0 \leq y \perp c - A(p)p \geq 0 \tag{1.4.14}$$

$$0 \leq p \perp b + A(p)^T y - d(p) \geq 0. \tag{1.4.15}$$

The economic interpretation of these conditions is as follows. Condition (1.4.14) states that activity levels are nonnegative and all activities yield nonpositive profits; moreover, activities with negative profits are not performed. The condition (1.4.15) states that prices are nonnegative and supplies must satisfy demands; moreover, excess supplies occur only in the case of free goods.

Clearly, the conditions (1.4.14) and (1.4.15) define the NCP (F) where

$$F(y,p) \equiv \begin{pmatrix} c - A(p)p \\ b + A(p)^T y - d(p) \end{pmatrix}.$$

An important special case arises when $A(p)$ is a constant matrix; this is the case of constant technology. In this case, the above NCP (F) becomes the KKT system of the VI (K, G), where

$$K \equiv \{ p \in \mathbb{R}^n_+ : c - Ap \geq 0 \}$$

and

$$G(p) \equiv b - d(p).$$

The latter VI is linearly constrained and is in terms of the price vector p only. The activity vector y consists of the multipliers to the constraint $c - Ap \geq 0$ that defines the polyhedron K.

There is yet another interpretation of the general equilibrium problem with constant technology that is often used in practice (e.g., in the PIES model). Consider the LP:

$$\begin{array}{ll} \text{minimize} & c^T y \\ \text{subject to} & A^T y \geq D, \quad y \geq 0, \end{array}$$

where D is a fixed vector. This linear program represents the supply side of the economy where D is a given demand vector. The constraints of this program stipulate that demands are being satisfied by some nonnegative economic activities; the objective function is the total activity cost. By linear programming duality, a vector y is optimal to the above LP if and only if there exists a vector of shadow prices p such that

$$\begin{array}{l} 0 \leq y \perp c - Ap \geq 0 \\ 0 \leq p \perp A^T y - D \geq 0. \end{array}$$

The demand side of the economy is described by the demand function $d(\pi)$ where π is the vector of market prices of the goods, and we have

$$D = b - d(\pi).$$

In this setting, the general equilibrium problem can be seen to be the problem of finding a set of market prices π such that the shadow prices p of the goods obtained from the solution of the supply side LP with D equal to $b - d(\pi)$ coincide with the market prices π.

There are various extensions of the general equilibrium problem. One such extension concerns the presence of taxes and/or subsidies on the economic goods. In this case, the NCP is defined by the modified function

$$\tilde{F}(y,p) \equiv \begin{pmatrix} c - C(p)p \\ b + A(p)^T y - d(p) \end{pmatrix},$$

where the matrix $C(p)$ is derived from the matrix $A(p)$ and incorporates the input/output taxes and/or subsidies. Further extensions of the basic model exist; see Section **1.9** for references.

A model of invariant capital stock

The next model is an economic growth model. Consider an economy with constant technology in which a capital stock invariant under optimization is sought; that is, an initial activity level is to be determined such that the maximization of the discounted sum of future utility flows over an infinite horizon can be achieved by reconstituting that activity level at the end of each period. The technology is given by two matrices A and B of the same dimension and a vector b:

$$\begin{aligned} A_{ij} &\equiv \text{amount of good } i \text{ used to operate activity } j \text{ at unit level} \\ B_{ij} &\equiv \text{amount of good } i \text{ produced by activity } j \text{ operated at unit level} \\ b_i &\equiv \text{amount of resource } i \text{ exogenously provided in each time period.} \end{aligned}$$

Note that $b_i < 0$ means that resource i is withdrawn for subsistence. The utility function $u(x)$ is assumed to be concave and continuously differentiable. Let x^t $(t = 1, 2, \ldots)$ denote the vector of activity levels in period t. The model then seeks a vector x so that $x^t = x$ $(t = 1, 2, \ldots)$ solves the problem $\mathrm{P}(Bx)$, where for a given vector b_0, $\mathrm{P}(b_0)$ denotes the problem of finding a sequence of activity levels $\{x^t\}_1^\infty$ in order to

$$\begin{array}{lll} \text{maximize} & \displaystyle\sum_{t=1}^{\infty} \rho^{t-1}\, u(x^t) & \\ \text{subject to} & Ax^1 \leq b_0 + b & \\ & Ax^t \leq Bx^{t-1} + b, & t = 2, 3, \ldots \\ & x^t \geq 0, & t = 1, 2, \ldots \end{array}$$

where $\rho \in (0,1)$ is the discount rate. The vector x computed by the model then provides an optimal capital stock invariant under discounted optimization. The result below shows how a vector can be obtained by solving a nonlinear complementarity problem.

1.4.5 Proposition. Suppose that B is a nonnegative matrix. If (x,y) satisfies the conditions below:

$$0 \leq (B-A)x+b \perp y \geq 0 \tag{1.4.16}$$

$$0 \leq -\nabla u(x)+(A^T-\rho B^T)y \perp x \geq 0, \tag{1.4.17}$$

then x is an invariant optimal capital stock.

Proof. Let $\{x^t\}_1^\infty$ be any feasible solution for the problem $\mathrm{P}(Bx)$. By the concavity of the utility function, we have

$$u(x^t) \leq u(x)+\nabla u(x)^T(x^t-x).$$

By condition (1.4.17) and the fact that $x^t \geq 0$, we derive

$$\nabla u(x)^T(x^t-x) \leq y^T(A-\rho B)(x^t-x).$$

Consequently, for any integer $s \geq 1$,

$$\sum_{t=1}^{s} \rho^{t-1} u(x^t) \leq \sum_{t=1}^{s} \rho^{t-1} u(x) + \sum_{t=1}^{s} \rho^{t-1} y^T(A-\rho B)(x^t-x);$$

rearrangement yields

$$\begin{aligned}\sum_{t=1}^{s}\rho^{t-1} u(x^t) &\leq \sum_{t=1}^{s}\rho^{t-1} u(x)+y^TA(x^1-x)\\ &+\sum_{t=1}^{s-1}\rho^t y^T(A(x^{t+1}-x)-B(x^t-x))-\rho^s y^TB(x^s-x),\end{aligned}$$

which implies by the feasibility of $\{x^t\}_1^\infty$ and condition (1.4.16),

$$\sum_{t=1}^{s} \rho^{t-1} u(x^t) \leq \sum_{t=1}^{s} \rho^{t-1} u(x) - \rho^s y^T B(x^s-x).$$

Since B is a nonnegative matrix, it follows that

$$\sum_{t=1}^{s} \rho^{t-1} u(x^t) \leq \sum_{t=1}^{s} \rho^{t-1} u(x) + \rho^s y^T Bx.$$

Passing to the limit $s \to \infty$, we deduce

$$\sum_{t=1}^{\infty} \rho^{t-1} u(x^t) \leq \sum_{t=1}^{\infty} \rho^{t-1} u(x),$$

which shows that the constant sequence $\{\tilde{x}^t\}$, where $\tilde{x}^t = x$ for all t, is optimal for the problem $\mathrm{P}(Bx)$. □

The two conditions (1.4.16) and (1.4.17) define the NCP (F), where

$$F(x, y) = \begin{pmatrix} -\nabla u(x) + (A^T - \rho B^T)y \\ b + (B - A)x \end{pmatrix}. \tag{1.4.18}$$

This NCP does not correspond to the KKT conditions of a constrained optimization problem due to the fact that $\rho < 1$. Nevertheless the function F has a special structure, which can be exploited to establish the existence of a solution to the model; see Exercise **2.9.30**.

1.4.5 Traffic equilibrium models

The purpose of a static traffic equilibrium model is to predict steady-state traffic flows in a congested traffic network. The node set of the network is denoted $\mathcal{N}$ and arc set is denoted $\mathcal{A}$. It is assumed that users of the network compete noncooperatively for the resources of the network in an attempt to minimize their travel costs, where the cost of travel along an arc $a \in \mathcal{A}$ is a nonlinear function $c_a(f)$ of the total flow vector f with components f_b for all $b \in \mathcal{A}$. Let $c(f)$ be the vector with components $c_a(f)$, $a \in \mathcal{A}$. There are two distinguished subsets of $\mathcal{N}$ that represent the set of origin nodes $\mathcal{O}$ and destination nodes $\mathcal{D}$, respectively. The set of origin-destination (OD) pairs is a given subset $\mathcal{W}$ of $\mathcal{O} \times \mathcal{D}$.

There are several descriptions of the static traffic equilibrium problem; one description is in terms of the flows on paths in the network, and another is in terms of multicommodity flows on arcs, where each commodity represents one OD pair. Both formulations are based on a certain equilibrium principle due to Wardrop that may be considered special cases of the Nash equilibrium concept.

The path formulation

For each $w \in \mathcal{W}$, let $\mathcal{P}_w$ denote the set of paths connecting the OD pair w and let $\mathcal{P}$ be the union of $\mathcal{P}_w$ for w ranging over all $w \in \mathcal{W}$. Figure 1.5 illustrates a simple network with two origins and three destinations. The paths between O1 and D2 that constitute the set $\mathcal{P}_{12}$ are in bold.

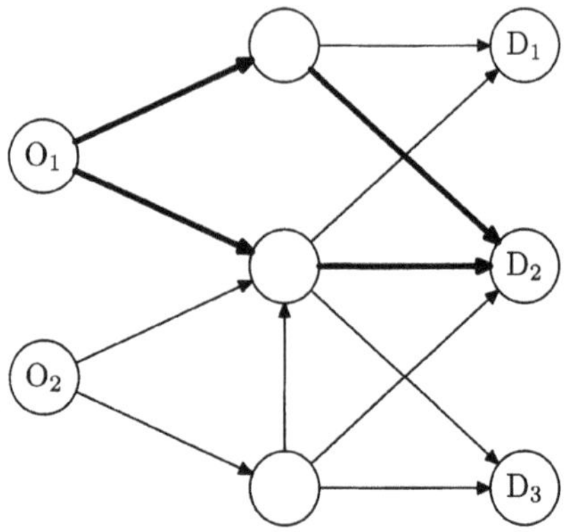

Figure 1.5: An illustration of the traffic equilibrium problem.

Let h_p denote the flow on path $p \in \mathcal{P}$ and let $C_p(h)$ be the travel cost on this path that is a function of the entire vector $h \equiv (h_p)$ of path flows. Let Δ be the arc-path incidence matrix with entries

$$\delta_{ap} \equiv \begin{cases} 1 & \text{if path } p \in \mathcal{P} \text{ traverses arc } a \in \mathcal{A} \\ 0 & \text{otherwise.} \end{cases}$$

Clearly, the vectors f and h are related by

$$f = \Delta h.$$

A common assumption on the path cost function $C(h) \equiv (C_p(h))$ is that it is *additive*; that is, for each $p \in \mathcal{P}$, $C_p(h)$ is the sum of the arc costs $c_a(f)$ on all the arcs a traversed by the path $p \in \mathcal{P}$. In vector notation, this assumption says

$$C(h) = \Delta^T c(f) = \Delta^T c(\Delta h). \tag{1.4.19}$$

Nevertheless, this assumption is not necessary for the formulation introduced below. On the demand side, for each $w \in \mathcal{W}$ a function $d_w(u)$ is given that represents the travel demand between the OD pair w, where $u \equiv (u_v)$ is the (unknown) vector of minimum travel costs between all OD pairs. This general case corresponds to an *elastic demand model*; it is in contrast to the *fixed demand model* where $d_w(u)$ is a constant for all $w \in \mathcal{W}$.

The Wardrop user equilibrium principle is a behavioral axiom that postulates the route choice of the network users. Specifically, the principle stipulates that users of the traffic network will choose the minimum cost path between each OD pair, and through this process the paths that are used (i.e., have positive flows) will have equal costs; moreover, paths with costs higher than the minimum will have no flow. Mathematically, this principle can be phrased succinctly as follows:

$$0 \leq C_p(h) - u_w \perp h_p \geq 0, \quad \forall\, w \in \mathcal{W} \text{ and } p \in \mathcal{P}_w; \tag{1.4.20}$$

moreover, the travel demand must be satisfied:

$$\sum_{p \in \mathcal{P}_w} h_p = d_w(u), \quad \forall\, w \in \mathcal{W}, \tag{1.4.21}$$

and the minimum travel costs must be nonnegative:

$$u_w \geq 0, \quad w \in \mathcal{W}. \tag{1.4.22}$$

The static traffic user equilibrium problem is to find a pair (h, u) of path flows and minimum travel costs, called a *traffic user equilibrium*, so that the above conditions (1.4.20)–(1.4.22) are satisfied. At first glance, these conditions are not quite in the form of a complementarity problem or variational inequality. Under a reasonable assumption on the travel cost and demand functions, the conditions (1.4.20)–(1.4.22) are indeed equivalent to an NCP (**F**) where the defining function **F** is given by:

$$\mathbf{F}(h, u) \equiv \begin{pmatrix} C(h) - \Omega^T u \\ \\ \Omega h - d(u) \end{pmatrix}, \tag{1.4.23}$$

where Ω is the (OD pair, path)-incidence matrix whose entries are

$$\omega_{wp} \equiv \begin{cases} 1 & \text{if } p \in \mathcal{P}_w \\ 0 & \text{otherwise.} \end{cases}$$

The following result makes this equivalence precise.

1.4.6 Proposition. Suppose that the travel cost and demand functions $C_p(h)$ and $d_w(u)$ are nonnegative and for each OD pair $w \in \mathcal{W}$,

$$\begin{aligned} &\left[\sum_{p \in \mathcal{P}_w} h_p C_p(h) = 0, \; h \equiv (h_p) \geq 0 \right] \\ &\qquad \Rightarrow \left[h_p = 0, \; \forall\, p \in \mathcal{P}_w \right]. \end{aligned} \tag{1.4.24}$$

The static traffic user equilibrium problem is equivalent to the NCP (**F**).

Proof. It suffices to show that every solution of the NCP must be a traffic user equilibrium. Let (h, u) solve the NCP (**F**). We only need to show the condition (1.4.21). Suppose for some $w \in \mathcal{W}$,

$$\sum_{p \in \mathcal{P}_w} h_p > d_w(u). \tag{1.4.25}$$

By the complementarity conditions, we must have $u_w = 0$ and

$$\sum_{p \in \mathcal{P}_w} h_p C_p(h) = u_w \sum_{p \in \mathcal{P}_w} h_p = 0.$$

Since $d_w(u)$ is nonnegative and each h_p is nonnegative, (1.4.25) implies that $h_q > 0$ for at least one $q \in \mathcal{P}_w$. Condition (1.4.24) implies

$$\sum_{p \in \mathcal{P}_w} h_p C_p(h) > 0.$$

But this is a contradiction. □

The sum on the left-hand side in (1.4.24) is the total travel cost between the OD pair w; it is called the *system cost* of this OD pair. Thus the assumption (1.4.24) stipulates that, if all travel costs are nonnegative, then the system cost of an OD pair must be positive unless there is no traffic flow between that OD pair.

Arc formulations

For networks of reasonable size, the *a priori* enumeration of all paths connecting elements of $\mathcal{W}$ is a prohibitive task, if there are many such elements. This imposes a heavy toll on the practical use of the path-flow formulation of the traffic equilibrium model. Nevertheless there are path generation techniques that utilize this formulation and generate the paths only if they are needed. There are alternative arc-flow formulations that circumvent the computational necessity of enumerating paths. One such formulation is possible when the demand function $d(u)$ is invertible and the path cost $C(h)$ is additive. In this case, the traffic user equilibrium problem can be formulated as a VI with the arc flow f as the exogenous variable and the path variables being treated endogenously. To present this formulation, we introduce the feasible set:

$$K \equiv \{ (f, d) : f = \Delta h \text{ and } d = \Omega h \text{ for some } h \geq 0 \}. \tag{1.4.26}$$

1.4.7 Proposition. Suppose that the path cost $C_p(h)$ satisfies (1.4.19) and $d(u)$ is an invertible function of u with inverse $\Phi(d)$. If (h, u) is a traffic user equilibrium flow-cost pattern, then (f, d), where $f \equiv \Delta h$ and $d \equiv d(u)$, is a solution of the VI $(K, \mathbf{G})$, where $\mathbf{G}(f, d) \equiv (c(f), -\Phi(d))$. Conversely, if $(f, d) \in \text{SOL}(K, \mathbf{G})$ and the inverse function Φ is nonnegative, then (h, u), where $u \equiv \Phi(d)$ and $h \geq 0$ is such that $f = \Delta h$ and $d = \Omega h$, is a traffic user equilibrium.

Proof. Using Proposition **1.1.3** and the fact that $f = \Delta h$ and $d = \Omega h$ for some $h \geq 0$ if $(f, d) \in K$, we may deduce that the VI $(K, \mathbf{G})$ is equivalent to the NCP $(\mathbf{H})$, where

$$\mathbf{H}(h) \equiv \Delta^T c(\Delta h) - \Omega^T \Phi(\Omega h),$$

in the sense that $(f, d) \in \text{SOL}(K, \mathbf{G})$ if and only if $h \geq 0$, where $f = \Delta h$ and $d = \Omega h$, solves the NCP $(\mathbf{H})$.

Based on the above observation, let (h, u) be a traffic user equilibrium and define $(f, d) \equiv (\Delta h, d(u))$. Then $u = \Phi(d)$ and $d = d(u) = \Omega h$; moreover, h clearly solves the NCP $(\mathbf{H})$. Consequently, $(f, d) \in \text{SOL}(K, \mathbf{G})$. Conversely, if $(f, d) \in \text{SOL}(K, \mathbf{G})$, let $h \geq 0$ be such that $f = \Delta h$ and $d = \Omega h$ and define $u \equiv \Phi(d)$. Then $u \geq 0$ and $d = d(u)$; thus $d(u) = \Omega h$. Moreover, since h solves the NCP $(\mathbf{H})$, we can easily verify that (h, u) is a traffic user equilibrium. □

In the fixed-demand traffic user equilibrium problem, each demand function $d_w(u)$ is a constant d_w. When the path cost $C(h)$ is also additive, then this equilibrium problem can be formulated as a VI in the arc flow vector f only. To derive this formulation, let $\mathcal{F}$ denote the set of *feasible flow patterns*; i.e.,

$$\mathcal{F} \equiv \{ f : f = \Delta h \text{ for some } h \geq 0 \text{ satisfying } \Omega h = d \}.$$

This set $\mathcal{F}$ is simply the canonical projection of the set K in (1.4.26) onto the set of arc flow vectors. For every feasible flow pattern $f \in \mathcal{F}$, there exists a path flow $h \geq 0$ such that $f = \Delta h$ and $\Omega h = d$. We say that $f \in \mathcal{F}$ *induces* a user equilibrium if there exists a cost vector u such that the pair (h, u) is a user equilibrium.

1.4.8 Proposition. Suppose that the path cost $C(h)$ is additive and that each arc cost $c_a(f)$ is a nonnegative function. Suppose that the traffic demand d is a fixed positive vector. A feasible flow pattern $f \in \mathcal{F}$ induces a user equilibrium if and only if $f \in \text{SOL}(\mathcal{F}, c)$.

Proof. Suppose f solves the VI $(\mathcal{F}, c)$. Then the path flow h must satisfy:

$$(h' - h)^T C(h) \geq 0, \quad \forall h' \geq 0 \text{ satisfying } \Omega h' = d.$$

This inequality defines the VI $(\mathcal{H}, C)$ where

$$\mathcal{H} \equiv \{ h' \geq 0 : \Omega h' = d \}.$$

Since $\mathcal{H}$ is a polyhedron and $h \in \text{SOL}(\mathcal{H}, C)$, it follows that there exists a vector u such that

$$0 \leq h \perp C(h) - \Omega^T u \geq 0.$$

We claim that u must be nonnegative. To show this, it suffices to show that for each $w \in \mathcal{W}$, $u_w = C_p(h)$ for at least one $p \in \mathcal{P}_w$. This is indeed true because

$$0 < d_w = \sum_{p \in \mathcal{P}_w} h_p;$$

thus $h_p > 0$ for at least one $p \in \mathcal{P}_w$. Hence by complementarity it follows that $u_w = C_p(h)$ as desired. Thus we have shown that if $f \in \text{SOL}(\mathcal{F}, c)$, then f induces a user equilibrium. The converse of this statement can be established by reversing the above argument. □

An extension of the traffic equilibrium problem with elastic demand can be obtained by replacing the OD demand function $d(u)$ with a more basic system of commodity supply and demand functions at each node of the network and a behavioral principle about the movement of goods between regions. This leads to the spatial price equilibrium problem, which has an NCP formulation that is closely related to the one in Proposition **1.4.6**. See Section **1.9** for notes and comments on the latter problem.

1.4.6 Frictional contact problems

Contact problems in mechanical structures provide a vast source where challenging CPs arise and whose solutions require advanced computational algorithms. Mathematically speaking, contact between two objects can be expressed very naturally as the complementary relation between the normal contact force and the gap between the contact surfaces: both quantities are sign restricted and one of them must be zero if the other is not zero. Recognizing this intuitive modeling device, we can easily obtain a host of CPs that describe a variety of contact problems in engineering mechanics.

Traditionally, contact problems from continuum mechanics are posed in infinite-dimensional functional spaces. In recent years, discrete and/or discretized contact problems giving rise to finite-dimensional CPs have gained tremendous momentum in research and applications as computational mechanicians are interested in solving the physical models numerically. In what follows, we describe a general mathematical model that provides a unified formulation for discrete contact problems of many kinds.

A time-incremental, discrete, unilateral contact problem with (generalized) Coulomb friction is the result of a time-space discretization and/or the numerical integration of a continuous-time contact model. In all such discrete contact models, there are four main components of the problem: geometry of contact, force equilibrium, normal contact conditions, friction laws and associated generalized directional conditions that govern the tangential contact forces. The geometry of a typical model is described in terms of certain nodal vectors of the body in contact, which we call *configurations*; these vectors are represented by $u \in \mathbb{R}^N$ for a positive integer N. The unilateral non-penetration constraints are defined by a finite collection of inequalities:

$$\psi_{in}(u) \leq 0, \quad i = 1, \ldots, n_c \tag{1.4.27}$$

where the integer n_c is the number of contact points and the subscript "n" refers to the normal direction on the contact surface. The displacement coordinates at the contact points that represent the "directions of generalized sliding" are described in terms of the functions

$$\psi_{ij}(u), \quad j = 1, \ldots, M_i; \ i = 1, \ldots, n_c,$$

where the integer M_i is the number of "sliding modes". There is a given reference configuration $u_{\text{ref}} \in \mathbb{R}^N$ that could be the result of solving a problem of the same kind at the preceding time step in a time-stepping numerical scheme. Typically, the functions ψ_{in} and ψ_{ij} are derived from some curvilinear coordinate systems of the contact surfaces; accordingly, we call these functions the normal and tangential coordinate functions, respectively.

The given external forces acting on the body are represented by a vector $f_{\text{ext}} \in \mathbb{R}^N$. The internal non-contact forces (e.g. stresses) are represented by a vector function $F : \mathbb{R}^N \to \mathbb{R}^N$. In many situations, F is a gradient map; e.g. $F(u) = \nabla\Theta(u)$, where $\Theta(u)$ is the potential energy as a function of the configuration u. The subsequent development does not rely on this special form of F. The force equilibrium equation is:

$$F(u) + \text{ contact forces } = f_{\text{ext}}.$$

At each contact node i, let p_{in} and p_{ij} for $j = 1, \ldots, M_i$ denote, respectively, the normal and generalized tangential contact forces defined relative to the local surface coordinate functions ψ_{in} and ψ_{ij}. These forces are constrained to lie in a "balanced" convex set with smooth boundary centered at the origin (possibly continuously dependent on the configuration). For simplicity, we assume that these force constraints are expressed by the following "abstract friction inequalities":

$$\phi_i(p_{in}, p_{it}, u) \leq 0, \quad i = 1, \ldots, n_c, \tag{1.4.28}$$

where p_{it} denotes the M_i-dimensional vector of the tangential forces:

$$p_{it} \equiv (p_{ij} : j = 1, \ldots, M_i)$$

and each ϕ_i is a vector function with components $\phi_{i\ell}$ for $\ell = 1, \ldots, m_i$ for some positive integer m_i that depends on the point of contact i. We label each constraint in (1.4.28) as an abstract friction inequality because it is an abstraction of certain commonly used friction laws in contact mechanics. For example, the well-known quadratic Coulomb friction laws can be described by the following functions:

$$\phi_{i\ell}(p_{in}, p_{it}, u) \equiv (p_{it})^T A_{i\ell}(u) p_{it} - \mu_{i\ell}(u)^2 p_{in}^2, \qquad (1.4.29)$$

where $A_{i\ell}(u)$ is a symmetric positive definite matrix and $\mu_{i\ell}(u)$ is a positive scalar (a friction coefficient), both dependent on the configuration u. Another class of friction functions $\phi_{i\ell}$ are obtained by replacing the quadratic friction laws by certain polygonal approximations. In these polygonal laws, the function $\phi_{i\ell}(p_{in}, p_{it}, u)$ is linear in (p_{in}, p_{it}). As an example of these polygonal friction laws, consider a planar law where $M_i = 1$ and

$$|\, p_{it} \,| \leq \mu_i \, p_{in},$$

where μ_i is a positive friction coefficient independent of u. For this example, we have $m_i = 2$ and

$$\phi_{i\ell}(p_{in}, p_{it}, u) \equiv \pm p_{it} - \mu_i \, p_{in}, \quad \ell = 1, 2. \qquad (1.4.30)$$

There are other polygonal friction laws, which we do not discuss.

The normal and tangential contact force are given by

$$\sum_{i=1}^{n_c} p_{in} \, \nabla \psi_{in}(u) \quad \text{and} \quad \sum_{i=1}^{n_c} \sum_{j=1}^{M_i} p_{ij} \, \nabla \psi_{ij}(u),$$

respectively. Thus, the force equilibrium equation becomes

$$F(u) + \sum_{i=1}^{n_c} p_{in} \, \nabla \psi_{in}(u) + \sum_{i=1}^{n_c} \sum_{j=1}^{M_i} p_{ij} \, \nabla \psi_{ij}(u) = f_{\text{ext}}. \qquad (1.4.31)$$

The Signorini contact conditions state that the normal contact forces are nontensile and at each contact point, such a force is positive only if there is actual contact. Written in the form of a complementarity system, these conditions are:

$$0 \leq p_{in} \perp \psi_{in}(u) \leq 0, \quad i = 1, \ldots, n_c. \qquad (1.4.32)$$

In addition to satisfying (1.4.28), the triple (p_{in}, p_{it}, u) is required to satisfy the "maximum dissipation principle". This principle generalizes the Coulomb point contact condition, which stipulates that the direction of friction force is directly opposed to the direction of sliding. Specifically, the maximum dissipation principle postulates that for each $i = 1, \ldots, n_c$, with (p_{in}, u) fixed, p_{it} solves the following maximization problem in the variable $q_{it} \equiv (q_{ij} : j = 1, \ldots, M_i)$:

$$\begin{aligned} &\text{maximize} \quad \sum_{j=1}^{M_i} q_{ij}\, \nabla\psi_{ij}(u)^T (\, u - u_{\text{ref}}\,) \\ &\text{subject to} \quad \phi_{i\ell}(p_{in}, q_{it}, u) \leq 0, \quad \ell = 1, \ldots, m_i. \end{aligned} \tag{1.4.33}$$

Notice that this is an optimization problem with a linear objective function but with possibly nonlinear constraints.

In principle, we have completed the mathematical formulation of the discrete frictional contact problem. Nevertheless, without some suitable conditions, the formulation is not very practical because it does not lend itself easily to analysis and computations. For one thing, we wish to write (1.4.33) in terms of its KKT conditions. In general, this is not possible. Consider the functions $\phi_{i\ell}$ given by (1.4.29). If $p_{in} = 0$, then the only feasible q_{it} is the zero vector; more seriously, all the constraints in (1.4.33) hold as equations. Thus no constraint qualification will hold and therefore the KKT conditions are not necessarily valid. Indeed, we need to resort to the more general Fritz-John necessary conditions; moreover, for various reasons, we wish to write these conditions with a particular multiplier for the objective function.

We make the following assumptions on the functions $\phi_{i\ell}$. These assumptions are all satisfied by (1.4.29), (1.4.30) and many other commonly used friction functions.

Properties of friction functions

For all pairs (p_{in}, u) with $p_{in} \geq 0$ and all $i = 1, \ldots, n_c$ and $\ell = 1, \ldots, m_i$,

(F1) $\phi_{i\ell}(p_{in}, p_{it}, u)$ is convex in p_{it};

(F2) $\phi_{i\ell}(p_{in}, 0, u) \leq 0$ with equality holding if and only if $p_{in} = 0$.

In addition, for all $i = 1, \ldots, n_c$,

(F3) for all vectors u, $\phi_i(0, p_{it}, u) \leq 0$ implies $p_{it} = 0$; and

(F4) there exists a positive scalar $\gamma_i \geq 1$ such that for all vectors u, $\phi_i(p_{in}, 0, u)$ is positively homogeneous of degree γ_i for $p_{in} \geq 0$; that is, for $p_{in} \geq 0$, $\phi_i(\tau p_{in}, 0, u) = \tau^{\gamma_i}\, \phi_i(p_{in}, 0, u), \quad \forall\, \tau \geq 0$.

Hypothesis (F1) implies that (1.4.33) is a concave maximization problem; hypothesis (F2) implies that $q_{it} = 0$ is a strictly feasible solution of (1.4.33) if $p_{in} > 0$; hypothesis (F3) states that if the normal force is equal to zero at a contact point, then the only admissible tangential force is zero (this is the "balanced" assumption on the set of feasible forces mentioned above); finally, hypothesis (F4) is quite mild and easily satisfied by such friction functions as (1.4.29) and (1.4.30). Under these four assumptions (F1)–(F4), it is not difficult to show that the maximization problem (1.4.33) is completely equivalent to its Fritz-John optimality conditions, which we can write as the following mixed complementarity system: for all $j = 1, \ldots, M_i$ and $\ell = 1, \ldots, m_i$,

$$\begin{aligned} (p_{in})^{\gamma_i} \nabla \psi_{ij}(u)^T (u - u_{\text{ref}}) - \sum_{\ell=1}^{m_i} \lambda_{i\ell} \frac{\partial \phi_{i\ell}(p_{in}, p_{it}, u)}{\partial p_{ij}} &= 0, \\ 0 \le \lambda_{i\ell} \perp \phi_{i\ell}(p_{in}, p_{it}, u) \le 0. \end{aligned} \tag{1.4.34}$$

Note the special multiplier $(p_{in})^{\gamma_i}$ in the first term of the first equation; this multiplier is justified by the hypotheses (F1)–(F4). If $\phi_i(p_{in}, \cdot, u)$ is a linear function for each fixed (p_{in}, u), (e.g. for polygonal friction laws such as (1.4.30)), then (1.4.33) is a linear program in the variable q_{it}. In this case, the Fritz-John conditions can be replaced by the KKT conditions where we simply drop the multiplier $(p_{in})^{\gamma_i}$ associated with the objective function.

In summary, we conclude that under the hypotheses (F1)–(F4), the discrete frictional contact problem is equivalent to the problem of finding a tuple $(p_{in}, p_{ij}, u, \lambda_{i\ell})$ for all $i = 1, \ldots, n_c$, $j = 1, \ldots, M_i$, and $\ell = 1, \ldots, m_i$, satisfying conditions (1.4.31), (1.4.32), and (1.4.34). Clearly, these conditions define a MiCP $(\mathbf{F})$, which contains some highly nonlinear functions in general. The form of the defining function $\mathbf{F}$ of this MiCP is as follows:

$$\mathbf{F}(u, \mathbf{p}_t, \mathbf{p}_n, \lambda) \equiv$$

$$\begin{pmatrix} F(u) + \sum_{i=1}^{n_c} p_{in} \nabla \psi_{in}(u) + \sum_{i=1}^{n_c} \sum_{j=1}^{M_i} p_{ij} \nabla \psi_{ij}(u) - f_{\text{ext}} \\ \left[(p_{in})^{\gamma_i} \nabla \psi_{ij}(u)^T (u - u_{\text{ref}}) - \sum_{\ell=1}^{m_i} \lambda_{i\ell} \frac{\partial \phi_{i\ell}(p_{in}, p_{it}, u)}{\partial p_{ij}} \right]_{i=1}^{n_c} \Bigg|_{j=1}^{M_i} \\ - \left[\psi_{in}(u) \right]_{i=1}^{n_c} \\ - \left[\phi_{i\ell}(p_{in}, p_{it}, u) \right]_{i=1}^{n_c} \Big|_{\ell=1}^{m_i} \end{pmatrix}.$$

If in addition for all $i = 1, \ldots, n_c$ and all pairs (p_{in}, u) with $p_{in} \ge 0$, $\phi_i(p_{in}, \cdot, u)$ is an affine function, then the same conclusion holds without the multiplier $(p_{in})^{\gamma_i}$ in (1.4.34) and the above function $\mathbf{F}$.

1.4.7 Elastoplastic structural analysis

The application of mathematical programming (particularly, complementarity) theory and methods to analyze elastoplastic structures dates back to the late 1960s and early 1970s. Largely due to the influence of Giulio Maier from the Italian school, quadratic programming and linear complementarity were the initial tools employed in dealing with this kind of structural analysis. Subsequently, along with the development of the NCP methodology and the availability of computer softwares, more sophisticated models were introduced to handle geometric nonlinearities, hardening and softening laws, as well as large-scale finite-element discretized structures.

In what follows, we consider a discretized, elastoplastic, semirigidly connected planar structure undergoing deformation when it is acted upon by external loads. The structure is discretized into M finite elements. For each element $m \in \{1, \ldots, M\}$, let Q^m, $q^m \in \mathbb{R}^3$ denote, respectively, the (unknown) vectors of generalized stresses and strains, and let $u^m \in \mathbb{R}^6$ denote the (unknown) vector of nodal displacements. Let $f^m_{\text{ext}} \in \mathbb{R}^6$ denote the vector of external forces acting on the m-th member element, which is of length L_m and at angle θ_m to some horizontal reference axis. Let

$$c_m \equiv \cos\theta_m \quad \text{and} \quad s_m \equiv \sin\theta_m.$$

It is convenient to introduce two matrices

$$A^m \equiv \begin{bmatrix} c_m & -s_m/L_m & -s_m/L_m \\ s_m & c_m/L_m & c_m/L_m \\ 0 & 1 & 0 \\ -c_m & s_m/L_m & s_m/L_m \\ -s_m & -c_m/L_m & -c_m/L_m \\ 0 & 0 & 1 \end{bmatrix} \quad \text{and} \quad A^m_\pi \equiv \begin{bmatrix} c_m & -s_m \\ s_m & c_m \\ 0 & 0 \\ -c_m & s_m \\ -s_m & -c_m \\ 0 & 0 \end{bmatrix}$$

and the auxiliary displacement vector

$$\delta^m \equiv (A^m_\pi)^T u^m = \begin{bmatrix} \delta^m_n \\ \delta^m_t \end{bmatrix}.$$

In terms of δ^m, we define the angle of member chord rotation and the length of the deformed member chord:

$$\rho_m \equiv \arctan \frac{\delta^m_t}{L_m - \delta^m_n} \quad \text{and} \quad L'_m \equiv \sqrt{(L_m - \delta^m_n)^2 + (\delta^m_t)^2}.$$

Let

$$c'_m \equiv \cos\rho_m \quad \text{and} \quad s'_m \equiv \sin\rho_m,$$

$$Z^m \equiv \left[\begin{array}{ccc} 1 - c'_m & -s'_m/L'_m & -s'_m/L'_m \\ s'_m & 1/L_m - c'_m/L'_m & 1/L_m - c'_m/L'_m \end{array} \right].$$

Finally, we define

$$C^m \equiv (\, A^m - A^m_\pi Z^m \,)^T.$$

Notice that through the auxiliary vector δ^m, the matrix C^m is a function of the displacement vector u^m.

Simple equilibrium considerations in the deformed state lead to the following balancing equation of forces and stresses:

$$f^m_{\text{ext}} = (\, C^m \,)^T Q^m. \tag{1.4.35}$$

The element kinematic relations between member deformations q^m and nodal displacements u^m can be written in the form:

$$(\, A^m \,)^T u^m = q^m + \left[\begin{array}{c} \delta^m_n - q^m_1 \\ \delta^m_t - \rho_m \\ \delta^m_t / L_m - \rho_m \end{array} \right]. \tag{1.4.36}$$

The latter two expressions are the static-kinematic equations of the structural model at each member element. We next describe the constitutive laws. These are summarized by the following equations:

$$q^m = e^m + p^m \tag{1.4.37}$$

$$Q^m = S^m e^m + R^m_b \tag{1.4.38}$$

$$p^m = N^m \lambda^m \tag{1.4.39}$$

$$0 \leq \lambda^m \perp \phi^m = (\, N^m \,)^T Q^m - H^m \lambda^m - R^m \leq 0. \tag{1.4.40}$$

A brief explanation of these equations is as follows. Equation (1.4.37) expresses the total element strains q^m as the sum of elastic e^m and plastic p^m strains. Equation (1.4.38) is the stress-strain relation augmented by the vector R^m_b, which accounts for the "bowing" effect; R^m_b is a 3-dimensional vector whose last two components are zero and the first component is a highly nonlinear function of the element axial force Q^m_1; S^m is an element elastic stiffness matrix; the entries of S^m are nonlinear functions of Q^m_1 and involve some constants pertaining to the semirigid connections at the two ends of the element. Thus (1.4.38) is in fact a nonlinear equation. In some applications, such as those involving small displacements, each S^m is symmetric positive semidefinite. Relations (1.4.39)–(1.4.40) describe the plastic holonomic (path-independent) constitutive laws, which define the plastic strains p^m as linear functions of the plastic multipliers λ^m, where N^m is a constant matrix containing the unit outward normals to the piecewise linearized yield hyperplanes; ϕ^m is the element yield function defined in terms of the element stress Q^m and the hardening (or softening) defined by $H^m\lambda^m + R^m$. The yield function is restricted to be nonpositive and complementary to the multiplier λ^m. Complementarity in this context stipulates that yielding can only occur at a stress point on the yield surface and implies that no local loading is allowed (i.e., $\phi_i^m < 0$ and $\lambda_i^m > 0$ is not acceptable). Finally, the matrix H^m can be positive or negative definite depending on whether there is hardening or softening in the element.

From equations (1.4.36)–(1.4.39), we can eliminate the variables q^m, e^m, Q^m, and p^m and substitute into the expressions (1.4.38) and (1.4.40); after some algebraic manipulation, we obtain the following MiCP for the member element m:

$$\begin{pmatrix} -\phi^m \\ 0 \end{pmatrix} = \begin{pmatrix} R^m \\ -f^m_{\text{ext}} \end{pmatrix} +$$

$$\begin{bmatrix} K^m_{\lambda\lambda} & K^m_{\lambda u} \\ K^m_{u\lambda} & K^m_{uu} \end{bmatrix} \begin{pmatrix} \lambda^m \\ u^m \end{pmatrix} + \begin{bmatrix} (N^m)^T \\ -(C^m)^T \end{bmatrix} (S^m R^m_q - R^m_b),$$

$$0 \leq \lambda^m \perp \phi^m \leq 0,$$

where

$$\begin{bmatrix} K^m_{\lambda\lambda} & K^m_{\lambda u} \\ K^m_{u\lambda} & K^m_{uu} \end{bmatrix} \equiv \begin{bmatrix} (N^m)^T S^m N^m + H^m & -(N^m)^T S^m C^m \\ -(C^m)^T S^m N^m & (C^m)^T S^m C^m \end{bmatrix}$$

and

$$R^m_q \equiv C^m u^m - q^m.$$

Although written in the form similar to that of a mixed LCP, the above system is actually highly nonlinear because of the nonlinear dependence of the matrix functions S^m and C^m on the variables u^m and Q^m_1.

The overall structural model is derived from the above element models and can be stated as the following MiCP:

$$\begin{pmatrix} -\mathbf{\Phi} \\ 0 \end{pmatrix} = \begin{pmatrix} R \\ -f_{\text{ext}} \end{pmatrix} + \begin{bmatrix} K_{\lambda\lambda} & K_{\lambda u} \\ K_{u\lambda} & K_{uu} \end{bmatrix} \begin{pmatrix} \mathbf{\Lambda} \\ \mathbf{u} \end{pmatrix} + \begin{bmatrix} N^T \\ -C^T \end{bmatrix} (SR_q - R_b)$$

$$0 \leq \mathbf{\Lambda} \perp \mathbf{\Phi} \leq 0, \tag{1.4.41}$$

where

$$K_{\lambda\lambda} \equiv N^T SN + H, \quad K_{u\lambda} = K^T_{\lambda u} \equiv -C^T SN, \quad K_{uu} \equiv C^T SC;$$

the matrices S, N, H are block diagonal matrices with diagonal blocks S^m, N^m, and H^m, respectively; the vectors $\mathbf{\Phi}$, $\mathbf{\Lambda}$, R, R_q, and R_b are concatenations of the element vectors ϕ^m, λ^m, R^m, R_q^m, and R_b^m, respectively. The vector $\mathbf{u}$ consists of the nodal displacements of the structure; the m-th element displacement vector u^m can be extracted from $\mathbf{u}$ according to the particular structure and the finite element discretization. The matrix C is assembled from the element matrices C^m, resulting in the vector $C\mathbf{u}$ being the concatenation of the element vectors $C^m u^m$. Due to physical properties, C has linearly independent columns.

The model (1.4.41) is highly complex; we are not aware of any formal study of the model's analytical properties. Existing methods for solving the model are rather ad hoc and lack a sound basis; in particular, there has been no attempt in applying the algorithms described in this book for solving the model.

In what follows, we describe a special elastoplastic structural problem with nonlinear hardening/softening that lies at the heart of stepwise holonomic elastic-plastic analysis. In essence, the problem describes the finite holonomic behavior of a discrete structural system and aims to find its response to a given monotonic external load. In the compact statement of the problem, the displacement components are eliminated from (1.4.41), resulting in a formulation that involves only the stress and strain quantities. The model further assumes a symmetric positive definite structural stiffness matrix S and a hardening/softening law described by the nonlinear function $\mathbf{H}(\mathbf{\Lambda})$. The resulting model can be formulated as the NCP:

$$0 \leq \mathbf{\Lambda} \perp -\mathbf{\Phi} = \mathbf{q} + \mathbf{H}(\mathbf{\Lambda}) + N^T Z N \mathbf{\Lambda} \geq 0, \tag{1.4.42}$$

where the constant vector $\mathbf{q}$ contains the plastic yield limits and the load history applied to the structure and Z is the matrix given by

$$Z \equiv S - SC(C^T SC)^{-1} C^T S.$$

Since C has linearly independent columns and S is symmetric positive definite, Z can be easily seen to be symmetric positive semidefinite. Furthermore, Z must be singular and its null space coincides with the column space of C^T. The hardening/softening function $\mathbf{H}$ is always nonnegative; it is separable in some applications. A large class of hardening laws $\mathbf{H}(\Lambda)$ is of the form:

$$\mathbf{H}(\mathbf{\Lambda}) \equiv H\mathbf{\Lambda}^k, \tag{1.4.43}$$

where H can be one of the following types: a positive diagonal matrix (Koiter's noninteracting hardening), a certain symmetric positive semidefinite matrix (Prager's kinematic hardening), or a special positive matrix (isotropic-type hardening); and $\mathbf{\Lambda}^k$ is the vector with components λ_i^k with $k \in (0, 1)$ being a given exponent. There are hardening laws of other types as well as softening laws. An example of a softening law is when $\mathbf{H}(\mathbf{\Lambda})$ is separable with each component $H_i(\lambda_i)$ being a nonnegative, convex, monotonically decreasing function of its argument. Typically, the softening laws lead to challenging NCPs whose properties are not fully understood. In particular, a satisfactory existence theory and provably convergent computational methods for the NCP (1.4.42) with a separable softening function $\mathbf{H}(\mathbf{\Lambda})$ as described are to date not known to be available.

Even more challenging is the problem of limit analysis where the following optimization problem has to be solved:

$$\begin{array}{ll} \text{maximize} & \mu \\ \text{subject to} & 0 \leq \mathbf{\Lambda} \perp \mu \mathbf{q} + \mathbf{H}(\mathbf{\Lambda}) + N^T Z N \mathbf{\Lambda} \geq 0. \end{array} \tag{1.4.44}$$

This is an instance of an MPEC, which will be discussed further in Subsection **1.4.10**. This problem is most difficult with a softening function $\mathbf{H}$ that could destroy many useful properties of the function

$$\Lambda \mapsto \mathbf{H}(\Lambda) + N^T Z N \Lambda,$$

such as its monotonicity.

1.4.8 Nonlinear obstacle problems

Many obstacle problems in mathematical physics naturally lead to complementarity problems and variational inequalities. In these contexts, the complementarity condition is intimately related to the notion of a free boundary that separates two physical states. Part of the objective of solving an obstacle problem is to determine this free boundary which, in essence, is equivalent to identifying the active constraints in the complementarity condition. Quadratic programming and LCP methods for solving some simple obstacle problems are well documented. In contrast, obstacle problems leading to (mixed) NCPs are less studied in the complementarity literature. Typically, these obstacle problems are formulated in infinite-dimensional spaces; their discretizations easily lead to finite-dimensional problems of very large size. The latter problems offer challenges for the NCP methods due to their nonlinearity and large size.

The von Kármán thin plate problem with obstacles is one which is not known to have been much investigated by numerical complementarity methods. In what follows, we give the formulation of this problem as a functional complementarity system. In this problem, the points of a thin elastic plate are given in a fixed, right-handed Cartesian coordinate system $0x_1x_2x_3$. The middle plane of the undeformed plate, which is assumed to have a constant thickness h, coincides with the $0x_1x_2$ plane. The points $(x_1, x_2, 0)$ of the undeformed plate constitute an open, bounded, connected subset $\Omega \subset \mathbb{R}^2$ with a Lipschitz boundary bd Ω. We denote by $u = (u_1, u_2)$ the horizontal and by ξ the vertical displacement of the point $x \in \Omega$. The plate is subjected to a distributed load $(0, 0, f)$, $f = f(x) \in \mathcal{L}^2(\Omega)$, per unit area of the middle surface. The shape of the obstacle is prescribed by a strictly concave function $\psi(x) \in \mathcal{H}^2(\Omega)$. An example of such a function is $\psi(x_1, x_2) \equiv -(x_1^2 + x_2^2)$. Let

$$K \equiv \frac{Eh^3}{12(1-\nu^2)}$$

be the plate's bending rigidity with E being Young's modulus of elasticity and $\nu \in (0, 0.5)$ the Poisson ratio.

In formulating the obstacle problem, we employ the Einstein summation convention with respect to a repeated index; i.e., we write $a_i b_{ij}$ to mean $\sum_i a_i b_{ij}$. Other notation used locally is as follows. For a real-valued function $\phi(x_1, x_2)$, $\phi_{,i} \equiv \partial\phi/\partial x_i$; $\partial\phi/\partial\mathbf{n} \equiv \nabla\phi^T\mathbf{n}$ is the directional derivative of ϕ in the direction of the outward unit normal vector $\mathbf{n}$ to $\partial\Omega$; for a vector-valued function $F(x_1, x_2)$, $F_{i,j} \equiv \partial F_i/\partial x_j$; finally $\Delta \equiv \sum_i \partial^2/\partial x_i^2$ is the Laplace operator, and $\Delta^2 \equiv \Delta\Delta$ is the biharmonic operator.

The von Kármán thin plate problem with obstacles is to find two functions $u(x_1, x_2) \equiv (u_1(x_1, x_2), u_2(x_1, x_2))$ and $\xi(x_1, x_2)$ satisfying the following system of partial differential equations and inequalities as well as the given boundary conditions:

$$K\,\Delta^2\xi - h\,(\sigma_{\alpha\beta}\xi_{,\beta})_{,\alpha} \geq f(x) \qquad \text{in } \Omega \tag{1.4.45}$$

$$(\sigma_{\alpha\beta})_{,\beta} = 0 \qquad \text{in } \Omega \tag{1.4.46}$$

$$\xi(x) \geq \psi(x) \qquad \text{in } \Omega \tag{1.4.47}$$

$$[K\,\Delta^2\xi - h\,(\sigma_{\alpha\beta}\xi_{,\beta})_{,\alpha}]\,[\xi(x) - \psi(x)] = 0 \qquad \text{in } \Omega \tag{1.4.48}$$

$$\psi(x),\ u_\alpha,\ \frac{\partial\psi}{\partial\mathbf{n}} \quad \text{all given} \qquad \text{on } \partial\Omega, \tag{1.4.49}$$

where $\sigma \equiv (\sigma_{\alpha\beta})$ and $\varepsilon \equiv (\varepsilon_{\alpha\beta})$ denote, respectively, the stress tensor and strain tensor in the plan of the plate that are linked by the elasticity tensor $C \equiv (C_{\alpha\beta\gamma\delta})$; i.e.,

$$\sigma_{\alpha\beta} = C_{\alpha\beta\gamma\delta}\left(\varepsilon_{\gamma\delta}(u) + \tfrac{1}{2}\,\psi_{,\gamma}\psi_{,\delta}\right)$$

with

$$\varepsilon_{\gamma\delta}(u) \equiv \tfrac{1}{2}\left(u_{\gamma,\delta} + u_{\delta,\gamma}\right) = \tfrac{1}{2}\left(\frac{\partial u_\gamma}{\partial x_\delta} + \frac{\partial u_\delta}{\partial x_\gamma}\right).$$

In general, each component $C_{\alpha\beta\gamma\delta}$ is a function in $\mathcal{L}^\infty(\Omega)$ and the tensor C satisfies the symmetry and ellipticity properties; i.e.,

$$C_{\alpha\beta\gamma\delta} = C_{\beta\alpha\gamma\delta} = C_{\gamma\delta\alpha\beta},$$

and there exists a constant $c > 0$ such that for all symmetric matrices $\varepsilon = (\varepsilon_{\alpha\beta})$,

$$C_{\alpha\beta\gamma\delta}\,\varepsilon_{\alpha\beta}\,\varepsilon_{\gamma\delta} \geq c\,\varepsilon_{\alpha\beta}\,\varepsilon_{\alpha\beta}.$$

For an isotropic plate, C is a constant tensor and has the form

$$C_{\alpha\beta\lambda\mu} = \frac{E}{2(1-\nu^2)}\left[\,(1-\nu)\,(\,\delta_{\alpha\lambda}\delta_{\beta\mu} + \delta_{\alpha\mu}\delta_{\beta\lambda} + 2\nu\,\delta_{alpha\beta}\delta_{\lambda\mu}\,)\,\right],$$

where $\delta_{\alpha\beta}$ is the standard Kronecker delta notation.

Upon discretization, the problem (1.4.45)–(1.4.49) leads to a large MiCP where the defining function is a multivariate cubic polynomial. A simplified one-dimensional version of the problem pertains to the computation of a scalar function $w(x)$ satisfying

$$\begin{aligned} w'''' - c\,w''\,(w')^2 &\geq f(x) & \forall\, x \in (0,L) \\ w(x) &\geq \psi(x) & \forall\, x \in (0,L) \\ [\,w'''' - c\,w''\,(w')^2 - f(x)\,]\,[\,w(x) - \psi(x)\,] &= 0 & \forall\, x \in (0,L) \end{aligned}$$

and given boundary conditions at $x = 0, L$, where c and L are two given positive scalars; and $f(x)$ and $\psi(x)$ are given univariate functions.

The obstacle Bratu problem is a model for nonlinear diffusion phenomena that take place, for example, in combustion and semiconductors. This is also an example of NCP that depends on a critical parameter, which has an important effect on the properties of the defining function. Posed in a function space, the problem is defined as follows. Given an bounded domain Ω in $\mathbb{R}^2$ and an obstacle function $\psi \geq 0$, determine a function $u(x_1, x_2)$ of two arguments satisfying

$$0 \leq \psi - u \perp \lambda e^u + \Delta u \geq 0, \quad \text{in } \Omega,$$

and $u = 0$ on $\partial\Omega$. With U being the unit square $[0,1]^2$ in the plane, a straightforward finite-difference discretization of the Laplace operator Δ by a regular mesh of N^2 points and size $h > 0$ leads to the NCP (F_λ), where

$$F_\lambda(x) \equiv A(x - \psi) + \lambda\,h^2\,e^{(\psi - x)}, \quad x \in \mathbb{R}^{N^2};$$

here A is a standard block tridiagonal matrix that corresponds to the discretization and $e^{\psi - x}$ is the N^2-vector with entries $e^{\psi_{ij} - x_{ij}}$. Although A is a symmetric positive definite matrix, values of the "eigen-parameter" λ have a direct effect on properties of F.

1.4.9 Pricing American options

The numerical valuation of options and derivative securities is of central importance in financial management. In essence an option is a financial instrumental that offers its owner the right (but not the obligation) to purchase or sell an underlying asset at a prescribed exercise price within a fixed period. An American option has the special feature that the option owner can exercise the option at any time before its expiry, whereas a European option does not have this early exercise feature. In 1973, Black and Scholes, under a set of market assumptions, developed for the first time an explicit formula for the valuation of a European option of a risky asset. Due to its enormous impact on financial economics, this work won Scholes (along with Merton) a Nobel Prize in Economics Science in 1997.

Unlike its European counterpart, the pricing of a vanilla (i.e., basic) American option is far more complicated. This is due to the fact that this pricing problem is intrinsically an optimal stopping problem of a martingale that can be formulated as a free boundary problem. Upon a time-state discretization, this free boundary problem becomes a large-scale, specially structure linear complementarity problem that can be solved by a host of effective direct and/or iterative methods. Many exotic options (i.e., those that are variations and/or extensions of the vanilla options) can be treated similarly by numerical LCP methods. In what follows, we describe the valuation of a vanilla American option in the presence of transaction costs. Depending on the assumption of the latter costs, we obtain either an LCP or an NCP after a time-state discretization.

Consider a risky asset whose random price pattern $S(t)$ as a function of time t is assumed to satisfy the following stochastic differential equation:

$$dS = (\mu S - D(S,t))dt + \sigma S \, dW$$

where μ and σ are positive constants with μ being the drift of the price process and σ being the "volatility" parameter; $D(S,t)$ is the dividend rate of the asset, and dW is a standard Wiener process with mean zero and variance dt. Let $r > 0$ denote the constant interest rate. Let $\Lambda(S,t)$ denote the given payoff function of an American option whose time of expiry is denoted T. Two most common forms of this function $\Lambda(S,t)$ are:

$$\Lambda(S,t) \equiv \max(E-S,0) \quad \text{for a call option}$$

$$\Lambda(S,t) \equiv \max(S-E,0) \quad \text{for a put option.}$$

The original Black-Scholes model assumes continuous rehedging and the absence of transaction costs in trade in the underlying assets. Under these two basic assumptions (and several others, most notably, the assumptions of no arbitrage and the constant asset volatility), it can be shown that the value $V(S,t)$ of the American option must satisfy a partial differential (PD) linear complementarity system defined by the basic Black-Scholes operator:

$$\mathcal{L}_{\mathrm{BS}} \equiv \frac{\partial}{\partial t} + \tfrac{1}{2}\sigma^2 S^2 \frac{\partial^2}{\partial S^2} + (rS - D(S,t))\frac{\partial}{\partial S} - r.$$

Numerous approaches have been suggested to handle the presence of transaction costs. Leland proposed one such approach that is based on a discrete hedging strategy. It is assumed that the asset price follows a discrete random walk given by

$$\delta S = (\mu S - D(S,t))\delta t + \sigma S \phi \sqrt{\delta t},$$

where $\delta t > 0$ is the time interval of rehedging and ϕ is a random number drawn from the standardized normal distribution. As a measure of the level of the transaction costs, the "gamma", which is the second partial derivative of $V(S,t)$ with respect to the asset price S, i.e., $V_{SS} \equiv \partial^2 V/\partial S^2$, provides a reasonable measure of the mishedging due to the discrete nature of the hedging strategy. A scalar-valued function $F(S, V_{SS})$ can then be introduced to model various transaction cost structures and hedging strategies. Examples of this function include the following. An extended Leland model in which the transaction cost is assumed to comprise of 3 components: a fixed component (k_1), a cost proportional to the number of shares traded (k_2N), and a cost proportional to the value traded (k_3NS), where k_1, k_2, and k_3 are positive constants and $N > 0$ is the number of shares traded to rehedge. Up to order $O(\sqrt{\delta t})$, the number N is equal to $\sqrt{\delta t}\sigma S|\phi V_{SS}|$; thus the expected number of share trades is approximated by $\sqrt{2\delta t/\pi}\sigma S|V_{SS}|$. This leads to a function F given by:

$$F(S,x) = \frac{k_1}{\delta t} + (k_2 + k_3 S)\sqrt{\frac{2}{\pi\,\delta t}}\,\sigma S\,|\,x\,|, \qquad (1.4.50)$$

Another cost function corresponds to a market practice model with the strategy of rehedging to the Black-Scholes value when the "delta" (i.e., $\partial V/\partial S$) moves outside of a "hedging bandwidth" of size ε/S, where ε is a positive function of S, t, as well as V and its derivatives. This yields a function F given by:

$$F(S,x) = \frac{\sigma^2}{\varepsilon}\left[k_1 + (k_2 + k_3 S)\frac{\sqrt{\varepsilon}}{S}\right] x^2. \tag{1.4.51}$$

A third function corresponds to a continuous time, small cost model that is based upon utility maximization:

$$F(S,x) = \\ \frac{e^{-r(T-t)}}{\gamma}\left(\frac{3k_3\gamma^2 S^4\sigma^3}{8\delta^2}\right)^{2/3}\left|x - \frac{e^{-r(T-t)}(\mu - r)}{\gamma S^2 \sigma^2}\right|^{4/3}, \tag{1.4.52}$$

where γ is a positive index of risk aversion and $\delta \equiv e^{-r(T-t)}$. Notice that the three functions $F(S,x)$ in (1.4.50), (1.4.51), and (1.4.52) are all nonnegative for all $S \geq 0$.

With the above setting and using an argument based on an optimally hedged portfolio, it can be shown that the option value $V(S,t)$ satisfies the following partial differential complementarity system: for all (S,t) in $(0,\infty) \times [0,T)$,

$$\begin{aligned} 0 &\leq V(S,t) - \Lambda(S,t), \\ 0 &\geq \mathcal{L}_{\mathrm{BS}}(V) - F(S, V_{SS}), \\ 0 &= [V(S,t) - \Lambda(S,t)][\mathcal{L}_{\mathrm{BS}}(V) - F(S, V_{SS})], \end{aligned} \tag{1.4.53}$$

and some suitable boundary conditions at $t = T$ and $S = 0$ and large values of S. The way the system (1.4.53) leads to a finite-dimensional LCP is through a finite-difference approximation of the time and state partial derivatives and the operator $\mathcal{L}_{\mathrm{BS}}$. To simplify the following discussion, we take the dividend function $D(S,t)$ to be a constant proportion of S; that is $D(S,t) = d\,S$ for some constant $d > 0$. We further assume that $r > d$.

For the discretization of the partial derivatives, we introduce a regular grid in the time-state space with grid sizes δt and δS and with the state variable S truncated so that

$$0 \leq S \leq N\,\delta S$$

where N is (large) positive integer. Let $M \equiv T/\delta t$ be the total number of time steps in the time discretization (the hedging interval δt is chosen such that M is an integer). We write: for $0 \leq n \leq N$ and $0 \leq m \leq M$,

$$V_n^m \equiv V(n\delta S, m\delta t) \quad \text{and} \quad \Lambda_n^m \equiv \Lambda(n\delta S, m\delta t).$$

The boundary conditions imply that V_0^m, V_N^m and V_n^M are given for all $m = 0, \dots, M$ and $n = 0, \dots, N$; moreover, the payoff values Λ_n^m are all given. Our goal is to calculate the discretized option values V_n^m for all $n = 1, \dots, N-1$ and $m = 0, \dots, M-1$. We employ the following finite-difference approximation of the time and state partial derivatives:

$$\frac{\partial V(S,t)}{\partial t} \approx \frac{V(S,t+\delta t) - V(S,t)}{\delta t},$$

$$\frac{\partial V(S,t)}{\partial S} \approx \theta \frac{V(S+\delta S,t) - V(S-\delta S,t)}{2\delta S} +$$

$$(1-\theta) \frac{V(S+\delta S,t+\delta t) - V(S-\delta S,t+\delta t)}{2\delta S},$$

and

$$\frac{\partial^2 V(S,t)}{\partial S^2} \approx \theta \frac{V(S+\delta S,t) - 2V(S,t) + V(S-\delta S,t)}{(\delta S)^2} +$$

$$(1-\theta) \frac{V(S+\delta S,t+\delta t) - 2V(S,t+\delta t) + V(S-\delta S,t+\delta t)}{(\delta S)^2},$$

where $\theta \in [0,1]$ is a given scalar. These approximations are of first order in time and second order in state, that is, $O(\delta t)$ and $O((\delta S)^2)$, where a function $f(x)$ is said to be $O(g(x))$ if

$$\limsup_{g(x)\to 0} \frac{|f(x)|}{|g(x)|} < \infty.$$

Thus,

$$\mathcal{L}_{\mathrm{BS}}(V) \approx -r\,V(S,t) + \frac{V(S,t+\delta t) - V(S,t)}{\delta t} +$$

$$\tfrac{1}{2}\sigma^2 S^2 \left[\theta \frac{V(S+\delta S,t) - 2V(S,t) + V(S-\delta S,t)}{(\delta S)^2} + \right.$$

$$\left. (1-\theta) \frac{V(S+\delta S,t+\delta t) - 2V(S,t+\delta t) + V(S-\delta S,t+\delta t)}{(\delta S)^2} \right] +$$

$$(r-d)\,S \left[\theta \frac{V(S+\delta S,t) - V(S-\delta S,t)}{2\delta S} + \right.$$

$$\left. (1-\theta) \frac{V(S+\delta S,t+\delta t) - V(S-\delta S,t+\delta t)}{2\delta S} \right].$$

This approximation is of the order $O(\delta t) + O((\delta S)^2)$. Thus by choosing δt to be $O((\delta S)^2)$, the partial differential operator $\mathcal{L}_{\rm BS}(V)$ is approximated by a finite-difference expression of the order $O((\delta S)^2)$.

After substituting the above finite-difference formulas into (1.4.53) and by an easy manipulation, we obtain a family of M finite-dimensional NCPs, each corresponding to a time step $t \equiv m\delta t$ and defined by the variable $\mathbf{V}^m \equiv (V_i^m)_{i=1}^{N-1}$ for $m = M-1, M-2 \ldots, 1, 0$. The initial condition of the PD LCP (1.4.53) is given at the terminal time T; thus, we need to solve the NCPs backward in time, starting at time $(M-1)\,\delta t$, using the given initial values $V(S,T)$ at time T and also the boundary values $V(0,t)$ at the zero boundary state $S = 0$ and $V(N\delta S, t)$ at the other boundary state $S = N\delta S$, which is a truncation of the true boundary state $S = \infty$. We proceed backward until we reach time $t = 0$. At each time t, the values $V(S, t+\delta t)$ computed at time $t + \delta t$ are part of the inputs to the NCP at time t. Define the following scalars: for $n = 1, \ldots, N-1$,

$$
\begin{aligned}
a_n &\equiv -\frac{\theta}{2}\left[\sigma^2 n^2 - (r-d)n\right], & a'_n &\equiv -\frac{1-\theta}{2}\left[\sigma^2 n^2 - (r-d)n\right],\\
b_n &\equiv \frac{1}{\delta t} + r + \theta\,\sigma^2 n^2, & b'_n &\equiv -\frac{1}{\delta t} + r + (1-\theta)\,\sigma^2 n^2,\\
c_n &\equiv -\frac{\theta}{2}\left[\sigma^2 n^2 + (r-d)n\right], & c'_n &\equiv -\frac{1-\theta}{2}\left[\sigma^2 n^2 + (r-d)n\right].
\end{aligned}
$$

The NCP at time step $t = m\delta t$ is of the form:

$$0 \leq \mathbf{V}^m - \boldsymbol{\Lambda}^m \perp \mathbf{q}^m + \mathbf{M}\mathbf{V}^m + \mathbf{F}(\mathbf{V}^m) \geq 0, \tag{1.4.54}$$

where $\mathbf{M}$ is the following $(N-1) \times (N-1)$ tridiagonal matrix:

$$\mathbf{M} \equiv \begin{bmatrix} b_1 & c_1 & 0 & 0 & \cdots & 0 \\ a_2 & b_2 & c_2 & 0 & \cdots & 0 \\ 0 & a_3 & b_3 & c_3 & & 0 \\ & & \ddots & \ddots & \ddots & \\ & & & a_{N-2} & b_{N-2} & c_{N-2} \\ & & & 0 & a_{N-1} & b_{N-1} \end{bmatrix}, \tag{1.4.55}$$

$\mathbf{q}^m \equiv \mathbf{M}'\mathbf{V}^{m+1}$, where $\mathbf{M}'$ is closely tied to the matrix $\mathbf{M}$:

$$\mathbf{M}' \equiv \begin{bmatrix} b_1' & c_1' & 0 & 0 & \cdots & 0 \\ a_2' & b_2' & c_2' & 0 & \cdots & 0 \\ 0 & a_3' & b_3' & c_3' & & 0 \\ & & \ddots & \ddots & \ddots & \\ & & & a_{N-2}' & b_{N-2}' & c_{N-2}' \\ & & & 0 & a_{N-1}' & b_{N-1}' \end{bmatrix},$$

and $\mathbf{F}^m(\mathbf{V}^m)$ is the $(N-1)$-dimensional vector function that is the finite-difference approximation of the function $F(S, V_{SS})$ at time $t = m\delta t$ and states $S = \delta S,\ 2\delta S,\ \cdots,\ (N-1)\delta S$. Since F is a nonnegative function, so is each $\mathbf{F}^m$. Observe that the matrix $\mathbf{M}$ is not symmetric. If

$$\frac{1}{\delta t} + r > |\, r - d \,|\, N,$$

then

$$b_n > |\, a_n \,| + |\, c_n \,|, \quad \forall\, n = 1, \ldots, N;$$

thus the matrix $\mathbf{M}$ is strictly row diagonally dominant, albeit not symmetric. Moreover, we may further restrict the time step δt in order for $\mathbf{M}$ (i.e. its symmetric part) to be positive definite. Such a condition can be easily derived and is omitted.

The functions $F(S, x)$ given by (1.4.50) and (1.4.52) contain the absolute value function, which is not differentiable in its argument. By means of a well-known substitution from linear programming, this absolute value function can be replaced by an additional complementarity relation, resulting in an alternative function that is smooth in its arguments. In what follows, we focus on the function given by (1.4.50) to illustrate this procedure. The resulting discretized problem is an LCP; the linearity of the complementarity problem is due to the piecewise linearity of the function $F(S, x)$ in the x argument.

For any scalar x, we can write

$$x = x^+ - x^- \quad \text{with} \quad |\, x \,| = x^+ + x^-,$$

where

$$(\, x^+, x^- \,) \geq 0 \quad \text{and} \quad (\, x^+ \,)(\, x^- \,) = 0. \tag{1.4.56}$$

Thus the function $F(S, x)$ given by (1.4.50) is equal to the function

$$G(S, x^+, x^-) = \frac{k_1}{\delta t} + (\, k_2 + k_3\, S\,) \sqrt{\frac{2}{\pi\, \delta t}}\, \sigma\, S\, (\, x^+ + x^- \,)$$

under the stipulation in (1.4.56). Note that G is linear in $x^{\pm}$. Making this substitution into (1.4.53), we obtain an equivalent formulation of this system as a partial differential LCP with $V(S,t)$, $W(S,t)$, and $Z(S,t)$ as the unknown functions: subject to the same initial and boundary conditions,

$$\begin{aligned} 0 &\leq V(S,t) - \Lambda(S,t), \\ 0 &\geq \mathcal{L}_{\mathrm{BS}}(V) - G(S,W,Z), \\ 0 &= [\, V(S,t) - \Lambda(S,t) \,] \, [\, \mathcal{L}_{\mathrm{BS}}(V) - G(S,W,Z) \,], \\ W(S,t) &= Z(S,t) - V_{SS} \\ 0 &\leq (\, W(S,t), Z(S,t) \,), \quad W(S,t)Z(S,t) = 0. \end{aligned} \tag{1.4.57}$$

With the substitution $W(S,t) = Z(S,t) - V_{SS}(S,t)$ in $G(S,W,Z)$, we obtain

$$G(S,W,Z) = \frac{k_1}{\delta t} + (\, k_2 + k_3 \, S \,) \sqrt{\frac{2}{\pi \, \delta t}} \, \sigma \, S \, (\, 2 \, Z - V_{SS} \,) \equiv \tilde{G}(S,Z,V).$$

By employing the same finite-difference formulas for the partial derivatives, the system (1.4.57) is approximated by a family of M finite-dimensional LCPs, with the m-th LCP being:

$$0 \leq \begin{pmatrix} \mathbf{V}^m \\ \mathbf{Z}^m \end{pmatrix} - \begin{pmatrix} \mathbf{\Lambda}^m \\ 0 \end{pmatrix} \perp \begin{pmatrix} \tilde{\mathbf{q}}^m \\ 0 \end{pmatrix} + \begin{bmatrix} \tilde{\mathbf{M}} & \mathbf{N} \\ \mathbf{Q} & \mathbf{I} \end{bmatrix} \begin{pmatrix} \mathbf{V}^m \\ \mathbf{Z}^m \end{pmatrix} \geq 0,$$

for some suitable constant vector $\tilde{\mathbf{q}}^m$ and matrices $\tilde{\mathbf{M}}$, $\mathbf{N}$, and $\mathbf{Q}$. Just like the matrix $\mathbf{M}$, the matrix $\tilde{\mathbf{M}}$ can be made strictly row diagonally dominant and/or positive definite by a suitable choice of the discretization steps. Moreover, the matrix $\mathbf{N}$ is nonnegative.

For the function $F(S,x)$ given by (1.4.52), we can eliminate the absolute value function in the same manner; but we still end up with an NCP due to the nonlinearity of F in the x argument.

The Black-Scholes operator $\mathcal{L}_{\mathrm{BS}}$ is an example of a diffusion-convection PD operator, with the term involving the second partial derivative with respect to the state S, $\partial^2/\partial S^2$, being the diffusion term and that involving the first partial derivative with respect to S, $\partial/\partial S$, being the convection term. In problems where the convection term dominates the diffusion term, (e.g., when the volatility σ is very small relative to the difference $r - d$), care must be taken in applying the finite-difference discretization to the partial derivative $\partial/\partial S$ in order to avoid solutions with oscillations due to numerical inaccuracies. In this regard, upwind schemes are quite popular in which the sign of the multiplicative coefficient of the convection term is taken into account in choosing the finite-difference approximation. There are some advanced upwind schemes that lead to nonlinear complementarity problems of the "implicit kind" defined as follows:

$$\begin{aligned} H(x, y, z) &= 0 \\ 0 \leq x \perp y &\geq 0, \end{aligned} \tag{1.4.58}$$

for some appropriate function $H : \mathbb{R}^{2n+m} \rightarrow \mathbb{R}^{n+m}$. In contrast to the MiCP in Definition **1.1.6**, where the relation between the complementarity variables is explicit; e.g., x and $y \equiv F(x)$ in the NCP (F), a complementarity problem of the form (1.4.58) distinguishes itself in that the relation between the complementary variables x and y is implicit in the function H. To highlight the latter feature, we call (1.4.58) an *implicit MiCP*.

The CP approach to pricing American options discussed above can be applied to many options of the exotic type and to options with multiple state variables. Options of the former type include Asian options in which the payoff at exercise depends on the price history of the asset. An example is a "lookback option" where the maximum or minimum of past asset price appears in the payoff function. Options with multiple state variables include options written on multiple assets and options with stochastic volatility. Suffices it to say that the CP approach provides a viable avenue for treating the pricing all these options with the early exercise feature.

1.4.10 Optimization with equilibrium constraints

Parametric VIs are problems where there is a parameter that is allowed to vary in a certain subset of an Euclidean space. To define these parametric problems formally, let $F : \mathcal{D} \times \mathcal{P} \subseteq \mathbb{R}^n \times \mathbb{R}^m \rightarrow \mathbb{R}^n$ be a given function of two arguments $(x, \mathbf{p})$; let $K : \mathbb{R}^m \rightarrow \mathcal{D}$ be a "multifunction" with values in $\mathcal{D}$; that is, for each $\mathbf{p} \in \mathbb{R}^m$, $K(\mathbf{p})$ is a (possibly empty) subset of $\mathcal{D}$. The *parametric family* of VIs is

$$\{ \text{VI } (K(\mathbf{p}), F(\cdot, \mathbf{p})) : \mathbf{p} \in \mathcal{P} \}$$

where $\mathbf{p}$ is the parameter varying in the given set $\mathcal{P}$.

Parametric VIs are a central ingredient in the class of **M**athematical **P**rograms with **E**quilibrium **C**onstraints, abbreviated as MPECs. These are constrained optimization problems that contain a family of parametric VIs as constraints:

$$\begin{array}{ll} \text{minimize} & \theta(x, y) \\ \text{subject to} & (x, y) \in Z \\ \text{and} & y \in \text{SOL}(K(x), F(\cdot, x)), \end{array} \tag{1.4.59}$$

where $\theta : \mathbb{R}^{n+m} \to \mathbb{R}$ is a given scalar-valued function and Z is a given subset in $\mathbb{R}^{n+m}$. The variables of this problem are of two types: x being the *design variable* and y the *state variable.* The respective roles played by these variables are quite different. A simple example of an MPEC is the limit analysis problem (1.4.44) in the elastoplastic structural application.

There are many applied contexts in which MPECs appear. In what follows, we briefly explain the "inverse problem" of an optimization problem, VI, or CP that naturally leads to an MPEC of the above type. In general every optimization problem, CP, or VI has its own inputs. For instance, in the LCP (q, M), the vector q and matrix M are the inputs; in the option pricing problem discussed in Subsection **1.4.9**, among the inputs are the volatility parameter σ, the interest rate r and other constants that are assumed given for this problem. By solving the VI/CP (or optimization problem), we obtain the *computed outputs.* In every applied context, these outputs have their own practical significance; typically, a set of *observed (or measured) outputs* is also available that may or may not be "consistent" with the computed outputs. For example, in the same option pricing problem, the computed outputs are the theoretical option prices, which are derived from a mathematical model, namely the model that is based on the Black-Scholes analysis. In reality, option prices are available from the stock markets and these market prices constitute the observed outputs. Obviously, due to many theoretical assumptions underlying the mathematical model, some of which are only approximations of reality, one would not expect the computed and observed outputs to be exactly the same. In general, given the discrepancy between these two kinds of "outputs", one can therefore pose the question of whether it is possible to determine a set of inputs so that the computed outputs (based on these revised inputs) will satisfy a certain minimization criterion with respect to the observed outputs. To formulate this inverse problem as the MPEC (1.4.59), simply let x be the unknown input vector and y be the computed output, $\theta(x, y)$ be whatever criterion one is interested in, and Z contain any additional constraints on the pair (x, y) that one deems important. Again using the option problem as an example, one may be interested in determining the volatility parameter (or more generally, a smooth surface of volatilities) so that the computed option prices will be as close to the market option prices as possible, where closeness is measured according to some distance function. The volatility parameters computed in this way are called *implied volatilities* to signify that they are implied by the market option values.

Engineering design problems are another major source of applications for MPECs; so is the classical Stackelberg leader-follower game; MPECs also include the bilevel optimization problems where the parametric VI corresponds to the first-order optimality conditions of an NLP. It is not possible to go into the details of these and other applications of MPECs; nor is it possible for us to explain the theory and algorithms of MPECs in this book. A main reason for introducing the MPEC is to give a strong motivation for the study of parametric VIs. Obviously, understanding how the solutions of the parametric VI $(K(x), F(\cdot, x))$ depend on the parameter x will be extremely useful for an in-depth examination of the MPEC.

1.4.11 CPs in SPSD matrices

To give an example of a CP (K, F) where the cone K is non-polyhedral, we mention the complementarity problem in symmetric positive semidefinite (SPSD) matrices. The linear version of this problem where F is an affine map was introduced by M. Kojima and his collaborators. The study of the CP in SPSD matrices is considerably more involved than the NCP; for one thing, such a study would necessarily entail a substantial background in matrix theory. For this reason, this book does not provide a comprehensive treatment of the CP in SPSD matrices; instead, we will rely on the exercises where the reader can find various results about such a CP.

The space of $n \times n$ real matrices can be identified with the Euclidean space $\mathbb{R}^{n^2}$. More precisely, the canonical mapping

$$\text{vec} : A \in \mathbb{R}^{n \times n} \mapsto \text{vec}(A) \in \mathbb{R}^{n^2},$$

where vec(A) is the n^2-dimensional vector whose components are the entries of A arranged row by row, defines an isomorphism between the two spaces $\mathbb{R}^{n\times n}$ and $\mathbb{R}^{n^2}$. For instance,

$$A \equiv \begin{bmatrix} 1 & 2 \\ 3 & 4 \end{bmatrix} \quad \text{and} \quad \text{vec}(A) = \begin{pmatrix} 1 \\ 2 \\ 3 \\ 4 \end{pmatrix}.$$

The usual scalar product of two vectors in $\mathbb{R}^{n^2}$ defines the *Frobenius (inner) product* of two matrices in $\mathbb{R}^{n\times n}$:

$$A \bullet B \equiv \text{vec}(A)^T \text{vec}(B) = \sum_{i,j=1}^{n} a_{ij} b_{ij} = \text{tr}(AB^T),$$

where "tr" denotes the trace of a matrix. Naturally, with this inner product, we can then say that two matrices A and B are perpendicular to mean

$A \bullet B = 0$; notationally, we continue to write $A \perp B$. The Frobenius inner product induces the Frobenius norm on the space of matrices:

$$\| A \|_F \equiv \sqrt{A \bullet A} = \sqrt{\sum_{i,j=1}^{n} a_{ij}^2}.$$

Let $\mathcal{M}^n$ denote the subspace of $\mathbb{R}^{n \times n}$ consisting of all the symmetric matrices of order n. We note that for any three symmetric matrices A, B, and C, we have

$$\text{tr}\,(ABC) = \sum_{i,j,k=1}^{n} a_{ik} b_{kj} c_{ij} = \text{tr}\,(BAC).$$

This commutability of the trace fails if the matrices are not symmetric. For instance, if we consider

$$A = \begin{bmatrix} 0 & 0 \\ 1 & 1 \end{bmatrix}, \quad B = \begin{bmatrix} 1 & 0 \\ 0 & 0 \end{bmatrix}, \quad \text{and} \quad C = \begin{bmatrix} 0 & 1 \\ 0 & 0 \end{bmatrix},$$

then we have

$$\text{tr}\,(ABC) = 1 \neq 0 = \text{tr}\,(BAC).$$

The above commutability property of the trace extends to any finite number of symmetric matrices.

Associated with the symmetric matrices, we can define an isomorphism between $\mathcal{M}^n$ and the Euclidean space $\mathbb{R}^{n(n+1)/2}$, which is of lower dimension than $\mathbb{R}^{n^2}$. Specifically, for each $A \in \mathcal{M}^n$, svec(A) is the $n(n+1)/2$-dimensional vector formed by stacking the lower triangular part of A, including the diagonal, row by row. For instance,

$$A \equiv \begin{bmatrix} 1 & 2 \\ 2 & 3 \end{bmatrix} \quad \text{and} \quad \text{svec}(A) = \begin{pmatrix} 1 \\ 2 \\ 3 \end{pmatrix}.$$

The operator svec has important roles to play in studying symmetric matrices. It is easy to see that

$$\| \text{svec}(A) \|_2 \leq \| A \|_F \leq \sqrt{2} \| \text{svec}(A) \|_2, \quad \forall\, A \in \mathcal{M}^n.$$

Let $\mathcal{M}^n_+ \subset \mathcal{M}^n$ denote the cone of all symmetric positive semidefinite matrices of order n. It is a known fact that like the nonnegative orthant of an Euclidean space, this cone $\mathcal{M}^n_+$ is "self-dual" in the space $\mathcal{M}^n$; that is $(\mathcal{M}^n_+)^* = \mathcal{M}^n_+$. The precise statement of this fact is known as Fejér's

Theorem and stated formally as Proposition **1.4.9** below. This proposition contains an additional assertion, which gives two equivalent conditions for two symmetric positive semidefinite matrices to be perpendicular to each other under the Frobenius inner product; see (1.4.60). In proving this equivalence, we need to use an elementary property of a symmetric positive semidefinite matrix; namely, if a diagonal entry of such a matrix is equal to zero, then the entire row and entire column of the matrix containing that entry must equal zero.

1.4.9 Proposition. Let $A \in \mathcal{M}^n$ be given. Then $A \bullet B \geq 0$ for all $B \in \mathcal{M}^n_+$ if and only if $A \in \mathcal{M}^n_+$. Moreover, if A and B both belong to $\mathcal{M}^n_+$, then

$$A \perp B \Leftrightarrow AB = 0 \Leftrightarrow AB + BA = 0. \tag{1.4.60}$$

Proof. Let $A \in \mathcal{M}^n$ be such that $A \bullet B$ is nonnegative for all $B \in \mathcal{M}^n_+$. Let $x \in \mathbb{R}^n$ be arbitrary. We have

$$x^T A x = \sum_{i,j=1}^n a_{ij} x_i x_j = A \bullet xx^T \geq 0,$$

where the last inequality holds because the rank-one matrix xx^T is clearly symmetric and positive semidefinite. Consequently, it follows that A is positive semidefinite.

Conversely, let A and B be any two matrices in $\mathcal{M}^n_+$. There exist an orthogonal matrix $S \in \mathbb{R}^n$ and a nonnegative diagonal matrix D such that $A = SDS^T$. Hence

$$\begin{aligned} A \bullet B &= \operatorname{tr}(AB) = \operatorname{tr}(SDS^T B) \\ &= \sum_{i=1}^n \sum_{j=1}^n s_{ij} d_{jj} (S^T B)_{ji} = \sum_{j=1}^n d_{jj} \sum_{i=1}^n s_{ij} (S^T B)_{ji} \\ &= \sum_{j=1}^n d_{jj} (S^T B S)_{jj}. \end{aligned}$$

Since $S^T BS$ is symmetric positive semidefinite, it has nonnegative diagonal entries. Moreover, since each d_{jj} is nonnegative, it follows that $A \bullet B$ is nonnegative. This completes the proof of the first assertion of the proposition.

To prove the second assertion, let A and B be two matrices belonging to $\mathcal{M}^n_+$. Suppose $A \perp B$. The above proof shows that

$$d_{jj} > 0 \Rightarrow (S^T BS)_{jj} = 0, \quad \forall j = 1, \ldots, n.$$

Since $S^T BS$ is symmetric positive semidefinite, the above further implies that the j-row and the j-column of $S^T BS$ must equal to zero. Consequently, we must have

$$DS^T BS = S^T BSD = 0.$$

In turn, this implies that $AB = BA = 0$ because S is an orthogonal, thus nonsingular, matrix.

Suppose $AB = 0$ then clearly $A \bullet B = 0$. The above proof then implies that $BA = 0$. Thus $AB + BA = 0$. Finally, if $AB + BA = 0$, then $A \bullet B$ is also equal to zero because $\operatorname{tr}(BA) = \operatorname{tr}(AB)$. This establishes the equivalences in (1.4.60). □

Let $F : A \in \mathcal{M}^n \mapsto F(A) \in \mathcal{M}^n$ be a mapping from $\mathcal{M}^n$ into itself. The complementarity problem in SPSD matrices is to find a matrix A satisfying

$$\mathcal{M}^n_+ \ni A \perp F(A) \in \mathcal{M}^n_+. \tag{1.4.61}$$

Proposition **1.4.9** offers several equivalent ways of formulating this problem (mainly the complementarity condition). For instance, (1.4.61) can be equivalently stated as:

$$A \in \mathcal{M}^n_+, \quad F(A) \in \mathcal{M}^n_+, \quad \text{and} \quad AF(A) = 0.$$

Moreover, if A is a solution of (1.4.61), then A and $F(A)$ commute; thus A and $F(A)$ can be simultaneously diagonalized by an orthogonal matrix. This observation allows one to convert the problem (1.4.61) into a mixed complementarity problem of the implicit kind; see Exercise **1.8.17**.

The significance of the CP in SPSD matrices lies in the fact that this problem has a fundamental connection to "semidefinite programs"; these are optimization problems whose unknowns include SPSD matrices. There are many applications of the latter optimization problems; their treatment is beyond the scope of this book. See the notes and comments for references.

The feasibility of the CP (1.4.61) can be determined by solving the following optimization problem:

$$\begin{array}{ll} \text{maximize} & \lambda_{\min}(F(A)) \\ \text{subject to} & A \in \mathcal{M}^n_+, \end{array} \tag{1.4.62}$$

where $\lambda_{\min}(B)$ denotes the smallest eigenvalue of the symmetric matrix B. We note that this eigenvalue function is concave in its argument. The following are two obvious facts:

- the CP (1.4.61) is feasible if and only if the optimal objective value of (1.4.62), which can be $+\infty$, is nonnegative; and

• the CP (1.4.61) is strictly feasible if and only if the optimal objective value of (1.4.62) is positive.

We further note that if F is an affine operator, then (1.4.62) is a concave maximization problem. In this case, the solution of (1.4.62), which itself is an instance of a semidefinite program, is amenable to a host of efficient interior point methods for convex programming. A detailed treatment of these methods for solving CPs of various kinds is presented in Chapter **11**.

1.5 Equivalent Formulations

In Subsection **1.3.1**, we have noted that a VI does not always arise as the stationary point problem of an optimization problem with the same feasible set. We have also observed in Section **1.1** that the VI includes as a special case the classical problem of solving systems of equations. It turns out that we can obtain "equivalent" formulations of a VI and a CP in terms of systems of equations and/or optimization problems of various kinds. Such formulations can be very beneficial for both analytical and computational purposes. Indeed, powerful theories from classical analysis of systems of equations can be applied to treat the VI/CP for proving the existence of solutions and for analyzing these solutions; efficient algorithms for solving equations and optimization problems can be borrowed and extended to solve the VI/CP. In this section, we present a preliminary foray into this vast subject of reformulations.

1.5.1 Equation reformulations of the NCP

To begin, let us consider the NCP (F). It is clear that a vector x solves the problem if and only if x, together with a $w \in \mathbb{R}^n$, is a solution of the system of *constrained equations*:

$$0 = H(x,w) \equiv \begin{pmatrix} w - F(x) \\ w \circ x \end{pmatrix} \qquad (1.5.1)$$

$$(x, w) \geq 0,$$

where $a \circ b$ denotes the *Hadamard product* of two vectors a and b; that is, $a \circ b$ is the vector whose i-component is $a_i b_i$. Note that the domain and range of the function H are both in $\mathbb{R}^{2n}$. Hence the system of equations $H(x,w) = 0$ has the same number of variables as equations; that is, it is a "square" system. The reason why we are interested in a square (as opposed to rectangular) system of equations will become clear subsequently when we introduce iterative methods for solving (1.5.1). In order to obtain such a square system, we have introduced the slack variable w and used the Hadamard (i.e., componentwise) product instead of the inner product $x^T w$. Moreover, in (1.5.1), there is the additional restriction that the variables are nonnegative.

C-functions

The previous, elementary equation reformulation of the NCP has the nice feature that the function involved inherits the same differentiability properties as F; for example, if F is continuously differentiable, then so is H. In the subsequent discussion, we introduce other equation formulations of the NCP. In several of these reformulations, the defining functions are not smooth even if F is. Thus, such algorithms as the traditional Newton method for solving smooth equations and many iterative methods for differentiable optimization problems fail to be directly applicable to these nonsmooth reformulations of the VI/CP. Nevertheless, as we see in later chapters, these nonsmooth formulations can still be put to use algorithmically; indeed, they offer a major avenue for solving the VI/CP.

The following definition captures the essence of all unconstrained equation formulations of complementarity problems of various kinds.

1.5.1 Definition. A function $\psi : \mathbb{R}^2 \to \mathbb{R}$ is called a *C-function*, where C stands for complementarity, if for any pair $(a, b) \in \mathbb{R}^2$,

$$\psi(a,b) = 0 \Leftrightarrow [\,(a,b) \geq 0 \text{ and } ab = 0\,];$$

equivalently, ψ is a C-function if the set of its zeros is the two nonnegative semiaxes. □

In essence, a C-function allows us to conveniently state the complementarity condition in a single equation. As we see below and also Chapter **9**, there are many C-functions. Perhaps the simplest one is the minimum function:

$$\psi_{\min}(a,b) = \min(a,b), \quad (a,b) \in \mathbb{R}^2.$$

It is trivial to verify that this is indeed a C-function. This function is not differentiable in its arguments (in the sense of Gâteau or Fréchet); indeed the nondifferentiable points lie on the line in the plane given by the equation $a - b = 0$. Nevertheless the function $\psi_{\min}$ has some distinct nonsmoothness properties that can be exploited. We will postpone the discussion of the latter properties until a later chapter.

In general, given any C-function ψ, we can immediately obtain an equivalent formulation of the NCP (F) as a system of equations:

$$0 = \mathbf{F}_\psi(x) \equiv \begin{pmatrix} \psi(x_1, F_1(x)) \\ \vdots \\ \psi(x_n, F_n(x)) \end{pmatrix}.$$

Applying this equivalence to the min function $\psi_{\min}$, we immediately obtain

$$x \text{ solves the NCP } (F) \Leftrightarrow \mathbf{F}_{\min}(x) \equiv \min(\,x,\, F(x)\,) = 0, \qquad (1.5.2)$$

where "min" is the componentwise minimum vector function. This system differs from the system (1.5.1) in several obvious ways. First, (1.5.2) is an unconstrained system, whereas (1.5.1) is constrained. Second, (1.5.2) involves the same variable x as the NCP (F), whereas (1.5.1) involves the additional variable w. Third, the function H in (1.5.1) retains the same smoothness property (if any) as the function F, whereas the function $\mathbf{F}_{\min}$ is definitely not smooth because of the min function.

The system (1.5.2) can be extended easily to a complementarity system defined by two or more functions with domain and range in different spaces. Specifically, consider two functions $G, H : \mathbb{R}^n \to \mathbb{R}^m$. Obviously, we have

$$[\,0 \leq G(x) \perp H(x) \geq 0\,] \Leftrightarrow \min(\,G(x),\, H(x)\,) = 0; \qquad (1.5.3)$$

thus a vector x solves the (generalized) NCP on the left side if and only if x is a zero of the min function on the right side. The problem (1.5.3), which we denote by CP (G, H), is an instance of a "vertical NCP"; see Section **1.6** for a formal definition. Applying the min function to the KKT system (1.3.5), we obtain the equivalent formulation of this MiCP as a system of nonsmooth equations:

$$\begin{aligned} \mathbf{L}(x, \mu, \lambda) &= 0 \\ h(x) &= 0 \\ \min(\,\lambda,\, -g(x)\,) &= 0. \end{aligned} \qquad (1.5.4)$$

As previously mentioned, the min function is not differentiable. A natural question to ask is whether there exist differentiable C-functions. The answer turns out to be affirmative. In what follows, we present some alternative C-functions that can be made as smooth as one desires. We begin with a large class of C-functions of this kind, which, with an appropriate choice of the key underlying function ζ, yields the min function. We attach the subscript "Man" to the class of C-functions in the following result.

This subscript, as well as others that we use subsequently, is derived from the name(s) of the author(s) who first proposed it. So "Man" stands for Mangasarian. We refer the reader to Section **1.9** for remarks on the history of C-functions.

1.5.2 Proposition. Let $\zeta : \mathbb{R} \to \mathbb{R}$ be any strictly increasing function with $\zeta(0) = 0$. The scalar function

$$\psi_{\text{Man}}(a,b) \equiv \zeta(|a-b|) - \zeta(b) - \zeta(a), \quad (a,b) \in \mathbb{R}^2.$$

is a C-function.

Proof. Suppose $\min(a,b) = 0$. Without loss of generality, we may assume that $a = 0$ and $b \geq 0$. Clearly we must have $\psi_{\text{Man}}(a,b) = 0$. Conversely, suppose $\psi_{\text{Man}}(a,b) = 0$. If $a < 0$, then since ζ is strictly increasing, we have

$$0 > \zeta(a) = \zeta(|a-b|) - \zeta(b) \geq -\zeta(b),$$

which implies $b > 0$. Consequently,

$$\zeta(|a-b|) > \zeta(b)$$

because $|a-b| = b - a > b$ and ζ is strictly increasing. Therefore we must have $a \geq 0$. Similarly, we must have $b \geq 0$. If both a and b are positive, then we can easily show

$$\zeta(|a-b|) < \zeta(a) + \zeta(b),$$

which contradicts $\psi_{\text{Man}}(a,b) = 0$. □

There are many choices for the function ζ. The simplest of these is $\zeta(t) \equiv t$. For this choice, we have

$$\psi_{\text{Man}}(a,b) = |a-b| - a - b = -2\min(a,b).$$

Thus the function ψ_{Man} is a constant multiple of the min function. Unlike the latter function, which is not differentiable, a choice like $\zeta(t) \equiv t^3$ yields a C-function $\psi_{\text{Man}}(a,b)$ that is twice continuously differentiable.

An important C-function, which plays a central role in the development of efficient algorithms for the solution of the NCP, is denoted by ψ_{FB} and usually referred to as the *Fischer-Burmeister (FB) C-function*:

$$\psi_{\text{FB}}(a,b) \equiv \sqrt{a^2+b^2} - (a+b) \quad \forall (a,b) \in \mathbb{R}^2.$$

This function is convex, differentiable everywhere except when (a,b) is equal to $(0,0)$; moreover $\psi_{\text{FB}}^2(a,b)$ is a continuously differentiable function on the entire plane. We summarize these properties of ψ_{FB} in the result below.

1.5.3 Proposition. The function $\psi_{\rm FB}$ is a C-function; moreover, $\psi_{\rm FB}^2$ is a continuously differentiable function on $\mathbb{R}^2$.

Proof. It is easy to see that if $\min(a,b) = 0$, then

$$\psi_{\rm FB}(a,b) = 0.$$

Conversely, suppose $\psi_{\rm FB}(a,b) = 0$. We have

$$a + b = \sqrt{a^2 + b^2}.$$

Squaring both sides yields $ab = 0$. If $a = 0$, then $b = \sqrt{b^2} \geq 0$; similarly, if $b = 0$, then $a = \sqrt{a^2} \geq 0$. Consequently, $\psi_{\rm FB}$ is a C-function. To prove the second assertion of the proposition, we note that obviously $\psi_{\rm FB}(a,b)$, and thus $\psi_{\rm FB}^2(a,b)$, is continuously differentiable at every $(a,b) \neq (0,0)$. It is easy to verify that $\psi_{\rm FB}^2(a,b)$ is also differentiable at $(a,b) = (0,0)$ with a zero gradient vector there. The continuity of the gradient function of $\psi_{\rm FB}^2$ is left as an exercise. □

There are many variants of the FB C-function based on which one can develop efficient numerical methods; see Chapter **9**. One such variant is

$$\psi_{\rm CCK}(a,b) \equiv \psi_{\rm FB}(a,b) - \tau \max(0,a) \max(0,b), \quad (a,b) \in \mathbb{R}^2,$$

for any scalar $\tau > 0$; the reader can verify that $\psi_{\rm CCK}$ has the same differentiability properties as the original function $\psi_{\rm FB}$.

Another class of C-functions can be defined as follows. For $m = 1, 2$, let $\mathbf{\Phi}_m$ denote the class of continuous functions $\phi : \mathbb{R}^m \to \mathbb{R}_+$ satisfying $\phi(t) = 0$ if and only if $t \leq 0$. There are many such functions; an example is

$$\phi(t_1,t_2) \equiv (\max(0,t_1)^2 + \max(0,t_2)^2)^r, \quad (t_1,t_2) \in \mathbb{R}^2,$$

for any positive integer r. This function is clearly continuously differentiable on $\mathbb{R}^2$. Based on the classes $\mathbf{\Phi}_1$ and $\mathbf{\Phi}_2$, we can introduce a class of C-functions.

1.5.4 Proposition. For any $\phi_1 \in \mathbf{\Phi}_1$ and $\phi_2 \in \mathbf{\Phi}_2$, the function

$$\psi_{\rm LTKYF}(a,b) \equiv \phi_1(ab) + \phi_2(-a,-b)$$

is a C-function.

Proof. This is elementary and thus omitted. □

The simplest example of a resulting equation reformulation of the NCP (F) that is derived from the LT family of C-functions is:

$$\max(0, x_i F_i(x)) + \max(0, -x_i) + \max(0, -F_i(x)) = 0, \qquad (1.5.5)$$

for $i = 1, \ldots, n$. A variant of this formulation is:

$$|\, x_i F_i(x) \,| + \max(\, 0, -x_i\,) + \max(\, 0, -F_i(x)\,) \,=\, 0, \quad i = 1, \ldots, n, \quad (1.5.6)$$

which is not derived from the family of C-functions in Proposition **1.5.4** because the absolute value function is not a member of $\mathbf{\Phi}_1$. Both of the above formulations are interesting because they possess a special "error bound" property that distinguishes them from the min or FB formulations; see Proposition **6.3.5**. Moreover, these formulations are fairly intuitive in view of the defining conditions of the NCP (F), which can be written as:

$$(\, x_i, F_i(x)\,) \geq 0, \quad x_i F_i(x) \,=\, 0, \quad \forall\, i \,=\, 1 \ldots, n.$$

In Exercise **1.8.24**, we introduce C-functions with more than two arguments and define the class of "sign-preserving" functions, to which many C-functions belong but which does not include the function $\psi_{\rm LTKYF}$.

1.5.2 Equation reformulations of the VI

The function $\mathbf{F}_{\min}$ defined in (1.5.2) can be written as follows:

$$\mathbf{F}_{\min}(x) \,=\, x - \max(\, 0, x - F(x)\,),$$

where "max" is the componentwise maximum operator. Recognizing that the max function: $z \mapsto z^+ \equiv \max(0, z)$ is the "Euclidean projector" onto the nonnegative orthant, which we denote $\Pi_{\mathbb{R}^n_+}$, we obtain

$$\mathbf{F}_{\min}(x) \,=\, x - \Pi_{\mathbb{R}^n_+}(x - F(x)).$$

This observation paves the way to the two equation reformulations of the general VI that we consider in this subsection. Both reformulations involve the Euclidean projector onto a closed convex set. Due to the central role of this operator in the study of the VI, we pause to give a formal introduction to the operator and establish some of its elementary properties. Advanced properties of the projection operator are covered in Chapter **4**.

The Euclidean projector

Let K be a closed convex subset of $\mathbb{R}^n$. It is well known from convex analysis that for every vector $x \in \mathbb{R}^n$ there exists a unique vector $\bar{x} \in K$ that is closest to x in the Euclidean norm; see Theorem **1.5.5** below for a proof. This closest vector $\bar{x}$ is called the *(Euclidean) projection* of x onto K and denoted $\Pi_K(x)$. The mapping $\Pi_K : x \mapsto \Pi_K(x)$ is called the *Euclidean projector onto* K. By definition, $\Pi_K(x)$ is the unique solution of the convex minimization problem in the variable y, where x is considered fixed:

$$\begin{array}{ll} \text{minimize} & \frac{1}{2}(y-x)^T(y-x) \\ \text{subject to} & y \in K. \end{array} \tag{1.5.7}$$

When K is a polyhedron, the above optimization problem is a strictly convex QP. When K is not polyhedral, computing the projection onto K is in general not a trivial task. We summarize the essential properties of the operator Π_K in Theorem **1.5.5** below. Figure 1.6 illustrates the projection of a point onto a closed convex set and the variational property of the projection.

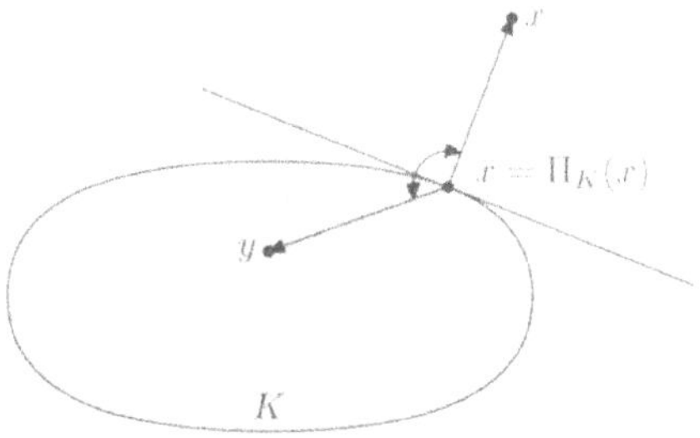

Figure 1.6: The projection of a point onto a closed convex set.

1.5.5 Theorem. Let K be a nonempty closed convex subset of $\mathbb{R}^n$. The following statements are valid.

(a) For each $x \in \mathbb{R}^n$, $\Pi_K(x)$ exists and is unique.

(b) For each $x \in \mathbb{R}^n$, $\Pi_K(x)$ is the unique vector $\bar{x} \in K$ satisfying the inequality:

$$(y-\bar{x})^T(\bar{x}-x) \geq 0, \quad \forall y \in K. \tag{1.5.8}$$

(c) For any two vectors u and v in $\mathbb{R}^n$,

$$(\Pi_K(u)-\Pi_K(v))^T(u-v) \geq \|\Pi_K(u)-\Pi_K(v)\|_2^2.$$

(d) As a function in x, $\Pi_K(x)$ is *nonexpansive*; that is for any two vectors u and v in $\mathbb{R}^n$,

$$\|\Pi_K(u)-\Pi_K(v)\|_2 \leq \|u-v\|_2;$$

thus Π_K is a globally Lipchitz continuous function on $\mathbb{R}^n$.

(e) The squared distance function

$$\rho(x) \equiv \tfrac{1}{2}\|x-\Pi_K(x)\|_2^2, \quad x \in \mathbb{R}^n,$$

is continuously differentiable in x; moreover,

$$\nabla \rho(x) = x - \Pi_K(x).$$

Proof. The optimization problem (1.5.7) has a strongly convex objective function; thus it has a unique optimal solution. This establishes part (a). The inequality (1.5.8) is simply the variational inequality associated with the optimization problem (1.5.7). Since the latter problem is a convex program, this VI characterizes the optimality of $\Pi_K(x)$ as the unique solution of (1.5.7). This establishes part (b). To prove part (c), let u and v be two arbitrary vectors in $\mathbb{R}^n$. By part (b), we have

$$(\Pi_K(u) - \Pi_K(v))^T(\Pi_K(v) - v) \geq 0$$

and

$$(\Pi_K(v) - \Pi_K(u))^T(\Pi_K(u) - u) \geq 0.$$

Adding these two inequalities and rearranging terms, we immediately obtain the desired inequality in part (c). The nonexpansiveness of Π_K claimed in part (d) is an easy consequence of part (c) and the Cauchy-Schwarz inequality. The global Lipschitz continuity of Π_K is an immediate consequence of its nonexpansiveness. Finally, to prove (e), it suffices to verify the limit

$$\lim_{h \to 0} \frac{2[\rho(x+h) - \rho(x) - (x - \Pi_K(x))^T h]}{\| h \|_2} = 0.$$

Write $\bar{x} \equiv \Pi_K(x)$ and $z \equiv \Pi_K(x+h)$. Also to simplify the notation somewhat, we omit the subscript "2" in the Euclidean norm $\| \cdot \|_2$. The numerator in the above limit is equal to

$$\begin{aligned}
&\| z - x - h \|^2 - \| \bar{x} - x \|^2 - 2 h^T(x - \bar{x}) \\
&\quad = \| z - x \|^2 - \| \bar{x} - x \|^2 - 2 h^T(z - \bar{x}) + \| h \|^2 \\
&\quad = (z - \bar{x})^T(z + \bar{x} - 2x - 2h) + \| h \|^2.
\end{aligned}$$

On the one hand, we have by parts (b) and (c),

$$\begin{aligned}
&(z - \bar{x})^T(z + \bar{x} - 2x - 2h) \\
&\quad = \| z - \bar{x} \|^2 + 2(z - \bar{x})^T(\bar{x} - x) - 2(z - \bar{x})^T h \\
&\quad \geq -2(z - \bar{x})^T h \geq -2 \| h \|^2.
\end{aligned}$$

On the other hand, we have by part (b),

$$\begin{aligned}
&(z - \bar{x})^T(z + \bar{x} - 2x - 2h) \\
&\quad = -\| z - \bar{x} \|^2 - 2(\bar{x} - z)^T(z - x - h) \\
&\quad \leq -\| z - \bar{x} \|^2 \leq 0.
\end{aligned}$$

Combining these expressions, we deduce

$$-\| h \|^2 \leq \| z - x - h \|^2 - \| \bar{x} - x \|^2 - 2 h^T (x - \bar{x}) \leq \| h \|^2.$$

Consequently the desired limit holds. □

Property (c) shows that the Euclidean projector Π_K is a "co-coercive" function. In general, a function $F : \mathbb{R}^n \to \mathbb{R}^n$ is *co-coercive* on a subset S of $\mathbb{R}^n$ if there exists a constant $c > 0$ such that

$$(F(x) - F(y))^T (x - y) \geq c \| F(x) - F(y) \|_2^2, \quad \forall x, y \in S.$$

Since the right-hand side is clearly nonnegative, we deduce that a co-coercive function must be *monotone*; that is,

$$(F(x) - F(y))^T (x - y) \geq 0, \quad \forall x, y \in S.$$

Conversely, every affine monotone function that is also symmetric must be co-coercive (on $\mathbb{R}^n$). This follows from the elementary matrix-theoretic fact that if M is a (nonzero) symmetric positive semidefinite matrix, then

$$x^T M x \geq (\lambda_{\max}(M))^{-1} \| M x \|_2^2, \quad \forall x \in \mathbb{R}^n,$$

where $\lambda_{\max}(M)$ denotes the largest eigenvalue of M. See Exercise **1.8.8** for more inequalities of this type. Monotone functions in general play an important role throughout the study of the VI/CP. A comprehensive treatment of this class of functions and its extensions begins in Section **2.3**.

When K is a closed convex cone, property (b) and Proposition **1.1.3** imply the following characterizing property of the projection: for all x in $\mathbb{R}^n$,

$$K \ni \Pi_K(x) \perp \Pi_K(x) - x \in K^*.$$

Since $K^{**} = K$, the above characterization implies that

$$\Pi_{(-K^*)}(x) = x - \Pi_K(x).$$

We have therefore recovered a well-known decomposition of $\mathbb{R}^n$: namely, given a closed convex cone K in $\mathbb{R}^n$,

$$x = \Pi_K(x) + \Pi_{-(K^*)}(x) \quad \forall x \in \mathbb{R}^n.$$

Finally, part (e) of Theorem **1.5.5** shows that the squared minimum distance function is continuously differentiable; furthermore, it follows that $\mathbf{I} - \Pi_K$, and thus Π_K, is a gradient map. See Exercise **2.9.13** for a consequence of the latter property. In general, neither the projection Π_K nor the distance function itself is differentiable in the sense of Gâteau or Fréchet.

1.5.6 Example. Let K be the (Euclidean) unit ball in the plane intersect with the nonnegative orthant; that is

$$K \equiv \{ (x_1, x_2) \in \mathbb{R}^2_+ : x_1^2 + x_2^2 \leq 1 \}.$$

For any vector $x \in \mathbb{R}^2$, the Euclidean projection $\Pi_K(x)$ is the unique solution to the following convex program:

$$\begin{array}{ll} \text{minimize} & \frac{1}{2} [(y_1 - x_1)^2 + (y_2 - x_2)^2] \\ \text{subject to} & y_1^2 + y_2^2 \leq 1 \\ \text{and} & y_1 \geq 0, \quad y_2 \geq 0. \end{array}$$

By letting $\lambda \geq 0$ be the multiplier of the quadratic constraint, we deduce that the projection of x is given by $\bar{x} = (\bar{x}_1, \bar{x}_2)$, where

$$\bar{x}_1 \equiv \max\left(0, \frac{1}{1+\lambda} x_1\right) \quad \text{and} \quad \bar{x}_2 \equiv \max\left(0, \frac{1}{1+\lambda} x_2\right).$$

Notice that λ is dependent on x. Consider the choice $x = (0, x_2)$ where $x_2 > 0$. From the above expression, we easily deduce

$$\bar{x}_1 = 0 \quad \text{and} \quad \bar{x}_2 = \min(1, x_2).$$

Clearly the function $\min(1, x_2)$ is not differentiable in x_2. The distance function is equal to

$$| x_2 - \min(1, x_2) | = \max(0, x_2 - 1),$$

which is not differentiable in x_2. Nevertheless the squared distance function is easily seen to be continuously differentiable in x_2. □

We consider a variant of the above example in which we vary the radius of the ball and examine the resulting projection as a function of the point and the radius. This example sheds some interesting light on the Euclidean projector onto a moving set. Indeed, the continuity and differentiability properties of such a projector are an important concern in the theory of the parametric VI/CP.

1.5.7 Example. For a scalar r, let $K(r)$ be the (Euclidean) ball in the plane with radius $r_+ \equiv \max(r, 0)$; that is

$$K(r) \equiv \left\{ (x_1, x_2) \in \mathbb{R}^2 : \sqrt{x_1^2 + x_2^2} \leq r_+ \right\}.$$

For $r \leq 0$, this ball reduces to the origin in the plane; thus the projection $\Pi_{K(r)}(x)$ is equal to $(0,0)$ for all $x \in \mathbb{R}^2$ and for all $r \leq 0$. For $r > 0$, it is easy to calculate $\Pi_{K(r)}(x)$. We summarize the result as follows:

$$\Pi_{K(r)}(x) = \min\left(\frac{r_+}{\sqrt{x_1^2 + x_2^2}}, 1 \right) x,$$

where $0/0$ is defined to be 1. It is not too difficult to show that this function is continuous at every $(r,x) \in \mathbb{R}^3$. What is not obvious is that the function actually possesses certain "semismoothness" property that is introduced in Chapter **7**. □

Instead of the Euclidean norm, we may use a vector norm induced by a symmetric positive definite matrix to define the projection operator. This results in a "skewed" projector, which leads to a generalization of the natural map of a VI. Specifically, for a given symmetric positive definite matrix $A \in \mathbb{R}^{n\times n}$, the A-norm defined on $\mathbb{R}^n$ is:

$$\| x \|_A \equiv \sqrt{x^T A x}, \quad \forall x \in \mathbb{R}^n.$$

Let $\Pi_{K,A}(x)$ denote the unique solution of the following strictly convex program in the variable y (for x fixed):

$$\begin{aligned} &\text{minimize} \quad \tfrac{1}{2}(y-x)^T A (y-x) \\ &\text{subject to} \quad y \in K. \end{aligned} \tag{1.5.9}$$

Since the objective function defines the squared A-norm of the vector $y - x$, $\Pi_{K,A}(x)$ is the point in the set K closest to the point x under the A-norm. See Figure 1.7 for an illustration of the skewed projection.

The variational characterization of $\Pi_{K,A}(x)$ is as follows: for every x in $\mathbb{R}^n$, $\Pi_{K,A}(x)$ is the unique vector $\bar{x}$ in K satisfying

$$(y - \bar{x})^T A(\bar{x} - x) \geq 0, \quad \forall y \in K. \tag{1.5.10}$$

Moreover, if u and v are two arbitrary vectors in $\mathbb{R}^n$, then we have

$$(\Pi_{K,A}(v) - \Pi_{K,A}(u))^T A(\Pi_{K,A}(u) - u) \geq 0$$

and

$$(\Pi_{K,A}(u) - \Pi_{K,A}(v))^T A(\Pi_{K,A}(v) - v) \geq 0.$$

Adding these two inequalities and rearranging terms, we deduce

$$\begin{aligned} &(\Pi_{K,A}(v) - \Pi_{K,A}(u))^T A(v-u) \geq \\ &\quad (\Pi_{K,A}(v) - \Pi_{K,A}(u))^T A(\Pi_{K,A}(v) - \Pi_{K,A}(u)). \end{aligned} \tag{1.5.11}$$

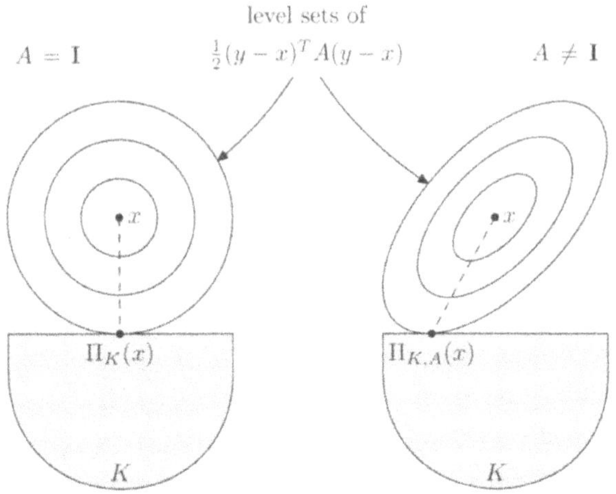

Figure 1.7: The skewed projection.

Consequently, if $\lambda_{\min}(A)$ denotes the smallest eigenvalue of A, which must be positive because A is symmetric positive definite, then on the one hand, the right-hand side of the above inequality yields

$$(\Pi_{K,A}(v) - \Pi_{K,A}(u))^T A(\Pi_{K,A}(v) - \Pi_{K,A}(u)) \geq \\ \lambda_{\min}(A) \| \Pi_{K,A}(v) - \Pi_{K,A}(u) \|_2^2,$$

which, together with (1.5.11), implies that the skewed projector $\Pi_{K,A}$ is also co-coercive. On the other hand, by the Cauchy-Schwarz inequality applied to the left-hand side of (1.5.11), we deduce

$$(\Pi_{K,A}(v) - \Pi_{K,A}(u))^T A(v - u) \leq \\ \| A \|_2 \| \Pi_{K,A}(v) - \Pi_{K,A}(u) \|_2 \| u - v \|_2.$$

Combining the last two inequalities, we obtain a property satisfied by the skewed projector that is similar to the nonexpansiness of the Euclidean projector; namely,

$$\| \Pi_{K,A}(v) - \Pi_{K,A}(u) \|_2 \leq \lambda_{\min}(A)^{-1} \| A \|_2 \| u - v \|_2.$$

Since A is symmetric positive definite, $\|A\|_2$ is equal to the maximum eigenvalue $\lambda_{\max}(A)$ of A. Since the ratio of the maximum eigenvalue of A to the minimum eigenvalue of A is equal to the condition number cond(A) of A, we therefore obtain

$$\| \Pi_{K,A}(v) - \Pi_{K,A}(u) \|_2 \leq \operatorname{cond}(A) \| u - v \|_2, \quad \forall u, v \in \mathbb{R}^n.$$

This inequality in turn implies that $\Pi_{K,A}$ is also a globally Lipschitz continuous function on $\mathbb{R}^n$, with a Lipschitz constant given by the condition number of A. For a direct demonstration of the nonexpansiveness of $\Pi_{K,A}$ under the A-norm, see Exercise **1.8.16**.

The skewed projector $\Pi_{K,A}$ plays an important role in the differentiability theory of the Euclidean projector Π_K when K is finitely representable by differentiable inequalities; details are given in Section **4.4**. Furthermore, analogously to the Euclidean projector, we show shortly that the skewed projector can be used to define equation reformulations of the VI.

The natural map and the normal map

Returning to the VI (K, F), we use the inequality (1.5.8) to establish the following result that gives an equivalent nonsmooth equation formulation of this problem.

1.5.8 Proposition. Let $K \subseteq \mathbb{R}^n$ be closed convex and $F : K \to \mathbb{R}^n$ be arbitrary. It holds that:

$$[\, x \in \text{SOL}(K, F) \,] \;\Leftrightarrow\; [\, \mathbf{F}_K^{\text{nat}}(x) \,=\, 0 \,],$$

where

$$\mathbf{F}_K^{\text{nat}}(v) \,\equiv\, v - \Pi_K(v - F(v)).$$

Proof. The defining inequality for the VI (K, F) is:

$$(\, y - x \,)^T F(x) \,\geq\, 0, \quad \forall\, y \,\in\, K,$$

which can be rewritten as:

$$(\, y - x \,)^T (\, x - (x - F(x)) \,) \,\geq\, 0, \quad \forall\, y \,\in\, K.$$

By (1.5.8), the last inequality is equivalent to:

$$x \,=\, \Pi_K(x - F(x)),$$

or equivalently $\mathbf{F}_K^{\text{nat}}(x) = 0$. □

From Proposition **1.5.8**, we can derive an alternative nonsmooth equation formulation of the VI.

1.5.9 Proposition. Let $K \subseteq \mathbb{R}^n$ be closed convex and $F : K \to \mathbb{R}^n$ be arbitrary. A vector x belongs to SOL(K, F) if and only if there exists a vector z such that $x = \Pi_K(z)$ and $\mathbf{F}_K^{\text{nor}}(z) = 0$, where

$$\mathbf{F}_K^{\text{nor}}(v) \,\equiv\, F(\Pi_K(v)) + v - \Pi_K(v).$$

Proof. Suppose that $x \in \text{SOL}(K, F)$. Then $x = \Pi_K(x - F(x))$ by Proposition **1.5.8**. Define $z \equiv x - F(x)$. Clearly $x = \Pi_K(z)$ and $\mathbf{F}_K^{\text{nor}}(z) = 0$. Conversely if the last two equations hold, then

$$z = x - F(x) \quad \text{and} \quad x = \Pi_K(x - F(x)).$$

Proposition **1.5.8** implies that x solves the VI (K, F). □

It is useful to clarify the two equation formulations of the VI (K, F):

$$\mathbf{F}_K^{\text{nat}}(x) = 0 \quad \text{and} \quad \mathbf{F}_K^{\text{nor}}(z) = 0. \tag{1.5.12}$$

The main difference between these two equations as far as their equivalence with the VI is concerned is that the former equation is formulated using the original variable of the VI whereas the latter equation is formulated via a change of variable: $x \equiv \Pi_K(z)$. This difference is made explicit in Propositions **1.5.8** and **1.5.9**. In general, the domain of definition of the function $\mathbf{F}_K^{\text{nat}}$ is the same as F, whereas the domain of definition of the function $\mathbf{F}_K^{\text{nor}}$ is always the entire space $\mathbb{R}^n$. This feature makes $\mathbf{F}_K^{\text{nor}}$ particularly attractive in computations, especially in the situation where the function F is not defined on the whole space $\mathbb{R}^n$; this is because when working with $\mathbf{F}_K^{\text{nat}}(x)$, we need to restrict the variable x to its appropriate domain whereas no such restriction is necessary when working with $\mathbf{F}_K^{\text{nor}}(z)$.

We call $\mathbf{F}_K^{\text{nat}}(x)$ and $\mathbf{F}_K^{\text{nor}}(z)$ the *natural map* and the *normal map* associated with the VI (K, F), respectively; we call the corresponding equations in (1.5.12) the *natural equation* and the *normal equation* . When specialized to the NCP (F), the normal map takes on the simple form:

$$\mathbf{F}_{\mathbb{R}_+^n}^{\text{nor}}(z) = F(z^+) - z^-,$$

where $z^- \equiv \max(0, -z) = z^+ - z$ is the nonpositive part of the vector z. Similarly, specialized to the KKT system (1.3.5), the normal map is given by (omitting the cone $K = \mathbb{R}^{n+\ell} \times \mathbb{R}_+^m$ for this MiCP):

$$\mathbf{F}^{\text{nor}}(x, \mu, z) = \begin{pmatrix} \mathbf{L}(x, \mu, z^+) \\ h(x) \\ g(x) + z^- \end{pmatrix}.$$

If F is the linear map $x \mapsto Mx$ for some square matrix M of order n, we write $\mathbf{M}_K^{\text{nat}}$ and $\mathbf{M}_K^{\text{nor}}$ for $\mathbf{F}_K^{\text{nat}}$ and $\mathbf{F}_K^{\text{nor}}$, respectively; that is

$$\mathbf{M}_K^{\text{nat}}(x) \equiv x - \Pi_K(x - Mx) \quad \text{and} \quad \mathbf{M}_K^{\text{nor}}(z) \equiv M\Pi_K(z) + z - \Pi_K(z).$$

As we mentioned earlier, we can use the skewed projector $\Pi_{K,A}$ in the definition of the natural map. Specifically, if $A \in \mathbb{R}^{n\times n}$ is symmetric and positive definite, we can set

$$\mathbf{F}^{\rm nat}_{K,A}(x) \equiv x - \Pi_{K,A}(x - A^{-1}F(x)),$$

which we call the *skewed natural map*. By the inequality (1.5.10) and similarly to Proposition **1.5.8**, we can show that

$$[\, x \in \mathrm{SOL}(K,F)\,] \Leftrightarrow [\, \mathbf{F}^{\rm nat}_{K,A}(x) = 0\,].$$

1.5.10 Example. If K is the rectangle (1.1.7), the projection $\Pi_K(x)$ admits a particularly simple form, which, with a change of notation, we denote as $\mathrm{mid}(a,b;x)$. Specifically, for a given pair of vectors a and b satisfying (1.1.8), the "mid function" $\mathrm{mid}(a,b;\cdot) : \mathbb{R}^n \to \mathbb{R}^n$ is defined by: for each $i = 1,\ldots,n$,

$$\mathrm{mid}(a,b;x)_i \equiv \Pi_{[a_i,b_i]}(x_i) = \begin{cases} a_i & \text{if } a_i > x_i \\ x_i & \text{if } a_i \le x_i \le b_i \\ b_i & \text{if } x_i > b_i. \end{cases}$$

It is trivial to see that $\Pi_K(x) = \mathrm{mid}(a,b;x)$. In terms of the mid function, the box constrained VI (K,F) is equivalent to the system of nonsmooth equations:

$$x - \mathrm{mid}(a,b;x - F(x)) = 0,$$

which is the analog of the natural min equation formulation of the NCP (F). Similarly, a simplified "normal equation" formulation for the same VI (K,F) is also possible.

We can define a generalization of a C-function that describes the implications in (1.2.4). There is also a natural extension of the FB function for the box constrained VI. These extensions provide alternative equation formulations for this VI that can be used profitably to design solution methods for solving the VI with simple bounds. Details of these formulations and the resulting algorithmic aspects are presented in Section **9.4**. □

A convenient feature of $\mathbf{F}^{\rm nor}_K$ is that if the original map F is translated by a constant vector q, then $\mathbf{F}^{\rm nor}_K$ is translated by the same constant. Based on this observation, the following result is easy to prove.

1.5.11 Proposition. Let K be a closed convex set in $\mathbb{R}^n$ and $F : K \to \mathbb{R}^n$ be arbitrary. The normal map $\mathbf{F}^{\rm nor}_K$ is a homeomorphism from $\mathbb{R}^n$ onto $\mathbb{R}^n$ if and only if the VI $(K, q+F)$ has a unique solution for all $q \in \mathbb{R}^n$.

Proof. Suppose that $\mathbf{F}_K^{\text{nor}}$ is a homeomorphism; it is then a bijection. For every vector q, the equation

$$q + F(\Pi_K(z)) + z - \Pi_K(z) = 0$$

has a unique solution. This establishes that the VI $(K, q+F)$ has a solution for all q. To show the uniqueness of such a solution, let x^1 and x^2 be two solutions of the VI $(K, q+F)$. It is then easy to show that

$$(x^1 - x^2)^T (F(x^1) - F(x^2)) \leq 0.$$

The vector $z^i \equiv x^i - q - F(x^i)$ is a zero of $\mathbf{F}_K^{\text{nor}}$ for $i = 1, 2$. By the injectivity of $\mathbf{F}_K^{\text{nor}}$, we have

$$x^1 - q - F(x^1) = x^2 - q - F(x^2),$$

which implies

$$x^1 - x^2 = F(x^1) - F(x^2).$$

Consequently, we must have $x^1 = x^2$, establishing the uniqueness of solution to the VI $(K, q+F)$.

Conversely, suppose that the VI $(K, q+F)$ has a unique solution for all vectors $q \in \mathbb{R}^n$. It follows easily that $\mathbf{F}_K^{\text{nor}}$ is surjective. To show that $\mathbf{F}_K^{\text{nor}}$ is injective, suppose $\mathbf{F}_K^{\text{nor}}(z^1) = \mathbf{F}_K^{\text{nor}}(z^2)$. With q denoting this common vector, it follows that $\Pi_K(z^1)$ and $\Pi_K(z^2)$ are both solutions of the VI $(K, -q+F)$. Therefore $\Pi_K(z^1) = \Pi_K(z^2)$. Since

$$q = F(\Pi_K(z^1)) + z^1 - \Pi_K(z^1) = F(\Pi_K(z^2)) + z^2 - \Pi_K(z^2),$$

we deduce $z^1 = z^2$, establishing that $\mathbf{F}_K^{\text{nor}}$ is bijective. Since F is continuous on K and Π_K is continous on $\mathbb{R}^n$, it follows that $\mathbf{F}_K^{\text{nor}}$ is continuous on $\mathbb{R}^n$. The continuity of the inverse $(\mathbf{F}_K^{\text{nor}})^{-1}$ follows from Proposition **2.1.12**. □

Unlike the normal map, the natural map $\mathbf{F}_K^{\text{nat}}$ does not have the convenient translational property. Nevertheless, it is possible to establish a result analogous to Proposition **1.5.11**; see Exercise **1.8.23**. For a polyhedral set K and a matrix M, it can be shown that $\mathbf{M}_K^{\text{nor}}$ is bijective if and only if $\mathbf{M}_K^{\text{nat}}$ is bijective. While the implication

$$\mathbf{M}_K^{\text{nor}} \text{ bijective} \Rightarrow \mathbf{M}_K^{\text{nat}} \text{ bijective}$$

is not hard to prove by combining Proposition **1.5.11** and Exercise **1.8.23**; the reverse implication

$$\mathbf{M}_K^{\text{nat}} \text{ bijective} \Rightarrow \mathbf{M}_K^{\text{nor}} \text{ bijective}$$

is not as easy as it looks. In fact, we are not aware of a simple proof. The latter implication follows from Theorem **4.3.2** whose proof requires some deep properties of the class of "piecewise affine maps", of which both $\mathbf{M}_K^{\rm nor}$ and $\mathbf{M}_K^{\rm nat}$ are members. For details, see Sections **4.2** and **4.3**.

1.5.3 Merit functions

So far, we have presented several reformulations of the VI and CP as a system of (unconstrained) equations. A different approach is to cast the VI/CP as a minimization problem. To illustrate this we start again with the NCP (F). It is clear that a vector x solves this problem if and only if x is a global minimizer of the optimization problem:

$$\begin{array}{ll} \text{minimize} & y^T F(y) \\ \text{subject to} & y \geq 0 \text{ and } F(y) \geq 0, \end{array} \tag{1.5.13}$$

and the optimum objective value $x^T F(x)$ is equal to zero. In this sense, we say that the optimization problem (1.5.13) is equivalent to the NCP and call the function $y^T F(y)$ a *merit function* for the NCP (F). Generalizing the above discussion we give the following definition of merit function.

1.5.12 Definition. A merit function for the VI (K, F) on a (closed) set $X \supseteq K$ is a nonnegative function $\theta : X \to \mathbb{R}_+$ such that $x \in \text{SOL}(K, F)$ if and only if $x \in X$ and $\theta(x) = 0$, that is, if and only if the solutions of the VI (K, F) coincide with the global solutions of the problem

$$\begin{array}{ll} \text{minimize} & \theta(y) \\ \text{subject to} & y \in X \end{array}$$

and the optimal objective value of this problem is zero. □

If $\text{SOL}(K, F)$ is empty, then either the global optimal value of θ over X is positive or θ has no global minima on X. Notice that unless F is affine, the feasible set of (1.5.13) is typically nonconvex. Furthermore, even for an LCP (q, M) where $F(y) \equiv q + My$, the objective function of (1.5.13) is not convex unless M is a positive semidefinite matrix. This raises the problem of finding "good" merit functions, where the exact meaning of "good" clearly depends on the use we have in mind for the merit function. In the remaining part of this section we first consider some basic types of merit functions and then illustrate two of their uses.

The equation reformulations presented above naturally lead to some merit functions for solving the VI/CP. Specifically, suppose that the system

$$H(x) = 0$$

is a reformulation of the VI/CP where H maps $\mathcal{D} \subseteq \mathbb{R}^n$ into $\mathbb{R}^n$. We can then associate the following scalar-valued function:

$$\theta(x) \equiv \| H(x) \|^r, \quad x \in \mathcal{D}, \tag{1.5.14}$$

where r is any positive integer, as a merit function for the VI/CP. Thus,

$$\theta_{\min}(x) \equiv \sum_{i=1}^{n} \min(x_i, F_i(x))^2$$

is a merit function for the NCP (F); more generally, if ψ is a C-function, then

$$\theta(x) \equiv \| \mathbf{F}_\psi(x) \|_2^2 = \sum_{i=1}^{n} \psi(x_i, F_i(x))^2$$

is a merit function for the same problem.

This kind of merit function can be very effective in developing algorithms for the NCP and for some of its extensions, but is typically not viable for a general VI. In this latter case it is usually preferable to use merit functions that are derived in a different way. We present here an alternative merit function for the VI that is the basis of all the merit functions discussed in Chapter **10** and that is not derived from an equation reformulation. This merit function, called the *gap function*, extends the reformulation (1.5.13) of the NCP that converts the latter problem into a constrained optimization problem.

Specifically, the *gap function* for the VI (K, F) is defined on the same domain $\mathcal{D}$ of F and is given by:

$$\theta_{\text{gap}}(x) \equiv \sup_{y \in K} F(x)^T(x - y), \quad x \in \mathcal{D} \supseteq K. \tag{1.5.15}$$

This is a nonnegative "extended-valued" function on K; $\theta_{\text{gap}}(x)$ is nonnegative for all $x \in K$; nevertheless it is possible for $\theta_{\text{gap}}(x)$ to be infinite for some x in K. In particular, when K is a cone, we have

$$\theta_{\text{gap}}(x) = \begin{cases} F(x)^T x & \text{if } F(x) \in K^* \\ +\infty & \text{otherwise.} \end{cases}$$

In general, we see that $x \in \text{SOL}(K, F)$ if and only if x is a global solution of the constrained gap minimization problem:

$$\begin{array}{ll} \text{minimize} & \theta_{\text{gap}}(z) \\ \text{subject to} & z \in K, \end{array} \tag{1.5.16}$$

and $\theta_{\text{gap}}(x) = 0$. Thus θ_{gap} is a merit function for the VI (K, F) on K.

When K is a cone, the gap program attains a particularly simple form:

$$\begin{array}{ll} \text{minimize} & x^T F(x) \\ \text{subject to} & x \in K \quad \text{and} \quad F(x) \in K^*. \end{array} \tag{1.5.17}$$

For $K = \mathbb{R}^n_+$, this problem is exactly (1.5.13). A noteworthy feature of the optimization problem (1.5.17) is that if F is a smooth function and K is "nice" (e.g., a polyhedron), (1.5.17) is a smooth optimization problem, though in general, neither the objective function is convex nor the feasible region is convex.

For a general VI (K, F), the gap function is defined as the "value function" of a parametric concave maximization problem with the linear objective function $y \mapsto F(x)^T(x - y)$; more precisely, $\theta_{\text{gap}}(x)$ is equal to the optimum objective value of the following optimization problem with variable y and parameter x:

$$\begin{array}{ll} \text{maximize} & F(x)^T(x - y) \\ \text{subject to} & y \in K. \end{array} \tag{1.5.18}$$

For each fixed but arbitrary x, this is a concave maximization problem in y. In the case of a polyhedral K (i.e., for a linearly constrained VI), the problem (1.5.18) becomes a linear program. Even in this case, the gap function is not differentiable.

If F is a "monotone" affine function given by $F(x) \equiv q + Mx$ where q is a given n-vector and M is a positive semidefinite matrix, the gap function $\theta_{\text{gap}}(x)$ is an extended-valued convex function and the gap program (1.5.16) becomes a convex minimization problem. The class of "monotone" VIs (including the nonlinear problems) is very important in applications. Many specialized results and algorithms exist for these VIs.

Merit functions and solution algorithms

Merit functions can be used in the design of numerical algorithms for solving the VI/CP. In particular, we can apply an iterative algorithm to minimize the merit function, with the hope of obtaining its global minimum. However, except in very rare cases, merit functions are typically not convex in their arguments; therefore we can not guarantee to obtain their global minima. At best, we can compute only a stationary point. Consequently, it is important to know when such a point will be a solution of the VI/CP: in fact this is one of the central issues of investigation in the study of merit functions. A classic result of this kind is available for the following QP formulation of the LCP (q, M):

$$\begin{array}{ll} \text{minimize} & x^T(q + Mx) \\ \text{subject to} & x \geq 0 \quad \text{and} \quad q + Mx \geq 0. \end{array} \tag{1.5.19}$$

Note that this is just the specialization of (1.5.13) when $F(x) \equiv q + Mx$. Indeed, it is known that the following two statements are equivalent:

(a) for every $q \in \mathbb{R}^n$ for which (1.5.19) is feasible, every stationary point of (1.5.19) solves the LCP (q, M), and

(b) M is a "row sufficient" matrix; that is, the following implication holds:

$$x \circ M^T x \leq 0 \Rightarrow x \circ M^T x = 0.$$

The reader can easily construct row sufficient matrices that are not positive semidefinite. For such a matrix M, (1.5.19) is a nonconvex quadratic program; yet all the stationary points (if they exist) of such a program are solutions of the LCP (q, M). For more discussion on this nonlinear programming approach to solving complementarity problems, see Exercise **1.8.13**.

The above discussion makes it clear that in general there must be conditions on the VI/CP in order for the (possibly constrained) stationary points of the merit functions of the VI/CP to be solutions of the problem in question; moreover, the merit functions of such a VI/CP need not be convex. For such a function, we can derive a necessary and sufficient condition for a stationary point of the function on the domain of minimization to be a zero of the function, thus a solution of the VI/CP. This condition is the underlying abstraction of all the "regularity" conditions that we will introduce subsequently when we discuss the convergence of descent algorithms for solving the VI/CP.

In order to provide a framework broad enough to handle a large family of (smooth and nonsmooth) merit functions, we consider an optimization problem of the following kind:

$$\begin{array}{ll} \text{minimize} & \theta(x) \\ \text{subject to} & x \in X, \end{array} \tag{1.5.20}$$

where the objective function $\theta : \mathcal{D} \supseteq X \to \mathbb{R}$ is defined on an open set $\mathcal{D}$ containing the feasible region X, which is a closed subset of $\mathbb{R}^n$. Since we are interested in functions θ that are not differentiable, we can not talk about a stationary point of (1.5.20) in terms of the gradient function of

θ, as defined in Subsection **1.3.1**. For our purpose here, assume that θ is directionally differentiable on X. A stationary point of (1.5.20) is then defined to be a feasible vector $x \in X$ such that

$$\theta'(x;d) \geq 0, \quad \forall\, d \in \mathcal{T}(x;X), \tag{1.5.21}$$

where $\theta'(x;d)$ is the directional derivative of θ at x along the direction d. More general concepts of stationarity can be defined for functions θ that are not directionally differentiable; the above definition using the directional derivative is sufficient for the discussion herein.

Under the above setting and assuming that θ is nonnegative (as in all the merit functions of the VI/CP that we have seen), the following simple result identifies a necessary and sufficient condition for a stationary point x of (1.5.20) to satisfy $\theta(x) = 0$.

1.5.13 Proposition. Let X be a nonempty closed subset of $\mathbb{R}^n$ and let $\theta : \mathcal{D} \supseteq X \to \mathbb{R}$ be nonnegative and directionally differentiable on X. A necessary and sufficient condition for a stationary point $x \in X$ to satisfy $\theta(x) = 0$ is that there exists a vector $d \in \mathcal{T}(x;X)$ such that

$$\theta(x) + \theta'(x;d) \leq 0. \tag{1.5.22}$$

Proof. If $\theta(x) = 0$, simply take d to be the zero vector; (1.5.22) holds trivially. Conversely, if a vector $d \in \mathcal{T}(x;X)$ satisfying (1.5.22) exists, then

$$0 \leq \theta(x) \leq \theta(x) + \theta'(x;d) \leq 0,$$

where the second inequality is due to (1.5.21). The above string of inequalities clearly implies $\theta(x) = 0$ as desired. □

We illustrate the above lemma with $X = \mathbb{R}^n$ and

$$\theta(x) \equiv \tfrac{1}{2} H(x)^T H(x),$$

where H is a continuously differentiable function from $\mathbb{R}^n$ into itself. The function θ is continuously differentiable in this case and we have

$$\theta'(x;d) = \sum_{i=1}^n H_i(x) \nabla H_i(x)^T d.$$

Thus

$$\theta(x) + \theta'(x;d) = \sum_{i=1}^n H_i(x) \left(\tfrac{1}{2} H_i(x) + \nabla H_i(x)^T d \right).$$

If the Jacobian matrix $JH(x)$ is nonsingular, by letting

$$d \equiv -\tfrac{1}{2} JH(x)^{-1} H(x),$$

we clearly have $\theta(x) + \theta'(x; d) = 0$. The nonsingularity of $JH(x)$ is a well-known condition in classical numerical analysis; it is the key condition required for the convergence of many Newton-type methods for solving the system of smooth equations $H(x) = 0$. We have shown here that this condition implies the one in Proposition **1.5.13**. See Exercise **1.8.25** for a further result where the nonsingularity of $JH(x)$ is not explicitly assumed.

Proposition **1.5.13** is applicable to a variety of merit functions of the VI/NCP. We postpone further discussion until Chapters **9** and **10**, where we will consider other important properties of merit functions such as smoothness and coerciveness.

Inexact solutions and error bounds

There is another major role that the merit functions play in the study of the VI/CP. This role grows out of the following practical consideration of an "inexact" (or approximate) solution of the VI/CP. Suppose there is a vector x that is known not to be a solution of the problem. Yet, we are interested in determining how close x is to "being a solution"; more specifically, we are interested in obtaining a quantitative measure of the violation of x with reference to the conditions that define the VI/CP. With a nonnegative merit function θ, such a measure can reasonably be prescribed by the quantity $\theta(x)$. Since the zeros of the merit function are the exact solutions of the problem, $\theta(x)$ is justifiably a sound measure of the inexactness of x being a non-solution. The theory that supports and clarifies this interpretation is known as "error bound study". Chapter **6** is devoted to the detailed investigation of this theory that is built on various merit functions.

In what follows, we make a very preliminary foray into this vast subject of error bounds with a further discussion about the concept of an approximate solution in terms of some simple merit functions. We begin by asking the following question. Suppose $\| \min(x, F(x)) \| \leq \varepsilon$ for some positive scalar $\varepsilon > 0$, can one say something about the vector x in terms of the NCP (F)? It turns out that one can say quite a few things by letting $r \equiv \min(x, F(x))$. It is trivial to observe that this is equivalent to

$$0 = \min(\, x - r,\; F(x) - r\,).$$

Writing this out in terms of the complementarity conditions, we obtain

$$0 \leq x - r \perp F(x) - r \geq 0,$$

or equivalently,

$$x \geq r, \quad F(x) \geq r,$$

and

$$x^T F(x) = r^T (x + F(x) - r).$$

Since r is a vector with presumably very small norm, we see that x satisfies approximately the defining conditions of the NCP (F) in the sense made precise by the above expressions. Alternatively, we can also say that the vector $y \equiv x - r$, which is a (small) perturbation of x (with ε being small), solves the perturbed NCP (G), where $G(z) \equiv F(z + r) - r$.

We can easily extend the above discussion to the VI (K, F) via the natural (or normal) map $\mathbf{F}_K^{\text{nat}}$. As before, suppose $\|\mathbf{F}_K^{\text{nat}}(x)\| \leq \varepsilon$. Letting

$$r \equiv \mathbf{F}_K^{\text{nat}}(x) = x - \Pi_K(x - F(x)),$$

we deduce that $x - r$ belongs to K and

$$(y - x + r)^T (F(x) - r) \geq 0, \quad \forall y \in K;$$

equivalently, the vector $y \equiv x - r$ solves the VI (K, G) with the same function G as in the case of the NCP. As an illustration, consider the case where K is a compact rectangle given by (1.1.7), where the bounds a_i and b_i are all finite. In this case, the equation

$$r = \mathbf{F}_K^{\text{nat}}(x) = x - \text{mid}(a, b; x - F(x))$$

is equivalent to $x \in r + K$ (i.e., $a_i + r_i \leq x_i \leq b_i + r_i$ for all i) and

$$\begin{aligned} x_i = a_i + r_i \quad &\Rightarrow \quad F_i(x) \geq r_i \\ a_i + r_i < x_i < b_i + r_i \quad &\Rightarrow \quad F_i(x) = r_i \\ x_i = b_i + r_i \quad &\Rightarrow \quad F_i(x) \leq r_i. \end{aligned}$$

Thus if in addition $|r_i| \leq \varepsilon$ for all i, then we have $x_i \in [a_i - \varepsilon, b_i + \varepsilon]$ and

$$\begin{aligned} |F_i(x)| \leq \varepsilon \quad & \text{if } x_i \in (a_i + \varepsilon,\, b_i - \varepsilon) \\ F_i(x) \geq -\varepsilon \quad & \text{if } |x_i - a_i| \leq \varepsilon \\ F_i(x) \leq \varepsilon \quad & \text{if } |x_i - b_i| \leq \varepsilon. \end{aligned}$$

The min function forms the basis for extensions to other merit functions for the NCP (F). This is accomplished via comparison results that give bounds between the min residual and the merit function in question. To illustrate the point, consider the FB merit function $\|\mathbf{F}_{\text{FB}}\|$, where

$$\mathbf{F}_{\text{FB}}(x) \equiv \begin{pmatrix} \psi_{\text{FB}}(x_1, F_1(x)) \\ \vdots \\ \psi_{\text{FB}}(x_n, F_n(x)) \end{pmatrix},$$

and suppose $\|\mathbf{F}_{\text{FB}}(x)\| \leq \varepsilon$. By Lemma **9.1.3**, whose proof is not difficult, it then follows that there exists a constant $c > 0$, depending only on the dimension n of the problem and the vector norm used, such that $\| \min(x, F(x))\| \leq c\varepsilon$. The above discussion now applies.

We end this preliminary discussion of error bounds by establishing a basic result relating various residual functions of a general VI (K, F). By now, we know that there are three equivalent ways to describe a solution of the VI (K, F) with a closed convex set K. Specifically, each of the following three statements is equivalent for a vector $x \in \mathbb{R}^n$ to be an element of SOL(K, F):

(a) $\mathbf{F}_K^{\text{nat}}(x) = 0$;

(b) $\mathbf{F}_K^{\text{nor}}(z) = 0$ and $x \equiv \Pi_K(z)$;

(c) $x \in K$ and $0 \in F(x) + \mathcal{N}(x; K)$.

Suppose now that $x \in K$ is an inexact solution of the VI (K, F). We are interested in establishing some quantitative relations between the following three (positive) quantities, all measured using the Euclidean norm: (a) $\|\mathbf{F}_K^{\text{nat}}(x)\|$, which is the residual of the natural equation evaluated at x, (b) $\|\mathbf{F}_K^{\text{nor}}(z)\|$, which is the residual of the normal equation evaluated at a vector z such that $\Pi_K(z) = x$ (or equivalently, $z \in \Pi_K^{-1}(x)$), and (c) $\text{dist}(-F(x), \mathcal{N}(x; K))$, which is the distance from $-F(x)$ to the normal cone $\mathcal{N}(x; K)$; this distance is by definition equal to:

$$\text{dist}(-F(x), \mathcal{N}(x; K)) \equiv \inf \{ \| F(x) + v \| \, : \, v \in \mathcal{N}(x; K) \}.$$

Since

$$z = \Pi_{\mathcal{T}(x;K)}(z) + \Pi_{\mathcal{N}(x;K)}(z), \quad \forall \, z \in \mathbb{R}^n,$$

it follows that

$$-F(x) - \Pi_{\mathcal{N}(x;K)}(-F(x)) = \Pi_{\mathcal{T}(x;K)}(-F(x)).$$

Hence

$$\text{dist}(-F(x), \mathcal{N}(x; K)) = \| \Pi_{\mathcal{T}(x;K)}(-F(x)) \|.$$

If $F(x) \equiv \nabla\theta(x)$ for a real-valued function θ, the vector $\Pi_{\mathcal{T}(x;K)}(-\nabla\theta(x))$ is traditionally called the *projected gradient* of the optimization problem:

$$\begin{array}{ll} \text{minimize} & \theta(y) \\ \text{subject to} & y \in K \end{array}$$

at the feasible vector $x \in K$. In general, we have the following result that connects the three measures (a), (b), and (c) of an inexact solution of a VI.

1.5.14 Proposition. Let K be a closed convex subset of $\mathbb{R}^n$ and F be a mapping from $\mathbb{R}^n$ into itself. For any vector $x \in K$,

$$\| \mathbf{F}_K^{\text{nat}}(x) \| \leq \text{dist}(-F(x), \mathcal{N}(x; K)) = \inf \{ \| \mathbf{F}_K^{\text{nor}}(z) \| : z \in \Pi_K^{-1}(x) \}.$$

Proof. Write $r \equiv \mathbf{F}_K^{\text{nat}}(x)$. We have $x - r = \Pi_K(x - F(x))$. Thus $x - r$ belongs to K and

$$(y - x + r)^T (F(x) - r) \geq 0, \quad \forall y \in K.$$

In particular, since x belongs to K, we deduce

$$r^T F(x) \geq r^T r.$$

Let $v \equiv \Pi_{\mathcal{N}(x;K)}(-F(x))$. It follows that

$$(y - x)^T v \leq 0, \quad \forall y \in K.$$

In particular, with y taken to be $x - r$, we obtain

$$0 \leq r^T v = r^T (v + F(x)) - r^T F(x);$$

thus $r^T F(x) \leq r^T (v + F(x))$. Consequently,

$$r^T r \leq r^T (v + F(x)),$$

which implies $\|r\| \leq \|v + F(x)\|$; i.e.,

$$\| \mathbf{F}_K^{\text{nat}}(x) \| \leq \text{dist}(-F(x), \mathcal{N}(x; K)).$$

Since

$$z \in \Pi_K^{-1}(x) \Leftrightarrow \Pi_K(z) = x \Leftrightarrow z - x \in \mathcal{N}(x; K),$$

it follows that

$$\text{dist}(-F(x), \mathcal{N}(x; K)) = \inf \left\{ \| \mathbf{F}_K^{\text{nor}}(z) \| : z \in \Pi_K^{-1}(x) \right\},$$

which is the claimed relation between the normal residual and the distance from $-F(x)$ to the normal cone $\mathcal{N}(x; K)$. □

In general, the inequality $\|\mathbf{F}_K^{\text{nat}}(x)\| \leq \text{dist}(-F(x), \mathcal{N}(x; K))$ can be strict; see Exercise **1.8.33**.

1.6 Generalizations

The VI/CP has many generalizations; we have seen some of these in the previous sections. The QVI is one such generalization; the vertical NCP is another; see (1.5.3). In this section, we introduce a few more extended problems; some of them will be revisited in later chapters and in exercises while others are included just to make connections with neighboring fields.

We consider two further generalizations of the VI. One is the "hemivariational inequality". This problem is defined as follows. Let $a(u,v)$ be a scalar function of two arguments $(u,v) \in \mathbb{R}^{2n}$ and let $\varphi(u)$ be a scalar function of one argument $u \in \mathbb{R}^n$. Let K be a nonempty subset of $\mathbb{R}^n$. The *hemivariational inequality*, also known as a variational inequality of the second kind, is to find a vector $x \in K$ such that

$$a(x, y-x) + \varphi(y) - \varphi(x) \geq 0, \quad \forall\, y \in K.$$

Clearly, this problem includes the VI and an optimization problem as special cases; the former corresponds to $a(u,v) \equiv v^T F(u)$ and $\varphi \equiv 0$ and the latter corresponds to $a \equiv 0$. If $a(u,\cdot)$ is linear in the second argument, as in the case of the VI, and if φ is directionally differentiable, then every solution x of the hemivariational inequality must be a solution of a quasi-variational inequality of a generalized type. Indeed, similar to (1.3.3), we can show that x must satisfy:

$$a(x, y-x) + \varphi'(x; y-x) \geq 0, \quad \forall\, y \in x + \mathcal{T}(x; K).$$

Another generalization of the VI that also includes the QVI (of the standard type) is defined as follows. Let K and F be two point-to-set maps defined on $\mathbb{R}^n$ and with values in the power set of $\mathbb{R}^n$; that is for every $x \in \mathbb{R}^n$, $K(x)$ and $F(x)$ are (possibly empty) subsets of $\mathbb{R}^n$. The *generalized quasi-variational inequality* defined by the pair (K, F) is to find two vectors x and v such that $x \in K(x)$, $v \in F(x)$ and

$$(y-x)^T v \geq 0, \quad \forall\, y \in K(x).$$

With F being a single-valued map, this generalized problem readily reduces to the QVI. An important source of the generalized QVI is a nonsmooth optimization problem. In Subsection **12.5.2**, we discuss a special case of the generalized QVI, called the *generalized* or *multivalued VI*, in which $K(x)$ is the same set for all x and F is a set-valued map.

Consider the optimization problem (1.3.1), where for simplicity we assume that the objective function θ is a convex function defined on $\mathbb{R}^n$, but the feasible region K is not necessarily convex. Let $\partial\theta(x)$ denote the *subdifferential* of θ at x; that is, $v \in \partial\theta(x)$ if and only if the *subgradient inequality* holds for all $y \in \mathbb{R}^n$,

$$\theta(y) - \theta(x) \geq v^T (y - x).$$

It is known from optimization theory that if x is locally optimal solution of (1.3.1), then there exists a vector $v \in \partial\theta(x)$ such that

$$(y - x)^T v \geq 0, \quad \forall y \in x + \mathcal{T}(x; K).$$

Thus, every local minimizer of (1.3.1) is a solution of the generalized QVI defined by the pair $(\mathcal{K}, \partial\theta)$, where

$$\mathcal{K}(v) \equiv \begin{cases} v + \mathcal{T}(v; K) & \text{if } v \in K \\ \emptyset & \text{otherwise} \end{cases}$$

Conversely, if in addition K is a convex set, then every solution of the same generalized (Q)VI is a globally optimal solution of (1.3.1).

Turning to the CP, we introduce a few of its generalizations. Let K be a cone in $\mathbb{R}^n$ and F be a mapping from $\mathbb{R}^{2n+m}$ into $\mathbb{R}^\ell$. We seek a triple of vectors $(x, y, z) \in \mathbb{R}^{2n+m}$ such that

$$\begin{aligned} F(x, y, z) &= 0 \\ K \ni x \perp y &\in K^*. \end{aligned} \tag{1.6.1}$$

This implicit CP is convenient for the development of "interior point methods"; see Chapter **11**. The vertical NCP (1.5.3) can be cast in this form too. Indeed it is not difficult to see that (1.5.3) is equivalent to:

$$\mathbf{F}(u, v, x) \equiv \begin{pmatrix} u - G(x) \\ v - H(x) \end{pmatrix} = 0$$

$$0 \leq u \perp v \geq 0.$$

We define another generalization of the NCP, called the *multi-vertical CP*, that is best described in terms of the min function. Specifically, for $i = 1, \ldots, r$, let $F^i : \mathcal{D} \subseteq \mathbb{R}^n \to \mathbb{R}^m$ be a finite family of vector valued functions defined on a common domain. The equation:

$$\min(F^1(x), \cdots, F^r(x)) = 0 \tag{1.6.2}$$

defines a complementarity problem of the generalized type. Here, complementarity is no longer between only two functions, but rather between r functions; namely, $F_j^1(x), \cdots, F_j^r(x)$. More precisely, at a solution $\bar{x}$ of the problem, all the vectors $F^i(\bar{x})$ are nonnegative and for each $j = 1, \ldots, m$ at least one of $F_j^1(\bar{x}), \cdots, F_j^r(\bar{x})$ must be zero. The above min formulation recasts this generalized CP as an equivalent nonsmooth equation. As such, one can extend easily the treatment of the standard NCP to this more general context. Furthermore, it is possible to extend the Fischer-Burmeister functional to this generalized CP; see Exercise **1.8.24** for a broad treatment of such a generalization.

1.7 Concluding Remarks

In this chapter, we have introduced the definitions of the VI and CP and their many variants. We have discussed the connection between these problems and several well-known problems, such as solving equations, optimization problems, and fixed point problems. We have also presented an extensive set of source problems from diverse subjects that can be formulated as VIs and/or CPs. Beginning in the next chapter, we undertake a comprehensive study of the VI/CP, covering such basic issues as existence of solutions, properties of solutions, sensitivity analysis, and theory of error bounds, as well as computational methods of various kinds. Regrettably, since each source problem has a very rich development of its own, a detailed treatment of each one is beyond the scope of this book. Instead, we will visit some of the source problems occasionally and use them to illustrate and motivate the general theory. Finally, some of the generalizations of the VI/CP presented in Section **1.6** can be treated by extending the techniques and results presented subsequently; others can not. Again, the scope of the book prevents us from deviating too far from its main concern, namely, the theory and methods for the basic VI and the CP.

1.8 Exercises

1.8.1 Let $F : K \to \mathbb{R}^n$ be a given mapping defined on the open set K. Show that the following four statements are equivalent.

(a) $x^* \in \mathrm{SOL}(K, F)$;

(b) $x^* \in \mathrm{SOL}(\mathrm{cl}\, K, F)$ and $x^* \in K$;

(c) $x^* \in K$ and $F(x^*) = 0$;

(d) there exists a closed set $K_0 \subset K$ such that $x^* \in \mathrm{SOL}(K_0, F) \cap \mathrm{int}\, K_0$.

1.8.2 Let $K \subseteq \mathbb{R}^n$ be a closed convex set and let x belong to K.

(a) Show that $\mathcal{N}(x; K)$ is a closed convex cone. Moreover $v \in \mathcal{N}(x; K)$ if and only if $x = \Pi_K(x + \tau v)$ for all $\tau \geq 0$. See Exercise **4.8.1**(a) for a characterization of a tangent vector in terms of the Euclidean projector.

(b) Suppose that $K = \{x \in \mathbb{R}^n : x^T a^i \leq b_i,\ i = 1, \ldots, m\}$ for some vectors $a^i \in \mathbb{R}^n$ and scalars b_i. Show that

$$\mathcal{N}(x; K) = \mathrm{pos}\{\, a^i : i \in \mathcal{I}(x)\,\},$$

where, we recall, $\mathcal{I}(x)$ denotes the set of indices of active constraints at x and it is understood that if $\mathcal{I}(x) = \emptyset$, then $\mathcal{N}(x; K) = \{0\}$.

(c) Suppose that $K = K_1 \times K_2$, where $K_1 \subseteq \mathbb{R}^{n_1}$ and $K_2 \subseteq \mathbb{R}^{n_2}$ are closed convex sets and $n_1 + n_2 = n$. Partition x accordingly as (x_1, x_2). Show that $\mathcal{N}(x; K) = \mathcal{N}(x_1; K_1) \times \mathcal{N}(x_2; K_2)$.

(d) Show that the normal cone depends only on the "local" structure of the set K, in the sense that, for every positive scalar δ,

$$\mathcal{N}(x; K) = \mathcal{N}(x; K \cap \operatorname{cl} \mathbb{B}(x, \delta)).$$

(See also Exercise **1.8.4**.)

(e) Show that properties analogous to (a),(c) and (d) also hold for the tangent cone.

(f) Consider the box constrained VI (K, F), where K is given by (1.1.7). Use part (d) to show (1.2.4).

1.8.3 Prove or disprove that if x is a unique optimal solution of the problem (1.2.1), then x is a unique solution of the VI (K, F).

1.8.4 Let S and T be two closed sets in $\mathbb{R}^n$. Show that if $x \in S \cap \operatorname{int} T$, then $\mathcal{T}(x; S \cap T) = \mathcal{T}(x; S)$.

1.8.5 Let $K(x) \equiv m(x) + \mathbb{R}^n_+$, where $m : \mathbb{R}^n \to \mathbb{R}^n$ is a vector-valued function. Show that a vector x is a solution of the QVI (K, F) if and only if x satisfies:

$$0 \leq x - m(x) \perp F(x) \geq 0. \tag{1.8.1}$$

This CP is a special case of the vertical CP with the two functions $\mathbf{I} - m$ and F.

1.8.6 Let $L(x, y)$ be a convex-concave saddle function. Show that the functions

$$\varphi(x) \equiv \sup\{\, L(x, v) \,:\, v \in Y \,\} \quad \text{and} \quad \psi(y) \equiv \inf\{\, L(u, y) \,:\, u \in X \,\}$$

are convex and concave, respectively.

1.8.7 Consider the minimization problem (1.3.1). We say that the objective function θ is *pseudo convex* on K if there exists an open convex set $\mathcal{D}$ containing K on which θ is continuously differentiable and such that for every x and y in $\mathcal{D}$,

$$(\, y - x \,)^T \nabla \theta(x) \geq 0 \;\Rightarrow\; \theta(y) \geq \theta(x).$$

(a) Show that if θ is convex on $\mathcal{D}$ and continuously differentiable there, then it is pseudo convex on the same set.

(b) Show that $\theta(x) = x^3 + x$ is pseudo convex but not convex on $\mathbb{R}$.

(c) Show that if θ is a pseudo convex function and K is convex set, then x is a global solution of problem (1.3.1) if and only if x is a solution of the stationary point problem, VI $(K, \nabla\theta)$.

(d) Give an example of a pair (q, M) such that the quadratic function $x^T(q + Mx)$ is pseudo convex but not convex on the nonnegative orthant $\mathbb{R}^n_+$.

1.8.8 Show that for any nonzero matrix $M \in \mathcal{M}^n_+$,

$$\frac{1}{\lambda^+_{\min}(M)} \| Mz \|^2 \geq z^T Mz \geq \frac{1}{\lambda_{\max}(M)} \| Mz \|^2, \quad \forall z \in \mathbb{R}^n,$$

where $\lambda^+_{\min}(M)$ is the smallest positive eigenvalue of M and $\lambda_{\max}(M)$ is the largest eigenvalue of M. With $M \equiv A^T EA$, where $E \in \mathcal{M}^m_+$ and $A \in \mathbb{R}^{m \times n}$, deduce that for every $z \in \mathbb{R}^n$,

$$\frac{1}{\lambda^+_{\min}(M)} \| Mz \|^2 \geq \frac{1}{\lambda_{\max}(E)} \| EAz \|^2$$

and

$$\frac{1}{\lambda^+_{\min}(E)} \| EAz \|^2 \geq \frac{1}{\lambda_{\max}(M)} \| Mz \|^2.$$

1.8.9 Consider the matrix given by (1.2.9). Throughout this exercise, M is not assumed symmetric. Let Z be any matrix whose columns form an orthonormal basis of the null space of C. We know that

$$\text{rank } Z + \text{rank } C = n.$$

Extend the basis Z through the orthogonal complement of the null space of C to an orthonormal basis P of $\mathbb{R}^n$. Thus $P = [Z \;\; W]$ for some matrix W whose columns form an orthonormal basis of the orthogonal complement of the null space of C. Moreover, P is an orthogonal matrix; i.e., $P^T P = \mathbf{I}_n$.

(i) Show that CW must have full column rank. Thus if C is of full row rank, then CW is a square, nonsingular matrix.

(ii) Suppose that C is of full row rank. Show that

$$\det \begin{bmatrix} M & C^T \\ -C & 0 \end{bmatrix} = (\det Z^T MZ)(\det CW)^2.$$

(Hint: pre- and post-multiply the matrix (1.2.9) by

$$\begin{bmatrix} P^T & 0 \\ 0 & \mathbf{I} \end{bmatrix}$$

and its transpose, respectively; use part (i), the Schur determinantal formula and some elementary matrix operations.)

(iii) Deduce from the above determinantal formula that the matrix (1.2.9) is nonsingular if and only if C has full row rank and $Z^T M Z$ is nonsingular. Give an appropriate interpretation of parts (ii) and (iii) when C is a square, nonsingular matrix.

1.8.10 For simplicity, we drop the matrix C in the MLCP (1.2.8); thus, consider the following problem:

$$\begin{aligned} 0 &= q + Mx + A^T \lambda \\ 0 &\leq \lambda \perp y \equiv b - Ax \geq 0. \end{aligned} \tag{1.8.2}$$

(i) Show that if there exists a set of rows B in A such that

$$\begin{bmatrix} M & B^T \\ -B & 0 \end{bmatrix}$$

is nonsingular, then by solving for the variables x and λ_β in (1.8.2), where β is the subset of $\{1, \ldots, m\}$ corresponding to the rows of B, one obtains an LCP in the variables y_β and $\lambda_{\bar\beta}$, where $\bar\beta$ is the complement of β in $\{1, \ldots, m\}$.

The following is a more practical approach of executing the conversion in part (i). Let Z be any matrix whose columns form an orthonormal basis of the null space of A. Extend the basis Z through the orthogonal complement of the null space of A to an orthonormal basis P of $\mathbb{R}^n$. Thus $P = [Z \;\; W]$ for some matrix W whose columns form an orthonormal basis of the orthogonal complement of the null space of A. Write

$$AW \equiv \begin{bmatrix} B \\ N \end{bmatrix},$$

where B is a nonsingular matrix. Why?

(ii) Suppose that the matrix $Z^T M Z$ is nonsingular. By considering the change of variables $x \equiv Py$, show that the MLCP (1.8.2) is equivalent to an LCP $(\mathbf{q}, \mathbf{M})$, where

$$\mathbf{M} \equiv \begin{bmatrix} \tilde{M} & -(NB^{-1})^T \\ NB^{-1} & 0 \end{bmatrix},$$

with

$$\begin{aligned}\tilde{M} \equiv\ & (WB^{-1})^T M W B^{-1} \\ & -(WB^{-1})^T M Z (Z^T M Z)^{-1} Z^T M W B^{-1}.\end{aligned} \tag{1.8.3}$$

The latter matrix $\tilde{M}$ is the Schur complement of $Z^T M Z$ in the matrix

$$\begin{bmatrix} Z^T M Z & Z^T M W B^{-1} \\ (WB^{-1})^T M Z & (WB^{-1})^T M W B^{-1} \end{bmatrix},$$

which is equal to

$$\begin{bmatrix} I & 0 \\ 0 & (B^{-1})^T \end{bmatrix} P^T M P \begin{bmatrix} I & 0 \\ 0 & B^{-1} \end{bmatrix}.$$

Deduce from the latter representation of $\tilde{M}$ that if M is positive semidefinite, then so is the matrix $\mathbf{M}$.

1.8.11 A special case of the frictional contact problem leads to an affine quasi-variational inequality (AQVI) with variable bounds (VB). To define this problem, let m and n be two positive integers. Let M be a square matrix of order $(n+m)$ partitioned as

$$M \equiv \begin{bmatrix} A & B \\ C & D \end{bmatrix},$$

where the principal submatrices A and D are of order $n \times n$ and $m \times m$, respectively, and the off-diagonal submatrices B and C are of order $n \times m$ and $m \times n$, respectively. Let q be an $(n+m)$-vector partitioned similarly,

$$q \equiv \begin{pmatrix} p \\ r \end{pmatrix},$$

where $p \in \mathbb{R}^n$ and $r \in \mathbb{R}^m$. Let K be a nonempty polyhedral convex set in $\mathbb{R}^n$. Let a and b be two m-vectors satisfying $a \leq 0 \leq b$ and $a < b$. Each index $i \in \{1, \ldots, m\}$ is associated with an index $j_i \in \{1, \ldots, n\}$. It is possible that $j_i = j_{i'}$ for two distinct indices i and i'. The AQVI/VB is to find a pair of vectors $z \equiv (x, y) \in K \times Q(x) \subset \mathbb{R}^{n+m}$ such that

$$(z' - z)^T (q + Mz) \geq 0, \quad \forall z' \in K \times Q(x),$$

where for each $x \in K$, $Q(x)$ is a compact rectangle in $\mathbb{R}^m$ given by

$$Q(x) \equiv \prod_{i=1}^{m} [\, a_i x_{j_i},\, b_i x_{j_i} \,].$$

In order for $Q(x)$ to be well defined, we assume that for simplicity that K is contained in the nonnegative orthant of $\mathbb{R}^n$.

(i) Write down the equivalent MLCP formulation of this problem. Suppose that the matrix M is positive definite. Must the defining matrix of the equivalent MLCP be positive semidefinite?

(ii) Show that a pair of vectors $(x, y) \in \mathbb{R}^{n+m}$ solves the AQVI/VB if and only if the conditions (a), (b), and (c) hold:

(a) x solves the AVI $(K, p + By, A)$;

(b) $y \in Q(x)$; and

(c) for all $i = 1, \ldots, m$, if $x_{j_i} \neq 0$, then

$$\begin{aligned} y_i &= a_i\, x_{j_i} &\Rightarrow\quad (\, r + Cx + Dy\,)_i &\geq 0 \\ a_i\, x_{j_i} < y_i &< b_i\, x_{j_i} &\Rightarrow\quad (\, r + Cx + Dy\,)_i &= 0 \\ y_i &= b_i\, x_{j_i} &\Rightarrow\quad (\, r + Cx + Dy\,)_i &\leq 0. \end{aligned}$$

1.8.12 Let a, b and r be given scalars with $r \geq 0$. Write $a \equiv d \cos\theta$ and $b \equiv d \sin\theta$. Consider the following simple minimization problem in two scalar variables x and y:

$$\begin{aligned} &\text{minimize} && ax + by \\ &\text{subject to} && \sqrt{x^2 + y^2} \leq r. \end{aligned}$$

Show that the unique solution to this problem is given by $x = \tau \cos\theta$ and $y = \tau \sin\theta$, where τ satisfies the equation

$$d + \min(\,0, r + \tau - d\,) + \max(\,0, -r + \tau - d\,) \;=\; 0.$$

This equation provides an effective way of formulating the 3-dimensional standard Coulomb friction law as a system of nonsmooth equations.

1.8.13 This exercise concerns the nonlinear programming approach to solving CPs. We start with the following QP associated with a *mixed, horizontal LCP* of the following type:

$$\begin{aligned} &\text{minimize} && x^T y \\ &\text{subject to} && Ax + By + Cz + q = 0 \\ &\text{and} && (\,x, y\,) \geq 0, \end{aligned} \tag{1.8.4}$$

where A and B are matrices of order $m \times n$, C is a matrix of order $m \times \ell$, and q is an m-vector. Let S denote the feasible region of the above program.

(a) Show that the objective function $x^T y$ is convex on S if and only if the following implication holds:

$$[(x^i, y^i, z^i) \in S,\ i = 1, 2] \Rightarrow (x^1 - x^2)^T(y^1 - y^2) \geq 0.$$

(b) Show that if $AB^T \in \mathbb{R}^{m \times m}$ is negative semidefinite on the null space of C^T, i.e., if

$$C^T v = 0 \Rightarrow v^T AB^T v \leq 0,$$

then every stationary point of (1.8.4) satisfies $x^T y = 0$. Prove this in two ways: one, verify that under the assumptions herein, Proposition **1.5.13** is applicable; two, apply the KKT conditions of (1.8.4).

The next part concerns an extension of the above treatment to the implicit MiCP (1.4.58), where $H : \mathbb{R}^{2n+m} \to \mathbb{R}^{n+m}$ is assumed to be continuously differentiable. Consider the nonlinear program

$$\begin{array}{ll} \text{minimize} & x^T y \\ \text{subject to} & H(x, y, z) = 0 \\ \text{and} & (x, y) \geq 0, \end{array} \tag{1.8.5}$$

which extends the quadratic program (1.8.4). Let (x^*, y^*, z^*) be a stationary point of (1.8.5). Suppose that (i) Abadie's CQ holds at (x^*, y^*, z^*) for the feasible region of (1.8.5), and (ii) the following implication holds:

$$\left.\begin{array}{r} J_z H(x^*, y^*, z^*)^T \lambda = 0 \\ (J_x H(x^*, y^*, z^*)^T \lambda) \circ (J_y H(x^*, y^*, z^*)^T \lambda) \geq 0 \end{array}\right\}$$

$$\Rightarrow (J_x H(x^*, y^*, z^*)^T \lambda) \circ (J_y H(x^*, y^*, z^*)^T \lambda) = 0.$$

Show that x^* is complementary to y^*.

1.8.14 Let E be an $m \times n$ matrix with full row rank. Let $K \subseteq \mathbb{R}^n$ be the null space of E.

(a) Show that the Euclidean projector Π_K is a linear transformation with the matrix representation

$$\Pi_K = \mathbf{I}_n - E^T(EE^T)^{-1}E.$$

(b) Let P denote the matrix representation of Π_K. Show that P is of the form (1.8.3) for some suitable choice of matrices M, W, Z, and B. Deduce from part (ii) of Exercise **1.8.10** that P is symmetric and positive semidefinite. (The positive semidefiniteness of P also follows from the monotonicity of the projector Π_K; see Theorem **1.5.5**.)

(c) Verify directly that $P^2 = P$. What is the geometric interpretation of this identity in connection with the projector Π_K. Use the nonexpansiveness property of Π_K to show that $\|P\|_2 \leq 1$.

(d) Give a geometric interpretation to the matrix $\tilde{P} \equiv \mathbf{I}_n - P$ and show that $\tilde{P}$ has the same algebraic properties as P; that is, $\tilde{P}$ is symmetric positive semidefinite, $\tilde{P}^2 = \tilde{P}$, and $\|\tilde{P}\|_2 \leq 1$.

(e) Let A be a symmetric positive definite matrix. Give a matrix representation of $\Pi_{K,A}$.

1.8.15 Design an "efficient" algorithm for computing the Euclidean projection of a vector onto the special polyhedral cone known as the *isotonic cone*:

$$\{ x \in \mathbb{R}^n : x_1 \leq x_2 \leq \cdots \leq x_n \}.$$

1.8.16 Let K be a closed convex set in $\mathbb{R}^n$ and A be a symmetric positive definite matrix of order n. Use (1.5.11) to show that the skewed projector is nonexpansive under the A-norm; i.e.,

$$\| \Pi_{K,A}(u) - \Pi_{K,A}(v) \|_A \leq \| u - v \|_A, \quad \forall u, v \in \mathbb{R}^n.$$

1.8.17 Let A and B be two SPSD matrices of order n. By means of the orthogonal decomposition of a SPSD matrix, show that $A \perp B$ if and only if there exist an orthogonal matrix P and two nonnegative diagonal matrices D_a and D_b such that

$$A = P D_a P^T, \quad B = P D_b P^T,$$

and $D_a D_b = 0$. Obviously, the diagonal entries of D_a and D_b are the eigenvalues of A and B, respectively; and the columns of P are the associated orthonormal eigenvectors. Thus, A and B are complementary if and only if they commute and with a suitable permutation, the eigenvalues of A are complementary to the eigenvalues of B. Use this fact to show the following two statements.

(a) For any matrix $A \in \mathcal{M}^n$, there exist an orthogonal matrix P and a diagonal matrix $D \equiv \operatorname{diag}(\lambda_1, \cdots, \lambda_n)$ such that

$$A = PDP^T \quad \text{and} \quad \Pi_{\mathcal{M}^n_+}(A) = PD_+P^T,$$

where $D_+ \equiv \operatorname{diag}((\lambda_1)_+, \cdots, (\lambda_n)_+)$.

(b) For a given mapping $\mathbf{F} : \mathcal{M}^n \to \mathcal{M}^n$, the complementarity problem (1.4.61) in SPSD matrices is equivalent to an implicit MiCP (1.4.58) for some appropriate mapping $H : \mathbb{R}^{2n+n^2} \to \mathbb{R}^{n+n^2}$.

1.8.18 Let $F : \mathcal{U} \subseteq \mathcal{M}^n \to \mathcal{M}^n$ be a given matrix function on an open set $\mathcal{U}$ of symmetric matrices of order n. Define

$$f : \operatorname{svec}(\mathcal{U}) \subseteq \mathbb{R}^{n(n+1)/2} \to \mathbb{R}^{n(n+1)/2}$$

by $f(\operatorname{svec}(A)) = \operatorname{svec}(F(A))$ for every A in $\mathcal{U}$. The Fréchet derivative of F at a matrix $A \in \mathcal{U}$ is a linear operator $F'(A; \cdot)$ from $\mathcal{M}^n$ into itself such that

$$\lim_{H(\in \mathcal{M}^n) \to 0} \frac{F(A+H) - F(A) - F'(A;H)}{\|H\|} = 0. \qquad (1.8.6)$$

(a) Assume that F is Fréchet differentiable at $A \in \mathcal{U}$. Show that

$$\operatorname{svec}(F'(A;H)) = Jf(\operatorname{svec}(A))\operatorname{svec}(H), \quad \forall H \in \mathcal{M}^n.$$

(b) Show by direct verification of the limit (1.8.6) that

(i) for $F(A) = A^2$, $F'(A;B) = AB + BA$ for any two matrices A and B in $\mathcal{M}^n$;

(ii) for $F(A) = A^{-1}$, $F'(A;B) = -A^{-1}BA^{-1}$ for any two matrices A and B in $\mathcal{M}^n$ with A nonsingular.

1.8.19 Let $\theta(X) = \log \det X$ for $X \in \mathcal{M}^n_{++}$.

(a) Show that $\nabla\theta(X) = X^{-1}$.

(b) Use the commutability property of the trace of products of symmetric matrices and part (ii) of Exercise **1.8.18**(b) to show that

$$B \bullet \nabla^2\theta(X)B \leq 0, \quad \forall B \in \mathcal{M}^n.$$

(c) Deduce from (b) that θ is a concave function on $\mathcal{M}^n_{++}$; hence

$$\det(\tau A + (1-\tau)B) \geq (\det A)^{\tau} (\det B)^{1-\tau},$$

for any two matrices A and B in $\mathcal{M}^n_{++}$ and scalar $\tau \in (0,1)$.

1.8.20 Show that the tangent cone $\mathcal{T}(A; \mathcal{M}^n_+)$ of a matrix $A \in \mathcal{M}^n_+$ is equal to the set of matrices $B \in \mathcal{M}^n$ that are positive semidefinite on the null space of A; equivalently, $B \in \mathcal{T}(A; \mathcal{M}^n_+)$ if and only if $B \in \mathcal{M}^n$ and for any matrix Z whose columns form an orthonormal basis of the null space of A, the matrix $Z^T BZ$ is positive semidefinite. This can be proved by convex analysis. In what follows, we ask the reader to attempt a direct matrix-theoretic proof. For the sufficiency part, let $A = PDP^T$ be an orthogonal decomposition of A, where P is an orthogonal matrix and D is a nonnegative diagonal matrix. Write $P \equiv [Z\ W]$, where the columns of

Z form an orthonormal basis of the null space of A and the columns of W are orthonormal eigenvectors corresponding to the positive eigenvalues of A. Similarly, write

$$D \equiv \begin{bmatrix} 0 & 0 \\ 0 & D_+ \end{bmatrix},$$

where D_+ is the diagonal matrix whose diagonal entries are the positive eigenvalues of A. For each $\varepsilon > 0$, define

$$C_\varepsilon \equiv \begin{bmatrix} \varepsilon Z^T B Z + \varepsilon^2 Z^T B W (D_+ + \varepsilon W^T B W)^{-1} W^T B Z & \varepsilon Z^T B W \\ \varepsilon W^T B Z & D_+ + \varepsilon W^T B W \end{bmatrix}.$$

To complete the proof, show that C_ε is positive semidefinite for all $\varepsilon > 0$ sufficiently small and

$$B = \lim_{\varepsilon \downarrow 0} \frac{P C_\varepsilon P^T - A}{\varepsilon}.$$

By noticing that

$$C_\varepsilon = P^T A P + \varepsilon P^T B P + \varepsilon^2 V(\varepsilon),$$

where

$$V(\varepsilon) \equiv \begin{bmatrix} Z^T B W (D_+ + \varepsilon W^T B W)^{-1} W^T B Z & 0 \\ 0 & 0 \end{bmatrix},$$

deduce that $B \in \mathcal{T}(A; \mathcal{M}^n_+)$ if and only if there exists $\bar{\varepsilon} > 0$ such that for every $\varepsilon \in [0, \bar{\varepsilon}]$ there exists $E(\varepsilon) \in \mathcal{M}^n_+$ satisfying (i) $A + \varepsilon B + \varepsilon^2 E(\varepsilon) \in \mathcal{M}^n_+$ and (ii) the limit of $E(\varepsilon)$ as $\varepsilon \downarrow 0$ exists.

1.8.21 The exercise gives two additional C-functions whose differentiability properties the reader is asked to study.

(a) Let κ be a positive constant. For $(a, b) \in \mathbb{R}^2$, define

$$\psi_{\mathrm{U}}(a,b) \equiv \begin{cases} \dfrac{a_+ \, b_+}{\kappa \left[1 - \exp\left(\dfrac{|a| + |b|}{\kappa} \right) \right]} - \sqrt{(a_-)^2 + (b_-)^2} & \text{if } (a,b) \neq 0 \\ 0 & \text{otherwise,} \end{cases}$$

where a_+ and a_- are, respectively, the nonnegative and nonpositive parts of the scalar a. Show that ψ_{U} is a locally Lipschitz continuous C-function. Study the differentiability properties of ψ_{U} and of ψ_{U}^2.

(b) Let η be any nonnegative scalar and set

$$\psi_{\mathrm{YYF}}(a,b) \equiv \frac{\eta}{2}(\,(ab)_+\,)^2 + \tfrac{1}{2}\,\psi_{\mathrm{FB}}(a,b)^2, \quad (a,b) \in \mathbb{R}^2.$$

Show that ψ_{YYF} is a differentiable C-function.

1.8.22 Show that the mid function $\mathrm{mid}(a,b;\cdot) : \mathbb{R}^n \to \mathbb{R}^n$ is directionally differentiable everywhere; compute its directional derivative at a point.

1.8.23 Let K be a closed convex set in $\mathbb{R}^n$ and $F : K \to \mathbb{R}^n$ be continuous. The natural map $\mathbf{F}_K^{\mathrm{nat}}$ is a bijection from $\mathbb{R}^n$ onto $\mathbb{R}^n$ if and only if the VI $(K, q + F_q)$ has a unique solution for all $q \in \mathbb{R}^n$, where $F_q(z) \equiv F(z - q)$.

1.8.24 A real-valued function $f : \mathbb{R}^n \to \mathbb{R}$ is *sign-preserving* if

$$[\,x \geq 0 \Rightarrow f(x) \geq 0\,] \quad \text{and} \quad [\,x \leq 0 \Rightarrow f(x) \leq 0\,].$$

A function f is *positively sign-preserving* if it is sign-preserving and satisfies

$$f(x) \geq 0 \Rightarrow x \geq 0.$$

A real-valued function $f : \mathbb{R}^n \to \mathbb{R}$ is a C-function if

$$f(x) = 0 \Leftrightarrow \min(\,x_1, \cdots, x_n\,) = 0.$$

Both the min and max functions are clearly sign-preserving. The min function is a positively sign-preserving C-function.

(a) Show that the negative of the FB functional ψ_{FB} and the negative of its variant ψ_{CCK} are positively sign-preserving functions of two arguments; i.e., the two functions:

$$(\,a,b\,) \mapsto a + b - \sqrt{a^2 + b^2}$$

and

$$(\,a,b\,) \mapsto a + b + \tau \max(0,a) \max(0,b) - \sqrt{a^2 + b^2},$$

where $\tau > 0$, are both positively sign-preserving.

(b) Show that if $f : \mathbb{R}^n \to \mathbb{R}$ and $\psi : \mathbb{R}^2 \to \mathbb{R}$ are (positively) sign-preserving, then so is $g : \mathbb{R}^{n+1} \to \mathbb{R}$ defined by

$$g(x,t) \equiv \psi(f(x),t), \quad \forall\,(\,x,t\,) \in \mathbb{R}^{n+1}.$$

Show that if f is a positively sign-preserving C-function and ψ is a C-function, then g is a C-function.

1.8.25 Let $H : \mathbb{R}^n \to \mathbb{R}^n$ be continuously differentiable. Show that x is a zero of H if and only if x is an unconstrained stationary point of $\frac{1}{2}H^T H$ and the linear equation

$$H(x) + JH(x)d = 0$$

has a solution in d.

1.8.26 The quadratic cone in $\mathbb{R}^{n+1}$ is defined as

$$\left\{ (x,t) \in \mathbb{R}^{n+1} : \sqrt{x^T Q x} \leq t \right\},$$

for some symmetric positive definite matrix Q of order n. This is the Coulomb friction cone with a unit friction coefficient; see (1.4.29).

(a) Show that for $Q = \mathbf{I}$, the resulting quadratic cone, which is called the *Lorentz cone*, is self-dual. See Figure 1.8.

(b) Use the result of part (a) to derive the dual cone for $Q \neq \mathbf{I}$.

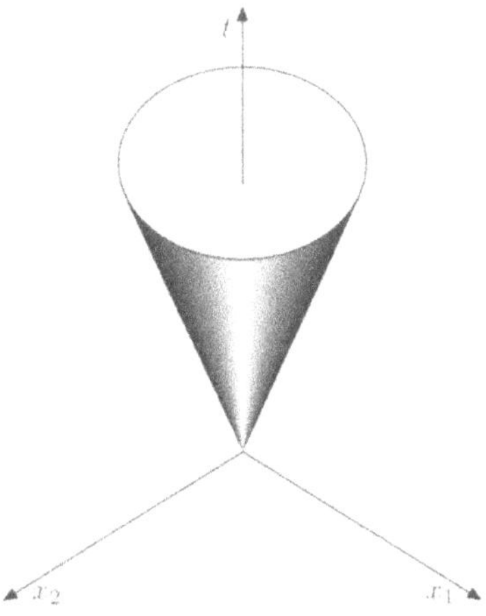

Figure 1.8: The Lorentz cone.

1.8.27 Let K be the Lorentz cone in $\mathbb{R}^{n+1}$:

$$\left\{ (x,t) \in \mathbb{R}^{n+1} : \sqrt{x^T x} \leq t \right\}.$$

(a) Show that

$$K \ni (x,t) \perp (y,s) \in K$$

if and only if

$$v \equiv t - \sqrt{x^T x} \geq 0, \quad u \equiv s - \sqrt{y^T y} \geq 0,$$

$$uv = ut = vs = 0,$$

and two nonnegative scalars ρ and σ, not both zero, exist such that $\rho x + \sigma y = 0$. Interpret this equivalence geometrically.

(b) Use (a) to show that the cone-CP:

$$K \ni (x, t) \perp F(x, t) \in K,$$

where $F : \mathbb{R}^{n+1} \to \mathbb{R}^{n+1}$, is equivalent to a mixed CP of the implicit kind (1.4.58) for some mapping $H : \mathbb{R}^{2+(3+n)} \to \mathbb{R}^{4+n}$.

1.8.28 Let K be the Lorentz cone in $\mathbb{R}^{n+1}$. Show that

$$\Pi_K(x, t) \equiv \begin{cases} \frac{1}{2}\left(1 + \frac{t}{\|x\|_2}\right)(x, \|x\|_2) & \text{if } |t| < \|x\|_2 \\ (x, t) & \text{if } \|x\|_2 \leq t \\ 0 & \text{if } \|x\|_2 \leq -t. \end{cases}$$

Deduce that

$$\text{dist}((x, t), K) = \begin{cases} \frac{f(x, t)_+}{\sqrt{2}} & \text{if } (x, t) \notin -K \\ \|(x, t)\|_2 \left(\in \left[\frac{f(x, t)_+}{\sqrt{2}}, f(x, t)_+\right]\right) & \text{if } (x, t) \in -K, \end{cases}$$

where $f(x, t) \equiv \|x\|_2 - t$. Consequently,

$$\frac{f(x, t)_+}{\sqrt{2}} \leq \text{dist}((x, t), K)) \leq f(x, t)_+, \quad (x, t) \in \mathbb{R}^{n+1}.$$

1.8.29 Let $F : K \to \mathbb{R}^n$ be Lipschitz continuous with modulus $L > 0$. Show that if $\tau \in (0, 1/L)$, then $x \in \text{SOL}(K, F)$ if and only if

$$x = \Pi_K(x - \tau F(\Pi_K(x - \tau F(x)))).$$

1.8.30 Let S be a closed set in $\mathbb{R}^m$ and let x^* be a vector in S. The cone of *attainable directions* of S at x^*, denoted $\mathcal{A}(x^*; S)$, consists of all vectors $d \in \mathbb{R}^m$ for which there exist a scalar $\bar{\varepsilon} > 0$ and a continuous function $x : [0, \bar{\varepsilon}] \to S$ such that

$$x(0) = x^* \quad \text{and} \quad \lim_{\tau \downarrow 0} \frac{x(\tau) - x^*}{\tau} = d.$$

Show that $\mathcal{A}(x^*; S) \subseteq \mathcal{T}(x^*; S)$ in general and that equality holds if S is either the Lorentz cone or the cone $\mathcal{M}^n_+$. (In Exercise **4.8.1**, it is shown that for any closed convex set S, the two cones $\mathcal{A}(x^*; S)$ and $\mathcal{T}(x^*; S)$ are always equal. The reader should attempt the present exercise without referring to the exercise in the later chapter.)

1.8.31 The exercise provides another CQ, known as the Kuhn-Tucker CQ and abbreviated as KTCQ, under which the KKT conditions of the VI (K, F) are necessary for a solution of this problem. The KTCQ implies Abadie's CQ.

(a) Let K be a finitely representable set given by (1.3.4) where every g_i and h_j are continuously differentiable. Let $x \in K$ be arbitrary. Show that

$$\mathcal{T}(x; K) \subseteq \mathcal{L}(x; K).$$

Thus the equality between the attainable cone $\mathcal{A}(x; K)$ and the linearization cone $\mathcal{L}(x; K)$ provides another CQ yielding the necessary satisfaction of the KKT conditions by any solution of the VI (K, F).

(b) We illustrate the above CQ with a special set. Let B be a symmetric positive definite matrix of order n and let

$$K \equiv \{ x \in \mathbb{R}^n_+ : x^T B x = 1 \}.$$

Let $x^* \in K$ be arbitrary. Show that every vector $v \in \mathcal{L}(x^*; K)$ is a tangent vector of K at x^* by considering the following curve

$$x(\tau) \equiv \frac{x^* + \tau v}{\sqrt{(x^* + \tau v)^T B (x^* + \tau v)}},$$

for $\tau > 0$ sufficiently small.

(c) Let K be given by (b). Show that if x is a solution of the VI (K, F), there exists a multiplier μ such that

$$0 \leq x \perp F(x) + \mu B x \geq 0.$$

This VI and its generalization, where $\mathbb{R}^n_+$ in the set K is replaced by an arbitrary closed convex cone C in $\mathbb{R}^n$, are particularly interesting because the defining set K is nonconvex. The special case where $F(x) \equiv Ax$ for some $n \times n$ matrix is a *constrained eigenvalue problem* defined by A on the cone C. We return to the latter problem in Exercise **2.9.28**.

1.8.32 Let C be a closed convex cone in $\mathbb{R}^n$. Let $x \in C$ with $\|x\| = 1$. Show that

$$\mathcal{T}(x; C \cap \operatorname{bd} \mathbb{B}(0,1)) = \mathcal{T}(x; C) \cap x^{\perp}.$$

Deduce that

$$\mathcal{N}(x; C \cap \operatorname{bd} \mathbb{B}(0,1)) = \mathcal{N}(x; C) + \mathbb{R}x.$$

(Hint: for the first equality, it suffices to verify that the right-hand intersection is contained in the left-hand tangent cone. Let d be an arbitrary vector in the right-hand set. Let $\{x^k\} \subset C$ be a sequence converging to x and $\{\tau_k\}$ be a sequence of positive scalars converging to zero such that

$$d = \lim_{k \to \infty} \frac{x^k - x}{\tau_k}.$$

Show that

$$0 = \lim_{k \to \infty} \frac{\dfrac{x^k}{\| x^k \|} - x - \tau_k\, d}{\tau_k}.$$

This completes the proof of the first equality. For the second equality, employ the following known fact from convex analysis: if S_1 and S_2 are two convex sets with S_1 polyhedral, then $(S_1 \cap S_2)^* = S_1^* + S_2^*$, provided that S_1 contains a relative interior point of S_2.)

1.8.33 Let K be the closed Euclidean ball in the plane with radius 1 and center at $(0, -1)$. Let x be the origin and $F(x) = (1, -1)$. Show that $\|\mathbf{F}_K^{\mathrm{nat}}(x)\|$ is strictly less than $\operatorname{dist}(-F(x), \mathcal{N}(x; K))$.

1.8.34 Let K be a closed convex subset of $\mathbb{R}^n$ and F be a continuous mapping from $\mathbb{R}^n$ into itself. Consider the following four statements.

(a) The natural map $\mathbf{F}_K^{\mathrm{nat}}$ is norm-coercive on K; i.e.,

$$\lim_{\substack{x \in K \\ \|x\| \to \infty}} \| \mathbf{F}_K^{\mathrm{nat}}(x) \| = \infty.$$

(b) For every bounded set S in $\mathbb{R}^n$, the union

$$\bigcup_{q \in S} \mathrm{SOL}(K, q + F)$$

is bounded.

(c) The normal map $\mathbf{F}_K^{\mathrm{nor}}$ is norm-coercive on $\mathbb{R}^n$; i.e.,

$$\lim_{\|z\| \to \infty} \| \mathbf{F}_K^{\mathrm{nor}}(z) \| = \infty \qquad (1.8.7)$$

(d) The distance function $\text{dist}(-F(x), \mathcal{N}(x;K))$ is coercive on K; i.e.,

$$\lim_{\substack{x \in K \\ \|x\| \to \infty}} \text{dist}(-F(x), \mathcal{N}(x;K)) = \infty. \tag{1.8.8}$$

Show that (a) $\Rightarrow$ (b) $\Rightarrow$ (c) $\Leftrightarrow$ (d). Show further that if F satisfies the property: for any two sequences $\{x^k\}$ and $\{y^k\}$ in K, the boundedness of $\{x^k - y^k\}$ implies the boundedness of $\{F(x^k) - F(y^k)\}$, then all four statements (a)–(d) are equivalent. Deduce from the latter result that

$$\lim_{\|z\| \to \infty} \| \mathbf{M}_K^{\text{nor}}(z) \| = \infty \Leftrightarrow \lim_{\substack{x \in K \\ \|x\| \to \infty}} \| \mathbf{M}_K^{\text{nat}}(x) \| = \infty.$$

See Exercise **10.5.12** for a continuation of this exercise.

1.9 Notes and Comments

The normal cone of a convex set is a fundamental geometric object in convex analysis. Several general references on this subject are [84, 217, 343, 746, 813]. The significance of this cone is due partly to its close connection with the first-order necessary conditions of a constrained optimization problem. In light of this well-known connection, it is no surprise that the normal cone continues to occupy a central role throughout the VI/CP theory. Although we do not utilize the normal cone of a nonconvex set in this book, the extended object has a similar, important role to play beyond convexity; see [127, 617, 752]. The terminology of a generalized equation (1.1.3) is coined by Robinson in [728].

The observation that a solution x of a VI (K, F) "solves its own program (1.2.1)" is due to Eaves [202], who used it to prove a version of Proposition **1.2.1** that pertained to the NCP with the nonnegative orthant $\mathbb{R}^n_+$ truncated by the hyperplane $d^T x \leq \eta$, where d is a positive vector and $\eta > 0$ is a positive scalar.

The papers by Cao and Ferris [104, 105], which are related to the earlier work by Eaves [207, 208], discussed how Lemke's pivotal algorithm can be extended to solve affine variational inequalities; see also [66, 446] for further extensions of this famous algorithm. Cao and Ferris employed a conversion scheme that is related to the conversion of the KKT system (1.2.8) into an LCP discussed in Section **1.2** and Exercise **1.8.10**.

The proof of Theorem **1.3.1** can be found in, e.g., [652], which is an excellent reference on the fundamentals of differentiable functions of several variables. For a discussion of the role of integrability in mathematical programming models of economic problems, see [110].

Quasi-variational inequalities were introduced by Bensoussan and Lions in a series of papers [51, 52, 53] in connection with the study of impulse control problems and their applications. The complementarity system in Exercise **1.8.5** was called the "implicit complementarity problem" by Capuzzo Dolcetta [107], who considered only the case where F is an affine function; see also [108, 627]. The paper [660] studies the existence and uniqueness of a solution to this problem. The term "implicit" was coined because of an implicit fixed-point perspective of the problem. Indeed, for each $z \in \mathbb{R}^n$, let $\Phi(z)$ denote the (possibly empty) solution set of the NCP:

$$0 \leq x - m(z) \perp F(x) \geq 0.$$

It is then clear that x is a solution of the CP (1.8.1) if and only if x is a fixed point of the set-valued map Φ.

The connection between (generalized) Nash games and (quasi-)VIs was recognized as early as 1974 by Bensoussan [49], who studied these problems with quadratic functionals in Hilbert spaces. Harker [330] revisited these problems in Euclidean spaces. Robinson [743, 744] discussed an application of a generalized Nash problem in a two-sided game model of combat. Kocvara and Outrata [439] described a class of QVIs with applications to engineering; see also [655]. Assuming that an associated VI map is single-valued, Outrata and Zowe [655] proposed using a Newton method for solving QVIs. De Luca and Maugeri [161] discussed the application of QVIs to traffic equilibria. Most recently, Fukushima and Pang [274] employ the QVI as a relaxation to multi-leader-follower games.

There are many general references on the theory and methods of nonlinear programming, [46, 59, 255, 563, 643, 695], to name a few. Among these, the classic by Mangasarian [563] remains an important source on the fundamentals of NLP theory. The book [895] treats optimization and variational problems as a subject in nonlinear functional analysis.

Constraint qualifications and the Karush-Kuhn-Tucker conditions have a long history in nonlinear programming. Chronologically, Karush obtained these conditions in his unpublished Master's Thesis [407], which was not known to Kuhn and Tucker when they published their seminal paper [465]. The latter paper is also the source for the Kuhn-Tucker CQ; see Exercise **1.8.31**. For further historical account of the KKT conditions, including Karush's work, see [464]. Abadie's CQ appears in [3].

When $L(x, y)$ is the scalar Lagrangian function of a constrained optimization problem, the inequality (1.4.4) is the well-known *weak duality* between a pair of primal and dual programs. In general, saddle problems have important applications in game theory. Indeed, the minimax characterization of a saddle point, Theorem **1.4.1**, is the foundation of game theory created by von Neumann [862]. With an eye toward applications to optimal control problems and stochastic programs, Rockafellar [747, 749] and Rockafellar and Wets [750, 751] identified the class of (extended) linear-quadratic programs as an important subclass of saddle problems. Specialized algorithms for solving the linear-quadratic programs can be found in [823, 904, 905].

Nash introduced his equilibrium concept in [634, 635] and proved the existence of equilibrium points in non-cooperative games based on Brouwer's fixed-point theorem. See [276] for a study of the uniqueness and stability of Nash equilibria. The Nash solution concept lies at the heart of all oligopolistic models of competition, among which the Nash-Cournot production/distribution problem is an important instance [650]. Inspired by the paper [628], which described a mathematical programming approach to compute a market equilibrium in a simple Nash-Cournot model, Harker [327] showed that the latter model can be solved much more simply by a VI approach; in fact, the model turns out to be equivalent to a highly structured convex quadratic program. Kolstad and Mathiesen [456] derived necessary and sufficient conditions for uniqueness of a Cournot equilibrium; in [457], the same authors examined convergence criteria of the SLCP algorithm for the computation of Nash-Cournot economic equilibria, based on the NCP formulation of the equilibrium problem. Application to homogeneous, segmented, and differentiated product markets was also discussed.

There is a substantial literature on using the VI/CP approach to study oligopolistic electricity models. Some recent references include [159, 347, 602, 673, 720, 791, 792, 869]. The model presented in Subsection **1.4.3** is based on the paper [869]. Hobbs [347] introduced electricity pricing models that include arbitrage and formulated them as LCPs; these models are extended and analyzed in two subsequent papers [602, 673] where interesting connections are established between models that distinguish themselves in how the firms handle arbitrage. Other gaming models of electricity markets exist; in particular, certain models lead to MPECs [109, 348].

The Markov perfect equilibrium problem is derived from a dynamic oligopolistic pricing problem in economics [590, 591]. The NCP formulation of the former problem is given in [787] where an infeasible interior-point method was proposed for computing such an equilibrium.

In the classic [866], the French economist Léon Walras published the first comprehensive mathematical analyses of general economic equilibrium. Based on the fundamental Walras' law of a pure exchange economy, Arrow and Debreu [18] established the existence of an equilibrium in an abstract economy that included production and consumption. This work, and the subsequent book [167], provided the rigorous foundation for the contemporary development of the field of mathematical economics. For his pioneering contributions, Debreu was awarded the 1983 Nobel Prize in economics. In the early days, the computation of a general equilibrium was largely influenced by Walras' *tantonement* process of repeated experiments. Scarf's paper [759] generated tremendous excitement with the fixed-point approach for computing equilibria. This line of research peaked about the time of publication of the PIES model whose contribution to the VI/CP was noted in the bird-eye overview in the preface of this book. Smale [790] defined a "global Newton method" for a C^2 function via an ordinary differential equation and showed that the method is closely related to a certain price adjustment scheme. More interesting, the method can be regarded as a differential version of Scarf's algorithm for computing economic equilibria.

In [581], Manne provides an overview of the formulation and solution of computable general equilibrium models. Our presentation of this problem in Subsection **1.4.4** follows the work of [594, 595, 754]. Applications of these models are plenty and can be found in the following sample of papers and books, some of which also report the numerical solution of the models by complementarity methods [180, 188, 213, 288, 388, 523, 582, 692, 693, 755, 783, 784]. The model of invariant capital stock is due to Hansen and Koopmans [325]. Dantzig and Manne [157] showed that the linear version of this problem with a linear utility function can be solved by Lemke's algorithm [490]; see also Jones [383].

In the seminal paper [868], Wardrop formulated two major behavioral principles that governed the users of a traffic network. The one that is the basis of the discussion in Subsection **1.4.5** is the user equilibrium principle. The other principle pertains to the system equilibrium and can be formulated as a nonlinear program whose objective function is the total system cost. These two principles form the foundation for the theoretical research of traffic theory. There is a large literature on the VI/CP approach to the traffic user equilibrium problem and the closely related spatial price equilibrium problem that deal with different aspects of these problems, including modelling, algorithms, and sensitivity analysis [1, 2, 20, 21, 58, 61, 116, 151, 153, 155, 229, 230, 233, 234, 241, 253, 254, 257, 258, 259, 264, 267, 268, 271, 277, 328, 329, 338, 384, 385, 486, 518, 519, 550, 583, 584, 631, 632, 642, 683, 702, 793, 794, 795, 796, 839]. Proposition **1.4.6** is a refinement of a result first established by Aashtiani and Magnanti [1], who also proved Proposition **2.2.14** by a direct application of Brouwer's fixed-point theorem. A result similar to Proposition **1.4.6** that provided the NCP formulation for the spatial price equilibrium problem was established by Friesz, Tobin, Smith, and Harker [268]. Patriksson's monograph [686] presents a comprehensive study of the traffic equilibrium problem and contains more than 1,000 references, many of which are works on the VI up to the mid-1990s. An important extension of the static traffic equilibrium problem is the dynamic problem. For references on the latter problem and its connection to a time-dependent VI; see [265, 266, 797].

Contact problems belong to the broad class of inequality problems in mechanics [346, 657, 658]. As early as the mid-1960s, the finite-dimensional contact problem was formulated as an LCP by Friedman and Chernina [263], who proposed a projected Gauss-Seidel method for its solution. With the influential work of Lötstedt [525, 526] for multi-rigid-body systems and Klarbring [412, 413, 414] for deformable bodies, it has become well known that complementarity methods offer a powerful approach for dealing with frictional contact problems of mechanical systems. Indeed, the monograph of Pfeiffer and Glocker [694] has made it quite clear that the mathematical concept of complementarity is synonymous with the physical phenomenon of contact. Half of this monograph is devoted to several complex mechanical systems where complementarity methods have played an important role in the numerical simulations. More generally, there are many compelling practical applications of constrained mechanical systems with frictional contact. The need to perform physical work in hazardous environments, the medical technology of minimally invasive computer assisted surgery, the desire to increase realism in virtual reality, and the automation of manufacturing processes provide some of the most prominent examples where frictional contact naturally arises. Generally speaking, rigid-body contact problems lead to ordinary differential complementarity systems that are a mixture of differential algebraic systems and complementarity conditions; elastic-body and elastoplastic contact problems typically involve partial differential operators to describe the continuum mechanics and complementarity conditions to describe the frictional contact. Time and space discretization of these problems in function spaces lead to the finite-dimensional problems, which are the main concern of Subsection **1.4.6**. The source of the discussion therein is the paper by Pang and Stewart [677].

There is a large literature on the complementarity approach to frictional contact problems, including several excellent surveys. For rigid-body systems, we mention [8, 9, 15, 16, 17, 37, 38, 89, 90, 160, 289, 290, 626, 679, 680, 681, 772, 806, 807, 809, 810, 811, 812, 843, 844, 845, 853]; in particular, the two surveys [89, 810] contain an extensive bibliography up to the year 2000. For elastic-body problems, Klarbring's Ph.D. thesis [412] and the subsequent papers [413, 414] were among the earliest to use mathematical programming, particularly complementarity, methods to study discrete and/or discretized frictional contact problems. Today the CP approach is well accepted in the mechanics community; selected references include [7, 14, 68, 69, 70, 124, 125, 126, 149, 150, 380, 381, 392, 409, 415, 416, 417, 418, 419, 420, 421, 422, 423, 424, 425, 426, 474, 475, 476, 487, 494, 511, 589, 707, 816, 817, 824]. Recent comprehensive surveys of discrete contact problems solvable by mathematical programming methods can be found in [418, 421].

The application of quadratic programming and linear complementarity to engineering plasticity, particularly, elastoplastic structural analysis, began with the seminal work of Giulio Maier [554, 555, 556]. The joint activities between these two disciplines in the 1970s led to the NATO conference on "Engineering Plasticity by Mathematical Programming", whose proceedings appeared as [130]. Kaneko made several important contributions to the piecewise linear complementarity approach to elastic-plastic analysis [394, 395]. Tin-Loi and his students and collaborators employed nonlinear complementarity methods to study semi-rigid frames [834, 836] and elastoplastic structures with nonlinear hardening/softening [832, 835, 837]; see [284, 560, 863, 864] for related work. The discussion in Subsection **1.4.7** is based on the work of Tin-Loi. Extending early works by Maier [557], Maier, Giannessi, and Nappi [559], and Nappi [633], Tin-Loi has recently joined forces with Maier and Bolzon to study parameter identification in fracture mechanics using the MPEC methodology [558]; see also [833]. Besides reporting on this MPEC application, the paper [558] also summarizes several topical problems in engineering mechanics involving the use of mathematical programming techniques.

Obstacle problems are a class of free boundary problems that have provided a rich source of applications for variational inequalities since the beginning of the field; see [35, 753]. The von Kármán thin plate problem with obstacles presented in Subsection **1.4.8** is based on the paper by Yau and Gao [887]; see [606] for a stability analysis of this problem. Chapter 7 in the monograph by Panagiotopoulos [657] is devoted to inequality problems in the theory of thin elastic plates, containing both the static and dynamic unilateral problems of von Kármán plates as well as the unilateral buckling problem that leads to an eigenvalue problem for variational inequalities. The obstacle Bratu problem is described in several articles [353, 604, 605, 611], where a continuation method for its solution was discussed. Numerical results with the application of finite-dimensional NCP methods for solving the discretized problem are reported in [279, 786].

There are other engineering applications of complementarity problems that are not covered in Section **1.4**, elasto-hydrodynamic lubrication being one. Beginning with Kostreva [459], who employed a block pivoting method [458] and solved an NCP, several authors [358, 644, 645, 646, 647, 648, 820] have employed MiCPs for solving mixed lubrication problems. A very recent application of the NCP to TCP/IP network with general topology is discussed in the paper [12]. Extending a logic-based approach [852], Westerberg and his colloborators used the complementarity condition to represent algebraic systems of disjunctive equations [717, 718, 719, 892] for the conditional modelling of various chemical engineering systems. In a similar vein, many engineering hybrid systems are studied via linear complementarity systems, which are combinations of ordinary differential equations with an LCP; see the Ph.D. thesis of Heemels [339] and the related papers [340, 341, 342, 858]. Indeed, linear complementarity systems are a new paradigm that demands careful investigation. A general form of a time-dependent, (nonlinear) VI system appears in Exercise **5.6.1**.

Originated from a paper of Cottle and Dantzig [140], the multi-vertical LCP, known traditionally as the generalized linear complementarity problem, i.e. the problem (1.6.2) where all the functions are affine, provides a useful mathematical model for the study of resistive piecewise-linear circuits [163, 859]. This affine problem has a generalization in a vector lattice known as the abstract linear order complementarity problem [83], which in turn has a further generalization [367, 825].

Published in 1973 [71], the Black-Scholes options pricing model has had a tremendous impact in the financial industry and has inspired intensive research in the pricing of financial derivatives of various kinds. The original Black-Scholes paper dealt only with European options. Building on the work of Bensoussan [50], Karatzas [406] formalized the problem of pricing American options as an optimal stopping problem of a stochastic process. Although not explicitly using the LCP framework, Brennan and Schwartz [85] are arguably the earliest authors who used an iterative LCP algorithm for solving the American option pricing problem. Jaillet, Lamberton, and Lapeyre [373] studied the problem of pricing American options by variational inequalities and discussed the numerical solution of the resulting finite-dimensional discretization of the problem by pivotal (such as Lemke's algorithm) and iterative LCP methods. In the monograph [870], Wilmott, Dewynne, and Howison provided a systematic treatment of the partial differential complementarity approach to the numerical resolution of the American option pricing problem. This approach has received increasing interest, as evidenced by the significant number of publications in recent years [131, 172, 173, 183, 261, 359, 674, 785, 861, 898, 912, 913, 914]. There are numerous papers that discuss non-PDE based methods for solving the American option pricing problem. A brief summary of the latter methods can be found in [88]; see also [86, 87]. The various models on options with transaction costs presented in Subsection **1.4.9** were introduced in the papers [158, 182, 352, 488]. The presentation in that subsection is based largely on the work of Huang and Pang [359, 674].

Extending the class of bilevel optimization problems, MPECs are a relatively new paradigm in mathematical programming. Detailed studies of MPECs can be found in [533, 654], which contain a large number of references. Many interesting engineering design applications are discussed in the latter reference and in [437, 438]; see also [122] for an application of MPEC in product positioning. Treatment of the implied volatility problem of an American option by the MPEC methodology is described in [360]; the related paper [132] uses standard NLP to calibrate the local volatility surface of an European option. There are many more studies on the MPEC, and it is not possible for us to mention all of them; some recent references include [622, 653, 671, 761, 765, 766, 767, 888, 889, 890]. Suffice it to say that this is a fast-growing, important topic that is in urgent need of much further research, since there remain many unresolved issues.

With important applications in engineering control and combinatorial optimization, semidefinite programs are another recent entry in mathematical programming that have received much attention. We refer the reader to Section 5.3 in the monograph [79] for a detailed, analytical treatment of these programs, including first- and second-order optimality conditions and stability and sensitivity analysis. Extending a semidefinite program, Kojima and his collaborators introduced the LCP in SPSD matrices in a series of papers [450, 451, 452, 453, 455]. Studies of the nonlinear version of this problem, which include the investigation of interior point methods and extended C-functions, can be found in [120, 613, 614, 616, 848, 881]. We refer to [64, 355, 356] for general references on matrix theory.

Mangasarian [564] introduced his family of C-functions ψ_{Man} with the intention of applying equation methods to solve NCPs. Soon after Fischer [247, 248, 249] introduced the FB C-function in the mathematical programming community, the usefulness of this function was very quickly recognized. Interested in obtaining C-functions that are convex, Luo and Tseng [544] introduced the function ψ_{LTKYF} and studied its role in solving

CPs. Specifically, Luo and Tseng focused directly on the merit function

$$\phi_1(x^T F(x)) + \sum_{i=1}^{n} \phi_2(-x_i, F_i(x)).$$

Kanzow, Yamashita, and Fukushima [397] refined the work of Luo and Tseng by considering the modified merit function

$$\sum_{i=1}^{n} [\, \phi_1(x_i F_i(x)) + \phi_2(-x_i, F_i(x)) \,].$$

The difference between the above two merit functions amounts to two points of view of the complementarity condition; the former corresponds to the inner-product formulation $x^T F(x) = 0$, whereas the latter corresponds to the Hadamard formulation $x \circ F(x) = 0$. The C-function ψ_{CCK} is introduced by Chen, Chen, and Kanzow [118].

Although the exact source of the function ψ_{FB} is not known to us, we can trace it to at least the Russian book by Rvachev [756], where it was introduced in the context of the "theory of R-functions." An extensive treatment of this theory is documented in the primer by Shapiro [779] who reported that Rvachev "first suggested the R-functions in 1963." According to this primer, "Informally, a real function is an R-function if it can change its property (sign) only when some of its arguments change the same property (sign)." The R-functions provide a fruitful methodology for the representation of rigid solids for use in computer graphics and visualization; see [780, 781]. Exercise **1.8.24**, which is inspired by these functions, allows one to construct C-functions in more than 2 arguments. The latter functions can then be used for solving piecewise smooth equations defined by functions that are the pointwise minimum of a finite number of C^1 functions, that is, for solving the vertical CP (1.6.2) with multiple functions. This area of research has yet to be fully investigated and deserves a careful study; the references [377, 396, 690, 782, 902] all consider the case of 2 functions.

For notes and comments on the Euclidean projector, see Section **4.9**. Minty [609] can be credited for introducing the map $\mathbf{F}_K^{\text{nor}}$; for this reason, this map is also known as the Minty map. As mentioned before, Eaves [202] uses the natural map $\mathbf{F}_K^{\text{nat}}$ to study the NCP. Aganagic [4] utilized the min map extensively in his Ph.D. dissertation. Neither Eaves nor Aganagic employed the term "natural map" for $\mathbf{F}_K^{\text{nat}}$. It was Mangasarian [571] who called $\| \min(x, q + Mx) \|$ the "natural residual" of the LCP (q, M). The term "normal map" was coined by Robinson [740, 741, 742], who obtained many of its deep properties.

In his Ph.D. thesis [135] (see also [137]), Cottle introduced the "positively bounded Jacobians" condition for a C^1 function F and showed that this condition along with an additional nondegeneracy condition implied that the NCP $(F+q)$ had a unique solution for all vectors $q \in \mathbb{R}^n$; see also the paper [137]. Unquestionably, this is the first existence and uniqueness result for the NCP. Cottle's proof of the result was constructive, based on a "principal pivoting" algorithm that solved nonlinear equations in each pivot. Megiddo and Kojima [601] gave an analytic proof of this classic result via the normal map $\mathbf{F}^{\text{nor}}_{\mathbb{R}^n_+}$ without the nondegeneracy condition. They called the property that the NCP $(F+q)$ has a unique solution for all vectors q the "globally uniquely solvable," abbreviated GUS, property. The proof of Megiddo and Kojima consists of first observing that the GUS property holds for the NCP (F) if and only if the normal map $\mathbf{F}^{\text{nor}}_{\mathbb{R}^n_+}$ is a homeomorphism from $\mathbb{R}^n$ onto itself and then showing that Cottle's condition of positively bounded Jacobians yields the desired property of the normal map. Proposition **1.5.11** generalizes the observation made by Megiddo and Kojima for the NCP to the VI. Extending Cottle's principal pivoting algorithm, Habetler and Kostreva [320] developed a direct algorithm for the NCP. The workhorse of these algorithms, which are "finite" under strong assumptions, is the solution of systems of nonlinear equations. This feature and the restrictive requirement for convergence are the main reasons why these early algorithms for the NCP are no longer in mode today.

When $F(x) \equiv q + Mx$, the minimization problem (1.5.13) was used extensively for the study of the LCP (q, M); see [136, 142, 143]. In particular, the class of row sufficient matrices was defined in the investigation of when the stationary points of the quadratic program (1.5.19) are solutions of the LCP (q, M). This investigation led to the discovery of the column sufficient matrices, which in turn play a central role in the convexity of the solution sets of the LCPs; for details, see [143]. Proposition **2.4.10** generalizes the results in the latter reference to the CP (K, F).

The origin of the primal gap function $\theta_{\text{gap}}(x)$ and the dual gap function $\theta_{\text{dual}}(x)$ can be traced to work by Zuhovisckii, Poljak, and Primak [908, 909, 910, 911] on n-person concave games. In these papers, the authors considered the saddle function in two arguments, $L(x, y) \equiv F(x)^T(x - y)$, and showed that under a monotonicity assumption on F, an equilibrium solution can be found by searching for a saddle point of L. The associated optimization problems (1.4.3) are precisely the primal gap program (1.5.16) and the dual gap program (2.3.15).

Auslender [27] formally introduced the gap functions for an optimization problem; Hearn [337] subsequently re-introduced the gap function $\theta_{\text{gap}}(x)$

for a convex program. Marcotte [583] and Marcotte and Dussault [585, 586] used the gap function θ_{gap} extensively to develop iterative descent algorithms for solving the monotone VI. Exercise **7.6.17** is inspired by Marcotte's work.

Proposition **1.5.13** appeared in a slightly different form in [665, 672]. This simple result was the basis for the original "s-regularity" concept that in turn led to the pointwise FB regularity in Definition **9.1.13**; see Section **7.7** for more discussion.

The projected gradient $\Pi_{\mathcal{T}(x;K)}(-\nabla\theta(x))$ was used by Calamai and Moré [103] and Burke and Moré [100] in their convergence analysis of the gradient projection method for solving a constrained optimization problem. Under very mild conditions, these authors showed that a sequence produced by the method must force the projected gradient to zero.

Hemivariational inequalities are discussed in the book [658]. Generalized VIs and CPs (K, F) in which K is a fixed set and F is a set-valued map were studied extensively by Fang and Peterson [228, 231, 232]. As a unification of the QVI and the generalized VI, Chan and Pang [112] defined the generalized QVI (K, F) in which both K and F are set-valued maps. Yao [884, 885, 886] studied the latter problem in much generality. The cited papers provide the basis for many subsequent papers that deal with various refinements of the basic existence results. A further generalization of the CP is the continuous complementarity problem in a measure space [13, 874]; this problem is potentially useful for the study of contact mechanics problems with impacts. The paper [304] studies the multivalued CP.

Mangasarian [562] introduced the class of pseudo convex functions (Exercise **1.8.7**). In her Ph.D. thesis [517], Lo developed a modified Lemke algorithm for solving the AQVI/VB (Exercise **1.8.11**) that pertains to the frictional contact problem. Exercise **1.8.12** is based on the paper [798], which proposes a nonsmooth model for the exact formulation of a 3-dimensional frictional contact problem. Parts (a) and (b) of Exercise **1.8.13** appear in [576]; the related paper [891] discusses an interior-point method of polynomial complexity for solving the extended LCP in this exercise. De Schutter and De Moor [164, 165, 166] study the extended LCP and its applications. Solodov [799] discusses some optimization reformulations of the extended LCP. The isotonic cone in Exercise **1.8.15** arises from ordered restricted statistical inference; for details see [721]. The idea of solving a semidefinite program as a standard nonlinear program is discussed extensively in the papers [93, 94]; this idea provides the motivation for Exercise **1.8.17**. The C-functions $\psi_{\text{U}}(a, b)$ and $\psi_{\text{YYF}}(a, b)$ in Exercise **1.8.21** are due to Ulbrich [854, 855, 856] and Yamada, Yamashita, and Fukushima [877], respectively. The emphasis of Ulbrich's work is on solving optimal control problems posed in functional spaces; the paper [877] proposes a derivative-free descent method for solving the (finite-dimensional) NCP. Together with the nonnegative orthant and the cone of symmetric positive semidefinite matrices, the Lorentz cone in Exercise **1.8.26** is one of several self-dual cones known as *symmetric cones*; for a comprehensive study of these cones, see [235]. The papers [272, 119] study, respectively, smoothing functions and smoothing methods for CPs on the Lorentz cone; the paper [54] analyzes in detail the approximation of the Lorentz cone by polyhedral cones. Such an approximation can be used to replace the quadratic Coulomb friction cone by a polygonal cone in a contact problem. This leads to an LCP formulation of the latter problem that can be solved by a modified Lemke algorithm; see [517] for details. The equivalence of the two limits was proved by Robinson [741], who called (1.8.8) K-properness of F and used this condition to study the homeomorphism property of a normal map on a polyhedron.

Chapter 2

Solution Analysis I

In this chapter, we begin an in-depth study of the VI/CP. Our first concern is the issue of the existence of a solution to the VI. There are several general approaches to obtain existence results for this problem. One is historical and has its origin in an infinite-dimensional setting. In this first approach, a basic existence theorem pertaining to the VI (K, F) with a compact convex set K and a continuous mapping F is obtained by a fixed-point theorem. From this theorem, extended results are derived by replacing the boundedness of the set K by refined conditions on F. These results can then be applied to various problem classes including the MiCP. The starting point of the alternative approach is to derive via a degree-theoretic argument an existence theorem for the VI (K, F) with a closed convex set K and a continuous map F under a key degree condition on the pair. This condition is then shown to hold for various special cases where the function F satisfies certain properties. Yet a third approach is via the demonstration that an equivalent (constrained or unconstrained) optimization problem of the VI/CP based on a certain merit function has an optimal solution, which can be shown to solve the VI/CP in question under additional conditions. In this and the next chapter, we do not consider the third approach, which is covered indirectly in the two algorithmic Chapters **9** and **10**.

Presumably the fixed-point approach is somewhat more elementary than the degree-theoretic approach, because the concept of degree, as witnessed from the development in Section **2.1**, requires a certain level of familiarity with nonlinear analysis. The proof of the basic theorem in each approach turns out to be quite simple. The fixed-point approach has the advantage that once a key extension of the basic theorem is obtained, the specialization of the extended result to various problem classes is not difficult. In contrast, the verification of the degree condition in the basic theorem using the degree-theoretic approach often requires artifices that lack an intuitive appeal, especially for those readers who are not well versed in degree arguments. A major advantage of the degree approach, however, is that the derived basic theorem asserts not only the existence of a solution to the VI (K, F) in question, but actually also to all VIs (L, G) where the pair $(L.G)$ is a "small perturbation" of (K, F). The latter assertion is a sensitivity result for the original VI (K, F). Furthermore, this additional conclusion is obtained as an immediate consequence of the degree condition and it requires no extra effort to be proven. In this sense, the degree result is more powerful than the fixed-point result. It is for this reason that, in spite of its advanced prerequisites, we adopt the degree-theoretic approach to the treatment of existence of a solution to the VI.

The other important topic of analysis is the property of the solution set of a VI/CP. Issues such as the uniqueness, convexity, compactness, and connectedness of solutions are covered in this and the next chapter; sufficient conditions are provided that ensure the validity of these useful properties. Since solution sensitivity is a vast topic in itself, we devote the entire Chapter **5** to a comprehensive treatment of this important subject.

2.1 Degree Theory and Nonlinear Analysis

This section presents several topics of nonlinear analysis: degree theory, a touch of the theory of homeomorphisms, elements of set-valued analysis, two celebrated fixed-point theorems, and a brief theory of contraction mappings. While degree theory is used in the next section, other topics are used in later chapters. Many results in this section are well known and presented without proof. Occasionally, some proofs are included in order to illustrate techniques that are relevant to the subsequent developments.

2.1.1 Degree theory

Degree theory is a classical mathematical tool that has diverse applications; it is particularly useful for the study of the existence of a solution to an equation $\Phi(x) = p$, where Φ is a continuous function defined on the closure of a bounded open subset Ω of $\mathbb{R}^n$ and takes values in $\mathbb{R}^n$ and p is a given vector not in $\Phi(\operatorname{bd}\Omega)$. In what follows, we review the basic definitions and results of degree theory, with the goal of laying down the background for the analytical results to be developed for the VI/CP.

Many approaches to the introduction of degree are possible. One approach is to first define the degree of a continuously differentiable mapping relative to a bounded open set and a "non-critical" value p. This definition is then extended by an approximation argument to continuous functions relative to critical values as well. Instead of following this illuminating but long approach, we adopt here an axiomatic point of view. From the axiomatic definition, numerous properties of the degree can be obtained. To present this definition of degree, let Γ be the collection of triples (Φ, Ω, p), where Ω is a bounded open subset of $\mathbb{R}^n$, Φ is a continuous mapping from $\mathrm{cl}\,\Omega$ into $\mathbb{R}^n$, and $p \notin \Phi(\mathrm{bd}\,\Omega)$.

The following definition identifies the degree as an integer-valued function with Γ as its domain. There are three axioms. Axiom (A1) simply postulates that the degree of the identity map relative to any bounded open set and value p belonging to this set is equal to unity. This is very reasonable because of the special property of this map. The next two axioms are useful for computing the degree in many situations. Axiom (A2) is an additive property of the degree as a function of its second argument; Axiom (A3) is called the *homotopy invariance principle* of the degree. It turns out that these three axioms alone will yield a host of important properties of the degree that have far-reaching consequences.

2.1.1 Definition. Let an integer $\deg(\Phi, \Omega, p)$ be associated with each triple (Φ, Ω, p) in the collection Γ. The function deg is called a *(topological) degree* if the following three axioms are satisfied:

(A1) $\deg(\mathbf{I}, \Omega, p) = 1$ if $p \in \Omega$;

(A2) $\deg(\Phi, \Omega, p) = \deg(\Phi, \Omega_1, p) + \deg(\Phi, \Omega_2, p)$ if Ω_1 and Ω_2 are two disjoint open subsets of Ω and $p \notin \Phi((\mathrm{cl}\,\Omega) \setminus (\Omega_1 \cup \Omega_2))$;

(A3) $\deg(H(\cdot, t), \Omega, p(t))$ is independent of $t \in [0,1]$ for any two continuous functions $H : \mathrm{cl}\,\Omega \times [0,1] \to \mathbb{R}^n$ and $p : [0,1] \to \mathbb{R}^n$ such that

$$p(t) \notin H(\mathrm{bd}\,\Omega, t), \qquad \forall\, t \in [0,1].$$

We call $\deg(\Phi, \Omega, p)$ the *degree* of Φ at p relative to Ω. If $p = 0$, we simply write $\deg(\Phi, \Omega)$ for $\deg(\Phi, \Omega, 0)$. □

There are two obvious questions that naturally arise: Does a degree exist? Can we have more than one degree? The answer to these questions is very neat and surprising: There exists one and only one degree. We do not prove this result. Instead, we are content with illustrating the main properties of the degree, particularly those that will allow us to actually compute the degree. The first such property is described in the following theorem, which elucidates the fundamental role of the degree throughout this book.

2.1.2 Theorem. Let Ω be a nonempty, bounded open subset of $\mathbb{R}^n$ and let $\Phi : \operatorname{cl}\Omega \to \mathbb{R}^n$ be a continuous function. Assume that $p \notin \Phi(\operatorname{bd}\Omega)$. If $\deg(\Phi, \Omega, p) \neq 0$, then there exists an $\bar{x} \in \Omega$ such that $\Phi(\bar{x}) = p$. Conversely, if $p \notin \operatorname{cl}\Phi(\Omega)$, then $\deg(\Phi, \Omega, p) = 0$. □

A main application of the above theorem is to establish the existence of a solution to a system of equations $\Phi(x) = p$, via the demonstration that $\deg(\Phi, \Omega, p)$ is well defined and nonzero. In this regard, Axiom (A3) offers a technique that is used repeatedly throughout the subsequent analysis. The basic idea is to construct a "homotopy function" $H : \operatorname{cl}\Omega \times [0,1] \to \mathbb{R}^n$ such that $H(x,0) \equiv \Phi(x)$ for all $x \in \operatorname{cl}\Omega$ and the degree of $(H(\cdot,1), \Omega, p)$ is known and presumably nonzero. If p does not belong to $H(\operatorname{bd}\Omega, t)$ for all $t \in [0,1]$, then by the homotopy invariance property, it follows immediately that the degree of the original triple (Φ, Ω, p) is equal to the degree of the auxiliary triple $(H(\cdot,1), \Omega, p)$; hence $\deg(\Phi, \Omega, p)$ is nonzero. In many applications, $H(\cdot,1)$ is the identity map and Axiom (A1) implies that $\deg(H(\cdot,1), \Omega, p)$ is equal to unity, provided that p belongs to Ω. More generally, by Axiom (A3), the computation of $\deg(\Phi, \Omega, p)$ reduces to the identification of the required homotopy function H that ensures the vector p does not belong to $H(\operatorname{bd}\Omega, t)$ for all $t \in [0,1]$.

The converse in Theorem **2.1.2** implies in particular that if Φ is a continuous function without a zero, then $\deg(\Phi, \Omega)$ is equal to zero for any nonempty, bounded open set Ω such that $\operatorname{cl}\Omega$ is contained in the domain of Φ. For an application of this observation, see Exercise **2.9.3**.

In the following proposition we collect several useful facts that facilitate the calculation of the degree of a continuous map. In part (c) of this proposition, the $\operatorname{dist}_\infty$ function to a closed set is defined in terms of the max-norm of vectors; that is, for a closed set $S \subseteq \mathbb{R}^n$,

$$\operatorname{dist}_\infty(x, S) \equiv \inf_{y \in S} \| x - y \|_\infty.$$

2.1.3 Proposition. Let Ω be a nonempty, bounded open subset of $\mathbb{R}^n$, and let Φ and Υ be continuous functions from $\operatorname{cl}\Omega$ to $\mathbb{R}^n$. Assume that $p \notin \Phi(\operatorname{bd}\Omega)$. The following properties hold:

(a) $\deg(\Phi, \Omega, p) = \deg(\Phi - p, \Omega)$;

(b) for any $a \in \mathbb{R}^n$, if $\Phi_a(x) \equiv \Phi(x + a)$, then

$$\deg(\Phi_a, \Omega - a, p) = \deg(\Phi, \Omega, p);$$

(c) $\deg(\Phi, \Omega, p) \ = \ \deg(\Upsilon, \Omega, p)$ if

$$\max_{x \in \mathrm{cl}\,\Omega} \| \Phi(x) - \Upsilon(x) \|_\infty \ < \ \mathrm{dist}_\infty(\, p, \Phi(\mathrm{bd}\,\Omega)\,);$$

(d) $\deg(\Phi, \Omega, p) = \deg(\Upsilon, \Omega, p)$ if $\Phi(x) = \Upsilon(x)$ for every $x \in \mathrm{bd}\,\Omega$;

(e) $\deg(\Phi, \Omega, p) = \deg(\Phi, \Omega, q)$ for every $q \in \mathbb{R}^n$ such that

$$\| p - q \|_\infty \ < \ \mathrm{dist}_\infty(p, \Phi(\mathrm{bd}\,\Omega));$$

(e) $\deg(\Phi, \Omega, p) = \deg(\Phi, \Omega, q)$ for every q belonging to the same connected component of $\mathbb{R}^n \setminus \Phi(\mathrm{bd}\,\Omega)$.

(f) $\deg(\Phi, \Omega, p) \ = \ \deg(\Phi, \Omega_1, p)$ for every open subset Ω_1 of Ω such that $p \notin \Phi(\Omega \setminus \Omega_1)$;

(g) if Ω_1 and Ω_2 are two disjoint open sets whose union is Ω, then

$$\deg(\Phi, \Omega, p) \ = \ \deg(\Phi, \Omega_1, p) + \deg(\Phi, \Omega_2, p);$$

(h) let $\Psi : \mathrm{cl}\,\Omega' \to \mathbb{R}^m$ be continuous, where Ω' is a nonempty, bounded open subset of $\mathbb{R}^m$; if $q \notin \Psi(\mathrm{bd}\,\Omega')$, then $\deg(\Phi \times \Psi, \Omega \times \Omega', (p, q))$ is well defined and

$$\deg(\Phi \times \Psi, \Omega \times \Omega', (p, q)) \ = \ \deg(\Phi, \Omega, p)\ \deg(\Psi, \Omega', q).$$

Part (c) says essentially that $\deg(\Phi, \Omega, p)$ is a continuous function of the first argument when we equip the linear space of continuous functions from the closure of Ω into $\mathbb{R}^n$ with the supremum norm:

$$\| \Phi \| \ \equiv \ \sup_{x \in \mathrm{cl}\,\Omega} \ \| \Phi(x) \|_\infty.$$

This property (c) of the degree is called the *nearness property*. Alternatively, in terms of a norm $\| \cdot \|$ other than the max-norm, there exists a constant $c > 0$ such that for every two functions Φ and Υ satisfying

$$\max_{x \in \mathrm{cl}\,\Omega} \| \Phi(x) - \Upsilon(x) \| \ < \ c\ \mathrm{dist}(\, p, \Phi(\mathrm{bd}\,\Omega)\,),$$

where the distance function $\mathrm{dist}(\cdot, S)$ to a subset S of $\mathbb{R}^n$ is induced by $\| \cdot \|$, we have $\deg(\Phi, \Omega, p) \ = \ \deg(\Upsilon, \Omega, p)$. For instance, if the Euclidean norm is used, then this constant c can be taken to be $1/7$.

Property (g) and property (h) are respectively the *domain decomposition property* and the *Cartesian product property* of the degree. Property (f) is called the *excision property* of the degree. This property leads to the concept of the "index" of a continuous function relative to a vector in the range of Φ with an "isolated" preimage. Specifically, suppose that $\bar{x} \in \Omega$ is an *isolated* p-point of Φ, i.e. $\Phi(\bar{x}) = p$ and there exists an open neighborhood $\Omega_1 \subseteq \Omega$ of $\bar{x}$ such that $\Phi^{-1}(p) \cap \operatorname{cl} \Omega_1 = \{\bar{x}\}$. We then have

$$\deg(\Phi, \Omega_1, p) = \deg(\Phi, \Omega_2, p) \tag{2.1.1}$$

for every open neighborhood $\Omega_2 \subseteq \Omega$ of $\bar{x}$ such that $\operatorname{cl} \Omega_2$ does not contain another point $\tilde{x}$ satisfying $\Phi(\tilde{x}) = p$. Indeed the excision property implies that

$$\deg(\Phi, \Omega_1, p) = \deg(\Phi, \Omega_1 \cup \Omega_2, p) = \deg(\Phi, \Omega_2, p).$$

Thus, (2.1.1) holds. This common degree is called the *index* of Φ at $\bar{x}$ and denoted by $\operatorname{ind}(\Phi, \bar{x})$. In summary, the index of a continuous function Φ is always well defined at a vector $\bar{x}$ in the domain Φ that is an isolated $\Phi(\bar{x})$-point.

The following result can be viewed as a generalization of Axiom (A1) in the definition of degree.

2.1.4 Proposition. Let Ω be a nonempty, bounded open subset of $\mathbb{R}^n$ and let $\Phi : \operatorname{cl} \Omega \to \mathbb{R}^n$ be a continuous injective mapping. For every $p \in \Phi(\Omega)$, $\deg(\Phi, \Omega, p) = \pm 1$. □

In particular, if Φ is a nonsingular affine map, say $\Phi(x) \equiv Ax + b$ for some nonsingular matrix $A \in \mathbb{R}^{n \times n}$ and a vector $b \in \mathbb{R}^n$, then for every bounded open subset Ω containing the vector $\bar{x} \equiv A^{-1}b$, we have

$$\deg(\Phi, \Omega) = \operatorname{ind}(\Phi, \bar{x}) = \pm 1 = \operatorname{sgn} \det A.$$

The fact that $\deg(\Phi, \Omega) = \operatorname{sgn} \det A$ does not follow from the discussion so far. In fact, this identity is the starting point of the alternative approach to define the degree that we mentioned before. For our purpose here, we take this as a known fact.

The above results can be generalized to a nonlinear function. Let Φ be continuously differentiable in a neighborhood of a vector $\bar{x}$. Assume that the Jacobian matrix $J\Phi(\bar{x})$ is nonsingular. By the inverse function theorem, it follows that Φ is a "local homeomorphism" at $\bar{x}$; that is, there exists an open neighborhood $\mathcal{N}$ of $\bar{x}$ such that the restriction

$$\Phi|_{\mathcal{N}} : \mathcal{N} \to \Phi(\mathcal{N})$$

is a homeomorphism; see Proposition **2.1.14**. With respect to any open neighborhood $\mathcal{N}'$ of $\bar{x}$ properly contained in $\mathcal{N}$, we have, with $p \equiv \Phi(\bar{x})$,

$$\deg(\Phi, \mathcal{N}', p) = \operatorname{ind}(\Phi, \bar{x}) = \operatorname{sgn} \det J\Phi(\bar{x}).$$

The following proposition is a further generalization of the latter result.

2.1.5 Proposition. Let Ω be a nonempty, bounded open subset of $\mathbb{R}^n$ and let Υ and Φ be two continuous functions defined on $\operatorname{cl}\Omega$. Suppose that $\bar{x}$ is a vector in Ω such that $\bar{x}$ is an isolated p-point of Φ and

$$\lim_{\bar{x}\neq x\to\bar{x}} \frac{\Phi(x)-\Upsilon(x)}{\|\, x-\bar{x}\,\|} = 0, \quad \text{and} \quad \liminf_{\bar{x}\neq x\to\bar{x}} \frac{\|\,\Upsilon(x)-\Upsilon(\bar{x})\,\|}{\|\, x-\bar{x}\,\|} > 0. \quad (2.1.2)$$

The following two statements hold:

(a) $\bar{x}$ is an isolated p-point of Υ;

(b) $\operatorname{ind}(\Phi,\bar{x})$ and $\operatorname{ind}(\Upsilon,\bar{x})$ are both well defined and equal.

Proof. The first limit in (2.1.2) implies that $\Upsilon(\bar{x}) = p$, by the continuity of Φ and Υ. Moreover, the second limit in (2.1.2) implies that $\bar{x}$ is an isolated p-point of Υ. Statement (a) therefore holds. Furthermore, both $\operatorname{ind}(\Phi,\bar{x})$ and $\operatorname{ind}(\Upsilon,\bar{x})$ are well defined. Let $\bar{\delta} > 0$ be such that $\bar{x}$ is the only vector in the closed Euclidean ball with center at $\bar{x}$ and radius $\bar{\delta}$ satisfying $\Phi(\bar{x}) = p$. We claim that there exists a constant $\gamma > 0$ such that for all $\delta \in (0,\bar{\delta})$,

$$\gamma\,\delta \leq \operatorname{dist}_\infty(p,\Phi(\operatorname{bd}\mathbb{B}(\bar{x},\delta))).$$

Suppose this claim is not valid. There exist sequences $\{\delta_k\} \subset (0,\bar{\delta})$ and $\{x^k\} \subset \operatorname{bd}\mathbb{B}(\bar{x},\delta_k)$ such that

$$\|\,\Phi(x^k)-\Phi(\bar{x})\,\|_\infty < \frac{1}{k}\,\delta_k.$$

Without loss of generality, we may assume that the sequence $\{\delta_k\}$ converges to some limit $\delta_\infty \in [0,\bar{\delta}]$. Since $\Phi^{-1}(p)\cap \operatorname{cl}\mathbb{B}(\bar{x},\bar{\delta}) = \{\bar{x}\}$, it follows that $\delta_\infty = 0$. We have

$$\Upsilon(x^k)-\Upsilon(\bar{x}) = \Upsilon(x^k)-\Phi(x^k)+\Phi(x^k)-\Phi(\bar{x}).$$

Dividing by $\|x^k-\bar{x}\| = \delta_k > 0$ and letting $k\to\infty$, we see that the left-hand term in the above expression is positive while the right-hand term approaches zero. This is a contradiction; hence the existence of the desired constant $\gamma > 0$ follows. By the first limit in (2.1.2), there exists an open neighborhood $\mathcal{N} \equiv \mathbb{B}(\bar{x},\delta)$ of $\bar{x}$, where $\delta \in (0,\bar{\delta})$, such that

$$\sup_{x\in\operatorname{cl}\mathcal{N}} \|\,\Phi(x)-\Upsilon(x)\,\|_\infty \leq \tfrac{1}{2}\gamma\,\delta < \operatorname{dist}_\infty(p,\Phi(\operatorname{bd}\mathbb{B}(\bar{x},\delta))).$$

By the nearness property of the degree, it follows that

$$\operatorname{ind}(\Phi,\bar{x}) = \deg(\Phi,\mathcal{N},p) = \deg(\Upsilon,\mathcal{N},p) = \operatorname{ind}(\Upsilon,\bar{x}).$$

This establishes (b). □

To understand the significance of Proposition **2.1.5**, let us examine the two limit conditions in (2.1.2). The first limit says that $\Upsilon(x)$ is a "first-order approximation" (FOA) of $\Phi(x)$ for x near $\bar{x}$; that is,

$$\Phi(x) = \Upsilon(x) + e(x)$$

where the difference function $e(x)$ is "little o" of $\|x - \bar{x}\|$. The second limit in (2.1.2) is a local growth condition of the approximating function Υ near $\bar{x}$; indeed, this limit is equivalent to the existence of a constant $c > 0$ such that for all x sufficiently near $\bar{x}$,

$$\Upsilon(x) = \Upsilon(\bar{x}) + h(x),$$

where $\|h(x)\| \geq c\|x - \bar{x}\|$. If $\Phi(x)$ is F-differentiable at $\bar{x}$, then

$$\Upsilon(x) \equiv \Phi(\bar{x}) + J\Phi(\bar{x})(x - \bar{x})$$

easily satisfies the first limit in (2.1.2); the second limit holds if $J\Phi(\bar{x})$ is nonsingular. For our purpose in this book, the most interesting application of Proposition **2.1.5** is when Φ is not differentiable but has an FOA that is "nonsingular" at $\bar{x}$; see Lemma **5.2.2** for a result of this kind.

Property (g) of Proposition **2.1.3** refers to a decomposition of the domain Ω. This property follows easily from Axiom (A2) of the degree. Using the domain decomposition property, we can establish the next result, which gives an identity to compute the degree in terms of the index.

2.1.6 Proposition. Let Ω be a nonempty, bounded open subset of $\mathbb{R}^n$ and let $\Phi : \operatorname{cl}\Omega \to \mathbb{R}^n$ be a continuous mapping. Let $p \notin \Phi(\operatorname{bd}\Omega)$ be such that $\Phi^{-1}(p)$ is finite. Then

$$\deg(\Phi, \Omega, p) = \sum_{x \in \Phi^{-1}(p)} \operatorname{ind}(\Phi, x).$$

Proof. Let $\{x^1, \cdots, x^s\}$ be the elements of $\Phi^{-1}(p)$ and let Ω_t $(t = 1, \cdots, s)$ be disjoint open neighborhoods of x^t all contained in Ω such that

$$\deg(\Phi, \Omega_t, p) = \operatorname{ind}(\Phi, x^t).$$

The desired identity for $\deg(\Phi, \Omega, p)$ follows from the excision property and the domain decomposition property of the degree. $\square$

If Φ is continuously differentiable on Ω and $p \notin \Phi(\operatorname{bd}\Omega)$, and if $J\Phi(x)$ is nonsingular at every $x \in \Phi^{-1}(p)$, then $\Phi^{-1}(p)$ is a finite set. This is left as an exercise for the reader to prove. A generalization of this remark is valid for certain nonsmooth functions, using the concept of some "generalized Jacobians" introduced in Chapter **7**.

More generally, suppose that the solution set of $\Phi(x) = p$ is nonempty and bounded, where Φ is defined and continuous on an open subset $\mathcal{D}$ of $\mathbb{R}^n$. Then $\deg(\Phi, \Omega, p)$ is independent of Ω as long as Ω is a bounded open set containing $\Phi^{-1}(p)$ and $\operatorname{cl}\Omega$ is contained in $\mathcal{D}$. This common degree can be thought of as an index of $\Phi^{-1}(p)$. The following result shows that this index can be calculated easily via a sequence of continuous functions that approximate Φ uniformly on compact sets.

2.1.7 Theorem. Let $\Phi : \operatorname{cl}\mathcal{D} \subseteq \mathbb{R}^n \to \mathbb{R}^n$ be a continuous function defined on the closure of the open set $\mathcal{D}$ and let $p \in \mathbb{R}^n$ be such that $\Phi^{-1}(p)$ is nonempty and compact. Let Ω be a bounded open set containing $\Phi^{-1}(p)$ such that $\operatorname{cl}\Omega \subseteq \operatorname{cl}\mathcal{D}$. For every sequence of continuous functions $\{\Phi^k\}$ converging uniformly to Φ on $\operatorname{cl}\Omega$,

$$\deg(\Phi, \Omega, p) = \lim_{k\to\infty} \deg(\Phi^k, \Omega, p); \qquad (2.1.3)$$

moreover, the limit is always reached after finitely many steps.

Proof. By part (a) of Proposition **2.1.3**, we can assume, without loss of generality, that $p = 0$. By the uniform convergence of $\{\Phi^k\}$ to Φ on $\operatorname{cl}\Omega$, and since $\operatorname{dist}_\infty(0, \Phi(\operatorname{bd}\Omega)) > 0$, there exists an integer $\bar{k}$ such that for all $k \geq \bar{k}$,

$$\max_{x\in\operatorname{cl}\Omega} \| \Phi^k(x) - \Phi(x) \|_\infty < \operatorname{dist}_\infty(0, \Phi(\operatorname{bd}\Omega)).$$

By part (c) of Proposition **2.1.3**, it follows that

$$\deg(\Phi^k, \Omega) = \deg(\Phi, \Omega),$$

establishing the desired limit (2.1.3) and the claim that this limit is reached after finitely many steps. □

We employ the results presented so far to establish a sufficient condition for the surjectivity of a nonlinear map.

2.1.8 Proposition. Let $\Phi : \mathbb{R}^n \to \mathbb{R}^n$ be continuous. If

$$\lim_{\|x\|\to\infty} \frac{x^T\Phi(x)}{\| x \|} = \infty. \qquad (2.1.4)$$

Then $\Phi(\mathbb{R}^n) = \mathbb{R}^n$.

Proof. Given $y \in \mathbb{R}^n$, define the homotopy:

$$H(x,t) \equiv tx + (1-t)\Phi(x) - y, \quad (x,t) \in \mathbb{R}^n \times [0,1].$$

Choose a scalar $r > \|y\|$ sufficiently large so that for all $x \in \text{bd}\,\mathbb{B}(0,r)$,

$$\frac{x^T\Phi(x)}{\|x\|} > \|y\|.$$

We have for all $t \in [0,1]$ and all $x \in \text{bd}\,\mathbb{B}(0,r)$

$$x^T H(x,t) \geq r\left[tr + (1-t)\frac{x^T\Phi(x)}{\|x\|} - \|y\|\right] > 0$$

Thus, for all $t \in [0,1]$, $H(\cdot,t)$ does not vanish on $\text{bd}\,\mathbb{B}(0,r)$. Hence by the homotopy invariance property of the degree, we have $\deg(\Phi, \mathbb{B}(0,r), y) = 1$. This implies that $\Phi(x) = y$ has a solution. □

The condition (2.1.4) postulates the growth of $\|\Phi(x)\|$ as $\|x\|$ becomes large. Since

$$\|\Phi(x)\| \geq \frac{x^T\Phi(x)}{\|x\|},$$

(2.1.4) implies the following weaker condition

$$\lim_{\|x\|\to\infty} \|\Phi(x)\| = \infty,$$

which is known as *norm-coercivity*. A sufficient condition for (2.1.4) to hold is that Φ is continuously differentiable on $\mathbb{R}^n$ and there exists a constant $c > 0$ such that

$$x^T J\Phi(y)x \geq c\|x\|^2, \quad \forall\, x, y \in \mathbb{R}^n.$$

To see that the latter condition implies (2.1.4), it suffices to note that, by the mean-value theorem in integral form, we have

$$\begin{aligned} x^T\Phi(x) &= x^T\Phi(0) + \int_0^1 x^T J\Phi(tx)x\,dt \\ &\geq x^T\Phi(0) + c\|x\|^2. \end{aligned}$$

For an application of Proposition **2.1.8**, see Exercise **2.9.19**. It is not difficult to see that a function Φ is norm-coercive on $\mathbb{R}^n$ if and only if the level set

$$\{x \in \mathbb{R}^n : \|\Phi(x)\| \leq \eta\}$$

is bounded for every scalar $\eta > 0$.

2.1.2 Global and local homeomorphisms

Homeomorphisms play an important role in the study of solution properties of the VI/CP. Indeed, many questions about the existence, uniqueness, and continuity of solutions to the VI/CP can be answered via the theory of these functions. For this reason, we give a concise definition of several kinds of homeomorphisms. To begin, we formally recall that a mapping $\Phi : S \to T$ is a homeomorphism from S onto T if Φ is continuous and bijective and the inverse $\Phi^{-1} : T \to S$ is also continuous. If S is an open set, the continuity of the inverse Φ^{-1} is an immediate consequence of the continuity and surjectivity of the mapping Φ itself; in this case T must be an open set too; see Theorem **2.1.11** and the proof of Proposition **2.1.12**.

2.1.9 Definition. A function $\Phi : \mathcal{D} \to \mathbb{R}^n$, where $\mathcal{D}$ is an open subset of $\mathbb{R}^n$, is said to be a

(a) *local homeomorphism* at a vector $x \in \mathcal{D}$ if there exists an open neighborhood $\mathcal{N} \subseteq \mathcal{D}$ of x such that the restricted map $\Phi|_{\mathcal{N}} : \mathcal{N} \to \Phi(\mathcal{N})$ is a homeomorphism; we call $(\Phi|_{\mathcal{N}})^{-1}$ a *local inverse* of Φ;

(b) *locally Lipschitz homeomorphism* at $x \in \mathcal{D}$ if there exists an open neighborhood $\mathcal{N} \subseteq \mathcal{D}$ of x such that the restricted map $\Phi|_{\mathcal{N}}$ is a homeomorphism and both $\Phi|_{\mathcal{N}}$ and $(\Phi|_{\mathcal{N}})^{-1}$ are Lipschitz continuous on their respective domains;

(c) *globally Lipschitz homeomorphism* if $\mathcal{D} = \mathbb{R}^n$ and Φ is a global homeomorphism from $\mathbb{R}^n$ onto itself and both Φ and Φ^{-1} are globally Lipschitz continuous on $\mathbb{R}^n$. □

The following classic result gives a necessary and sufficient condition for a continuous function from $\mathbb{R}^n$ into itself to be a global homeomorphism.

2.1.10 Theorem. Let $\Phi : \mathbb{R}^n \to \mathbb{R}^n$ be continuous. For Φ to be a global homeomorphism from $\mathbb{R}^n$ onto $\mathbb{R}^n$, it is necessary and sufficient for Φ to be norm-coercive and everywhere a local homeomorphism. □

We next present conditions for Φ to be a local homeomorphism and a locally Lipschitz homeomorphism, respectively. As a preliminary result, we state a version of the *domain invariance theorem* in nonlinear analysis; this theorem asserts that a continuous, injective function maps open sets onto open sets. In general, a function that maps open sets onto open sets is called an *open mapping*. Thus, we also refer the following result as the "open mapping theorem".

2.1.11 Theorem. If $\Phi : \mathcal{D} \to \mathbb{R}^n$ is continuous and injective in the open subset $\mathcal{D}$ of $\mathbb{R}^n$, then $\Phi(\mathcal{D})$ is open. □

Based on the above theorem, it is not difficult to establish the following simple local homeomorphism result.

2.1.12 Proposition. If $\Phi : \mathcal{D} \to \mathbb{R}^n$ is continuous and injective on the open set $\mathcal{D} \subseteq \mathbb{R}^n$, then Φ is a local homeomorphism at every point of $\mathcal{D}$.

Proof. Let x be an arbitrary vector in $\mathcal{D}$ and let $\mathcal{N}$ be an open neighborhood of x such that $\mathrm{cl}\,\mathcal{N}$ is contained in $\mathcal{D}$. It suffices to show that the inverse of the restricted map $\Phi|_{\mathcal{N}}$ is continuous on $\Phi(\mathcal{N})$, which is an open set. Let $\{x^k\}$ be a sequence of vectors in $\mathcal{N}$ such that $\{\Phi(x^k)\}$ converges to a vector $v \in \Phi(\mathcal{N})$. We need to show that $\{x^k\}$ converges to $\Phi^{-1}(v)$. The sequence $\{x^k\}$, being contained in $\mathcal{N}$, is bounded. Every accumulation point x^∞ of $\{x^k\}$ belongs to $\mathrm{cl}\,\mathcal{N}$, which is a subset of $\mathcal{D}$. Since Φ is continuous on $\mathcal{D}$, we have $\Phi(x^\infty) = v$. Thus every accumulation point of $\{x^k\}$ is equal to the preimage of v; since this preimage is unique, it follows that $\{x^k\}$ converges to $\Phi^{-1}(v)$. □

Using the above proposition, we can establish the following characterization of a locally Lipschitz homeomorphism.

2.1.13 Corollary. A function $\Phi : \mathcal{D} \to \mathbb{R}^n$, where $\mathcal{D}$ is an open subset of $\mathbb{R}^n$, is a locally Lipschitz homeomorphism at a vector $x \in \mathcal{D}$ if and only if Φ is locally Lipchitz continuous at x and there exist an open neighborhood $\mathcal{N}$ of x and a constant $c > 0$ such that

$$\| \Phi(y) - \Phi(z) \| \geq c \| y - z \|, \quad \forall y, z \in \mathcal{N}. \tag{2.1.5}$$

Proof. The "only if" statement is straightforward. For the converse, note that (2.1.5) implies that Φ is injective in $\mathcal{N}$. The proof of Proposition **2.1.12** then implies that $\Phi|_{\mathcal{N}}$ is a homeomorphism, and thus a Lipschitz homeomorphism by assumption and (2.1.5). □

If Φ is strongly Fréchet differentiable at a vector x, then the nonsingularity of the Jacobian matrix $J\Phi(x)$ provides a necessary and sufficient condition for Φ to be a locally Lipschitz homeomorphism near x. This is the assertion of Proposition **2.1.14** below. As a reminder, we recall that the strong Fréchet differentiability of Φ at x means that $JF(x)$ exists and

$$\lim_{\substack{u \neq v \\ (u,v) \to 0}} \frac{\Phi(x+u) - \Phi(x+v) - J\Phi(x)(u-v)}{\| u - v \|} = 0. \tag{2.1.6}$$

A sufficient condition for Φ to have a strong Fréchet derivative at x is that Φ is continuously differentiable in a neighbhorhood of x.

2.1.14 Proposition. Let $\Phi : \mathcal{D} \to \mathbb{R}^n$ be strongly Fréchet differentiable at a point x in the open subset $\mathcal{D}$ of $\mathbb{R}^n$. The Jacobian matrix $J\Phi(x)$ is nonsingular if and only if Φ is a locally Lipschitz homeomorphism at x.

Proof. Since Φ is strongly Frécht differentiable at x, it is locally Lipchitz continuous there. If $J\Phi(x)$ is nonsingular, it follows easily from the limit (2.1.6) that there exists a neighborhood $\mathcal{N}$ of x and a constant $c > 0$ such that (2.1.5) holds. Thus Φ is a locally Lipschitz homeomorphism at x by Corollary **2.1.13**.

Conversely, suppose that Φ is a locally Lipschitz homeomorphism at x but that there exists a nonzero vector $v \in \mathbb{R}^n$ such that $J\Phi(x)v = 0$. Since Φ^{-1} is Lipschitz continuous in a neighborhood of $\Phi(x)$, it follows that there exists a constant $L > 0$ such that for all $\tau > 0$ sufficiently small,

$$\tau \, \| \, v \, \| \, \leq \, L \, \| \, \Phi(x + \tau v) - \Phi(x) \, \|.$$

(This is just (2.1.5).) Dividing by τ and letting $\tau \downarrow 0$, we deduce that $v = 0$, which is a contradiction. □

Proposition **2.1.14** can be extended to a nondifferentiable function. Exercise **3.7.6** gives one such extension for a B-differentiable function (see **3.1.2** for the definition of B-differentiability); there, the necessary and sufficient condition for the locally Lipschitz homeomorphism property is expressed in terms of the B-derivative. In what follows, we present a characterization for a locally Lipschitz continuous function to be a locally Lipschitz homeomorphism in terms of a set of "directional derivative like" vectors. Specifically, for a given function Φ that is locally Lipschitz continuous at a vector x in its domain, define $\Delta\Phi(x)$ to be the set of vectors $d^\infty \in \mathbb{R}^n$ that are limits of points

$$d^k \equiv \frac{\Phi(x^k) - \Phi(y^k)}{\| \, x^k - y^k \, \|}, \quad k = 1, 2, \ldots,$$

where $\{x^k\}$ and $\{y^k\}$ are arbitrary sequences converging to x and $x^k \neq y^k$ for each k. By the locally Lipschitz continuity of Φ, the set $\Delta\Phi(x)$ is clearly nonempty. The proposition below is an immediate consequence of Corollary **2.1.13**. In spite of its simplicity, the extent to which the set $\Delta\Phi(x)$ can be used remains to be investigated.

2.1.15 Proposition. A locally Lipschitz continuous function Φ from $\mathbb{R}^n$ into itself is a locally Lipschitz homeomorphism at $x \in \mathbb{R}^n$ if and only if the set $\Delta\Phi(x)$ does not contain the origin. □

2.1.3 Elementary set-valued analysis

A point-to-set map, also called a multifunction or a set-valued map, is a map Φ from $\mathbb{R}^n$ into the power set of $\mathbb{R}^n$, i.e., for every $x \in \mathbb{R}^n$, $\Phi(x)$ is a (possibly empty) subset of $\mathbb{R}^n$. Whereas the term "point-to-set map" is more conventional, the other two terms, "set-valued map" and "multifunction", are more contemporary. In this book, we use these three terms interchangeably. For notational convenience, we use the usual functional notation $\Phi : \mathbb{R}^n \to \mathbb{R}^n$ to denote such a map, but we will clarify the context by specifying that Φ is a set-valued map. The domain of Φ, denoted $\operatorname{dom}\Phi$, the range of Φ, denoted $\operatorname{ran}\Phi$, and the graph of Φ, denoted $\operatorname{gph}\Phi$, are, respectively, the sets:

$$\operatorname{dom}\Phi \equiv \{ x \in \mathbb{R}^n : \Phi(x) \neq \emptyset \}$$

$$\operatorname{ran}\Phi \equiv \bigcup_{x \in \operatorname{dom}\Phi} \Phi(x)$$

$$\operatorname{gph}\Phi \equiv \{ (x, y) \in \mathbb{R}^{2n} : y \in \Phi(x) \}.$$

These are illustrated in Figure 2.1. The following definition contains several classical concepts relevant to a set-valued map.

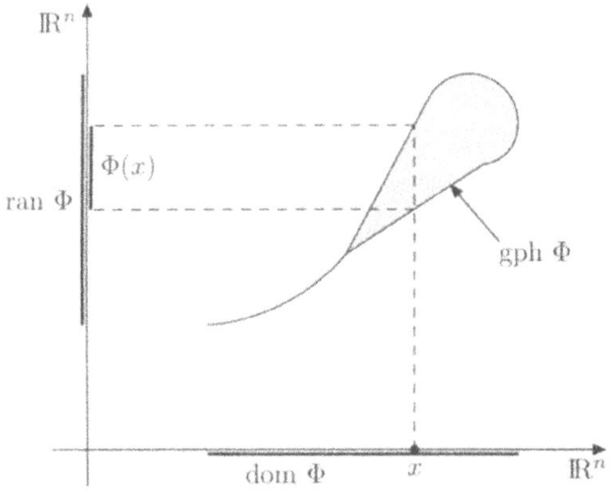

Figure 2.1: Illustration of a set-valued map.

2.1.16 Definition. A set-valued map $\Phi : \mathbb{R}^n \to \mathbb{R}^n$ is said to be

(a) *closed* at a point $\bar{x}$ if

$$\left. \begin{array}{r} \{ x^k \} \to \bar{x} \\ y^k \in \Phi(x^k) \quad \forall\, k \\ \{ y^k \} \to \bar{y} \end{array} \right\} \Rightarrow \bar{y} \in \Phi(\bar{x});$$

(b) *locally bounded* at a point $\bar{x}$ if there exists an open neighborhood $\mathcal{N}$ of $\bar{x}$ such that the set

$$\bigcup_{x \in \mathcal{N} \cap \operatorname{dom} \Phi} \Phi(x)$$

is bounded;

(c) *lower semicontinuous* at a point $\bar{x}$ if for every open set $\mathcal{U}$ such that $\Phi(\bar{x}) \cap \mathcal{U} \neq \emptyset$, there exists an open neighborhood $\mathcal{N}$ of $\bar{x}$ such that, for each $x \in \mathcal{N}$, $\Phi(x) \cap \mathcal{U} \neq \emptyset$;

(d) *upper semicontinuous* at a point $\bar{x}$ if for every open set $\mathcal{V}$ containing $\Phi(\bar{x})$, there exists an open neighborhood $\mathcal{N}$ of $\bar{x}$ such that, for each $x \in \mathcal{N}$, $\mathcal{V}$ contains $\Phi(x)$;

(e) *continuous* at a point $\bar{x}$ if Φ is both lower and upper semicontinuous at $\bar{x}$;

(f) *closed* on a set S if Φ is closed at every point in S; if $S = \mathbb{R}^n$, we simply say that Φ is closed;

(g) (lower, upper semi)continuous on a set S if Φ is (respectively, lower, upper semi)continuous at every point in S. □

The following are useful facts that can be proved easily from the above definitions. Let $\Phi : \mathbb{R}^n \to \mathbb{R}^n$ be a given set-valued map.

• If Φ is closed at $\bar{x}$, then $\Phi(\bar{x})$ is a closed set.

• Suppose $\Phi(\bar{x})$ is a closed set. If Φ is upper semicontinuous at $\bar{x}$, then Φ is closed at $\bar{x}$.

• Conversely, if Φ is closed and locally bounded at $\bar{x}$, then Φ is upper semicontinuous at $\bar{x}$.

• A set-valued map Φ is closed if and only if its graph is a closed set.

• If Φ is lower semicontinuous at $\bar{x} \in \operatorname{dom}(\Phi)$, then $\bar{x} \in \operatorname{int}(\operatorname{dom} \Phi)$.

• If Φ is lower semicontinuous at $\bar{x}$, then for every $y \in \Phi(\bar{x})$ and every sequence $\{x^k\}$ converging to $\bar{x}$, there exists sequence of vectors $\{y^k\}$ converging to y such that $y^k \in \Phi(x^k)$ for all k; in other words, elements in $\Phi(\bar{x})$ are approximable by elements in $\Phi(x)$ with x sufficiently close to $\bar{x}$.

• If Φ is upper semicontinuous at $\bar{x}$, then for every scalar $\varepsilon > 0$, there exists an open neighborhood $\mathcal{N}$ of $\bar{x}$ such that

$$x \in \mathcal{N} \Rightarrow \Phi(x) \subseteq \Phi(\bar{x}) + \mathbb{B}(0, \varepsilon).$$

In Definition **2.1.16**, we have defined the lower semicontinuity, upper semicontinuity, and continuity of a set-valued map in the conventional way.

Using contemporary set-valued analysis, we give equivalent definitions of these continuity properties in terms of the concepts of liminf, limsup, and limit of set-valued maps.

Specifically, let $\Phi : \mathbb{R}^n \to \mathbb{R}^n$ be a set-valued map. Suppose that Φ is *closed valued*; that is, $\Phi(x)$ is a closed set for all $x \in \operatorname{dom}\Phi$. Thus, for every $x \in \operatorname{dom}\Phi$, the distance function $\operatorname{dist}(\cdot, \Phi(x))$ is well defined, where

$$\operatorname{dist}(z, \Phi(x)) \equiv \inf_{y \in \Phi(x)} \| y - z \|.$$

The *liminf* (or *inner limit*) of $\Phi(y)$ as y tends to x, denoted $\liminf_{y \to x} \Phi(y)$, is defined to be the set of vectors $z \in \mathbb{R}^n$ such that

$$\lim_{y \to x} \operatorname{dist}(z, \Phi(y)) = 0.$$

Similarly, the *limsup* (or *outer limit*) of $\Phi(y)$ as y tends to x, denoted $\limsup_{y \to x} \Phi(y)$, is defined to the set of vectors $z \in \mathbb{R}^n$ such that

$$\liminf_{y \to x} \operatorname{dist}(z, \Phi(y)) = 0.$$

Clearly, for every $x \in \mathbb{R}^n$,

$$\liminf_{y \to x} \Phi(y) \subseteq \Phi(x) \subseteq \limsup_{y \to x} \Phi(y).$$

If equality holds between the liminf and limsup sets, we write the common set as $\lim_{y \to x} \Phi(y)$ and call it the limit of $\Phi(y)$ as y tends x. Notice that if $\liminf_{y \to x} \Phi(y)$ is nonempty, then $\Phi(y)$ must be nonempty for all y sufficiently close to x; thus x must be an element of $\operatorname{int}(\operatorname{dom}\Phi)$. In contrast, if $\limsup_{y \to x} \Phi(y)$ is nonempty, then there exists a sequence of vectors $\{y^k\} \subset \operatorname{dom}\Phi$ converging to x.

The following result shows how the above set-valued functional limits are related to the previously defined set-valued continuity concepts.

2.1.17 Proposition. Let $\Phi : \mathbb{R}^n \to \mathbb{R}^n$ be a closed-valued multifunction. The following statements are valid.

(a) Φ is lower semicontinuous at x if and only if

$$\liminf_{y \to x} \Phi(y) = \Phi(x).$$

(b) Φ is closed at x if and only if

$$\limsup_{y \to x} \Phi(y) = \Phi(x).$$

(c) If Φ is continuous at x, then

$$\liminf_{y \to x} \Phi(y) = \limsup_{y \to x} \Phi(y). \tag{2.1.7}$$

Conversely if (2.1.7) holds and Φ is locally bounded at x, then Φ is continuous at x.

Proof. Suppose that Φ is lower semicontinuous at x. It suffices to show

$$\Phi(x) \subseteq \liminf_{y \to x} \Phi(y).$$

Let z be an arbitrary element of $\Phi(x)$. For every scalar $\varepsilon > 0$, there exists an open neighborhood $\mathcal{N}$ of x such that $\Phi(y) \cap \mathbb{B}(z, \varepsilon) \neq \emptyset$ for all $y \in \mathcal{N}$. Let v be a common element of $\Phi(y)$ and $\mathbb{B}(z, \varepsilon)$. We have

$$\operatorname{dist}(z, \Phi(y)) \leq \| z - v \| \leq \varepsilon.$$

Hence,

$$\lim_{y \to x} \operatorname{dist}(z, \Phi(y)) = 0. \tag{2.1.8}$$

Conversely, suppose $\Phi(x) = \liminf_{y \to x} \Phi(y)$. To show that Φ is lower semicontinuous at x, let $\mathcal{U}$ be any open set such that $\Phi(x) \cap \mathcal{U}$ is nonempty. Let z be a common element of $\Phi(x)$ and $\mathcal{U}$. Then (2.1.8) holds. Without loss of generality, we may assume that $\mathcal{U}$ is an open ball centered at z and with radius $\varepsilon > 0$. There exists an open neighborhood $\mathcal{N}$ of x such that

$$\operatorname{dist}(z, \Phi(y)) < \varepsilon, \quad \forall y \in \mathcal{N}.$$

Thus, for all $y \in \mathcal{N}$, $\Phi(y)$ contains at least one point belonging to $\mathcal{U}$. Hence Φ is lower semicontinuous at x. The proof of statement (b) is fairly easy. Statement (c) follows from (a) and (b) and the fact that if a set-valued map is locally bounded at a point, then the map is closed at this point if and only if it is upper semicontinuous there. □

The above concepts of the pointwise inner limit, outer limit, and limit of a set-valued map can be modified, respectively, to define the inner limit, outer limit, and limit of a sequence of closed sets in $\mathbb{R}^n$. Thus we can speak about "set convergence" using these set limits in a way similar to the pointwise continuity of a set-valued map. Nevertheless, we do not use such terminology in the rest of the book.

2.1.4 Fixed-point theorems

Degree theory can be used to prove many different kinds of fixed-point theorems. We report below a famous fixed-point theorem that was instrumental for proving the first existence result for a VI. By definition, a *fixed point* of a function $\Phi : \mathcal{D} \subseteq \mathbb{R}^n \to \mathbb{R}^n$ is a point $x \in \mathcal{D}$ such that $\Phi(x) = x$. In other words a fixed point of a function is a point that is mapped onto itself by the given function. Note that every equation $\Phi(x) = 0$ can be cast as the problem of finding a fixed point of the function $x \mapsto x + \Phi(x)$.

The following is the celebrated Brouwer fixed-point theorem for a continuous map.

2.1.18 Theorem. Let $C \subset \mathbb{R}^n$ be a nonempty convex compact set. Every continuous function $\Phi : C \to C$ has a fixed point in C.

Proof. We prove this theorem for the case where C is the closed unit Euclidean ball $\operatorname{cl}\mathbb{B}(0,1)$ in $\mathbb{R}^n$. The proof illustrates how the homotopy argument can be used as we have mentioned before. Assume by contradiction that Φ has no fixed point in the ball $\operatorname{cl}\mathbb{B}(0,1)$. Define the homotopy function:

$$H(x,t) \equiv x - t\,\Phi(x), \quad \forall\,(x,t) \in \operatorname{cl}\mathbb{B}(0,1) \times [0,1].$$

For $t = 0$, we have $H(x,t) = x$; that is, $H(\cdot,0)$ is the identity map. By Axiom (A1) of the degree, $\deg(H(\cdot,0),\mathbb{B}(0,1)) = 1$. Assume $H(x,t) = 0$ for some (x,t) in the domain of H. It follows that $x = t\Phi(x)$, which implies $\|x\| = t\|\Phi(x)\|$. If $x \in \operatorname{bd}\mathbb{B}(0,1)$, the last equality implies $t = 1$. Therefore $x = \Phi(x)$, contradicting the non-existence of a fixed point of Φ. Consequently, we have $0 \notin H(\operatorname{bd}\mathbb{B}(0,1),t)$ for all $t \in [0,1]$. By the homotopy invariance property of the degree, we deduce

$$1 = \deg(H(\cdot,0),\mathbb{B}(0,1)) = \deg(H(\cdot,1),\mathbb{B}(0,1)).$$

Consequently, $H(\cdot,1)$ has a zero in $\mathbb{B}(0,1)$; equivalently, Φ has a fixed point in $\mathbb{B}(0,1)$. This is a contradiction.

The above proof serves as the basis for proving the general case of a convex compact set C. This generalization is based on a topological argument showing that every nonempty convex compact set is homeomorphic to a possibly lower-dimensional closed unit Euclidean ball. Since this argument is not essential to our analysis, we omit the details. □

Another celebrated fixed-point theorem is due to Kakutani; it pertains to a point-to-set map. Formally, a fixed point of a set-valued map Φ is a vector x such that $x \in \Phi(x)$. The following is Kakutani's fixed-point theorem, which is an extension of Theorem **2.1.18** to a point-to-set map.

2.1.19 Theorem. Let $C \subset \mathbb{R}^n$ be a nonempty, convex compact set. Let $\Phi : C \to C$ be a set-valued map such that for each $x \in C$, $\Phi(x)$ is a nonempty closed convex subset of C. If Φ is closed on C, then Φ has a fixed point. □

Fixed points of set-valued maps are clearly relevant to the QVI. Indeed, a necessary condition for the QVI (K, F) to have a solution is for the multifunction K to have a fixed point. Moreover, Kakutani's fixed-point theorem can be used to prove a basic existence theorem for a VI on a compact convex set; see Exercise **2.9.10**.

Just like point-to-point functions, there are extensions of the continuity concepts for set-valued maps to Lipschitz continuity. We do not define these extended concepts here but will do so when they actually arise from the applied contexts; see the discussion following Corollary **3.2.5** where Lipschitz continuity is introduced for a particular set-valued map.

2.1.5 Contractive mappings

Without formally giving the definition, we have shown in Theorem **1.5.5** that the Euclidean projector is a nonexpansive map with respect to the ℓ_2-norm. In what follows, we introduce an important subclass of such maps. Throughout this subsection, we let $\| \cdot \|$ be a given vector norm (not necessarily the Euclidean norm). Let $G : \Omega \subseteq \mathbb{R}^n \to \mathbb{R}^n$ be a given mapping, where Ω is a nonempty closed set. We say that G is a *contraction* with respect to the norm $\| \cdot \|$ if there exists a constant $\eta \in (0, 1)$ such that

$$\| G(x) - G(y) \| \leq \eta \| x - y \| \tag{2.1.9}$$

for all x and y in Ω; the constant η is called a *contraction constant* of G. Clearly, every contraction is nonexpansive and thus continuous. More interestingly, every contraction G must have a unique fixed point in Ω; see Theorem **2.1.21** below. The computation of the unique fixed point of a contraction can be accomplished by a simple fixed-point iteration, which we introduce in the following iterative algorithm.

Fixed-Point Contraction Algorithm (FPCA)

2.1.20 Algorithm.

Data: $x^0 \in \Omega$.

Step 0: Set $k = 0$.

Step 1: If $x^k = G(x^k)$ stop.

Step 2: Set $x^{k+1} \equiv G(x^k)$ and $k \leftarrow k + 1$; go to Step 1.

The convergence of the above algorithm is stated in the following result, which is referred to as the Banach fixed-point theorem.

2.1.21 Theorem. Let $G : \Omega \subseteq \mathbb{R}^n \to \Omega$ be a contraction with constant $\eta \in (0,1)$, where $\Omega \neq \emptyset$ is a closed set. The three statements below hold.

(a) The map G has a unique fixed point x^* in Ω.

(b) For any starting point x^0 belonging to Ω, Algorithm **2.1.20** generates a sequence $\{x^k\}$ converging to x^*.

(c) For any sequence $\{x^k\}$ given in (b),

$$\| x^k - x^* \| \le \frac{\eta^k}{1-\eta} \| x^0 - G(x^0) \|, \quad \forall\, k \ge 1.$$

Proof. Since G maps Ω into itself, the sequence $\{x^k\}$ is well defined. If this sequence converges, its limit must be a fixed point of G in Ω. Since G is a contraction, such a fixed point must be unique. Therefore, it suffices to show that the sequence $\{x^k\}$ in (b) converges and the error inequality in (c) holds. We have

$$\| x^{k+1} - x^k \| = \| G(x^k) - G(x^{k-1}) \| \le \eta \| x^k - x^{k-1} \|,$$

for every k, which implies

$$\| x^{k+1} - x^k \| \le \eta^k \| x^1 - x^0 \|$$

for every k. It follows that if i is any positive integer, we can write

$$\begin{aligned} \| x^{k+i} - x^k \| &\le \sum_{m=1}^{i} \| x^{k+m} - x^{k+m-1} \| \\ &\le \eta^k (1 + \eta + \cdots + \eta^{i-1}) \| x^1 - x^0 \| \\ &\le \frac{\eta^k}{1-\eta} \| x^1 - x^0 \|. \end{aligned}$$

This shows that the sequence $\{x^k\}$ is a Cauchy sequence and therefore converges to a limit x^*. Passing to the limit $i \to \infty$ in the above string of inequalities, we obtain (c) readily. □

The contraction property of G is essential for the sequence $\{x^k\}$ produced by Algorithm **2.1.20** to converge. For example, if $G(x) = x^2$ and $\Omega = [1, \infty)$, $x = 1$ is the only fixed point of G. Yet if we apply Algorithm **2.1.20** with any x^0 different from 1 a sequence of points diverging to infinity will be generated. Furthermore, Theorem **2.1.21** is false if G is nonexpansive but not contractive. A simple example is $G(x) = x + 1$ and $\Omega = \mathbb{R}$. With any x^0, the sequence $\{x^k\}$ produced by Algorithm **2.1.20** clearly diverges.

2.2 Existence Results

Based on the expression (1.5.12), which identifies two equation reformulations of the VI (K, F), we can easily prove the following fundamental existence result.

2.2.1 Theorem. Let $K \subseteq \mathbb{R}^n$ be closed convex and $F : \mathcal{D} \supseteq K \to \mathbb{R}^n$ be continuous on the open set $\mathcal{D}$. Let $\mathbf{F}_K^{\text{nat}}$ and $\mathbf{F}_K^{\text{nor}}$ denote respectively the natural map and normal map of the pair (K, F). The following two statements hold:

(a) if there exists a bounded open set $\mathcal{U}$ satisfying $\text{cl}\,\mathcal{U} \subseteq \mathcal{D}$ and such that $\deg(\mathbf{F}_K^{\text{nat}}, \mathcal{U})$ is well defined and nonzero, then the VI (K, F) has a solution in $\mathcal{U}$;

(b) if there exists a bounded open set $\mathcal{U}'$ such that $\deg(\mathbf{F}_K^{\text{nor}}, \mathcal{U}')$ is well defined and nonzero, then the VI (K, F) has a solution x such that $x - F(x) \in \mathcal{U}'$.

Proof. Both statements follow immediately from the property of the degree. Indeed, if $\deg(\mathbf{F}_K^{\text{nat}}, \mathcal{U})$ is well defined and nonzero, then $\mathbf{F}_K^{\text{nat}}$ has a zero in $\mathcal{U}$; but such a zero is also a solution of the VI (K, F) by Proposition **1.5.8**. In case (b), $\mathbf{F}_K^{\text{nor}}$ has a zero, say z, in $\mathcal{U}'$. By Proposition **1.5.9**, $x \equiv \Pi_K(z) \in \text{SOL}(K, F)$. Since $0 = \mathbf{F}_K^{\text{nor}}(z) = F(x) + z - x$, it follows that $x - F(x) \in \mathcal{U}'$ as asserted. □

It is natural to ask the question of whether the degree condition in statement (a) of Theorem **2.2.1** implies the degree condition in statement (b) or vice versa. These two degree conditions pertain to different maps, the former to the natural map $\mathbf{F}_K^{\text{nat}}$ and the latter to the normal map $\mathbf{F}_K^{\text{nor}}$. The somewhat different conclusions about the solutions to the VI in the respective statements reflect the distinct domains of definition of these two maps. Presently, we do not have an answer to the question.

By the above theorem, the task of establishing the existence of a solution to the VI (K, F) reduces to verifying the degree condition in either statement (a) or (b) of Theorem **2.2.1**. To facilitate the application of this theorem, we recall a classic theorem in topology that says every continuous function defined on a closed set $\mathbb{R}^n$ has a continuous extension to the entire space $\mathbb{R}^n$. This result, known as the Tietze-Urysohn Extension Theorem, is valid in general metric spaces; for our purpose here, we state the result as a lemma on Euclidean spaces.

2.2.2 Lemma. Let $\Phi : S \subseteq \mathbb{R}^n \to \mathbb{R}^m$ be a continuous function defined on the nonempty closed set S. A continuous extension $\bar{\Phi} : \mathbb{R}^n \to \mathbb{R}^m$ exists such that $\bar{\Phi}(x) = \Phi(x)$ for all $x \in S$. □

Consider the VI (K, F), where F is a continuous function from the subset K of $\mathbb{R}^n$ into $\mathbb{R}^n$. Suppose that K is closed. If $\bar{F} : \mathbb{R}^n \to \mathbb{R}^n$ denotes a continuous extension of F as stipulated by Lemma **2.2.2**, then $\text{SOL}(K, F) = \text{SOL}(K, \bar{F})$. Based on this observation, we next give a very broad sufficient condition under which the VI (K, F) has a solution. The proof of the solution existence in the following Proposition **2.2.3** is based on a homotopy argument that is very standard in many results of this kind. Namely, we "homotopize" a suitable map associated with the VI (e.g., either the natural map or the normal map) with the identity map that is known to have a degree of plus one. More generally, we can use a "simple" map with known nonzero degree as the auxiliary map to define the homotopy. In the remainder of this section, we will establish our results using the homotopy between the natural map $\mathbf{F}_K^{\text{nat}}$ and the identity map.

2.2.3 Proposition. Let $K \subseteq \mathbb{R}^n$ be closed convex and $F : K \to \mathbb{R}^n$ be continuous. Consider the following statements:

(a) There exists a vector $x^{\text{ref}} \in K$ such that the set

$$L_< \equiv \{ x \in K : F(x)^T (x - x^{\text{ref}}) < 0 \}$$

is bounded (possibly empty).

(b) There exist a bounded open set Ω and a vector $x^{\text{ref}} \in K \cap \Omega$ such that

$$F(x)^T (x - x^{\text{ref}}) \geq 0 \quad \forall x \in K \cap \text{bd}\,\Omega. \tag{2.2.1}$$

(c) The VI (K, F) has a solution.

It holds that (a) $\Rightarrow$ (b) $\Rightarrow$ (c). Moreover, if the set

$$L_\leq \equiv \{ x \in K : F(x)^T (x - x^{\text{ref}}) \leq 0 \},$$

which is nonempty and larger than $L_<$, is bounded, then $\text{SOL}(K, F)$ is nonempty and compact.

Proof. Assume the conditions in (a). Let Ω be a bounded open set containing $\{x^{\text{ref}}\} \cup L_<$. Since Ω is open and contains $L_<$, we must have $L_< \cap \text{bd}\,\Omega = \emptyset$. Consequently (2.2.1) holds and (b) follows.

Assume (b) holds. Let $\bar{F} : \mathbb{R}^n \to \mathbb{R}^n$ be the Tietze-Urysohn continuous extension of F. Since $\bar{F}$ and F agree on K, we have

$$\bar{F}(x)^T (x - x^{\text{ref}}) \geq 0 \quad \forall x \in K \cap \text{bd}\,\Omega.$$

For simplicity, we drop the bar on $\bar{F}$ and assume that F is a continuous mapping defined on the entire space $\mathbb{R}^n$. To show that SOL(K, F) is nonempty, we argue by contradiction. Suppose this solution set is empty. Since the zeros (if any) of $\mathbf{F}_K^{\text{nat}}$ coincide with the solutions of the VI (K, F) and the latter problem has no solutions by assumption, we have $(\mathbf{F}_K^{\text{nat}})^{-1}(0) \cap \text{bd}\,\Omega = \emptyset$; thus $\deg(\mathbf{F}_K^{\text{nat}}, \Omega)$ is well defined. We claim that this degree is nonzero. Consider the homotopy:

$$H(x,t) \equiv x - \Pi_K(t(x - F(x)) + (1-t)x^{\text{ref}}), \quad (x,t) \in \text{cl}\,\Omega \times [0,1].$$

We have $H(x,0) = x - x^{\text{ref}}$; since $x^{\text{ref}} \in \Omega$, it follows that $\deg(H(\cdot,0),\Omega)$ is well defined and equal to unity. Furthermore, $H(x,1) = \mathbf{F}_K^{\text{nat}}(x)$. We now show that if $H(x,t) = 0$ for some $(x,t) \in \text{cl}\,\Omega \times (0,1)$, then $x \notin \text{bd}\,\Omega$. Assume $H(x,t) = 0$ for some $0 < t < 1$. Without loss of generality, we may assume $x \neq x^{\text{ref}}$. Since $0 = H(x,t)$, by the definition of H, we have $x \in K$ and

$$(y - x)^T[\,x - t(x - F(x)) - (1-t)x^{\text{ref}}\,] \geq 0, \quad \forall\, y \in K.$$

In particular, for $y = x^{\text{ref}}$, we deduce

$$(x^{\text{ref}} - x)^T[\,tF(x) + (1-t)(x - x^{\text{ref}})\,] \geq 0$$

which implies

$$(x^{\text{ref}} - x)^T F(x) \geq \frac{1-t}{t} \,\|\, x - x^{\text{ref}} \,\|_2^2 > 0$$

where the last inequality holds because $t \in (0,1)$ and $x \neq x^{\text{ref}}$. Thus x does not belong to $\text{bd}\,\Omega$. Consequently, by the homotopy invariance property of the degree, we deduce that

$$\deg(\mathbf{F}_K^{\text{nat}}, \Omega) = \deg(H(\cdot,1),\Omega) = \deg(H(\cdot,0),\Omega) = 1.$$

By Theorem **2.2.1**, it follows that SOL(K, F) is nonempty. A contradiction is obtained; thus (b) $\Rightarrow$ (c). If the set $L_{\leq}$ is bounded, then so is $L_{<}$; hence SOL(K, F) is nonempty. To show the compactness of SOL(K, F), it suffices to note that SOL(K, F) is a subset of $L_{\leq}$. □

2.2.4 Remark. The argument of replacing F with its continuous extension $\bar{F}$ appears in several results in this chapter that involve the application of a degree-theoretic proof. □

Condition (b) of Proposition **2.2.3**, which is illustrated in Figure **2.2**, is referred to as the *nonnegativity property at infinity*. This condition is an abstract generalization of the opposite-sign condition in the well-known intermediate value theorem for a scalar function of one variable. Indeed consider the very simple case where $n = 1$ and $K = \mathbb{R}$. With Ω taken to be a bounded open interval, say (a, b), and x^{ref} to be an arbitrary point in this interval, the condition (2.2.1) stipulates that $F(a)$ and $F(b)$ have opposite signs. In this case, the proposition implies that F has a zero in the interval $[a, b]$. We have thus recovered an elementary result in calculus.

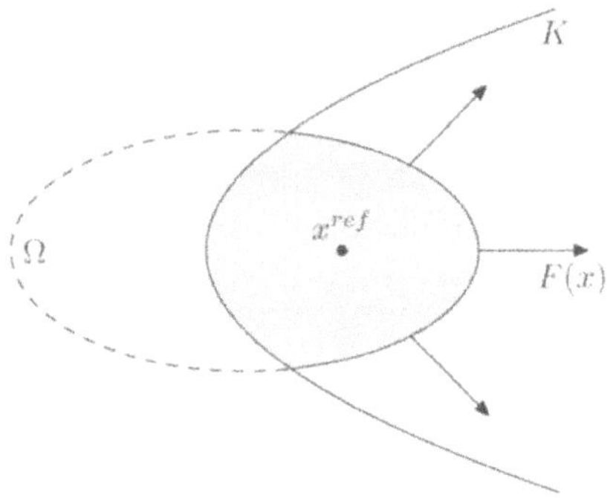

Figure 2.2: Condition (b) of Proposition **2.2.3**.

Proposition **2.2.3** has many special cases. The first special case is the following fundamental result whose proof is very simple.

2.2.5 Corollary. Let $K \subseteq \mathbb{R}^n$ be compact convex and let $F : K \to \mathbb{R}^n$ be continuous. The set $\text{SOL}(K, F)$ is nonempty and compact.

Proof. The set $L_{\leq}$ is obviously compact for every choice of $x^{\text{ref}} \in K$. □

It follows from Corollary **2.2.5** that the box constrained VI (K, F), where K is a compact rectangle and F is continuous, always has a solution. Instead of the compactness assumption in Corollary **2.2.5**, we can assume certain conditions on the function F to establish the same conclusion about $\text{SOL}(K, F)$. The next result is another consequence of Proposition **2.2.3**, which allows for an unbounded set K, and is equally simple to prove.

2.2.6 Corollary. Let $K \subseteq \mathbb{R}^n$ be closed convex and let $F : K \to \mathbb{R}^n$ be continuous. If there exists a vector x^{ref} in K such that

$$F(x)^T(x - x^{\text{ref}}) \geq 0, \quad \forall x \in K.$$

then the VI (K, F) has a solution.

Proof. The assumption clearly implies that the set $L_{<}$ is empty. □

Another class of problems that allow for an unbounded set K is the class of "coercive" VIs (K, F), where the map F satisfies a certain growth property as $x \in K$ grows in norm. There are different coercivity conditions; we have seen one of these in (2.1.4). The following condition is a weakening of (2.1.4): for some $x^{\text{ref}} \in K$ and $\zeta \geq 0$,

$$\liminf_{\substack{x \in K \\ \|x\| \to \infty}} \frac{F(x)^T(x - x^{\text{ref}})}{\|x\|^\zeta} > 0. \tag{2.2.2}$$

Equivalently, this condition postulates the existence of a constant $c > 0$ such that for all $x \in K$ with sufficiently large norm,

$$F(x)^T(x - x^{\text{ref}}) \geq c\|x\|^\zeta.$$

Based on this observation, the following result is easily proved.

2.2.7 Proposition. Let $K \subseteq \mathbb{R}^n$ be closed convex and $F : K \to \mathbb{R}^n$ be continuous. If there exist a vector $x^{\text{ref}} \in K$ and a scalar $\zeta \geq 0$ such that (2.2.2) holds, then the VI (K, F) has a nonempty compact solution set.

Proof. It suffices to note that the coercivity condition (2.2.2) implies the boundedness of the set $L_\leq$ in Proposition **2.2.3**. □

The next result identifies a necessary and sufficient condition for the VI (K, F) to have a solution. The proof of the result turns out to be quite elementary. Traditionally, the proposition below, along with Corollary **2.2.5**, was the basis for deriving all known existence results for the VI.

2.2.8 Proposition. Let $K \subseteq \mathbb{R}^n$ be closed convex and $F : K \to \mathbb{R}^n$ be continuous. The set SOL(K, F) is nonempty if and only if there exists a closed set $E \subseteq \mathbb{R}^n$ with $\operatorname{int} E \neq \emptyset$ such that the "restricted" VI $(K \cap E, F)$ has a solution in $\operatorname{int} E$.

Proof. If the VI (K, F) has a solution x, simply take E to be a closed Euclidean ball with that solution as the center. It is trivial to see that x must solve the VI $(K \cap E, F)$. Thus the necessary condition holds. Conversely, suppose that a set E with the prescribed property exists. Let x^* be an element of SOL$(K \cap E, F) \cap \operatorname{int} E$, which exists by assumption. Let $y \in K$ be arbitrary. Since K is convex and $x^* \in \operatorname{int} E$, it follows that for some $\tau \in (0, 1)$, the vector $x^* + \tau(y - x^*) \in K \cap E$. Therefore, since $x^* \in$ SOL$(K \cap E, F)$, we have $\tau(y - x^*)^T F(x^*) \geq 0$. Consequently $x^* \in$ SOL(K, F) as desired. □

Using the above proposition, we can obtain a generalization of Corollary **2.2.6**; see Exercise **2.9.8**.

2.2.1 Applications to source problems

To illustrate the existence results obtained so far, we apply them to some of the source problems of the VI/CP discussed in Section **1.4**. We begin with the Nash equilibrium problem formulated as the VI described in Proposition **1.4.2**. Specializing Corollary **2.2.5** to this VI, we obtain the following existence result for the Nash equilibrium problem.

2.2.9 Proposition. Let each K_i be a compact convex subset of $\mathbb{R}^{n_i}$ and each θ_i be continuously differentiable. Suppose that for each fixed tuple $\tilde{\mathbf{x}}^i$, the function $\theta_i(x^i, \tilde{\mathbf{x}}^i)$ is convex in x^i. Then the set of Nash equilibrium tuples is nonempty and compact.

Proof. According to Proposition **1.4.2**, the Nash equilibrium problem is equivalent to the VI $(\mathbf{K}, \mathbf{F})$, where

$$\mathbf{K} \equiv \prod_{i=1}^{N} K_i \quad \text{and} \quad \mathbf{F}(\mathbf{x}) \equiv (\nabla_{x^i} \theta_i(\mathbf{x}))_{i=1}^{N}.$$

Since each K_i is compact and convex, hence so is $\mathbf{K}$; moreover, $\mathbf{F}$ is continuous. Consequently, by Corollary **2.2.5**, the desired conclusion follows. □

A special case of the Nash problem is the saddle problem. Thus the following corollary is immediate.

2.2.10 Corollary. Let X and Y be compact convex subsets of $\mathbb{R}^n$ and $\mathbb{R}^m$ respectively. Let $L : X \times Y \to \mathbb{R}$ be a continuously differentiable convex-concave saddle function. The set of saddle points of the triple (L, X, Y) is nonempty and compact. Moreover,

$$\min_{x \in X} \max_{y \in Y} L(x,y) = \max_{y \in Y} \min_{x \in X} L(x,y).$$

Proof. Let $K_1 \equiv X$, $K_2 \equiv Y$, $\theta_1(x,y) \equiv L(x,y)$, and $\theta_2(x,y) \equiv -L(x,y)$. Apply Proposition **2.2.9** to deduce the first assertion of the corollary. The second assertion follows from Theorem **1.4.1**. □

The next application concerns the Walrasian equilibrium problem formulated as the NCP (F) with

$$F(y,p) \equiv \begin{pmatrix} c - A(p)p \\ b + A(p)^T y - d(p) \end{pmatrix},$$

where $A(p)$ is the input-output matrix of the economy and $d(p)$ is the demand function of the commodities. Typically, the latter function $d(p)$ is obtained via a utility maximization problem subject to a budget constraint of the type $p^T d \le p^T b$. Invariably, this function satisfies the so-called Walras law: that is, $p^T d(p) = p^T b$ for all price vectors p in the domain of the function $d(p)$; cf. the function in (1.4.13). In general, the demand function $d(p)$ is not defined on all of $\mathbb{R}^n_+$; e.g., the function in (1.4.13) is not defined at the origin; in this case, further assumptions are needed to ensure the existence of a Walrasian equilibrium. In order to simplify our treatment of this model, we assume that $d(p)$ is a continuous function defined on $\mathbb{R}^n_+$ and that it satisfies Walras' law.

2.2.11 Proposition. Suppose that the functions $A(p)$ and $d(p)$ are continuous and that $p^T d(p) = d^T b$ for all $p \ge 0$. If $c \ge 0$, then there exists a Walrasian equilibrium.

Proof. The function F of the Walarsian NCP is clearly continuous; moreover, we have

$$F(y,p)^T \begin{pmatrix} y \\ p \end{pmatrix} = c^T y \ge 0,$$

for all $y \ge 0$. Thus taking x^{ref} to be the zero vector, we may apply Corollary **2.2.6** to the NCP (F), which is equivalent to the VI defined on the nonnegative orthant. □

The NCP formulation of the American option pricing problem with transaction costs:

$$0 \le \mathbf{V}^m - \mathbf{\Lambda}^m \perp \mathbf{q}^m + \mathbf{M}\mathbf{V}^m + \mathbf{F}(\mathbf{V}^m) \ge 0.$$

and the NCP formulation of the elastoplastic structural problem with hardening laws:

$$0 \le \mathbf{\Lambda} \perp \mathbf{q} + N^T Z N \mathbf{\Lambda} + \mathbf{H}(\mathbf{\Lambda}) \ge 0,$$

have a common property: the defining function is the sum of an affine function and a nonlinear function, both with special properties. The affine summand is of the form $q + Mx$, where M is either positive definite (in the option problem with a proper choice of the discretization steps) or symmetric positive semidefinite (in the structural problem); thus M is *positive semidefinite plus* in both cases (i.e., M is positive semidefinite and $x^T M x = 0 \Rightarrow Mx = 0$). The nonlinear summand is a nonnegative function in the option problem; the same summand is nonnegative on the nonnegative orthant in the structural problem with certain hardening laws. The next result establishes the existence of a solution to an NCP with the defining function satisfying the above conditions.

2.2.12 Proposition. Let $h : \mathbb{R}^n_+ \to \mathbb{R}^n$ be a continuous function; let $M \in \mathbb{R}^{n \times n}$ be a positive semidefinite matrix, and let $q \in \mathbb{R}^n$ be arbitrary. The NCP

$$0 \leq x \perp q + Mx + h(x) \geq 0 \tag{2.2.3}$$

has a nonempty bounded solution set under either one of the following two conditions:

(a) $x^T h(x)$ is nonnegative on $\mathbb{R}^n_+$, M is positive semidefinite plus, and there exists a vector z such that $q + Mz > 0$;

(b) h satisfies the coercivity condition:

$$\lim_{\substack{x \geq 0 \\ \|x\| \to \infty}} \frac{x^T h(x)}{\|x\|} = \infty.$$

Proof. For (a), we show that the set

$$\{ x \geq 0 : x^T (q + Mx + h(x)) \leq 0 \}$$

is bounded. Since $x^T h(x)$ is nonnegative for $x \geq 0$, the above set is bounded if the larger set

$$\{ x \geq 0 : x^T (q + Mx) \leq 0 \}$$

is bounded. Assume for contradiction that the latter set is unbounded. There exists a sequence $\{x^k\} \subseteq \mathbb{R}^n_+$ such that

$$(x^k)^T (q + Mx^k) \leq 0 \quad \forall k,$$

$$\lim_{k \to \infty} \| x^k \| = \infty, \quad \text{and} \quad \lim_{k \to \infty} \frac{x^k}{\| x^k \|} = d$$

for some nonzero vector $d \geq 0$. Since M is positive semidefinite, it is not difficult to show, by a normalization followed by a limiting argument that d satisfies $d^T M d = 0$ and $q^T d \leq 0$. By the plus property of M, it follows that $Md = M^T d = 0$. Moreover, since d is nonzero and nonnegative, we have

$$0 \geq d^T q > -d^T M z = 0,$$

which is a contradiction. By Proposition **2.2.3**, it follows that the NCP (2.2.3) has a nonempty bounded solution set. To prove (b), we note that the coercivity condition on h and the positive semidefiniteness of M implies

$$\lim_{\substack{x \geq 0 \\ \|x\| \to \infty}} \frac{x^T (q + Mx + h(x))}{\|x\|} = \infty.$$

Therefore (b) follows from Proposition **2.2.7** with $x^{\text{ref}} = 0$ and $\zeta = 1$. □

2.2.13 Remark. If M is a positive semidefinite plus matrix, the above proof shows that if there exists a vector z satisfying $q + Mz > 0$, then the LCP (q, M) has a nonempty bounded solution set. The converse turns out to be true too and this is an immediate consequence of Corollary **2.4.5**. □

Proposition **2.2.12** has not taken full advantage of the Cartesian product structure of the nonnegative orthant. This topic is the subject of Section **3.5**; see in particular Proposition **3.5.6** that pertains to the NCP of the option problem.

We next consider the traffic equilibrium problem formulated using the path flow variables as the NCP $(\mathbf{F})$, where $\mathbf{F}$ is defined by (1.4.23):

$$\mathbf{F}(h, u) \equiv \begin{pmatrix} C(h) - \Omega^T u \\ \Omega h - d(u) \end{pmatrix}.$$

Under some reasonable assumptions on the path cost function $C(h)$ and the travel demand function $d(u)$, the following result establishes the existence of a traffic user equilibrium. The proof of this result is essentially an application of Proposition **2.2.8**, but exploits the special structure of the above function $\mathbf{F}$.

2.2.14 Proposition. If each $C_p(h)$ is a nonnegative continuous function and each $d_w(u)$ is a continuous function bounded above for all $u \geq 0$, then the NCP $(\mathbf{F})$ with the above function $\mathbf{F}$ has a solution.

Proof. Choose two scalars c_1 and c_2 such that $c_2 > c_1 > 0$ and

$$c_1 > \max_{w \in \mathcal{W}} \max_{u \geq 0} d_w(u) \quad \text{and} \quad c_2 > \max_{w \in \mathcal{W}} \max_{p \in \mathcal{P}_w} \max_{0 \leq h \leq c_1 \mathbf{1}} C_p(h).$$

Let E be the rectangular box:

$$E \equiv \{ (h, u) \geq 0 : h \leq c_1 \mathbf{1}, \; u \leq c_2 \mathbf{1} \}.$$

The VI $(E, \mathbf{F})$ has a solution because E is a compact convex set and $\mathbf{F}$ is a continuous function. If (h^*, u^*) denotes this solution, then there exist multipliers λ and μ such that

$$\begin{aligned} 0 \leq h^* \quad &\perp \quad C(h^*) - \Omega^T u^* + \lambda \geq 0 \\ 0 \leq u^* \quad &\perp \quad \Omega h^* - d(u^*) + \mu \geq 0 \\ 0 \leq \lambda \quad &\perp \quad c_1 \mathbf{1} - h^* \geq 0 \\ 0 \leq \mu \quad &\perp \quad c_2 \mathbf{1} - u^* \geq 0. \end{aligned}$$

The proof will be completed if we can show λ and μ are both equal to zero. Suppose for the sake of contradiction that $\lambda_p > 0$ for some $p \in \mathcal{P}_w$ with $w \in \mathcal{W}$. By complementarity, we have $h_p^* = c_1$. By the definition of c_1, it follows that

$$\sum_{p' \in \mathcal{P}_w} h_{p'}^* - d_w(u^*) + \mu_w \geq h_p^* - d_w(u^*) + \mu_w > 0.$$

Thus by complementarity again, we have $u_w^* = 0$. Moreover, we have, since $h_p^* = c_1 > 0$,

$$0 = C_p(h^*) - u_w^* + \lambda_p = C_p(h^*) + \lambda_p > 0,$$

because $C_p(h^*)$ is nonnegative and λ_p is positive. This contradiction yields $\lambda = 0$.

Similarly, if $\mu_w > 0$ for some $w \in \mathcal{W}$, then $u_w^* = c_2$. By the definition of c_2, we have

$$u_w^* > C_p(h^*), \quad \forall\, p \in \mathcal{P}_w.$$

But this contradicts

$$0 \leq C_p(h^*) - u_w^* + \lambda_p = C_p(h^*) - u_w^*$$

for all $p \in \mathcal{P}_w$. □

2.3 Monotonicity

In this section, we introduce several "monotonicity" properties of vector functions that are naturally satisfied by the gradient maps of convex functions of the respective kinds. Indeed, the class of monotone vector functions play a similarly important role in the VI/CP as the class of convex functions in optimization. In particular, the existence and uniqueness of a solution to the VI can be established under the strong monotonicity defined below; see Theorem **2.3.3**. Monotone functions are first defined after the proof of Theorem **1.5.5** as a class of functions to which the Euclidean projector belongs. Formally, we have the following definition.

2.3.1 Definition. A mapping $F : K \subseteq \mathbb{R}^n \to \mathbb{R}^n$ is said to be

(a) *pseudo monotone* on K if for all vectors x and y in K,

$$(x - y)^T F(y) \geq 0 \Rightarrow (x - y)^T F(x) \geq 0;$$

(b) *monotone* on K if

$$(F(x) - F(y))^T (x - y) \geq 0, \quad \forall\, x, y \in K;$$

(c) *strictly monotone* on K if

$$(F(x) - F(y))^T(x - y) > 0, \quad \forall x, y \in K \text{ and } x \neq y;$$

(d) *ξ-monotone* on K for some $\xi > 1$ if there exists a constant $c > 0$ such that

$$(F(x) - F(y))^T(x - y) \geq c\| x - y \|^\xi, \quad \forall x, y \in K. \tag{2.3.1}$$

(e) *strongly monotone* on K if there exists a constant $c > 0$ such that

$$(F(x) - F(y))^T(x - y) \geq c\| x - y \|^2, \quad \forall x, y \in K,$$

i.e., if F is 2-monotone on K. □

In principle, one could define ξ-monotonicity for any positive scalar ξ. Nevertheless the case where $\xi \leq 1$ is not very interesting because no particularly meaningful result can be obtained. Thus when we talk about ξ-monotonicity, we always take ξ to be greater than one. It is clear that every strictly monotone function must be injective; moreover, among the above monotonicity properties, the following relations hold:

strongly monotone	$\Rightarrow$	ξ monotone	$\Rightarrow$	strictly monotone	$\Rightarrow$	monotone
						$\Downarrow$
						pseudo monotone.

Moreover, for an affine map $F(x) \equiv Ax + b$ and with $K = \mathbb{R}^n$, where A is an $n \times n$ matrix, not necessarily symmetric, and b is an n-vector, we have

strongly monotone	$\Leftrightarrow$	ξ monotone	$\Leftrightarrow$	strictly monotone	$\Leftrightarrow$	A is positive definite

and

monotonicity $\Leftrightarrow$ A is positive semidefinite.

More generally, if F is a continuously differentiable function defined on an open convex set, we have the following connection between the above monotonicity properties and the positive semidefiniteness of the Jacobian matrices of F. The reader is asked to supply the proof in Exercise **2.9.16**.

2.3.2 Proposition. Let $F : \mathcal{D} \subseteq \mathbb{R}^n \to \mathbb{R}^n$ be continuously differentiable on the open convex set $\mathcal{D}$. The following statements are valid.

(a) F is monotone on $\mathcal{D}$ if and only if $JF(x)$ is positive semidefinite for all x in $\mathcal{D}$;

(b) F is strictly monotone on $\mathcal{D}$ if $JF(x)$ is positive definite for all x in $\mathcal{D}$;

(c) F is strongly monotone on $\mathcal{D}$ if and only if $JF(x)$ is uniformly positive definite for all x in $\mathcal{D}$; i.e., there exists a constant $c' > 0$ such that

$$y^T JF(x) y \geq c' \| y \|^2, \quad \forall y \in \mathbb{R}^n.$$

for all $x \in \mathcal{D}$. □

Clearly, every ξ-monotone function for some $\xi > 1$ satisfies (2.2.2) with x^{ref} being any vector in K. Indeed, for every fixed vector y, (2.3.1) implies

$$\lim_{\substack{x \in K \\ \|x\| \to \infty}} \frac{F(x)^T (x - y)}{\|x\|^{\frac{\xi+1}{2}}} = \infty.$$

For such a monotone map, we can establish the existence and uniqueness of a solution to the VI. In addition to formally stating this result, the following theorem also asserts that the VI (K, F) can have at most one solution if F is strictly monotone on K and that for a Lipschitz continuous ξ-monotone function, there exists a computable upper bound on the distance from an arbitrary vector in a given domain to the unique solution of the VI in terms of the norm of the natural map; cf. part (c) of the theorem below.

2.3.3 Theorem. Let $K \subseteq \mathbb{R}^n$ be closed convex and $F : K \to \mathbb{R}^n$ be continuous.

(a) If F is strictly monotone on K, the VI (K, F) has at most one solution.

(b) If F is ξ-monotone on K for some $\xi > 1$, the VI (K, F) has a unique solution x^*.

(c) If F is defined, Lipschitz continuous, and ξ-monotone on a set $\Omega \supseteq K$ for some $\xi > 1$, then there exists a constant $c' > 0$ such that for every vector $x \in \Omega$,

$$\| x - x^* \| \leq c' \| \mathbf{F}_K^{\text{nat}}(x) \|^{\frac{1}{\xi - 1}},$$

where x^* is the unique solution of the VI (K, F).

Proof. Suppose that F is strictly monotone on K. If $x \neq x'$ are two distinct solutions of the VI (K, F), we have, for all $y \in K$,

$$(y - x)^T F(x) \geq 0 \quad \text{and} \quad (y - x')^T F(x') \geq 0.$$

Substituting $y = x'$ into the first inequality and $y = x$ into the second inequality, we deduce,

$$(x' - x)^T F(x) \geq 0 \quad \text{and} \quad (x - x')^T F(x') \geq 0.$$

Adding these two inequalities, we deduce

$$(x' - x)^T (F(x') - F(x)) \leq 0.$$

This inequality contradicts the strict monotonicity of F, thus establishing statement (a).

If F is ξ-monotone on K for some $\xi > 1$, the existence of a solution to the VI (K, F) follows from Proposition **2.2.3** and the observation noted before the statement of the proposition; the uniqueness of the solution follows from part (a).

To establish (c), let $c > 0$ be such that (2.3.1) holds. For a given vector $x \in \Omega$, write $r \equiv \mathbf{F}_K^{\rm nat}(x)$. We have

$$x - r = \Pi_K(x - F(x)).$$

By the variational characterization of the projection, it follows that

$$(y - x + r)^T (F(x) - r) \geq 0, \quad \forall y \in K.$$

In particular, with $y = x^*$, we obtain

$$(x^* - x + r)^T (F(x) - r) \geq 0.$$

Since $x^* \in \mathrm{SOL}(K, F)$ and $x - r \in K$, we have

$$(x - r - x^*)^T F(x^*) \geq 0.$$

Adding the two inequalities and rearranging terms, we deduce

$$(x - x^*)^T (F(x) - F(x^*)) \leq r^T (F(x) - F(x^*)).$$

By the ξ-monotonicity of F on Ω, the left-hand side is not smaller than $c\|x - x^*\|^\xi$ while the right-hand side is not greater than $L\|r\| \, \|x - x^*\|$, where $L > 0$ is a Lipschitz constant of F on Ω. Consequently,

$$\| x - x^* \| \leq (c^{-1} L)^{\frac{1}{\xi - 1}} \| r \|^{\frac{1}{\xi - 1}}.$$

With $c' \equiv (c^{-1} L)^{\frac{1}{\xi - 1}}$, part (c) follows. □

The strict monotonicity of F on K is in general not sufficient for the VI (K, F) to have a solution. This is easily illustrated by the scalar equation

$e^t = 0$ which has no zero on the real line. Interestingly, if F is pseudo monotone on K, then the three statements (a), (b), and (c) in Proposition **2.2.3** are equivalent. Therefore, we have a necessary and sufficient condition for a pseudo monotone VI to have a solution; the resulting condition is different from that in Proposition **2.2.8**, which does not rely on the pseudo monotonicity of F.

2.3.4 Theorem. Let $K \subseteq \mathbb{R}^n$ be closed convex and $F : K \to \mathbb{R}^n$ be continuous. Assume that F is pseudo monotone on K. The three statements (a), (b), and (c) in Proposition **2.2.3** are equivalent.

Proof. It suffices to prove that (c) implies (a). If the VI (K, F) has a solution, let x^{ref} be any such solution. The pseudo monotonicity of F on K then easily implies that the set $L_<$ is empty. □

The next result, Theorem **2.3.5**, shows that the solution set of a pseudo monotone VI is always convex and gives a sufficient condition for such a VI to have a nonempty bounded solution set. The latter condition involves the recession cone of K. To prepare for the result, we quickly summarize the key recession properties of a set in $\mathbb{R}^n$. By definition, a *recession direction* of a set X (not necessarily convex) is a vector d such that for some vector $x \in X$, the ray $\{x + \tau d : \tau \geq 0\}$ is contained in X. The set of all recession directions of X is denoted X_∞ and called the *recession cone* of X. Clearly, if X_∞ contains a nonzero vector, then X is unbounded. If X is a closed and convex set, then $x + \tau d \in X$ for all $x \in X$, all $d \in X_\infty$, and $\tau \geq 0$; moreover, in this case, X is bounded if and only if $X_\infty = \{0\}$. If X is a closed cone, then $X = X_\infty$. If X is a polyhedral set, say,

$$X = \{x \in \mathbb{R}^n : Ax \leq b\},$$

for some matrix $A \in \mathbb{R}^{m \times n}$ and vector $b \in \mathbb{R}^m$, then

$$X_\infty = \{d \in \mathbb{R}^n : Ad \leq 0\}.$$

For clarity of notation, we caution the reader that we write $\text{int}(X_\infty)^*$ to mean $\text{int}((X_\infty)^*)$; i.e., the interior of $(X_\infty)^*$.

2.3.5 Theorem. Let $K \subseteq \mathbb{R}^n$ be closed convex and $F : K \to \mathbb{R}^n$ be continuous. Assume that F is pseudo monotone on K.

(a) The solution set SOL(K, F) is convex.

(b) If there exists a vector $x^{\text{ref}} \in K$ satisfying $F(x^{\text{ref}}) \in \text{int}(K_\infty)^*$, then SOL$(K, F)$ is nonempty, convex, and compact.

Proof. Let F be pseudo monotone on K. We claim that

$$\mathrm{SOL}(K,F) = \bigcap_{y \in K} \{ x \in K : F(y)^T(y-x) \geq 0 \}. \tag{2.3.2}$$

Indeed, if $x \in \mathrm{SOL}(K,F)$, then

$$F(x)^T(y-x) \geq 0, \quad \forall y \in K.$$

By the pseudo monotonicity of F on K, this implies

$$F(y)^T(y-x) \geq 0, \quad \forall y \in K;$$

thus x belongs to the right-hand set in (2.3.2). Conversely, suppose x belongs to the latter set. Let $z \in K$ be arbitrary. The vector

$$y \equiv \tau x + (1-\tau) z$$

belongs to K for all $\tau \in [0,1]$. Thus we have

$$F(\tau x + (1-\tau)z)^T(z-x) \geq 0$$

for all $\tau \in (0,1)$. Letting $\tau \to 1$ yields

$$F(x)^T(z-x) \geq 0, \quad \forall z \in K.$$

Hence $x \in \mathrm{SOL}(K,F)$, thus establishing the identity (2.3.2). Since for each fixed but arbitrary $y \in K$, the set

$$\{ x \in K : F(y)^T(y-x) \geq 0 \}$$

is convex and since the intersection of any number of convex sets is convex, it follows that $\mathrm{SOL}(K,F)$ is convex. Thus statement (a) is proved.

To establish statement (b), it suffices to show that if the vector x^{ref} exists with the prescribed property, then the set

$$L_{\leq} \equiv \{ x \in K : F(x)^T(x - x^{\mathrm{ref}}) \leq 0 \}$$

is bounded. The pseudo monotonicity of F on K implies

$$L_{\leq} \subseteq \{ x \in K : F(x^{\mathrm{ref}})^T(x^{\mathrm{ref}} - x) \geq 0 \}. \tag{2.3.3}$$

The set in the right side is a closed convex set. If it is unbounded, then it must have a nonzero recession direction; that is, there exists a nonzero vector $d \in K_\infty$ such that $F(x^{\mathrm{ref}})^T d \leq 0$. Since $F(x^{\mathrm{ref}}) \in \mathrm{int}(K_\infty)^*$, it follows that for some scalar $\delta > 0$, $F(x^{\mathrm{ref}}) - \delta d \in (K_\infty)^*$. Thus we have

$$0 \leq d^T(F(x^{\mathrm{ref}}) - \delta d) \leq -\delta d^T d < 0,$$

where the last inequality is due to the fact that $d \neq 0$. This contradiction shows that the set on the right side of (2.3.3), thus $L_{\leq}$, is bounded. □

Earlier, we have labelled condition (b) of Proposition **2.2.3** as a nonnegativity property at infinity. Theorem **2.3.4** shows that this property is necessary and sufficient for a pseudo monotone VI to have a solution. Part (b) of Theorem **2.3.5** provides a sufficient condition for the pair (K, F) to be "positive at infinity", thus for the VI to have bounded solutions.

In general, for a convex set K, we have

$$F(x) \in (K_\infty)^*, \quad \forall x \in \mathrm{SOL}(K, F);$$

that is, the inclusion

$$F(\mathrm{SOL}(K, F)) \subseteq (K_\infty)^* \tag{2.3.4}$$

holds in general. Condition (b) in Theorem **2.3.5** requires that

$$F(K) \cap \mathrm{int}(K_\infty)^* \neq \emptyset. \tag{2.3.5}$$

This condition has a natural interpretation in the case where K is a convex cone so that the VI (K, F) is equivalent to the CP (K, F). Indeed since $K_\infty = K$ when K is a convex cone, the condition (2.3.5) is precisely the strict feasibility of the CP (K, F). Thus, it follows immediately from Theorem **2.3.5** that if the pseudo monotone CP (K, F) is strictly feasible, then it has a nonempty, convex compact solution set. It turns out that the converse of this result also holds, provided that K^* has a nonempty interior to start with; for details, see Theorem **2.4.4**. More discussion about the existence of solutions to the CP is given in Subsection **2.4.2**.

For a convex set K, we clearly have

$$\mathrm{SOL}(K, F)_\infty \subseteq K_\infty.$$

By duality and (2.3.4), we deduce

$$F(\mathrm{SOL}(K, F)) \subseteq (\,\mathrm{SOL}(K, F)_\infty\,)^*.$$

The next result identifies an additional property satisfied by the solutions to a pseudo monotone VI, which sharpens the above inclusion. The proof of the following result is very simple. For an application of the result to a convex-concave linear-quadratic program, see Exercise **2.9.18**.

2.3.6 Proposition. Let $F : K \to \mathbb{R}^n$ be pseudo monotone on the convex set $K \subseteq \mathbb{R}^n$. For any two solutions x^1 and x^2 in $\mathrm{SOL}(K, F)$, it holds that

$$(\,x^1 - x^2\,)^T F(x^1) \,=\, (\,x^1 - x^2\,)^T F(x^2) \,=\, 0; \tag{2.3.6}$$

consequently,

$$(x^1 - x^2)^T (F(x^1) - F(x^2)) = 0 \tag{2.3.7}$$

and $F(\mathrm{SOL}(K,F)) \subseteq (\mathrm{SOL}(K,F)_\infty)^\perp$.

Proof. Since x^1 and x^2 both are solutions of the VI, we have

$$(x^1 - x^2)^T F(x^2) \geq 0 \quad \text{and} \quad (x^2 - x^1)^T F(x^1) \geq 0.$$

By the pseudo monotonicity of F on K, these inequalities imply,

$$(x^1 - x^2)^T F(x^1) \geq 0 \quad \text{and} \quad (x^2 - x^1)^T F(x^2) \geq 0,$$

respectively. Consequently, equalities hold throughout, establishing (2.3.6). The equality (2.3.7) follows easily from (2.3.6). Finally, if $d \in \mathrm{SOL}(K,F)_\infty$ and $x \in \mathrm{SOL}(K,F)$, then we have $x + d \in \mathrm{SOL}(K,F)$; thus $d^T F(x) = 0$. This establishes that $F(\mathrm{SOL}(K,F))$ is a subset of the orthogonal complement of $\mathrm{SOL}(K,F)_\infty$. □

Proposition **2.3.6** has an interesting consequence when F is a monotone gradient map. Although the following result remains valid when F is not necessarily differentiable, to simplify the proof, we state and prove the result for a continuously differentiable F.

2.3.7 Corollary. Let $K \subseteq \mathbb{R}^n$ be closed convex and $F : \mathcal{D} \supset K \to \mathbb{R}^n$ be continuously differentiable on the open set $\mathcal{D}$. If $JF(x)$ is symmetric for all $x \in K$ and F is monotone on K, then $F(\mathrm{SOL}(K,F))$ is a singleton.

Proof. Let x^1 and x^2 be two solutions of the VI (K,F). By the integral form of the mean value theorem for vector functions, we have

$$F(x^1) - F(x^2) = \int_0^1 JF(x^1 + t(x^2 - x^1))(x^1 - x^2)\, dt$$

Hence, by (2.3.7), we deduce

$$0 = (x^1 - x^2)^T \int_0^1 JF(x^1 + t(x^2 - x^1))\, dt\, (x^1 - x^2).$$

Since F is monotone, the matrix $JF(x^1 + t(x^2 - x^1))$ is positive semidefinite for all $t \in [0,1]$; moreover, it is symmetric by assumption. Consequently, the above equality implies

$$0 = \int_0^1 JF(x^1 + t(x^2 - x^1))\, dt\, (x^1 - x^2),$$

which in turn implies $F(x^1) = F(x^2)$. Consequently, $F(\mathrm{SOL}(K,F))$ is a singleton as claimed. □

In the context of the convex program:

$$\begin{array}{ll} \text{minimize} & \theta(x) \\ \text{subject to} & x \in K, \end{array} \tag{2.3.8}$$

where $\theta : \mathcal{D} \supseteq K \to \mathbb{R}$ is a twice continuously differentiable convex function defined on the open convex set $\mathcal{D}$ and K is closed convex, Corollary **2.3.7** implies that $\nabla\theta(x)$ is a constant on the set of optimal solutions of this program. Based on this fact, we can establish a representation of the optimal solution set of (2.3.8), denoted S_{opt}, in terms of the gradient of the objective function.

2.3.8 Corollary. Let K be a closed convex set in $\mathbb{R}^n$ and θ be a twice continuously differentiable convex function defined on an open set containing K. If S_{opt} is nonempty, then, for any $\bar{x} \in S_{\text{opt}}$,

$$S_{\text{opt}} = \{ x \in K : \nabla\theta(x) = \nabla\theta(\bar{x}), \ \nabla\theta(\bar{x})^T(x - \bar{x}) = 0 \}. \tag{2.3.9}$$

Proof. Let $\bar{S}$ denote the set on right-hand side of (2.3.9). By Proposition **2.3.6** and Corollary **2.3.7**, S_{opt} is a subset of $\bar{S}$. To show equality between these two sets, let x belong to $\bar{S}$. By the gradient inequality of a convex function, we have

$$\theta(\bar{x}) - \theta(x) \geq \nabla\theta(x)^T(\bar{x} - x) = \nabla\theta(\bar{x})^T(\bar{x} - x) = 0.$$

Hence $\theta(x) = \theta(\bar{x})$ and x therefore belongs to S_{opt}. □

2.3.1 Plus properties and F-uniqueness

Corollary **2.3.7** provides a sufficient condition for the set $F(\text{SOL}(K,F))$ to be a singleton. This is probably the next best thing to have if $\text{SOL}(K,F)$ contains multiple elements. This property is important enough to warrant a name. Formally, we say that the $\text{SOL}(K,F)$ is *F-unique* if $F(\text{SOL}(K,F))$ is at most a singleton. In what follows, we define several classes of functions that yield this property. One of these is the class of co-coercive functions that are first defined after Theorem **1.5.5** and which include the Euclidean projector. For convenience, we repeat the definition of this class.

2.3.9 Definition. A mapping $F : K \subseteq \mathbb{R}^n \to \mathbb{R}^n$ is said to be

(a) *pseudo monotone plus* on K if it is pseudo monotone on K and for all vectors x and y in K,

$$[(x - y)^T F(y) \geq 0 \text{ and } (x - y)^T F(x) = 0] \Rightarrow F(x) = F(y);$$

(b) *monotone plus* on K if it is monotone on K and for all vectors x and y in K,

$$(x - y)^T (F(x) - F(y)) = 0 \Rightarrow F(x) = F(y);$$

(c) *co-coercive* on K if there exists a constant $c > 0$ such that

$$(F(x) - F(y))^T (x - y) \geq c \| F(x) - F(y) \|^2, \quad \forall x, y \in K.$$

(d) *(strictly, strongly) monotone composite* if a function $G : \mathbb{R}^m \to \mathbb{R}^m$, a matrix $A \in \mathbb{R}^{m \times n}$, and a vector $b \in \mathbb{R}^n$ exist such that

$$F(x) \equiv A^T G(Ax) + b, \quad \forall x \in K \tag{2.3.10}$$

and G is (strictly, strongly) monotone on the range of A;

(e) *L-(strictly, strongly) monotone composite* if the function G in (d) is Lipschitz continuous and (strictly, strongly) monotone on the range of A. □

By the same proof as in Proposition **2.3.6**, we obtain the following result, which does not require a further proof.

2.3.10 Corollary. Let $F : K \to \mathbb{R}^n$ be pseudo monotone plus on the convex set $K \subseteq \mathbb{R}^n$. The solution set SOL(K, F) is F-unique. □

Before exploring consequences of the F-uniqueness of solutions to the VI, we first clarify the interrelations between the various function classes in Definition **2.3.9**. Each of the functional properties (b)–(e) is invariant under a constant translation of the function. It is not difficult to deduce that if F is monotone plus on K, then F is pseudo monotone plus on K. In essence, the proof of Corollary **2.3.7** has shown that if F is a continuously differentiable, symmetric, monotone function on an open convex set, then it must be monotone plus there. If F is a strictly monotone composite function on K, then it is monotone plus there. If F is co-coercive on K, then it is monotone and nonexpansive, and thus Lipschitz continuous, on K. The converse turns out to be true also if F is a gradient map; i.e., a monotone, Lipschitz continuous gradient map on an open convex set must be co-coercive there; see Exercise **2.9.25**. If F is an L-strongly monotone composite function on K, then it is co-coercive there. To see this, let F be given by (2.3.10), where G is Lipschitz continuous and strongly monotone on the range of A. Let $c > 0$ and $L > 0$ be, respectively, a strong monotonicity and a Lipschitz constant of G. We have, on the one hand,

$$(F(x) - F(y))^T (x - y) = (Ax - Ay)^T (G(Ax) - G(Ay)) \geq c \| Ax - Ay \|^2;$$

on the other hand,

$$\| F(x) - F(y) \|^2 = \| A^T (G(Ax) - G(Ay)) \|^2 \leq \| A^T \|^2 L^2 \| Ax - Ay \|^2.$$

Consequently, we deduce

$$(F(x) - F(y))^T (x - y) \geq c \| A^T \|^{-2} L^{-2} \| F(x) - F(y) \|^2.$$

In particular, every Lipschitz continuous, strongly monotone function is co-coercive. Moreover, if $F : \mathbb{R}^n \to \mathbb{R}^n$ is co-coercive, then so is $E^T F \circ E$ for any $n \times m$ matrix E. The following diagram summarizes the above discussion.

L-strongly monotone composite		strictly monotone composite		
$\Downarrow$		$\Downarrow$		symmetric
co-coercive	$\Rightarrow$	monotone plus	$\Leftarrow$	monotone
$\Uparrow$		$\Downarrow$		
Lipschitz, symmetric, monotone		pseudo monotone plus	$\Rightarrow$	F-uniqueness of solutions.

In particular, for a Lipschtiz continuous and symmetric function,

$$\text{co-coercive} \Leftrightarrow \text{monotone plus} \Leftrightarrow \text{monotone}.$$

In Exercise **2.9.24**, the reader is asked to show that if F is an affine map, then F is co-coercive on $\mathbb{R}^n$ if and only if it is monotone plus, and if and only if it is "affine strongly (strictly) monotone composite".

Another class of co-coercive functions is obtained by considering the unique solution of a strongly monotone VI $(K, F - q)$ with a varying vector q. Such a VI includes the one arising from the Euclidean projection Π_K onto a closed convex set K, which corresponds to F being the identity function. A generalization of the following result to a broader class of VIs can be found in Exercise **3.7.34**. See also Exercise **2.9.17** that pertains to a monotone VI $(K, F - q)$ where F is not strongly monotone.

2.3.11 Proposition. Let $K \subseteq \mathbb{R}^n$ be closed convex and $F : K \to \mathbb{R}^n$ be continuous and strongly monotone. If $x(q)$ denotes the unique solution of the VI $(K, F - q)$, then $x(q)$ is co-coercive on $\mathbb{R}^n$.

Proof. The proof is an easy extension of that for the Euclidean projector; see part (c) of Theorem **1.5.5**. □

In general, if $F(\mathrm{SOL}(K,F))$ is a singleton, then the single element of $F(\mathrm{SOL}(K,F))$ yields a simple description of $\mathrm{SOL}(K,F)$.

2.3.12 Proposition. Suppose that $F(\mathrm{SOL}(K,F))$ is a singleton $\{w\}$. It holds that

$$\mathrm{SOL}(K,F) \,=\, F^{-1}(w) \cap \operatorname{argmin} \{ x^T w \,:\, x \in K \}.$$

If in addition K is polyhedral, then the KKT multiplier set $\mathcal{M}(x)$ is a polyhedron that is independent of $x \in \mathrm{SOL}(K,F)$

Proof. Let S denote the set on the right side of the displayed expression. Let $x \in \mathrm{SOL}(K,F)$. We have $F(x) = w$; thus $x \in F^{-1}(w)$. Moreover,

$$(x' - x)^T F(x) \geq 0, \quad \forall\, x' \in K.$$

Hence $\mathrm{SOL}(K,F) \subseteq S$. Conversely, let $x \in S$. Then $F(x) = w$ and it follows easily that $x \in \mathrm{SOL}(K,F)$. Hence $\mathrm{SOL}(K,F) = S$. To prove the second assertion of the proposition, write

$$K \equiv \{ x \in \mathbb{R}^n \,:\, Ax \leq b \}$$

for some matrix A and vector b of appropriate dimension. We claim that for every $x \in \mathrm{SOL}(K,F)$, $\mathcal{M}(x)$ is the optimal solution set of the linear program:

$$\begin{array}{ll} \text{minimize} & b^T \lambda \\ \text{subject to} & w + A^T \lambda = 0 \\ \text{and} & \lambda \geq 0, \end{array} \tag{2.3.11}$$

which is obviously polyhedral and independent of x. The dual of the above LP is

$$\begin{array}{ll} \text{minimize} & y^T w \\ \text{subject to} & y \in K. \end{array} \tag{2.3.12}$$

Suppose $\lambda \in \mathcal{M}(x)$ for some $x \in \mathrm{SOL}(K,F)$. Clearly x is an optimal solution of (2.3.12), λ is feasible to (2.3.11), and $\lambda \perp Ax - b$. It follows from LP duality that λ is an optimal solution of (2.3.11). Conversely, if λ is an optimal solution of the latter LP and $x \in \mathrm{SOL}(K,F)$, then by LP complementary slackness, we must have $\lambda \perp Ax - b$, establishing that $\lambda \in \mathcal{M}(x)$. $\square$

Notice that $\mathrm{SOL}(K,F)$ is not asserted to be polyhedral in the above proposition, even though the multiplier set must be polyhedral (assuming that K is polyhedral). If F is a strictly monotone composite function on

K, SOL(K, F) has a slightly different representation, which can be used to show that if K is polyhedral, then so is SOL(K, F). The latter conclusion is interesting because it does not require F to be affine; nor does it require G to be Lipschitz continuous. The polyhedrality of SOL(K, F) can be used to obtain an error bound for the solutions to the VIs of this class; see Theorem **6.2.8**.

2.3.13 Corollary. Let F be given by (2.3.10) for some G that is strictly monotone on the range of A. If SOL$(K, F) \neq \emptyset$, then $A(\text{SOL}(K, F))$ is a singleton. With $v \in \mathbb{R}^m$ denoting the single element of $A(\text{SOL}(K, F))$, it holds that

$$\text{SOL}(K, F) = A^{-1}(v) \cap \operatorname{argmin} \{ x^T w : x \in K \},$$

where w is the single element of $F(\text{SOL}(K, F))$. If in addition K is a polyhedron, then so is SOL(K, F).

Proof. Under the given assumptions on F, it holds that $F(x) = F(y)$ if and only if $Ax = Ay$. Moreover, $A(\text{SOL}(K, F))$ must be a singleton (this property does not require G to be Lipschitz continuous, nor does it require G to be strongly monotone; the strict monotonicity of G suffices). The desired representation of SOL(K, F) in terms of the vector v follows readily. If K is a polyhedron, then the argmin set in the representation of SOL(K, F), being the optimal solution set of a linear program, is a polyhedron; hence so is SOL(K, F). □

We can establish a result analogous to part (c) of Theorem **2.3.3** for a co-coercive function.

2.3.14 Proposition. Let $F : \Omega \supseteq K \to \mathbb{R}^n$ be co-coercive on Ω with constant $c > 0$. Suppose SOL$(K, F) \neq \emptyset$ and let w be the unique element of $F(\text{SOL}(K, F))$. For every $x \in \Omega$,

$$\| F(x) - w \| \leq c^{-1} \| \mathbf{F}_K^{\text{nat}}(x) \|.$$

Proof. This follows easily from the proof of part (c) of Theorem **2.3.3** by using the co-coercivity of F. □

2.3.2 The dual gap function

In addition to providing a proof to the convexity of the solution set of a pseudo monotone VI, the identity (2.3.2) leads to a "dual gap function" that is worth mentioning. Recalling the function $\theta_{\mathrm{gap}}(x)$ defined by (1.5.15), we define

$$\theta_{\mathrm{dual}}(x) \equiv \inf_{y \in K} F(y)^T(y-x), \quad x \in \mathbb{R}^n. \tag{2.3.13}$$

In order to distinguish the two functions θ_{gap} and θ_{dual}, we call the former the *primal gap function* and the latter the *dual gap function*. The following inequalities clearly hold between these functions:

$$-\infty \leq \theta_{\mathrm{dual}}(x) \leq 0 \leq \theta_{\mathrm{gap}}(x) \leq \infty, \quad \forall x \in K;$$

in particular, the dual gap function is also extended-valued in the sense that $\theta_{\mathrm{dual}}(x)$ could equal $-\infty$ for some $x \in \mathbb{R}^n$. There are several important differences between $\theta_{\mathrm{gap}}(x)$ and $\theta_{\mathrm{dual}}(x)$. First, $\theta_{\mathrm{gap}}(x)$ is defined on the same domain of $F(x)$, whereas $\theta_{\mathrm{dual}}(x)$ is defined for all $x \in \mathbb{R}^n$. Second, the function $\theta_{\mathrm{dual}}(x)$ is always concave. To see this, consider the hypograph of θ_{dual}, which is equal to the set:

$$\bigcap_{y \in K} \{(x,\eta) \in \mathbb{R}^n \times \mathbb{R} : F(y)^T(y-x) \geq \eta\}.$$

This set is the intersection of a family of halfplanes in $\mathbb{R}^{n+1}$, hence it is convex. Third, similar to (1.5.18), the evaluation of $\theta_{\mathrm{dual}}(x)$ requires solving the following optimization problem in the variable y and parameterized by x:

$$\begin{array}{ll} \text{minimize} & F(y)^T(y-x) \\ \text{subject to} & y \in K. \end{array} \tag{2.3.14}$$

Unlike its counterpart (1.5.18), the program (2.3.14) is in general a nonconvex program. The exceptional case is when F is an affine monotone function; in this case, the objective of (2.3.14) is a convex quadratic function in y. If in addition K is polyhedral (thus the VI (K,F) becomes a monotone AVI), then (2.3.14) becomes a parametric convex quadratic program.

In summary, the primal gap function $\theta_{\mathrm{gap}}(x)$ is in general neither convex nor concave and its evaluation requires solving convex programs; in contrast, the dual gap function $\theta_{\mathrm{dual}}(x)$ is always concave but its evaluation requires solving nonconvex programs (except for monotone AVIs). See Exercise **10.5.3** for an interesting differentiability property of θ_{dual} in the case of a compact convex set K and a pseudo monotone plus function F.

Similar to (1.5.16), we can introduce a *dual gap program*:

$$\begin{array}{ll} \text{maximize} & \theta_{\mathrm{dual}}(x) \\ \text{subject to} & x \in K. \end{array} \tag{2.3.15}$$

Since θ_{dual} is a concave function, this becomes a concave maximization problem, which in principle is easier to solve than (1.5.16); but the hidden difficulty in solving (2.3.15) for a nonlinear function $F(x)$ lies in the evaluation of $\theta_{\text{dual}}(x)$. In spite of this computational hardship, (2.3.15) provides a useful theoretical link between a pseudo monotone VI and a concave maximization problem. This link is made precise in the following result, which is an immediate consequence of Theorem **2.3.5** and the above discussion.

2.3.15 Proposition. Let $K \subseteq \mathbb{R}^n$ be closed convex and $F : K \to \mathbb{R}^n$ be continuous. If F is pseudo monotone on K, then $x \in \text{SOL}(K, F)$ if and only if x is a global maximizer of (2.3.15) and $\theta_{\text{dual}}(x) = 0$. □

The primal gap function θ_{gap} and the dual gap function θ_{dual} can be seen to be the pair of optimization problems associated with the saddle function (see Subsection **1.4.1**):

$$L(x, y) \equiv F(x)^T(x - y);$$

For a generalization of this point of view, see Exercise **2.9.22**.

2.3.3 Boundedness of solutions

Based on the dual gap program, it is possible to develop an alternative theory of the existence of solutions to a pseudo monotone VI. In particular, a necessary and sufficient condition for such a VI to have a nonempty compact solution set can be obtained via convex analysis. In what follows, we present this condition and establish the theorem below as a consequence of our development so far, without relying on the dual gap program. The reader can easily verify that the assumption in part (b) of Theorem **2.3.5** is a special case of the condition (2.3.16) below.

2.3.16 Theorem. Let $K \subseteq \mathbb{R}^n$ be closed convex and $F : K \to \mathbb{R}^n$ be continuous. If F is pseudo monotone on K, then SOL(K, F) is nonempty and bounded if and only if

$$K_\infty \cap [-(F(K)^*)] = \{0\}. \tag{2.3.16}$$

Proof. Suppose that SOL(K, F) is nonempty and bounded. Let d be an arbitrary vector belonging to the left-hand set in (2.3.16). Since $d \in K_\infty$, it follows that $x + \tau d \in K$ for all $\tau \geq 0$ and $x \in K$. Moreover, we have $d^T F(y) \leq 0$ for all $y \in K$. For every $x \in \text{SOL}(K, F)$, by the representation (2.3.2) of SOL(K, F), we deduce

$$F(y)^T[y - (x + \tau d)] \geq 0$$

for every $y \in K$ and $\tau \geq 0$. Thus d is a recession direction of SOL(K, F). Since this set is bounded, we must have $d = 0$. This establishes the "only if" statement.

Conversely, if (2.3.16) holds, then by using the fact that every unbounded closed convex set must have a nonzero recession direction and by reversing the above argument, we see that it suffices to show the existence of a solution to the VI (K, F). For this demonstration, we use a homotopy argument similar to the proof of (b) $\Rightarrow$ (c) in Proposition **2.2.3**. As in this proof, we may assume for the sake of contradiction that SOL(K, F) is empty. Recalling Remark **2.2.4**, we may further assume that F is defined and continuous on the entire space $\mathbb{R}^n$. Let $a \in K$ be an arbitrary vector and consider the homotopy:

$$H(x,t) \equiv x - \Pi_K(\, t(x - F(x)) + (1-t)a\,), \quad (x,t) \in \mathbb{R}^n \times [0,1].$$

We claim that the set of zeros:

$$\bigcup_{t \in [0,1]} H(\cdot,t)^{-1}(0)$$

is bounded. Once this claim is established, by the same homotopy argument used in Proposition **2.2.3**, a contradiction will be obtained and the proof of the theorem will be completed. So assume that the displayed union is unbounded. Thus there exist a sequence of scalars $\{t_k\} \subset [0,1]$ and a sequence of vectors $\{x_k\}$ such that

$$\lim_{k \to \infty} \| x^k \| = \infty \quad \text{and} \quad H(x^k, t_k) = 0 \quad \forall\, k.$$

Without loss of generality, we may assume that each t_k is positive. By the definition of the homotopy map H, we have $x^k \in K$ for all k; moreover, for all $y \in K$,

$$(\, y - x^k\,)^T [\, t_k\, F(x^k) + (1-t_k)\,(\, x^k - a\,)] \geq 0,$$

which implies

$$(\, y - x^k\,)^T F(x^k) \geq -\frac{1-t_k}{t_k}\,(\, y - x^k\,)^T (\, x^k - a\,).$$

For each fixed but arbitrary $y \in K$, the right-hand side in the above expression is nonnegative for all k sufficiently large. Thus, by the pseudo-monotonicity of F on K, we have

$$(\, y - x^k\,)^T F(y) \geq 0$$

for all k sufficiently large. Since $\{x^k\}$ is an unbounded sequence in K, which is closed and convex, it follows that every accumulation point of the normalized sequence $\{x^k/\|x^k\|\}$ must be a nonzero element of K_∞. Let d be any such point. Then $d \in K_\infty$ and

$$d^T F(y) \leq 0, \quad \forall y \in K.$$

This contradicts the expression (2.3.16). □

From the proof of Theorem **2.3.16**, we can establish the following result, which has important implications in the stability analysis of the VI. As always, $\mathbf{F}_K^{\text{nat}}$ denotes the natural map of the pair (K, F).

2.3.17 Proposition. Let $K \subseteq \mathbb{R}^n$ be closed convex and $F : \text{cl}\,\mathcal{D} \to \mathbb{R}^n$ be continuous, where $\mathcal{D}$ is an open set containing K. If F is pseudo monotone on K and SOL(K, F) is nonempty and bounded, then $\deg(\mathbf{F}_K^{\text{nat}}, \Omega) = 1$ for every bounded open set Ω containing SOL(K, F) such that $\text{cl}\,\Omega \subseteq \text{cl}\,\mathcal{D}$.

Proof. For simplicity, we may assume that $\mathcal{D} = \mathbb{R}^n$. Let a be an arbitrary vector in K. By the proof of Theorem **2.3.16**, there exists a bounded open set $\mathcal{U}$ containing SOL$(K, F) \cup \{a\}$ such that

$$\deg(\mathbf{F}_K^{\text{nat}}, \mathcal{U}) = 1.$$

By the excision property of the degree, $\mathcal{U}$ can be replaced by every bounded open set Ω containing SOL(K, F). □

In Exercise **2.9.30**, the reader is asked to use Proposition **2.3.17** to show the existence of a solution to the NCP (F) obtained from the model of invariant capital stock in Proposition **1.4.5**. Note however that the mapping F in (1.4.18) is not monotone, but it is the sum of a monotone function and a nonnegative linear function if the matrix B is nonnegative.

2.4 Monotone CPs and AVIs

Refined existence results and sharpened solution properties can be established for the CP (K, F), which corresponds to the VI (K, F) with K being a cone. The existence results all assume the feasibility (or its equivalent) of the CP. This is a distinguishing feature that is absent in the case of the VI defined on a general convex set. Since the CP is defined on a cone, properties of closed convex cones play an important role. For this reason, we devote the next subsection to discuss these properties.

2.4.1 Properties of cones

In the last part of the proof of Theorem **2.3.5**, we have used a consequence of the assumption $F(x^{\text{ref}}) \in \text{int}(K_\infty)^*$ that pertains to a general property of the elements in the interior of the dual of an arbitrary closed convex cone. This property turns out to characterize the elements in the interior of such a dual cone. In what follows, we generalize this property to the relative interior of the dual of a closed convex cone. Specifically, given an arbitrary closed convex cone C, we establish several equivalent conditions for a vector to be a relative interior point of C^*. In Section **3.4**, we employ these conditions to define an important property of a solution to a VI.

We recall some elementary properties of closed convex cones in $\mathbb{R}^n$. Let C be a closed convex cone in $\mathbb{R}^n$. The intersection $C \cap (-C)$ is a linear subspace in $\mathbb{R}^n$, called the *lineality space* of C. The lineality space of C is the largest subspace contained in C. For any two subsets S_1 and S_2 of $\mathbb{R}^n$ both containing the origin, we have

$$(S_1 + S_2)^* \, = \, S_1^* \cap S_2^*.$$

Thus if C_1 and C_2 are two closed convex cones such that the sum $C_1^* + C_2^*$ is closed, then we must have

$$(C_1 \cap C_2)^* \, = \, C_1^* + C_2^*.$$

Based on this fact, we can show that the orthogonal complement of the lineality space of C is equal to the linear hull of C^*, denoted $\text{lin}\, C^*$; that is,

$$\text{lin}\, C^* \, = \, (C \cap (-C))^\perp.$$

This holds because

$$(C \cap (-C))^\perp \, = \, (C \cap (-C))^* \, = \, C^* - C^* \, = \, \text{lin}\, C^*.$$

There are four equivalent conditions in Proposition **2.4.1**, which concerns a given closed convex cone C in $\mathbb{R}^n$. The first condition (a) is just the statement of a vector v being in the relative interior of the dual cone C^*, which we denote $\text{ri}\, C^*$. The second statement (b) is an angle condition; it asserts that a relative interior vector of C^* must make an acute angle with every vector common to C and the linear hull of C^* (since C^* is a convex cone, its linear hull coincides with its affine hull). This statement is essentially the one used in the proof of Theorem **2.3.5** when C^* is of full dimension so that its relative interior becomes its topological interior. The third statement (c) shows that such a relative interior vector is related to the boundedness of certain "truncations" of the cone C intersected with the hull $\text{lin}\, C^*$. The fourth and last condition (d) is the dual of condition (a). The notation $v^\perp$ therein refers to the orthogonal complement of the vector v; that is $v^\perp$ is the linear subspace consisting of vectors that are perpendicular to v.

2.4.1 Proposition. Let C be a closed convex cone in $\mathbb{R}^n$. The following four statements are equivalent:

(a) $v \in \text{ri}\, C^*$;

(b) for all nonzero vectors $x \in C \cap \text{lin}\, C^*$, $v^T x > 0$;

(c) for every scalar $\eta > 0$, the set

$$S(v, \eta) \equiv \{ x \in C \cap \text{lin}\, C^* \, : \, v^T x \leq \eta \}$$

is bounded;

(d) $v \in C^*$ and the intersection $C \cap v^\perp$ is a linear subspace.

If any of the above holds, then $C \cap v^\perp$ is equal to the lineality space of C. □

Proof. (a) ⇒ (b). The proof is very similar to the argument used at the end of the proof of Theorem **2.3.5**. Suppose $v \in \text{ri}\, C^*$. Assume for contradiction that there exists a nonzero vector $x \in C \cap \text{lin}\, C^*$ satisfying $v^T x \leq 0$. The vector $v - \delta x$ belongs to C^* for all $\delta > 0$ sufficiently small. We obtain a contradiction as in the proof of Theorem **2.3.5**.

(b) ⇒ (c). If the set $S(v, \eta)$ is unbounded for some $\eta > 0$, then there exists an unbounded sequence $\{x^k\} \subset S(v, \eta)$. The normalized sequence $\{x^k / \|x^k\|\}$ must have at least one accumulation point and every such point must be a nonzero vector $u \in C \cap \text{lin}\, C^*$ satisfying $v^T u \leq 0$. This contradicts (b).

(c) ⇒ (a). First assume that $\text{lin}\, C^* = \mathbb{R}^n$ so that $\text{ri}\, C^* = \text{int}\, C^*$. Let v be such that $S(v, \eta)$ is bounded for all $\eta > 0$. We claim that $v \in C^*$. Indeed if there exists a nonzero element u in C such that $v^T u \leq 0$, then τu belongs to $S(v, \eta)$ for all positive scalars τ and η. This contradicts the boundedness of $S(v, \eta)$. Consequently, $v^T x > 0$ for all nonzero $x \in C$; in particular $v \in C^*$. Suppose that $v \notin \text{int}\, C^*$. For every scalar $\delta > 0$, there exist vectors $y(\delta) \in \mathbb{R}^n$ and $x(\delta) \in C$ satisfying

$$\| v - y(\delta) \|_2 \leq \delta \quad \text{and} \quad y(\delta)^T x(\delta) < 0. \tag{2.4.1}$$

The last inequality implies $x(\delta) \neq 0$ for all $\delta > 0$. We claim that the vector $u(\delta) \equiv \delta^{-1} x(\delta) / \|x(\delta)\|_2$ belongs to $S(v, 1)$. Indeed since C is a cone, $u(\delta) \in C$. Moreover,

$$v^T u(\delta) = \delta^{-1} (v - y(\delta))^T \frac{x(\delta)}{\| x(\delta) \|_2} + \delta^{-1} y(\delta)^T \frac{x(\delta)}{\| x(\delta) \|_2}.$$

By the Cauchy-Schwarz inequality and (2.4.1), the first summand in the right-hand expression is not greater than unity and the second summand is negative. Consequently $u(\delta) \in S(v, 1)$ for all $\delta > 0$. Clearly $\|u(\delta)\|_2 = \delta^{-1}$. This contradicts the boundedness of $S(v, 1)$.

Consider the general case where $\operatorname{lin} C^*$ is a proper subset of $\mathbb{R}^n$. We can apply what has just been proved to deduce that under the assumption of (c), we must have $v \in \operatorname{int}(C \cap \operatorname{lin} C^*)^*$. Note that

$$(C \cap \operatorname{lin} C^*)^* = C^* + (C \cap (-C))$$

because the right-hand sum is a closed set. Hence there exists a neighborhood $\mathcal{N}$ of v such that

$$\mathcal{N} \subseteq C^* + (C \cap (-C)).$$

To complete the proof, it suffices to verify that

$$[C^* + (C \cap (-C))] \cap \operatorname{lin} C^* \subseteq C^*.$$

Let $x \in C^*$ and $y \in C \cap (-C)$ be such that $x + y \in \operatorname{lin} C^*$. Since $\operatorname{lin} C^*$ is the orthogonal complement of $C \cap (-C)$, we have

$$0 = y^T (x + y) = y^T y,$$

which implies that $y = 0$. This completes the proof of the equivalence between the three statements (a), (b), and (c).

(a) $\Leftrightarrow$ (d). Suppose that $v \in \operatorname{ri} C^*$. Let $x \in C \cap v^\perp$. We claim that $-x$ belongs to C. Since $C = C^{**}$, it suffices to show that for every $y \in C^*$, $x^T y \leq 0$. There exists $\delta > 0$ such that $v - \delta y \in C^*$. Since $x \in C$, we have

$$0 \leq x^T (v - \delta y) = -\delta x^T y,$$

which implies $x^T y \leq 0$. Consequently, $C \cap v^\perp$ is a linear subspace. Conversely, suppose that $C \cap v^\perp$ is a linear subspace. It suffices to show that if x is a nonzero vector in $C \cap \operatorname{lin} C^*$, then $x^T v > 0$. Since $v \in C^*$ by assumption, we have $x^T v \geq 0$. If $x^T v = 0$, then $x \in C \cap v^\perp$. So $x \in -C$. Thus x belongs to $C \cap (-C)$. But x also belongs to $\operatorname{lin} C^*$, which is equal to the orthogonal complement of $C \cap (-C)$, we obtain a contradiction to the assumption that x is nonzero.

Finally, to prove the last assertion of the proposition, suppose that $C \cap v^{\perp}$ is a linear subspace. Since the lineality space of C is the largest linear subspace contained in C, it suffices to show that if $x \in C \cap (-C)$, then $x^T v = 0$. But this is clear because $v \in C^*$. □

Existence results for nonlinear CPs often require that the defining cone and its dual possess two important properties, which we define below.

2.4.2 Definition. A cone $C \subseteq \mathbb{R}^n$ is *pointed* if $C \cap (-C) = \{0\}$. A set $S \subseteq \mathbb{R}^n$ is *solid* if $\operatorname{int} S \neq \emptyset$. □

Clearly, the nonnegative orthant is both pointed and solid. Moreover, so is any simplicial cone, i.e., a cone that is equal to pos(A) for some nonsingular matrix A. The Lorentz cone and the cone of symmetric positive semidefinite matrices are non-polyhedral, pointed, and solid cones. The lemma below asserts that a closed convex cone is solid (pointed) if and only if its dual is pointed (solid).

2.4.3 Proposition. Let C be a closed convex cone in $\mathbb{R}^n$. The following two statements hold.

(a) If C is solid, then its dual C^* is pointed.

(b) If C is pointed, then its dual C^* is solid.

Thus C is solid (pointed) if and only if C^* is pointed (solid).

Proof. Suppose that C is solid. To show that C^* is pointed, let d be an element of $C^* \cap (-C^*)$. Then $d^T x = 0$ for all $x \in C$. Since C is closed and convex, $C = C^{**}$. Since C is solid, $\operatorname{int} C$ is nonempty. Let y be an interior point of $C = C^{**}$. By Proposition **2.4.1**, if $d \neq 0$, then $d^T y > 0$. But this is not possible. Thus $d = 0$ and C^* is pointed.

Suppose that C is pointed. To show that C^* has a nonempty interior, let k be the maximum number of linearly independent vectors in C^* and let $\{y^1, \ldots, y^k\}$ denote k such vectors. If $k = n$, then C^* contains the simplicial cone generated by these $k = n$ vectors. Since the latter cone has a nonempty interior, so does C^*. If $k < n$, then the system

$$p^T y^i = 0, \quad i = 1, \ldots, k$$

has a nonzero solution $p \in \mathbb{R}^n$. By the maximality of k, $p^T y = 0$ for all $y \in C^*$. Thus $p \in C^{**} \cap (-C^{**})$. Since $C^{**} = C$, it follows that $0 \neq p \in C \cap (-C)$; this contradicts the pointedness of C.

Finally, to establish the last assertion of the lemma, suppose that C^* is pointed. By part (b), it follows that C^{**} is solid. Since C is a closed convex cone, we have $C^{**} = C$. Thus C is solid. Similarly, we can show that if C^* is solid, then C must be pointed. □

Exercise **2.9.26** contains further properties of a closed convex cone. Not adopted in the book, the term "proper cone" is sometimes used in the literature to mean a closed convex cone that is both solid and pointed.

2.4.2 Existence results

By combining Theorem **2.3.5**, Theorem **2.3.16**, and Proposition **2.3.17**, we obtain the following result, which provides several necessary and sufficient conditions for a pseudo monotone CP to have a nonempty compact solution set. Among these conditions is the strict feasibility of the CP.

2.4.4 Theorem. Let K be a closed convex cone in $\mathbb{R}^n$ and let F be a continuous map from K into $\mathbb{R}^n$ that is pseudo monotone on K. The following three statements are equivalent.

(a) The CP (K, F) is strictly feasible.

(b) The dual cone K^* has a nonempty interior and the CP (K, F) has a nonempty compact solution set.

(c) The dual cone K^* has a nonempty interior and

$$K \cap [-(F(K)^*)] = \{0\}. \tag{2.4.2}$$

Proof. (a) ⇒ (b). If the CP (K, F) is strictly feasible, then clearly $\operatorname{int} K^*$ is nonempty. Moreover, by Theorem **2.3.5**, it follows that $\operatorname{SOL}(K, F)$ is nonempty and compact.

(b) ⇒ (c). This follows from Theorem **2.3.16** by noting that $K = K_\infty$ because K is a convex cone.

(c) ⇒ (a). Without loss of generality, we may assume that F is defined and continuous on $\mathbb{R}^n$. By Proposition **2.3.17**, $\deg(\mathbf{F}_K^{\text{nat}}, \Omega) = 1$ for every bounded open set Ω containing $\operatorname{SOL}(K, F)$. Let q be an arbitrary vector in $\operatorname{int} K^*$. For any fixed but arbitrary scalar $\varepsilon > 0$, let

$$\mathbf{F}_K^{\varepsilon,\text{nat}}(x) \equiv x - \Pi_K(x - F(x) + \varepsilon\, q)$$

be the natural map associated with the perturbed CP:

$$K \ni x \perp F(x) - \varepsilon\, q \in K^*. \tag{2.4.3}$$

It follows that for all $\varepsilon > 0$ sufficiently small,

$$\deg(\mathbf{F}_K^{\varepsilon,\text{nat}}, \Omega) = \deg(\mathbf{F}_K^{\text{nat}}, \Omega) = 1.$$

Thus the CP (2.4.3) has a solution, say x. Since $q \in \operatorname{int} K^*$, x must belong to K and $F(x)$ to $\operatorname{int} K^*$. Hence the CP (K, F) is strictly feasible. □

When F is an affine map, we can further specialize the condition (2.4.2) and obtain the following result, which does not require K to be polyhedral.

2.4.5 Corollary. Let $F(x) \equiv q + Mx$ be an affine map that is pseudo monotone on a closed convex cone K in $\mathbb{R}^n$. The following three statements are equivalent.

(a) There exists a vector x in K such that $q + Mx$ is in $\operatorname{int} K^*$.

(b) The dual cone K^* has a nonempty interior and the CP (K, q, M) has a nonempty compact solution set.

(c) The dual cone K^* has a nonempty interior and the implication below holds:

$$[\, d \in K, \; M^T d \in -K^*, \; q^T d \leq 0 \,] \; \Rightarrow \; d = 0.$$

Proof. It suffices to show that (2.4.2) is equivalent to the implication in part (c). Clearly the former condition is equivalent to the following implication:

$$\left. \begin{array}{r} d \in K \\ (\, q + Mx \,)^T d \leq 0, \; \forall \, x \in K \end{array} \right\} \; \Rightarrow \; d = 0.$$

It is now a simple observation to note that this implication is further equivalent to the desired implication in part (c). □

We have mentioned at the end of Subsection **1.4.11** that when K is the cone $\mathcal{M}^n_+$ of $n \times n$ symmetric positive semidefinite matrices, the strict feasibility of the CP $(K, q, M) = (\mathcal{M}^n_+, q, M)$ can be determined by solving a semidefinite program. When K is a polyhedral cone, the strict feasibility of the CP (K, q, M) can be checked by linear programming. Instead of presenting the details of how this can accomplished, which we leave as an exercise for the reader, we consider the LCP (q, M), which corresponds to $K = \mathbb{R}^n_+$. It is not difficult to verify that the LCP (q, M) is strictly feasible if and only if the following piecewise linear program:

$$\begin{array}{ll} \text{maximize} & \displaystyle\min_{1 \leq i \leq n} \, (\, q + Mx \,)_i \\[2ex] \text{subject to} & x \geq 0 \end{array}$$

has a positive optimum objective value (that could be $+\infty$). This is a concave maximization problem that can be easily converted into a standard linear program by well-known techniques.

From Theorem **2.4.4**, one is led to ask whether the feasibility of a pseudo monotone CP is sufficient for its solvability. The example below shows that even for a monotone NCP, feasibility is not sufficient for solvability. This example also illustrates an interesting property of a feasible monotone NCP.

2.4.6 Example. Consider the NCP (F) where

$$F(x_1, x_2) \equiv \begin{pmatrix} 2x_1x_2 - 2x_2 + 1 \\ -x_1^2 + 2x_1 - 1 \end{pmatrix}, \quad (x_1, x_2) \in \mathbb{R}^2.$$

The Jacobian matrix of F is given by:

$$JF(x_1, x_2) \equiv \begin{bmatrix} 2x_2 & 2x_1 - 2 \\ 2 - 2x_1 & 0 \end{bmatrix},$$

which is skew-symmetric thus positive semidefinite for all (x_1, x_2) with $x_2 \geq 0$. Thus F is monotone on $\mathbb{R}^2_+$. It is easy to show that the feasible region of the NCP (F) is the set

$$\{ (x_1, x_2) \in \mathbb{R}^2 : x_1 = 1, \; x_2 \geq 0 \}.$$

Nevertheless the NCP (F) has no solution. Although no exact solution exists, this NCP has ε-solutions in the following sense. For each scalar $\varepsilon \in (0, 1)$, consider the positive vector $x^\varepsilon \equiv (1 - \varepsilon, 1/(2\varepsilon))$. It is easy to verify that

$$\lim_{\varepsilon \to 0} \min(x^\varepsilon, F(x^\varepsilon)) = 0.$$

Thus for every $\varepsilon > 0$, there exists a vector $\tilde{x}$ satisfying

$$\| \min(\tilde{x}, F(\tilde{x})) \| \leq \varepsilon.$$

Such a vector $\tilde{x}$ is called an ε-(approximate) solution of the NCP (F). □

The above example raises two questions. The first question is: when does the feasibility of a monotone CP (K, F) imply its solvability? Originated from the study of the LCP, this question is practically significant because it aims at identifying the situations in which it suffices to check the feasibility of the CP in order to infer its solvability. Notice that Theorem **2.4.4** does not quite address the question raised here because the result there involves the boundedness of the solutions. The second question is: does every feasible, monotone NCP have ε-solutions? The answer to the second question turns out to be affirmative. A constructive proof exists, which is based on an iterative algorithm that actually computes such an approximate solution; see part (a) of Proposition **11.5.8**.

There are two important situations where the first question raised above has an affirmative answer: (a) when K is polyhedral and F is affine, (i.e., for an affine CP) and (b) when F is strictly monotone on K (but not affine) and K is pointed (but non-polyhedral). We first deal with the affine CP. The proof of the following result makes use of the fundamental Frank-Wolfe Theorem in quadratic programming, which states that a quadratic function bounded below on a polyhedral set attains its minimum there.

2.4.7 Theorem. Let K be a polyhedral cone in $\mathbb{R}^n$ and F be a monotone affine map from $\mathbb{R}^n$ into itself. The CP (K, F) is solvable if and only if it is feasible.

Proof. Since K is a polyhedral cone, $K^{**} = K$. Let us write

$$F(x) \equiv q + Mx, \quad \forall x \in \mathbb{R}^n,$$

for some n-vector q and $n \times n$ positive semidefinite matrix M. Consider the gap program (1.5.17), which we write as:

$$\begin{array}{ll} \text{minimize} & x^T(q + Mx) \\ \text{subject to} & x \in K \quad \text{and} \quad q + Mx \in K^*. \end{array}$$

This is a convex quadratic program whose objective function is bounded below by zero on the feasible region of the program. Therefore by the Frank-Wolfe theorem the program has an optimal solution x^*, which along with a multiplier $\lambda \in \mathbb{R}^n$ must satisfy the following complementarity system:

$$K \ni x^* \perp v \equiv q + (M + M^T)x^* - M^T\lambda \in K^*,$$

$$K^* \ni q + Mx^* \perp \lambda \in K^{**} = K.$$

From these conditions, we deduce

$$\begin{aligned} 0 &\geq (x^* - \lambda)^T v \\ &= (x^* - \lambda)^T[q + Mx^* + M^T(x^* - \lambda)] \geq (q + Mx^*)^T x^*, \end{aligned}$$

where we have used the positive semidefiniteness of M in the last inequality. Consequently, $(q + Mx^*)^T x^* = 0$ and $x^* \in \mathrm{SOL}(K, F)$. □

The treatment of the nonlinear CP requires Proposition **2.4.3**. Using the proposition, we can prove the following result for a strictly monotone nonlinear CP.

2.4.8 Theorem. Let K be a pointed, closed, and convex cone in $\mathbb{R}^n$ and let $F : K \to \mathbb{R}^n$ be a continuous map. Consider the following statements:

(a) F is strictly monotone on K and $\text{FEA}(K, F) \neq \emptyset$;

(b) F is strictly monotone on K and the CP (K, F) is strictly feasible;

(c) the CP (K, F) has a unique solution.

It holds that (a) $\Leftrightarrow$ (b) $\Rightarrow$ (c).

Proof. It suffices to show that (a) $\Rightarrow$ (b). By Proposition **2.4.3**, there exists a vector $w \in \text{int}\, K^*$. Since $\text{lin}\, K^* = \mathbb{R}^n$, Proposition **2.4.1**(b) implies that $w^T x > 0$ for all nonzero x in K. Let $z \in \text{FEA}(K, F)$ be an arbitrary feasible vector. For every scalar $\eta > 0$, the set

$$S_\eta \equiv \{ x \in z + K : (x - z)^T w = \eta \},$$

which must be nonempty and convex, is compact. Indeed, if S_η is unbounded, then there exists a sequence $\{x^k\} \subset S_\eta$ such that

$$\lim_{k \to \infty} \| x^k \| = \infty \quad \text{and} \quad \lim_{k \to \infty} \frac{x^k}{\| x^k \|} = v.$$

It is easily seen that v belongs to K and satisfies $v^T w = 0$, which is a contradiction. Hence the VI (S_η, G) has a solution, which we denote $x(\eta)$, where G is the function F shifted by the vector $F(z)$; that is,

$$G(x) \equiv F(x) - F(z) \quad \forall x \in K.$$

Thus

$$(y - x(\eta))^T (F(x(\eta)) - F(z)) \geq 0, \quad \forall y \in S_\eta.$$

Clearly we have

$$\lim_{\eta \to \infty} \| x(\eta) \| = \infty.$$

Thus $x(\eta) \neq z$ for all η sufficiently large. By the strict monotonicity of F, we have for all $y \in S_\eta$ and η sufficiently large,

$$\begin{aligned} &(y - z)^T (F(x(\eta)) - F(z)) = \\ &(y - x(\eta))^T (F(x(\eta)) - F(z)) + (x(\eta) - z)^T (F(x(\eta)) - F(z)) > 0. \end{aligned}$$

Fix an $\eta > 0$ sufficiently large. Let u be an arbitrary nonzero vector in K. Since $w \in \text{int}\, K^*$, we have $w^T u > 0$. The vector

$$y \equiv z + \frac{\eta}{w^T u} u$$

belongs to S_η. Consequently, we deduce $u^T (F(x(\eta)) - F(z)) > 0$, which implies, since $F(z) \in K^*$ by the definition of z,

$$u^T F(x(\eta)) > u^T F(z) \geq 0.$$

Thus $F(x(\eta)) \in \text{int}\, K^*$; the vector $x(\eta)$ is a desired strictly feasible solution of the CP (K, F). □

2.4.9 Remark. By the same proof, it is possible to show that if K is as stated in Theorem **2.4.8**, F is continuous and monotone on K, and there exists a vector $z \in \text{FEA}(K, F)$ such that the set

$$\{ x \in \mathbb{R}^n : K \ni x - z \perp F(x) - F(z) \in K^* \}$$

is bounded, then the CP (K, F) has a nonempty, convex, and compact solution set. □

2.4.3 Polyhedrality of the solution set

In the rest of this section, we establish some further properties of the solutions to monotone CPs and AVIs. Besides being of independent interest, these properties are useful for the demonstration of the polyhedrality of the solution set of a monotone AVI. This demonstration is contained in two main results, Theorems **2.4.13** and **2.4.15**. We begin by establishing a result that sharpens Proposition **2.3.6** for a CP.

2.4.10 Proposition. Let K be a convex cone in $\mathbb{R}^n$ and $F : K \to \mathbb{R}^n$ be a given map. The following two statements are valid.

(a) If F is pseudo monotone on K, then for any two solutions x^1 and x^2 in $\text{SOL}(K, F)$, it holds that

$$x^1 \perp F(x^2). \tag{2.4.4}$$

(b) If F is affine, $\text{SOL}(K, F)$ is a convex set if and only if (2.4.4) holds for any x^1 and x^2 in $\text{SOL}(K, F)$.

Proof. Part (a) follows readily from Proposition **2.3.6**. Indeed from equation (2.3.6), we easily obtain (2.4.4) because $x^1 \perp F(x^1)$.

To show (b), it suffices to establish the "if" statement. Let x and y be any two solutions of the CP (K, F). We need to verify that the vector $z(\tau) \equiv \tau x + (1 - \tau) y$ belongs to $\text{SOL}(K, F)$ for every scalar $\tau \in (0, 1)$. Clearly $z(\tau)$ is in K by the convexity of K. Moreover, since F is affine and K^* is convex, we have $F(z(\tau)) = \tau F(x) + (1 - \tau) F(y) \in K^*$. Finally,

$$z(\tau)^T F(z(\tau)) =$$
$$\tau^2 x^T F(x) + \tau(1 - \tau) [\, x^T F(y) + y^T F(x) \,] + (1 - \tau)^2 y^T F(y) = 0$$

by (2.4.4) and the fact that x and y are both solutions of the CP (K, F). □

The property (2.4.4) is referred to as "cross complementarity (or cross orthogonality)" between any pair of solutions to a CP. The noteworthy point of part (b) in Proposition **2.4.10** is that this property characterizes the convexity of the solution set of a CP (K, F) with an affine F that is not assumed monotone. The following simple LCP (q, M) shows that it is possible for the solution set $\mathrm{SOL}(q, M)$ to be convex even if M is not positive semidefinite.

2.4.11 Example. Consider the nonmonotone LCP in 2 variables:

$$0 \leq \begin{pmatrix} x_1 \\ x_2 \end{pmatrix} \perp \begin{pmatrix} 0 \\ -1 \end{pmatrix} + \begin{bmatrix} 0 & 1 \\ 1 & 0 \end{bmatrix} \begin{pmatrix} x_1 \\ x_2 \end{pmatrix} \geq 0.$$

The solutions of this LCP are: $(x_1, 0)$ for all $x_1 \geq 1$. Thus the solution set is a convex ray but the matrix M is clearly not positive semidefinite. □

In Subsection **1.5.3**, we have defined a row sufficient matrix and recalled one of its roles in LCP theory. Interestingly, the transpose of such a matrix has an important role to play in the convexity of the solution set of LCPs. Specifically, for a given matrix $M \in \mathbb{R}^{n \times n}$, the LCP (q, M) has a convex (possibly empty) solution set for all vectors $q \in \mathbb{R}^n$ if and only if M is "column sufficient", i.e. if and only if M^T is row sufficient:

$$x \circ Mx \leq 0 \Rightarrow x \circ Mx = 0.$$

When F is an affine monotone map, we can obtain an alternative representation of the solution set of the CP (K, F).

2.4.12 Lemma. Let K be a convex cone in $\mathbb{R}^n$ and $F(x) \equiv q + Mx$, where $q \in \mathbb{R}^n$ and $M \in \mathbb{R}^{n \times n}$ is positive semidefinite. For any $y \in \mathrm{SOL}(K, q, M)$, it holds that

$$\begin{aligned} \mathrm{SOL}(K, q, M) = \{ & x \in K : q + Mx \in K^*, \\ & (M^T + M)(x - y) = 0, \ q^T(x - y) = 0 \}. \end{aligned} \tag{2.4.5}$$

Proof. By Proposition **2.3.6**, we have, for each $x \in \mathrm{SOL}(K, q, M)$,

$$(x - y)^T M (x - y) = 0.$$

By the positive semidefiniteness of M, it follows that

$$(M^T + M)(x - y) = 0,$$

which easily implies

$$x^T M x = x^T \left(\frac{M + M^T}{2} \right) y = y^T M y.$$

Consequently, we have

$$q^T x = -x^T M x = -y^T M y = q^T y.$$

This establishes that SOL(K, F) is contained in the right-hand set in the expression (2.4.5). To show the reverse inclusion, let x belong to the latter set. It suffices to show $x^T(q + Mx) = 0$. Since $q^T(x - y) = 0$ and $(M^T + M)(x - y) = 0$ by the above argument, we deduce

$$x^T(q + Mx) = y^T(q + My) = 0$$

because $y \in$ SOL(K, F). This establishes (2.4.5) as desired. □

Lemma **2.4.12** shows that for a monotone CP (K, q, M), the scalar $q^T x$ and the vector $(M + M^T)x$ are constants for all $x \in$ SOL(K, q, M). An immediate consequence of the expression (2.4.5) is that the solution set of a monotone affine CP (K, q, M), where K is a polyhedral cone, is polyhedral. We extend this conclusion to the monotone AVI (K, q, M), where K is a polyhedral set.

2.4.13 Theorem. Let K be a polyhedral set in $\mathbb{R}^n$ and $F(x) \equiv q + Mx$, where $q \in \mathbb{R}^n$ and $M \in \mathbb{R}^{n \times n}$ is positive semidefinite. The solution set of the AVI (K, q, M) is polyhedral.

Proof. Let K be given by

$$K \equiv \{ x \in \mathbb{R}^n : Cx = d, \ Ax \leq b \},$$

for some given matrices $C \in \mathbb{R}^{\ell \times n}$ and $A \in \mathbb{R}^{m \times n}$ and vectors $d \in \mathbb{R}^\ell$ and $b \in \mathbb{R}^m$. By Proposition **1.2.1**, a vector x is a solution of the AVI (K, q, M) if and only if there exists a pair of multipliers $(\mu, \lambda) \in \mathbb{R}^{\ell+m}$ such that the triple (x, μ, λ) is a solution to the augmented MLCP (1.2.3) whose defining matrix $\mathbf{Q}$, given by (1.2.7), is positive semidefinite if M is so. For ease of reference, the latter MLCP is repeated below:

$$\begin{aligned} 0 &= q + Mx + C^T \mu + A^T \lambda \\ 0 &= d - Cx \\ 0 &\leq \lambda \perp b - Ax \geq 0. \end{aligned} \tag{2.4.6}$$

By the aforementioned observation, the solution set of the latter augmented MLCP (being a special monotone affine CP) is polyhedral. Since the solution set of the AVI (K, q, M) is the image of the latter polyhedral set under the canonical projection:

$$\begin{pmatrix} x \\ \mu \\ \lambda \end{pmatrix} \in \mathbb{R}^{n+\ell+m} \mapsto x \in \mathbb{R}^n, \tag{2.4.7}$$

it follows that the solution set of the AVI (K, q, M) is also polyhedral. □

We next proceed to establish a linear inequality representation of the solution set of the monotone AVI (K, q, M). It turns out that this is not as straightforward as that for the monotone CP (K, q, M); cf. Lemma **2.4.12**. We first establish a useful property of the solutions of a monotone VI (K, q, M).

2.4.14 Lemma. Let K be a closed convex set in $\mathbb{R}^n$ and let M be an $n \times n$ positive semidefinite matrix. Assume that SOL(K, q, M) is nonempty. There exist a vector $d \in \mathbb{R}^n$ and a (nonnegative) scalar σ such that for all x in SOL(K, q, M),

$$(M + M^T)x = d \quad \text{and} \quad x^T M x = \sigma.$$

If K is polyhedral given by

$$K \equiv \{ x \in \mathbb{R}^n : Ax \leq b \},$$

and if (x, λ) is any KKT pair of the AVI (K, q, M), then

$$\sigma + q^T x + b^T \lambda = 0.$$

Proof. By Proposition **2.3.6**, we know that $(x^1 - x^2)^T M (x^1 - x^2) = 0$ for any two solutions x^1 and x^2 of the VI (K, q, M). Since M is positive semidefinite, this implies that $(M + M^T)x^1 = (M + M^T)x^2$. Hence the existence of the vector d follows. From the last equality, we may deduce $(x^1)^T M x^1 = (x^2)^T M x^2$, which establishes the existence of the scalar σ. If (x, λ) is a KKT pair, then we have

$$\begin{aligned} 0 &= q + Mx + A^T \lambda \\ 0 &\leq \lambda \perp b - Ax \geq 0, \end{aligned}$$

which easily yields $0 = q^T x + x^T M x + b^T \lambda = q^T x + \sigma + b^T \lambda$. □

In the rest of this section, we focus on the AVI (K, q, M). Recall that the gap function is the extended-value function $\theta_{\text{gap}} : \mathbb{R}^n \to \mathbb{R} \cup \{\infty\}$ defined by

$$\theta_{\text{gap}}(x) \equiv \sup_{y \in K} (q + Mx)^T (x - y), \quad \forall x \in \mathbb{R}^n.$$

cf. (1.5.15). We can write

$$\theta_{\text{gap}}(x) = x^T(q + Mx) - \omega(x),$$

where

$$\omega(x) \equiv \inf_{y \in K} y^T(q + Mx),$$

is the optimal objective value of a linear program parameterized by x. Note that $\omega(x)$ is also extended-valued in that it is possible for $\omega(x)$ to equal $-\infty$ for some $x \in \mathbb{R}^n$. Of particular interest is the effective domain of ω, that is the set

$$\Omega \equiv \{ x \in \mathbb{R}^n : \omega(x) > -\infty \}.$$

Since K is polyhedral, we can write

$$K \equiv \operatorname{conv} E + \operatorname{pos} H$$

where E and H are finite sets and $\operatorname{conv} E$ denotes the convex hull of E and $\operatorname{pos} H$ is the conical hull of H. With this representation, we have $K_\infty = \operatorname{pos} H$; moreover, it is easy to see that $\omega(x)$ is finite if and only if $(q + Mx)$ belongs to $(K_\infty)^*$. Thus,

$$\begin{aligned} \Omega &= \{ x \in \mathbb{R}^n : q + Mx \in (K_\infty)^* \} \\ &= \{ x \in \mathbb{R}^n : y^T(q + Mx) \geq 0 \text{ for all } y \in H \}; \end{aligned}$$

the second equality represents Ω in terms of a finite system of linear inequalities. Consequently, Ω is a polyhedron. By (2.3.4), it follows that Ω contains $\text{SOL}(K, q, M)$.

Next, we consider the set

$$\Omega' \equiv \{ x \in \mathbb{R}^n : \omega(x) - (\sigma + q^T x) \geq 0 \},$$

which is a subset of Ω. This set Ω' is also polyhedral and has the representation

$$\Omega' = \{ x \in \Omega : v^T(q + Mx) \geq \sigma + q^T x \text{ for all } v \in E \}. \tag{2.4.8}$$

Note that if E is empty, or equivalently, if K is a polyhedral cone, then $x \in \Omega \Leftrightarrow \omega(x) = 0$. In this case, the representation for Ω' reduces to

$$\Omega' = \{ x \in \Omega : 0 \geq \sigma + q^T x \}.$$

In what follows, we adopt the convention that if E is empty, any term involving a vector in this empty set is interpreted as zero. The convention enables us to treat this special case as a part of the general framework.

With the above preparation, we may state and prove the promised polyhedral representation of the solution set of the monotone AVI (K, q, M), which generalizes (2.4.5) for the case of a polyhedral cone K.

2.4.15 Theorem. Let K be a polyhedron in $\mathbb{R}^n$ and let M be an $n \times n$ positive semidefinite matrix. Assume that SOL(K, q, M) is nonempty. Let d and σ be the two invariants associated with the solutions of the AVI (K, q, M) (see Lemma **2.4.14**). It holds that

$$\text{SOL}(K, q, M) = \{ x \in K \cap \Omega' : (M + M^T)x = d \}.$$

Proof. We first show the inclusion

$$\text{SOL}(K, q, M) \subseteq \{ x \in K \cap \Omega' : (M + M^T)x = d \}.$$

Let $x \in \text{SOL}(K, q, M)$. If suffices to verify that $x \in \Omega'$. As noted above, we have $x \in \Omega$. Moreover, using the fact that $\sigma = x^T Mx$ and the inequality

$$(y - x)^T (q + Mx) \geq 0,$$

which holds for all y in K, it follows that $x \in \Omega'$. This establishes the desired inclusion.

To prove the reverse inclusion, let $x \in K \cap \Omega'$ satisfy $(M + M^T)x = d$. By the definition of d, we have

$$(M + M^T)x = (M + M^T)\bar{x}, \quad \forall \bar{x} \in \text{SOL}(K, q, M),$$

which implies $x^T Mx = \bar{x}^T M\bar{x} = \sigma$. Let $y \in K$ be arbitrary. Since $x \in \Omega'$, we have

$$y^T (q + Mx) \geq \sigma + q^T x = x^T (q + Mx),$$

which shows that $x \in \text{SOL}(K, q, M)$. □

The above discussion raises the question about the structure of the solution set of the AVI (K, q, M), for a matrix M that is not positive semidefinite. The proof of Theorem **2.4.13** suggests that we may examine this set via the solution set of the associated KKT system (2.4.6), which is an MLCP. The details are presented in the next section that is devoted to a comprehensive treatment of the VI (K, q, M), in which K is not necessarily polyhedral. Subsection **2.5.2** treats the AVI; in particular, Theorem **2.5.15** identifies the structure of this affine problem.

2.5 The VI (K, q, M) and Copositivity

The importance of the VI (K, q, M) stems from a "semi-linearization" of the VI (K, F) where the nonlinear function F is approximated by a first-order Taylor expansion near a given point. Specifically, if x^0 is a given vector, the linearization of F near x^0 is given by

$$x \mapsto F(x^0) + JF(x^0)(x - x^0).$$

Replacing F by the above affine function in the VI (K, F), we obtain the VI $(K, q, JF(x^0))$, where $q \equiv F(x^0) - JF(x^0)x^0$. As we see in Section **7.3**, the semi-linearization idea lies at the heart of a Newton method for solving the VI (K, F).

In this section, we study the VI (K, q, M) under certain copositivity properties of M on the recession cone K_∞. Of particular emphasis herein are the CP (K, q, M) where K is a convex cone and the AVI (K, q, M) where K is a polyhedron. We begin with a formal definition of various copositivity concepts.

2.5.1 Definition. Let C be a cone in $\mathbb{R}^n$. A matrix $M \in \mathbb{R}^{n \times n}$ is said to be

(a) *copositive* on C if

$$x^T M x \geq 0, \quad \forall x \in C;$$

(b) *copositive star* on C if M is copositive on C and

$$[\, C \ni x \perp Mx \in C^* \,] \Rightarrow -M^T x \in C^*;$$

(c) *copositive plus* on C if M is copositive on C and

$$[\, x^T M x = 0, \; x \in C \,] \Rightarrow (M + M^T)x = 0;$$

(d) *strictly copositive* on C if

$$x^T M x > 0, \quad \forall x \in C \setminus \{0\}.$$

If M is copositive plus on C, then M is copositive star on C. It is also clear that if M is a positive semidefinite matrix, then M is copositive plus on every cone in $\mathbb{R}^n$. The converse of the latter statement is false, because for example every strictly copositive matrix is copositive plus and every positive matrix is strictly copositive on the nonnegative orthant but there are obviously positive matrices that are not positive semidefinite.

As a tool for studying the VI (K, q, M), we introduce several basic point sets associated with this problem. These sets have their origin from LCP theory where they form the basis for the definitions of various matrix classes. The development in this section is largely motivated by this theory. The sets defined below are particularly relevant when we deal with the VI (K, q, M) with a fixed pair (K, M) but with q being arbitrary. Specifically, given the pair (K, M), where K is an arbitrary subset of $\mathbb{R}^n$ and M is an arbitrary $n \times n$ matrix, we define

$$\begin{aligned} \mathcal{R}(K,M) &\equiv \{ q \in \mathbb{R}^n : \mathrm{SOL}(K,q,M) \neq \emptyset \} \\ \mathcal{D}(K,M) &\equiv (K_\infty)^* - MK \\ \mathcal{K}(K,M) &\equiv \mathrm{SOL}(K_\infty, 0, M). \end{aligned}$$

The first set $\mathcal{R}(K, M)$ consists of all vectors q for which the VI (K, q, M) has a solution. We call $\mathcal{R}(K, M)$ the *VI range* of the pair (K, M). It is easy to see that

$$-MK \subseteq \mathcal{R}(K, M). \tag{2.5.1}$$

Understanding the VI range $\mathcal{R}(K, M)$ is clearly useful because it depicts the set of all constant vectors for which the VI is solvable, for a given pair (K, M). In particular, it would be of interest to have a constructive procedure to characterize those pairs (K, M) for which this range is the entire space $\mathbb{R}^n$. This is not an easy task in general.

The third set $\mathcal{K}(K, M)$ and its dual have a close connection to the VI range $\mathcal{R}(K, M)$. The set $\mathcal{K}(K, M)$ consists of all solutions of the homogeneous CP defined on the recession cone K_∞ and by the matrix M; that is, a vector v belongs to $\mathcal{K}(K, M)$ if and only if

$$K_\infty \ni v \perp Mv \in (K_\infty)^*.$$

We call $\mathcal{K}(K, M)$ the *VI kernel* of the pair (K, M). This kernel is always a closed cone, albeit not necessarily convex.

The middle set $\mathcal{D}(K, M)$ is always convex, provided that K is closed and convex. We call $\mathcal{D}(K, M)$ the *VI domain* of the pair (K, M). To understand why we use the term "domain" to describe this set, we consider two cases. The first case is when K is a cone (not necessarily polyhedral). In this case, it is easy to see that q belongs to $\mathcal{D}(K, M)$ if and only if the CP (K, q, M) is feasible; i.e., if and only if $\mathrm{FEA}(K, F) \neq \emptyset$, where $F(x) \equiv q + Mx$. The second case is when K is polyhedral and has the external representation (1.2.2). We have

$$K_\infty = \{ d \in \mathbb{R}^n : Cd = 0, \ Ad \leq 0 \},$$

which implies

$$(K_\infty)^* = C^T \mathbb{R}^\ell - A^T \mathbb{R}^m_+.$$

Thus,

$$\mathcal{D}(K, M) = C^T \mathbb{R}^\ell - A^T \mathbb{R}^m_+ - MK; \tag{2.5.2}$$

therefore, $\mathcal{D}(K, M)$ is a polyhedron, provided that K is so. The KKT system of the AVI (K, q, M) is given by (1.2.3):

$$0 = q + Mx + C^T\mu + A^T\lambda$$

$$0 = d - Cx$$

$$0 \leq \lambda \perp b - Ax \geq 0.$$

From (2.5.2), we see that $\mathcal{D}(K, M)$ coincides with the set of all vectors q for which the above KKT system is feasible (but not necessarily solvable) as an MLCP. In summary, the consideration of these two cases inspires naming the set $\mathcal{D}(K, M)$ the VI domain of the pair (K, M).

The next lemma shows that if M is copositive star on K_∞, then the VI kernel $\mathcal{K}(K, M)$ is equal to $K_\infty \cap (-MK_\infty)^*$, and hence, is convex. Rather than using K_∞, we present this lemma assuming that K is a convex cone (thus $K_\infty = K$).

2.5.2 Lemma. Let K be a convex cone in $\mathbb{R}^n$ and let M be an $n \times n$ matrix copositive on K. It holds that

$$\mathcal{K}(K, M) \supseteq \{ v \in K : -M^T v \in K^* \} = K \cap (-MK)^*. \tag{2.5.3}$$

Moreover, equality holds throughout (2.5.3) if and only if M is copositive star on K.

Proof. The equality in (2.5.3) is obvious. If M is copositive on K, then for each vector $v \in K \cap (-MK)^*$, we must have $v^T Mv = 0$; therefore, such a vector v is a global minimizer of the problem:

$$\begin{array}{ll} \text{minimize} & x^T Mx \\ \text{subject to} & x \in K. \end{array}$$

By the variational principle for this problem and the fact that K is a convex cone, we deduce $(M + M^T)v \in K^*$, which implies

$$Mv \in -M^T v + K^* \subseteq K^* + K^* = K^*.$$

Hence, v belongs to the CP kernel $\mathcal{K}(K, M)$, thus establishing (2.5.3). It is clear that M is copositive star on K if and only if M is copositive and

$$\mathcal{K}(K, M) \subseteq \{ v \in K : -M^T v \in K^* \}.$$

Thus the last conclusion of the lemma follows immediately. □

The VI kernel $\mathcal{K}(K, M)$ plays an important role in the boundedness of solutions of the VIs (K, q, M) for all q. This role is elucidated in two results, Propositions **2.5.3** and **2.5.6**.

2.5.3 Proposition. Let K be a closed convex set in $\mathbb{R}^n$ and let M be an $n \times n$ matrix. It holds that

$$\bigcup_{q \in \mathbb{R}^n} (\,\mathrm{SOL}(K, q, M)\,)_\infty \subseteq \mathcal{K}(K, M). \tag{2.5.4}$$

Moreover, if $\mathcal{K}(K, M) = \{0\}$, then for all $q \in \mathbb{R}^n$, $\mathrm{SOL}(K, q, M)$ is bounded (possibly empty).

Proof. Let $q \in \mathbb{R}^n$ be arbitrary. Suppose that d is a recession direction of the set $\mathrm{SOL}(K, q, M)$. Thus there exists a vector $x \in \mathbb{R}^n$ such that

$$x + \tau\, d \in \mathrm{SOL}(K, q, M) \quad \forall\, \tau \geq 0.$$

Consequently we have, for all $\tau \geq 0$, $x + \tau\, d \in K$ and

$$(\,y - x - \tau\, d\,)^T (\,q + Mx + \tau\, Md\,) \geq 0, \quad \forall\, y \in K.$$

Hence $d \in K_\infty$ and $d^T Md \leq 0$. Letting $y \equiv x + 2\tau\, d$ and $\tau > 0$ be sufficiently large, we deduce $d^T Md \geq 0$ and therefore $d^T Md = 0$. For an arbitrary vector $d' \in K_\infty$, we have $y \equiv x + \tau\, d + d' \in K$; thus

$$(\,d'\,)^T (\,q + Mx + \tau\, Md\,) \geq 0.$$

Since this holds for all $\tau \geq 0$, we must have

$$(\,d'\,)^T Md \geq 0.$$

Since this holds for all $d' \in K_\infty$, we have therefore shown that d satisfies

$$K_\infty \ni d \perp Md \in (\,K_\infty\,)^*;$$

thus, $d \in \mathcal{K}(K, M)$. This establishes the inclusion (2.5.4). Suppose that $\mathcal{K}(K, M) = \{0\}$ and, for the sake of contradiction, that for some $q \in \mathbb{R}^n$, $\mathrm{SOL}(K, q, M)$ contains a sequence $\{x^k\}$ with $\|x^k\|$ tending to ∞. It is not difficult to verify that any limit point of the normalized sequence $\{x^k/\|x^k\|\}$ is a nonzero element in the kernel $\mathcal{K}(K, M)$. □

Clearly, if M is strictly copositive on K_∞, then $\mathcal{K}(K, M) = \{0\}$. In general, we say that the pair (K, M) is an R_0 *pair* if $\mathcal{K}(K, M) = \{0\}$. If (K, M) is an R_0 pair then, since $\mathrm{SOL}(K, q, M)$ is bounded, it follows that if $\mathrm{SOL}(K, q, M)$ is nonempty, then for every bounded open set $\mathcal{U}$ containing $\mathrm{SOL}(K, q, M)$, $\deg(\mathbf{F}^{\mathrm{nat}}_K, \mathcal{U})$ is well defined and independent of $\mathcal{U}$, where $\mathbf{F}^{\mathrm{nat}}_K$ is the natural map associated with the VI (K, q, M). A similar statement can be made for the normal map $\mathbf{F}^{\mathrm{nor}}_K$ associated with the same problem. In particular, when $\mathrm{SOL}(K,q,M)$ is a singleton, say $\{x^*\}$, it follows that $\mathrm{ind}(\mathbf{F}_K^{\mathrm{nat}}, x^*)$ and $\mathrm{ind}(\mathbf{F}_K^{\mathrm{nor}}, z^*)$, where $z^* \equiv x^* - (q + Mx^*)$, are both well defined.

Elements of $\mathrm{SOL}(K,q,M)_\infty$ are called *solution rays* of the VI (K,q,M). The following proposition provides a necessary and sufficient condition for a vector to be a solution ray of the VI (K,q,M).

2.5.4 Proposition. A vector $d \in \mathbb{R}^n$ is a solution ray of the VI (K,q,M) if and only if there exists $x \in \mathrm{SOL}(K,q,M)$ such that

(a) $d \in \mathcal{K}(K,M)$,

(b) $d^T(q + Mx) = 0$, and

(c) $x^T M d \leq y^T M d$ for all $y \in K$.

Proof. Suppose that d is a solution ray of the VI (K,q,M). As in the proof of Proposition **2.5.3**, we deduce that $d \in \mathcal{K}(K,M)$; moreover, there exists a solution $x \in \mathrm{SOL}(K,q,M)$ such that, for all $\tau \geq 0$, $x + \tau d \in K$ and

$$(y - x - \tau d)^T(q + Mx + \tau Md) \geq 0, \quad \forall y \in K. \tag{2.5.5}$$

Thus it suffices to show that (b) and (c) hold. Since $d \perp Md$, letting $y = x$ and $\tau = 1$ yields $d^T(q + Mx) \leq 0$. Moreover, letting $y = x + 2d$ and $\tau = 1$ yields $d^T(q + Mx) \geq 0$. Hence $d \perp q + Mx$ and (b) holds. Since

$$d^T(q + Mx) = d^T M d = 0,$$

(2.5.5) becomes

$$(y - x)^T(q + Mx + \tau Md) \geq 0, \quad \forall \tau \geq 0 \text{ and } y \in K.$$

Thus $(y - x)^T M d \geq 0$ for all $y \in K$; hence (c) holds.

Conversely, suppose $x \in \mathrm{SOL}(K,q,M)$ exists such that (a), (b) and (c) hold. It is easy to see that for all $\tau \geq 0$, $x + \tau d \in K$ and (2.5.5) holds. Thus d is a solution ray of the VI (K,q,M). □

Based on the above proposition, we can identify a large class of VIs (K,q,M) that have no solution rays. First a definition. We say that the pair (K,M) has the *sharp property* if

$$\left.\begin{array}{r} d \in \mathcal{K}(K,M) \\ x \in \operatorname{argmin}_{z \in K} z^T M d \end{array}\right\} \Rightarrow d^T M x \geq 0. \tag{2.5.6}$$

2.5.5 Proposition. Let K be a closed convex set in $\mathbb{R}^n$ and M be an $n \times n$ matrix. The sharp property holds for the pair (K, M) under any one of the following five conditions:

(a) $\mathcal{K}(K, M) = \{0\}$;

(b) M is positive semidefinite and for every $d \in \mathcal{K}(K, M)$, there exists $y \in K$ such that $d^T M y \geq 0$ (in turn, the latter condition holds trivially if K contains the origin);

(c) K contains the origin and

$$d \in \mathcal{K}(K, M) \Rightarrow (M + M^T)d \in (\operatorname{pos} K)^*;$$

(d) K contains the origin and M is copositive on $\operatorname{pos} K$;

(e) K is a cone and M is symmetric.

If (K, M) has the sharp property, then the VI (K, q, M) has no solution rays for all vectors $q \in \operatorname{int} \mathcal{K}(K, M)^*$.

Proof. The last statement of the proposition is easy because q belongs to $\operatorname{int} \mathcal{K}(K, M)^*$ if and only if $q^T d > 0$ for all $d \in \mathcal{K}(K, M) \setminus \{0\}$, by Proposition **2.4.1**.

We now show that the pair (K, M) has the sharp property under any one of the conditions (a)–(e). This is clear for (a). Let (d, x) be a pair of vectors satisfying the left-hand conditions in the sharp condition (2.5.6). If (b) holds, then we must have $(M + M^T)d = 0$. Hence by the definition of x and the particular vector y associated with d, we obtain

$$d^T M x = -x^T M d \geq -y^T M d = d^T M y \geq 0.$$

Hence the right-hand side in (2.5.6) holds.

If (c) holds, then since K contains the origin, we must have $x^T M d \leq 0$. Since $x \in K \subseteq \operatorname{pos} K$, it follows from the assumption that

$$0 \leq x^T (M + M^T) d \leq d^T M x,$$

which is the desired right-hand side in (2.5.6).

If (d) holds, since $d^T M d = 0$ and M is copositive on $\operatorname{pos} K$, which contains $K_\infty \ni d$, it follows that $(M + M^T)d \in (\operatorname{pos} K)^*$. The above proof of part (c) applies and (2.5.6) holds. (Actually, part (d) is a special case of part (c).)

Finally, part (e) is also a special case of part (c) because with K being a closed convex cone, we have $K = K_\infty = \operatorname{pos} K$. □

2.5.1 The CP (K, q, M)

If K is a cone, we call $\mathcal{R}(K, M)$, $\mathcal{D}(K, M)$, and $\mathcal{K}(K, M)$ the CP range, the CP domain, and the CP kernel of the pair (K, M), respectively. This terminology is consistent with our usage of the prefix CP throughout the book. Note that the CP range, CP domain, and CP kernel are all cones, although only $\mathcal{D}(K, M)$ is convex in general. If K is a closed convex cone, Proposition **2.5.3** can be sharpened.

2.5.6 Proposition. Let K be a closed convex cone in $\mathbb{R}^n$ and M be an $n \times n$ matrix. It holds that

$$\bigcup_{q \in \mathbb{R}^n} (\,\mathrm{SOL}(K, q, M)\,)_\infty = \mathcal{K}(K, M). \tag{2.5.7}$$

Moreover, $\mathcal{K}(K, M) = \{0\}$ if and only if there exists a constant $c > 0$ such that for all $q \in \mathbb{R}^n$,

$$\|\, x \,\| \leq c \|\, q \,\|, \quad \forall\, x \in \mathrm{SOL}(K, q, M).$$

Proof. With K being a convex cone, we have $K = K_\infty$ so that

$$\mathcal{K}(K, M) = \mathrm{SOL}(K, 0, M) \subseteq \mathrm{SOL}(K, 0, M)_\infty.$$

Consequently (2.5.7) must hold by (2.5.4). To prove the second assertion of the proposition, suppose first that $\mathcal{K}(K, M) = \{0\}$ but no such constant c exists. There exist sequences $\{q^\nu\}$ and $\{x^\nu\}$ such that $q^\nu \neq 0$, x^ν is nonzero and belongs to $\mathrm{SOL}(K, q^\nu, M)$ for each ν, and

$$\lim_{\nu \to \infty} \frac{\|\, x^\nu \,\|}{\|\, q^\nu \,\|} = \infty.$$

The normalized sequence $\{x^\nu / \|x^\nu\|\}$ must have at least one nonzero accumulation point; it is not hard to verify that such a point must be a solution of the homogeneous CP $(K, 0, M)$, thus belongs to $\mathcal{K}(K, M)$. This is a contradiction.

Conversely, if the constant $c > 0$ with the desired property exists, then it follows easily that $\mathcal{K}(K, M) = \mathrm{SOL}(K, 0, M) = \{0\}$. □

The second part of the above proposition says that for a closed convex cone K, (K, M) is an R_0 pair if and only if the solutions of the CP (K, q, M) are uniformly bounded for all q belonging to a bounded set. We have a special terminology when K is the nonnegative orthant. Specifically, a matrix $M \in \mathbb{R}^{n \times n}$ is said to be an *R_0 matrix* if the homogeneous LCP $(0, M)$ has the zero vector as the unique solution. A proper subclass of the class of R_0 matrices consists of the *nondegenerate matrices*, which are real square matrices whose principal minors are nonzero.

If M is symmetric and copositive on a convex cone K, the R_0 property of the pair (K, M) is equivalent to the strict copositivity of M on K. We state and prove this observation in the result below, which is useful in the context of second-order conditions of an NLP.

2.5.7 Proposition. Let M be a symmetric, copositive matrix on a convex cone K. The pair (K, M) has the R_0 property if and only if M is strictly copositive on K.

Proof. Clearly, it suffices to show that if (K, M) is an R_0 pair, then M is strictly copositive on K. Assume not. Then there exists a nonzero vector $v \in K$ such that $v^T M v = 0$. By the proof of Lemma **2.5.2** and the symmetry of M, it follows that $Mv \in K^*$; thus $v \in \mathcal{K}(K, M)$. This contradicts the R_0 property of the pair (K, M). □

2.5.8 Remark. The symmetry of M is essential for Proposition **2.5.7** to hold. A counterexample without the symmetry assumption is provided by the matrix

$$M \equiv \begin{bmatrix} 1 & -1 \\ 1 & 0 \end{bmatrix},$$

which is copositive, but not strictly copositive, on the nonnegative orthant $\mathbb{R}^2_+$. It is easy to verify that the homogeneous LCP $(0, M)$ has a unique solution; thus M is an R_0 matrix. □

Neither Proposition **2.5.3** nor Proposition **2.5.6** asserts the existence of a solution to the CP. We next address this existence issue. First we give a definition. If C is a closed convex cone in $\mathbb{R}^n$ and M is an $n \times n$ matrix and if (C, M) is an R_0 pair, then $\text{ind}(\mathbf{M}_C^{\text{nat}}, 0)$ and $\text{ind}(\mathbf{M}_C^{\text{nor}}, 0)$ are both well defined. We call the former the *natural index* of the pair (C, M) and the latter the *normal index* of the pair (C, M) and denote these indices by $\text{ind}\,\mathbf{M}_C^{\text{nat}}$ and $\text{ind}\,\mathbf{M}_C^{\text{nor}}$, respectively. The fundamental role of these special indices is described in the next result.

2.5.9 Proposition. Let K be a closed convex cone in $\mathbb{R}^n$ and M be an $n \times n$ matrix. Suppose that (K, M) is an R_0 pair. If either the natural index or the normal index of the pair (K, M) is nonzero, then the CP (K, q, M) has a nonempty and bounded solution set for all vectors $q \in \mathbb{R}^n$.

Proof. We assume that the natural index is nonzero. The proof is the same if the normal index is nonzero. Let $q \in \mathbb{R}^n$ be arbitrary. Only the nonemptiness of SOL(K, q, M) requires a proof. Let $F(x) \equiv q + Mx$, since

$$\| \mathbf{M}_K^{\text{nat}}(x) - \mathbf{F}_K^{\text{nat}}(x) \| \leq \| q \|,$$

it follows from the nearness property of the degree that for any open ball $\mathbb{B}$ centered at the origin, $\deg(\mathbf{F}_K^{\text{nat}}, \mathbb{B})$ is nonzero for all q with $\|q\|$ sufficiently small. Therefore the CP (K, q, M) has a solution in $\mathbb{B}$ for all such vectors q. Since K is a cone, by a simple scaling argument, it follows that the CP (K, q, M) has a solution for all vectors $q \in \mathbb{R}^n$. □

In the next result, we provide two sufficient conditions for the natural index and the normal index of an R_0 pair (K, M) to be nonzero. These conditions are: (A) M is copositive on K, and (B) there exists a particular vector q^* belonging to $\operatorname{int} K^*$ such that SOL(K, q^*, M) is a singleton; we show further that under either one of these two assumptions, the natural map associated with the CP (K, q, M) has a well-defined degree on every bounded open set containing SOL(K, q, M) and this degree is equal to one. Before presenting this result, we remark that if K is a cone, then the zero vector is always an element of SOL(K, q, M) for all vectors $q \in K^*$; if, in addition, M is strictly copositive on K, then SOL$(K, q, M) = \{0\}$ for all $q \in K^*$

2.5.10 Theorem. Let K be a closed convex cone in $\mathbb{R}^n$ and M be an $n \times n$ matrix. Suppose that (K, M) is an R_0 pair. If either one of the following two assumptions holds:

(A) M is copositive on K, or

(B) there exists a vector $q^* \in \operatorname{int} K^*$ such that SOL$(K, q^*, M) = \{0\}$,

then the following two statements hold for all vectors $q \in \mathbb{R}^n$:

(a) the set SOL(K, q, M) is nonempty and bounded; and

(b) for every bounded open set $\mathcal{U}$ containing SOL(K, q, M), $\deg(\mathbf{F}_K^{\text{nat}}, \mathcal{U})$ is well defined and equal to one, where $\mathbf{F}_K^{\text{nat}}$ is the natural map associated with the CP (K, q, M).

Moreover, under either (A) or (B), both the natural index and the normal index of the pair (K, M) are equal to 1.

Proof. Let q be an arbitrary vector. The boundedness of SOL(K, q, M), if nonempty, is an immediate consequence of the R_0 property of the pair (K, M). We claim that there exists a bounded open set $\mathcal{U}$ containing SOL(K, q, M) such that $\deg(\mathbf{F}_K^{\text{nat}}, \mathcal{U})$ is well defined and equal to one. Once this claim is established, it then follows that $\mathbf{F}_K^{\rm nat}$ has a zero in $\mathcal{U}$, thus SOL(K, q, M) is nonempty. Moreover, by the excision property of the degree, it follows that $\mathcal{U}$ can be replaced by any bounded open set containing SOL(K, q, M).

First assume condition (B). Consider the homotopy:

$$H(x,t) \equiv x - \Pi_K(x - (tq + (1-t)q^* + Mx)), \quad (x,t) \in \mathbb{R}^n \times [0,1].$$

Clearly, $H(\cdot, t)$ is the natural map associated with the CP $(K, q(t), M)$, where $q(t) \equiv tq + (1-t)q^*$. Since the line segment $\{q(t) : t \in [0,1]\}$ is bounded and (K, M) is an R_0 pair, Proposition **2.5.6** implies that

$$\bigcup_{t \in [0,1]} \mathrm{SOL}(K, q(t), M) = \bigcup_{t \in [0,1]} H(\cdot, t)^{-1}(0) \tag{2.5.8}$$

is bounded. Let $\mathcal{U}$ be a bounded open set containing this union of solutions. By the homotopy invariance property of the degree, we have

$$\deg(\mathbf{F}_K^{\rm nat}, \mathcal{U}) = \deg(H(\cdot,1), \mathcal{U}) = \deg(H(\cdot,0), \mathcal{U})$$

and this common degree is independent of the set $\mathcal{U}$ as long as it contains the union (2.5.8). To complete the proof, we show that $\deg(H(\cdot,0), \mathcal{U})$ is equal to unity. Indeed, $H(\cdot, 0)$ is equal to the natural map associated with the CP (K, q^*, M). By assumption, the latter CP has a unique solution, namely, the zero vector. Since q^* belongs to $\operatorname{int} K^*$, there exists a neighborhood $\mathcal{Q}$ of q^* such that q' remains in $\operatorname{int} K^*$ for all $q' \in \mathcal{Q}$. Thus, there exists a neighborhood $\mathcal{N}$ of the origin such that for all $x \in \mathcal{N}$, $q^* - x + Mx$ belongs to $\mathcal{Q} \subseteq K^*$; or equivalently, $x - q^* - Mx \in -K^*$ for all $x \in \mathcal{N}$. It follows that

$$H(x,0) = x - \Pi_K(x - q^* - Mx) = x, \quad \forall\, x \in \mathcal{N},$$

because the projection onto K of every vector in $-K^*$ is equal to the origin. That is, the map $H(\cdot, 0)$ is equal to the identity map in the neighborhood $\mathcal{N}$. Consequently, by the excision property of the degree, we have

$$1 = \deg(H(\cdot,0), \mathcal{N}) = \deg(H(\cdot,0), \mathcal{U}),$$

as desired.

Next assume condition (A). Consider the homotopy

$$H(x,t) \equiv x - \Pi_K(tx - (q + tMx)), \quad (x,t) \in \mathbb{R}^n \times [0,1].$$

It is easy to see that for all $t \in [0,1]$, $H(\cdot, t)$ is the natural map associated with the CP $(K, q, tM + (1-t)\mathbf{I})$. Moreover, $H(x, 0)$ is a translation of the identity map. We claim that

$$\bigcup_{t\in[0,1]} \mathrm{SOL}(K,q,tM+(1-t)\mathbf{I}) = \bigcup_{t\in[0,1]} H(\cdot,t)^{-1}(0) \tag{2.5.9}$$

is bounded. Assume by contradiction that there exist a sequence $\{x^k\}$ of vectors and a sequence $\{t_k\}$ of scalars in $[0,1]$ such that

$$\lim_{k\to\infty} \| x^k \| = \infty, \qquad \lim_{k\to\infty} \frac{x^k}{\| x^k \|} = d$$

for some nonzero vector d, and that for every k, x^k is a solution of the CP $(K,q,t_kM+(1-t_k)\mathbf{I})$. Without loss of generality, we may assume that $\{t_k\}$ converges to a scalar t_∞ in $[0,1]$. It is easy to show that the vector d must be a solution of the homogeneous CP $(K,0,t_\infty M+(1-t_\infty)\mathbf{I})$. By the copositivity of M and the R_0 property of the pair (K,M), the latter CP has the origin as the unique solution. This contradiction establishes the desired boundedness of the union in (2.5.9). Since $H(\cdot,0)$ is a translation of the identity map, it follows easily that $\deg(H(\cdot,0),\mathcal{U})$, and thus $\deg(\mathbf{F}_K^{\mathrm{nat}},\mathcal{U})$, is equal to one for any bounded open set $\mathcal{U}$ containing (2.5.9).

Finally, the natural index of the pair (K,M) is equal to one by taking q to be the zero vector. By using the normal map in defining the homotopies in the above proof, we can similarly show that the normal index of the pair (K,M) is also equal to one. □

Theorem **2.5.10** yields the existence and boundedness of solutions to the CP (K,q,M) for all vectors q if M is copositive on K and (K,M) is an R_0 pair. The next result generalizes this conclusion by establishing that the CP (K,q,M) with a copositive M has bounded solutions if q belongs to the interior of the dual cone of the CP kernel $\mathcal{K}(K,M)$. We state the latter assertion in terms of a set-theoretic inclusion (2.5.10).

2.5.11 Proposition. Let K be a closed convex cone in $\mathbb{R}^n$; let M be an $n\times n$ matrix copositive on K. It holds that

$$\operatorname{int}\mathcal{K}(K,M)^* \subseteq \mathcal{R}(K,M). \tag{2.5.10}$$

Moreover, for all $q\in\operatorname{int}\mathcal{K}(K,M)^*$, the set $\mathrm{SOL}(K,q,M)$ is compact.

Proof. Let $q\in\operatorname{int}\mathcal{K}(K,M)^*$ be given. By the homotopy argument employed several times already, in particular, by considering the homotopy

$$H(x,t) \equiv x - t\Pi_K(x-(q+Mx)), \quad (x,t)\in\mathbb{R}^n\times[0,1],$$

it suffices to show that

$$\bigcup_{t \in [0,1]} H(\cdot, t)^{-1}(0)$$

is bounded, which will then imply that the VI (K, q, M) has a solution and establish the inclusion (2.5.10). Assume for the sake of contradiction that there exists a sequence $\{x^k\}$ such that, for some nonzero vector $d \in K$,

$$\lim_{k \to \infty} \| x^k \| = \infty, \quad \lim_{k \to \infty} \frac{x^k}{\| x^k \|} = d, \quad \text{and} \quad H(x^k, t_k) = 0 \quad \forall k.$$

Without loss of generality, we may assume that each t_k is positive. Let $\tau_k \equiv (1 - t_k)/t_k$. By the definition of the homotopy map H and the fact that K is a cone, we deduce, for each k,

$$K \ni x^k \perp \tau_k x^k + q + M x^k \in K^*.$$

Since

$$0 = (x^k)^T (\tau_k x^k + q + M x^k),$$

dividing by $\|x^k\|^2$ and passing to the limit $k \to \infty$, we deduce, by the copositivity of M on K, that $d^T M d = 0$ and

$$\lim_{k \to \infty} \tau_k = 0.$$

Moreover, we have

$$0 \geq q^T x^k,$$

which implies $q^T d \leq 0$. Since K^* is a cone, we have, for all k,

$$\tau_k \frac{x^k}{\| x^k \|} + \frac{q}{\| x^k \|} + M \frac{x^k}{\| x^k \|} \in K^*;$$

letting $k \to \infty$, we deduce that $Md \in K^*$. Summarizing, we have shown that the vector d satisfies the following conditions:

$$K \ni d \perp Md \in K^* \quad \text{and} \quad q^T d \leq 0.$$

But this contradicts the assumption that $q \in \operatorname{int} \mathcal{K}(K, M)^*$. □

Proposition **2.5.11** has an important consequence, which is not difficult to prove.

2.5.12 Corollary. Let K be a closed convex cone in $\mathbb{R}^n$ and M be an $n \times n$ matrix copositive on K. If $\operatorname{int} \mathcal{K}(K, M)^*$ is nonempty and $\mathcal{R}(K, M)$ is closed, then

$$\mathcal{K}(K, M)^* \subseteq \mathcal{R}(K, M).$$

Proof. By taking closures on both sides in the expression (2.5.10), we immediately obtain the desired conclusion. □

The two assumptions in the above corollary deserves further discussion. The following lemma gives several sufficient conditions for $\operatorname{int} \mathcal{K}(K, M)^*$ to be nonempty.

2.5.13 Lemma. Let K be a closed convex set in $\mathbb{R}^n$ and M be an $n \times n$ matrix. Consider the following statements:

(a) K has an extreme point;

(b) K contains no line;

(c) K_∞ is pointed;

(d) $\operatorname{int}(K_\infty)^*$ is nonempty;

(e) $\operatorname{int} \mathcal{K}(K, M)^*$ is nonempty.

It holds that (a) $\Rightarrow$ (b) $\Leftrightarrow$ (c) $\Rightarrow$ (d) $\Rightarrow$ (e).

Proof. Since $\mathcal{K}(K, M)$ is a subset of K_∞, it follows by duality that $\mathcal{K}(K, M)^*$ is a superset of $(K_\infty)^*$. Thus, if $(K_\infty)^*$ has a nonempty interior, then so does $\mathcal{K}(K, M)^*$. This establishes (d) $\Rightarrow$ (e). In turn, if K_∞ is pointed, then $\operatorname{int} K^*$, and thus $\operatorname{int} \mathcal{K}(K, M)^*$, is nonempty, by part (b) of Proposition **2.4.3**. This establishes (c) $\Rightarrow$ (d). If K_∞ is not pointed, there exists a nonzero vector v such that $x \pm \tau v$ belongs to K for all vectors $x \in K$ and all scalars $\tau \geq 0$. Thus through every point in K passes an entire line. Consequently, either (a) or (b) $\Rightarrow$ (c). Conversely, if K contains a line, this line must be of the form $\{x \pm \tau v : \tau \geq 0\}$ for some $x \in K$ and some nonzero vector $v \in \mathbb{R}^n$. The vector v shows that K_∞ is not pointed. □

Unlike the kernel, the VI range $\mathcal{R}(K, M)$ is not always closed. In essence, this range is closed if and only if for every convergent sequence of vectors $\{q^k\}$ with limit q^∞ and for which the individual VI (K, q^k, M) is solvable for each k, the limiting VI (K, q^∞, M) is also solvable. Subsequently, we show that $\mathcal{R}(K, q, M)$ is closed if K is a polyhedron; see Theorem **2.5.15**. We first give an example to show that $\mathcal{R}(K, M)$ is not necessarily closed if K is not a polyhedron. For such an example, we take K to be the cone of symmetric positive definite matrices of order 2.

2.5.14 Example. We consider the LCP in SPSD matrices:

$$\mathcal{M}^n_+ \ni A \perp Q + \mathcal{L}(A) \in \mathcal{M}^n_+.$$

Let $n = 2$,

$$Q \equiv \begin{bmatrix} 0 & 1 \\ 1 & 0 \end{bmatrix} \notin \mathcal{M}^2_+$$

and $\mathcal{L} : \mathcal{M}^2 \to \mathcal{M}^2$ be the linear operator defined by

$$\mathcal{L}\begin{bmatrix} a & b \\ b & d \end{bmatrix} \equiv \begin{bmatrix} a & b \\ b & 0 \end{bmatrix}, \quad \forall \begin{bmatrix} a & b \\ b & d \end{bmatrix} \in \mathcal{M}^2.$$

It is easy to see that $\mathcal{L}$ is a monotone operator; that is,

$$A \bullet \mathcal{L}(A) = a^2 + b^2 \geq 0, \quad \forall\, A = \begin{bmatrix} a & b \\ b & d \end{bmatrix} \in \mathcal{M}^2.$$

The CP $(\mathcal{M}^2_+, Q, \mathcal{L})$ has no solution because

$$A \perp Q + \mathcal{L}(A) \;\Rightarrow\; A = \begin{bmatrix} 0 & 0 \\ 0 & d \end{bmatrix} \quad \text{for some scalar } d \geq 0;$$

thus,

$$Q + \mathcal{L}(A) \equiv \begin{bmatrix} 0 & 1 \\ 1 & 0 \end{bmatrix} \notin \mathcal{M}^2_+.$$

Consequently, $\mathrm{SOL}(\mathcal{M}^2_+, Q, \mathcal{L}) = \emptyset$. Nevertheless we have

$$Q^k \equiv \begin{bmatrix} -1/k & 1 \\ 1 & 0 \end{bmatrix} \to Q$$

and for all k,

$$\begin{bmatrix} 1/k & -1 \\ -1 & k \end{bmatrix} \in \mathrm{SOL}(\mathcal{M}^2_+, Q, \mathcal{L}).$$

Consequently, $\mathcal{R}(\mathcal{M}^2_+, \mathcal{L})$ is not closed. □

2.5.2 The AVI (K, q, M)

The next result shows that if K is a polyhedron, then the AVI range $\mathcal{R}(K, M)$ is closed. This result also establishes the long awaited property of the solution set of the AVI (K, q, M).

2.5.15 Theorem. Let K be a polyhedron in $\mathbb{R}^n$ and M be an $n \times n$ matrix. The following two statements are valid.

(a) For every vector $q \in \mathbb{R}^n$, the set $\mathrm{SOL}(K, q, M)$ is the union of finitely many polyhedra in $\mathbb{R}^n$.

(b) The AVI range $\mathcal{R}(K, M)$ is the union of finitely many polyhedra in $\mathbb{R}^n$; thus it is closed.

Proof. Adopting the notation in Theorem **2.4.13**, we know that the set $\text{SOL}(K, q, M)$ is the image of the solution set of the mixed linear complementarity system (2.4.6) under the canonical projection (2.4.7) of $\mathbb{R}^{n+\ell+m}$ onto $\mathbb{R}^n$. Thus if we can establish that the latter solution set is the union of finitely many polyhedra (in $\mathbb{R}^{n+\ell+m}$), then the same will hold for $\text{SOL}(K, q, M)$. Indeed, the solution set of the MLCP (2.4.6) is equal to

$$\bigcup_{\alpha} \mathcal{P}_\alpha,$$

where the union ranges over all subsets α of $\{1, \ldots, m\}$. With $\bar{\alpha}$ denoting the complement of α in $\{1, \ldots, m\}$, we can write

$$\begin{aligned} \mathcal{P}_\alpha \equiv \{ (x, \mu, \lambda) \in \mathbb{R}^{n+\ell+m} \quad & 0 = q + Mx + C^T\mu + (A_{\alpha\cdot})^T\lambda_\alpha \\ & 0 = d - Cx \\ & 0 = (b - Ax)_\alpha, \quad \lambda_\alpha \geq 0 \\ & 0 \leq (b - Ax)_{\bar{\alpha}}, \quad \lambda_{\bar{\alpha}} = 0 \}. \end{aligned}$$

Since each $\mathcal{P}_\alpha$ is clearly polyhedral, the first claim (a) is established. To establish the second claim (b), let

$$K_\alpha \equiv \{ x \in \mathbb{R}^n : Cx = d, \ (Ax = b)_\alpha, \ (Ax \leq b)_{\bar{\alpha}} \}$$

be the α-face of K. It is then easy to see that

$$\mathcal{R}(K, M) = \bigcup_{\alpha} Q_\alpha$$

where

$$Q_\alpha \equiv -MK_\alpha - C\mathbb{R}^\ell - (A_{\alpha\cdot})^T\mathbb{R}_+^{|\alpha|}.$$

Since each Q_α is a polyhedron, (b) follows. □

We call each of the polyhedra whose union is $\text{SOL}(K, q, M)$ a *piece* of this solution set. In essence, Theorem **2.5.15** states that the solution set of an AVI is *piecewise polyhedral.* An immediate consequence of this piecewise polyhedrality is that the solution set of an AVI is bounded if and only if there exists no nonzero solution ray of the problem.

2.5.16 Corollary. The solution set of an AVI (K, q, M) is bounded if and only if there exists no nonzero solution ray.

Proof. It suffices to show that if there exists no nonzero solution ray, then SOL(K, q, M) is bounded. Indeed, if SOL(K, q, M) is unbounded, then one of its polyhedral pieces is unbounded. Being an unbounded polyhedral set, this piece must contain a ray emanating from a certain vector in the piece. The direction of the ray furnishes a nonzero solution ray of the AVI (K, q, M). □

The polyhedrality of the solution set of an AVI can be characterized in several equivalent ways. These characterizations are presented in the following result, which refines several results established in the previous subsections: Proposition **2.3.6** that pertains to a monotone VI, Proposition **2.4.10** that pertains to a monotone CP, and Theorem **2.4.13** that pertains to a monotone AVI.

2.5.17 Theorem. The following four statements are equivalent for the AVI (K, q, M).

(a) The solution set SOL(K, q, M) is convex.

(b) For any two solutions x^1 and x^2 in SOL(K, q, M),

$$(x^1)^T(q + Mx^1) = (x^2)^T(q + Mx^1).$$

(c) For any two solutions x^1 and x^2 in SOL(K, q, M),

$$(x^1 - x^2)^T M (x^1 - x^2) = 0.$$

(d) The solution set SOL(K, q, M) is polyhedral.

Moreover if any one of the above conditions hold and if K is given by (1.2.2), then SOL(K, q, M) is equal to

$$\{ x \in K : -(q + Mx) \in C^T \mathbb{R}^\ell + \text{pos}(A_{\beta\cdot})^T, \ (Ax = b)_\beta \} \qquad (2.5.11)$$

where

$$\beta \equiv \{ i : (Ax = b)_i \quad \forall x \in \text{SOL}(K, q, M) \}.$$

Proof. (a) ⇒ (b). The proof is similar to that of Proposition **2.4.10**. Let x^1 and x^2 be two solutions of the AVI (K, q, M). Write $w^i \equiv q + Mx^i$ for $i = 1, 2$. Since SOL(K, q, M) is convex, we have for every $\tau \in [0, 1]$,

$$\begin{aligned} 0 &\leq [x^1 - (\tau x^1 + (1 - \tau) x^2)]^T [q + M(\tau x^1 + (1 - \tau) x^2)] \\ &= (1 - \tau)(x^1 - x^2)^T [\tau w^1 + (1 - \tau) w^2], \end{aligned}$$

which yields

$$(x^1 - x^2)^T [\tau w^1 + (1 - \tau) w^2] \geq 0$$

for all $\tau \in (0,1)$. Letting $\tau \to 1$ we deduce

$$(x^1 - x^2)^T w^1 \geq 0.$$

Hence equality must hold because $x^1 \in \text{SOL}(K, q, M)$ and $x^2 \in K$. This establishes (b).

(b) $\Rightarrow$ (c). Adding the two equations:

$$(x^1 - x^2)^T (q + M x^1) = 0 \quad \text{and} \quad (x^2 - x^1)^T (q + M x^2) = 0$$

easily yield (c).

(c) $\Rightarrow$ (d). It suffices to verify that $\text{SOL}(K, q, M)$ is equal to (2.5.11). By the equivalence of the AVI and its KKT system, we have $x \in \text{SOL}(K, q, M)$ if and only if there exists (μ, λ) such that (x, λ, μ) satisfies (2.4.6). Condition (c) implies that if x^1 and x^2 are two solutions of the AVI (K, q, M) and (μ^1, λ^1) and (μ^2, λ^2) are the corresponding multipliers, then

$$(\lambda^1)^T (b - A x^2) = 0 = (\lambda^2)^T (b - A x^1).$$

This is the cross complementarity of the KKT system as an MLCP. In particular we deduce that for every triple (x, λ, μ) satisfying (2.4.6), we must have

$$\lambda_i = 0 \quad \forall\, i \not\in \beta.$$

Based on observation, we can easily establish the desired representation of $\text{SOL}(K, q, M)$.

(d) $\Rightarrow$ (a). This requires no proof. □

2.5.3 Solvability in terms of feasibility

In Subsection **2.4.2**, we have asked the question of when the feasibility of a CP implies its solvability and given two situations where an affirmative answer exists. In the present subsection, we consider the related question of when the range of a pair (K, M) is equal to its domain. In fact, we embed this question in a slightly broader context for technical reasons.

Specializing the inclusion (2.3.4) to the affine map $F(x) = q + Mx$, we deduce that if $q \in \mathcal{R}(K, M)$, then for all $x \in \text{SOL}(K, q, M)$, we have

$$q + Mx \in (K_\infty)^*,$$

which implies, since x belongs to K,

$$q \in (K_\infty)^* - MK.$$

As we have mentioned in the proof of Lemma **2.5.13**, by dualizing the inclusion $\mathcal{K}(K, M) \subseteq K_\infty$, we obtain

$$(K_\infty)^* \subseteq (\mathcal{K}(K, M))^*.$$

Consequently, we have established, for a pair (K, M) with K being a closed convex set,

$$\mathcal{R}(K, M) \subseteq \operatorname{conv}(\mathcal{R}(K, M)) \subseteq \mathcal{D}(K, M) \subseteq (\mathcal{K}(K, M))^* - MK. \quad (2.5.12)$$

Simple examples easily demonstrate that each of these inclusions is proper. The main goal of this subsection is to derive conditions on the pair (K, M) in order for equalities to hold throughout the above expression.

We note that a necessary condition for the VI range $\mathcal{R}(K, M)$ to be equal to the VI domain $\mathcal{D}(K, M)$ is that the former is a convex set. It turns out that if K is a convex cone, the convexity of the CP range $\mathcal{R}(K, M)$ is sufficient for equalities to hold throughout (2.5.12), provided that the pair (K, M) satisfies the assumptions of Corollary **2.5.12**.

2.5.18 Proposition. Let K be a closed convex cone in $\mathbb{R}^n$ and M be an $n \times n$ matrix copositive on K. Suppose that $\operatorname{int} \mathcal{K}(K, M)^*$ is nonempty and $\mathcal{R}(K, M)$ is closed. Equalities hold throughout (2.5.12) if and only if $\mathcal{R}(K, M)$ is convex.

Proof. By the above remark, it suffices to establish the "if" statement. Suppose that $\mathcal{R}(K, M)$ is convex. Recalling (2.5.1), we deduce from Corollary **2.5.12** that

$$\mathcal{R}(K, M) = \mathcal{R}(K, M) - MK \supseteq \mathcal{K}(K, M) - MK \supseteq \mathcal{R}(K, M),$$

where the first equality holds because $\mathcal{R}(K, M)$ and $-MK$ are both convex cones and the former contains the latter. Hence, the first and last set in (2.5.12) are equal; thus equalities hold throughout this expression. □.

Under the assumptions of Proposition **2.5.18**, it follows that the set of q's for which the CP (K, q, M) is feasible coincides with the set of q's for which the CP (K, q, M) is solvable, provided that the pair (K, M) has a convex CP range.

The remainder of this section deals with the case where K is a polyhedral set. In this case, the equality between the AVI range $\mathcal{R}(K, M)$ and the AVI domain $\mathcal{D}(K, M)$ means that the set of all vectors q for which the AVI (K, q, M) is solvable coincides with the set of all vectors q for which the KKT of the AVI (K, q, M) is feasible as an MLCP. The equality between the first and last set in (2.5.12) are inspired by results from the LCP where a great deal is known about the relation between the LCP kernel SOL$(0, M)$ and the LCP range of a copositive matrix M.

The proof of the main result, Theorem **2.5.20**, turns out to be not as easy as the case of a cone K; cf. Proposition **2.5.18**. In fact, in addition to requiring K to be polyhedral, we need to further assume that M is copositive plus on the recession cone K_∞. Under the latter assumption, we first show that the last inclusion in (2.5.12) holds as an equality.

2.5.19 Lemma. Let K be a polyhedral set in $\mathbb{R}^n$ and M be an $n \times n$ matrix. If M is copositive plus on K_∞ then

$$\mathcal{D}(K, M) = (\mathcal{K}(K, M))^* - MK.$$

Proof. Applying Lemma (2.5.2) to K_∞, we have

$$\mathcal{K}(K, M) = K_\infty \cap (-MK_\infty)^*.$$

Dualizing this equality and using the polyhedrality of K, we deduce

$$\mathcal{K}(K, M)^* = (K_\infty)^* - MK_\infty.$$

Since $K + K_\infty = K$, we have

$$\begin{aligned} \mathcal{K}(K, M)^* - MK &= (K_\infty)^* - M(K + K_\infty) \\ &= (K_\infty)^* - MK = \mathcal{D}(K, M), \end{aligned}$$

as desired. □

We state and prove the final result of this section.

2.5.20 Theorem. Let K be a polyhedral set in $\mathbb{R}^n$ and M be an $n \times n$ matrix copositive plus on K_∞. Consider the following statements:

(a) there exists an extreme point c of K such that M is copositive on the conical hull of $K - c$;

(b) int $\mathcal{K}(K, M)^*$ is nonempty and there exists a vector $q^* \in \mathcal{R}(K, M)$ such that SOL(K, q^*, M) is bounded and for some bounded open set containing SOL(K, q^*, M), $\deg(\mathbf{F}_K^{\text{nat}}, \mathcal{U})$ is nonzero, where $\mathbf{F}_K^{\text{nat}}$ is the natural map associated with the AVI (K, q^*, M);

(c) equalities hold throughout (2.5.12).

It holds that (a) $\Rightarrow$ (b) $\Rightarrow$ (c).

Proof. Suppose that (a) holds. Lemma **2.5.13** implies that int $\mathcal{K}(K, M)^*$ is nonempty. Indeed, by the proof of this lemma, we can establish that

$\text{int}\,\mathcal{T}(c; K)^*$ is nonempty. More directly, since c is extreme point of K, the cone $\mathcal{T}(c; K)$ must be pointed. Hence by Proposition **2.4.3**, the dual cone $\mathcal{T}(c; K)^*$ must be solid; that is, $\text{int}\,\mathcal{T}(c; K)^*$ is nonempty. Let d be an arbitrary vector in the latter interior and define $q^* \equiv d - Mc$. We claim that for every vector q' that is sufficiently close to q^*, $\text{SOL}(K, q', M) = \{c\}$. Indeed, if q' is such a vector, then $d' \equiv q' + Mc$ is sufficiently close to d. Since d belongs to the interior of $\mathcal{T}(c; K)$, by choosing a suitable neighborhood N of q^*, it follows that the vector d' also belongs to $\text{int}\,\mathcal{T}(c; K)^*$. For any such vector d', we have, since $y - c \in \mathcal{T}(c; K)$ for all $y \in K$,

$$0 \leq (y-c)^T d' = (y-c)^T (q' + Mc).$$

Thus $c \in \text{SOL}(K, q', M)$. To see that c is the only solution of such an AVI (K, q', M), let c' be another solution. Similar to the proof of Proposition **2.3.6**, we can show that, using the copositivity of M on the conical hull of $K - c$

$$0 = (c' - c)^T (q' + Mc) = (c' - c)^T d'.$$

Since $d' \in \text{int}\,\mathcal{T}(c; K)^*$ and $c' - c$ is an element of $\mathcal{T}(c; K)$, we have

$$(c' - c)^T d' > 0$$

for $c' \neq c$. Consequently, we must have $\text{SOL}(K, q', M) = \{c\}$.

We next claim that for all vectors y with $\|y\|$ sufficiently small, the equation $\mathbf{F}_K^{\text{nat}}(x) = y$ has at most one solution. Since the latter equation holds if and only if

$$0 = x - y - \Pi_K(x - y + y - q^* - M(x - y) - My),$$

we see that $x \in (\mathbf{F}_K^{\text{nat}})^{-1}(y)$ if and only if $x - y \in \text{SOL}(K, q', M)$, where $q' \equiv q^* - y + My$. Since $\|y\|$ is sufficiently small, it follows that q' is sufficiently close to q^*. Hence there exists an $\varepsilon > 0$ such that for all $\|y\| < \varepsilon$, the set $(\mathbf{F}_K^{\text{nat}})^{-1}(y)$ is either empty or a singleton. Since $\mathbf{F}_K^{\text{nat}}(c) = 0$, there exists an open neighborhood $\mathcal{U}$ of c such that $\|\mathbf{F}_K^{\text{nat}}(x)\| < \varepsilon$ for all $x \in \mathcal{U}$. The restricted map $\mathbf{F}_K^{\text{nat}} : \mathcal{U} \to \mathbb{R}^n$ is injective; consequently, since $0 \in \mathbf{F}_K^{\text{nat}}(\mathcal{U})$, it follows that $\deg(\mathbf{F}_K^{\text{nat}}, \mathcal{U})$ is well defined and nonzero. Therefore, we have established (a) $\Rightarrow$ (b).

To prove (b) $\Rightarrow$ (c), it suffices to show $\mathcal{D}(K, M) \subseteq \mathcal{R}(K, M)$. In turn, since $\mathcal{R}(K, M)$ is a closed set by Theorem **2.5.15**, we only need to show

$$\emptyset \neq \text{int}\,\mathcal{D}(K, M) \subseteq \mathcal{R}(K, M).$$

The nonemptiness of $\text{int}\,\mathcal{D}(K, M)$ is not difficult because we have

$$\text{int}\,\mathcal{K}(K, M)^* - MK \subseteq \text{int}(\,\mathcal{K}(K, M)^* - MK\,) = \text{int}\,\mathcal{D}(K, M),$$

where the first inclusion holds by an easy argument and the second equality holds by Lemma **2.5.19**. Since $\operatorname{int} \mathcal{K}(K, M)^*$ is nonempty by assumption, it follows that $\operatorname{int} \mathcal{D}(K, M)$ is nonempty.

Let $q \in \operatorname{int} \mathcal{D}(K, M)$ be arbitrary and q^* be as prescribed by assumption (b). By the invariance property of the degree, it follows that for all bounded open sets $\mathcal{U}$ containing $\mathrm{SOL}(K, q^*, M)$, $\deg(\mathbf{F}_K^{\mathrm{nat}}, \mathcal{U})$ is well defined and nonzero. Consider the homotopy:

$$H(x,t) \equiv x - \Pi_K(x - tq - (1-t)q^* - Mx), \quad (x,t) \in \mathbb{R}^{n+1}.$$

By the homotopy invariance property of the degree, provided that we can show

$$\bigcup_{t \in (0,1]} H(\cdot, t)^{-1}(0) \tag{2.5.13}$$

is bounded, then we must have $\deg(H(\cdot,1), \mathcal{U})$ is well defined and nonzero for some bounded open set $\mathcal{U}$ containing the above union of zero sets. Since $H(x,1)$ is the natural map associated with the AVI (K, q, M), it follows that this problem must have a solution; therefore, q belongs to $\mathcal{R}(K, M)$ as desired.

Consequently, the last thing we need to establish is the boundedness of the union (2.5.13). Suppose for the sake of contradiction that this union is not bounded. Then there exist a sequence of scalars $t_k \in (0, 1]$, a sequence of vectors $\{x^k\}$, and a nonzero vector d such that

$$\lim_{k \to \infty} \| x^k \| = \infty, \quad \lim_{k \to \infty} \frac{x^k}{\| x^k \|} = d, \quad \text{and} \quad H(x^k, t_k) = 0, \quad \forall\, k.$$

The last equation is equivalent to $x^k \in \mathrm{SOL}(K, q^k, M)$, where

$$q^k \equiv t_k\, q + (1 - t_k)\, q^*.$$

Similar to the proof of Proposition **2.5.11**, we can show that d belongs to $\mathcal{K}(K, M)$. Thus $-M^T d = Md$ because M is copositive plus on K_∞.

For simplicity, let us assume that K is given by

$$K \equiv \{ x \in \mathbb{R}^n : Ax \leq b \}$$

for some $m \times n$ matrix A and m-vector b. For each k, define the active index set at x^k:

$$\alpha_k \equiv \{ i : (Ax^k = b)_i \}.$$

Since there are only finitely many such index sets α_k and there are infinitely many indices k, by working with an appropriate subsequence of $\{x^k\}$, we may assume without loss of generality that all the index sets α_k are the same; let α denote this common index set. Thus

$$(Ax^k = b)_\alpha \quad \forall\, k.$$

Hence we must have $(Ad = 0)_i$ for all i belonging to α. In terms of this index set, the KKT system of the AVI (K, q^k, M) can be written as:

$$0 = q^k + Mx^k + \sum_{i \in \alpha} \lambda_i^k \, (A_{i\cdot})^T \tag{2.5.14}$$

for some

$$\lambda_i^k \geq 0, \quad \forall\, i \in \alpha.$$

Equivalently, this system says that

$$-(q^k + Mx^k) \in \text{pos}(A_{\alpha\cdot})^T.$$

Since the right-hand set is a polyhedral cone, it follows that

$$-Md \in \text{pos}(A_{\alpha\cdot})^T.$$

Hence there exists $\lambda_\alpha^\infty \geq 0$ such that

$$M^T d = -Md = \sum_{i \in \alpha} \lambda_i^\infty \, (A_{i\cdot})^T.$$

From (2.5.14), we obtain

$$0 = d^T(q^k + Mx^k) = t_k \, d^T(q + Mx^k) + (1 - t_k) \, d^T(q^* + Mx^k). \tag{2.5.15}$$

We claim that

$$d^T(q + Mx^k) > 0 \quad \text{and} \quad d^T(q^* + Mx^k) \geq 0. \tag{2.5.16}$$

By assumption, $q \in \text{int}\, \mathcal{D}(K, M) = \text{int}(\mathcal{K}(K, M)^* - MK)$. Thus, there exist a scalar $\varepsilon > 0$ and a vector $y \in K$ such that the vector

$$p \equiv q - \varepsilon\, d + My$$

belongs to $\mathcal{K}(K, M)^*$. Hence $d^T p \geq 0$; moreover,

$$\begin{aligned}
& d^T(q + Mx^k) \\
&= d^T[\, p + \varepsilon\, d + M(x^k - y)\,] > d^T M(x^k - y) \\
&= (y - x^k)^T Md = \sum_{i \in \alpha} \lambda_i^\infty \, A_{i\cdot}(x^k - y) \\
&= \sum_{i \in \alpha} \lambda_i^\infty \, (b - Ay)_i \geq 0.
\end{aligned}$$

In a similar way, since $q^* \in \mathcal{R}(K,M) \subseteq \mathcal{K}(K,M)^* - MK$, we can deduce $d^T(q^* + Mx^k) \geq 0$. We have therefore established the claim (2.5.16). But this contradicts (2.5.15) because $t_k \in (0,1]$. □

2.5.21 Remark. At the beginning of the proof of Theorem **2.5.20**, we have established a fact for an arbitrary affine pair (K,M) that does not require any property of the matrix M. Namely, if K is a polyhedron with an extreme point, then $\mathcal{R}(K,M)$ contains an open set. See Proposition **6.3.17** for an important consequence of this fact. □

2.6 Further Existence Results for CPs

While the existence results in Subsection **2.4.2** for the CP (K,F) all require F to be at least pseudo monotone, those in Section **2.5** assume that F is affine. In what follows, using Theorem **2.5.10**, we obtain a sufficient condition for the CP (K,F) to have a solution without any monotonicity assumption on the nonlinear function F.

2.6.1 Theorem. Let K be a closed convex cone in $\mathbb{R}^n$; let F be a continuous map from K into $\mathbb{R}^n$. If there exists a copositive matrix $E \in \mathbb{R}^{n\times n}$ on K such that (K,E) is an R_0 pair and the union

$$\bigcup_{\tau>0} \mathrm{SOL}(K, F + \tau E)$$

is bounded, then the CP (K,F) has a solution.

Proof. As we have done so several times already, we may assume that F is defined and continuous on the entire space $\mathbb{R}^n$. Assume for the sake of contradiction that $\mathrm{SOL}(K,F)$ is empty. According to Theorem **2.2.1**, it suffices to show that $\deg(\mathbf{F}_K^{\mathrm{nat}}, \mathcal{U})$ is well defined and nonzero for some bounded open set $\mathcal{U}$. Consider the homotopy:

$$H(x,t) \equiv x - \Pi_K(x - tF(x) - (1-t)Ex), \quad (x,t) \in \mathbb{R}^n \times [0,1].$$

Clearly, $H(\cdot,t)$ is the natural map associated with the CP $(K, tF+(1-t)E)$. For $t \in (0,1]$, the latter CP is equivalent to the CP $(K, F+\tau E)$, where $\tau \equiv (1-t)/t$. By assumption, the union of the solution sets of these CPs for all $\tau > 0$ is bounded; for $\tau = 0$, the CP (K,F) is assumed to have no solution. Since (K,E) is an R_0 pair, the CP $(K,0,E)$ has a unique solution, namely, the origin. Consequently, the union

$$\bigcup_{t\in[0,1]} H(\cdot,t)^{-1}(0)$$

is bounded. It suffices to take $\mathcal{U}$ to be a bounded open set containing the latter union and apply Theorem **2.5.10** and the homotopy invariance property of the degree to deduce

$$\deg(\mathbf{F}_K^{\rm nat},\mathcal{U}) \,=\, \deg(H(\cdot,1),\mathcal{U}) \,=\, \deg(H(\cdot,0),\mathcal{U}) \,=\, \deg(\mathbf{E}_K^{\rm nat},\mathcal{U}) \,=\, 1,$$

where $\mathbf{E}_K^{\rm nat}(x) \equiv x - \Pi_K(x - Ex)$ is the natural map associated with the homogeneous CP $(K, 0, E)$. □

Theorem **2.6.1** has several interesting consequences. The first consequence is an "alternative theorem" for the existence of a solution to a CP.

2.6.2 Corollary. Let K be a closed convex cone in $\mathbb{R}^n$; let F be a continuous map from K into $\mathbb{R}^n$. Either the CP (K, F) has a solution or there exist an unbounded sequence of vectors $\{x^k\}$ and a sequence of positive scalars $\{\tau_k\}$ such that, for every k,

$$K \ni x^k \perp F(x^k) + \tau_k\, x^k \in K^*. \tag{2.6.1}$$

Proof. It suffices to let E be the identity matrix in Theorem **2.6.1**. □

Another consequence of Theorem **2.6.1** pertains to a co-coercive CP. To motivate the corollary, we note that, by Theorem **2.4.4**, if F is a continuous pseudo monotone map and K is a pointed, closed, convex cone, then the CP (K, F) has a nonempty compact solution set if and only if the problem is strictly feasible. It is not difficult to show that if $F(x) \equiv q + Mx$ for some symmetric positive semidefinite matrix M and an arbitrary vector q, then the strict feasibility of the LCP (q, M) is equivalent to the existence of a vector u, not necessarily nonnegative, such that $q + Mu > 0$. The corollary below extends this result for an LCP to a co-coercive CP on a pointed, closed, convex cone.

2.6.3 Corollary. Let K be a pointed, closed, convex cone in $\mathbb{R}^n$ and let $F : \mathbb{R}^n \to \mathbb{R}^n$ be a continuous map. If F is co-coercive on $\mathbb{R}^n$, then the CP (K, F) has a nonempty compact solution set if and only if there exists a vector $u \in \mathbb{R}^n$ satisfying $F(u) \in \operatorname{int} K^*$.

Proof. Only the sufficiency requires a proof. Let $q \equiv F(u) \in \operatorname{int} K^*$ and consider the CP:

$$K \ni x \perp -q + F(x) \in K^*.$$

If this problem has a solution then the CP (K, F) is clearly strictly feasible. Assume that the above CP has no solution. By Corollary **2.6.2**, there exist a sequence of vectors $\{x^k\}$ and a sequence of positive scalars $\{\tau_k\}$ such that

$$\lim_{k\to\infty} \| x^k \| = \infty$$

and for every k,

$$K \ni x^k \perp -q + F(x^k) + \tau_k x^k \in K^*.$$

We claim that

$$\lim_{k\to\infty} \tau_k x^k = 0. \tag{2.6.2}$$

Let $c > 0$ be a co-coercive constant of F on $\mathbb{R}^n$. For every k, we have

$$\begin{aligned} & c \| F(x^k) - F(u) \|^2 \\ & \leq (x^k - u)^T (F(x^k) - F(u)) \\ & = -u^T (F(x^k) - F(u)) + (x^k)^T (F(x^k) - q + \tau_k x^k) - \tau_k \| x^k \|^2 \\ & \leq \| u \| \| F(x^k) - F(u) \|. \end{aligned}$$

This implies that

$$\| F(x^k) - F(u) \| \leq c^{-1} \| u \|, \quad \forall k.$$

Moreover,

$$\tau_k \| x^k \|^2 \leq \| u \| \| F(x^k) - F(u) \| - c \| F(x^k) - F(u) \|^2.$$

Since $\{\|x^k\|\}$ tends to infinity and the right-hand expression is bounded, it follows that (2.6.2) holds. This implies, since $q \in \operatorname{int} K^*$, that $q - \tau_k x^k$ belongs to $\operatorname{int} K^*$ for all k sufficiently large. Consequently, the strictly feasibility of the CP (K, F) follows. □

The next consequence of Theorem **2.6.1** requires the norm-coercivity of the natural map $\mathbf{F}_K^{\text{nat}}(x)$ on K. Specifically, we assume that

$$\lim_{\substack{x \in K \\ \|x\| \to \infty}} \| \mathbf{F}_K^{\text{nat}}(x) \| = \infty. \tag{2.6.3}$$

This condition is equivalent to the inf-compactness of the merit function $\|\mathbf{F}_K^{\text{nat}}(x)\|$ on K; i.e., for every scalar η, the level set

$$\{ x \in K : \| \mathbf{F}_K^{\text{nat}}(x) \| \leq \eta \}$$

is bounded. Either condition implies that the minimization problem:

$$\begin{aligned} & \text{minimize} && \| \mathbf{F}_K^{\text{nat}}(x) \| \\ & \text{subject to} && x \in K \end{aligned}$$

attains its finite minimum. By assuming an additional copositivity condition on the function F on K, we can ensure that this minimum value is zero, thus $\operatorname{SOL}(K, F)$ is nonempty. In fact, the conclusion of the following corollary pertains to the CP $(K, q + F)$ for all vectors $q \in \mathbb{R}^n$.

2.6.4 Corollary. Let K be a closed convex cone in $\mathbb{R}^n$; let $F : K \to \mathbb{R}^n$ be a continuous map. Suppose that (2.6.3) holds and

$$x^T (F(x) - F(0)) \geq 0, \quad \forall x \in K.$$

The CP $(K, q + F)$ has a nonempty compact solution set for all $q \in \mathbb{R}^n$.

Proof. First consider the case $q = 0$. Clearly, SOL(K, F) must be bounded by the coercivity of $\|\mathbf{F}_K^{\mathrm{nat}}(x)\|$ on K. To show that this solution set is nonempty, we proceed by contradiction. Suppose that the CP (K, F) has no solution. There exist a sequence of vectors $\{x^k\}$ and a sequence of positive scalars $\{\tau_k\}$ such that

$$\lim_{k \to \infty} \| x^k \| = \infty$$

and for every k,

$$K \ni x^k \perp F(x^k) + \tau_k x^k \in K^*.$$

The latter implies

$$x^k - \Pi_K(x^k - F(x^k) - \tau_k x^k) = 0, \quad \forall k. \tag{2.6.4}$$

Similar to the proof of Corollary **2.6.3**, we have for every k,

$$\begin{aligned} 0 &\leq (x^k)^T (F(x^k) - F(0)) \\ &= (x^k)^T (-\tau_k x^k - F(0)), \end{aligned}$$

which implies that

$$\tau_k \| x^k \| \leq \| F(0) \|.$$

We have, by (2.6.4) and the non expansiveness of the Euclidean projector,

$$\begin{aligned} \infty &= \lim_{k \to \infty} \| \mathbf{F}_K^{\mathrm{nat}}(x^k) \| \\ &= \lim_{k \to \infty} \| \Pi_K(x^k - F(x^k) - \tau_k x^k) - \Pi_K(x^k - F(x^k)) \| \\ &\leq \| F(0) \|; \end{aligned}$$

this is clearly a contradiction.

For an arbitrary nonzero vector $q \in \mathbb{R}^n$, we can apply what we have just proved to the function $\tilde{F} \equiv q + F$. To make sure that the argument is applicable, it suffices to verify that the function

$$x \mapsto x - \Pi_K(x - F(x) - q)$$

is norm-coercive on K. But this is obvious because the difference between this function and $\mathbf{F}_K^{\mathrm{nat}}(x)$ is bounded in norm by the constant $\|q\|$ for all x. □

The norm-coercivity of the natural map $\mathbf{F}_K^{\text{nat}}$ plays an important role in the convergence analysis of iterative descent algorithms for solving the VI (K, F). See Exercise **1.8.34** for a preliminary treatment of the coercivity issue and Chapters **9** and **10** for details of the algorithmic implications. In what follows, we show that if $F(x) \equiv q + Mx$ is an affine map, the mentioned coercivity condition is equivalent to the R_0 property of the pair (K, M).

2.6.5 Proposition. Let K be a closed convex cone in $\mathbb{R}^n$ and M be a matrix in $\mathbb{R}^{n \times n}$. The following four statements are equivalent.

(a) For all vectors $q \in \mathbb{R}^n$, the natural map $\mathbf{F}_K^{\text{nat}}$ of the CP (K, q, M) is norm-coercive on K (or $\mathbb{R}^n$).

(b) For some vector $q \in \mathbb{R}^n$, the natural map $\mathbf{F}_K^{\text{nat}}$ of the CP (K, q, M) is norm-coercive on K (or $\mathbb{R}^n$)

(c) The natural map $\mathbf{M}_K^{\text{nat}}$ of the pair (K, M) is norm-coercive on K (or $\mathbb{R}^n$).

(d) (K, M) is an R_0 pair.

Proof. By the last part of the proof of Corollary **2.6.4**, it is clear that (a), (b), and (c) are equivalent and that (c) implies (d). It remains to show that (d) implies (c). Suppose that (K, M) is an R_0 pair but there exists a sequence of vectors $\{x^k\}$ and a scalar η such that $\|\mathbf{M}_K^{\text{nat}}(x^k)\| \leq \eta$ for all k and

$$\lim_{k \to \infty} \| x^k \| = \infty.$$

Without loss of generality, we may assume that

$$\lim_{k \to \infty} \frac{x^k}{\| x^k \|} = d$$

for some nonzero vector d. By a normalization followed by a limiting argument and by the cone property of K, it is not difficult to show that d is a solution of the homogeneous CP $(K, 0, M)$. This contradicts the R_0 property of the pair (K, M). Therefore $\mathbf{M}_K^{\text{nat}}$ is norm-coercive on $\mathbb{R}^n$; hence it must be norm-coercive on K too. □

In view of the above proposition, we see that Corollary **2.6.4** is a nonlinear extension of part (a) of Theorem **2.5.10**. In Proposition **9.1.27**, we derive a necessary and sufficient condition for the min map $\min(x, F(x))$ to be norm-coercive on $\mathbb{R}^n$.

2.7 A Frictional Contact Problem

We consider a discrete linear elastic, small displacement, planar contact problem under a standard Coulomb friction law. Although being a much simplified version of the model described in Subsection **1.4.6**, the problem discussed in this section is an important realization of the family of discrete frictional contact problems; furthermore, the line of treatment employed herein can be extended to deal with the full model where the detailed analysis becomes highly technical but not particularly illuminating, due to the nonlinearities of the functions involved. The treatment herein illustrates how some of the existence results established in the previous sections can be used to prove the solvability of a practical engineering model.

Mathematically, the problem treated here can be formulated as follows. Given a symmetric positive semidefinite stiffness matrix $M \in \mathbb{R}^{N \times N}$, two $n_c \times N$ matrices C_n and C_t whose rows are, respectively, the normal and tangential vectors at the contact points on the contact surface that is assumed planar, an N-vector of external force f^{ext}, an n_c-vector of initial gap distance g, a reference N-vector u^{ref} of displacements, and a friction coefficient $\mu > 0$, find a displacement vector $u \in \mathbb{R}^{n_c}$, two contact force vectors p_n and p_t, both n_c-dimensional, and two dual variables $\lambda_t^{\pm}$ such that the following conditions hold:

$$Mu + C_n^T p_n + C_t^T p_t = f^{\text{ext}} \tag{2.7.1}$$

$$0 \leq p_n \perp g - C_n u \geq 0 \tag{2.7.2}$$

$$C_t(u - u^{\text{ref}}) = \lambda_t^+ - \lambda_t^- \tag{2.7.3}$$

$$\left.\begin{array}{l} 0 \leq \lambda_t^+ \;\perp\; p_t - \mu p_n \leq 0 \\ 0 \leq \lambda_t^- \;\perp\; -p_t - \mu p_n \leq 0 \end{array}\right\}. \tag{2.7.4}$$

The first of these equations, (2.7.1), is the force equilibrium equation; the second of these equations, (2.7.2), is the Signorini normal contact condition; and the last three equations, (2.7.3) and (2.7.4), are Coulomb's friction law and the maximum dissipative law of the tangential contact forces. Observe that there exist $\lambda_t^{\pm}$ such that (2.7.3) and (2.7.4) hold if and only if for all $i = 1, \ldots, n_c$,

$$p_{it} \begin{cases} = +\mu p_{in} & \text{if } C_{it}(u - u^{\text{ref}}) > 0 \\ \in [-\mu p_{in}, +\mu p_{in}] & \text{if } C_{it}(u - u^{\text{ref}}) = 0 \\ = -\mu p_{in} & \text{if } C_{it}(u - u^{\text{ref}}) < 0. \end{cases} \tag{2.7.5}$$

It is easy to see that the latter condition is necessary for (2.7.3) and (2.7.4). Conversely, if (2.7.5) holds, then one choice for $\lambda_t^\pm$ is

$$\lambda_t^+ = \max(0, C_t(u - u^{\text{ref}})), \quad \text{and} \quad \lambda_t^- = \max(0, -C_t(u - u^{\text{ref}})).$$

The model (2.7.1)–(2.7.4) is thus equivalent to (2.7.1), (2.7.2), and (2.7.5) with the latter three conditions involving only the triple (u, p_n, p_t).

The case of small friction

We consider the simplest case under the following assumptions:

(A) M is positive definite;

(B) the columns of the matrix $[\, C_n^T \quad C_t^T \,]$ are linearly independent;

(C) the friction coefficient μ is "sufficiently small".

We will quantify the magnitude of μ subsequently. Our immediate goal is to show that under the assumptions (A), (B), and (C), the frictional contact problem presented above can be formulated as a copositive LCP $(\mathbf{q}, \mathbf{M})$ with $\mathbf{q}$ being an element of the dual of the LCP kernel of $\mathbf{M}$. Thus, by Corollary **2.5.12**, the friction problem has a solution; more importantly, this solution can be computed by the well-known Lemke almost complementary pivotal method.

We eliminate the (free) variable p_t by introducing the slack variables:

$$v^+ \equiv \mu\, p_n - p_t \quad \text{and} \quad v^- \equiv \mu\, p_n + p_t.$$

The first equation yields

$$p_t = \mu\, p_n - v^+;$$

substituting this expression in the model equations (2.7.1)–(2.7.4), we obtain

$$\begin{aligned}
&Mu + (\, C_n - \mu\, C_t\,)^T p_n - C_t^T v^+ - f^{\text{ext}} = 0 \\
&0 \le p_n \quad \perp \quad g - C_n u \ge 0 \\
&0 \le v^+ \quad \perp \quad \lambda_t^+ = C_t(\, u - u^{\text{ref}}\,) + \lambda_t^- \ge 0 \\
&0 \le \lambda_t^- \quad \perp \quad v^- = 2\mu\, p_n - v^+ \ge 0.
\end{aligned}$$

Since M is assumed to be positive definite, we may use the first equation to solve for the (free) variable u, obtaining

$$u = -M^{-1}\left[\,(\, C_n - \mu\, C_t\,)^T p_n - C_t^T v^+ - f^{\text{ext}}\,\right];$$

we can then use this expression to eliminate u from the model. This results in an equivalent LCP $(\mathbf{q}, \mathbf{M})$ formulation of the model, where

$$\mathbf{q} \equiv \begin{pmatrix} g - C_n M^{-1} f^{\text{ext}} \\ C_t (M^{-1} f^{\text{ext}} - u^{\text{ref}}) \\ 0 \end{pmatrix}$$

and

$$\mathbf{M} \equiv \begin{bmatrix} C_n M^{-1} (C_n - \mu C_t)^T & C_n M^{-1} C_t^T & 0 \\ C_t M^{-1} (C_n - \mu C_t)^T & C_t M^{-1} C_t^T & \mathbf{I} \\ 2 \mu \mathbf{I} & -\mathbf{I} & 0 \end{bmatrix}.$$

2.7.1 Proposition. Under assumptions (A) and (B), there exists a scalar $\bar{\mu} > 0$ such that for all $\mu \in [0, \bar{\mu}]$, the matrix $\mathbf{M}$ is copositive on $\mathbb{R}^{3n_c}_+$; moreover, it holds that

$$[0 \leq \mathbf{x} \perp \mathbf{M}\mathbf{x} \geq 0] \Rightarrow \mathbf{q}^T \mathbf{x} \geq 0.$$

Hence, the LCP $(\mathbf{q}, \mathbf{M})$, and thus the frictional contact model (2.7.1)–(2.7.4), has a solution.

Proof. We can write

$$\mathbf{M} = \mathbf{M}_1 - \mu \mathbf{M}_2 + \mathbf{M}_3,$$

where

$$\mathbf{M}_1 \equiv \begin{bmatrix} C_n M^{-1} C_n^T & C_n M^{-1} C_t^T & 0 \\ C_t M^{-1} C_n^T & C_t M^{-1} C_t^T & 0 \\ 0 & 0 & 0 \end{bmatrix},$$

$$\mathbf{M}_2 \equiv \begin{bmatrix} C_n M^{-1} C_t^T & 0 & 0 \\ C_t M^{-1} C_t^T & 0 & 0 \\ 0 & 0 & 0 \end{bmatrix}, \quad \text{and} \quad \mathbf{M}_3 \equiv \begin{bmatrix} 0 & 0 & 0 \\ 0^T & 0 & \mathbf{I} \\ 2\mu \mathbf{I} & -\mathbf{I} & 0 \end{bmatrix}.$$

By assumptions (A) and (B), it follows that the matrix

$$\begin{bmatrix} C_n M^{-1} C_n^T & C_n M^{-1} C_t^T \\ C_t M^{-1} C_n^T & C_t M^{-1} C_t^T \end{bmatrix}$$

is positive definite. Thus there exists a scalar $\bar{\mu} > 0$ such that the matrix

$$\begin{bmatrix} C_n M^{-1} C_n^T & C_n M^{-1} C_t^T \\ C_t M^{-1} C_n^T & C_t M^{-1} C_t^T \end{bmatrix} - \mu \begin{bmatrix} C_n M^{-1} C_t^T & 0 \\ C_t M^{-1} C_t^T & 0 \end{bmatrix}$$

remains positive definite for all $\mu \in [0, \bar{\mu}]$. Thus for all these μ's, $\mathbf{M}_1 - \mu\mathbf{M}_2$ is positive semidefinite. It is easy to see that $\mathbf{M}_3$ is copositive. Consequently, the matrix $\mathbf{M}$ is copositive for all $\mu \in [0, \bar{\mu}]$. Suppose $\mathbf{x}$ is an element of the LCP kernel of $\mathbf{M}$. It follows that

$$\mathbf{x}^T(\mathbf{M}_1 - \mu\mathbf{M}_2)\mathbf{x} = 0,$$

which implies $\mathbf{x}_1 = \mathbf{x}_2 = 0$, where $\mathbf{x}_1$ and $\mathbf{x}_2$ denote the first two component blocks of $\mathbf{x}$ partitioned according to $\mathbf{M}$. By the definition of $\mathbf{q}$, we deduce $\mathbf{q}^T\mathbf{x} = 0$ as claimed. □

We remark that the matrix $\mathbf{M}$ is not copositive star (and thus not copositive plus) on $\mathbb{R}^{3n_c}_+$ for $\mu > 0$.

The case of arbitrary friction

Next, we drop assumption (C) and allow the friction coefficient μ to be any positive scalar. We further relax assumption (B) and instead assume that the following implication holds:

$$\left.\begin{aligned} C_t^T p_t + C_n^T p_n &= 0 \\ p_n &\geq 0 \end{aligned}\right\} \Rightarrow (p_t, p_n) = 0; \tag{2.7.6}$$

or equivalently, there exists a scalar $\delta > 0$ such that

$$\|(p_t, p_n)\| \leq \delta \| C_t^T p_t + C_n^T p_n \|, \quad \forall (p_t, p_n) \text{ with } p_n \geq 0. \tag{2.7.7}$$

The previous analysis is no longer applicable and we need to resort to the MLCP formulation (2.7.1)–(2.7.4). The affine map that defines this MLCP is:

$$\mathbf{F}(u, p_t, p_n, \lambda_t^{\pm}) \equiv \begin{pmatrix} Mu + C_t^T p_t + C_n^T p_n - f^{\text{ext}} \\ -C_t(u - u^{\text{ref}}) + \lambda_t^+ - \lambda_t^- \\ g - C_n u \\ \mu p_n - p_t \\ \mu p_n + p_t \end{pmatrix},$$

with (u, p_t) being the free variables and $(p_n, \lambda_t^{\pm})$ the nonnegative variables. There are two ways to establish the solvability of this MLCP: One is to apply Corollary **2.5.12**; and the other is to apply Theorem **2.6.1**. In what follows, we illustrate how to apply the latter theorem. For this purpose, we first establish the following lemma.

2.7.2 Lemma. Under assumption (A) and (2.7.6), there exists a constant $\eta > 0$ such that for all scalars $\tau > 0$ and all triples $\mathbf{x} \equiv (u, p_t, p_n, \lambda_t^{\pm})$ satisfying the following conditions:

$$Mu + \tau u + C_t^T p_t + C_n^T p_n = f^{\text{ext}} \tag{2.7.8}$$

$$\tau p_t - C_t(u - u^{\text{ref}}) = -\lambda_t^+ + \lambda_t^- \tag{2.7.9}$$

$$\left.\begin{array}{lll} 0 \leq p_n & \perp & g - C_n u + \tau p_n \geq 0 \\ 0 \leq \lambda_t^+ & \perp & \mu p_n - p_t + \tau \lambda_t^+ \geq 0 \\ 0 \leq \lambda_t^- & \perp & \mu p_n + p_t + \tau \lambda_t^- \geq 0 \end{array}\right\}, \tag{2.7.10}$$

we have $\|\mathbf{x}\| \leq \eta$.

Proof. We first derive some general properties of an arbitrary solution $\mathbf{x}$ of (2.7.8)–(2.7.10) corresponding to a given positive τ. By (2.7.7), it follows from (2.7.8) that there exists a scalar $c > 0$, independent of τ, such that

$$\|(p_t, p_n)\| \leq c[\|f^{\text{ext}}\| + (1+\tau)\|u\|]. \tag{2.7.11}$$

From (2.7.10), we have

$$p_n^T(g - C_n u + \tau p_n) = 0$$

$$(\lambda_t^+)^T(\mu p_n - p_t + \tau \lambda_t^+) = 0$$

$$(\lambda_t^-)^T(\mu p_n + p_t + \tau \lambda_t^-) = 0.$$

Premultiplying (2.7.8) by u^T, (2.7.9) by p_t, and using the above equalities, we deduce

$$\begin{aligned} u^T M u + \tau[u^T u + p_n^T p_n + p_t^T p_t + (\lambda_t^+)^T(\lambda_t^+) + (\lambda_t^-)^T(\lambda_t^-)] \\ + \mu p_n^T(\lambda_t^+ + \lambda_t^-) = u^T f^{\text{ext}} - p_n^T g - p_t^T C_t u^{\text{ref}}. \end{aligned} \tag{2.7.12}$$

Consequently, we obtain, for some constant $c' > 0$,

$$\begin{aligned} \lambda_{\min}(M)\|u\|^2 + \tau\|\mathbf{x}\|^2 &\leq c'[\|u\| + \|(p_n, p_t)\|] \\ &\leq c'[\|u\| + c(\|f^{\text{ext}}\| + (1+\tau)\|u\|)]. \end{aligned}$$

We also note that, for each $i = 1, \ldots, n_c$, $\lambda_{it}^+ \lambda_{it}^- = 0$. If not, then for some i both λ_{it}^+ and λ_{it}^- are positive. This implies

$$\mu p_{in} - p_{it} + \tau \lambda_{it}^+ = 0 \quad \text{and} \quad \mu p_{in} + p_{it} + \tau \lambda_{it}^- = 0.$$

Adding these two equations yields a contradiction.

Assume for the sake of contradiction that the lemma is false; i.e., no such constant η exists. There exist then a sequence of positive scalars $\{\tau_k\}$ and a sequence of tuples $\{\mathbf{x}^k \equiv (u^k, p_t^k, p_n^k, \lambda_t^{k,\pm})\}$ such that

$$\lim_{k\to\infty} \| \mathbf{x}^k \| = \infty$$

and, for each k, $\mathbf{x}^k$ is a solution of the MLCP (2.7.8)–(2.7.10) corresponding to τ_k. We claim that $\{\tau_k\}$ must tend to zero as $k \to \infty$. Assume for contradiction that

$$\limsup_{k\to\infty} \tau_k > 0.$$

We have, for every k,

$$\tau_k \| \mathbf{x}^k \|^2 \leq c' [\| u^k \| + c(\| f^{\text{ext}} \| + (1 + \tau_k) \| u^k \|)].$$

Dividing by $\|\mathbf{x}^k\|^2$ and taking the limsup as $k \to \infty$, we see that the left-hand side tends to a positive limit whereas the right-hand side tends to zero. This contradiction establishes our claim about $\{\tau_k\}$. We have, for every k,

$$\lambda_{\min}(M) \| u^k \|^2 \leq c' [\| u^k \| + c(\| f^{\text{ext}} \| + (1 + \tau_k) \| u^k \|)].$$

Dividing by $\|\mathbf{x}^k\|^2$ and taking the limit as $k \to \infty$, we obtain

$$\lim_{k\to\infty} \frac{u^k}{\| \mathbf{x}^k \|} = 0.$$

From (2.7.11), we deduce

$$\lim_{k\to\infty} \frac{(p_t^k, p_n^k)}{\| \mathbf{x}^k \|} = 0.$$

Since, for every k,

$$\tau_k p_t^k - C_t(u^k - u^{\text{ref}}) = -\lambda_t^{k,+} + \lambda_t^{k,-},$$

dividing by $\|\mathbf{x}^k\|$, letting $k \to \infty$, and using the fact that $\lambda_t^{k,+}$ and $\lambda_t^{k,-}$ are complementary for each k, we deduce that

$$\lim_{k\to\infty} \frac{\lambda_t^{k,\pm}}{\| \mathbf{x}^k \|} = 0.$$

In summary, we have shown that

$$\lim_{k\to\infty} \frac{\mathbf{x}^k}{\| \mathbf{x}^k \|} = 0,$$

which is a contradiction. □

Combining the above lemma and Theorem **2.6.1**, we immediately obtain the following existence result, which requires no proof.

2.7.3 Proposition. Under assumption (A) and (2.7.6), the friction problem (2.7.1)–(2.7.4) has a solution. □

The semicoercive case

We relax condition (A) and assume that the stiffness matrix M is only semidefinite. By Proposition **2.7.3**, it follows that for every $\varepsilon > 0$, there exists a tuple

$$\mathbf{x}(\varepsilon) \equiv (\, u(\varepsilon), p_t(\varepsilon), p_n(\varepsilon), \lambda_t^{\pm}(\varepsilon)\,)$$

satisfying the system (2.7.8)–(2.7.10) with M perturbed by $\varepsilon\,\mathbf{I}$; that is, $\mathbf{x}(\varepsilon)$ solves the MLCP $(\mathbf{F}_\varepsilon)$, where

$$\mathbf{F}_\varepsilon(u, p_t, p_n, \lambda_t^{\pm}) \equiv \begin{pmatrix} Mu + \varepsilon\, u + C_t^T p_t + C_n^T p_n - f^{\text{ext}} \\ -C_t(\, u - u^{\text{ref}}\,) + \lambda_t^+ - \lambda_t^- \\ g - C_n u \\ \mu\, p_n - p_t \\ \mu\, p_n + p_t \end{pmatrix},$$

If we can show that

$$\lim_{\varepsilon \downarrow 0} \varepsilon\, u(\varepsilon) = 0, \tag{2.7.13}$$

then, since the MLCP range of a matrix is a closed set, it follows that the MLCP $(\mathbf{F})$, and thus the frictional model (2.7.1)–(2.7.4), has a solution. Toward establishing (2.7.13), we can show, similar to (2.7.12), that for each $\varepsilon > 0$,

$$u(\varepsilon)^T(\, M + \varepsilon\,\mathbf{I}\,)\, u(\varepsilon) \le u(\varepsilon)^T f^{\text{ext}} - p_n(\varepsilon)^T g - p_t(\varepsilon)^T C_t u^{\text{ref}}. \tag{2.7.14}$$

Let X denote the set of triples (u, p_t, p_n) satisfying:

$$0 \le p_n \perp g - C_n u \ge 0$$

and for each $i = 1, \ldots, n_c$,

$$p_{it} \begin{cases} = +\mu\, p_{in} & \text{if } C_{it}(u - u^{\text{ref}}) > 0 \\ \in [\, -\mu\, p_{in},\, +\mu\, p_{in}\,] & \text{if } C_{it}(u - u^{\text{ref}}) = 0 \\ = -\mu\, p_{in} & \text{if } C_{it}(u - u^{\text{ref}}) < 0. \end{cases}$$

In terms of the alternate formulation (2.7.1), (2.7.2), and (2.7.5), we see that the set X serves like the "feasible set" of the friction problem; that is, every solution triple (u, p_t, p_n) of the problem must be an element of X. Making use of this set X, we establish the following existence result for the semicoercive frictional contact problem.

2.7.4 Proposition. Let M be a positive semidefinite matrix. Suppose that (2.7.6) holds. If the linear function

$$(u, p_t, p_n) \mapsto u^T f^{\text{ext}} - p_n^T g - p_t^T C_t u^{\text{ref}}$$

is bounded above on the set X, then the frictional contact problem (2.7.1)–(2.7.4) has a solution.

Proof. Continuing the argument from above, we note that the triple $(u(\varepsilon), p_t(\varepsilon), p_n(\varepsilon))$ must be an element of X. Thus by (2.7.14), we deduce the existence of a scalar $\gamma > 0$ such that

$$u(\varepsilon)^T (M + \varepsilon \mathbf{I}) u(\varepsilon) \leq \gamma,$$

for all $\varepsilon > 0$. Since M is positive semidefinite, we obtain

$$\varepsilon \| u(\varepsilon) \|_2^2 \leq \gamma$$

for all $\varepsilon > 0$. This implies the desired limit (2.7.13). □

The limit (2.7.13) is not sufficient to complete the analysis for the full model described in Subsection **1.4.6**, which is defined by a MiCP with a nonlinear function $\mathbf{F}$. We actually need the trajectory $\{u(\varepsilon)\}$ to be bounded. Thus the bounded-above condition introduced in Proposition **2.7.4** has to be strengthened in order to ensure the boundedness of $\{u(\varepsilon)\}$. We omit the details.

2.8 Extended Problems

We close this chapter with a brief discussion of some existence results for two generalizations of the basic VI and CP. The first extended problem to be discussed herein is the QVI (K, F) where $K : \mathbb{R}^n \to \mathbb{R}^n$ is a set-valued map and $F : \mathbb{R}^n \to \mathbb{R}^n$ is a point-to-point function. Throughout the discussion, we assume that for each $x \in \operatorname{dom} K$, $K(x)$ is a closed convex subset of $\mathbb{R}^n$; that is, K is a closed-valued and convex-valued multifunction (with $K(x)$ being empty for $x \notin \operatorname{dom} K$). Similar to the natural map of the standard VI, we can define the natural map of this QVI:

$$\mathbf{F}_K^{\text{nat}}(x) \equiv x - \Pi_{K(x)}(x - F(x)), \quad x \in \operatorname{dom} K. \tag{2.8.1}$$

Extending Proposition **1.5.8**, we can establish the following equivalent formulation of the QVI as a system of nonsmooth equation.

2.8.1 Proposition. Let $K : \mathbb{R}^n \to \mathbb{R}^n$ be a closed-valued and convex-valued point-to-set map. Let $F : \mathbb{R}^n \to \mathbb{R}^n$ be a given point-to-point function. A vector x solves the QVI (K, F) if and only if x is a zero of the natural map $\mathbf{F}_K^{\rm nat}$ of the pair (K, F) defined in (2.8.1). □

In order to utilize the equation (2.8.1) to establish the existence of a solution to the QVI (K, F), our first order of business is to ensure that the natural map $\mathbf{F}_K^{\rm nat}$ is continuous. Of course, this is not a trivial matter due the dependence of K on the variable x. For this purpose, we establish the following result that provides a necessary and sufficient condition for the parameter dependent projector $\Pi_{K(x)}(y)$ to be a continuous function of the two arguments (x, y). No special form of the set $K(x)$ is assumed.

2.8.2 Lemma. Let $K : \mathbb{R}^p \to \mathbb{R}^n$ be a closed-valued and convex-valued point-to-set map. Let $\bar{x} \in \operatorname{dom} K$ be given. The map $\Phi(x, y) \equiv \Pi_{K(x)}(y)$ is continuous at $(\bar{x}, y)$ for all $y \in \mathbb{R}^n$ if and only if

$$\lim_{x \to \bar{x}} K(x) = K(\bar{x}). \tag{2.8.2}$$

Proof. Suppose that Φ is continuous at $(\bar{x}, y)$ for all $y \in \mathbb{R}^n$. This implies in particular that for all x sufficiently close to $\bar{x}$, $K(x)$ is nonempty; thus $\Pi_{K(x)}$ is well defined. We first show that

$$\limsup_{x \to \bar{x}} K(x) \subseteq K(\bar{x}).$$

Let z be a vector belonging to the left-hand set. There exists a sequence $\{x^k\}$ converging to $\bar{x}$ such that

$$\lim_{k \to \infty} \operatorname{dist}(z, K(x^k)) = 0.$$

Let $y^k \equiv \Pi_{K(x^k)}(z)$ for each k. We have

$$z = \lim_{k \to \infty} y^k = \lim_{k \to \infty} \Pi_{K(x^k)}(z) = \Pi_{K(\bar{x})}(z),$$

where the last equality is by the continuity of $\Pi_{K(x)}(z)$ at $\bar{x}$ for z fixed. Consequently, z belongs to $K(\bar{x})$. We next show

$$K(\bar{x}) \subseteq \liminf_{x \to \bar{x}} K(x).$$

Let $v \in K(\bar{x})$ be arbitrary. We have

$$v = \Pi_{K(\bar{x})}(v) = \lim_{x \to \bar{x}} \Pi_{K(x)}(v).$$

Since

$$\operatorname{dist}(v, K(x)) = \| v - \Pi_{K(x)}(v) \|,$$

and the right-hand side approaches zero as x tends to $\bar{x}$, it follows that v belongs to $\liminf_{x \to \bar{x}} K(x)$. Consequently (2.8.2) holds.

Conversely, suppose that (2.8.2) holds. In particular, it follows that $K(x)$ is nonempty for all x sufficiently close to $\bar{x}$. Let $\{(x^k, y^k)\}$ be an arbitrary sequence of vector pairs converging to $(\bar{x}, y^\infty)$. Without loss of generality, we may assume that $K(x^k)$ is nonempty for all k. It suffices to show that

$$\Pi_{K(\bar{x})}(y^\infty) = \lim_{k \to \infty} \Pi_{K(x^k)}(y^k).$$

Let v be an arbitrary element of $K(\bar{x})$. By the lower semicontinuity of K at $\bar{x}$, there exists for each k a vector $v^k \in K(x^k)$ such that the sequence $\{v^k\}$ converges to v. We have

$$\| \Pi_{K(x^k)}(y^k) - y^k \| \leq \| v^k - y^k \|.$$

The right-hand term converges to $\|v - y^\infty\|$ as $k \to \infty$. Thus the above inequality implies that the sequence $\{\Pi_{K(x^k)}(y^k)\}$ is bounded; moreover, every accumulation point z of this sequence must satisfy:

$$\| z - y^\infty \| \leq \| v - y^\infty \|, \quad \forall\, v \in K(\bar{x})$$

and z must belong to $K(\bar{x})$. Consequently, z is the (unique) projection of y^∞ onto $K(\bar{x})$. Hence, the sequence $\{\Pi_{K(x^k)}(y^k)\}$ has a unique accumulation point that is the Euclidean projection of y^∞ onto the set $K(\bar{x})$. Thus, $\Pi_{K(x)}(y)$ is continuous at $(\bar{x}, y^\infty)$. □

Based on the above lemma and degree theory, we can state an existence result for the QVI that extends part (a) of Theorem **2.2.1**. The proof of the extended result follows immediately from Proposition **2.8.1** and Lemma **2.8.2**.

2.8.3 Theorem. Let $K : \mathbb{R}^n \to \mathbb{R}^n$ be a closed-valued and convex-valued point-to-set map. Let $F : \mathbb{R}^n \to \mathbb{R}^n$ be a continuous function. Suppose there exists a bounded open set $\Omega \subset \mathbb{R}^n$ such that (2.8.2) holds at every $\bar{x} \in \operatorname{cl}\Omega$ and $\deg(\mathbf{F}_K^{\text{nat}}, \Omega)$ is well defined and nonzero. Then the QVI (K, F) has a solution in Ω. □

As an application of this theorem, we state a special existence result for the QVI that extends the corresponding existence result for the VI described in Proposition **2.2.3**.

2.8.4 Corollary. Let $K : \mathbb{R}^n \to \mathbb{R}^n$ be a closed-valued and convex-valued point-to-set map. Let $F : \mathbb{R}^n \to \mathbb{R}^n$ be a continuous function. Suppose there exist a bounded open set $\Omega \subset \mathbb{R}^n$ and a vector $x^{\text{ref}} \in \Omega$ such that

(a) K is nonempty-valued and (2.8.2) holds at every $\bar{x} \in \operatorname{cl}\Omega$;

(b) x^{ref} belongs to $K(x)$ for every $x \in \operatorname{cl}\Omega$; and

(c) the following holds:

$$\{ x \in K(x) : (x - x^{\text{ref}})^T F(x) < 0 \} \cap \operatorname{bd}\Omega = \emptyset.$$

The QVI (K, F) has a solution.

Proof. For the sake of contradiction, assume that the QVI (K, F) has no solution. Define the homotopy:

$$H(x,t) \equiv x - \Pi_{K(x)}(t(x - F(x)) + (1-t)x^{\text{ref}}), \quad (x,t) \in \operatorname{cl}\Omega \times [0,1].$$

This is a continuous homotopy with

$$H(x,0) = x - \Pi_{K(x)}(x^{\text{ref}}) = x - x^{\text{ref}}, \quad \forall\, x \in \operatorname{cl}\Omega$$

and $H(x,1) = \mathbf{F}_K^{\text{nat}}(x)$ for all $x \in \operatorname{cl}\Omega$. We claim that

$$\bigcup_{t \in (0,1)} H(\cdot,t)^{-1}(0) \cap \operatorname{bd}\Omega = \emptyset.$$

Suppose $H(x,t) = 0$ for some $(x,t) \in \operatorname{bd}\Omega \times (0,1)$. Then $x \neq x^{\text{ref}}$; moreover, $x \in K(x)$ and

$$(y - x)^T [x - t(x - F(x)) - (1-t)x^{\text{ref}}] \geq 0, \quad \forall\, y \in K.$$

In particular, for $y = x^{\text{ref}}$, we deduce

$$(x^{\text{ref}} - x)^T [tF(x) + (1-t)(x - x^{\text{ref}})] \geq 0,$$

which implies

$$(x^{\text{ref}} - x)^T F(x) \geq \frac{1-t}{t} \| x - x^{\text{ref}} \|_2^2 > 0$$

where the last inequality holds because $t \in (0,1)$ and $x \neq x^{\text{ref}}$. Thus x does not belong to $\operatorname{bd}\Omega$ by condition (c). Consequently, by the homotopy invariance property of the degree, we readily deduce that $\deg(\mathbf{F}_K^{\text{nat}}, \Omega)$ is well defined and equal to unity. Thus the QVI (K, F) has a solution. This is a contradiction. □

Corollary **2.8.4** provides the basis for deriving a host of existence results for the QVI. As it presently stands, conditions (a), mainly the lower semicontinuity of the multifunction K, and condition (b) are fairly restrictive. Condition (c) will easily hold under certain assumptions such as the boundedness of the range of K or the coerciveness of F on this range. Compared to the development of the VI, that of the QVI is still very much in its infancy. The further pursuance of the QVI is beyond the scope of this book. See Exercise **4.8.14** for a further existence result for a QVI constrained by convex inequalities.

We turn to another extended problem. Consider the vertical CP defined by a pair of functions (F, G):

$$0 \leq F(x) \perp G(x) \geq 0;$$

this problem is equivalent to the nonsmooth equation:

$$H(x) \equiv \min(F(x), G(x)) = 0.$$

From the point of view of analysis, this problem is most interesting when neither F nor G is a homeomorphism. Indeed, suppose one of them, say F, is a homeomorphism with inverse F^{-1}. Letting $y \equiv F(x)$ so that $x \equiv F^{-1}(y)$, the CP (F, G) is clearly equivalent to the standard NCP:

$$0 \leq y \perp G \circ F^{-1}(y) \geq 0.$$

In what follows, we present an existence result for the CP (F, G) that requires one of the two functions to be injective but not bijective.

2.8.5 Proposition. Let F and G be two continuous functions from $\operatorname{cl}\mathcal{D}$, where $\mathcal{D}$ is an open subset of $\mathbb{R}^n$, into $\mathbb{R}^n$ with G being injective. Suppose that there exists a vector $u \in \operatorname{cl}\mathcal{D}$ satisfying $G(u) \geq 0$ and there are positive constants c, L, and γ such that for all vectors $x \in \operatorname{cl}\mathcal{D}$ satisfying $G(x) \geq 0$ and $\|x\| \geq c$, it holds that

(a) $\|G(x) - G(u)\| \leq L\|x - u\|$;

(b) $\max\limits_{1 \leq i \leq n} (F_i(x) - F_i(u))(G_i(x) - G_i(u)) \geq \gamma\|x - u\|^2$.

The CP (F, G) has a solution.

Proof. Assume for the sake of contradiction that the CP (F, G) has no solution. Define the homotopy: for all $(x, t) \in \operatorname{cl}\mathcal{D} \times [0, 1]$,

$$H(x, t) \equiv \min\{G(x), t\,F(x) + (1 - t)\,(G(x) - G(u))\}.$$

We have $H(x, 1) = \min(F(x), G(x))$ and

$$H(x, 0) = \min(G(x), G(x) - G(u)) = G(x) - G(u)$$

because $G(u)$ is nonnegative. Since G is injective, it follows that

$$\deg(H(\cdot,0),\Omega) = \pm 1$$

for every bounded open set $\Omega \subseteq \mathcal{D}$ containing u. By the homotopy invariance property of the degree, it remains to show that

$$\bigcup_{t\in(0,1)} H(\cdot,t)^{-1}(0)$$

is bounded. Assume otherwise. Let $\{t_k\} \subset (0,1)$ be a sequence of scalars and $\{x^k\} \subset \operatorname{cl}\mathcal{D}$ be a sequence of vectors such that

$$\lim_{k\to\infty} \| x^k \| = \infty \quad \text{and} \quad H(x^k, t_k) = 0 \quad \forall\, k.$$

Thus, we have

$$0 \leq G(x^k) \perp t_k F(x^k) + (1 - t_k)(G(x^k) - G(u)) \geq 0.$$

Hence, for each index i,

$$0 \geq (G_i(x^k) - G_i(u))\,[\,t_k F_i(x^k) + (1 - t_k)\,(G_i(x^k) - G_i(u))\,],$$

which implies

$$\begin{aligned} 0 &\geq (G_i(x^k) - G_i(u))\,F_i(x^k) \\ &= (G_i(x^k) - G_i(u)\,(F_i(x^k) - F_i(u)) + (G_i(x^k) - G_i(u))\,F_i(u). \end{aligned}$$

Consequently,

$$\| G(x^k) - G(u) \| \, \| F(u) \| \geq \max_{1\leq i\leq n} (G_i(x^k) - G_i(u))(F_i(x^k) - F_i(u)).$$

The left-hand term is bounded above by $L\,\|x^k - u\|\,\|F(u)\|$ whereas the right-hand term is bounded below by $\gamma\,\|x^k - u\|^2$. This is a contradiction because $\|x^k\|$ is unbounded. □

An alternative approach to deal with the vertical CP (F,G) is to consider it as a special case of the following implicit CP:

$$H(u,v,x) \equiv \begin{pmatrix} u - F(x) \\ v - G(x) \end{pmatrix} = 0$$

$$0 \leq u \perp v \geq 0.$$

An extensive theory for the latter CP is presented in Section **11.4**, which can be used to obtain alternative existence results for the CP (F,G); see in particular part (b) of Corollary **11.4.19**.

2.9 Exercises

2.9.1 Let $\Phi : \Omega \subseteq \mathbb{R}^n \to \mathbb{R}^n$ be continuously differentiable, where Ω is open and bounded. Let $p \notin \Phi(\mathrm{bd}\,\Omega)$ be a given vector. Show that if $J\Phi(x)$ is nonsingular at every $x \in \Phi^{-1}(p)$, then $\Phi^{-1}(p)$ is a finite set.

2.9.2 Let

$$F(x_1, x_2) \equiv \begin{pmatrix} \min(x_1^2 + x_2, |x_2|) \\ x_1 \end{pmatrix}, \qquad (x_1, x_2) \in \mathbb{R}^2.$$

Show that the origin is the unique zero of F and that the index of F at the origin is equal to negative one. Does F have a Fréchet derivative at the origin? (Hint: to show the index assertion, homotopize F with a suitable linear map with a negative determinant.)

2.9.3 Let $F : \mathbb{R}^n \to \mathbb{R}^n_+$ be a continuous function with the origin as its unique zero. Show that $\mathrm{ind}(F, 0) = 0$. (Hint: homotopize F with a suitable constant map; use the converse of Theorem **2.1.2** and the homotopy invariance of degree.)

2.9.4 Let $F : [a, b] \to \mathbb{R}$ be a continuous function. Show that

$$\deg(F, (a,b)) = \begin{cases} 0 & \text{if } F(a)F(b) > 0 \\ -1 & \text{if } F(a) > 0 \text{ and } F(b) < 0 \\ 1 & \text{if } F(a) < 0 \text{ and } F(b) > 0. \end{cases}$$

2.9.5 A function $F : \mathcal{D} \subseteq \mathbb{R}^n \to \mathbb{R}^n$ is said to be (a) *inverse isotone* if

$$F(x) \leq F(y) \Rightarrow x \leq y;$$

(b) an *S function* (or S_0 function) if for every $x \in \mathcal{D}$, there exists $y \in \mathcal{D}$ such that $x \neq y \geq x$ and $F(y) > F(x)$ (or $F(y) \geq F(x)$). Show that

(a) if F is inverse isotone on $\mathcal{D}$, then it must be injective there;

(b) if $\mathcal{D}$ is open and convex, F is F-differentiable on $\mathcal{D}$, and each component function F_i is convex, then F is inverse isotone on $\mathcal{D}$ if and only if $JF(x)$ is invertible and has a nonnegative inverse for all $x \in \mathcal{D}$;

(c) if $\mathcal{D}$ is open and if F is inverse isotone on $\mathcal{D}$, then F is an S function on $\mathcal{D}$. (Hint: use the domain invariance theorem.)

2.9.6 Let $F : \mathcal{D} \subseteq \mathbb{R}^n \to \mathbb{R}^n$ be a local homeomorphism on the open set $\mathcal{D}$. Suppose that $F + \tau\mathbf{I}$ is injective on $\mathcal{D}$ for all $\tau > 0$. Use Proposition **2.1.6** to show that F is injective on $\mathcal{D}$.

2.9.7 Let $K \subseteq \mathbb{R}^n$ be closed convex and $F : K \to \mathbb{R}^n$ be continuous. Show that either the VI (K, F) has a solution or for every $x^{\text{ref}} \in K$ there exists an unbounded sequence $\{x^k\} \subset K$ such that $F(x^k)^T(x^k - x^{\text{ref}}) < 0$ for all k.

2.9.8 This is a classical existence result for the VI that can be proved by Proposition **2.2.8**. Let $K \subseteq \mathbb{R}^n$ be closed convex and $F : K \to \mathbb{R}^n$ be continuous. Suppose that there exists a bounded subset D of K such that for every $x \in K \setminus D$, a vector $y \in D$ exists satisfying

$$F(x)^T(x - y) \geq 0.$$

Show that the VI (K, F) has a solution. (Hint: let E be a sufficiently large closed Euclidean ball such that D is contained in the interior of E.)

2.9.9 Let K be a closed convex set such that $\text{int}(K_\infty)^*$ is nonempty. Let $F : \mathbb{R}^n \to \mathbb{R}^n$ be continuous.

(a) Suppose that $\deg(\mathbf{F}_K^{\text{nat}}, \mathcal{U})$ is well defined and nonzero for some bounded open set $\mathcal{U}$. Show that (2.3.5) holds.

(b) Show that either a vector $x^{\text{ref}} \in K$ exists with $F(x^{\text{ref}}) \in \text{int}(K_\infty)^*$ or for every $v \in \text{int}(K_\infty)^*$ and every $x^{\text{ref}} \in K$ an unbounded sequence $\{x^k\} \subset K$ exists such that $(F(x^k) - v)^T(x^k - x^{\text{ref}}) < 0$ for all k.

2.9.10 Use (i) Brouwer fixed-point Theorem **2.1.18** and (ii) Kakutani fixed-point Theorem **2.1.19** to prove Corollary **2.2.5**. (Hint: for (i), use the natural map $\mathbf{F}_K^{\text{nat}}$; for (ii), consider the set-valued map $\Phi : K \to K$, where for each $x \in K$, $\Phi(x) \equiv \text{ argmin } \{y^T F(x) : y \in K\}$.)

2.9.11 Let K be a closed convex subset of $\mathbb{R}^n$, $F : K \to \mathbb{R}^n$ be continuous and $\varphi : K \to \mathbb{R}$ be convex and continuous. Consider the hemivariational inequality of finding a vector $x \in K$ satisfying

$$F(x)^T(y - x) + \varphi(y) - \varphi(x) \geq 0, \quad \forall y \in K.$$

(a) Use Kakutani's fixed-point theorem to show that if K is bounded, then a solution exists.

(b) Suppose that φ is coercive on K and there exist a vector $x^{\text{ref}} \in K$ and a scalar $\eta < 0$ such that the set

$$L_< \equiv \{x \in K : F(x)^T(x - x^{\text{ref}}) < \eta\}$$

is bounded (possibly empty). Show that a solution to the hemivariational inequality exists. (Hint: for every scalar $t > 0$, consider the hemivariational inequality on the truncated set $K \cap \text{cl}\,\mathbb{B}(0, t)$. Use part (a) and take $t \to \infty$.)

(c) Suppose that φ is coercive on K and there exist a vector $x^{\text{ref}} \in K$ and a scalar $\zeta \geq 0$ such that (2.2.2) holds. Show that a solution to the hemivariational inequality exists.

2.9.12 The Euclidean projection onto a closed but nonconvex set defines a set-valued map. More precisely, if S is a closed subset of $\mathbb{R}^n$, then for every $x \in \mathbb{R}^n$,

$$\Pi_S(x) \equiv \{ y \in S : \| x - y \| = \operatorname{dist}(x, S) \}.$$

Let $K : \mathbb{R}^n \to \mathbb{R}^n$ be a closed-valued point-to-set map that is continuous at $x^* \in \operatorname{dom} K$. Show that for every $y^* \in \mathbb{R}^n$,

$$\limsup_{(x,y) \to (x^*,y^*)} \Pi_{K(x)}(y) = \Pi_{K(x^*)}(y^*).$$

2.9.13 Let K be a closed convex subset of $\mathbb{R}^n$.

(a) Use Theorem **1.5.5** to show that $\|x - \Pi_K(x)\|^2$ and $\|x\|^2 - \|x - \Pi_K(x)\|^2$ are convex functions.

(b) Use (a) to show that $\mathbf{I} - \Pi_K$ is a co-coercive function. (This can also be verified directly as in part (c) of Theorem **1.5.5**.)

(c) Show that if $G : \mathbb{R}^m \to \mathbb{R}^n$ is an affine function, then $\operatorname{dist}(G(y), K)^2$ is a convex function on $\mathbb{R}^m$.

2.9.14 Employ Proposition **2.2.8** to prove the nonemptiness assertion in Proposition **2.2.7**.

2.9.15 A set-valued map $\Phi : \mathbb{R}^n \to \mathbb{R}^n$ is said to be (strongly) monotone on $\Omega \subseteq \operatorname{dom} \Phi$ if there exists a constant $c(>) \geq 0$ such that

$$(x - y)^T (u - v) \geq c \| x - y \|^2$$

for all x and y in Ω, and all u in $\Phi(x)$ and v in $\Phi(y)$. Let $F : X \to \mathbb{R}^n$ be a point-to-point function. The inverse of F is a set-valued map with domain $F(X)$. Show that F is co-coercive on X if and only if F^{-1} is strongly monotone on $F(X)$.

2.9.16 Prove Proposition **2.3.2**.

2.9.17 This exercise generalizes Proposition **2.3.11** by relaxing the strong monotonicity assumption. Let $F : K \to \mathbb{R}^n$ be continuous and monotone on the closed convex set $K \subseteq \mathbb{R}^n$. Let $\Phi(q)$ denote the (possibly empty) solution set of the VI $(K, F - q)$. Show that Φ is a monotone set-valued map on its domain. Show further that if F is strictly monotone, then Φ is a monotone plus single-valued map on its domain.

2.9.18 Let X and Y be closed convex subsets of $\mathbb{R}^n$ and $\mathbb{R}^m$ respectively. Let

$$L(x,y) = p^T x + q^T y + \tfrac{1}{2} x^T P x + x^T R y - \tfrac{1}{2} y^T Q y, \quad (x,y) \in \mathbb{R}^{n+m},$$

where the matrices P and Q are symmetric positive semidefinite. Show that if (x^1, y^1) and (x^2, y^2) are two pairs of saddle points of the triple (L, X, Y), then

$$P x^1 = P x^2 \quad Q y^1 = Q y^2,$$

and

$$(x^1 - x^2)^T (p + R y^1) = (y^1 - y^2)^T (q + R^T x^1) = 0.$$

Show further that if X and Y are polyhedra with the representation:

$$X \equiv \{ x \in \mathbb{R}^n : Ax \leq b \}$$

and

$$Y \equiv \{ y \in \mathbb{R}^m : Cy \leq d \},$$

the set of tuples (p, q, b, d) for which the saddle problem (L, X, Y) has a solution is a polyhedral cone (with the matrices A, C, P, R, and Q fixed).

2.9.19 Let K be a closed convex set and $F : K \to \mathbb{R}^n$ be a continuous mapping such that for some constant $c \in (0,1)$,

$$\| F(x) - x \| \leq c \| x \|, \quad \forall x \in K.$$

By applying Proposition **2.1.8** to the normal map $\mathbf{F}_K^{\text{nor}}$, show that the VI $(K, q + F)$ has a solution for all vectors $q \in \mathbb{R}^n$.

2.9.20 Let K be a closed convex set in $\mathbb{R}^n$ and let $M \equiv A^T E A$, where E is a symmetric positive semidefinite $m \times m$ matrix and A is an arbitrary $m \times n$ matrix.

(a) Show that for every $q \in \mathcal{R}(K, M)$ (the VI range of the pair (K, M)), if x^1 and x^2 are any two solutions of the VI (K, q, M), $EAx^1 = EAx^2$. Let $\tilde{w}(q)$ denote the common vector EAx for any solution x of the VI (K, q, M).

(b) Suppose that K is a polyhedral cone. Show that a constant $c > 0$ exists satisfying

$$\| \tilde{w}(q) \| \leq c \| q \|, \quad \forall q \in \mathcal{R}(K, M).$$

Deduce that the function $\tilde{w} : \mathcal{R}(K, M) \to \mathbb{R}^n$ is continuous on its domain. (See Exercise **5.6.15** for a strengthening of this continuity property that holds for a polyhedral set K.)

2.9.21 Let K be the Cartesian product of two polyhedra $K_1 \subseteq \mathbb{R}^{n_1}$ and $K_2 \subset \mathbb{R}^{n_2}$ with K_2 being compact. Let F be defined by: for (x, y) in $\mathbb{R}^{n_1+n_2}$,

$$F(x,y) \equiv \begin{pmatrix} q \\ r \end{pmatrix} + \begin{bmatrix} M_{11} & M_{12} \\ M_{21} & M_{22} \end{bmatrix} \begin{pmatrix} x \\ y \end{pmatrix} + \begin{pmatrix} 0 \\ h(y) \end{pmatrix},$$

where $(q, r) \in \mathbb{R}^{n_1+n_2}$, $h : \mathbb{R}^{n_2} \to \mathbb{R}^{n_2}$ is a continuous function and the matrix

$$\begin{bmatrix} M_{11} & M_{12} \\ M_{21} & M_{22} \end{bmatrix} \in \mathbb{R}^{(n_1+n_2)\times(n_1+n_2)}$$

is positive semidefinite (not necessarily symmetric). Assume that for all $\hat{y} \in K_2$, the AVI $(K, F^{\hat{y}})$ has a solution, where

$$F^{\hat{y}}(x,y) \equiv \begin{pmatrix} q \\ r + h(\hat{y}) \end{pmatrix} + \begin{bmatrix} M_{11} & M_{12} \\ M_{21} & M_{22} \end{bmatrix} \begin{pmatrix} x \\ y \end{pmatrix}, \qquad (x,y) \in \mathbb{R}^{n_1+n_2}.$$

Show that the VI (K, F) has a solution. (Hint: apply Kakutani's fixed-point theorem to the set-valued map $\Phi : K_2 \to K_2$ defined as follows. For each $\hat{y} \in K_2$, $\Phi(\hat{y})$ consists of all vectors $y \in K_2$ for which there exists a vector $x \in K_1$ such that the pair (x, y) solves the VI $(K, F^{\hat{y}})$.)

2.9.22 Let K be a closed convex set in $\mathbb{R}^n$, $f : \mathcal{D} \supset K \to \mathbb{R}$ be a convex, continuously differentiable function on the open set $\mathcal{D}$, and $F : K \to \mathbb{R}^n$ be continuous. Define the saddle function

$$\tilde{L}(x,y) \equiv f(x) - f(y) + [\,F(x) - \nabla f(x)\,]^T (\,x - y\,).$$

Show that if (x^*, y^*) is a saddle point of L on the set $K \times K$, then x^* is a solution of the VI (K, F). (Hint: use Theorem **1.4.1**.)

2.9.23 Let K be a closed convex subset of $\mathbb{R}^n$ and $F : K \to \mathbb{R}^n$ be a given function. Consider the dual gap function

$$\theta_{\text{dual}}(x) \equiv \inf_{y \in K} F(y)^T(\,y - x\,), \quad x \in \mathbb{R}^n.$$

For every $x \in \mathbb{R}^n$ let $\Lambda(x)$ denote the (possibly empty) argmin of the minimization problem defining $\theta_{\text{dual}}(x)$. Suppose that F is pseudo monotone plus on K. Show that if $x^* \in \text{SOL}(K, F)$, then $F(y)$ is equal to $F(x^*)$ for every $y \in \Lambda(x^*)$. Deduce that $\Lambda(x^*) = \text{SOL}(K, F)$. See Exercise **10.5.3** for further properties of the dual gap function with a compact set K.

2.9.24 Let $M \in \mathbb{R}^{n\times n}$ and $F(x) \equiv Mx$. Show that the following statements are equivalent.

(a) F is co-coercive on $\mathbb{R}^n$.

(b) $M = E^T A E$ for some matrices $E \in \mathbb{R}^{m \times n}$ and $A \in \mathbb{R}^{m \times m}$ for some positive integer m, where A is positive definite. (This is the "affine strongly monotone composite" property.)

(c) M is positive semidefinite and $Mx = 0$ whenever $x^T M x = 0$. (This is the "positive semidefinite plus" property.)

2.9.25 Let $F : C \subseteq \mathbb{R}^n \to \mathbb{R}^n$ be continuously differentiable on the open convex set C.

(a) Show that F is co-coercive on C if and only if there exists a constant $\eta > 0$ such that

$$y^T JF(x) y \geq \eta \| JF(x) y \|^2, \quad \forall x \in C \text{ and } \forall y \in \mathbb{R}^n.$$

(b) Show that if F is a gradient function, then F is co-coercive on C if and only if F is monotone and Lipschitz continuous on C.

2.9.26 Let C be a closed convex cone in $\mathbb{R}^n$.

(a) Show that C is solid if and only if it is *reproducing*, i.e., if and only if $\mathbb{R}^n = C - C$. (Hint: if C is solid, $C - C$ is a full-dimensional linear subspace. The converse can be proved by showing that C^* is pointed if $\mathbb{R}^n = C - C$ and by applying part (b) of Proposition **2.4.3**.)

(b) Show that $C \cap C^* = \{0\}$ if and only if C is a subspace.

(c) Show that C contains C^* if and only if for every $x \in \mathbb{R}^n$ there exist u and v in C such that $x = u - v$ and $u \perp v$.

2.9.27 Let $F : \mathbb{R}^n \to \mathbb{R}^n$ be a monotone mapping. Suppose that $\hat{x}$ is a strictly feasible vector of the NCP (F). Show that every solution x^* of the NCP (F) satisfies

$$x_i^* \leq \frac{\hat{x}^T F(\hat{x})}{F_i(\hat{x})}, \quad \forall i = 1, \ldots, n.$$

This gives a simple bound on the solutions of a strictly feasible, monotone NCP.

2.9.28 Let F be a continuous function from $\mathbb{R}^n$ into itself. Let B be a symmetric positive definite matrix of order n.

(a) Show that if F is a gradient map, then there exists a pair (x, μ) such that $x^T B x = 1$ and

$$0 \leq x \perp F(x) + \mu B x \geq 0 \tag{2.9.1}$$

(b) Suppose that $F(x) \equiv Ax$ for some matrix $A \in \mathbb{R}^{n \times n}$. Show that a pair (x, μ) satisfying $x^T Bx = 1$ and (2.9.1) exists if and only if $A + \mu B$ is not an R_0 matrix.

(c) Show that if $F(x) \equiv Ax$ for some matrix A, there exist only finitely many μ's satisfying (2.9.1), even though there may be infinitely many x's satisfying the same condition. Finally, show that if A is a copositive matrix on the nonnegative orthant, then any μ satisfying (2.9.1) must be nonpositive.

2.9.29 Let $K \equiv \mathbb{R}^n_+ \cap \operatorname{bd} \mathbb{B}(0,1)$. Although K is nonconvex, the gap function of the VI (K, F):

$$\theta_{\text{gap}}(x) \equiv \sup_{y \in K} F(x)^T (x - y)$$

is well defined on the domain of F because K is a compact set. Show that

$$\theta_{\text{gap}}(x) = x^T F(x) - \begin{cases} \min\limits_{1 \leq i \leq n} F_i(x) & \text{if } F(x) \geq 0 \\ -\| \max(0, -F(x)) \| & \text{if } F(x) \not\geq 0. \end{cases}$$

Verify directly that $\theta_{\text{gap}}(x)$ is continuous, assuming that F is continuous.

2.9.30 Consider the NCP (F), where

$$F(x, y) = \begin{pmatrix} -\nabla u(x) + (A^T - \rho B^T) y \\ b + (B - A) x \end{pmatrix},$$

with $u : \mathbb{R}^n \to \mathbb{R}$ being a concave and continuously differentiable function and $\rho \in (0, 1)$. This NCP is the one obtained from the model of invariant capital stock in Proposition **1.4.5**. Assume that B is nonnegative. Show that if (a) there exists a vector $\bar{y} \geq 0$ such that $(A - B)^T \bar{y} > 0$ and (b)

$$\left.\begin{array}{r} (A - \rho B)^T y \geq 0 \\ b^T y \leq 0 \\ y \geq 0 \end{array}\right\} \Rightarrow y = 0,$$

then the NCP (F) has a solution. (Hint: assume for the sake of contradiction that the NCP (F) has no solution. For each $t \in [0, 1]$, define

$$F^t(x, y) = \begin{pmatrix} -\nabla u(x) + (A^T - B^T) y \\ b + (B - A) x \end{pmatrix} + \begin{pmatrix} t(1 - \rho) B^T y \\ 0 \end{pmatrix}$$

and let $H(x, y, t)$ be the natural (i.e., min) map associated the NCP (F^t). Show that for $t = 0$, the NCP (F^0) is the KKT formulation of the concave maximization problem:

$$\begin{array}{ll} \text{maximize} & u(x) \\ \text{subject to} & b + (B - A)x \geq 0 \\ \text{and} & x \geq 0. \end{array}$$

Use assumptions (a) and (b) to show that the feasible region of this optimization problem is nonempty and bounded and that the set

$$\bigcup_{t \in [0,1)} H(\cdot, \cdot, t)^{-1}(0)$$

is bounded. Use Proposition **2.3.17** and a degree-theoretic argument to complete the proof.) Show further that the solution set of the NCP (F) is bounded.

2.9.31 Let P be a polyhedron in $\mathbb{R}^n$, q be an n-vector and M be an $n \times n$ symmetric matrix (not necessarily positive semidefinite). Consider the quadratic program:

$$\begin{array}{ll} \text{minimize} & \theta(x) \equiv q^T x + \frac{1}{2} x^T M x \\ \text{subject to} & x \in P. \end{array}$$

Assume that this program attains its finite minimum value. Let $S \subseteq \mathbb{R}^n$ denote the set of stationary points of the program.

(a) Show that S is the union of finitely many polyhedra.

(b) Show that θ attains finitely many values on S.

(c) For each stationary value $\eta \in \theta(S)$, show that $\theta^{-1}(\eta) \cap S$ is the union of finitely many polyhedra by obtaining a characterization of a stationary point with θ-value equal to η.

(d) Deduce from part (c) that the set of global minima of a (possibly nonconvex) quadratic program is the union of finitely many polyhedra. Prove that the same is true for the set of local minima.

2.9.32 Consider the affine CP $(\mathcal{M}^n_+, Q, \mathbf{L})$ in SPSD matrices:

$$\mathcal{M}^n_+ \ni A \perp Q + \mathbf{L}(A) \in \mathcal{M}^n_+,$$

where $Q \in \mathcal{M}^n$ is a given symmetric matrix and $\mathbf{L} : \mathcal{M}^n \to \mathcal{M}^n$ is given linear operator. It follows from Theorem **2.5.10**(B) that if (i) $\mathbf{L}$ is an R_0

operator, i.e., if the homogenous CP $(\mathcal{M}^n_+, 0, \mathbf{L})$ has a unique solution, and (ii) there exists a matrix $E \in \mathcal{M}^n_{++}$ such that the CP $(\mathcal{M}^n_+, E, \mathbf{L})$ also has a unique solution, then the CP $(\mathcal{M}^n_+, Q, \mathbf{L})$ has a solution for all $Q \in \mathcal{M}^n$.

(a) Suppose that $\mathbf{L}$ satisfies the following implication:

$$[\, X \in \mathcal{M}^n_+, \, X\mathbf{L}(X) + \mathbf{L}(X)X \in -\mathcal{M}^n_+ \,] \Rightarrow X = 0. \tag{2.9.2}$$

Show that the CP $(\mathcal{M}^n_+, Q, \mathbf{L})$ has a solution for all $Q \in \mathcal{M}^n$.

(b) Suppose that $\mathbf{L}$ satisfies the *Jordan P property*:

$$X\mathbf{L}(X) + \mathbf{L}(X)X \in -\mathcal{M}^n_+ \Rightarrow X = 0. \tag{2.9.3}$$

Assume further that for any two solutions X^1 and X^2 of the CP $(\mathcal{M}^n_+, Q, \mathbf{L})$, the matrix $X^1W^2 + X^2W^1$ is positive semidefinite (but not necessarily symmetric), where $W^i \equiv Q + \mathbf{L}(X^i)$ for $i = 1, 2$. Show that $\mathrm{SOL}(\mathcal{M}^n_+, Q, \mathbf{L})$ is a singleton.

2.9.33 Another consequence of Theorem **2.5.10**(B) is the following. Let K be a closed convex cone in $\mathbb{R}^n$. Let $M \in \mathbb{R}^{n\times n}$ be given. Suppose there exists a vector $q^* \in \mathrm{int}\, K^*$ such that

$$\bigcup_{t\geq 0} \mathrm{SOL}(K, t\,q^*, M) = \{\, 0 \,\}.$$

Show that there exists a neighborhood $\mathcal{U} \subset \mathbb{R}^{n\times n}$ of M such that for every $M' \in \mathcal{U}$ and every $q \in \mathbb{R}^n$, the CP (K, q, M') has a nonempty bounded solution set.

2.9.34 Consider the set-valued map $K : \mathbb{R} \to \mathbb{R}^n$ defined by

$$K(\tau) \equiv \mathrm{cl}\, \mathbb{B}(0, \tau_+); \quad \tau \in \mathbb{R}_+.$$

Show that K is continuous on $\mathbb{R}$ by using only Definition **2.1.16**.

More generally, let $f : \mathbb{R}^{n+m} \to \mathbb{R}$ be a continuous function. Suppose that $f(\cdot, \bar{y})$ is a convex function and a vector $\bar{x} \in \mathbb{R}^n$ exists satisfying $f(\bar{x}, \bar{y}) < 0$. Show that the set-valued map $K : \mathbb{R}^m \to \mathbb{R}^n$ defined by

$$K(y) \equiv \{\, x \in \mathbb{R}^n \, : \, f(x, y) \leq 0 \,\}, \quad y \in \mathbb{R}^m$$

is continuous at $\bar{y}$.

2.9.35 Let $K : \mathbb{R}^m \to \mathbb{R}^n$ be a continuous, closed-valued, convex-valued, set-valued map. Let $F : \mathbb{R}^{m+n} \to \mathbb{R}^n$ be continuous. Show that the set-valued map $x \in \mathbb{R}^m \mapsto \mathrm{SOL}(K(x), F(x, \cdot)) \subset \mathbb{R}^n$ is closed.

2.10 Notes and Comments

Degree theory, homeomorphisms, fixed-point theorems, and contraction mappings are well-studied topics in classical mathematical analysis; there are many excellent references, including [170, 260, 369, 516, 652, 896]. The definition of degree given in Subsection **2.1.1** is based on [170]. The cited references contain the omitted proofs of the degree-theoretic results in Subsection **2.1.1**. The global homeomorphism Theorem **2.1.10** was proved by Palais [656]. The necessary part of Proposition **2.1.14** is the classical inverse function theorem for a smooth map. This characterization of the locally Lipschitz homeomorphism property of a strongly F-differentiable function and its generalization to a strongly B-differentiable function, Exercise **3.7.6**, both appeared in [762]. Proposition **2.1.15**, which is an inverse-function theorem for a locally Lipschitz continuous function, is due to Kummer [467]; see also the related papers [468, 469] for inverse-function-type theorems in the class of $C^{1,1}$ optimization problems; the latter are optimization problems whose objective and constraint functions are C^1 with locally Lipschitz continuous gradients.

A classic reference on set-valued mappings is Berge [55]. A contemporary treatment of these mappings can be found in the book by Aubin and Frankowska [25]. The fundamental role played by set-valued mappings in variational analysis is well documented in the treatise by Rockafellar and Wets [752]. This outstanding book presents in a coherent manner a unified mathematical theory that is most relevant for the sensitivity and stability study of a host of variational problems in finite- and infinite-dimensional spaces. The book by Aubin and Ekeland [24] is a good reference for many analytical tools that we use throughout our work.

A complete proof of Brouwer's fixed-point Theorem **2.1.18** can be found in several of the references cited above; see, e.g., [652]. A proof of Kakutani's fixed-point theorem is available in many references; see, e.g., [813, 841]. See [203] for a constructive approach to computing a Kakutani fixed point. The monograph [81] presents an excellent exposition of the application of fixed-point theorems in economics and game theory. For a more complete mathematical treatment of the latter theory, see [22]. With particular reference to VIs/CPs, Villar [860] deals with the existence of many economic problems using operator theorems. Not used in our book, the Eilenberg-Montgomery fixed-point theorem [215] deals with contractible set-valued maps; Chan and Pang [112] employed the latter theorem in the study of the generalized QVI. Incidentally, it would be of interest to understand the contractability of the solution set of a VI/CP.

The Banach fixed-point Theorem **2.1.21** is classical, and there are many generalizations. Particularly worthy of note are Opial's extension for a nonexpansive map under an asymptotical regularity condition [651] and Dunn's extension to an averaging scheme [198]. These extended nonexpansive fixed-point results are the basis for much of the work of Magnanti and Perakis [552, 553] on the unified convergence analysis of fixed-point iterative methods for solving VIs.

The Tietze-Urysohn Extension Lemma **2.2.2** is a classic result in topology; see, e.g., [197]. The paper by Hartman and Stampacchia [333] is the source for many of the results in the beginning part of Section **2.2**, including Corollary **2.2.5** and Proposition **2.2.8**. According to the bibliographic notes in [410], the (Brouwer) fixed-point based proof of the cited corollary (Exercise **2.9.10**) is due to Brezis. The use of coercivity conditions for the study of nonlinear operator equations apparently began with Browder [92] and Minty [610]. Moré [623] documented the important role of coercivity conditions in the existence of solutions to the NCP. For a contemporary study of non-coercive variational problems in function spaces using recession analysis, see [292]; the related paper [256] studies the existence of solutions in generalized non-coercive equilibrium problems.

Proposition **2.2.9** was proved by Nash [634, 635]. The existence result of a saddle point, Corollary **2.2.10**, was proved first by von Neumann [862]. For extensions of this result to abstract spaces, see [227, 788]. There are also extensions to noncompact sets; see, e.g., Section 6.4 in [813]. Our treatment of the existence of a Walrasian equilibrium of an economy, Proposition **2.2.11**, is a very crude simplification of the general theory. The literature on this topic is enormous; even a partial documentation is beyond the scope of the notes and comments herein. Proposition **2.2.12** and the related Proposition **3.5.6** are drawn from [674, 835]. See Section **1.9** for more notes and comments on the applied problems in Subsection **2.2.1**.

Kachurovskii [393] was apparently the first to note that gradients of differentiable convex functions are monotone maps, and he coined the term "monotonicity" for this property. Beginning with his seminal work on monotone networks [608], Minty [609] studied monotone set-valued maps in their full generality and established many fundamental results in this landmark paper. As early as 1960, Zarantonello [893] employed strong monotonicity to establish contractive properties of iterative schemes for solving functional equations. Such a contractive approach was very useful in the convergence analysis of many iterative algorithms for solving monotone VIs; see Chapter **12** for details. The proof of the basic Proposition **2.3.2** can be found in many places; see, e.g., [652]. Theorem **2.3.4** gives for the first time a necessary and sufficient condition for a pseudo monotone VI (K, F) on a general closed convex set to have a solution. Cohen and Chaplais [129] introduce a "strongly nested monotone" function, which is shown by Marcotte and Wynter [587] to have an important role to play in multi-class network equilibrium problems.

Extending the gradient of a differentiable pseudo convex function, Karamardian [401] defined a pseudo monotone vector function and established that a scalar function is pseudo convex if and only if its gradient is a pseudo monotone vector function. Inspired by this paper, Karamardian and Schaible [404] introduced several generalized monotonicity concepts, including "strict pseudomonotonicity", "strong pseudomonoticity", and "quasimonotonicity". For each type of generalized monotone maps, these authors established a relation to a corresponding type of generalized convex functions. In [405], Karamardian, Schaible, and Crouzeix investigated differentiable generalized monotone maps and obtained characterizations of these maps in terms of their Jacobian matrices. Zhu and Marcotte [906] introduced more generalized monotonicity classes. Crouzeix [147] presents a survey of generalized convexity and generalized monotonicity. Research activities in this area continue to expand; strong emphasis is placed on connections to VIs and other topics in mathematical programming.

The equality (2.3.2), which establishes the convexity of the solution set of a pseudo monotone VI, is a particular case of a result commonly referred to as Minty's Lemma. Mangasarian [569] was the first to establish the constancy of the gradient of the objective function and the resulting representation of the optimal solution set of a differentiable convex program (Corollary **2.3.8**). He also extended this result to a nonsmooth objective function. Further study of this topic for nondifferentiable programs can be found in Burke and Ferris [98].

Extending the symmetric positive semidefinite matrices, the class of monotone composite functions and the associated VIs were the subject of study in several papers by Luo and Tseng [537, 538, 539, 540, 541] and Tseng [847, 850]. Although not explicitly stated in their present form, Proposition **2.3.12** and its Corollary **2.3.13** are implicit in the work of these authors. In the context of the LCP (q, M), the F-uniqueness property of $\mathrm{SOL}(q, M)$ is called w-uniqueness. It is known [142] that for a given matrix $M \in \mathbb{R}^{n \times n}$, w-uniqueness holds for the LCP (q, M) for all $q \in \mathbb{R}^n$ if and only if M is "column adequate"; i.e., for all $\alpha \subseteq \{1, \ldots, n\}$,

$$\det M_{\alpha\alpha} = 0 \Rightarrow M_{\cdot\alpha} \text{ has linearly independent columns.}$$

Column adequate matrices must be column sufficient.

In the literature on Bregman functions (see Subsection **12.7.2** for this theory and Section **12.9** for related notes and comments), a *paramonotone* property is defined for set-valued maps, which when specialized to a function becomes the monotone plus property in part (b) of Definition **2.3.9**; see [111]. We feel that the terminology "monotone plus" is more appropriate for the said property for the following two reasons. First, the terminology is consistent with the term "copositive plus" associated with matrices that was coined in late 1960s (see the notes below for the class of copositive matrices). Second, the prefix "para" means "akin to" and is less descriptive than the adjective "plus", which refers to the additional implication in the monotonicity property.

Originated from the inverses of strongly monotone set-valued maps (Exercise **2.9.15**), co-coercive functions have been called different names in the literature. Golshtein uses the term "inverse strong monotonicity", which he introduced in 1975 (see [296]). Several authors from the French school of G. Cohen [220, 219, 592] attributed the co-coercivity property to Dunn [199] and called it the "Dunn property". Dunn showed that "in Hilbert spaces the gradient maps of convex functionals with uniformly bounded continuous second Fréchet derivatives satisfy monotonicity conditions that insure that some convex combination of the identity $\mathbf{I}$ and $\mathbf{I} - \nabla f$ is either strictly contractive or at worst nonexpansive". Eckstein and Bertsekas [214] and Lemaire [489] used the term "firmly nonexpansive" for "co-coercive". Magnanti and Perakis [551, 552, 553] used the term "strongly F-monotone function" for a co-coercive function; presumably, the letter "F" was intended to mean "functional". We follow in this book the usage of Tseng [846] and Zhu and Marcotte [907]. The papers [553, 907] give a detailed account of the role of co-coercivity in the convergence of iterative methods for solving monotone VIs. Recently, Fischer, Jeyakumar, and Luc [250] introduce a pointwise variant of the co-coercivity property, which they call directional co-monotonicity. A function H mapping $\mathbb{R}^n$ into itself is *co-monotone* at $x \in \mathbb{R}^n$ along the direction $u \in \mathbb{R}^n$ if a constant $c > 0$ exists, which depends on the pair (x, u), such that

$$(H(x + \tau u) - H(x))^T u \geq c \| H(x + \tau u) - H(x) \|,$$

for all $\tau > 0$ sufficiently small. Unlike co-copositivity, the right-hand norm in the above expression is not squared. The authors of [250] used co-monotonicity to derive a derivative-free algorithm for solving an NCP (F) with F continuous but not necessarily locally Lipschitz continuous. Incidentally, the co-monotonicity here is distinct from the co-monotonicity of a function of two arguments on a set that we introduce in Definition **11.4.12**.

The concavity of the dual gap function $\theta_{\text{dual}}(x)$ was profitably exploited by several authors for multiple uses. Hearn, Lawphongpanich, and Nguyen [338] employed this function to define a convex programming formulation of the asymmetric traffic equilibrium problem. Nguyen and Dupuis [642] proposed a special cutting plane method for solving the dual gap program (2.3.15) with application to the latter traffic problem. (Cutting plane methods for solving VIs are not covered in this book. For references, see [148, 175, 293, 546, 547, 612].) Crouzeix [146] used the dual gap function to establish Theorem **2.3.16** for a generalized VI. Based on the theory of weak sharp minima of Burke and Ferris [99], Marcotte and Zhu [588] derived error bounds for monotone VIs in terms of the dual gap function. Exercise **2.9.23** is drawn from the latter reference. Dietrich [185] defines a smooth dual gap function for a class of QVIs.

As mentioned in Section **1.9**, the primal and dual gap functions, θ_{gap} and θ_{dual}, can be obtained from a saddle function. Taking this point of view, Auchmuty [26] introduced the saddle function $\tilde{L}(x,y)$ and obtained the saddle result in Exercise **2.9.22**. Based on the latter result, we can define the "generalized primal gap function" $\tilde{\theta}_{\text{gap}}(x)$ and the "generalized dual gap function" $\tilde{\theta}_{\text{dual}}(x)$ as follows:

$$\tilde{\theta}_{\text{gap}}(x) \equiv \sup_{y \in K} \tilde{L}(x,y), \quad x \in K,$$

$$\tilde{\theta}_{\text{dual}}(y) \equiv \inf_{x \in K} \tilde{L}(x,y), \quad y \in K.$$

A unified study of these generalized gap functions is given by Larsson and Patriksson [485]. This study also touches upon the closely related "auxiliary problem principle" due to Cohen [128].

Properties of convex cones can be found in many books [57, 315, 746, 813]. Proposition **2.4.3** is due to Krein and Rutman [461]; part (a) of Exercise **2.9.26** is due to Krasnoselskii [460]; part (b) is due to Gaddum [280]; and part (c) is due to Haynsworth and Hoffman [334].

The existence of a solution to a strictly feasible, monotone NCP was proved by Moré [624], who also asked the question of whether a feasible monotone NCP is solvable. Megiddo [599] constructed the NCP in Example **2.4.6** to give Moré's question a negative answer. The existence and boundedness of solutions to a pseudo monotone, strictly feasible CP was proved in Karamardian [401]. Our proof of Theorem **2.4.4** is different. The bound given in Exercise **2.9.27** for the solutions to a strictly feasible, monotone NCP is obtained by Mangasarian and McLinden [575]. The proof of Theorem **2.4.8** is based on that of Karamardian's original result, which is stated in Remark **2.4.9**. For $K = \mathbb{R}^n_+$, Theorem **2.4.7** was proved in the classic paper of Cottle [136]. The Frank-Wolfe Theorem of quadratic programming, on which the proof of Theorem **2.4.7** is based, appeared in [262]. There are several interesting generalizations of the Frank-Wolfe Theorem that are worth mentioning. Terlaky [830] studied ℓ_p programming and established in particular a Frank-Wolfe type result for convex quadratic-quadratic programs. Luo and Zhang [545] studied convex and nonconvex quadratic constraint systems and rediscovered Terlaky's extended Frank-Wolfe result. Unbeknownst to these authors, the latter result had already been proved in a 1977 book written in Russian by Belousov [47]. Presently, the most general Frank-Wolfe type theorem is that proved by Belousov and Klatte [48], who showed that if the objective function is a convex polynomial bounded below on a constraint set defined by a system of convex polynomial inequalities, then the objective function attains its constrained minimum on the set.

The representation Theorem **2.4.15** appeared in [306]. The latter reference also studied solution rays for the AVI, introduced the sharp property (2.5.6), and established Proposition **2.5.5**. The main Theorem **2.5.20** in Subsection **2.5.3** is also obtained in this reference.

First studied by Motzkin in an unpublished report in 1952, copositive matrices play an important role in LCP theory; see [142] for a brief historical account of these matrices. In particular, the term "copositive plus" was coined in [139], and the class of LCPs with copositive plus matrices was used by Lemke [490] to demonstrate the finite termination of his well-known algorithm. Gowda [297] introduced the class of copositive star matrices and established Lemma **2.5.2**. Defined in [306], the VI range, VI domain, and VI kernel of a pair (K, M) are the respective generalizations of the LCP range, LCP domain, and LCP kernel of a matrix M [142]. Qi and Jiang [700] study the range set of a VI. Copositive CPs in abstract spaces were studied in two papers [298, 309], which were primarily concerned with extending results for the finite-dimensional LCP to infinite-dimensional spaces. Example **2.5.14** is due to Gowda in a private communication dated December 15, 1998.

Smith [796] introduced and used "exceptional sequences" to study the existence of solutions to the NCP (F) and applied the results to the spatial price equilibrium problem. Specifically, an unbounded sequence $\{x^k\}$ of vectors is said to be *exceptional* with respect to F if there exists a sequence of positive scalars $\{\tau_k\}$ such that for every k, x^k is a solution of the NCP $(F + \tau_k \mathbf{I})$. Extending Smith's exceptional sequences, Isac, Bulavski, and Kalashnikov [365] defined an exceptional family of elements for a continuous function; see also [364, 366]. Zhao and Han [899] and Zhao, Han, and Qi [900] defined an exceptional family of elements for a pair (K, F), where the set K is assumed to be convex and finitely representable and satisfy the generalized Slater CQ. These authors obtained existence results for the CP/VI under the assumption that no such exceptional families of elements exist. In the case of the CP, the concept defined by Isac, Bulavski, and Kalashnikov is very close to the sequence in Corollary **2.6.2**. Exercise **2.9.7** can be used to define an exceptional family of elements of a pair (K, F) without any restriction on the representation of the closed convex set K. It should be noted that all these definitions of exceptional sequences are special considerations of an original idea of Eaves [202], who studied the NCP by truncating the nonnegative orthant. Proposition **2.2.3** and its consequences, particularly Proposition **2.2.8**, are refinements of Eaves' idea. The condition in Exercise **2.9.8** is due to Karamardian [400]; Isac [363] generalizes this existence result to an infinite-dimensional space. Under the blanket assumption that the defining set K is convex and finitely representable and satisfies the generalized Slater condition, Zhao and Han [899] essentially proved Theorem **2.3.4**. In our development, there is no assumption on K other than its closedness and convexity.

Applying the concept of an exceptional sequence to the function $F - \mu \mathbf{I}$ for a positive scalar μ, Zhao and Li [901] obtained various interesting results pertaining to the strict feasibility of the NCP (F). Their work has inspired the extension to more general sets K, resulting in Corollaries **2.6.3** and **2.6.4**. Exercise **2.9.9** deals with the strict feasibility of a VI in the sense of (2.3.5).

The development in Section **2.7** is largely based on the reference [425]. The extension to treat the nonlinear model discussed in Subsection **1.4.6** can be found in [677]. The fundamental role of a copositive LCP in discrete frictional contact problems was first recognized in [844]. The fact that the dual of the LCP kernel of a copositive matrix M is contained in the LCP range of M, which is the key to the proof of Proposition **2.7.1**, has since been used several times; see [15, 16, 809].

The degree-based treatment of the QVI in Section **2.8** follows that of the VI in the earlier sections. Thus, while Corollary **2.8.4** is related to known results in the literature, such as [112], Theorem **2.8.3** appears herein for the first time. Proposition **2.8.5** is drawn from [682].

Inverse isotone functions and S functions were introduced by Moré and Rheinboldt [625], which is the source for Exercises **2.9.5** and **2.9.6**. Exercise **2.9.21** appears in [673]. The equivalence of (a) and (c) in Exercise **2.9.24** is proved by Zhu and Marcotte [907]; that of (b) and (c) is proved by Luo and Tseng [535]. The fact that the objective function of a quadratic program attains finitely many values on the set of stationary points, part (b) of Exercise **2.9.31**, was noted by Luo and Luo [528], who used the result to obtain an error bound for a quadratic inequality system.

Gowda and Song [310], Gowda and Parthasarathy [307], and, most recently, Parthasarathy, Sampangi Raman, and Sriparna [685] and the Ph.D. thesis of Song [801] study LCPs in SPSD matrices and investigate the relationships between various fundamental solution properties of these problems, particularly the GUS property. Exercise **2.9.32** is based on the work of [310], which introduces (2.9.2) and the Jordan P property (2.9.3) for linear operators defined on the space $\mathcal{M}^n$. There remain many open questions about this class of generalized LCPs. At this time, besides those that are derived from optimization problems, interesting applications of LCPs in SPSD matrices await to be discovered.

Exercise **2.9.28** pertains to the "eigenvalue complementarity problem"; see [134, 705, 769] for more discussion of this problem, including its application to elastic frictional contact systems. The paper [705] focuses on the "symmetric" problem where $F(x) \equiv -Ax$ for a symmetric matrix A and the multiplier μ is required to be positive. In particular it is shown that if A and B are symmetric matrices with B positive definite, then the problem

$$0 \leq x \perp (\mu B - A)x \geq 0$$

has a nonzero solution x with $\mu > 0$ if and only if a nonnegative vector $\hat{x}$ exists satisfying $\hat{x}^T A\hat{x} > 0$.

Chapter 3

Solution Analysis II

This chapter is a continuation of Chapter **2**. Specifically, we present further properties of the solutions to the VI/CP. We begin this chapter with a study of the class of B(ouligand)-differentiable functions, these being locally Lipschtiz continuous functions that have directional derivatives. This class of nonsmooth functions plays a central role in the rest of the book. We next present a detailed exposition of constraint qualifications of finitely representable sets. In order to establish a key continuity property of the KKT multipliers associated with a VI, we introduce a fundamental "error bound" of a polyhedral set first proved by Hoffman (see Lemma **3.2.3**). Armed with these background materials, we study in Section **3.3** the local uniqueness of a solution to the VI/CP. For a VI with a finitely representable set (including the NCP), the latter property is closely tied to the R_0 property of a suitable "reduced" affine problem. When the solution in question happens to be "nondegenerate" (a concept defined in Section **3.4**), the local uniqueness property is implied by the nonsingularity of a certain basic matrix.

Another major topic discussed in this chapter is the class of VIs whose defining sets are equal to the Cartesian products of sets of lower dimensions; see Section **3.5**. Two prominent examples of this class of VIs are the NCPs and the box constrained problems. For such VIs, much of the theory for the general VI can be refined; analogous results can be established under weaker assumptions. For instance, a generalization of monotonicity, known as the P_0 property, can be defined that is tailored to the Cartesian product structure of the defining set. It turns out that the P_0 property is the basis for establishing the connectedness of the solution set of a non-monotone VI; see Section **3.6**. The latter property is clearly a generalization of the convexity of the solution set, a topic treated in the previous chapter.

3.1 Bouligand Differentiable Functions

The class of locally Lipschitz continuous functions is the foundation of the differentiability theories of nonsmooth functions developed throughout the book. For this reason, it is useful to state a famous theorem due to Rademacher. First, we recall from measure theory that a subset $A \subset \mathbb{R}^n$ is called *negligible* if for every $\varepsilon > 0$ there is a family of boxes $\{B^k\}$ with n-dimensional volume $\varepsilon_k > 0$ such that

$$A \subset \bigcup_{k=1}^{\infty} B^k \quad \text{and} \quad \sum_{k=1}^{\infty} \varepsilon_k < \varepsilon.$$

Clearly, every subset of a negligible set is negligible.

3.1.1 Theorem. Let $F : \mathcal{D} \subseteq \mathbb{R}^m$ be a locally Lipschitz continuous function defined on the open set $\mathcal{D}$ in $\mathbb{R}^n$. Let Ω be the subset of $\mathcal{D}$ consisting of points where F is Fréchet differentiable. The set $\mathcal{D} \setminus \Omega$ is negligible; thus Ω is dense in $\mathcal{D}$; i.e., $\mathcal{D} \subset \operatorname{cl} \Omega$. □

If Φ is a F(réchet)-differentiable function at x, then the linear function:

$$y \mapsto \Phi(x) + J\Phi(x)(y - x)$$

provides a good approximation of Φ in an open neighborhood of x. Since we are interested in broadening the theory to a nondifferentiable function Φ, we wish to introduce an important class of nonsmooth functions that is useful not only here but also in subsequent developments. To motivate this definition, recall that by definition a function Φ is directionally differentiable at x along a direction d if the following limit

$$\lim_{\tau \downarrow 0} \frac{\Phi(x + \tau d) - \Phi(x)}{\tau}$$

exists; this limit, denoted $\Phi'(x; d)$, is called the directional derivative of Φ at x along d. Thus,

$$\lim_{\tau \downarrow 0} \frac{\Phi(x + \tau d) - \Phi(x) - \tau \Phi'(x; d)}{\tau} = 0. \tag{3.1.1}$$

Observe that $\Phi'(x; \tau d) = \tau \Phi'(x; d)$ for all scalars $\tau \geq 0$; that is, $\Phi'(x; \cdot)$ is positively homogeneous in the second argument. If Φ is directionally differentiable at x along all directions, then we say that Φ is directionally differentiable at x. It is well known from elementary multivariate calculus that there are functions which are directionally differentiable at a point without being continuous there. In essence, the class of nonsmooth functions introduced below consists of the locally Lipschitz continuous functions that are also directionally differentiable.

3.1.2 Definition. A function $\Phi : \mathcal{D} \subseteq \mathbb{R}^n \to \mathbb{R}^m$ defined on the open set $\mathcal{D}$ is said to be *B(ouligand)-differentiable* at a vector $x \in \mathcal{D}$ if Φ is Lipschitz continuous in a neighborhood of x and directionally differentiable at x. If Φ is B-differentiable at x, we call the directional derivative $\Phi'(x; d)$ the *B-derivative* of Φ at x along d. The B-derivative $\Phi'(x; \cdot)$ is *strong* if the error function

$$\mathbf{e}(y) \equiv \Phi(y) - \Phi(x) - \Phi'(x; y - x)$$

satisfies

$$\lim_{\substack{y^1 \neq y^2 \\ (y^1, y^2) \to (x, x)}} \frac{\mathbf{e}(y^1) - \mathbf{e}(y^2)}{\| y^1 - y^2 \|} = 0.$$

In this case, we say that Φ is *strongly B-differentiable* at x. We say that Φ is B-differentiable near x if Φ is B-differentiable at every point in a certain neighborhood of x. □

Clearly, if $\Phi : \mathcal{D} \to \mathbb{R}^m$ is locally Lipschitz continuous and has an F-derivative at a vector $x \in \mathcal{D}$, then Φ is B-differentiable at x. If Φ has a strong F-derivative at x; that is, if

$$\lim_{\substack{y^1 \neq y^2 \\ (y^1, y^2) \to (x, x)}} \frac{\Phi(y^1) - \Phi(y^2) - J\Phi(x)(y^1 - y^2)}{\| y^1 - y^2 \|} = 0,$$

then Φ has a strong B-derivative there. In general, if Φ is B-differentiable at x, then the B-derivative $\Phi'(x; \cdot)$ is a globally Lipschitz continuous function in the second argument; moreover, $\Phi'(x; \cdot)$ is strong if and only if the error function $\mathbf{e}$ has a strong F-derivative at x that is identically equal to zero. If F is B-differentiable in a neighborhood of x, the B-derivative $\Phi'(\cdot; \cdot)$ is generally not continuous in the first argument; it can be shown (see Exercise **3.7.2**) that if for every fixed direction d, $\Phi'(\cdot; d)$ is continuous in the first argument at x, then Φ is F-differentiable at x.

An important consequence of the B-differentiability is that the limit of the directional derivative (3.1.1) is uniform on compact sets of directions. This consequence is formally stated in terms of a limit condition that strengthens the directional derivative limit (3.1.1).

3.1.3 Proposition. Let $\Phi : \mathcal{D} \to \mathbb{R}^m$ be B-differentiable at a vector x in the open set $\mathcal{D} \subseteq \mathbb{R}^n$. The following uniform limit holds:

$$\lim_{\substack{y \neq x \\ y \to x}} \frac{\Phi(y) - \Phi(x) - \Phi'(x; y - x)}{\| y - x \|} = 0. \tag{3.1.2}$$

Proof. Assume for the sake of contradiction that the desired limit fails to hold. There exists a sequence of vectors $\{y^k\}$ converging to x such that $y^k \neq x$ for all k and

$$\liminf_{k \to \infty} \frac{\Phi(y^k) - \Phi(x) - \Phi'(x; y^k - x)}{\| y^k - x \|} > 0.$$

Without loss of generality, we may assume that there exists a nonzero vector d such that

$$\lim_{k \to \infty} \frac{y^k - x}{\| y^k - x \|} = d.$$

Let

$$\tau_k \equiv \| y^k - x \| \quad \text{and} \quad d^k \equiv \frac{y^k - x}{\tau_k}.$$

As k tends to ∞, $\{\tau_k\}$ converges to zero and $\{d^k\}$ converges to d. Writing

$$\Phi(y^k) = \Phi(x + \tau_k d^k) = \Phi(x + \tau_k d) + \Phi(x + \tau_k d^k) - \Phi(x + \tau_k d),$$

we have

$$\frac{\Phi(y^k) - \Phi(x) - \Phi'(x; y^k - x)}{\| y^k - x \|} = [\, \Phi'(x; d) - \Phi'(x; d^k) \,]$$
$$+ \frac{\Phi(x + \tau_k d) - \Phi(x) - \tau_k \, \Phi'(x; d)}{\tau_k} + \frac{\Phi(x + \tau_k d^k) - \Phi(x + \tau_k d)}{\tau_k}.$$

As k tends to ∞, the difference within the square bracket approaches zero, by the continuity of the directional derivative $\Phi'(x; \cdot)$ in the second argument; the first fraction in the second line of the above equation approaches zero, by the definition of the directional derivative; and the second fraction approaches zero, by the local Lipschitz continuity of Φ at x. Consequently, the desired limit (3.1.2) holds. □

It follows from (3.1.2) that if the B-derivative $\Phi'(x; \cdot)$ is linear in the second argument, then Φ is F-differentiable at x. We illustrate the B-differentiability using the min function of two arguments.

3.1.4 Example. Consider the function of 2 variables:

$$\Phi : (a, b) \in \mathbb{R}^2 \mapsto \min(a, b) \in \mathbb{R}.$$

This function is continuously differentiable at all points (a, b) for which $a \neq b$. At a point (a, a), the min function has a strong B-derivative given by

$$\Phi'((a,a);(c,d)) = \min(c,d) = \Phi(c,d), \quad \forall\,(c,d) \in \mathbb{R}^2.$$

In fact, the error function $\mathbf{e}(c, d)$ at such a pair (a, a) is identically equal to zero because

$$\mathbf{e}(c,d) = \Phi(c,d) - \Phi(a,a) - \Phi'((a,a);(c-a,d-a)) = 0,$$

where the last equality holds by an easy substitution. □

Next, we demonstrate the B-differentiability of the function in Example **1.5.7**. It turns out that this demonstration is not totally trivial. In some cases, the calculation of the directional derivatives is omitted; the reader can either calculate them directly or use the chain rule in Proposition **3.1.6** to simplify the calculation.

3.1.5 Example. Let

$$\Phi(r,x_1,x_2) \equiv \min\left(\frac{r_+}{\sqrt{x_1^2+x_2^2}},\, 1\right) x, \qquad (r,x_1,x_2) \in \mathbb{R}^3.$$

We claim that this function is B-differentiable everywhere on $\mathbb{R}^3$. First we show that Φ is globally Lipschitz continuous on $\mathbb{R}^3$. (The continuity of Φ can be proved by combining Lemma **2.8.2** and Exercise **2.9.34**; the proof of the Lipschitz continuity of Φ given below exploits the special structure of Φ.) Specifically, we need to show the existence of a constant $L > 0$ such that for any two triples (c_i, a_i, b_i) for $i = 1, 2$ we have

$$\|\,\Phi(c_1,a_1,b_1) - \Phi(c_2,a_2,b_2)\,\| \le L\,\|\,(c_1,a_1,b_1) - (c_2,a_2,b_2)\,\|. \qquad (3.1.3)$$

Let (c_i, a_i, b_i) for $i = 1, 2$ be two arbitrary triples. We divide the argument into several cases. If either $(a_1, b_1) = (0, 0)$ or $(a_2, b_2) = (0, 0)$, then (3.1.3) holds with $L = 1$. Similarly, if $c_1 \le 0$ or $c_2 \le 0$, the same inequality also holds with the same constant L. So we can assume that $(a_1, b_1) \neq (0, 0)$, $(a_2, b_2) \neq (0, 0)$, and c_1 and c_2 are both positive. There are four remaining cases to consider:

(i) $c_i \ge \sqrt{a_i^2 + b_i^2}$ for $i = 1, 2$;

(ii) $c_1 \ge \sqrt{a_1^2 + b_1^2}$ and $c_2 < \sqrt{a_2^2 + b_2^2}$;

(iii) $c_1 < \sqrt{a_1^2 + b_1^2}$ and $c_2 \ge \sqrt{a_2^2 + b_2^2}$;

(iv) $c_i < \sqrt{a_i^2 + b_i^2}$ for $i = 1, 2$.

In case (i), the inequality (3.1.3) also holds with $L = 1$. Case (ii) and (iii) are similar, so it suffices to prove (ii). Assume (ii). We have

$$\begin{aligned}
&\Phi(c_1, a_1, b_1) - \Phi(c_2, a_2, b_2) \\
&= \begin{pmatrix} a_1 \\ b_1 \end{pmatrix} - \frac{c_2}{\sqrt{a_2^2 + b_2^2}} \begin{pmatrix} a_2 \\ b_2 \end{pmatrix} \\
&= \begin{pmatrix} a_1 - a_2 \\ b_1 - b_2 \end{pmatrix} + \left(1 - \frac{c_2}{\sqrt{a_2^2 + b_2^2}} \right) \begin{pmatrix} a_2 \\ b_2 \end{pmatrix}.
\end{aligned}$$

Thus

$$\begin{aligned}
&\| \Phi(c_1, a_1, b_1) - \Phi(c_2, a_2, b_2) \| \\
&\leq \left\| \begin{pmatrix} a_1 - a_2 \\ b_1 - b_2 \end{pmatrix} \right\| + \left(1 - \frac{c_2}{\sqrt{a_2^2 + b_2^2}} \right) \left\| \begin{pmatrix} a_2 \\ b_2 \end{pmatrix} \right\| \\
&= \sqrt{(a_1 - a_2)^2 + (b_1 - b_2)^2} + \sqrt{a_2^2 + b_2^2} - c_2 \\
&\leq \sqrt{(a_1 - a_2)^2 + (b_1 - b_2)^2} + \sqrt{a_2^2 + b_2^2} - \sqrt{a_1^2 + b_1^2} + c_1 - c_2 \\
&\leq 2 \sqrt{(a_1 - a_2)^2 + (b_1 - b_2)^2} + | c_1 - c_2 |
\end{aligned}$$

Hence the inequality (3.1.3) holds with $L = 4$.

Finally, consider the case (iv). We have

$$\Phi(c_1, a_1, b_1) - \Phi(c_2, a_2, b_2) = \frac{c_1}{\sqrt{a_1^2 + b_1^2}} \begin{pmatrix} a_1 \\ b_1 \end{pmatrix} - \frac{c_2}{\sqrt{a_2^2 + b_2^2}} \begin{pmatrix} a_2 \\ b_2 \end{pmatrix}.$$

Letting $x^i \equiv (a_i, b_i)$ for $i = 1, 2$, we have

$$\begin{aligned}
&\| \Phi(c_1, a_1, b_1) - \Phi(c_2, a_2, b_2) \|^2 \\
&= c_1^2 - 2 \frac{c_1}{\sqrt{a_1^2 + b_1^2}} \frac{c_2}{\sqrt{a_2^2 + b_2^2}} (a_1 a_2 + b_1 b_2) + c_2^2 \\
&= (c_1 - c_2)^2 + 2 \frac{c_1}{\sqrt{a_1^2 + b_1^2}} \frac{c_2}{\sqrt{a_2^2 + b_2^2}} (\| x^1 \| \| x^2 \| - (x^1)^T x^2) \\
&\leq (c_1 - c_2)^2 + \| x^1 - x^2 \|^2.
\end{aligned}$$

Hence (3.1.3) holds with $L = 1$.

Having established the Lipschitz continuity of Φ, we next show that Φ is directionally differentiable everywhere on $\mathbb{R}^3$. Depending on the given triple $z^* \equiv (r^*, x_1^*, x_2^*)$, we divide the analysis into several cases.

1. $0 < r^* < \sqrt{(x_1^*)^2 + (x_2^*)^2}$. In a suitable neighborhood of z^*, we have

$$\Phi(r, x_1, x_2) = \frac{r}{\sqrt{x_1^2 + x_2^2}} x;$$

thus Φ is continuously differentiable in a neighborhood of z^*.

2. $0 < \sqrt{(x_1^*)^2 + (x_2^*)^2} < r^*$. In a suitable neighborhood of z^*, we have

$$\Phi(r, x_1, x_2) = x;$$

again, Φ is continuously differentiable in a neighborhood of z^*.

3. $0 < \sqrt{(x_1^*)^2 + (x_2^*)^2} = r^*$. In a suitable neighborhood of z^*, we have

$$\Phi(r, x_1, x_2) \equiv \min\left(\frac{r}{\sqrt{x_1^2 + x_2^2}}, 1 \right) x;$$

it is not difficult to show that Φ is directionally differentiable in such a neighborhood because it is straightforward to manipulate the fraction.

4. $r^* < 0$. The function Φ is identically equal to zero in a suitable neighborhood of z^*; thus the continuous differentiability of Φ in such a neighborhood follows trivially.

5. $r^* = 0 < \sqrt{(x_1^*)^2 + (x_2^*)^2}$. It is not difficult to show that in a suitable neighborhood of z^*,

$$\Phi(r, x_1, x_2) = \frac{r_+}{\sqrt{x_1^2 + x_2^2}} x.$$

By Proposition **3.1.6**, it follows that Φ is directionally differentiable in such a neighborhood.

6. $(r^*, x_1^*, x_2^*) = (0, 0, 0)$. By the positive homogeneity of Φ, we have, for any triple (c, a, b),

$$\Phi'((0,0,0); (c, a, b)) = \Phi(c, a, b).$$

Consequently, Φ is Lipschitz continuous and directionally differentiable everywhere on $\mathbb{R}^3$. □

The composition of two B-differentiable functions has some interesting properties. First, it remains B-differentiable; there is a chain rule. Moreover, it is possible to establish that the B-derivative of the composite map is strong under an appropriate restriction.

3.1.6 Proposition. Let $\mathcal{D}$ and $\mathcal{D}'$ be open sets in $\mathbb{R}^n$ and $\mathbb{R}^m$ respectively. Let $\Phi : \mathcal{D} \to \mathbb{R}^m$ and $\Psi : \mathcal{D}' \to \mathbb{R}^p$ be B-differentiable at $x \in \mathcal{D}$ and $\Phi(x) \in \mathcal{D}'$ respectively. Suppose that $\Phi(\mathcal{D}) \subseteq \mathcal{D}'$. The following two statements hold.

(a) The composite map $\Gamma \equiv \Psi \circ \Phi : \mathcal{D} \to \mathbb{R}^p$ is B-differentiable at x; moreover

$$\Gamma'(x; d) = \Psi'(\Phi(x); \Phi'(x; d)), \quad \forall\, d \in \mathbb{R}^n.$$

(b) If Ψ is strongly F-differentiable at $\Phi(x)$ and Φ has a strong B-derivative at x, then Γ has a strong B-derivative at x.

Proof. We prove (b) only; the following proof is applicable to (a) with a minor modification. It suffices to show

$$\lim_{\substack{y^1 \neq y^2 \\ (y^1, y^2) \to (x,x)}} \frac{\mathbf{e}_\Gamma(y^1) - \mathbf{e}_\Gamma(y^2)}{\| y^1 - y^2 \|} = 0, \tag{3.1.4}$$

where

$$\mathbf{e}_\Gamma(y) \equiv \Psi(\Phi(y)) - \Psi(\Phi(x)) - \Psi'(\Phi(x); \Phi'(x; y - x)).$$

Since Ψ is F-differentiable at $v \equiv \Phi(x)$, thus $\Psi'(v; \cdot)$ is linear in the second argument, we have, for $i = 1, 2$,

$$\mathbf{e}_\Gamma(y^i) = \mathbf{e}_\Psi(\Phi(y^i)) + J\Psi(\Phi(x))\mathbf{e}_\Phi(y^i),$$

where

$$\mathbf{e}_\Psi(u) \equiv \Psi(u) - \Psi(v) - J\Psi(v)(u - v), \quad \forall\, u \in \mathbb{R}^m,$$

and

$$\mathbf{e}_\Phi(y) \equiv \Phi(y) - \Phi(x) - \Phi'(x; y - x), \quad \forall\, y \in \mathbb{R}^n.$$

Since Ψ has a strong F-derivative at v, we have

$$\lim_{\substack{u^1 \neq u^2 \\ (u^1, u^2) \to (v,v)}} \frac{\mathbf{e}_\Psi(u^1) - \mathbf{e}_\Psi(u^2)}{\| u^1 - u^2 \|} = 0;$$

hence

$$\lim_{\substack{y^1 \neq y^2 \\ (y^1, y^2) \to (x,x)}} \frac{\mathbf{e}_\Psi(\Phi(y^1)) - \mathbf{e}_\Psi(\Phi(y^2))}{\| y^1 - y^2 \|} =$$

$$\lim_{\substack{y^1 \neq y^2 \\ (y^1, y^2) \to (x,x)}} \frac{\mathbf{e}_\Psi(\Phi(y^1)) - \mathbf{e}_\Psi(\Phi(y^2))}{\| \Phi(y^1) - \Phi(y^2) \|} \, \frac{\| \Phi(y^1) - \Phi(y^2) \|}{\| y^1 - y^2 \|} = 0,$$

where the last equality holds because Φ is locally Lipschitz continuous at x. Similarly, since Φ has a strong B-derivative at x, we have

$$\lim_{\substack{y^1 \neq y^2 \\ (y^1, y^2) \to (x,x)}} \frac{\mathbf{e}_\Phi(y^1) - \mathbf{e}_\Phi(y^2)}{\| y^1 - y^2 \|} = 0.$$

Combining the last two expressions immediately yields the desired limit (3.1.4). □

It is important to note that in part (b) of the above proposition, the order of composition is important; more precisely, if Ψ has a strong B-derivative at $\Phi(x)$ and Φ has a strong F-derivative at x, the composite map $\Psi \circ \Phi$ does not necessarily have a strong B-derivative at x.

3.1.7 Example. Let

$$\Psi(a,b) \equiv \min(a,b), \quad \forall\,(a,b) \in \mathbb{R}^2,$$

and

$$\Phi(x_1,x_2) \equiv (\, x_1^3,\, x_2\,), \quad \forall\,(\, x_1, x_2\,) \in \mathbb{R}^2.$$

Then

$$\Gamma(x_1,x_2) \equiv \Psi \circ \Phi(x_1,x_2) = \min(\, x_1^3,\, x_2\,).$$

We have $\Gamma(0,0) = 0$ and

$$\Gamma'((0,0);(v_1,v_2)) = \min(\,0, v_2\,), \quad \forall\,(\, v_1, v_2\,) \in \mathbb{R}^2;$$

thus the error function $\mathbf{e}(y)$ is given by

$$\mathbf{e}(y_1,y_2) = \Gamma(y_1,y_2) - \Gamma'((0,0);(y_1,y_2)) = \min(\, y_1^3, y_2\,) - \min(\,0, y_2\,).$$

For $\varepsilon > 0$ let

$$\begin{pmatrix} y_1^1 \\ y_2^1 \end{pmatrix} \equiv \begin{pmatrix} -\varepsilon \\ -0.5\,\varepsilon^3 \end{pmatrix} \quad \text{and} \quad \begin{pmatrix} y_1^2 \\ y_2^2 \end{pmatrix} \equiv \begin{pmatrix} -\varepsilon \\ \varepsilon^4 \end{pmatrix}.$$

We have

$$\frac{\mathbf{e}(y_1^1,y_2^1) - \mathbf{e}(y_1^2,y_2^2)}{\sqrt{(y_1^1 - y_1^2)^2 + (y_2^1 - y_2^2)^2}} = \frac{.5\,\varepsilon^3}{\sqrt{0.25\varepsilon^6 + \varepsilon^7 + \varepsilon^8}},$$

which implies

$$\lim_{\varepsilon \downarrow 0} \frac{\mathbf{e}(y_1^1,y_2^1) - \mathbf{e}(y_1^2,y_2^2)}{\sqrt{(y_1^1 - y_1^2)^2 + (y_2^1 - y_2^2)^2}} = 1 \neq 0.$$

Thus, Γ is not strongly B-differentiable at the origin. □

Proposition **3.1.6** and Example **3.1.7** illustrate an important structural difference in terms of directional differentiation between the natural map $\mathbf{F}_K^{\text{nat}}$ and the normal map $\mathbf{F}_K^{\text{nor}}$:

$$\mathbf{F}_K^{\text{nat}}(x) \equiv x - \Pi_K(x - F(x)) \quad \text{and} \quad \mathbf{F}_K^{\text{nor}}(z) \equiv F(\Pi_K(z)) + z - \Pi_K(z).$$

Consider the case where K is the nonnegative orthant, which yields

$$\Pi_{\mathbb{R}^n_+}(x) = \max(0, x), \quad \forall x \in \mathbb{R}^n.$$

By Example **3.1.4**, Π_K has a strong B-derivative everywhere on $\mathbb{R}^n$. The natural map $\mathbf{F}_K^{\text{nat}}$ is of the type in Example **3.1.7**; thus it is generally not possible for $\mathbf{F}_K^{\text{nat}}$ to have a strong B-derivative, even if F is very smooth. The normal map $\mathbf{F}_K^{\text{nor}}$ is of the type in part (b) of Proposition **3.1.6**, provided that F is continuously differentiable. In this case, $\mathbf{F}_K^{\text{nor}}$ has a strong B-derivative. Consequently, in situations where the strong B-differentiability of the nonsmooth formulation of the VI is needed (such as in the study of the strong stability of an isolated solution to a linearly constrained VI; see Theorem **5.3.17**), we need to employ the normal map but not the natural map.

3.2 Constraint Qualifications

In this section, we present a detailed summary of various CQs used in this book. All these CQs imply Abadie's CQ introduced in Subsection **1.3.1**, but each of them has additional consequences, therefore making them useful for particular purposes.

Throughout this section, we assume that K is a finitely representable set given by (1.3.4):

$$K \equiv \{ x \in \mathbb{R}^n : h(x) = 0, \; g(x) \leq 0 \}, \tag{3.2.1}$$

with $h : \mathbb{R}^n \to \mathbb{R}^\ell$ and $g : \mathbb{R}^n \to \mathbb{R}^m$ being continuously differentiable. We do not assume that K is convex unless otherwise stated. By definition, the **M**angasarian **F**romovitz **C**onstraint **Q**ualification, abbreviated as MFCQ, holds at a vector $x \in K$ if

(a) the gradients

$$\{ \nabla h_j(x) : j = 1, \ldots, \ell \} \tag{3.2.2}$$

are linearly independent, and

(b) there exists a vector $v \in \mathbb{R}^n$ such that

$$\nabla h_j(x)^T v = 0, \quad \forall j = 1, \ldots, \ell,$$
$$\nabla g_i(x)^T v < 0, \quad \forall i \in \mathcal{I}(x),$$

where

$$\mathcal{I}(x) \equiv \{ i : g_i(x) = 0 \}$$

is the index set of active constraints at x. For a pair (x, λ) satisfying $x \in K$ and

$$0 \leq \lambda \perp g(x) \leq 0, \tag{3.2.3}$$

define the index sets:

$$\alpha \equiv \{ i \in \mathcal{I}(x) : \lambda_i > 0 \} \quad \text{and} \quad \beta \equiv \{ i \in \mathcal{I}(x) : \lambda_i = 0 \}.$$

The **S**trict **M**angasarian **F**romovitz **C**onstraint **Q**ualification, abbreviated as SMFCQ, holds at such a pair (x, λ) if

(a) the gradients

$$\{ \nabla h_j(x) : j = 1, \ldots, \ell \} \cup \{ \nabla g_i(x) : i \in \alpha \} \tag{3.2.4}$$

are linearly independent, and

(b) there exists a vector $v \in \mathbb{R}^n$ such that

$$\begin{aligned} \nabla h_j(x)^T v &= 0, \quad \forall j = 1, \ldots, \ell, \\ \nabla g_i(x)^T v &= 0, \quad \forall i \in \alpha \\ \nabla g_i(x)^T v &< 0, \quad \forall i \in \beta. \end{aligned}$$

For completeness, we also introduce the **L**inear **I**ndependence **C**onstraint **Q**ualification at x, which postulates that the vectors

$$\{ \nabla g_i(x) : i \in \mathcal{I}(x) \} \cup \{ \nabla h_j(x) : j = 1, \ldots, \ell \}$$

are linearly independent.

Whereas both the MFCQ and the LICQ are well defined at an arbitrary vector x belonging to K, the SMFCQ is well defined at a pair (x, λ) satisfying $x \in K$ and (3.2.3). In particular, we can speak of the latter CQ at a KKT triple (x, μ, λ) of a VI (K, F). By a theorem of the alternative, it is not difficult to show that the SMFCQ holds at a pair (x, λ) if and only if

$$\left.\begin{aligned} \sum_{j=1}^{\ell} d\mu_j \nabla h_j(x) + \sum_{i \in \mathcal{I}(x)} d\lambda_i \nabla g_i(x) &= 0 \\ d\lambda_i \geq 0 \quad \forall i \in \beta \end{aligned}\right\} \tag{3.2.5}$$

$$\Rightarrow d\mu_j = d\lambda_i = 0, \quad \forall j = 1, \ldots, \ell; \; i \in \mathcal{I}(x).$$

Similarly, the MFCQ holds at $x \in K$ if and only if

$$\left.\begin{aligned} \sum_{j=1}^{\ell} d\mu_j \nabla h_j(x) + \sum_{i \in \mathcal{I}(x)} d\lambda_i \nabla g_i(x) &= 0 \\ d\lambda_i \geq 0 \quad \forall i \in \mathcal{I}(x) \end{aligned}\right\} \tag{3.2.6}$$

$$\Rightarrow d\mu_j = d\lambda_i = 0, \quad \forall j = 1, \ldots, \ell; \; i \in \mathcal{I}(x).$$

Further properties of these CQs are summarized in the result below. In particular, it follows from this result that

$$\text{LICQ} \Rightarrow \text{SMFCQ} \Rightarrow \text{MFCQ} \Rightarrow \text{Abadie CQ}.$$

3.2.1 Proposition. Let K be defined by (3.2.1) with all h_j and g_i being continuously differentiable in a neighborhood of a vector $x \in K$. Let F be a continuous mapping from K into $\mathbb{R}^n$.

(a) If the SMFCQ holds at a pair (x, λ), where λ satisfies (3.2.3), then the MFCQ holds at x.

(b) If $x \in \text{SOL}(K, F)$, then the MFCQ holds at x if and only if $\mathcal{M}(x)$ is nonempty and bounded.

(c) Let $(\mu, \lambda) \in \mathcal{M}(x)$ be given. The SMFCQ holds at (x, λ) if and only if $\mathcal{M}(x) = \{(\mu, \lambda)\}$.

(d) If the MFCQ holds at x, then there exist a neighborhood $\mathcal{N}$ of x and a scalar $c > 0$ such that the MFCQ holds at every x' in $K \cap \mathcal{N}$ and, for every x' belonging to $\text{SOL}(K, F) \cap \mathcal{N}$, $\mathcal{M}(x')$ is nonempty and

$$\| (\mu', \lambda') \| \leq c, \quad \forall (\mu', \lambda') \in \mathcal{M}(x').$$

Proof. It is clear that if the implication (3.2.5) holds, then so does the implication (3.2.6). Hence (a) follows.

Since x is a solution of the VI (K, F), it is a minimizer of the nonlinear program:

$$\begin{array}{ll} \text{minimize} & y^T F(x) \\ \text{subject to} & h(y) = 0, \quad g(y) \leq 0. \end{array}$$

As such, statement (b) follows from well-known NLP theory.

Suppose the SMFCQ holds at the KKT triple (x, μ, λ). If (μ', λ') is another pair of KKT multipliers belonging to $\mathcal{M}(x)$, we have

$$\sum_{j=1}^{\ell} (\mu_j - \mu_j') \nabla h_j(x) + \sum_{i \in \alpha} (\lambda_i - \lambda_i') \nabla g_i(x) - \sum_{i \in \beta} \lambda_i' \nabla g_i(x) = 0.$$

By (3.2.5), it follows that

$$\begin{aligned} \mu_j &= \mu_j', \quad \forall j = 1, \ldots, \ell \\ \lambda_i &= \lambda_i' \quad \forall i \in \alpha \\ \lambda_i' &= 0, \quad \forall i \in \beta. \end{aligned}$$

Thus (μ, λ) is unique. Conversely, suppose that $\mathcal{M}(x)$ is the singleton $\{(\mu, \lambda)\}$. If the SMFCQ fails to hold, let $(d\mu, d\lambda_{\mathcal{I}(x)})$ be a nonzero tuple satisfying the left-hand side of the implication (3.2.5). It is then easy to verify that $(\mu, \lambda) + \tau(d\mu, d\lambda)$ belongs to $\mathcal{M}(x)$ for all $\tau > 0$ sufficiently small. This is a contradiction. Consequently, the SMFCQ holds. This establishes statement (c) of the lemma.

Suppose that the MFCQ holds at x. We claim that there exists an open neighborhood $\mathcal{N}$ of x such that the MFCQ remains valid at every $x' \in K \cap \mathcal{N}$. Since the gradients (3.2.2) are linearly independent and each h_j is continuously differentiable near x, it follows that there exists a open neighborhood $\mathcal{N}$ of x such that the gradients

$$\{ \nabla h_j(x') : j = 1, \ldots, \ell \}$$

remain linearly independent for all $x' \in \mathcal{N}$. Suppose there exists a sequence of vectors $\{x^k\} \subset K$ converging to x such that for each k, condition (b) in the MFCQ fails to hold at x^k. By the equivalent implication (3.2.6) and the linear independence of the vectors $\{\nabla h_j(x^k) : j = 1, \ldots, \ell\}$, it follows that, for each k, there exist vectors $d\mu^k$ and $d\lambda^k$ with $d\lambda^k \neq 0$ such that

$$\sum_{j=1}^{\ell} d\mu_j^k \, \nabla h_j(x^k) + \sum_{i \in \mathcal{I}(x^k)} d\lambda_i^k \, \nabla g_i(x^k) = 0, \tag{3.2.7}$$

$$d\lambda_i^k \geq 0, \quad \forall \, i \in \mathcal{I}(x^k),$$

and $d\lambda_i^k = 0$ for all $i \notin \mathcal{I}(x^k)$. Since there are only finitely many index subsets of $\{1, \ldots, m\}$ and since $\mathcal{I}(x^k)$ is a subset of $\mathcal{I}(x)$ for all k sufficiently large, we may assume without loss of generality, by working with an appropriate subsequence of $\{(d\mu^k, d\lambda^k)\}$ if necessary, that there exists an index set $\mathcal{J} \subseteq \mathcal{I}(x)$ such that $\mathcal{I}(x^k) = \mathcal{J}$ for all k. By a standard normalizing argument, we can establish that the sequence $\{(d\mu^k, d\lambda^k)\}$ is bounded and every accumulation point of this sequence yields a contradiction to the validity of the MFCQ at x. Indeed, if $\{(d\mu^k, d\lambda^k) : k \in \kappa\}$ is a subsequence such that

$$\lim_{k (\in \kappa) \to \infty} \| \, (\, d\mu^k, d\lambda^k \,) \, \| = \infty$$

and

$$\lim_{k (\in \kappa) \to \infty} \frac{(\, d\mu^k, d\lambda^k \,)}{\| \, (\, d\mu^k, d\lambda^k \,) \, \|} = (\, d\mu^\infty, d\lambda^\infty \,),$$

for some nonzero pair $(d\mu^\infty, d\lambda^\infty)$. Dividing (3.2.7) by $\|(d\mu^k, d\lambda^k)\|$ and letting $k (\in \kappa) \to \infty$, we deduce

$$\sum_{j=1}^{\ell} d\mu_j^\infty \, \nabla h_j(x) + \sum_{i \in \mathcal{J}} d\lambda_i^\infty \, \nabla g_i(x) = 0,$$

which contradicts the MFCQ at x. Consequently, it follows that the MFCQ continues to hold in a suitable neighborhood of x. Hence, for every x' in SOL(K,F) that is sufficiently near x, $\mathcal{M}(x')$ is nonempty.

Finally, the existence of the scalar c that bounds all the multiplier pairs in $\mathcal{M}(x')$ can be proved by a similar contradiction. It suffices to note that each pair $(\mu',\lambda') \in \mathcal{M}(x')$, where $x' \in \text{SOL}(K,F)$, satisfies

$$F(x') + \sum_{j=1}^{\ell} \mu_j' \nabla h_j(x') + \sum_{i \in \mathcal{I}(x')} \lambda_i' \nabla g_i(x') = 0;$$

moreover, provided that x' is sufficiently close to x, $\mathcal{I}(x')$ is contained in $\mathcal{I}(x)$. Thus, a similar normalizing argument will produce the desired contradiction if no such scalar c exists. We omit the details. □

We may consider the multiplier map

$$\mathcal{M} : x' \in \text{SOL}(K,F) \mapsto \mathcal{M}(x') \subset \mathbb{R}^{\ell} \times \mathbb{R}^m_+$$

as a multifunction. It turns out that if the MFCQ holds at a solution x of the VI (K,F), then this multifunction is nonempty-valued in a neighborhood of x and upper semicontinuous at x. We formally state this property in the result below.

3.2.2 Proposition. Assume the same setting as in Proposition **3.2.1**. If the MFCQ holds at $x \in \text{SOL}(K,F)$, then for every scalar $\varepsilon > 0$, there exists a neighborhood $\mathcal{N}$ of x such that for every $x' \in \text{SOL}(K,F) \cap \mathcal{N}$,

$$\emptyset \neq \mathcal{M}(x') \subseteq \mathcal{M}(x) + \text{cl}\,\mathbb{B}(0,\varepsilon). \tag{3.2.8}$$

Hoffman's error bound for polyhedra

The proof of Proposition **3.2.2** relies on a fundamental property of a polyhedron known as an "error bound". A full treatment of this topic for more general sets is presented in Chapter **6**. For now, we present a preliminary foray into this vast topic by restricting the set of interest to a polyhedron in $\mathbb{R}^n$ represented by

$$P \equiv \{ x \in \mathbb{R}^n : Cx = d,\ Ax \leq b \},$$

where C and A are given $\ell \times n$ and $m \times n$ matrices respectively and d and b are ℓ and m-vectors respectively. Define the *residual function* $r : \mathbb{R}^n \to \mathbb{R}_+$ of P to be the scalar function:

$$r(x) \equiv \| Cx - d \| + \| \max(0, Ax - b) \|, \quad \forall x \in \mathbb{R}^n.$$

Clearly, $r(x) = 0$ if and only if $x \in P$. For a vector $x \notin P$, $r(x)$ provides a quantitative measure of the violation by x of the constraints that define P. It is natural to wonder how this measure is related to the distance from x to P. Since $r(x)$ is clearly a Lipschitz continuous function on $\mathbb{R}^n$, there exists a constant $c' > 0$ such that

$$c' r(x) \leq \operatorname{dist}(x, P), \quad \forall x \in \mathbb{R}^n.$$

The following result is the celebrated Hoffman error bound for polyhedra; it shows that $\operatorname{dist}(x, P)$ is bounded above by a multiplicative constant times $r(x)$ for all $x \in \mathbb{R}^n$.

3.2.3 Lemma. For any matrices $C \in \mathbb{R}^{\ell \times n}$ and $A \in \mathbb{R}^{m \times n}$, there exists a constant $c > 0$ such that for all vectors $d \in \mathbb{R}^\ell$ and $b \in \mathbb{R}^m$ for which $P \equiv \{x \in \mathbb{R}^n : Cx = d, Ax \leq b\}$ is nonempty,

$$\operatorname{dist}(x, P) \leq c[\,\|\,Cx - d\,\| + \|\,\max(\,0, Ax - b\,)\,\|\,], \quad \forall x \in \mathbb{R}^n. \tag{3.2.9}$$

Proof. For simplicity, linear equalities are omitted from the representation of P to simplify the notation. Let b be such that P is nonempty. Consider the projection problem in Euclidean norm:

$$\begin{array}{ll} \text{minimize} & \frac{1}{2}(\,y - x\,)^T(\,y - x\,) \\ \text{subject to} & Ay \leq b. \end{array}$$

There exists a vector $\lambda \in \mathbb{R}^m$ such that

$$\Pi_P(x) - x = -A^T\lambda$$

$$0 \leq \lambda \perp b - A\Pi_P(x) \geq 0.$$

We may choose λ such that the row vectors

$$\{\,A_{i\cdot} \,:\, i \in \operatorname{supp}(\lambda)\,\}$$

are linearly independent. Let $\mathcal{J} \equiv \operatorname{supp}(\lambda)$. Thus,

$$B \equiv A_{\mathcal{J}\cdot}(\,A_{\mathcal{J}\cdot}\,)^T$$

is a symmetric positive definite matrix. We have

$$0 = b_{\mathcal{J}} - A_{\mathcal{J}\cdot}\Pi_P(x) = b_{\mathcal{J}} - A_{\mathcal{J}\cdot}x + B\lambda_{\mathcal{J}},$$

which implies,

$$(\,\lambda_{\mathcal{J}}\,)^T B\lambda_{\mathcal{J}} = (\,\lambda_{\mathcal{J}}\,)^T(\,A_{\mathcal{J}\cdot}x - b_{\mathcal{J}}\,) \leq (\,\lambda_{\mathcal{J}}\,)^T \max(0, A_{\mathcal{J}\cdot}x - b_{\mathcal{J}}\,),$$

where the last inequality holds because λ is nonnegative. Hence if $\delta > 0$ denotes the smallest eigenvalue of B, we deduce

$$\| \lambda \| \leq \delta^{-1} \| \max(0, Ax - b) \|.$$

Consequently, we obtain

$$\| \Pi_P(x) - x \| \leq \delta^{-1} \| A^T \| \| \max(0, Ax - b) \|.$$

To define the desired constant $c > 0$, let $\mathcal{F}$ denote the family of submatrices of A whose rows are linearly independent rows of A and $\mathcal{B}$ be the collection of symmetric positive definite matrices B such that $B = A_{\mathcal{I}\cdot}(A_{\mathcal{I}\cdot})^T$ for some member $A_{\mathcal{I}\cdot}$ in $\mathcal{F}$. For each member $B \in \mathcal{B}$, let $\delta(B)$ denote the smallest eigenvalue of B. Define

$$c \equiv \max_{B \in \mathcal{B}} \delta(B)^{-1} \| A^T \|, \tag{3.2.10}$$

which is independent of b. From the above proof and this definition of c, the desired inequality (3.2.9) follows easily. □

3.2.4 Remark. The expression (3.2.10) gives a computable, albeit combinatorially impractical, formula for the multiplicative constant in the inequality (3.2.9). The sensitivity of this constant as the matrix A is perturbed is a subject that has received considerable amount of research. Of particular interest is the question whether this constant will remain bounded in a neighborhood of A. Needless to say, when an alternative norm is used in (3.2.10), the expression for the constant will be different, yet it will remain independent of b. □

In what follows, we consider distance bounds from solutions of varying polyhedra to a base polyhedron. For this purpose, we write

$$P(A, b) \equiv \{ x \in \mathbb{R}^n : Ax \leq b \}$$

to denote the dependence of this polyhedron on the pair (A, b). The following result is one of Lemma **3.2.3**'s many consequences. There are two parts in this result. The first part pertains to solutions of $P(A, b')$, where only the right-hand vector b' is different; the second part pertains to solutions of $P(A', b')$, where both the matrix A' and the right-hand vector b' are different.

3.2.5 Corollary. Let the matrix $A \in \mathbb{R}^{m \times n}$ be given. The following two statements are valid.

(a) There exists a constant $L > 0$ such that for every vector $b \in \mathbb{R}^m$ for which $P(A, b)$ is nonempty and for every $x \in P(A, b')$,

$$\operatorname{dist}(x, P(A, b)) \leq L \| b - b' \|.$$

(b) For every bounded subset S of $\mathbb{R}^n$, there exists a constant $L > 0$ such that for every vector $b \in \mathbb{R}^m$ for which $P(A, b)$ is nonempty and for every $x \in S \cap P(A', b')$,

$$\operatorname{dist}(x, P(A, b)) \leq L [\| A - A' \| + \| b - b' \|] \tag{3.2.11}$$

Proof. By Lemma **3.2.3**, there exists a constant $c > 0$, dependent on A only, such that for every $b \in \mathbb{R}^m$ for which $P(A, b)$ is nonempty,

$$\operatorname{dist}(x, P(A, b)) \leq c \| \max(0, Ax - b) \|, \quad \forall x \in \mathbb{R}^n.$$

Let S be a bounded set and $x \in S \cap P(A', b')$ be given. Since

$$Ax - b = (A'x - b') + (A - A')x - (b - b')$$

and $A'x - b'$ is nonpositive, it follows that

$$\begin{aligned} \| \max(0, Ax - b) \| &\leq \| (A - A')x - (b - b') \| \\ &\leq \| A - A' \| \| x \| + \| b - b' \|. \end{aligned}$$

Since $\|x\|$ is bounded, with an obvious choice of the constant L, (3.2.11) follows easily. It is also clear that if $A' = A$, $\|x\|$ does not need to be bounded. This observation provides the proof of statement (a). □

In set notation, the inequality (3.2.11) implies that

$$\begin{aligned} &S \cap P(A', b') \\ &\subseteq P(A, b) + L [\| A - A' \| + \| b - b' \|] \operatorname{cl} \mathbb{B}(0, 1). \end{aligned} \tag{3.2.12}$$

This inclusion can be interpreted in terms of certain Lipschitz property of polyhedra. First consider the case $A' = A$, which is part (a) of Corollary **3.2.5**. For a given matrix $A \in \mathbb{R}^{m \times n}$, $P(A, \cdot)$ defines a multifunction from $\mathbb{R}^m$ into $\mathbb{R}^n$. Part (a) of the above corollary says that this set-valued map is *Lipschitz continuous* on its domain; that is, there exists a constant $L > 0$ such that for every b and b' in $\operatorname{dom} P(A, \cdot)$,

$$P(A, b') \subseteq P(A, b) + L \| b - b' \| \operatorname{cl} \mathbb{B}(0, 1).$$

Similarly, by varying A also, $P : (A, b) \mapsto P(A, b)$ defines a multifunction from $\mathbb{R}^{n \times n} \times \mathbb{R}^m \to \mathbb{R}^n$. Part (b) of Corollary **3.2.5** implies that this set-valued map is *strongly pseudo upper Lipschitz continuous* around every pair

(A, b) in its domain in the sense that for every scalar $\rho > 0$, there exists a scalar $L > 0$ such that the inclusion (3.2.12) holds with $S \equiv \mathrm{cl}\,\mathbb{B}(0,\rho)$ for all (A', b'). The adverb "strongly" is used to emphasize the fact that the scalar L is the same for all scalars ρ and that there is no restriction on the pair (A', b'); the adjective "pseudo" refers to the fact that the compact set S is needed in the left-hand set of the inclusion (3.2.12); "upper" refers to the fact that (A, b) appears in the upper set, i.e., the right-hand set in (3.2.12). Such a generalized Lipschitz property of multifunctions is very common in many set-valued maps that arise from the solution sets of parameter dependent inequality systems (including VIs and CPs) when the parameter is subject to variations.

Specializing $P(A, b)$ to the set of multipliers $\mathcal{M}(x)$, we can provide the proof of Proposition **3.2.2**.

Proof of Proposition 3.2.2. By part (d) of Proposition **3.2.1**, there exist a neighborhood $\mathcal{N}$ of x and a bounded set S such that

$$\bigcup_{x' \in \mathcal{N} \cap \mathrm{SOL}(K,F)} \mathcal{M}(x') \subseteq S.$$

Let $L > 0$ be the constant associated with the polyhedron $\mathcal{M}(x)$ and the set S, as asserted by Corollary **3.2.5**. For every $x' \in \mathcal{N} \cap \mathrm{SOL}(K, F)$, every pair (μ', λ') in $\mathcal{M}(x')$ satisfies

$$F(x') + \sum_{j=1}^{\ell} \mu_j' \nabla h_j(x') + \sum_{i \in \mathcal{I}(x)} \lambda_i' \nabla g_i(x') = 0$$

$$\lambda_i' \geq 0, \quad \forall i \in \mathcal{I}(x),$$

thus it follows that

$$\mathrm{dist}((\mu', \lambda'), \mathcal{M}(x)) \leq L \left[\| F(x) - F(x') \| + \right.$$

$$\left. \sum_{j=1}^{\ell} \| \nabla h_j(x') - \nabla h_j(x) \| + \sum_{i \in \mathcal{I}(x)} \| \nabla g_i(x') - \nabla g_i(x) \| \right].$$

Let $\varepsilon > 0$ be given. We may restrict the neighborhood $\mathcal{N}$ such that for all $x' \in \mathcal{N}$, the right-hand side in the above inequality is no greater than ε, by continuity of the functions involved. This is enough to established the desired inclusion (3.2.8). □

An important consequence of Proposition **3.2.2** occurs when $\mathcal{M}(x)$ is a singleton. We present this special result in the following corollary. The proof of the corollary is immediate from the proposition and is omitted.

3.2.6 Corollary. Assume the setting of Proposition **3.2.1**. If $\mathcal{M}(x)$ is the singleton $\{(\mu, \lambda)\}$, then

$$\lim_{x' \to x} \sup \{ \| (\mu', \lambda') - (\mu, \lambda) \| : (\mu', \lambda') \in \mathcal{M}(x') \} = 0.$$

The discussion of the CQs so far has not assumed the convexity of the set K. If K is convex, we have the following result, which shows that the MFCQ holding at a given point of K implies the MFCQ holding at every point of K.

3.2.7 Proposition. Let each $g_i : \mathbb{R}^n \to \mathbb{R}$ be convex and continuously differentiable for $i = 1, \ldots, m$ and let $h : \mathbb{R}^n \to \mathbb{R}^\ell$ be affine. Let K be given by (3.2.1). The MFCQ holds at a point in K if and only if the MFCQ holds at every point in K. In particular, if $\ell = 0$ so that

$$K \equiv \{ x \in \mathbb{R}^n : g(x) \leq 0 \},$$

then the MFCQ holding at any point in K is equivalent to the existence of a vector $\hat{x}$ satisfying $g(\hat{x}) < 0$.

Proof. Suppose that the MFCQ holds at a vector $\bar{x}$ in K. Let $\bar{v}$ satisfy

$$\nabla h_j(\bar{x})^T \bar{v} = 0, \quad \forall j = 1, \ldots, \ell,$$
$$\nabla g_i(\bar{x})^T \bar{v} < 0, \quad \forall i \in \mathcal{I}(\bar{x}).$$

It is easy to verify that for all $\tau > 0$ sufficiently small, the vector $z \equiv \bar{x} + \tau \bar{v}$ satisfies $h(z) = 0$ and $g(z) < 0$. Let x be an arbitrary vector in K. Since each h_j is affine, it suffices to verify condition (b) in the MFCQ. We claim that the vector $v \equiv z - x$ satisfy this required condition. Since h is affine, we clearly have $\nabla h_j(x)^T v = 0$ for all j. Since g_i is convex, we have, for every $i \in \mathcal{I}(x)$, by the gradient inequality of a convex function,

$$0 > g_i(z) \geq g_i(x) + \nabla g_i(x)^T (z - x) = \nabla g_i(x)^T v$$

as desired. The last assertion of the proposition is trivial. □

The existence of a vector $\hat{x}$ satisfying $g(\hat{x}) < 0$ is commonly referred to as the *Slater CQ* for a set K defined by only convex inequalities $g_i(x) \leq 0$, where each g_i is convex. Notice that the Slater CQ is well defined without g being differentiable. Unlike the MFCQ, the Slater CQ is somewhat restrictive because it requires K to be convex and equality free. More generally, when linear equalities are present in the representation of the convex set K, we say that the *generalized Slater CQ* holds if there exists $\hat{x} \in K$ such that $g(\bar{x}) < 0$. In this case, if g is in addition continuously differentiable, then by removing the linearly dependent equality constraints in K (which are linear by assumption), we obtain an equivalent representation of K such that the MFCQ holds at all feasible vectors in K. A consequence of this observation is that if K is a convex set defined by differentiable inequalities and linear equalities and satisfying the generalized Slater CQ, then the set of KKT multipliers $\mathcal{M}(x)$ is nonempty at all solutions x of the VI (K, F).

In what follows, we introduce two more CQs, one of which is a generalization of the LICQ and includes the case of polyhedral constraints, while the other is a unification of several of the CQs studied herein.

We say that the **C**onstant **R**ank **C**onstraint **Q**ualification, abbreviated as CRCQ, holds at a vector $x \in K$ if there exists a neighborhood $\mathcal{N}$ of x such that for every pair of index subsets $\mathcal{I}' \subseteq \mathcal{I}(x)$ and $\mathcal{J}' \subseteq \{1, \ldots, \ell\}$, the family of gradient vectors

$$\{ \nabla g_i(x') : i \in \mathcal{I}' \} \cup \{ \nabla h_j(x') : j \in \mathcal{J}' \}$$

has the same rank, which depends on $(\mathcal{I}', \mathcal{J}')$, for all $x' \in K \cap \mathcal{N}$. It is easy to see that the CRCQ holds at all vectors in K if K is polyhedral; i.e., if h and g are both affine. Clearly, the CRCQ is implied by the LICQ. It is nontrivial to show that the CRCQ implies the KTCQ defined in Exercise **1.8.31**. As in the proof of part (b) of Proposition **3.2.1**, it therefore follows that if the CRCQ holds at a solution x of the VI (K, F), then $\mathcal{M}(x)$ is nonempty although it can be unbounded.

We say that the **S**equentially **B**ounded **C**onstraint **Q**ualification, abbreviated as SBCQ, holds at $x \in \text{SOL}(K, F)$ if for every sequence of vectors $\{x^k\} \subset \text{SOL}(K, F)$ converging to x, there exists an integer $\bar{k} > 0$ such that for all $k \geq \bar{k}$, a pair of multipliers $(\mu^k, \lambda^k) \in \mathcal{M}(x^k)$ exists such that the sequence $\{(\mu^k, \lambda^k)\}$ is bounded. This CQ implies that $\mathcal{M}(x')$ is nonempty for all $x' \in \text{SOL}(K, F)$ sufficiently close to x. The following result shows that the SBCQ is a very broad constraint qualification.

3.2.8 Lemma. Let $h : \mathbb{R}^n \to \mathbb{R}^\ell$ and $g : \mathbb{R}^n \to \mathbb{R}^m$ be continuously differentiable in a neighborhood of a solution $x \in \text{SOL}(K, F)$, where K is given by (3.2.1). If either the MFCQ or the CRCQ holds at x, then so does the SBCQ.

Proof. By part (d) of Proposition **3.2.1**, it follows that the MFCQ implies the SBCQ. Suppose that the CRCQ holds at x. Let $\{x^k\} \subset \text{SOL}(K, F)$ be an arbitrary sequence converging x. For all k sufficiently large, $\mathcal{I}(x^k)$ is a subset of $\mathcal{I}(x)$. Hence the CRCQ remains valid at x^k for all k sufficiently large; therefore, $\mathcal{M}(x^k)$ is nonempty for all such k.

Assume for the sake of contradiction that the SBCQ fails to hold at x. There exists an infinite subset κ of $\{1, 2, \ldots\}$ such that every sequence $\{(\mu^k, \lambda^k) : k \in \kappa\}$, where the pair (μ^k, λ^k) belongs to $\mathcal{M}(x^k)$ for each $k \in \kappa$, satisfies

$$\lim_{k(\in\kappa)\to\infty} \| (\mu^k, \lambda^k) \| = \infty.$$

For each $k \in \kappa$ sufficiently large, choose a pair $(\mu^k, \lambda^k) \in \mathcal{M}(x^k)$ such that the gradients

$$\{ \nabla h_j(x^k) : j \in \operatorname{supp}(\mu^k) \} \cup \{ \nabla g_i(x^k) : i \in \operatorname{supp}(\lambda^k) \}$$

are linearly independent. There must exist an infinite subset κ' of κ such that the pair $(\operatorname{supp}(\mu^k), \operatorname{supp}(\lambda^k))$ is the same for all $k \in \kappa'$. Let $(\mathcal{J}', \mathcal{I}')$ denote this common pair of index sets. Since the CRCQ holds at x, it follows that the limiting gradients

$$\{ \nabla h_j(x) : j \in \mathcal{J}' \} \cup \{ \nabla g_i(x) : i \in \mathcal{I}' \}$$

must be linearly independent. Thus the subsequence $\{(\mu^k, \lambda^k) : k \in \kappa'\}$ is bounded. This is a contradiction. Consequently, the CRCQ implies the SBCQ. □

The CRCQ can be described in terms of the continuity of certain set-valued maps defined by the active gradients. In order to simplify the notation, we omit the equalities in the set K; thus in the following discussion, we let

$$K \equiv \{ x \in \mathbb{R}^n : g(x) \leq 0 \}. \tag{3.2.13}$$

For a given vector $y \in \mathbb{R}^n$ and index set $\mathcal{K} \subseteq \{1, \ldots, m\}$ with cardinality s, let $R(y, \mathcal{K})$ denote the linear subspace of $\mathbb{R}^s$ spanned by the columns of the matrix:

$$Jg_{\mathcal{K}}(y) \equiv (\nabla g_i(y)^T : i \in \mathcal{K}) \in \mathbb{R}^{s \times n},$$

and let $N(y, \mathcal{K})$ be the orthogonal complement of $R(y, \mathcal{K})$ in $\mathbb{R}^s$. Thus

$$N(y, \mathcal{K}) \equiv \left\{ \lambda_{\mathcal{K}} \in \mathbb{R}^s : \sum_{i \in \mathcal{K}} \lambda_i \nabla g_i(y) = 0 \right\}.$$

For any vector x and any index set $\mathcal{K}$, we have

$$\liminf_{y \to x} N(y, \mathcal{K}) \subseteq \limsup_{y \to x} N(y, \mathcal{K}) = N(x, \mathcal{K}),$$

$$\mathcal{R}(x, \mathcal{K}) = \liminf_{y \to x} R(y, \mathcal{K}) \subseteq \limsup_{y \to x} R(y, \mathcal{K}).$$

The following proposition shows that equalities holding throughout these inclusions at $x \in K$ for all index set $\mathcal{K} \subseteq \mathcal{I}(x)$ is equivalent to the CRCQ holding at x. Thus, the CRCQ holds at x if and only if for every $\mathcal{K} \subseteq \mathcal{I}(x)$, the set-valued maps

$$N(\cdot, \mathcal{K}) \,:\, y \mapsto N(y, \mathcal{K}) \quad \text{and} \quad R(\cdot, \mathcal{K}) \,:\, y \mapsto R(y, \mathcal{K})$$

are both closed and lower semicontinuous at x.

3.2.9 Proposition. Let K be given by (3.2.13) where each g_i is twice continuously differentiable. Let $x \in K$ be arbitrary. The following statements are equivalent.

(a) The CRCQ holds at x.

(b) For any nonempty subset $\mathcal{K}$ of $\mathcal{I}(x)$,

$$\lim_{y \to x} N(y, \mathcal{K}) \;=\; N(x, \mathcal{K}).$$

(c) For any nonempty subset $\mathcal{K}$ of $\mathcal{I}(x)$,

$$\lim_{y \to x} R(y, \mathcal{K}) \;=\; R(x, \mathcal{K}).$$

Proof. (a) $\Rightarrow$ (b). It suffices to show that

$$N(x, \mathcal{K}) \subseteq \liminf_{y \to x} N(y, \mathcal{K}); \tag{3.2.14}$$

that is, for every $\lambda_{\mathcal{K}} \in N(x, \mathcal{K})$ and every sequence $\{y^k\}$ converging to x, there exists for every k a vector $\lambda_{\mathcal{K}}^k$ in $N(y^k, \mathcal{K})$ such that the sequence $\{\lambda_{\mathcal{K}}^k\}$ converges to $\lambda_{\mathcal{K}}$. Let

$$\{\, \nabla g_i(x) \,:\, i \,\in\, \mathcal{B} \,\}, \tag{3.2.15}$$

where $\mathcal{B} \subseteq \mathcal{K}$, be a set of maximally linearly independent vectors among the gradients $\{\nabla g_i(x) : i \in \mathcal{K}\}$. Let B denote the matrix with (3.2.15) as its columns and $\bar{B}$ the matrix with

$$\{\, \nabla g_i(x) \,:\, i \,\in\, \bar{\mathcal{B}} \,\}$$

as its columns, where $\bar{\mathcal{B}} \equiv \mathcal{K} \setminus \mathcal{B}$. By the maximality of $\mathcal{B}$, it follows that each of the vectors in the latter family is a linear combination of the columns of the matrix B. More precisely, with $C \equiv (B^T B)^{-1} B^T \bar{B}$, we have $\bar{B} = BC$. It follows that $\lambda_{\mathcal{B}} = -C\lambda_{\bar{\mathcal{B}}}$. By the CRCQ, it follows that for all k sufficiently large, the gradients

$$\{\, \nabla g_i(y^k) \,:\, i \,\in\, \mathcal{B} \,\} \tag{3.2.16}$$

are also linearly independent and each vector in the family

$$\{ \nabla g_i(y^k) \, : \, i \in \bar{\mathcal{B}} \} \tag{3.2.17}$$

is a linear combination of the vectors in the family (3.2.16). With B^k and $\bar{B}^k$ denoting, respectively, the matrix with the vectors in (3.2.16) and (3.2.17) as its columns, it follows that for all k sufficiently large, we have

$$\bar{B}^k = B^k C^k,$$

where $C^k \equiv ((B^k)^T B^k)^{-1} (B^k)^T \bar{B}^k$. Clearly, $C^k \to C$. Define

$$\lambda^k_{\bar{\mathcal{B}}} \equiv \lambda_{\bar{\mathcal{B}}} \quad \text{and} \quad \lambda^k_{\mathcal{B}} \equiv -C^k \lambda^k_{\bar{\mathcal{B}}}.$$

Clearly, we have

$$\sum_{i \in \mathcal{K}} \lambda^k_i \, \nabla g(y^k) = 0$$

and the sequence $\{\lambda^k_{\mathcal{K}}\}$ converges to $\lambda_{\mathcal{K}}$. Thus (b) follows.

(b) $\Rightarrow$ (a). Suppose that the CRCQ fails at x. There exists a subset $\mathcal{K}$ of $\mathcal{I}(x)$ and a sequence $\{y^k\}$ converging to x such that for every k, the gradients

$$\{ \nabla g_i(y^k) \, : \, i \in \mathcal{K} \}$$

are linearly independent, but the limiting gradients

$$\{ \nabla g_i(x) \, : \, i \in \mathcal{K} \}$$

are linearly dependent. This easily contradicts (b).

(b) $\Rightarrow$ (c). Let $s \equiv |\mathcal{K}|$. It suffices to show that

$$\limsup_{y \to x} R(y, \mathcal{K}) \subseteq R(x, \mathcal{K}).$$

Let $\{y^k\}$ be a sequence converging to x; let a sequence $\{\lambda^{R,k}_{\mathcal{K}}\}$ converge to some vector $\lambda^\infty_{\mathcal{K}}$, where each $\lambda^{R,k}_{\mathcal{K}}$ is an element in $R(y^k, \mathcal{K})$. Since $\mathbb{R}^s$ is the orthogonal sum of $R(x, \mathcal{K})$ and $N(x, \mathcal{K})$, there exist orthogonal vectors $\bar{\lambda}^R_{\mathcal{K}}$ in $R(x, \mathcal{K})$ and $\bar{\lambda}^N_{\mathcal{K}}$ in $N(x, \mathcal{K})$ such that

$$\lambda^\infty_{\mathcal{K}} = \bar{\lambda}^{R,\infty}_{\mathcal{K}} + \bar{\lambda}^{N,\infty}_{\mathcal{K}}. \tag{3.2.18}$$

By (b), there exists a sequence $\{\lambda^{N,k}_{\mathcal{K}}\}$ converging to $\bar{\lambda}^{N,\infty}_{\mathcal{K}}$ such that $\lambda^{N,k}_{\mathcal{K}}$ belongs to $N(y^k, \mathcal{K})$ for every k. Since

$$(\lambda^{N,k}_{\mathcal{K}})^T \lambda^{R,k}_{\mathcal{K}} = 0,$$

by letting $k \to \infty$, it follows that

$$(\bar{\lambda}_{\mathcal{K}}^{N,\infty})^T \lambda_{\mathcal{K}}^{\infty} = 0.$$

Consequently, by (3.2.18), it follows that $\bar{\lambda}_{\mathcal{K}}^{N,\infty} = 0$. Thus, $\lambda_{\mathcal{K}}^{\infty}$ belongs to $\mathcal{R}(x, \mathcal{K})$ as desired.

(c) $\Rightarrow$ (b). Again, it suffices to show (3.2.14). Let $\lambda_{\mathcal{K}}$ be arbitrary element of $N(x, \mathcal{K})$. Let $\{y^k\}$ be an arbitrary sequence converging to x. For each k, let

$$\lambda_{\mathcal{K}}^{N,k} \equiv \Pi_{N_k}(\lambda_{\mathcal{K}}) \quad \text{and} \quad \lambda_{\mathcal{K}}^{R,k} \equiv \Pi_{R_k}(\lambda_{\mathcal{K}}),$$

where $N_k \equiv N(y^k, \mathcal{K})$ and $R_k \equiv R(y^k, \mathcal{K})$. Both sequences $\{\lambda_{\mathcal{K}}^{R,k}\}$ and $\{\lambda_{\mathcal{K}}^{N,k}\}$ are bounded. Without loss of generality, we may assume that

$$\lim_{k \to \infty} \lambda_{\mathcal{K}}^{R,k} = \bar{\lambda}_{\mathcal{K}}^{R,\infty} \quad \text{and} \quad \lim_{k \to \infty} \lambda_{\mathcal{K}}^{N,k} = \bar{\lambda}_{\mathcal{K}}^{N,\infty}$$

for some $\bar{\lambda}_{\mathcal{K}}^{R,\infty}$ and $\bar{\lambda}_{\mathcal{K}}^{N,\infty}$. By (c), it follows that the $\bar{\lambda}_{\mathcal{K}}^{R,\infty}$ must belong to $R(x, \mathcal{K})$. It is easy to see that $\bar{\lambda}_{\mathcal{K}}^{N,\infty}$ must belong to $N(x, \mathcal{K})$. Consequently,

$$\bar{\lambda}_{\mathcal{K}}^{R,\infty} = \lambda_{\mathcal{K}} - \bar{\lambda}_{\mathcal{K}}^{N,\infty} \in N(\bar{x}, \mathcal{K}) \cap R(x, \mathcal{K}) = \{0\}.$$

This shows that the entire sequence $\{\lambda_{\mathcal{K}}^{R,k}\}$ converges to zero. Thus,

$$\lambda_{\mathcal{K}} = \lim_{k \to \infty} \lambda_{\mathcal{K}}^{N,k},$$

as desired. □

More discussion on the various CQs can be found in Exercises **3.7.7**, **3.7.9**, and **3.7.10**.

3.3 Local Uniqueness of Solutions

It follows from Theorem **2.3.3** that the strict monotonicity of the map F on K ensures the global uniqueness of the solution to the VI (K, F), if this solution exists. Often, the local uniqueness of a solution is also of interest. Indeed, this property is central to the local convergence analysis of "Newton-type" iterative methods for solving the VI, especially in establishing their fast rate of convergence, as we see in a later chapter. In this section, we undertake a comprehensive study of the local uniqueness property of a solution to a VI/CP. We begin by formally introducing the local uniqueness concept in the following definition.

3.3.1 Definition. A solution x of the VI (K, F) is said to be *locally unique*, or *isolated*, if there exists a neighborhood $\mathcal{N}$ of x such that

$$\text{SOL}(K, F) \cap \mathcal{N} = \{x\}.$$

It is in general possible for a VI to have multiple solutions each of which is isolated. For instance, if the matrix M is nondegenerate, then a known result in LCP theory says that for every vector q, the LCP (q, M) must have a finite number (possibly zero) of solutions all of which must therefore be isolated. In general, it would be of interest to identify a class of VIs for which the global uniqueness of a solution (assuming that it exists) is implied by the existence of a locally unique solution to the problem. This issue is treated in Section **3.5**.

Due to the local nature of the isolatedness concept, it is possible to obtain sufficient conditions for a given solution to a VI to be locally unique in terms of an appropriate "local approximation" of the pair (K, F) near the solution. For F, we use the B-derivative as a pointwise local approximation. In the next subsection, we deal with the local approximation of K at a given solution of the VI (K, F).

3.3.1 The critical cone

The local approximation is obtained via the tangent cone $\mathcal{T}(x; K)$ to K at a vector $x \in K$ and the orthogonal complement of the vector $F(x)$. Specifically, we define the *critical cone* of the pair (K, F) at $x \in K$ as:

$$\mathcal{C}(x; K, F) \equiv \mathcal{T}(x; K) \cap F(x)^{\perp};$$

we call elements of the critical cone $\mathcal{C}(x; K, F)$ *critical vectors* of the pair (K, F) at x. See Figure 3.1 for an illustration of the critical cone. Presumably, these critical vectors provide a potential source for the existence of distinct solutions of the VI (K, F) that are arbitrarily close to a given solution $x \in \text{SOL}(K, F)$. Therefore, in order for x to be locally unique, it is imperative that such critical vectors be well behaved.

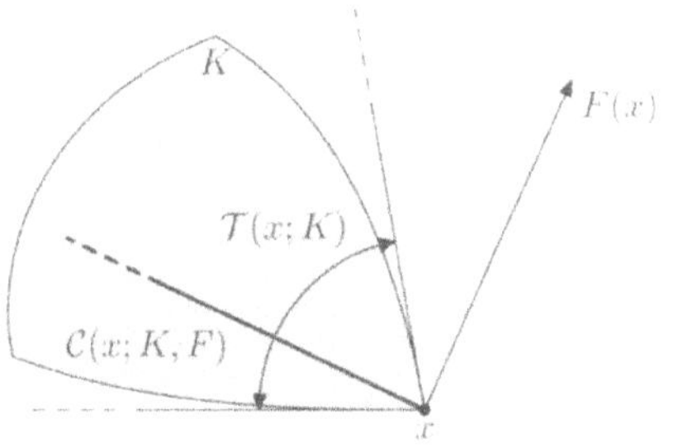

Figure 3.1: Illustration of the critical cone.

In what follows, we discuss properties of the critical cone and its dual. These properties require no conditions on the function F. First, the cone

$\mathcal{C}(x; K, F)$ can be represented in a slightly different form for a vector x in $\text{SOL}(K, F)$. Indeed for such a solution x, a vector v is a critical vector of the pair (K, F) if and only if $v \in \mathcal{T}(x; K)$ and $F(x)^T v \le 0$. The justification of this equivalence (mainly the weakening of the equality $F(x)^T v = 0$ to the inequality $F(x)^T v \le 0$) is as follows: if $v \in \mathcal{T}(x; K)$, then there exist a sequence $\{x^k\} \subset K$ and a sequence of positive scalars $\{\tau_k\}$ converging to zero such that

$$v = \lim_{k \to \infty} \frac{x^k - x}{\tau_k}.$$

We have for each k,

$$F(x)^T (x^k - x) \ge 0;$$

thus dividing by τ_k and letting k tend to infinity, we deduce $F(x)^T v \ge 0$. Consequently we have

$$\mathcal{T}(x; K) \subseteq F(x)^*, \tag{3.3.1}$$

where $F(x)^*$ is the dual cone of the singleton $\{F(x)\}$, i.e., the set of all vectors in $\mathbb{R}^n$ whose scalar products with $F(x)$ are nonnegative. From this fact, the alternative description of the elements of $\mathcal{C}(x; K, F)$ follows readily.

We apply the above observation to the case where K is finitely representable by differentiable equalities and inequalities satisfying a constraint qualification, including the case where K is polyhedral. In this case, the elements of the dual cone $\mathcal{C}(x; K, F)^*$ can be described in two equivalent ways. One way is in terms of the active gradients at x and $F(x)$ and the other way is in terms certain multipliers without involving $F(x)$. The latter description is key to the proof of Proposition **3.3.7** in which K is a polyhedron. Specifically, let K be given by (1.3.4); that is,

$$K \equiv \{ x \in \mathbb{R}^n : h(x) = 0, \; g(x) \le 0 \}, \tag{3.3.2}$$

with $h : \mathbb{R}^n \to \mathbb{R}^\ell$ and $g : \mathbb{R}^n \to \mathbb{R}^m$ being continuously differentiable vector functions. Suppose that $x \in K$ satisfies the Abadie CQ so that

$$\begin{aligned} \mathcal{T}(x; K) \equiv \{ v \in \mathbb{R}^n : \;\; & v^T \nabla h_j(x) = 0, \quad \forall j = 1, \ldots, \ell \\ & v^T \nabla g_i(x) \le 0, \quad \forall i \in \mathcal{I}(x) \}, \end{aligned}$$

where $\mathcal{I}(x)$ is the active index set at x. Under this setting and using the above alternative description of the critical vectors, we see that for a given $x \in \text{SOL}(K, F)$,

$$\begin{aligned} \mathcal{C}(x; K, F) = \{ v \in \mathbb{R}^n : \;\; & v^T F(x) \le 0 \\ & v^T \nabla h_j(x) = 0, \; \forall j = 1, \ldots, \ell, \\ & v^T \nabla g_i(x) \le 0, \; \forall i \in \mathcal{I}(x) \}; \end{aligned} \tag{3.3.3}$$

thus by duality, it follows that a vector u belongs to $\mathcal{C}(x; K, F)^*$ if and only if there exist multipliers $(\eta, \delta, \theta) \in \mathbb{R}^{\ell+|\mathcal{I}(x)|+1}$ such that

$$0 = u + \sum_{j=1}^{\ell} \eta_j \nabla h_j(x) + \sum_{i \in \mathcal{I}(x)} \delta_i \nabla g_i(x) + \theta F(x),$$

$$\delta_i \geq 0, \quad \forall i \in \mathcal{I}(x), \quad \text{and} \quad \theta \geq 0.$$

Consequently,

$$-\mathcal{C}(x; K, F)^* = \operatorname{pos}\{ \{ F(x) \} \cup \{ \nabla g_i(x) \}_{i \in \mathcal{I}(x)} \} \cup \operatorname{lin}\{ \nabla h_j(x) \}_{j=1}^{\ell}.$$

The above descriptions of $\mathcal{C}(x; K, F)$ and its dual do not involve the multipliers in $\mathcal{M}(x)$ but involve $F(x)$. Alternatively, equivalent descriptions of $\mathcal{C}(x; K, F)$ and its dual can be obtained based on any pair of multipliers $(\mu, \lambda) \in \mathcal{M}(x)$ without using $F(x)$ explicitly. For convenience, we restate the KKT system as follows:

$$\begin{aligned} 0 &= F(x) + \sum_{j=1}^{\ell} \mu_j \nabla h_j(x) + \sum_{i=1}^{m} \lambda_i \nabla g_i(x), \\ 0 &= h(x), \\ 0 \leq \lambda &\perp g(x) \leq 0. \end{aligned} \tag{3.3.4}$$

Let $(\mu, \lambda) \in \mathcal{M}(x)$ be an arbitrary pair of KKT multipliers. We introduce three fundamental index sets associated with the triple (x, μ, λ):

$$\begin{aligned} \alpha &\equiv \{ i : \lambda_i > 0 = g_i(x) \} \\ \beta &\equiv \{ i : \lambda_i = 0 = g_i(x) \} \\ \gamma &\equiv \{ i : \lambda_i = 0 > g_i(x) \} \end{aligned}$$

and call these the *strongly active*, *degenerate*, and *inactive* set of (x, μ, λ), respectively. Notice that $\alpha \cup \beta = \mathcal{I}(x)$ and α is simply the *support* of the multiplier λ. If the degenerate set β is empty, then we say that the triple (x, μ, λ) is *nondegenerate*; otherwise, we say that the triple is *degenerate*. Clearly, (x, μ, λ) is a nondegenerate KKT triple if and only if $\lambda - g(x) > 0$. The latter condition is called the *strict complementarity condition* between the multiplier λ and the constraint $g(x)$. It is equivalent to the condition that

$$\lambda_i > 0 \Leftrightarrow g_i(x) = 0.$$

See Section **3.4** for more discussion about nondegenerate solutions of VIs in general.

The following lemma gives an alternate description of $\mathcal{C}(x; K, F)$ in terms of the above three index sets; based on this alternative description, a corresponding description of $\mathcal{C}(x; K, F)^*$ can be derived accordingly.

3.3.2 Lemma. Let K be given by (3.3.2) with each h_j and g_i being continuously differentiable. Suppose that the Abadie CQ holds at x in SOL(K, F). For any $(\mu, \lambda) \in \mathcal{M}(x)$, it holds that

$$\begin{aligned} \mathcal{C}(x; K, F) = \{ v \in \mathbb{R}^n : \; & v^T \nabla h_j(x) = 0, \;\; \forall j = 1, \ldots, \ell, \\ & v^T \nabla g_i(x) = 0, \;\; \forall i \in \alpha, \\ & v^T \nabla g_i(x) \leq 0, \;\; \forall i \in \beta \}. \end{aligned} \tag{3.3.5}$$

Consequently, a vector u belongs to $\mathcal{C}(x; K, F)^*$ if and only if there exist multipliers $d\mu$, $d\lambda_\alpha$, and $d\lambda_\beta$, with $d\lambda_\beta \geq 0$ such that

$$0 = u + \sum_{j=1}^{\ell} d\mu_j \, \nabla h_j(x) + \sum_{i \in \alpha \cup \beta} d\lambda_i \, \nabla g_i(x).$$

Proof. Let $\mathcal{C}'$ denote the right-hand set in (3.3.5). If $v \in \mathcal{C}'$, then clearly $v \in \mathcal{T}(x; K)$. Premultiplying the first equation in (3.3.4) by v^T, we deduce that $v^T F(x) = 0$. Thus v belongs to $\mathcal{C}(x; K, F)$. The converse can be proved easily. Finally, the assertion about the dual $\mathcal{C}(x; K, F)^*$ is straightforward. □

3.3.3 Remark. The index sets (α, β, γ) depend on the KKT pair (μ, λ), each yielding a different, yet equivalent, representation of the critical cone $\mathcal{C}(x; K, F)$. □

When K is $\mathbb{R}^n_+$, i.e. for the NCP (F), the above representations of the critical cone $\mathcal{C}(x; K, F)$ simplify significantly. Specifically, let x be a given solution of the NCP (F). In the notation of the KKT system (3.3.4), we have $g(x) = -x$ and $\lambda = F(x)$. Thus specializing the above derivation, we deduce that the strongly active, degenerate, and inactive sets associated with x are as follows:

$$\begin{aligned} \alpha &= \{ i : x_i = 0 < F_i(x) \} \\ \beta &= \{ i : x_i = 0 = F_i(x) \} \\ \gamma &= \{ i : x_i > 0 = F_i(x) \} = \text{supp}(x). \end{aligned}$$

A word of caution on the terminology is needed. The labeling of "strongly active" and "inactive" sets, α and γ respectively, is with reference to the vector x; with reference to the vector $F(x)$, α and γ become the inactive and strongly active sets, respectively. Throughout, we continue to refer to these index sets with x as the reference vector.

In terms of the above index sets, it is not hard to verify that the critical cone of the pair $(\mathbb{R}^n_+, F)$ at x comprises precisely those vectors $v \in \mathbb{R}^n$ for which

$$[v_i \geq 0 \quad \forall\, i \in \beta] \quad \text{and} \quad [v_i = 0 \quad \forall\, i \in \alpha];$$

the dual of the critical cone is therefore the cone of all vectors $u \in \mathbb{R}^n$ for which

$$[u_i \geq 0 \quad \forall\, i \in \beta] \quad \text{and} \quad [u_i = 0 \quad \forall\, i \in \gamma];$$

These remarks are useful subsequently when we derive the special results for the NCP.

3.3.2 Conditions for local uniqueness

We present a series of results having to do with the local uniqueness of a solution to a VI/CP. The first result pertains to a general VI (K, F) where no special structure is assumed on the set K. In fact K is not even required to be convex. (See Exercise **3.7.12** for a sharper result if K is convex.) In the following proposition, we assume that $F : \mathcal{D} \to \mathbb{R}^n$, where $\mathcal{D}$ is an open set containing K, is B-differentiable at a solution $x \in \text{SOL}(K, F)$. This assumption implies that we can write

$$F(y) = F(x) + F'(x; y - x) + \mathbf{e}(y - x),$$

where the error function $\mathbf{e}$ satisfies

$$\lim_{v \to 0} \frac{\mathbf{e}(v)}{\| v \|} = 0.$$

3.3.4 Proposition. If F is B-differentiable at a solution x of the VI (K, F) and

$$v^T F'(x; v) > 0, \quad \forall\, v \in \mathcal{C}(x; K, F) \setminus \{0\},$$

then x is locally unique.

Proof. Assume for the sake of contradiction that x is not locally unique. There exists a sequence $\{x^k\} \subset \text{SOL}(K, F)$ converging to x with $x^k \neq x$ for all k. We claim that every accumulation point of the normalized sequence

$$\left\{ \frac{x^k - x}{\| x^k - x \|} \right\}$$

is a critical vector of the VI (K, F) at x. Indeed, let v be any such point. Then $v \in \mathcal{T}(x; K)$. Let κ be the infinite subset of indices such that

$$\lim_{k(\in\kappa)\to\infty} \frac{x^k - x}{\| x^k - x \|} = v.$$

For each k, we have

$$(x^k - x)^T F(x) \geq 0 \quad \text{and} \quad (x - x^k)^T F(x^k) \geq 0.$$

Dividing both inequalities by $\|x^k - x\|$ and letting $k(\in \kappa) \to \infty$, we deduce $v^T F(x) = 0$. Thus $v \in \mathcal{C}(x; K, F)$. By the B-differentiability of F at x, we have

$$\begin{aligned} 0 &\leq (x - x^k)^T F(x^k) \\ &= (x - x^k)^T [F(x) + F'(x; x^k - x) + o(\| x^k - x \|)] \\ &\leq (x - x^k)^T [F'(x; x^k - x) + o(\| x^k - x \|)]. \end{aligned}$$

Dividing by $\|x^k - x\|^2$ and letting $k(\in \kappa) \to \infty$, we deduce $v^T F'(x; v) \leq 0$. But this is a contradiction because v is clearly nonzero. □

3.3.5 Remark. If F is F-differentiable at the solution x with the Jacobian matrix $JF(x)$ (thus $F'(x; v) = JF(x)v$ for all $v \in \mathbb{R}^n$), Proposition **3.3.4** says that if $JF(x)$ is strictly copositive on the critical cone $\mathcal{C}(x; K, F)$, then x is a locally unique solution of the VI (K, F). □

For a VI with a polyhedral set, the sufficient condition for the local uniqueness of a solution can be substantially weakened. In particular, for an AVI, a necessary and sufficient condition for the local uniqueness of a solution can be obtained. The key to these refined results is due to two special properties of the tangent cone of a polyhedral set, which we state in the following lemma.

3.3.6 Lemma. Let K be a polyhedron in $\mathbb{R}^n$ and let $x \in K$ be given. The following two statements are valid.

(a) The tangent cone of K at $x \in K$ coincides with the feasible cone of K at x; that is, $d \in \mathcal{T}(x; K)$ if and only if $x + \tau d \in K$ for all $\tau > 0$ sufficiently small.

(b) There exists a neighborhood $\mathcal{N}$ of x such that $\mathcal{T}(x; K) \subseteq \mathcal{T}(y; K)$ for all $y \in K \cap \mathcal{N}$.

Proof. Write

$$K \equiv \{ x \in \mathbb{R}^n : Cx = d, \; Ax \leq b \}$$

for some matrices C and A and vectors d and b. We have

$$\mathcal{T}(x; K) = \{ d \in \mathbb{R}^n : Cd = 0, \; A_{i \cdot} d \leq 0, \; \forall i \in \mathcal{I}(x) \}.$$

Assertion (a) follows easily from this representation and the linearity of the constraints. Assertion (b) follows from the simple observation that $\mathcal{I}(y) \subseteq \mathcal{I}(x)$ for all y sufficiently close to x. □

The following result pertains to the local uniqueness of a solution to a linearly constrained VI, in particular, the AVI.

3.3.7 Proposition. Let K be a polyhedron and let $x \in \mathrm{SOL}(K, F)$ be given. Assume that F is B-differentiable at x. Consider the following three statements:

(a) the homogeneous CP:

$$\mathcal{C}(x; K, F) \ni v \perp F'(x; v) \in \mathcal{C}(x; K, F)^* \tag{3.3.6}$$

has $v = 0$ as the unique solution;

(b) the implication below holds:

$$\left. \begin{array}{r} x + d \in K, \quad d \perp F(x), \quad d \neq 0 \\ (F(x) + F'(x; d))^T (y - x) \geq 0, \quad \forall y \in K \end{array} \right\} \tag{3.3.7}$$

$$\Rightarrow \; d^T F'(x; d) > 0;$$

(c) x is locally unique.

It holds that (a) ⇔ (b) ⇒ (c). If in addition F is affine, then all three statements are equivalent.

Proof. (a) ⇒ (b). Let d be a vector satisfying the left-hand condition in (3.3.7). Clearly $d \in \mathcal{C}(x; K, F)$. Moreover, for every $d' \in \mathcal{C}(x; K, F)$, we have $d' \perp F(x)$ and $y \equiv x + \tau d' \in K$ for all $\tau > 0$ sufficiently small. Consequently it follows that $F'(x; d)^T d' \geq 0$. Thus $F'(x; d) \in \mathcal{C}(x; K, F)^*$. Moreover we have $F'(x; d)^T d \geq 0$. By condition (a), it follows that $F'(x; d)^T d > 0$ because $d \neq 0$. This establishes (b).

(b) ⇒ (a). Suppose for the sake of contradiction that the CP (3.3.6) has a nonzero solution v. We claim that for all $\tau > 0$ sufficiently small, the vector $d \equiv \tau v$ satisfies the left-hand condition in (3.3.7). If suffices to verify that for all $\tau > 0$ sufficiently small,

$$(F(x) + \tau F'(x; v))^T (y - x) \geq 0, \quad \forall y \in K.$$

Since K is polyhedral, we represent it as:

$$K = \{ x \in \mathbb{R}^n : Cx = d, \; Ax \leq b \}.$$

Since $F'(x; v) \in \mathcal{C}(x; K, F)^*$, there exists a triple $(\eta, \delta, \theta) \in \mathbb{R}^{\ell + |\mathcal{I}(x)| + 1}$ such that

$$0 = F'(x; v) + C^T \eta + (A_{\mathcal{I} \cdot})^T \delta + \theta F(x),$$

$$\delta_i \geq 0, \quad \forall i \in \mathcal{I}(x), \quad \text{and} \quad \theta \geq 0.$$

For every $y \in K$, we deduce

$$(y - x)^T [\theta F(x) + F'(x, v)] \geq 0.$$

Since $(y - x)^T F(x) \geq 0$, we obtain

$$(y - x)^T \left[F(x) + \frac{1}{1 + \theta} F'(x, v) \right] \geq 0.$$

It follows that for all nonnegative scalars $\tau \leq 1/(1 + \theta)$,

$$(y - x)^T (F(x) + \tau F'(x, v)) \geq 0.$$

This establishes our claim; therefore (b) implies (a).

(a) $\Rightarrow$ (c). We proceed as in the proof of Proposition **3.3.4**. Assume that x is not locally unique. Let $\{x^k\}$ be a sequence of solutions converging to x and $x^k \neq x$ for all k. Without loss of generality we may assume that

$$\lim_{k \to \infty} \frac{x^k - x}{\| x^k - x \|} = v \in \mathcal{T}(x; K) \setminus \{ 0 \}.$$

Similar to the proof of the previous proposition, we can show that

$$v \perp F(x) \quad \text{and} \quad v^T F'(x; v) \leq 0.$$

It remains to show that $F'(x; v) \in \mathcal{C}(x; K, F)^*$. Let $d' \in \mathcal{C}(x; K, F)$. By part (b) of Lemma **3.3.6**, it follows that $d' \in \mathcal{T}(x^k; K)$ for all k sufficiently large. Consequently, by (3.3.1), we have

$$\begin{aligned} 0 &\leq F(x^k)^T d' \\ &= [F(x) + F'(x; x^k - x) + o(\| x - x^k \|)]^T d' \\ &= [F'(x; x^k - x) + o(\| x - x^k \|)]^T d'. \end{aligned}$$

Dividing by $\|x^k - x\|$ and letting $k \to \infty$, we obtain $F'(x;v)^T d' \geq 0$. This completes the proof that (a) $\Leftrightarrow$ (b) $\Rightarrow$ (c).

Finally, we show that if in addition F is affine, then (c) $\Rightarrow$ (b). Assume that x is locally unique. Suppose that there exists a vector d satisfying the left-hand condition of (3.3.7) but $d^T F'(x;d) \leq 0$. We must have for all $\tau \in [0,1]$,

$$(y - x)^T [F(x) + \tau F'(x;d)] \geq 0, \quad \forall y \in K.$$

Since F is affine, we have

$$F(x + \tau d) = F(x) + \tau F'(x;d).$$

Consequently, it follows that

$$(y - x - \tau d)^T F(x + \tau d) \geq 0.$$

Thus $x + \tau d \in \mathrm{SOL}(K, F)$ for all $\tau \in [0,1]$. Since d is nonzero, we obtain a contradiction to the local uniqueness of x. □

In Exercise **3.7.13**, the reader is asked to demonstrate that the implication (a) $\Rightarrow$ (b) in Proposition **3.3.7** remains valid for a non-polyhedral set K if the tangent cone $\mathcal{T}(\cdot; K)$ is lower semicontinuous at x. In what follows, we give an immediate corollary of this proposition that does not require a proof.

3.3.8 Corollary. A solution x of the AVI (K, q, M) is isolated if and only if (C, M) is an R_0 pair, where C is the critical cone of the AVI at x. □

To prepare for the specialization of Proposition **3.3.7** to the NCP and for other related results (see e.g. Corollary **5.3.20**), we first recall the matrix concept of a Schur complement and discuss some important matrix-theoretic facts. Let A be a matrix of order $m \times n$ and let B be any nonsingular submatrix of A. There exist permutation matrices P and Q of order m and n respectively such that PAQ can be written in the partitioned form

$$PAQ = \begin{bmatrix} B & C \\ D & E \end{bmatrix}$$

for appropriate matrices C, D, and E. The Schur complement of B in A, which we denote A/B, is the matrix $E - DB^{-1}C$. If M is a square matrix of order n and L is a subset of $\{1, \ldots, n\}$ such that M_{LL} is nonsingular, the Schur complement of M_{LL} in M is the matrix

$$M/M_{LL} \equiv M_{JJ} - M_{JL}(M_{LL})^{-1} M_{LJ}.$$

where J is the complement of L in $\{1, \ldots, n\}$. If J is empty, it is understood that M/M_{LL} is vacuous and any condition imposed on this Schur complement is automatically satisfied. It is easy to see that if J' is a subset of J, then $(M/M_{LL})_{J'J'}$ is the Schur complement of M_{LL} in the following matrix

$$\begin{bmatrix} M_{LL} & M_{LJ'} \\ M_{J'L} & M_{J'J'} \end{bmatrix},$$

which itself is a principal submatrix of M. It follows that the principal submatrices of the Schur-complement M/M_{LL} are given by M_{II}/M_{LL} for all index sets I such that $L \subseteq I \subseteq L \cup J$.

An important determinantal formula, known as the *Schur determinantal formula*, states that if M_{LL} is nonsingular, then the determinant of M is equal to the product of the determinant of M_{LL} and the Schur complement M/M_{LL}; that is,

$$\det M = \det M_{LL} \det M/M_{LL}.$$

A trivial consequence of this formula is that if M_{LL} is nonsingular, then M is nonsingular if and only if the Schur complement M/M_{LL} is nonsingular; moreover, if $\det M_{LL}$ is positive and if either M or M/M_{LL} is nonsingular, then $\det M$ and $\det M/M_{LL}$ have the same nonzero sign. It follows from the former consequence that if M_{QQ} is a nonsingular submatrix of M_{LL}, then M_{LL}/M_{QQ} is well defined, and nonsingular if M_{LL} is nonsingular. A classical result, known as the *quotient formula for the Schur complement*, states that if M_{QQ} is a nonsingular principal submatrix of the nonsingular M_{LL}, then

$$M/M_{LL} = (M/M_{QQ})/(M_{LL}/M_{QQ}). \tag{3.3.8}$$

One feature to notice is how this formula formally resembles the algebraic rule used for simplifying a complex fraction; that is, the M_{QQ} in the right-hand "quotient" simply "cancels" itself, yielding the left-hand Schur complement.

The next result is the specialization of Proposition **3.3.7** to the NCP defined by a F-differentiable map. We recall the index sets β (the degenerate set) and γ (the support of x) introduced at the end of Subsection **3.3.1**.

3.3.9 Corollary. Suppose that F is continuously differentiable in an open neighborhood of a solution x of the NCP (F). Consider the following four statements:

(a) the principal submatrix

$$J_\gamma F_\gamma(x) \tag{3.3.9}$$

is nonsingular and the Schur complement:

$$J_\beta F_\beta(x) - J_\gamma F_\beta(x)(\, J_\gamma F_\gamma(x)\,)^{-1} J_\beta F_\gamma(x) \tag{3.3.10}$$

is nondegenerate;

(b) for every index subset γ' of $\{1, \ldots, n\}$ satisfying

$$\gamma \subseteq \gamma' \subseteq \gamma \cup \beta,$$

the principal submatrix

$$JF(x)_{\gamma'\gamma'}$$

is nonsingular;

(c) the MLCP in the variable (v_γ, v_β):

$$J_\gamma F_\gamma(x)\, v_\gamma + J_\beta F_\gamma(x)\, v_\beta = 0$$

$$0 \leq v_\beta \perp J_\gamma F_\beta(x)\, v_\gamma + J_\beta F_\beta(x)\, v_\beta \geq 0$$

has $(v_\gamma, v_\beta) = (0,0)$ as the unique solution;

(d) x is locally unique.

It holds that (a) $\Leftrightarrow$ (b) $\Rightarrow$ (c) $\Rightarrow$ (d). Moreover, (c) and (d) are equivalent if F is affine.

Proof. (a) $\Leftrightarrow$ (b). This follows easily from the Schur determinantal formula. Indeed suppose (a) holds. Let $\gamma' = \gamma \cup \beta'$ for some subset β' of β. Since the submatrix (3.3.9) is nonsingular, the Schur determinantal formula yields

$$\det JF(x)_{\gamma'\gamma'} = \det JF(x)_{\gamma\gamma} \det \left(J_{\beta'} F_{\beta'}(x) - J_\gamma F_{\beta'}(x)\, (\, J_\gamma F_\gamma(x)\,)^{-1} J_{\beta'} F_\gamma(x) \right).$$

The matrix

$$J_{\beta'} F_{\beta'}(x) - J_\gamma F_{\beta'}(x)\, (\, J_\gamma F_\gamma(x)\,)^{-1} J_{\beta'} F_\gamma(x)$$

is clearly a principal submatrix of the Schur complement (3.3.10); since the latter matrix is nondegenerate, (b) follows. Conversely, if (b) holds, then the matrix (3.3.9) must be nonsingular. Reversing the above argument shows that the Schur complement (3.3.10) is nondegenerate; thus (a) follows.

(b) $\Rightarrow$ (c). Suppose the displayed MLCP has a nonzero solution (v_γ, v_β). Let β' be the subset of β consisting of those indices $i \in \beta$ for which $v_i > 0$.

The subvector $v_{\gamma'} \equiv (v_\gamma, v_{\beta'})$, where $\gamma' \equiv \gamma \cup \beta'$, is nonzero and satisfies the homogeneous system of linear equations:

$$JF(x)_{\gamma'\gamma'} v_{\gamma'} = 0.$$

This contradicts (b). Thus (c) holds.

(c) $\Rightarrow$ (d). By invoking the special structure of the critical cone associated with an NCP, we deduce that the CP (3.3.6) reduces to the displayed MLCP. Thus this implication is an immediate consequence of Proposition **3.3.7**. Moreover, if F is affine, then (c) and (d) are equivalent by the same proposition. □

Notice that the functions F_i for i in the strongly active set α of a solution x plays no role in the local uniqueness of x. The intuitive explanation is that for any solution x' of the NCP (F) that is near x, x_i' must remain equal to zero for all such indices i. Property (a) or equivalently (b) in Corollary **3.3.9** is sufficiently important to deserve a name.

3.3.10 Definition. Let x be a solution of the NCP (F). We call x a *b-regular* solution if x satisfies the property (a), or equivalently (b), in Corollary **3.3.9**. We also call (3.3.9) the *basic matrix* of x. □

Although every b-regular solution of an NCP must be locally unique, the converse is not always true. In fact, if x is a locally unique solution of the NCP (F), it is not even necessary for the basic matrix of x to be nonsingular. This is illustrated by the following LCP in three variables.

3.3.11 Example. Consider the LCP (q, M) with data:

$$q = \begin{pmatrix} 0 \\ -1 \\ 1 \end{pmatrix} \quad \text{and} \quad M = \begin{bmatrix} 0 & -1 & -1 \\ 1 & 0 & 0 \\ -1 & 0 & 0 \end{bmatrix}.$$

This LCP has a unique solution $x = (1, 0, 0)$, which yields $\gamma = \{1\}$ and $\beta = \{2, 3\}$. Since $m_{11} = 0$ the matrix (3.3.9) is therefore singular. □

If the basic matrix (3.3.9) of the solution x is nonsingular, then the homogeneous MLCP in part (c) of Corollary **3.3.9** has a unique solution if and only if the homogeneous LCP $(0, M_{\text{dgn}})$ has a unique solution, i.e., if and only if M_{dgn} is an R_0 matrix, where

$$M_{\text{dgn}} \equiv J_\beta F_\beta(x) - J_\gamma F_\beta(x) \, (\, J_\gamma F_\gamma(x) \,)^{-1} J_\beta F_\gamma(x).$$

In particular, if x is a nondegenerate solution of the NCP (F) and if $J_\gamma F_\gamma(x)$ is nonsingular, then x is locally unique.

3.3.3 Local uniqueness in terms of KKT triples

Corollary **3.3.9** can easily be extended to an MiCP. We omit the details but will use the extension to establish the next result, which pertains to the VI (K, F) with K being finitely representable.

3.3.12 Theorem. Consider the VI (K, F) where K is defined by (3.3.2) with F being continuously differentiable and all h_j and g_i being twice continuously differentiable in a neighborhood of a solution x of the VI. Suppose that x satisfies the SBCQ and that for every $(\mu, \lambda) \in \mathcal{M}(x)$, the homogeneous linear complementarity system:

$$\mathcal{C}(x; K, F) \ni v \perp J_x\mathbf{L}(x, \mu, \lambda)\, v \in \mathcal{C}(x; K, F)^*, \tag{3.3.11}$$

has $v = 0$ as the unique solution. The following two statements hold.

(a) The solution x is locally unique.

(b) A triple (x, μ, λ) is a locally unique solution of the KKT system (3.3.4) if and only if $\mathcal{M}(x) = \{(\mu, \lambda)\}$.

Proof. To establish (a), assume for the sake of contradiction that x is not a locally unique solution of the VI (K, F). Let $\{x^k\} \subset \mathrm{SOL}(K, F) \setminus \{x\}$ be a sequence of solutions converging to x. Without loss of generality, we may assume that

$$\lim_{k\to\infty} \frac{x^k - x}{\| x^k - x \|} = v \neq 0$$

for some vector v, which can be shown to belong to $\mathcal{C}(x; K, F)$; see the proof of Proposition **3.3.4**. Since x satisfies the SBCQ, it follows that for all k sufficiently large, there exists $(\mu^k, \lambda^k) \in \mathcal{M}(x^k)$ such that the sequence $\{(\mu^k, \lambda^k)\}$ converges to some (μ, λ) belonging to $\mathcal{M}(x)$. We have, for every k sufficiently large,

$$\mathbf{L}(x^k, \mu^k, \lambda^k) = 0, \tag{3.3.12}$$

$$0 \leq \lambda^k \perp g(x^k) \leq 0,$$

and $\mathcal{I}(x^k) \subseteq \mathcal{I}(x)$. Thus $\lambda_i^k = 0$ for all $i \notin \mathcal{I}(x)$ and all k sufficiently large.

We claim that for all k sufficiently large,

$$\lambda_i^k\, v^T \nabla g_i(x) = 0, \quad \forall\, i. \tag{3.3.13}$$

Suppose this is not true; then there exists an infinite index set κ and an index i such that for all $k \in \kappa$,

$$\lambda_i^k\, v^T \nabla g_i(x) \neq 0.$$

Hence $\lambda_i^k > 0$, which implies

$$0 \,=\, g_i(x^k) \,=\, g_i(x) + \nabla g_i(x)^T(\,x^k - x\,) + o(\|\,x^k - x\,\|).$$

Thus $i \in \mathcal{I}(x^k)$, which yields $g_i(x) = 0$. Consequently for the specified index i, we deduce that for all $k \in \kappa$ sufficiently large,

$$0 \,=\, \nabla g_i(x)^T(\,x^k - x\,) + o(\|\,x^k - x\,\|).$$

Dividing by $\|x^k - x\|$ and letting $k(\in \kappa) \to \infty$, we obtain $v^T \nabla g_i(x) = 0$, which is a contradiction. Thus our claim is established.

We can write

$$F(x^k) \,=\, F(x) + JF(x)\,(\,x^k - x\,) + \mathbf{e}_F(x^k - x),$$

$$\nabla h_j(x^k) \,=\, \nabla h_j(x) + \nabla^2 h_j(x)\,(\,x^k - x\,) + e_j^h(x^k - x), \quad \forall\, j \,=\, 1, \ldots, \ell,$$

$$\nabla g_i(x^k) \,=\, \nabla g_i(x) + \nabla^2 g_i(x)\,(\,x^k - x\,) + e_i^g(x^k - x), \quad \forall\, i \,=\, 1, \ldots, m,$$

where the error functions $\mathbf{e}_F$, e_j^h, and e_i^g all satisfy

$$\lim_{v \to 0} \frac{e(v)}{\|\,v\,\|} \,=\, 0.$$

Substituting these expressions into (3.3.12), we deduce

$$\mathbf{L}(x, \mu^k, \lambda^k) + J_x\mathbf{L}(x, \mu^k, \lambda^k)\,(\,x^k - x\,) + \mathbf{e}^k \,=\, 0, \tag{3.3.14}$$

where $\mathbf{e}^k$ satisfies

$$\lim_{k \to \infty} \frac{\mathbf{e}^k}{\|\,x^k - x\,\|} \,=\, 0.$$

Since $\lambda_i^k = 0$ for all $i \notin \mathcal{I}(x)$, the equation (3.3.14) implies that

$$J_x\mathbf{L}(x, \mu^k, \lambda^k)\,(\,x^k - x\,) + \mathbf{e}^k \,\in\, \mathcal{C}(x; K, F)^*$$

for all k sufficiently large. Since $\mathcal{C}(x; K, F)$ is a polyhedral cone, it follows that

$$J_x\mathbf{L}(x, \mu, \lambda)\,v \,\in\, \mathcal{C}(x; K, F)^*.$$

Premultiplying (3.3.14) by v^T, using (3.3.13) and noting that $v^T \nabla h_j(x)$ is equal to zero for all j, we deduce

$$v^T[\,J_x\mathbf{L}(x, \mu^k, \lambda^k)\,(\,x^k - x\,) + \mathbf{e}^k\,] \,=\, 0.$$

Normalizing this equation by $\|x^k - x\|$ and letting $k \to \infty$, we conclude that $v^T J_x\mathbf{L}(x, \mu, \lambda)\,v = 0$. Thus the assumption that the system (3.3.11) has a unique solution yields a contradiction. Consequently, x must be a locally unique solution of the VI (K, F).

To prove statement (b), assume first that $\mathcal{M}(x)$ is a singleton with its sole element being (μ, λ). Considering the KKT system (3.3.4) as a MiCP defined by the mapping (1.3.8) and using the extension of Corollary **3.3.9** to a MiCP, we may deduce that (x, μ, λ) is a locally unique KKT triple, provided we can show that the following MLCP has $(u, d\mu, d\lambda_{\mathcal{I}(x)}) = (0, 0, 0)$ as the unique solution:

$$\begin{aligned}
0 &= J_x\mathbf{L}(x, \mu, \lambda)\, u + \sum_{j=1}^{\ell} d\mu_j\, \nabla h_j(x) + \sum_{i \in \mathcal{I}(x)} d\lambda_i\, \nabla g_i(x), \\
0 &= \nabla h_j(x)^T u, \quad \forall\, j = 1, \ldots, \ell, \\
0 &= \nabla g_i(x)^T u, \quad \forall\, i \in \alpha, \\
0 &\geq \nabla g_i(x)^T u \perp d\lambda_i \geq 0, \quad \forall\, i \in \beta.
\end{aligned} \tag{3.3.15}$$

We proceed to show the last statement. Let $(u, d\mu, d\lambda_{\mathcal{I}(x)})$ satisfy the system (3.3.15). By the alternative description of $\mathcal{C}(x; K, F)$ given in Lemma **3.3.2**, it is easy to see that u satisfies (3.3.11). By assumption, we must have $u = 0$. Since (μ, λ) is the unique KKT multiplier, it follows that the SMFCQ holds at (x, μ, λ); this easily implies that $(d\mu, d\lambda_{\mathcal{I}(x)})$ must equal to $(0, 0)$.

For the converse, if (x, μ, λ) is a locally unique KKT triple, then since $\mathcal{M}(x)$ is a polyhedron, it follows that $\mathcal{M}(x)$ must be a singleton. □

3.3.13 Remark. The unique solvability of the homogeneous linear complementarity system (3.3.11) is equivalent to the R_0 property of the pair $(\mathcal{C}(x; K, F), J_x\mathbf{L}(x, \mu, \lambda))$. □

It is natural to ask the following question: for a VI (K, F) where K is given by (3.3.2), if x is a solution of the VI (K, F) and (x, μ, λ) is a locally unique KKT triple, is x necessarily an isolated solution of the VI? In essence, Theorem **3.3.12** shows that the answer is affirmative, under the assumption that $(\mathcal{C}(x; K, F), J_x\mathbf{L}(x, \mu, \lambda))$ is an R_0 pair. The following result shows that the answer remains affirmative even without this R_0 assumption.

3.3.14 Proposition. Consider the VI (K, F) where K is defined by (3.3.2) with F being continuous and all h_j and g_i being continuously differentiable near a solution x of the VI. If (x, μ, λ) is a locally unique KKT triple, then x is a locally unique solution of the VI.

Proof. Since (x, μ, λ) is a locally unique KKT triple, the last part of the proof of Theorem **3.3.12** shows that $\mathcal{M}(x)$ is a singleton. (This does not require the R_0 property of the pair $(\mathcal{C}(x; K, F), J_x\mathbf{L}(x, \mu, \lambda))$.) Thus the MFCQ holds at x. If $\{x^k\} \subset \text{SOL}(K, F)$ is an arbitrary sequence of solutions converging to x, then for all k sufficiently large, $\mathcal{M}(x^k)$ is nonempty; moreover, every sequence of multipliers $\{(\mu^k, \lambda^k)\}$, where $(\mu^k, \lambda^k) \in \mathcal{M}(x^k)$ for each k, must converge to the unique multiplier (μ, λ) of the KKT system at the limit point x. Since (x, μ, λ) is a locally unique KKT triple, it follows that (x^k, μ^k, λ^k) must coincide with (x, μ, λ) for all k sufficiently large; in particular, we have $x^k = x$ for all k sufficiently large. Consequently, x is a locally unique solution of the VI (K, F) □

Under the assumptions of Theorem **3.3.12**, we can strengthen part (a) of this theorem; in particular, we get a quantitative error bound from any triple (x', μ, λ) that is near the set $\{x\} \times \mathcal{M}(x)$ to this set in terms of the residual of the KKT system. See Proposition **6.2.7** for details. Thus the R_0 assumption in Theorem **3.3.12** yields more than the local uniqueness of the solution $x \in \text{SOL}(K, F)$. The converse of Proposition **3.3.14** is easily seen to be false, even if K is convex. In what follows, we present an example to illustrate that it is possible for the VI (K, F) to have a locally unique solution x for which there are multiple KKT pairs in $\mathcal{M}(x)$.

3.3.15 Example. Let K be the closed unit ball in the plane represented with a redundant constraint:

$$K \equiv \{ (x_1, x_2) \in \mathbb{R}^2 : x_1^2 + x_2^2 \le 1, \; x_1 \le 1 \}.$$

Let $F(x_1, x_2)$ be any continuously differentiable function satisfying:

$$F(1,0) = (-1,0) \quad \text{and} \quad \frac{\partial F_2(1,0)}{\partial x_2} > 0;$$

e.g. $F(x_1, x_2) \equiv (-x_1 e^{x_2}, x_2)$, which is non-integrable and non-monotone. It is then easy to see that $(1,0)$ is a solution of the VI (K, F) because the following KKT system is satisfied at this vector:

$$\begin{aligned} F_1(x_1, x_2) + 2\lambda_1 x_1 + \lambda_2 &= 0 \\ F_2(x_1, x_2) + 2\lambda_1 x_2 &= 0 \\ 0 \le \lambda_1 \perp x_1^2 + x_2^2 - 1 &\le 0 \\ 0 \le \lambda_2 \perp x_1 - 1 &\le 0. \end{aligned}$$

The multiplier set $\mathcal{M}(1,0)$ is the line segment in the plane joining the two points $(1/2, 0)$ and $(0, 1)$. Since this is certainly a compact set, the MFCQ holds at $(1,0)$. The tangent cone of K at this vector is the half-plane: $x_1 \leq 0$. Thus the critical cone $\mathcal{C}$ of the pair (K,F) at $(1,0)$ is the x_2 axis. The dual $\mathcal{C}^*$ of the latter cone is therefore the x_1 axis. For any $\lambda \in \mathcal{M}(1,0)$, consider a vector $v \in \mathbb{R}^2$ satisfying (3.3.11). With $\mathcal{C}$ and $\mathcal{C}^*$ being the x_2 and x_1 axis, respectively, we deduce $v_1 = 0$ and

$$\left(\frac{\partial F_2(1,0)}{\partial x_2} + \lambda_1 \right) v_2 = 0.$$

Since the sum of the two terms within the parentheses is positive, it follows that $v_2 = 0$. Consequently, the assumptions of Theorem **3.3.12** are satisfied. Thus $(1,0)$ is a locally unique solution of the VI (K,F). With $F(x_1,x_2) \equiv (-x_1 e^{x_2}, x_2)$, the reader can verify this conclusion directly, via the KKT system. With the latter function, the VI (K,F) has another solution, namely, $(0,0)$. □

3.3.4 Local uniqueness theory in NLP

As noted in Subsection **1.3.1**, if K is a convex set, the VI $(K, \nabla\theta)$ is the stationary point problem of the NLP of minimizing θ on the set K. Thus the local uniqueness results derived previously can be applied to infer the local uniqueness of a stationary point of such an NLP. In what follows, we present a brief local uniqueness theory of a standard NLP without assuming the convexity of its feasible set.

We begin by repeating some basic material about the NLP:

$$\begin{array}{ll} \text{minimize} & \theta(x) \\ \text{subject to} & h(x) = 0, \; g(x) \leq 0, \end{array} \tag{3.3.16}$$

where θ and each g_i and h_j are twice continuously differentiable real-valued functions. Let

$$K \equiv \{ x \in \mathbb{R}^n : h(x) = 0, \; g(x) \leq 0 \}$$

denote the (possibly nonconvex) feasible set of (3.3.16). By definition, a stationary point of (3.3.16) is a feasible vector $x \in K$ such that

$$d^T \nabla\theta(x) \geq 0, \quad \forall \, d \in \mathcal{T}(x;K).$$

Such a point is not necessarily a solution of the VI $(K, \nabla\theta)$. The KKT system of (3.3.16) is the MiCP:

$$\begin{aligned} 0 &= \nabla\theta(x) + \sum_{j=1}^{\ell} \mu_j \nabla h_j(x) + \sum_{i=1}^{m} \lambda_i \nabla g_i(x), \\ 0 &= h(x), \\ 0 &\leq \lambda \perp g(x) \leq 0. \end{aligned} \tag{3.3.17}$$

Let $\mathcal{M}_{\mathrm{NLP}}(x)$ denote the set of KKT multipliers (μ, λ) satisfying the above system (3.3.17). If $x \in K$ satisfies the Abadie CQ, then x is a stationary point of the NLP (3.3.16) if and only if $\mathcal{M}_{\mathrm{NLP}}(x)$ is nonempty.

We introduce two definitions that are the main topic of discussion in what follows.

3.3.16 Definition. A stationary point x of the NLP (3.3.16) is *isolated* or *locally unique* if there exists a neighborhood $\mathcal{N}$ of x such that x is the only stationary point of (3.3.16) that lies in $\mathcal{N}$. □

The next definition distinguishes several types of local minimizers of the NLP (3.3.16). The goal of the discussion herein is to clarify the relation between these types of local minimizers.

3.3.17 Definition. A local minimizer x of the NLP (3.3.16) is said to be

(a) *isolated* if there exists a neighborhood $\mathcal{N}$ of x such that (3.3.16) has no other local minimizer within $\mathcal{N}$;

(b) *strict* if there exists a neighborhood $\mathcal{N}$ of x such that for all feasible vectors x' in $\mathcal{N}$ distinct from x, $\theta(x') > \theta(x)$;

(c) *strong* if there exist a neighborhood $\mathcal{N}$ of x and a constant $c > 0$ such that for all feasible vectors x' in $\mathcal{N}$,

$$\theta(x') \geq \theta(x) + c \, \| x - x' \|^2.$$

It is clear that every local minimizer that is an isolated stationary point must be an isolated local minimizer. Moreover, every strong local minimizer is strict; but the converse is not true. Every isolated local minimizer must be strict; again, the converse is not true. The following counterexamples illustrate the distinction between these various types of local minimizers. See also Figure 3.2.

3.3.18 Example. Consider the following simple equality-constrained nonlinear program in one variable:

$$\begin{array}{ll} \text{minimize} & x^4 \\ \text{subject to} & x^6 \sin(1/x) = 0, \end{array}$$

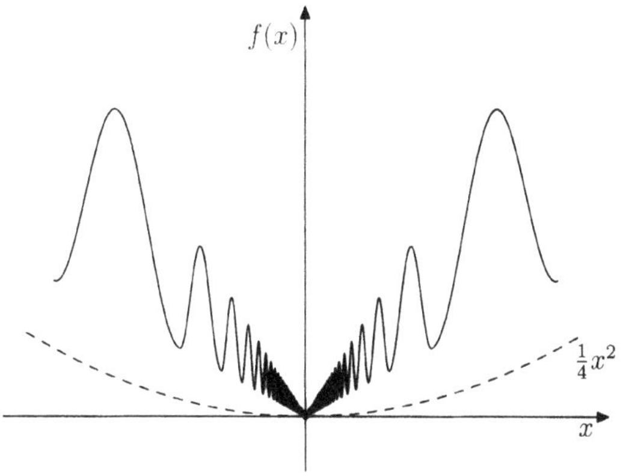

Figure 3.2: A strong minimum that is not isolated.

where $0 \sin(1/0)$ is defined to be 0. For this nonlinear program, $x = 0$ is the unique global minimizer; thus it is a strict local minimizer, but it is not strong because there cannot exist any positive constant c such that $x^4 \geq cx^2$ in any neighborhood of 0. Moreover, $x = 0$ is not an isolated minimizer of the NLP because for every positive integer k, $x_k \equiv (k\pi)^{-1}$ is an isolated feasible solution and thus an isolated local minimizer of the NLP.

Consider the slightly modified nonlinear program:

$$\begin{array}{ll} \text{minimize} & x^2 \\ \text{subject to} & x^6 \sin(1/x) \,=\, 0. \end{array}$$

We see that $x = 0$ is a strong, unique global minimum but every feasible solution is a local minimizer. □

The scalar Lagrangian function of the NLP (3.3.16) is:

$$L(x, \mu, \lambda) \equiv \theta(x) + \mu^T h(x) + \lambda^T g(x).$$

The critical cone of this NLP at a feasible vector x is:

$$\mathcal{C}(x; K, \nabla\theta) \equiv \mathcal{T}(x; K) \cap \nabla\theta(x)^{\perp}.$$

For each pair of KKT multipliers $(\mu, \lambda) \in \mathcal{M}_{\text{NLP}}(x)$, consider the following homogeneous quadratic program:

$$\begin{array}{ll} \text{minimize} & \frac{1}{2} v \nabla^2_{xx} L(x, \mu, \lambda) v \\ \text{subject to} & v \in \mathcal{C}(x; K, \nabla\theta). \end{array} \tag{3.3.18}$$

It is easy to see that the pair $(\mathcal{C}(x; K, \nabla\theta), \nabla^2_{xx} L(x, \mu, \lambda))$ has the R_0 property if and only if $v = 0$ is the unique stationary point of (3.3.18). Moreover,

$\nabla^2_{xx}L(x, \mu, \lambda)$ is (strictly) copositive on $\mathcal{C}(x; K, \nabla\theta)$ if and only if $v = 0$ is a (unique) global minimum of (3.3.18).

The strict copositivity of $\nabla^2_{xx}L(x, \mu, \lambda)$ on $\mathcal{C}(x; K, \nabla\theta)$ for some (μ, λ) in $\mathcal{M}_{\text{NLP}}(x)$ coincides with the "second-order sufficiency condition" in NLP. Since $\nabla^2_{xx}L(x, \mu, \lambda)$ is a symmetric matrix, it follows from Proposition **2.5.7** that $\nabla^2_{xx}L(x, \mu, \lambda)$ is strictly copositive on $\mathcal{C}(x; K, \nabla\theta)$ if and only if this matrix is copositive on the critical cone and $(\nabla^2_{xx}L(x, \mu, \lambda), \mathcal{C}(x; K, \nabla\theta))$ is an R_0 pair. If $\mathcal{M}_{\text{NLP}}(x)$ is the singleton $\{(\mu, \lambda)\}$, then the copositivity of $\nabla^2_{xx}L(x, \mu, \lambda)$ on $\mathcal{C}(x; K, \nabla\theta)$ coincides with the classical "second-order necessity condition" in NLP.

Similar to Theorem **3.3.12**, we can prove the following result.

3.3.19 Proposition. Let θ, g_i and h_j be twice continuously differentiable functions. Suppose that the stationary point x of (3.3.16) satisfies the MFCQ and that for every $(\mu, \lambda) \in \mathcal{M}_{\text{NLP}}(x)$, the QP (3.3.18) has $v = 0$ as the unique stationary point. Then x is an isolated stationary point of (3.3.16).

Proof. Assume for the sake of contradiction that $\{x^\nu\}$ is a sequence of stationary points of (3.3.16) each distinct from x that converges to x. Without loss of generality, we may assume that each x^ν satisfies the MFCQ. By following the same proof of Theorem **3.3.12**, we can obtain a contradiction to the assumption about the QP (3.3.18). □

The following corollary follows easily from Proposition **3.3.19** and an aforementioned fact.

3.3.20 Corollary. Let θ, g_i, and h_j be twice continuously differentiable functions. Suppose that the stationary point x of (3.3.16) satisfies the MFCQ and that for every $(\mu, \lambda) \in \mathcal{M}_{\text{NLP}}(x)$, $\nabla^2_{xx}L(x, \mu, \lambda)$ is strictly copositive on $\mathcal{C}(x; K, \nabla\theta)$. Then x is an isolated, strong local minimizer of (3.3.16).

Proof. By a classical result in NLP, the strict copositivity of $\nabla^2_{xx}L(x, \mu, \lambda)$ on $\mathcal{C}(x; K, \nabla\theta)$ is sufficient for x to be a strong local minimizer of (3.3.16). Proposition **3.3.19** implies that x is an isolated stationary point. These two properties of x in turn implies that x is an isolated local minimizer of (3.3.16). □

We return in Section **7.4.1** with a further result about a local minimizer of an NLP that complements Corollary **3.3.20**; see Proposition **7.4.12**.

3.3.5 A nonsmooth-equation approach

There is an alternative way of deriving several of the above results, namely, via a local analysis of nonsmooth equations. Here, we present a preliminary treatment of the latter theory and postpone the full treatment until Section **5.2**. This nonsmooth-equation approach is useful for dealing with some stability issues of the VI/CP; see Section **5.3**. In what follows, we present a local uniqueness result for a general B-differentiable equation and specialize the result to the vertical CP (1.5.3) that involves two functions F and G:

$$0 \leq F(x) \perp G(x) \geq 0. \tag{3.3.19}$$

See Proposition **2.8.5** for an existence result for this CP, which we denoted (F, G). Further specializations of the local uniqueness theory presented herein to systems like

$$F(x, y) = 0$$
$$0 \leq x \perp y \geq 0,$$

can be similarly derived. The following proposition is an immediate consequence of part (a) of Proposition **2.1.5**. We give a direct proof that is quite simple.

3.3.21 Proposition. Let $H : \mathcal{D} \subseteq \mathbb{R}^n \to \mathbb{R}^m$, with $\mathcal{D}$ open, be continuous. Suppose that $H(x) = 0$ and H is B-differentiable at $x \in \mathcal{D}$. If

$$H'(x; v) = 0 \Rightarrow v = 0, \tag{3.3.20}$$

then x is an isolated zero of H.

Proof. Assume for the sake of contradiction that $\{x^\nu\}$ is a sequence of zeros of H converging to x with $x^\nu \neq x$ for every ν. By using the B-differentiability of H, it is easy to show that every accumulation point of the normalized sequence $\{(x^\nu - x)/\|x^\nu - x\|\}$ must be a nonzero vector v satisfying $H'(x; v) = 0$. This is a contradiction to the assumption. □

The converse of the above proposition is easily seen to be false. For instance, consider the scalar function $H(t) = t^2$, which has a unique zero, namely, $t = 0$. But the implication (3.3.20) does not hold because $H'(0; \cdot)$ is identically equal to zero. Subsequently, we show in Theorem **5.2.12** that if x is a "stable zero" of H (see Definition **5.2.3**) then (3.3.20) holds; see also Remark **5.2.13**.

Proposition **3.3.21** can be immediately specialized to the natural map or the normal map of the VI (K, F), provided that the Euclidean projector Π_K is well defined, and more importantly, B-differentiable. Unfortunately,

Π_K is not necessarily B-differentiable for an arbitrary closed convex set K. Section **4.4** addresses this issue in detail. For now, we derive a consequence of Proposition **3.3.21** for the map H defined by the pointwise minimum of two B-differentiable functions:

$$H(x) \equiv \min(F(x), G(x)), \quad \forall x \in \mathcal{D},$$

where $F, G : \mathcal{D} \subseteq \mathbb{R}^n \to \mathbb{R}^m$ are two B-differentiable maps. For every vector $x \in \mathcal{D}$, it is easy to verify that, for all $i = 1, \ldots, n$, (cf. Example **3.1.4** and Proposition **3.1.6**)

$$H_i'(x;v) = \begin{cases} F_i'(x;v) & \text{if } F_i(x) < G_i(x) \\ \min(F_i'(x;v), G_i'(x;v)) & \text{if } F_i(x) = G_i(x) \\ G_i'(x;v) & \text{if } F_i(x) > G_i(x). \end{cases}$$

With this expression, we can easily prove the following result, the first part of which is an immediate corollary of Proposition **3.3.21**.

3.3.22 Corollary. Let $F, G : \mathcal{D} \subseteq \mathbb{R}^n \to \mathbb{R}^m$ be two B-differentiable maps defined on the open set $\mathcal{D}$. Let x be a solution of the complementarity problem (3.3.19). If the system:

$$\begin{aligned} F_i'(x;v) &= 0, & \forall i \text{ such that } F_i(x) &= 0 < G_i(x) \\ \min(F_i'(x;v), G_i'(x;v)) &= 0, & \forall i \text{ such that } F_i(x) &= 0 = G_i(x) \\ G_i'(x;v) &= 0, & \forall i \text{ such that } F_i(x) &> 0 = G_i(x) \end{aligned}$$

has a unique solution $v = 0$, then x is an isolated solution of the CP (3.3.19). Conversely, if x is an isolated solution of (3.3.19) and if F and G are affine, then the above system has $v = 0$ as the unique solution.

Proof. It suffices to prove the converse. If F and G are affine and if v is any solution of the displayed system, then $x + \tau v$ remains a solution of (3.3.19) for all $\tau > 0$ sufficiently small. Consequently, the converse of the corollary follows easily. □

Corollary **3.3.22** can be further specialized if $m = n$ and the functions F and G are both F-differentiable. To present this specialized result, we introduce a concept that generalizes that of a principal submatrix to the case of a pair of matrices of the same order.

3.3.23 Definition. Let A and B be two $m \times n$ matrices. An $m \times n$ matrix C is said to be a *row (column) representative matrix* of the pair (A, B) if each row (column) $C_{i\cdot}$ $(C_{\cdot i})$ of C is equal to either $A_{i\cdot}$ $(A_{\cdot i})$ or $B_{i\cdot}$ (or $B_{\cdot i}$). □

We have the following consequence of Corollary **3.3.22**.

3.3.24 Corollary. Let $F, G : \mathcal{D} \subseteq \mathbb{R}^n \to \mathbb{R}^n$ be two F-differentiable maps defined on the open set $\mathcal{D}$. Let x be a solution of the complementarity system (3.3.19). If every row representative matrix M of the pair $(JF(x), JG(x))$ with the property that

$$M_{i\cdot} = \begin{cases} \nabla F_i(x)^T & \forall\, i \text{ such that } F_i(x) = 0 < G_i(x) \\ \nabla G_i(x)^T & \forall\, i \text{ such that } F_i(x) > 0 = G_i(x), \end{cases} \tag{3.3.21}$$

is nonsingular, then x is locally unique. □

Representative matrices M of $(JF(x), JG(x))$ satisfying (3.3.21) differ from each other in those row $M_{i\cdot}$ that correspond to the degenerate components of the solution x of (3.3.19); in turn, these indices i are such that $F_i(x) = G_i(x) = 0$. If G is the identity function, we recover the implication (b) $\Rightarrow$ (d) in Corollary **3.3.9**.

3.4 Nondegenerate Solutions

In this section, we introduce a special property of a solution to the VI/CP under which much of the local analysis of this problem can be simplified significantly. By definition, if $\bar{x}$ is a solution of the VI (K, F), then $-F(\bar{x})$ belongs to the normal cone $\mathcal{N}(\bar{x}; K)$ of K at $\bar{x}$. By further restricting $-F(\bar{x})$, we arrive at the following definition. See Figure 3.3 for an illustration of the concept.

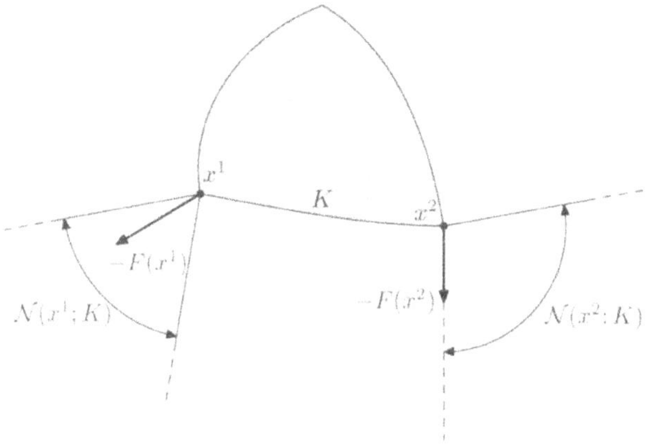

Figure 3.3: x^1 is a nondegenerate solution while x^2 is a degenerate solution.

3.4.1 Definition. A solution x of the VI (K, F) is *nondegnerate* if $-F(x)$ is a relative interior normal vector of K at x; i.e., $-F(x) \in \operatorname{ri} \mathcal{N}(x; K)$. A *degenerate solution* is one that is not nondegenerate. □

The following result is an immediate corollary of Proposition **2.4.1**; no proof is needed. The result gives an equivalent geometric description of a nondegenerate solution.

3.4.2 Proposition. Let K be a closed convex set in $\mathbb{R}^n$ and $F : \mathbb{R}^n \to \mathbb{R}^n$ be given. Let $x \in \mathrm{SOL}(K, F)$. The following two statements are equivalent.

(a) x is nondegenerate.

(b) The critical cone $\mathcal{C}(x; K, F)$ is a linear subspace.

In this case, the critical cone $\mathcal{C}(x; K, F)$ must equal the lineality space of the tangent cone $\mathcal{T}(x; K)$. □

As a simple example of Definition **3.4.1**, consider the case where K is the nonnegative orthant, so that the VI (K, F) becomes the NCP (F). It then follows that a solution x is nondegenerate if and only if $x + F(x)$ is a positive vector, that is, if and only if the degenerate set β is empty. This is easily proved by noticing that the normal cone $\mathcal{N}(x; \mathbb{R}^n_+)$ is equal to the cone generated by the negative coordinate vectors in $\mathbb{R}^n$ corresponding to the index set $\mathcal{I}(x)$. By Corollary **3.3.9**, if x is a nondegenerate solution of the NCP (F), then x is isolated if the matrix $J_\gamma F_\gamma(x)$ is nonsingular, where γ is the support of x.

Nondegenerate solutions are very special and they do not always exist. The NCP (F) with F being the identity my easily illustrates the nonexistence of a nondegenerate solution. The existence of a nondegenerate solution to a VI/CP provides a special property of the problem that often leads to useful consequences. For example, see Corollary **5.4.14** for an important role of a nondegenerate solution in the study of parametric VIs.

When K is finitely represented by a system of convex inequalities, a further characterization of the nondegeneracy of a solution is possible in terms of the KKT multipliers of the VI (K, F). Let us write

$$K \equiv \{ x \in \mathbb{R}^n : g(x) \leq 0 \} \tag{3.4.1}$$

where each $g_i : \mathbb{R}^n \to \mathbb{R}$ is convex and continuously differentiable. If the Abadie CQ holds at a vector $x \in K$, then we have

$$\mathcal{T}(x; K) = \{ d \in \mathbb{R}^n : \nabla g_i(x)^T d \leq 0, \ \forall i \in \mathcal{I}(x) \},$$

where $\mathcal{I}(x)$ is the index set of binding constraints at x. With this representation of $\mathcal{T}(x; K)$, it follows that a vector v belongs to $\mathrm{ri}\,\mathcal{N}(x; K)$ if and only if there exists $\lambda_{\mathcal{I}(x)} > 0$ such that

$$v + \sum_{i \in \mathcal{I}(x)} \lambda_i \nabla g_i(x) = 0;$$

see Exercise **3.7.15**. Based on this observation, the following result does not require a proof.

3.4.3 Corollary. Let K be given by (3.4.1), where each $g_i : \mathbb{R}^n \to \mathbb{R}$ is convex and continuously differentiable. If the Abadie CQ holds at a vector $x \in \mathrm{SOL}(K, F)$, then the following two statements are equivalent.

(a) x is nondegenerate.

(b) There exists a KKT multiplier λ such that $\lambda - g(x) > 0$. □

Corollary **3.4.3** can be used to obtain a necessary and sufficient condition for a monotone AVI to have a nondegenerate solution; in turn, the existence of such a solution induces a "global error bound" of a certain kind for such an AVI; see Section **6.4** for details. The above corollary motivates the following definition. Let K be a finitely representable, convex subset of $\mathbb{R}^n$. We say that a solution x of the VI (K, F) is *strongly nondegenerate* if x is nondegenerate, $\mathcal{M}(x)$ is nonempty, and for every $\lambda \in \mathcal{M}(x)$, $\lambda - g(x) > 0$. We have the following characterization of a strongly nondegenerate solution of the VI, which shows that strong nondegeneracy is equivalent to nondegeneracy plus LICQ.

3.4.4 Proposition. Let K be given by (3.4.1), where each $g_i : \mathbb{R}^n \to \mathbb{R}$ is convex and continuously differentiable. A solution $x \in \mathrm{SOL}(K, F)$ is strongly nondegenerate if and only if the set of active gradients:

$$\{\, \nabla g_i(x) \,:\, i \in \mathcal{I}(x) \,\}$$

is linearly independent and x is nondegenerate.

Proof. It is known that the linear independence of the active gradients implies that $\mathcal{M}(x)$ is a singleton (see Section **3.2** for a detailed discussion of this CQ). Hence the sufficiency part of the proposition is obvious. Conversely, suppose that x is strongly nondegenerate but the active gradients are dependent. There exist scalars η_i, $i \in \mathcal{I}(x)$, not all zero, such that

$$\sum_{i \in \mathcal{I}(x)} \eta_i \nabla g_i(x) = 0.$$

Take any $\lambda \in \mathcal{M}(x)$. By the strong nondegeneracy of x, we have $\lambda_i > 0$ for all $i \in \mathcal{I}(x)$. By choosing an $\varepsilon > 0$ appropriately, the scalars

$$\lambda_i(\varepsilon) \equiv \lambda_i - \varepsilon\,\eta_i, \quad i \in \mathcal{I}(x)$$

remain nonnegative and equal to zero for at least one $i \in \mathcal{I}(x)$. For this choice of ε, the vector $\lambda(\varepsilon)$ (with $\lambda_i(\varepsilon) \equiv 0$ for $i \notin \mathcal{I}(x)$) remains an element of $\mathcal{M}(x)$ but clearly violates the strict complementarity condition; hence a contradiction. □

3.5 VIs on Cartesian Products

The MiCPs (and NCPs in particular) and box constrained VIs are special VIs where the defining sets are the Cartesian products of closed, one-dimensional intervals, i.e., closed rectangles. The Nash equilibrium problem discussed in Subsection **1.4.2** provides another example of a VI whose defining set is the Cartesian product of lower-dimensional sets. For such a VI, which we call a *partitioned VI*, it is natural to consider a weakening of many of the properties that we have introduced in Chapter **2** so that they become consistent with the Cartesian structure of the underlying set.

Throughout this section, we consider a set $K \subseteq \mathbb{R}^n$ given by:

$$K = \prod_{\nu=1}^{N} K_\nu, \tag{3.5.1}$$

where N is a positive integer and each K_ν is a subset of $\mathbb{R}^{n_\nu}$ with

$$\sum_{\nu=1}^{N} n_\nu = n.$$

Consistent with this structure of K, we write

$$\mathbb{R}^n = \prod_{\nu=1}^{N} \mathbb{R}^{n_\nu};$$

we also partition and represent all vectors in $\mathbb{R}^n$ in component blocks accordingly; thus for a vector $x \in K$, we write $x = (x_\nu)$, where each x_ν belongs to K_ν.

For a partitioned VI (K, F) where K has the Cartesian structure (3.5.1), an existence result analogous to Proposition **2.2.3** can be obtained. The difference between the result below and the latter proposition is that the conditions (a), (b), and (c) are weakened as a consequence of the product structure of K.

3.5.1 Proposition. Let K be given by (3.5.1) where each $K_\nu \subseteq \mathbb{R}^{n_\nu}$ is closed and convex; let $F : K \to \mathbb{R}^n$ be continuous. Consider the following statements:

(a) There exists a vector $x^{\mathrm{ref}} \in K$ such that the set

$$L'_< \equiv \{ x \in K : F_\nu(x)^T(x_\nu - x_\nu^{\mathrm{ref}}) < 0 \ \forall \nu \text{ such that } x_\nu \neq x_\nu^{\mathrm{ref}} \}$$

is bounded.

(b) There exist a bounded open set $\Omega \subset \mathbb{R}^n$ and a vector $x^{\mathrm{ref}} \in K \cap \Omega$ such that for every $x \in K \cap \mathrm{bd}\,\Omega$, an index ν exists such that $x_\nu \neq x_\nu^{\mathrm{ref}}$ and

$$F_\nu(x)^T(x_\nu - x_\nu^{\mathrm{ref}}) \geq 0.$$

(c) The VI (K, F) has a solution.

It holds that (a) $\Rightarrow$ (b) $\Rightarrow$ (c). Moreover, if the set

$$L'_\leq \equiv \{ x \in K : \max_{1 \leq \nu \leq N} F_\nu(x)^T(x_\nu - x_\nu^{\mathrm{ref}}) \leq 0 \},$$

which contains $L'_<$, is bounded, then SOL(K, F) is nonempty and compact.

Proof. We proceed as in the proof of Proposition **2.2.3**. Indeed the only part of that proof which requires a modification is to show, under (b),

$$[\, H(x,t) = 0, \quad t \in (0,1) \,] \Rightarrow x \notin \mathrm{bd}\,\Omega.$$

Let (x, t) satisfy the left-hand condition. Assume for the sake of contradiction that x belongs to $\mathrm{bd}\,\Omega$. By the definition of H, we have $x \in K$ and

$$(y - x)^T[\, t\, F(x) + (1-t)(x - x^{\mathrm{ref}})\,] \geq 0, \quad \forall y \in K. \tag{3.5.2}$$

Since $x \in K \cap \mathrm{bd}\,\Omega$, by (b), there must exist an index ν such that $x_\nu \neq x_\nu^{\mathrm{ref}}$ and $F_\nu(x)^T(x_\nu - x_\nu^{\mathrm{ref}}) \geq 0$. Define a vector $y \equiv (y_\nu) \in K$ as follows: for each $\nu' \in \{1, \ldots, N\}$,

$$y_{\nu'} \equiv \begin{cases} x_\nu^{\mathrm{ref}} & \text{if } \nu' = \nu \\ x_\nu & \text{otherwise} \end{cases} \tag{3.5.3}$$

we deduce from (3.5.2) that

$$(x_\nu^{\mathrm{ref}} - x_\nu)^T[\, t\, F_\nu(x) + (1-t)(x_\nu - x_\nu^{\mathrm{ref}})\,] \geq 0.$$

This is a contradiction. □

Conditions (a), (b), and (c) in Proposition **3.5.1** are weaker than their counterparts in Proposition **3.5.1**. The reason is due to the identity:

$$x^T y = \sum_{\nu=1}^{N} (x_\nu)^T y_\nu,$$

which clearly implies that if $x \in L'_<$, then either $x = x^{\text{ref}}$ or

$$F(x)^T (x - x^{\text{ref}}) < 0.$$

Thus the set $L'_<$ in Proposition **3.5.1** is contained in the set $L_<$ in Proposition **2.2.3** union $\{x^{\text{ref}}\}$; hence, the boundedness of $L_<$ implies the boundedness of $L'_<$. Consequently, if condition (a) in Proposition **2.2.3** holds, then so does condition (a) in Proposition **3.5.1**. A similar argument applies to condition (b) and (c).

The proof of Proposition **3.5.1** employs a technique that is pertinent to the Cartesian structure of K. Namely, this structure of K enables the independent choice of the components of the special vector y defined by (3.5.3). We use this idea again several times throughout this section.

3.5.1 Semicopositive matrices

We can specialize Proposition **3.5.1** to the CP (K, q, M). For this purpose, we introduce the following generalized copositivity concepts associated with a cone K having the Cartesian product structure (3.5.1).

3.5.2 Definition. Let each K_ν be a cone in $\mathbb{R}^{n_\nu}$ and K be a subset of $\mathbb{R}^n$ defined by (3.5.1). An $n \times n$ matrix M is said to be

(a) *semicopositive* on K if for every nonzero vector x in K, there exists an index $\nu \in \{1, \ldots, N\}$ such that

$$x_\nu \neq 0 \quad \text{and} \quad (x_\nu)^T (Mx)_\nu \geq 0;$$

(b) *strictly semicopositive* on K if

$$\max_{1 \leq \nu \leq N} (x_\nu)^T (Mx)_\nu > 0, \quad \forall x \in K \setminus \{ 0 \}.$$

When K is the nonnegative orthant of $\mathbb{R}^n$, the (strict) semicopositivity on K is meant with K represented as the Cartesian product of n semi-infinite intervals $[0, \infty)$. □

When $N = 1$, (strict) semicopositivity becomes (strict) copositivity. If M is semicopositive on K, then $M + \varepsilon \mathbf{I}$ is strictly semicopositive on K for every $\varepsilon > 0$. In order for this statement to be true, it is essential that in the definition of semicopositivity, the index ν is such that x_ν is nonzero. As a counterexample, consider the matrix

$$M \equiv \begin{bmatrix} 1 & 0 \\ 0 & -1 \end{bmatrix},$$

which satisfies

$$\max(\, x_1(Mx)_1,\, x_2(Mx)_2\,) \geq 0, \quad \forall\,(\, x_1, x_2\,) \in \mathbb{R}^2;$$

yet for $\varepsilon \in (0,1)$, the matrix $M + \varepsilon\,\mathbf{I}_2$ is not strictly (semi)copositive on the nonnegative orthant $\mathbb{R}^2_+ = \mathbb{R}_+ \times \mathbb{R}_+$. A simple way to see the latter statement is to recognize that the second diagonal entry of $M + \varepsilon\mathbf{I}$ is negative for all $\varepsilon > 0$ sufficiently small.

If each K_ν is a cone and K is given by (3.5.1), and if M is (strictly) copositive on K, then M is (strictly) semicopostivie on K; the converse is not true. As a counterexample, consider the matrix

$$M \equiv \begin{bmatrix} 1 & -3 \\ 1 & 0 \end{bmatrix},$$

which is easily seen to be semicopositive on $\mathbb{R}^2_+ = \mathbb{R}_+ \times \mathbb{R}_+$ but not copositive on $\mathbb{R}^2_+$. When each K_ν is equal to the nonnegative real line, a semicopositive matrix M on $K = \mathbb{R}^n_+$ is traditionally called a *semimonotone* matrix in the LCP literature. We feel that the term "monotone" is being abused in this context and is not consistent with the above remark for the case $N = 1$; indeed, this case suggests that Definition **3.5.2** really refers to a generalization of the copositivity concept, which itself is somewhat different from monotonicity. Thus we prefer to depart from tradition and use the terminology "semicopositivity" instead of "semimonotonicity".

We have the following corollary of Proposition **3.5.1**. This corollary generalizes Theorem **2.5.10**, which corresponds to the case $N = 1$.

3.5.3 Corollary. Let K be given by (3.5.1) where each $K_\nu \subseteq \mathbb{R}^{n_\nu}$ is a closed convex cone. Let M be an $n \times n$ matrix. Consider the following three statements:

(a) M is strictly semicopositive on K;

(b) M is semicopositive on K, (K, M) is an R_0 pair, and $\mathcal{R}(K, M)$ is closed;

(c) for all vectors $q \in \mathbb{R}^{n \times n}$, the CP (K, q, M) has a nonempty bounded solution set.

It holds that (a) $\Rightarrow$ (b) $\Rightarrow$ (c). Moreover, if M is semicopositive on K and (K, M) is an R_0 pair, then both the natural index and the normal index of the pair (K, M) are equal to 1.

Proof. The way we prove the implications is to first show that (a) implies (c); we next use this to show that (b) implies (c). Finally, (a) $\Rightarrow$ (b) is proved as follows. If M is strictly semicopositive on K, then it is semicopositive on K. To see that (K, M) is an R_0 pair, let x be a nonzero element of the CP kernel $\mathcal{K}(K, M)$. On the one hand, by the strict semicopositivity of M on K, there exists an index $\nu \in \{1, \ldots, N\}$ such that $(x_\nu)^T(Mx)_\nu > 0$. On the other hand, by the Cartesian cone structure of K and the fact that $x \in \mathcal{K}(K, M)$, we must have $(x_\nu)^T(Mx)_\nu = 0$. Consequently, the CP kernel must consist of the zero vector alone. Thus (K, M) is an R_0 pair. Finally since (a) implies (c), it follows that if M is strictly semicopositive on K, then $\mathcal{R}(K, M)$ is the whole space $\mathbb{R}^n$ and is thus closed.

Suppose that (a) holds. Let $F(x) \equiv q + Mx$, where $q \in \mathbb{R}^n$ is arbitrary. Since K is a cone, with x^{ref} chosen to be the origin, which is an element of K, it is easy to show that the set $L'_{\leq}$ is bounded. Therefore, by Proposition **3.5.1**, statement (c) holds.

Suppose that (b) holds. As mentioned before, for every $\varepsilon > 0$, the matrix $M + \varepsilon\mathbf{I}$ is strictly semicopositive on K. Thus for every vector q, the CP $(K, q, M + \varepsilon\,\mathbf{I})$ has a solution. Fix an arbitrary vector q and let $x(\varepsilon)$ be a solution of the latter CP. It is easy to show that, since (K, M) is an R_0 pair, for every sequence $\{\varepsilon_k\}$ of positive scalars converging to zero, the corresponding sequence of solutions $\{x(\varepsilon_k)\}$ must be bounded. We have $q + \varepsilon_k\, x^k \in \mathcal{R}(K, M)$ for all k. Since the CP range is closed by assumption, it follows that $q \in \mathcal{R}(K, M)$; that is $\text{SOL}(K, q, M)$ is nonempty. The boundedness of this solution set follows from the R_0 property of the pair (K, M). Finally, to prove the index assertion, it suffices to homotopize the natural map or the normal map of the homogeneous CP $(K, 0, M)$ as in the proof of Theorem **2.5.10**. The details are not repeated. □

3.5.4 Remark. In the above corollary, statement (b) does not imply statement (a). To present a counterexample, consider the positive semidefinite matrix

$$M \equiv \begin{bmatrix} 1 & -1 \\ 1 & 0 \end{bmatrix},$$

which is easily seen to be an R_0 matrix but not strictly semicopositive on $\mathbb{R}^2_+ = \mathbb{R}_+ \times \mathbb{R}_+$. Thus the pair $(\mathbb{R}^2_+, M)$ satisfies (b) but fails (a). □

We illustrate the above theory by mentioning a realization of a strictly semicopositive matrix on the nonnegative orthant in an application of the

LCP. Indeed, in the problem of pricing American options with transaction costs described in Subsection **1.4.9**, we encountered a partitioned matrix of the following form:

$$S \equiv \begin{bmatrix} A & B \\ Q & \mathbf{I} \end{bmatrix},$$

where A is positive definite and B is nonnegative (the property of Q is not essential for the discussion here). The following simple lemma establishes the strict semicopositivity property of the matrix S with a broader class of matrices A.

3.5.5 Lemma. Let A be a strictly semicopositive $n \times n$ matrix on $\mathbb{R}^n_+$ and let B be a nonnegative matrix of order $n \times m$. The above matrix S is strictly semicopositive on $\mathbb{R}^{n+m}_+$.

Proof. Let $(x, y) \in \mathbb{R}^{n+m}_+$ be a nonzero vector. If x is nonzero, then by the strict semicopositivity of A and the nonnegativity of B, it follows that there exists an index $i \in \{1, \ldots, n\}$ such that

$$x_i\,(\,Ax\,)_i + x_i\,(\,By\,)_i > 0.$$

If x is zero, then y must be nonzero; so there exists an index $j \in \{1, \ldots, m\}$ such that $y_j^2 > 0$. This establishes that S is strictly semicopositive on $\mathbb{R}^{n+m}_+$. □

A consequence of the above lemma is that the discretized LCP arising from the American options problem with the transaction cost function (1.4.50) must have a solution; moreover, such a solution can be computed by Lemke's algorithm.

More generally, for the American option pricing problem with nonlinear transaction costs, the discretization of the differential operators leads to the NCP:

$$0 \leq \mathbf{V}^m - \mathbf{\Lambda}^m \perp \mathbf{q}^m + \mathbf{M}\mathbf{V}^m + \mathbf{F}(\mathbf{V}^m) \geq 0, \tag{3.5.4}$$

where the function $\mathbf{F}$ is nonnegative; see the original discussion in Subsection **1.4.9** for the notation. In Subsection **2.2.1**, we have noted that if the matrix $\mathbf{M}$ is positive definite, then the existence results of Section **2.2** can be applied to establish the existence of a solution to (3.5.4). In order to allow for more relaxed discretization steps that do not necessarily yield a positive definite matrix $\mathbf{M}$, we establish below an existence result for the above NCP, assuming that $\mathbf{M}$ is strictly semicopositive on $\mathbb{R}^{N-1}_+$, where $N - 1$ is the order of the matrix $\mathbf{M}$.

3.5.6 Proposition. Suppose that $\mathbf{M}$ is strictly semicopositive on $\mathbb{R}^{N-1}_+$ and $\mathbf{F}$ is a nonnegative function. The NCP (3.5.4) has a nonempty bounded solution set for all Λ^m and $\mathbf{q}^m$.

Proof. Let $\mathbf{x} \equiv \mathbf{V}^m - \mathbf{\Lambda}^m$. The NCP (3.5.4) is clearly equivalent to:

$$0 \leq \mathbf{x} \perp \mathbf{q}' + \mathbf{M}\mathbf{x} + \tilde{\mathbf{F}}(\mathbf{x}) \geq 0,$$

where $\mathbf{q}' \equiv \mathbf{q}^m + \mathbf{M}\mathbf{\Lambda}^m$ and $\tilde{\mathbf{F}}(\mathbf{x}) \equiv \mathbf{F}(\mathbf{x} + \mathbf{\Lambda}^m)$. With $\mathbf{x}^{\text{ref}}$ taken to be the origin, by Proposition **3.5.1**, it suffices to verify that the set

$$\left\{ \mathbf{x} \geq 0 : \max_{1 \leq \nu \leq N-1} x_\nu [\, (\mathbf{q}' + \mathbf{M}\mathbf{x})_\nu + \tilde{\mathbf{F}}_\nu(\mathbf{x}) \,] \leq 0 \right\}$$

is bounded. By the nonnegativity of the function $\tilde{\mathbf{F}}$, this set is a subset of

$$\left\{ \mathbf{x} \geq 0 : \max_{1 \leq \nu \leq N-1} x_\nu [\, (\mathbf{q}' + \mathbf{M}\mathbf{x})_\nu \,] \leq 0 \right\},$$

which must be bounded, by the strict semicopositivity of $\mathbf{M}$ on the nonnegative orthant. □

3.5.7 Remark. Although this concept has not been formally defined, the (nonlinear) function

$$\mathbf{x} \mapsto \mathbf{q}' + \mathbf{M}\mathbf{x} + \tilde{\mathbf{F}}(\mathbf{x})$$

is an instance of a "strictly semicopositive" function on the nonnegative orthant. See Exercise **3.7.31** for a formal definition of such a function. □

3.5.2 P properties

Monotonicity and copositivity are related to each other in an obvious way. Definition **3.5.2** introduces the matrix-theoretic concept of semicopositivity as a weakening of copositivity when the underlying cone K has the Cartesian structure (3.5.1). In a similar fashion, we introduce a weakening of functional monotonicity by exploiting the same Cartesian structure.

3.5.8 Definition. Let K be given by (3.5.1). A map $F : K \to \mathbb{R}^n$ is said to be

(a) a P_0 *function* on K if for all pairs of distinct vectors x and y in K, there exists $\nu \in \{1, \ldots, N\}$ such that

$$x_\nu \neq y_\nu \quad \text{and} \quad (\, x_\nu - y_\nu \,)^T (\, F_\nu(x) - F_\nu(y) \,) \geq 0;$$

(b) a $P_*(\sigma)$ *function* on K for some $\sigma > 0$ if for all x and y in K,

$$(x-y)^T(F(x)-F(y)) \geq -\sigma \sum_{\nu \in I_+(x,y)} (x_\nu - y_\nu)^T(F_\nu(x)-F_\nu(y)),$$

where

$$I_+(x,y) \equiv \{\nu : (x_\nu - y_\nu)^T(F_\nu(x) - F_\nu(y)) > 0\}$$

(if $I_+(x,y)$ is empty, the summation in the right-hand side of the above expression is defined to be zero);

(c) a *P function* on K if for all pairs of distinct vectors x and y in K,

$$\max_{1 \leq \nu \leq N} (x_\nu - y_\nu)^T(F_\nu(x) - F_\nu(y)) > 0;$$

(d) a *uniformly P function* on K if there exists a constant $\mu > 0$ such that for all pairs of vectors x and y in K,

$$\max_{1 \leq \nu \leq N} (x_\nu - y_\nu)^T(F_\nu(x) - F_\nu(y)) \geq \mu \| x - y \|_2^2.$$

Unless otherwise stated, when K is a rectangle in $\mathbb{R}^n$, each of the above P properties is meant with K represented as the Cartesian product of the n intervals that define K. □

Just like the definition of a semicopositive matrix, we cannot use the following as the definition of a P_0 function: for all pairs of vectors x and y in K,

$$\max_{1 \leq \nu \leq N} (x_\nu - y_\nu)^T(F_\nu(x) - F_\nu(y)) \geq 0. \tag{3.5.5}$$

In other words, it is essential to require the index ν in the definition of a P_0 function to be such that $x_\nu \neq y_\nu$.

Clearly, every uniformly P function must be a P function, which in turn must be a P_0 function. Moreover, if F is a P_0 function on K, then $F + \varepsilon \mathbf{I}$ is a P function on K for every $\varepsilon > 0$. It is clear that a monotone map must be a $P_*(\sigma)$ function for all $\sigma > 0$ and that a $P_*(\sigma)$ function for any $\sigma > 0$ must be a P_0 function. Moreover, if F is a $P_*(\sigma)$ function on K, then the following implication holds for all x, y in K:

$$[(x_\nu - y_\nu)^T(F_\nu(x) - F_\nu(y)) \leq 0 \quad \forall \nu]$$
$$\Rightarrow [(x_\nu - y_\nu)^T(F_\nu(x) - F_\nu(y)) = 0 \quad \forall \nu].$$

If $F(x) \equiv Mx$ for some $n \times n$ matrix M, the following statements hold:

(a) if $K - K$ is closed, then F is a P function on K if and only if it is uniformly P on K;

(b) if each K_ν is a cone and if F is P_0 function on K, then M is semicopositive on K;

(c) if each K_ν is a cone and if F is P function on K, then M is strictly semicopositive on K;

(d) if each K_ν is the real line, then F is a $\mathrm{P}_*(\sigma)$ function for some $\sigma > 0$ if and only if M is column sufficient. (The proof of this result is nontrivial; see Section **3.8**.)

If F is a nonlinear function, the following diagram summarizes the relations between monotonicity and P properties:

strongly monotone	$\Rightarrow$	strictly monotone	$\Rightarrow$	monotone
$\Downarrow$		$\Downarrow$		$\Downarrow$
uniformly P	$\Rightarrow$	P	$\Rightarrow$	P_0.

The functional P properties can be used to define analogous matrix-theoretic properties by taking F to be a linear function. Specifically, let a matrix $M \in \mathbb{R}^{n\times n}$ be given and let $F(x) \equiv Mx$. We say that

(a) M is a P_0 matrix if F is a P_0 function on $\mathbb{R}^n$; and

(b) M is a P matrix if F is a P function on $\mathbb{R}^n$.

The classes of P and P_0 matrices play an important role throughout the mathematical sciences. There are many equivalent definitions of these matrices. For our purpose, we note the following characterization. A matrix $M \in \mathbb{R}^{n\times n}$ is P (P_0) if and only if all principal minors of M are positive (nonnegative).

In Proposition **2.3.2**, we have seen that for a continuously differentiable function F defined on an open convex set, the monotonicity of F is equivalent to the positive semidefiniteness of the Jacobian matrix $JF(x)$ for all x in the domain; and the strict monotonicity of F is implied by the positive definiteness of the Jacobian matrix $JF(x)$ for all x in the domain. Thus it is reasonable to expect that similar statements hold for a P_0 and for a P function. We formally state and prove these statements in Proposition **3.5.9**, which assumes that K is a rectangle. The proof of this result makes use of the classical Zorn's lemma, which states that every nonempty partially ordered set such that every totally ordered subset has a lower bound must have a minimal element. It is not known whether parts (a) and (b) of the following proposition remain valid if K is not a rectangle.

3.5.9 Proposition. Let $F : \Omega \supset K \to \mathbb{R}^n$ be continuously differentiable on the open set Ω containing the given rectangle K (not necessarily open or closed).

(a) If $JF(x)$ is a P matrix for all $x \in K$, then F is a P function on K.

(b) If $JF(x)$ is a P_0 matrix for all $x \in K$, then F is a P_0 function on K.

(c) If K is open and F is a P_0 function on K, then $JF(x)$ is a P_0 matrix for all $x \in K$.

(d) If K is open and F is a uniformly P function on K, then there exists a constant $c > 0$ such that for all $x \in K$,

$$\max_{1 \le i \le n} y_i(JF(x)y)_i \ge c \| y \|^2, \quad \forall y \in \mathbb{R}^n.$$

Proof. We first prove part (c). Let K be an open rectangle and let $x \in K$ be arbitrary. For every $v \in \mathbb{R}^n$, we have

$$JF(x)v = \lim_{\tau \downarrow 0} \frac{F(x + \tau v) - F(x)}{\tau},$$

Thus, for every i,

$$v_i(JF(x)v)_i = \lim_{\tau \downarrow 0} \frac{v_i(F(x + \tau v) - F(x))_i}{\tau}.$$

If $v \neq 0$, then for every $\tau > 0$ sufficiently small, there exists an index i for which v_i is nonzero and $v_i(F(x + \tau v) - F(x))_i$ is nonnegative (this is where the openness of K is needed to ensure that $x + \tau v$ belongs to K). Since there are only finitely many indices, it follows that there must be an index i for which $v_i \neq 0$ and $v_i(JF(x)v)_i$ is nonnegative. Hence $JF(x)$ is a P_0 matrix. Thus (c) follows. Part (d) can be proved in a similar way; the details are omitted.

We next prove (a). First assume that K is closed. By means of an inductive hypothesis (on the validity of the assertion for $n - 1$), we may assume that for any two vectors $x \neq y$ in K with $x_i = y_i$ for at least one component i, we have

$$\max_{1 \le i \le n} (x - y)_i(F_i(x) - F_i(y)) > 0.$$

For any fixed vector $y \in K$, we show that the set

$$K_y \equiv \{ x \in K : F(x) \le F(y), x > y \}.$$

is empty. Assume the contrary. We argue that K_y is closed. Let $\{x^k\} \subset K_y$ be any sequence converging to some vector x. By the closedness of K, we have $x \in K$. Moreover, by continuity, $F(x) \leq F(y)$ and $x \geq y$. By the inductive hypothesis, we have either $x = y$ or $x > y$. If $x = y$, then by the F-differentiability of F, we have

$$\lim_{k \to \infty} \frac{F(x^k) - F(y) - JF(y)(x^k - y)}{\| x^k - y \|} = 0. \tag{3.5.6}$$

Since $JF(y)$ is a P matrix, there exists a constant $c > 0$ such that for every k,

$$\max_{1 \leq i \leq n} (x^k - y)_i [JF(y)(x^k - y)]_i \geq c \| x^k - y \|^2 > 0.$$

Since $x^k > y$, it follows that some component of $JF(y)(x^k - y)$ must be greater than $c\|x^k - y\|$. Hence in view of the limit (3.5.6), some component of $F(x^k) - F(y)$ must be positive. This contradicts the fact that x^k is in K_y. Consequently, we must have $x > y$; hence K_y is closed.

Applying Zorn's lemma to K_y, which is a partially ordered set with the usual componentwise order and is bounded below by y, we deduce that K_y has a minimal element u; that is, $u \in K_y$ and

$$[x \in K_y, \text{ and } x \leq u] \Rightarrow x = u.$$

Since $JF(u)$ is a P matrix, there is a vector $h < 0$ such that

$$0 > JF(u)h = \lim_{\tau \downarrow 0} \frac{F(u + \tau h) - F(u)}{\tau}.$$

Hence for $\tau > 0$ sufficiently small, we must have $F(u + \tau h) < F(u) \leq F(y)$ and $y < u + \tau h$. Thus $u + \tau h$ belongs to K_y. But this contradicts the minimality of u. Consequently, the set K_y is empty.

Let x and y be any two vectors in K such that $x_i \neq y_i$ for all i. Suppose that

$$(x - y)_i (F(x) - F(y))_i < 0, \quad \forall i.$$

Define a diagonal matrix D with diagonal entries

$$d_i \equiv \begin{cases} 1 & \text{if } x_i > y_i \\ -1 & \text{if } x_i < y_i \end{cases}$$

and a mapping $G : D^{-1}K \to \mathbb{R}^n$ by $G(z) \equiv DF(Dz)$. For any $z \in D^{-1}K$, $JG(z) = DJF(Dz)D$ is again a P matrix. Moreover, by construction of D, we have $D^{-1}x > D^{-1}y$. as well as $G(D^{-1}x) \leq G(D^{-1}y)$, which contradicts the emptiness of the set $\tilde{K}_{\tilde{y}}$ defined with respect to the pair $(\tilde{K}, G)$, where $\tilde{K} \equiv D^{-1}K$ and $\tilde{y} \equiv D^{-1}y$. This contradiction establishes part (a) with K closed.

Consider a general rectangle, not necessarily closed. Suppose that $JF(x)$ is a P matrix for all x in K but F is not a P function. There exist two distinct vectors x and y in K such that

$$\max_{1 \leq i \leq n} (x_i - y_i)(F_i(x) - F_i(y)) \leq 0.$$

Let K' be a closed subrectangle contained in K that contains the two vectors x and y. By what has just been proved, we know that F is a P function on K'. This immediately yields a contradiction. Consequently part (a) is valid without any closedness assumption on K.

Suppose that $JF(x)$ is a P_0 matrix for all x in K. Let $x \neq y$ be two distinct vectors in K. Let $\{\varepsilon_k\}$ be a sequence of positive scalars converging to zero. Since by part (a), $F + \varepsilon_k \mathbf{I}$ is a P function for every k, it follows that for every k, there exists an index i such that $x_i \neq y_i$

$$(x - y)_i(F_i(x) - F_i(y)) + \varepsilon_k (x_i - y_i)^2 > 0.$$

Since there are only finitely many indices, there must be an i such that the above inequality holds for infinitely many k and $x_i \neq y_i$. Passing to the limit $k \to \infty$ in the inequality completes the proof of (b). □

The converse of part (a) of the above proposition is false. A simple counterexample is the classic strictly monotone function $F(t) \equiv t^3$ for $t \in \mathbb{R}$, which is a P function with a zero derivative at $t = 0$. It might be conjectured that if F is a P function with a nonsingular Jacobian everywhere, then the Jacobian would be a P matrix. The following example shows that even this is false. Let $F : \mathbb{R}^2 \to \mathbb{R}^2$ be given by

$$F(x_1, x_2) \equiv \begin{pmatrix} x_1^3 - x_2 \\ x_1 + x_2^3 \end{pmatrix}, \quad (x_1, x_2) \in \mathbb{R}^2.$$

It is readily verified that F is a P function and that its Jacobian,

$$JF(x_1, x_2) = \begin{pmatrix} 3x_1^2 & -1 \\ 1 & 3x_2^2 \end{pmatrix},$$

is nonsingular for every $x \in \mathbb{R}^2$. However, $JF(0)$ is not a P matrix.

The next result addresses the existence and uniqueness of a solution to the VI (K, F) when F is a P function on K.

3.5.10 Proposition. Let K be given by (3.5.1).

(a) If F is a P function on K, then the VI (K, F) has at most one solution.

(b) If each K_ν is closed convex and F is a continuous uniformly P function on K, then the VI (K,F) has a unique solution.

Proof. Suppose x^1 and x^2 are two distinct solutions. For each index $\nu = 1,\ldots,N$, define the vector $z \equiv (z^{\nu'}) \in K$ as follows: for $\nu' = 1,\ldots,N$,

$$z_{\nu'} \equiv \begin{cases} x^1_\nu & \text{if } \nu' = \nu \\ x^2_{\nu'}, & \text{if } \nu' \neq \nu. \end{cases}$$

Thus

$$0 \le F(x^2)^T(z - x^2) = F_\nu(x^2)^T(x^1_\nu - x^2_\nu).$$

Similarly, we can deduce

$$0 \le F_\nu(x^1)^T(x^2_\nu - x^1_\nu).$$

Adding these two inequalities, we obtain

$$0 \ge (F_\nu(x^1) - F_\nu(x^2))^T(x^1_\nu - x^2_\nu) \quad \forall \nu = 1,\ldots,N.$$

This contradicts the P property of F on K; thus statement (a) holds.

If F is a uniformly P function on K, then the set $L'_\le$ in Proposition **3.5.1** must be bounded for all x^{ref} in K. Thus statement (b) holds. □

In what follows, we present a necessary and sufficient condition for a partitioned VI with a P_0 function to have a solution. This result extends Theorem **2.3.4**, which pertains to a monotone VI.

3.5.11 Theorem. Let K be given by (3.5.1), with each $K_\nu \subseteq \mathbb{R}^{n_\nu}$ being closed and convex. Let $F : K \to \mathbb{R}^n$ be a continuous P_0 function on K. The three conditions (a), (b), and (c) in Proposition **3.5.1** are equivalent; hence, for the VI (K,F) to have a solution, it is necessary and sufficient that a vector $x^{\text{ref}} \in K$ exists for which the set $L'_<$ in the proposition is bounded.

Proof. It suffices to prove that (c) implies (a). Let x^{ref} be a solution of the VI (K,F). Using the Cartesian structure of K, we can show that

$$(x_\nu - x^{\text{ref}})^T F_\nu(x^{\text{ref}}) \ge 0, \quad \forall x_\nu \in K_\nu, \quad \forall \nu.$$

If $x \neq x^{\text{ref}}$, there exists ν such that $x_\nu \neq x^{\text{ref}}_\nu$ and

$$(x_\nu - x^{\text{ref}})^T F_\nu(x) \ge (x_\nu - x^{\text{ref}})^T F_\nu(x^{\text{ref}}).$$

Hence the set $L'_<$ is empty. □

In principle, it is possible to refine the proof of Theorem **2.3.16** and establish a solution boundedness result for the VI (K, F) when F is P_0 function on K. Instead of following this approach, we refer the reader to Theorem **5.5.15** for such a result, where a necessary and sufficient condition for a VI of the P_0 type to have a nonempty bounded solution set is obtained in terms of a simple degree condition.

The next result establishes the relation between solution boundedness and strict feasibility of a CP of the P_0 type; a consequence of the result is that these two properties are equivalent for a CP defined by a $P_*(\sigma)$ function.

3.5.12 Theorem. Let K be given by (3.5.1), where each K_ν is a pointed, closed, convex cone in $\mathbb{R}^{n_\nu}$. Let $F : K \to \mathbb{R}^n$ be a continuous P_0 function on K. The following two statements are valid.

(a) If SOL(K, F) is nonempty and compact, then the CP (K, F) is strictly feasible.

(b) Conversely, if F is a $P_*(\sigma)$ function on K for some $\sigma > 0$ and if the CP (K, F) is strictly feasible, then SOL(K, F) is nonempty and compact.

Proof. Suppose SOL(K, F) is nonempty and compact. By part (d) of Theorem **3.6.6**, the CP $(K, q + F)$ has a solution for all vectors q with sufficiently small norm. Let q be such that $-q$ belongs to int K^*. It follows that for $\varepsilon > 0$ sufficiently small, there exists a vector $x \in K$ satisfying $F(x) \in -\varepsilon q + K^*$ The strict feasibility of the CP (K, F) thus follows.

Conversely, assume the conditions in (b). Let $x^{\text{ref}} \in K$ be such that $F(x^{\text{ref}}) \in \text{int}\, K^*$. We claim that the set $L'_{\leq}$ is bounded. Let x be an arbitrary vector in this set. We have $x \in K$ and

$$F_\nu(x)^T(x_\nu - x_\nu^{\text{ref}}) \leq 0, \quad \forall\, i = 1, \ldots, n.$$

For an index ν in the set $I_+(x, x^{\text{ref}})$, we have

$$(x_\nu - x_\nu^{\text{ref}})^T(F_\nu(x) - F_\nu(x^{\text{ref}})) > 0,$$

which implies,

$$(x_\nu - x_\nu^{\text{ref}})^T F_\nu(x^{\text{ref}}) < 0.$$

Since $F_\nu(x^{\text{ref}}) \in \text{int}\, K_\nu^*$, the set

$$\{ z_\nu \in K_\nu : z_\nu^T F_\nu(x^{\text{ref}}) \leq \eta \}$$

is bounded for every scalar η. Therefore there exists a constant $c > 0$ such that for every $x \in L'_{\leq}$,

$$\| x_\nu \| \leq c, \quad \forall\, \nu \in I_+(x, x^{\text{ref}}).$$

By the $\mathrm{P}_*(\sigma)$ property, we have

$$(1+\sigma) \sum_{\nu \in I_+(x,x^{\mathrm{ref}})} (x_\nu - x_\nu^{\mathrm{ref}})^T (F_\nu(x) - F_\nu(x^{\mathrm{ref}}))$$

$$+ \sum_{\nu \notin I_+(x,x^{\mathrm{ref}})} (x_\nu - x_\nu^{\mathrm{ref}})^T (F_\nu(x) - F_\nu(x^{\mathrm{ref}})) \geq 0,$$

which implies

$$\sum_{\nu \notin I_+(x,x^{\mathrm{ref}})} (x_\nu - x_\nu^{\mathrm{ref}})^T F_\nu(x^{\mathrm{ref}}) \leq -(1+\sigma) \sum_{\nu \in I_+(x,x^{\mathrm{ref}})} (x_\nu - x_\nu^{\mathrm{ref}})^T F_\nu(x^{\mathrm{ref}}).$$

Since the right-hand side is bounded and each $F_i(x^{\mathrm{ref}})$ belongs to int K_ν^*, it follows that x_ν must also be bounded for all $\nu \notin I_+(x, x^{\mathrm{ref}})$. Thus our claim holds. Therefore, by Proposition **3.5.1**, the CP (K, F) has a nonempty compact solution set. □

We give an example to show that the converse of part (a) of the above theorem is false; thus the $\mathrm{P}_*(\sigma)$ property in essential for part (b) to hold.

3.5.13 Example. Consider the LCP (q, M) with

$$q \equiv \begin{pmatrix} -1 \\ 1 \end{pmatrix} \quad \text{and} \quad M \equiv \begin{bmatrix} 0 & 1 \\ 0 & 1 \end{bmatrix}.$$

Clearly M is a P_0 matrix and the LCP (q, M) is strictly feasible. But this LCP has no solution. The homogeneous LCP $(0, M)$ is also strictly feasible but has unbounded solutions of the form $(x_1, 0)$ for all $x_1 \geq 0$. □

For a further result on the existence of solutions to an NCP (F), see Corollary **9.1.31**. In the rest of this subsection, we take up an issue related to the "regularization" of a VI of the P_0 type. The discussion below provides the background for an iterative algorithm for solving such a VI; see Section **12.2**.

In generalizing the monotonicity property to the P_0 property, we loose an important consequence of the former property. Namely, if F is monotone on K, then for every scalar $\varepsilon > 0$, the *Tikhonov map*:

$$F_\varepsilon \equiv F + \varepsilon \mathbf{I}$$

is strongly monotone on K. Nevertheless, if F is merely a P_0 function on K, then although F_ε must be a P function on K, the map F_ε is not necessarily a uniformly P function on K. The following example illustrates this "loss of uniformity".

3.5.14 Example. Define the function:

$$F(x_1, x_2) \equiv \begin{pmatrix} 0 \\ -e^{-x_1} \end{pmatrix}, \quad (x_1, x_2) \in \mathbb{R}^2.$$

It is easy to verify that this is a P_0 function on $\mathbb{R}^2$. We demonstrate that the map F_ε is not uniformly P on $\mathbb{R}^2$. For any fixed but arbitrary $\mu > 0$, choose a constant $c \geq 2$ satisfying

$$\frac{\varepsilon^2}{\mu}(c-1)^2 - \sqrt{\frac{\varepsilon}{\mu}}(e^c - e) \leq \varepsilon.$$

This is always possible because the second term on the left-hand side is negative and decreases exponentially with c. Multiplying both sides in the above inequality by $(c-1)^2$, we obtain

$$\frac{\varepsilon^2}{\mu}(c-1)^4 - \sqrt{\frac{\varepsilon}{\mu}}(c-1)^2(e^c - e) \leq \varepsilon(c-1)^2.$$

Define the points $x = (x_1, x_2)$ and $y \equiv cx$, where

$$x_1 \equiv 1 \quad \text{and} \quad x_2 \equiv \sqrt{\frac{\varepsilon}{\mu}}(c-1).$$

Omitting some obvious details, we obtain

$$\begin{aligned} \mu \| x - y \|_2^2 &\geq \frac{\mu}{(c-1)^2} \| x - y \|_2^2 = \mu(x_1^2 + x_2^2) > \varepsilon(c-1)^2 \\ &= \max\left\{ \varepsilon(c-1)^2, \frac{\varepsilon^2}{\mu}(c-1)^4 - \sqrt{\frac{\varepsilon}{\mu}}(c-1)^2(e^c - e) \right\} \\ &= \max_{i=1,2} (x_i - y_i)(F_{\varepsilon,i}(x) - F_{\varepsilon,i}(y)). \end{aligned}$$

Consequently, for every scalar $\mu > 0$, we can always find two points x and y that violate the uniformly P condition of F_ε with μ as the constant. Thus F_ε is not uniformly P for any $\varepsilon > 0$. □

Although F_ε is not necessarily a uniformly P function, the VI (K, F_ε) still possesses a unique solution, provided that F is a P_0 function on K. This is the assertion of the next result.

3.5.15 Theorem. Let K be given by (3.5.1) where each K_ν is closed convex. Let $F : K \to \mathbb{R}^n$ be a continuous P_0 function on K. The VI (K, F_ε) has a unique solution for every $\varepsilon > 0$.

Proof. Let $\varepsilon > 0$ be given. Since F_ε is a P function on K, it suffices to show the existence of a solution to the VI (K, F_ε). Fix an arbitrary vector $a \in K$ and consider the homotopy:

$$H(x,t) \equiv x - \Pi_K(\, t(x - F_\varepsilon(x)) + (1-t)a\,), \quad (x,t) \in K \times [0,1].$$

As in the proof of Theorem **2.3.16**, we show that the set of zeros:

$$\bigcup_{t \in [0,1]} H(\cdot,t)^{-1}(0)$$

is bounded. Proceeding as in the previous proof, we assume for the sake of contradiction that there exist a sequence of scalars $\{t_k\} \subset (0,1]$ and a sequence of vectors $\{x_k\} \subset K$ such that

$$\lim_{k\to\infty} \| x^k \| = \infty, \quad H(x^k, t_k) = 0 \quad \forall\, k,$$

and for all $y \in K$,

$$(\, y - x^k\,)^T [\, t_k\, F_\varepsilon(x^k) + (1 - t_k)\,(\, x^k - a\,)\,] \geq 0,$$

which implies

$$(\, y - x^k\,)^T [\, F(x^k) + \varepsilon\, x^k\,] \geq -\frac{1-t_k}{t_k}\,(\, y - x^k\,)^T (\, x^k - a\,).$$

By the Cartesian structure of K, we deduce that for each $\nu = 1, \ldots, N$ and for all $y_\nu \in K_\nu$,

$$(\, y_\nu - x^k_\nu\,)^T [\, F_\nu(x^k) + \varepsilon\, x^k\,] \geq -\frac{1-t_k}{t_k}\,(\, y_\nu - x^k_\nu\,)^T (\, x^k_\nu - a_\nu\,). \quad (3.5.7)$$

Define two complementary index sets:

$$\alpha \equiv \{\, \nu : \{\, x^k_\nu \,\} \text{ is bounded} \,\} \quad \text{and} \quad \gamma \equiv \{\, \nu : \{\, x^k_\nu \,\} \text{ is unbounded} \,\}.$$

Notice that $\gamma \neq \emptyset$. By working with an appropriate subsequence of $\{x^k\}$ if necessary, we may assume without loss of generality that for each $\nu \in \gamma$,

$$\lim_{k\to\infty} \| x^k_\nu \| = \infty. \quad (3.5.8)$$

For each $k = 1, 2, \ldots$, define the vector $y^k \equiv (y^k_\nu) \in K$ where for each $\nu \in \{1, \ldots, N\}$,

$$y^k_\nu \equiv \begin{cases} x^k_\nu & \text{if } k \in \alpha \\ a_\nu & \text{if } k \in \gamma \end{cases}$$

The sequence $\{y^k\}$ is bounded. Moreover $x^k \neq y^k$ for all k sufficiently large. By the P_0 property of F, it follows that for each k, there exists an index $\nu \in \{1, \ldots, N\}$ such that

$$y^k_\nu \neq x^k_\nu \quad \text{and} \quad (\, y^k_\nu - x^k_\nu \,)^T (\, F_\nu(y^k) - F_\nu(x^k) \,) \geq 0.$$

This index ν (which depends on k) must belong to γ. By (3.5.7) and the definition of y^k_ν, we obtain

$$(\, a_\nu - x^k_\nu \,)^T [\, F_\nu(y^k) + \varepsilon\, x^k_\nu \,] \geq -\frac{1 - t_k}{t_k} (\, a_\nu - x^k_\nu \,)^T (\, x^k_\nu - a_\nu \,) \geq 0.$$

We have shown that for each k, there exists an index $\nu \in \gamma$ such that

$$(\, a_\nu - x^k_\nu \,)^T [\, F_\nu(y^k) + \varepsilon\, x^k_\nu \,] \geq 0.$$

In particular, there is an index ν such that the above inequality holds for infinitely many k's. Since ε is positive and the sequence $\{y^k\}$ is bounded, this contradicts (3.5.8). □

3.6 Connectedness of Solutions

So far, we have discussed several important solution properties of the VI and the CP; these include global and local uniqueness, convexity, boundedness, and regularization. In this section, we present a theory that generalizes the theory of solution convexity. Specifically, we investigate the connectedness of the solution set of a VI. Although the main theory is applicable to a broader context, our focus is on the VI (K, F) where K is the Cartesian product of lower-dimensional sets and F is a continuous P_0 function. For this analysis, we continue to adopt the setting of the last section. At the end of the present section, we show that in the special case of the LCP, the P_0 property is in some sense necessary for solution connectedness; see Theorem **3.6.4** for details.

The theory developed herein is based on a nonsmooth equation formulation of the VI. Although either the natural or the normal equation can be used for this purpose, we choose to employ the natural map $\mathbf{F}^{\text{nat}}_K$ associated with the pair (K, F); this choice has an advantage and a disadvantage. The advantage is that it is more direct and simplifies the proofs somewhat. The disadvantage is that we need to assume the P_0 property of F globally (outside of K) due to a change of variables; see Lemma **3.6.1** below. Throughout this section, the statement "F is a P_0 function on $\mathbb{R}^n$" means that for all pairs of distinct vectors x and y in $\mathbb{R}^n$, there exists $\nu \in \{1, \ldots, N\}$ such that

$$x_\nu \neq y_\nu \quad \text{and} \quad (\, x_\nu - y_\nu \,)^T (\, F_\nu(x) - F_\nu(y) \,) \geq 0.$$

We know that the zeros of the map $\mathbf{F}_K^{\text{nat}}$ are precisely the solutions of the VI (K, F). More generally, there is a one-to-one correspondence between the solutions of the equation

$$\mathbf{F}_K^{\text{nat}}(x) = q$$

and a "translated" VI. In order to make this statement precise, we define the map

$$F_q(y) \equiv F(q+y) - q, \quad \forall\, y \in \mathbb{R}^n.$$

3.6.1 Lemma. For every vector $q \in \mathbb{R}^n$, a vector $x \in (\mathbf{F}_K^{\text{nat}})^{-1}(q)$ if and only if the vector $x - q \in \text{SOL}(K, F_q)$.

Proof. We have

$$\begin{aligned} x \in (\mathbf{F}_K^{\text{nat}})^{-1}(q) \quad &\Leftrightarrow \quad 0 = (x-q) - \Pi_K(x - F(x)) \\ &\Leftrightarrow \quad 0 = (x-q) - \Pi_K(x - q - F_q(x-q)). \end{aligned}$$

Thus the desired equivalence follows readily. □

The following result is an immediate consequence of Proposition **3.5.10** and Lemma **3.6.1**.

3.6.2 Corollary. Let K be given by (3.5.1) where each K_ν is closed convex and let $F : \mathbb{R}^n \to \mathbb{R}^n$ be a continuous P_0 function on $\mathbb{R}^n$. For every $\varepsilon > 0$ the perturbed natural map $\mathbf{F}_{\varepsilon,K}^{\text{nat}}$ is injective, where

$$\mathbf{F}_{\varepsilon,K}^{\text{nat}}(x) \equiv x - \Pi_K(x - \varepsilon\, x - F(x)), \quad \forall\, x \in \mathbb{R}^n.$$

Proof. For every vector $q \in \mathbb{R}^n$, the map

$$F_{\varepsilon,q}(y) \equiv \varepsilon\,(q+y) + F(q+y) - q, \quad y \in \mathbb{R}^n,$$

is clearly a P function on K. To establish the corollary, let

$$\mathbf{F}_{\varepsilon,K}^{\text{nat}}(x^1) = \mathbf{F}_{\varepsilon,K}^{\text{nat}}(x^2)$$

and denote this common vector by q. By Lemma **3.6.1**, it follows that $x^i - q \in \text{SOL}(K, F_{\varepsilon,q})$ for $i = 1, 2$. In turn by Proposition **3.5.10**(a), we must have $x^1 = x^2$ as desired. □

3.6.1 Weakly univalent functions

Clearly, the family of injective maps $\{\mathbf{F}_{\varepsilon,K}^{\text{nat}}\}$ converges pointwise to $\mathbf{F}_K^{\text{nat}}$. Moreover, the convergence is uniform on bounded sets; in fact, by the nonexpansiveness of the Euclidean projector Π_K, we have, for all x in the domain of F,

$$\| \mathbf{F}^{\rm nat}_{\varepsilon,K}(x) - \mathbf{F}^{\rm nat}_K(x) \| \leq \varepsilon \| x \|,$$

from which the uniform convergence is obvious if $\|x\|$ is bounded. Hence, under the assumptions of Corollary **3.6.2**, the natural map $\mathbf{F}^{\rm nat}_K$ can be approximated by a sequence of injective maps, and the approximation is uniform on bounded sets. This kind of a map has a special name. See Figure 3.4 for an illustration of such a map.

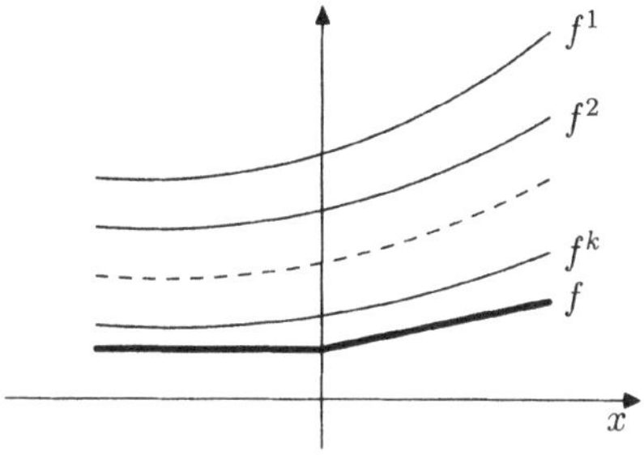

Figure 3.4: A weakly univalent function.

3.6.3 Definition. A mapping $f : \mathcal{D} \subseteq \mathbb{R}^n \to \mathbb{R}^n$ is said to be *weakly univalent* on its domain if it is continuous and there exists a sequence of univalent (i.e., continuous and injective) functions $\{f^\nu\}$ from $\mathcal{D}$ into $\mathbb{R}^n$ such that $\{f^\nu\}$ converges to f uniformly on bounded subsets of $\mathcal{D}$. □

Corollary **3.6.2** shows that if F is a P_0 function on $\mathbb{R}^n$, then the natural map $\mathbf{F}^{\rm nat}_K$ is weakly univalent. This observation provides the key for the applicability of the following result that unifies and extends several famous theorems in nonlinear analysis. We say that a subset E of another set F is *clopen* in F if E is both open and closed in F. We recall that a subset W of $\mathbb{R}^n$ is connected if the only subsets of W that are clopen in W are $\emptyset$ and W itself.

3.6.4 Theorem. Let $\mathcal{D}$ be a subset of $\mathbb{R}^n$ with nonempty interior and $f : \mathcal{D} \to \mathbb{R}^n$ be weakly univalent. Suppose that there is a nonempty subset E of $f^{-1}(0) \cap \operatorname{int}\mathcal{D}$ such that E is compact and clopen in $f^{-1}(0)$. The following statements are then valid:

(a) $f^{-1}(0) = E$;

(b) $f^{-1}(0)$ is nonempty, connected, and compact;

(c) for every bounded open set $U \supset f^{-1}(0)$, $\deg(f, U) = \pm 1$.

Proof. We write $S \equiv f^{-1}(0)$ and $\Omega \equiv \operatorname{int} \mathcal{D}$. We first establish the existence of a bounded open set U such that

$$E \subset U \subset \operatorname{cl} U \subset \Omega \quad \text{and} \quad (S \setminus E) \cap \operatorname{cl} U = \emptyset. \tag{3.6.1}$$

For this purpose, define $F \equiv \Omega^c \cup \operatorname{cl}(S \setminus E)$, which is clearly a closed set. Since $E \subseteq \Omega$, $E \cap \Omega^c = \emptyset$. Moreover, we must have $E \cap \operatorname{cl}(S \setminus E) = \emptyset$. Indeed if there is a vector $x \in E \cap \operatorname{cl}(S \setminus E)$, then by the openness of E in S it follows that there exists a neighborhood $\mathcal{N}$ of x such that $\mathcal{N} \cap S \subseteq E$. But since $x \in \operatorname{cl}(S \setminus E)$ there exists a sequence $\{x^\nu\} \subset S \setminus E$ converging to x. It follows that for all ν sufficiently large, $x^\nu \in \mathcal{N} \cap S$ which implies $x^\nu \in E$; a contradiction.

In summary, we have therefore shown that $F \cap E = \emptyset$. Since E and F are disjoint closed sets with E bounded, it follows that there exists a bounded open set U such that

$$E \subset U \subset \operatorname{cl} U \subset F^c = \Omega \cap (\operatorname{cl}(S \setminus E))^c.$$

Clearly, this implies (3.6.1). Noting that $S \cap \operatorname{bd} U = \emptyset$, we deduce that 0 does not belong to $f(\operatorname{bd} U)$; thus $\operatorname{dist}(0, f(\operatorname{bd} U)) > 0$.

Let $x^* \in E$. For the sake of contradiction suppose there is a vector $\bar{x} \in S \setminus E$. Let $\{f^\nu\}$ be a sequence of univalent functions converging to f uniformly on bounded subsets of $\mathcal{D}$. Define

$$G^\nu(x) \equiv f^\nu(x) - f^\nu(x^*) \quad \text{and} \quad H^\nu(x) \equiv f^\nu(x) - f^\nu(\bar{x}).$$

Each G^ν is a univalent function on D and its unique zero is x^*, which lies in E, it follows that $\deg(G^\nu, U) = \pm 1$ for all ν. Since f^ν converges to f uniformly on $\operatorname{cl} U$ and $f(x^*) = 0 = f(\bar{x})$, it follows that for all ν sufficiently large,

$$\sup_{x \in \operatorname{cl} U} \| G^\nu(x) - f(x) \|_\infty < \operatorname{dist}_\infty(0, f(\operatorname{bd} U))$$

and

$$\sup_{x \in \operatorname{cl} U} \| H^\nu(x) - f(x) \|_\infty < \operatorname{dist}_\infty(0, f(\operatorname{bd} U)).$$

Consequently, by the nearness property of degree, we have

$$\deg(H^\nu, U) = \deg(f, U) = \deg(G^\nu, U) = \pm 1.$$

Thus H^ν has a zero in U; but since the unique zero of H^ν is $\bar{x}$, which lies in $S \setminus E$, we obtain a contradiction to (3.6.1). This establishes that $S = E$ and statement (a) is proved. Moreover, the above argument shows that $\deg(f, U) = \pm 1$ for at least one bounded open subset U containing $f^{-1}(0)$.

By the excision property of the degree, it follows that $\deg(f, U) = \pm 1$ for all such subsets U. This establishes (c). (Incidentally, (c) also follows from Theorem **2.1.7**.)

For statement (b) it suffices to show that S is connected. Let W be a nonempty subset of S that is clopen in S. With W playing the role of the set E, the above proof shows that W must equal to S. Consequently S is connected. □

There are many consequences of the above theorem. We state several of these in the corollary below. See Figure 3.5 for an illustration of the result.

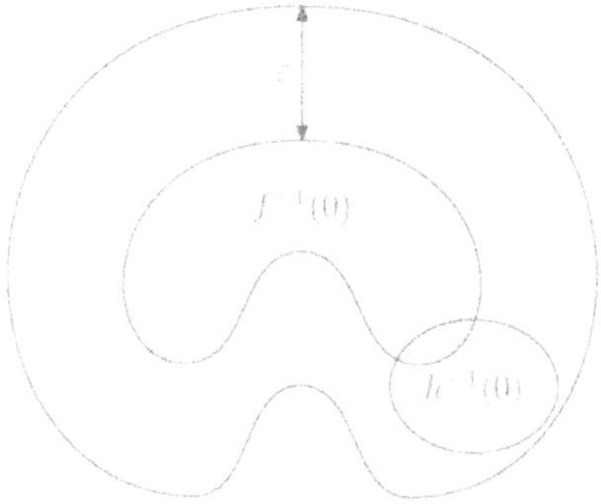

Figure 3.5: Illustration of Corollary **3.6.5**.

3.6.5 Corollary. Let $f : \mathbb{R}^n \to \mathbb{R}^n$ be a weakly univalent function. Suppose $f^{-1}(0) \neq \emptyset$. The following statements are valid:

(a) if $f^{-1}(0)$ is compact, then it is connected;

(b) if $f^{-1}(0)$ contains an isolated vector, then $f^{-1}(0)$ is a singleton;

(c) if $f^{-1}(0)$ is compact, then for every $\varepsilon > 0$ there exists $\delta > 0$ such that for every weakly univalent function $h : \mathbb{R}^n \to \mathbb{R}^n$ satisfying

$$\sup \{ \| h(x) - f(x) \| : x \in \operatorname{cl}(f^{-1}(0) + \mathbb{B}(0, \varepsilon)) \} \leq \delta, \qquad (3.6.2)$$

we have

$$\emptyset \neq h^{-1}(0) \subseteq f^{-1}(0) + \mathbb{B}(0, \varepsilon)$$

and $h^{-1}(0)$ is connected; thus the level set

$$\{ x \in \mathbb{R}^n : \| f(x) \| \leq \delta \}$$

is compact.

Proof. In case (a), take E to be $f^{-1}(0)$. Theorem **3.6.4** immediately implies that $f^{-1}(0)$ is connected.

In case (b), let x^* be an isolated vector in $f^{-1}(0)$. Let $\mathcal{N}$ be a neighborhood of x^* such that $\mathcal{N} \cap f^{-1}(0) = \{x^*\}$. Take E to be $\{x^*\}$. Then E is clopen in $f^{-1}(0)$. So the same theorem implies $f^{-1}(0) = \{x^*\}$.

To prove (c), write $\Omega \equiv f^{-1}(0) + \mathbb{B}(0, \varepsilon)$. By part (c) of Theorem **3.6.4**, we have $\deg(f, \Omega) = \pm 1$. Let $\delta \equiv \frac{1}{2}\,\mathrm{dist}_\infty(0, f(\mathrm{bd}\,\Omega))$. If h is a weakly univalent function satisfying (3.6.2), then $\deg(h, \Omega) = \pm 1$ by the nearness property of the degree. It follows that $h^{-1}(0) \cap \Omega \neq \emptyset$ and $h^{-1}(0) \cap \mathrm{bd}\,\Omega = \emptyset$. Thus

$$h^{-1}(0) \cap \Omega = h^{-1}(0) \cap \mathrm{cl}\,\Omega$$

is a nonempty, compact, clopen subset of $h^{-1}(0)$. By Theorem **3.6.4**, we must have $h^{-1}(0) \cap \Omega = h^{-1}(0)$, which implies $h^{-1}(0) \subseteq \Omega$. The connectedness of $h^{-1}(0)$ also follows from this theorem. To complete the proof of part (c), let x satisfy $\| f(x) \| \leq \delta$. Let $q \equiv f(x)$ and define $h \equiv f - q$. It follows that $x \in f^{-1}(0) + \mathbb{B}(0, \varepsilon)$. Since the latter set is bounded, thus so is the δ-level set of $\| f \|$. □

Applying Corollary **3.6.5** to the natural map $\mathbf{F}_K^{\mathrm{nat}}$ associated with a P_0 function F and a closed convex set K having a Cartesian structure, we obtain the following result for the partitioned VI (K, F).

3.6.6 Theorem. Let K be given by (3.5.1), where each K_ν is a closed convex set in $\mathbb{R}^{n_\nu}$. Let $F : \mathbb{R}^n \to \mathbb{R}^n$ be a continuous P_0 function on $\mathbb{R}^n$. Suppose that $\mathrm{SOL}(K, F) \neq \emptyset$. The following four statements are valid:

(a) if $\mathrm{SOL}(K, F)$ is compact, then it is connected;

(b) if the VI (K, F) has an isolated solution, then it has a unique solution;

(c) if $\mathrm{SOL}(K, F)$ is compact, there exists $\eta > 0$ such that the level set

$$\{ x \in \mathbb{R}^n : \| \mathbf{F}_K^{\mathrm{nat}}(x) \| \leq \eta \}$$

is bounded;

(d) if $\mathrm{SOL}(K, F)$ is compact, there exists $\varepsilon > 0$ such that for all vectors $q \in \mathbb{R}^n$ with $\| q \| < \varepsilon$, the VI $(K, q + F)$ has a nonempty compact solution set.

Proof. By Corollary **3.6.5** applied to the natural map $\mathbf{F}_K^{\mathrm{nat}}$, which is weakly univalent by Corollary **3.6.2**, parts (a), (b), and (c) follow readily. To prove part (d), it suffices to observe that

$$\| \mathbf{F}_K^{\mathrm{nat}}(x) - (x - \Pi_K(x - F(x) - q)) \| \leq \| q \|$$

for all $x \in \mathbb{R}^n$, therefore with $\|q\|$ sufficiently small, part (d) also follows easily from Corollary **3.6.5**. □

3.6.7 Remark. In essence, Theorem **3.6.6** holds more generally for a VI (K, F) such that the natural map is weakly univalent. But the question of whether there is a property of F besides the P_0 property that will yield the weakly univalence of $\mathbf{F}_K^{\text{nat}}$ does not have an answer at this time. □

Part (d) of Theorem **3.6.6** is a kind of stability result for the solution set of a VI of the P_0 type. For an extension of this result, see Theorem **5.5.15**. A 3-dimensional LCP shows that the boundedness assumption in statement (a) of Theorem **3.6.6** cannot be removed.

3.6.8 Example. Let

$$q \equiv \begin{pmatrix} -1 \\ 0 \\ 0 \end{pmatrix} \quad \text{and} \quad M \equiv \begin{bmatrix} 0 & 1 & 1 \\ 0 & 0 & 1 \\ 0 & 0 & 0 \end{bmatrix}.$$

It is easy to see that M is P_0 matrix. Nevertheless SOL(q, M) is not connected because it has two connected components (see Figure 3.6):

$$\begin{aligned} \text{SOL}(q, M) &= \{ (0, 0, x_3) : x_3 \geq 1 \} \cup \{ (x_1, 0, 1) : x_1 \geq 0 \} \\ &\cup \{ (0, x_2, 0) : x_2 \geq 1 \} \cup \{ (x_1, 1, 0) : x_1 \geq 0 \}. \end{aligned}$$

The set SOL(q, M) is clearly unbounded. Notice that the matrix M is nonnegative, hence semicopositive on $\mathbb{R}^3_+$. □

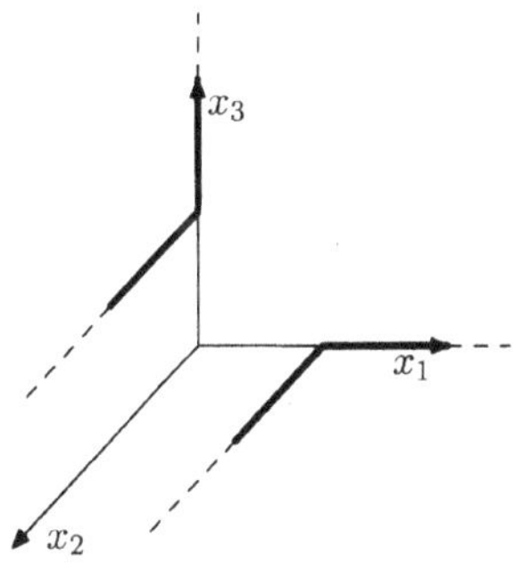

Figure 3.6: Solution set of an LCP.

Specializing Theorem **3.6.6** to the LCP (q, M) we may conclude that if M is a P_0 matrix and if this LCP has a nonempty bounded solution set, then the latter set must be connected. Thus if M is an R_0 and P_0 matrix, then $\mathrm{SOL}(q, M)$ is connected for all vectors q. This statement has a converse, which asserts that the P_0 property is necessary when one is interested in the connectedness of the solutions to all LCPs defined by a fixed matrix M. Thus, the P_0 property of an R_0 matrix M is characterized by the connectedness property of $\mathrm{SOL}(q, M)$ for all q.

3.6.9 Proposition. Let M be an $n \times n$ R_0 matrix. The following two statements are equivalent:

(a) M is a P_0 matrix;

(b) for all n-vectors q, the LCP (q, M) has a connected solution set.

Proof In view of the above remarks, it suffices to show (b) implies (a). Suppose that $\det M_{\alpha\alpha} < 0$ for some proper subset α of $\{1, \ldots, n\}$ with complement $\bar{\alpha}$. Define a vector $q \equiv (q_\alpha, q_{\bar{\alpha}})$ as follows:

$$q_\alpha \equiv -M_{\alpha\alpha} x_\alpha \quad \text{and} \quad q_{\bar{\alpha}} \equiv v_{\bar{\alpha}} - M_{\bar{\alpha}\alpha} x_\alpha,$$

where x_α is an arbitrary positive vector and $v_{\bar{\alpha}}$ is such that $q_{\bar{\alpha}}$ is positive. For this vector q, the LCP (q, M) has $(x_\alpha, 0)$ as a solution. This solution is nondegenerate with support α. Since $M_{\alpha\alpha}$ is nonsingular, Corollary **3.3.9** implies that this solution is locally unique. Since $\mathrm{SOL}(q, M)$ is by assumption connected, it follows that $\mathrm{SOL}(q, M)$ is a singleton.

Consider the homotopy:

$$H(z, t) \equiv t\,q + (1-t)\,\mathbf{1}_n + Mz^+ - z^-, \quad (z, t) \in \mathbb{R}^n \times [0, 1],$$

where $\mathbf{1}_n$ is the n-vector of all ones. The function $H(\cdot, t)$ is simply the normal map associated with the LCP $(tq + (1-t)\mathbf{1}_n, M)$. Since M is an R_0 matrix, the solutions of this family of LCPs are uniformly bounded. Let $\mathcal{U}$ be a bounded open subset of $\mathbb{R}^n$ containing all these solutions. By the homotopy invariance property of the degree, we have

$$\deg(H(\cdot, 0), \mathcal{U}) = \deg(H(\cdot, 1), \mathcal{U}).$$

We now argue that the left-hand degree is equal to 1 while the right-hand degree is equal to -1. Indeed, $H(\cdot, 0)$ is the normal map of the LCP $(\mathbf{1}_n, M)$ that has $x = 0$ as the unique solution. By Corollary **3.3.9**, this solution is locally unique. Since $\mathrm{SOL}(\mathbf{1}_n, M)$ is by assumption connected, it follows that $\mathrm{SOL}(\mathbf{1}_n, M)$ is a singleton. Consequently, $z = -\mathbf{1}_n$ is the unique zero of the map $H(\cdot, 0)$. Moreover, in a neighborhood $\mathcal{N}$ of this zero, the map $H(\cdot, 0)$ is equal to the identity map (it suffices to choose $\mathcal{N}$ so that all vectors in $\mathcal{N}$ are negative). Thus we have $H(z,0) = z$ for all $z \in \mathcal{N}$. Since $\mathcal{U}$ is a bounded open subset containing $-\mathbf{1}_n$, which is the unique zero of $H(\cdot,0)$, it follows that

$$\det(H(\cdot,0),\mathcal{U}) = \deg(H(\cdot,\mathcal{N}) = 1.$$

The map $H(\cdot,1)$ is the normal map of the LCP (q,M), which has a unique solution x with support α. It follows that $z \equiv (x_\alpha, -v_{\bar{\alpha}})$ is the unique zero of the map $H(\cdot,1)$. Moreover, in a suitable neighborhood of this zero, we have

$$H(z,1) = q + \begin{bmatrix} M_{\alpha\alpha} & 0 \\ M_{\bar{\alpha}\alpha} & \mathbf{I}_{|\bar{\alpha}|} \end{bmatrix} z.$$

Hence as above, we may deduce that the degree of $H(\cdot,1)$ on $\mathcal{U}$ is equal to the degree of the $H(\cdot,1)$ in this neighborhood where $H(\cdot,1)$ is the above affine map. Consequently,

$$\deg(H(\cdot,1),\mathcal{U}) = \operatorname{sgn}\det \begin{bmatrix} M_{\alpha\alpha} & 0 \\ M_{\bar{\alpha}\alpha} & \mathbf{I}_{|\bar{\alpha}|} \end{bmatrix} = \operatorname{sgn}\det M_{\alpha\alpha} < 0.$$

This is a contradiction. □

3.7 Exercises

3.7.1 Let $f : \mathbb{R} \to \mathbb{R}$ be B-differentiable at $x \in \mathbb{R}$. Let $f'(x+)$ and $f'(x-)$ denote the right and left derivative of f at x, respectively. Show that

$$f'(x;y) = f'(x+)y_+ - f'(x-)y_-,$$

where $y_\pm \equiv \max(0, \pm y)$.

3.7.2 Let $\Phi : \mathcal{D} \to \mathbb{R}^m$ be B-differentiable in a neighborhood of a vector x in the open set $\mathcal{D}$. Suppose that for every fixed direction d, $\Phi'(\cdot;d)$ is continuous in the first argument at x. Show that Φ has a strong F-derivative at x. Conversely, show that if Φ has a strong F-derivative at x, the directional derivative Φ' is continuous at (x,v) for every $v \in \mathbb{R}^n$.

3.7.3 Let $\theta : \mathbb{R}^2 \to \mathbb{R}$ be given by

$$\theta(x_1,x_2) \equiv \begin{cases} 0 & \text{if } x_1 \leq \sqrt{|x_2|} \text{ or if } x_2 = 0 \\ 1 & \text{otherwise} \end{cases}$$

Show that θ is directionally differentiable at the origin but not even continuous there. Consider now the modified function

$$\theta(x_1, x_2) \equiv \begin{cases} 0 & \text{if } x_1 \leq \sqrt{|x_2|} \\ x_1 & \text{otherwise} \end{cases}$$

Show that θ is continuous and directionally differentiable at the origin but the directional derivative at the origin is not continuous in the second argument.

3.7.4 Let $F : \mathbb{R}^{n+m} \to \mathbb{R}^\ell$ be locally Lipschitz continuous at (x^0, y^0) in $\mathbb{R}^{n+m}$.

(a) Assume that F is B-differentiable at (x^0, y^0). Show that $F(\cdot, y^0)$ is B-differentiable at x^0 and $F(x^0, \cdot)$ is B-differentiable at y^0. Let $F'_x((x^0, y^0); u)$ denote the B-derivative of $F(\cdot, y^0)$ at x^0 along the direction $u \in \mathbb{R}^n$ and let $F'_y((x^0, y^0); v)$ denote the B-derivative of $F(x^0, \cdot)$ at y^0 along the direction $v \in \mathbb{R}^m$. Give an example to show that the sum formula

$$F'((x^0, y^0); (u, v)) = F'_x((x^0, y^0); u) + F'_y((x^0, y^0); v) \qquad (3.7.1)$$

does not necessarily hold.

(b) Suppose that $F(\cdot, y^0)$ and $F(x^0, \cdot)$ are B-differentiable at x^0 and at y^0, respectively, and that

$$\lim_{(x,y) \to (x^0, y^0)} \frac{F(x, y) - F(x^0, y) - F'_x((x^0, y^0); x - x^0)}{\| x - x^0 \|} = 0.$$

Show that F is B-differentiable at (x^0, y^0) and (3.7.1) holds.

(c) Suppose in addition to the assumptions in (b) the following limit also holds

$$\lim_{(x,y) \to (x^0, y^0)} \frac{F(x, y) - F(x, y^0) - F'_y((x^0, y^0); y - y^0)}{\| y - y^0 \|} = 0.$$

Show that F is strongly B-differentiable at (x^0, y^0).

3.7.5 Let $F : \mathbb{R}^n \to \mathbb{R}^n$ be a co-coercive function that is directionally differentiable at a vector x. Show that the B-derivative $F'(x; \cdot)$ is co-coercive on $\mathbb{R}^n$.

3.7.6 Let $\Phi : \mathcal{D} \subseteq \mathbb{R}^n \to \mathbb{R}^n$, with $\mathcal{D}$ open, be B-differentiable at $x \in \mathcal{D}$.

(a) Show that $\Phi'(x; \cdot)$ is a globally Lipschitz homeomorphism on $\mathbb{R}^n$ if and only if $\Phi'(x; \cdot)$ is a locally Lipschitz homeomorphism at the origin.

(b) Suppose that Φ is a locally Lipschitz homeomorphism at x. Show that the B-derivative $\Phi'(x;\cdot)$ is a globally Lipschitz homeomorphism on $\mathbb{R}^n$. Show further that the local inverse of Φ is B-differentiable at $\Phi(x)$ and the B-derivative of this inverse is equal to the inverse of the function $\Phi'(x;\cdot)$. (Hint: use Corollary **2.1.13** to prove the first assertion and directly verify the limit

$$\lim_{v \to 0} \frac{\Phi^{-1}(y+v) - x - z}{\| v \|} = 0,$$

where $y \equiv \Phi(x)$ and $\Phi'(x;z) = v$, which establishes the second assertion.)

(c) Conversely, show that if Φ is strongly B-differentiable at x and the B-derivative $\Phi'(x;\cdot)$ is a locally Lipschitz homeomorphism at the origin, then Φ is a locally Lipschitz homeomorphism at x.

3.7.7 Let K be given by (3.2.1). Write

$$F(x) \equiv \begin{pmatrix} h(x) \\ g(x) \end{pmatrix} \in \mathbb{R}^{\ell+m}.$$

(a) Let $x \in K$ and let L denote the lineality space of the tangent cone $\mathcal{T}(-g(x);\mathbb{R}^m_+)$. Show that the LICQ holds at $x \in K$ if and only if $JF(x)(\mathbb{R}^n) + \{0\}^\ell \times L = \mathbb{R}^{\ell+m}$.

(b) Show that the MFCQ holds at $x \in K$ if and only if

$$0 \in \operatorname{int}(\, F(x) + JF(x)(\mathbb{R}^n) + \{0\}^\ell \times \mathbb{R}^m_+ \,).$$

(c) Show that the SMFCQ holds at a pair (x,λ), where $x \in K$ and λ satisfies (3.2.3), if and only if

$$0 \in \operatorname{int}(\, F(x) + JF(x)(\mathbb{R}^n) + \{0\}^\ell \times S \,),$$

where

$$S \equiv \{ y \in \mathbb{R}^m_+ \,:\, \lambda^T(\, y + g(x)\,) = 0 \}.$$

(d) Suppose that $\ell = 0$ and each g_i is a convex function. Show that the Slater CQ holds if and only if $0 \in \operatorname{int}(g(\mathbb{R}^n) + \mathbb{R}^n_+)$.

3.7.8 Consider the following abstract constraint system:

$$0 \in F(x) + S, \quad x \in C,$$

where $F : \mathbb{R}^n \to \mathbb{R}^m$ is continuously differentiable, C is a closed convex set in $\mathbb{R}^n$, and S is a closed convex set in $\mathbb{R}^m$. Let x^0 be a given feasible vector and let L_x and L_y denote the lineality space of the tangent cone $\mathcal{T}(x^0; C)$ and $\mathcal{T}(-F(x^0); S)$, respectively. Consider several generalized constraint qualifications:

(a) $JF(x^0)(L_x) + L_y = \mathbb{R}^m$;

(b) $0 \in \operatorname{int}(F(x^0) + JF(x^0)(C - x^0) + S)$;

(c) $0 \in \operatorname{int}(F(C) + S)$;

(d) $F(C) \cap (-\operatorname{int} S) \neq \emptyset$ if $\operatorname{int} S \neq \emptyset$.

Show that (a) implies (b). Moreover, if F is S-convex, that is, if

$$\tau F(x) + (1 - \tau) F(y) - F(\tau x + (1 - \tau)y) + S \subseteq S$$

for all x and y in $\mathbb{R}^n$ and all $\tau \in [0, 1]$, then (b) and (c) are equivalent. Finally, if $\operatorname{int} S$ is nonempty, then (c) and (d) are equivalent.

3.7.9 Consider the convex program:

$$\begin{array}{ll} \text{minimize} & \frac{1}{2} x_1^2 + \frac{1}{2} (x_2 - 1)^2 \\ \text{subject to} & x_2 \leq 0 \\ & -x_1 - x_2 \leq 0 \\ \text{and} & \frac{1}{2} [(x_1 - 1)^2 + (x_2 - 1)^2] \leq 1. \end{array}$$

This program has $x^* \equiv (0, 0)$ as its unique optimal solution. Show that $(1, 0, 0)$ is the unique KKT multiplier of the problem. Thus the SMFCQ holds. Show nevertheless that the CRCQ fails to hold at x^*.

3.7.10 A CQ that is implied by either the SMFCQ or the CRCQ but not conversely is the Weak Constant Rank Constraint Qualification (WCRCQ) defined as follows. Let K be given by (3.2.1). For a given $x \in K$, define

$$\mathcal{J}(x) \equiv \{ i \in \mathcal{I}(x) : \exists (\mu, \lambda) \in \mathcal{M}(x) \text{ with } \lambda_i > 0 \}.$$

The WCRCQ is said to hold at $x \in K$ if there exists a neighborhood $\mathcal{N}$ of x such that for every pair of index subsets $\mathcal{I}' \subseteq \mathcal{J}(x)$ and $\mathcal{J}' \subseteq \{1, \ldots, \ell\}$, the family of gradient vectors

$$\{ \nabla g_i(x') : i \in \mathcal{I}' \} \cup \{ \nabla h_j(x') : j \in \mathcal{J}' \}$$

has the same rank, which depends on the pair $(\mathcal{I}', \mathcal{J}')$, for all $x' \in \mathcal{N} \cap K$.

(a) Show that either the CRCQ at a vector $x \in K$ or the SMFCQ at a pair (x, λ), where $x \in K$ and $(\mu, \lambda) \in \mathcal{M}(x)$ for some μ, implies the WCRCQ at x.

(b) Consider a modification of the program in Exercise **3.7.9**:

$$\begin{array}{ll} \text{minimize} & \frac{1}{2}\,x_1^2 + \frac{1}{2}\,(\,x_2 - 1\,)^2 + \frac{1}{2}\,x_3^2 \\ \text{subject to} & e^{x_2} - x_3 \,\leq\, 1 \\ & x_2 + x_3 \,\leq\, 0 \\ & x_3 \,\geq\, 0 \\ & -x_1 - x_2 \,\leq\, 0 \\ \text{and} & \frac{1}{2}\,[\,(\,x_1 - 1\,)^2 + (\,x_2 - 1\,)^2\,] \,\leq\, 1. \end{array}$$

Show that the WCRCQ and the MFCQ, but not the CRCQ, hold at the unique optimal solution $x^* \equiv (0,0,0)$; show further that there are nonunique multipliers.

3.7.11 Consider the linear program:

$$\begin{array}{ll} \text{minimize} & c^T x \\ \text{subject to} & x \,\in\, P, \end{array}$$

where P is a polyhedron in $\mathbb{R}^n$ and $c \in \mathbb{R}^n$. Suppose that this program has a nonempty optimal solution set, which we denote S. Use Hoffman's Lemma **3.2.3** and an obvious polyhedral representation of S to show that there exists a scalar $\rho > 0$ such that, for any $\bar{x} \in S$,

$$\operatorname{dist}(x, S) \,\leq\, \rho\, c^T(\,x - \bar{x}\,), \quad \forall\, x \,\in\, P.$$

Optimization problems that have a bound like the above one are said to have "weak sharp minima". See Exercise **3.7.14** for an NLP having such minima and Sections **6.4** and **6.5** for more discussion on this topic.

3.7.12 Show that under the assumption of Proposition **3.3.4**, if K is convex, then there exist a constant $\eta > 0$ and a neighborhood $\mathcal{U}$ of x such that for all $y \in K \cap \mathcal{U}$,

$$\|\, x - y \,\| \,\leq\, \eta \operatorname{dist}(-F(y); \mathcal{N}(y; K)).$$

See Corollary **5.1.8** for an extension of this result.

3.7.13 Let F be B-differentiable at a solution x of the VI (K, F).

(a) Show that x is locally unique if

$$\left.\begin{array}{r} v \,\in\, \mathcal{C}(x; K, F) \\ F'(x; v) \,\in\, \left(\displaystyle\liminf_{\substack{y \in \mathrm{SOL}(K,F) \\ y \to x}} \left(\, \mathcal{T}(y; K) \,\cap\, F(x)^{\perp} \,\right) \right)^{*} \\ v^T F'(x; v) \,\leq\, 0 \end{array}\right\} \;\Rightarrow\; v \,=\, 0.$$

(b) Suppose that

$$\mathcal{C}(x; K, F) \subseteq \liminf_{\substack{y \in \mathrm{SOL}(K,F) \\ y \to x}} \left(\mathcal{T}(y; K) \cap F(x)^{\perp} \right)$$

If the homogeneous CP (3.3.6) has $v = 0$ as the unique solution, show that x is locally unique.

3.7.14 Let K be a closed convex set in $\mathbb{R}^n$ and let F be a mapping from K into $\mathbb{R}^n$.

(a) Show that $x \in \mathrm{SOL}(K, F)$ is nondegenerate if and only if there exists $\eta > 0$ such that

$$v^T F(x) \geq \eta \| v \|, \quad \forall v \in \mathcal{T}(x; K) \cap \mathrm{lin}\, \mathcal{T}(x; K)^*.$$

(b) Let $x \in \mathrm{SOL}(K, F)$ and suppose that $\mathcal{T}(x; K)$ is pointed. Show that the inequality in (a) is equivalent to

$$(y - x)^T F(x) \geq \eta \| y - x \|, \quad \forall y \in K.$$

(c) Let $x \in K$ be arbitrary. Show that if $\mathcal{T}(x; K)$ is pointed then x must be an extreme point of K and that the converse holds if K is polyhedral. (We have used the latter fact in proving Theorem **2.5.20**.)

(d) Suppose that F is monotone on K. Deduce that if the VI (K, F) has a nondegenerate solution x such that $\mathcal{T}(x; K)$ is pointed, then a constant $c > 0$ exists such that

$$\mathrm{dist}(y, \mathrm{SOL}(K, F)) \leq c\, \theta_{\mathrm{gap}}(y), \quad \forall y \in K.$$

This result is an error bound for the monotone VI with the gap function as the residual, under the assumption that a nondegenerate solution yielding a pointed tangent cone exists.

(e) Consider the convex program:

$$\begin{array}{ll} \text{minimize} & \theta(x) \\ \text{subject to} & x \in K, \end{array}$$

where θ is a C^1 convex function on an open convex set containing K. Let S_{opt} denote the optimal solution set of the above program. Suppose that S_{opt} contains a nondegenerate optimal solution x (in the sense that $-\nabla\theta(x) \in \mathrm{ri}\, \mathcal{N}(x; K)$) such that $\mathcal{T}(x; K)$ is pointed. Show that a constant $c > 0$ exists such that

$$\mathrm{dist}(y, S_{\mathrm{opt}}) \leq c\, (\, \theta(y) - \theta_{\min}\,), \quad \forall y \in K,$$

where $\theta_{\min}$ is the minimum objective value of the program. Deduce that this holds if K is polyhedral and S_{opt} contains a nondegenerate optimal solution that is an extreme point of K.

3.7.15 Let C be the polyhedral cone generated by the finite family of vectors $\{a^i : i = 1, \ldots, k\}$ in $\mathbb{R}^n$. Show that $v \in \operatorname{ri} C$ if and only if v is equal to a positive combination of the generators.

3.7.16 Let K be given by (3.5.1) where each K_ν is closed convex and let $F : \mathbb{R}^n \to \mathbb{R}^n$ be a continuous differentiable function on an open set containing a solution $x \in \text{SOL}(K, F)$.

(a) Show that

$$\mathcal{C}(x; K, F) = \prod_{\nu=1}^{N} (\mathcal{T}(x_\nu; K_\nu) \cap F_\nu(x)^\perp),$$

which shows that $\mathcal{C}(x; K, F)$ is the Cartesian product of the cones $\mathcal{T}(x_\nu; K_\nu) \cap F_\nu(x)^\perp$ for $\nu = 1, \ldots, N$.

(b) Show that if $JF(x)$ is strictly semicopositive on $\mathcal{C}(x; K, F)$, then x is locally unique.

3.7.17 A function $\Phi : \mathbb{R}^n \to \mathbb{R}^m$ is said to *H-differentiable* at x^* if there exists a nonempty subset $T(x^*)$ of matrices of order $n \times m$ such that for every sequence $\{x^k\}$ converging to x^*, a subsequence $\{x^k : k \in \kappa\}$ and a matrix $A \in T(x^*)$ exist satisfying

$$\lim_{k(\in\kappa)\to\infty} \frac{\Phi(x^k) - \Phi(x^*) - A(x^k - x^*)}{\| x^k - x^* \|} = 0.$$

(a) Show that Φ is H-differentiable at x^* if and only if positive constants δ and η exist such that

$$\| \Phi(x) - \Phi(x^*) \| \leq \eta \| x - x^* \|, \quad \forall x \in \mathbb{B}(x^*, \delta).$$

(b) Consider the linearly constrained VI (K, F) where F is H-differentiable at a solution $x^* \in \text{SOL}(K, F)$. Show that if for all $A \in T(x^*)$, the pair $(\mathcal{C}(x^*; K, F), A)$ is R_0, then x^* is locally unique.

3.7.18 Let $f : \mathbb{R}^n \to \mathbb{R}$ be continuously differentiable with $\nabla f(x) \geq 0$ for all $x \in \mathbb{R}^n$. Let $g : \mathbb{R}^n \to \mathbb{R}^n$ be such that $g_i(x) = f(x)$ for all i. Show that g is a P_0 function on $\mathbb{R}^n$ that is not monotone in general. (Hint: show that $Jg(x)$ is a P_0 matrix.)

3.7.19 Let $f : \mathbb{R}^{n+m} \to \mathbb{R}$ and $g : \mathbb{R}^{n+m} \to \mathbb{R}$ be two continuously differentiable function. Suppose that for some scalar $c > 0$,

$$\frac{\partial f(x,y)}{\partial x_i} \geq c, \quad \forall\, i$$

and

$$\frac{\partial g(x,y)}{\partial y_j} \geq c, \quad \forall\, j$$

for all $(x,y) \in \mathbb{R}^{n+m}$. Suppose further that for some scalar $c' > 0$,

$$\min(\,\|\, J_y f(x,y)\,\|,\ \|\, J_x g(x,y)\,\|\,) \leq c'\eta, \quad \forall\,(\,x,y\,) \in \mathbb{R}^{n+m}.$$

Show that there exists $\bar{\eta} > 0$, which depends on c and c', such that for all $\eta \in (0, \bar{\eta})$, the function $h : \mathbb{R}^{n+m} \to \mathbb{R}^{n+m}$ defined by

$$h_i(x,y) \equiv \begin{cases} f(x,y) & \text{for } i = 1, \ldots, n \\ g(x,y) & \text{for } i = n+1, \ldots, n+m, \end{cases}$$

is a P_0 function on $\mathbb{R}^{n+m}$. (Hint: show that $Jh(x,y)$ is a P_0 matrix.)

3.7.20 Let $F : \mathbb{R}^n \to \mathbb{R}^n$ be a P function, i.e.,

$$\max_{1 \leq i \leq n}\,(\,x_i - y_i\,)\,(\,F_i(x) - F_i(y)\,) > 0, \quad \forall\, x \neq y \in \mathbb{R}^n.$$

(a) Show that for every i and every vector $x \in \mathbb{R}^n$, the one-dimensional function $f_i : \mathbb{R} \to \mathbb{R}$ defined by

$$f_i(t) \equiv F_i(x_1, \ldots, x_{i-1}, t, x_{i+1}, \ldots, x_n), \quad t \in \mathbb{R}$$

is increasing. Thus the "diagonal functions" of a P function are increasing.

(b) Let F be continuous and uniform P. Show that for every $q \in \mathbb{R}^n$, there exists a vector $x \in \mathbb{R}^n_+$ such that $F(x) \geq q$.

3.7.21 Let X be the Cartesian product of n intervals X_i for $i = 1, \ldots, n$, each of which is not necessarily open or closed. A function $F : X \to \mathbb{R}^n$ is said to be a *Z function* on X if for every pair of indices $i \neq j$ and for every vector $x \in X$, the one-dimensional function $f_{ij} : X_i \to \mathbb{R}$ defined by

$$f_{ij}(t) \equiv F_j(x_1, \ldots, x_{i-1}, t, x_{i+1}, \ldots, x_n), \quad t \in X_i$$

is nonincreasing. These f_{ij} are the "off-diagonal functions" of F. Thus a Z function has nonincreasing off-diagonal functions; for this reason, a Z function is also called an *off-diagonal antitone* function. It is easy to show that if each X_i is an open interval and F is F-differentiable on X, then F is a Z function on X if and only if $JF(x)$ is a *Z matrix* for every $x \in X$, that is, the off-diagonal entries of $JF(x)$ are all nonpositive.

Let X be an open rectangle containing $\mathbb{R}^n_+$ and let $F : X \to \mathbb{R}^n$ be a continuous Z function. Suppose that the NCP (F) is feasible.

(a) Show that if x^1 and x^2 are two feasible solutions of the NCP (F), then so is $\min(x^1, x^2)$.

(b) Show that the feasible region of the NCP (F) contains a vector $\bar{x}$ such that $\bar{x} \leq x$ for all feasible vectors x. Moreover, $\bar{x}$ is an optimal solution of the minimization problem

$$\begin{array}{ll} \text{minimize} & p^T x \\ \text{subject to} & F(x) \geq 0, \quad x \geq 0, \end{array}$$

for any positive vector p. Such a vector $\bar{x}$ must necessarily be unique and is called the *least element* of the feasible set of the NCP (F).

(c) Show that the least feasible element of the NCP (F) must be a solution of the NCP (F).

(d) Show that the least-element solution $\bar{x}$ of the NCP (F) must also be the unique least-norm feasible solution; that is, if x is any other feasible solution, then $\|\bar{x}\| < \|x\|$.

3.7.22 This exercise extends Exercise **3.7.21** to the VI (K, F) where K is a closed rectangle in $\mathbb{R}^n$. Write

$$K \equiv \prod_{i=1}^{n} K_i,$$

where each K_i is one of the following four types of closed intervals:

$$[a_i, \infty), \ (-\infty, b_i], \ [a_i, b_i], \ (\infty, \infty).$$

Let F be a continuous Z function on K. Define the closed set

$$S \equiv \bigcap_{i=1}^{n} \{ x \in K : \ \text{either } x_i = b_i \text{ or } F_i(x) \geq 0 \}.$$

The statement $x_i = b_i$ in the above expression is interpreted as vacuously false if K_i is not bounded above. In particular, if $K = \mathbb{R}^n_+$, the set S becomes the feasible region of the NCP (F).

(a) Show that if x^1 and x^2 are two elements of S, then so is $\min(x^1, x^2)$.

(b) Suppose that S is nonempty and bounded below. Show that S contains a least element x^*; moreover, x^* solves the VI (K, F).

(c) If K is contained in $\mathbb{R}^n_+$, show that the least element of S (assuming that it exists) must be the unique least-norm solution of the VI (K, F).

3.7.23 Recall the concept of an inverse isotone function in Exercise **2.9.5**. Let $F : K \to \mathbb{R}^n$ be a Z function on the rectangle $K \subseteq \mathbb{R}^n$. Show that F is a P function on K if and only if it is inverse isotone on K.

3.7.24 Consider the function $F : \mathbb{R}^2 \to \mathbb{R}^2$ given by

$$F(x^1, x^2) \equiv \begin{pmatrix} x_1 + x_2^2 \\ x_2 \end{pmatrix}, \quad (x_1, x_2) \in \mathbb{R}^2.$$

Show that, for every $x \in \mathbb{R}^2$, all principal minors of $JF(x)$ are equal to 1 (thus $JF(x)$ is a P matrix), that the NCP $(F + q)$ has a unique solution for every $q \in \mathbb{R}^2$, but that F is not a uniformly P function.

3.7.25 Let A and B be two SPSD matrices of order n. Suppose that $A \perp B$. By Exercise **1.8.17**, there exist an orthogonal matrix P and two nonnegative diagonal matrices D_a and D_b such that

$$A = PD_aP^T, \quad B = PD_bP^T,$$

and $D_aD_b = 0$. Write $P \equiv [Z\ W]$, where the columns of Z form an orthonormal basis of the null space of A and the columns of W are orthonormal eigenvectors corresponding to the positive eigenvalues of A. Show that a symmetric matrix C belongs to $\mathcal{T}(A; \mathcal{M}^n_+) \cap B^\perp$ if and only if Z^TCZ is positive semidefinite and, after a suitable rearrangement of its rows and same columns, is of the form

$$\begin{bmatrix} 0 & 0 \\ 0 & (Z^TCZ)_{\beta\beta} \end{bmatrix},$$

where β is the set of indices i such that $(D_a)_{ii} = (D_b)_{ii} = 0$; that is, β is the "degenerate index set" of the complementary pair (A, B). Verify that the matrix

$$C \equiv \mathbf{I} - Z_{\cdot\alpha}(Z_{\cdot\alpha})^T,$$

where α is the set of indices i such that $(D_a)_{ii} = 0 < (D_b)_{ii}$, belongs to $\mathcal{T}(A; \mathcal{M}^n_+) \cap B^\perp$. (Hint: use the fact that Z^TZ is the identity matrix of order $|\alpha| + |\beta|$.)

3.7.26 Let $F : \mathcal{M}^n \to \mathcal{M}^n$ be a continuously differentiable matrix function. Let A^* be a given solution of the CP in SPSD matrices (1.4.61):

$$\mathcal{M}^n_+ \ni A \perp F(A) \in \mathcal{M}^n_+. \tag{3.7.2}$$

Show that if $(\mathcal{C}(A^*;\mathcal{M}^n_+,F), JF(A^*))$ is an R$_0$ pair, then A^* is a locally unique solution of the (3.7.2). (Hint: suppose that $\{A^k\} \subset \mathcal{M}^n_+$ is a sequence of solutions of (3.7.2) that converges to A^*. Suppose further that $A^k \neq A$ for all k and

$$\lim_{k\to\infty} \frac{A^k - A^*}{\| A^k - A^* \|} = B$$

exists. Show that B is a solution of the CP $(\mathcal{C}(A^*;\mathcal{M}^n_+,F), JF(A^*))$. The R$_0$ property of the latter pair then yields a contradiction. As in the proof of (a) $\Rightarrow$ (c) in Proposition **3.3.7**, it suffices to show that $JF(A^*)B$ belongs to $\mathcal{C}(A^*;\mathcal{M}^n_+,F)^*$. Let V be an arbitrary matrix in $\mathcal{C}(A^*;\mathcal{M}^n_+,F)$. Similar to Exercise **1.8.20**, show that for every k sufficiently large, there exists $\bar{\varepsilon}_k > 0$ such that for every $\varepsilon \in [0,\bar{\varepsilon}_k]$ there exists $E_k(\varepsilon) \in \mathcal{M}^n_+$ satisfying (i) $V_k(\varepsilon) \equiv A^k + \varepsilon V + \varepsilon^2 E_k(\varepsilon) \in \mathcal{M}^n_+$ and (ii) the limit of $E_k(\varepsilon)$ as $\varepsilon \downarrow 0$ exists. We have

$$\begin{aligned} 0 &\leq V_k(\varepsilon) \bullet F(A^k) \\ &= \varepsilon\,(\,V \bullet F(A^k) + \varepsilon\, E_k(\varepsilon) \bullet F(A^k)\,). \end{aligned}$$

The reader can now complete the proof.)

Show further that if F is affine and if A^* is a locally unique solution of (3.7.2), then $(\mathcal{C}(A^*;\mathcal{M}^n_+,F), JF(A^*))$ is an R$_0$ pair.

3.7.27 Let A^* be a solution of the CP (3.7.2) in SPSD matrices. Use Exercise **3.7.25**) to show that A^* is a nondegenerate solution if and only if there exist an orthogonal matrix P and two nonnegative diagonal matrices D_a and D_b such that

$$A = PD_aP^T, \quad F(A) = PD_bP^T,$$

$D_aD_b = 0$ and $D_a + D_b$ is a positive diagonal matrix; i.e., the degenerate index set β is empty. In this case, the critical cone $\mathcal{C}(A^*;\mathcal{M}^n_+,F)$ consists of all matrices $C \in \mathcal{M}^n$ such that $Z^TCZ = 0$.

3.7.28 Let $F : \mathbb{R}^n \to \mathbb{R}^n$ be continuous. Suppose that for all $\varepsilon > 0$ sufficiently small, the perturbed NCP $(F + \varepsilon\mathbf{I})$ has a solution, which may not be unique. Show that if there exists a vector $\bar{x} \geq 0$ such that $F(\bar{x}) \geq 0$ and the set

$$\{\, x \in \mathbb{R}^n_+ \,:\, (\,x - \bar{x}\,) \circ (\,F(x) - F(\bar{x})\,) \leq \bar{x} \circ F(\bar{x}) + \tau\,\mathbf{1}_n \,\}$$

is bounded for all $\tau > 0$ sufficiently small, then

$$\limsup_{\varepsilon \downarrow 0} \sup \{ \| x(\varepsilon) \| : x(\varepsilon) \text{ solves NCP } (F + \varepsilon \mathbf{I}) \} < \infty.$$

Show also that the NCP (F) has a nonempty compact solution set.

3.7.29 This exercise shows that it is possible to test in finite time if a given matrix is semicopositive or copositive on the nonnegative orthant, albeit not necessarily efficiently. Such a finite test is not obvious from the respective definition, which involves the continuum of all nonnegative vectors. First a definition. A matrix $M \in \mathbb{R}^{n \times n}$ is said to be an S_0 matrix if the system of linear inequalities

$$Mx \geq 0, \quad 0 \neq x \geq 0$$

has a solution. Clearly, linear programming can be used to test whether a matrix M is S_0. Moreover, M is an S_0 matrix if and only if the function $F(x) \equiv Mx$ is an S_0 function defined in Exercise **2.9.5**.

Let M be an arbitrary matrix in $\mathbb{R}^{n \times n}$. Show that the following four statements are equivalent.

(a) M is semicopositive on $\mathbb{R}^n_+$.

(b) Every principal submatrix of M is S_0.

(c) Every principal submatrix of M^T is S_0.

(d) M^T is semicopositive on $\mathbb{R}^n_+$.

If M is symmetric, show that M is copositive on $\mathbb{R}^n_+$ if and only if M is semicopositive on $\mathbb{R}^n_+$. Finally, give finite procedures to test if a matrix, not necessarily symmetric, is semicopositive or copositive on the nonnegative orthant.

3.7.30 Let C_ν be a pointed, closed convex cone in $\mathbb{R}^{n_\nu}$ for $\nu = 1, \ldots, N$; let

$$C \equiv \prod_{\nu=1}^{N} C_\nu \quad \text{and} \quad n \equiv \sum_{\nu=1}^{N} n_\nu.$$

Let M be a symmetric matrix in $\mathbb{R}^{n \times n}$. Show that M is (strictly) copositive on C if and only if M is (strictly) semicopositive on C.

3.7.31 A function $F : \mathbb{R}^n_+ \to \mathbb{R}^n$ is said to be *semicopositive* on $\mathbb{R}^n_+$ if for every nonzero vector $x \geq 0$, there exists an index i such that $x_i > 0$ and $F_i(x) \geq F_i(0)$. The function F is said to *strongly semicopositive* on $\mathbb{R}^n_+$ if there exists a constant $c > 0$ such that

$$\max_{1 \leq i \leq n} x_i \, (F_i(x) - F_i(0)) \geq c \| x \|^2, \quad \forall x \in \mathbb{R}^n_+.$$

Show that if F is continuous and strongly semicopositive on $\mathbb{R}^n_+$, then the NCP (F) has a nonempty compact solution set. Does the NCP necessarily have a unique solution?

Suppose that F is continuous and semicopositive. Prove or disprove that the NCP $(F + \varepsilon\mathbf{I})$ has a solution for all $\varepsilon > 0$.

3.7.32 Let K be a closed convex set in $\mathbb{R}^n$ and $A : K \to \mathbb{R}^{n \times n}$ be a continuous map. Let $x \in K$ be given. Consider the VI (K, F), where

$$F(y) \equiv A(y)(\, y - x\,), \qquad y \in K.$$

Show that if $A(x)$ is strictly copositive on the tangent cone $\mathcal{T}(x; K)$, then x is a locally unique solution of the VI (K, F).

3.7.33 Consider the multiplier map $\mathcal{M} : \mathbb{R}^n \to \mathbb{R}^{\ell+m}$ associated with a VI defined by a finitely representable set. Assume that $\mathcal{M}(\bar{x})$ is a singleton. Show that $\mathcal{M}$ is continuous at $\bar{x}$. Prove or disprove that $\mathcal{M}(x)$ is a singleton for all x sufficiently near $\bar{x}$.

3.7.34 Part (b) of this exercise generalizes Proposition **2.3.11**, while parts (c) and (d) generalize Exercise **2.9.17**. Let K be given by (3.5.1) where each K_ν is closed convex and let $F : K \to \mathbb{R}^n$ be a continuous. Show that

(a) if F is uniformly P on K, the normal map $\mathbf{F}^{\rm nor}_K$ of the pair (K, F) is a global homeomorphism on $\mathbb{R}^n$ (cf. Proposition **1.5.11**);

(b) if F is uniformly P on K, the unique solution of the VI $(K, F - q)$ is a co-coercive function of q;

(c) if F is a $\mathrm{P}_*(\sigma)$ function on K for some $\sigma > 0$, then for any two vectors q and q' and any solutions $x(q)$ and $x(q')$ of the VI $(K, F - q)$ and the VI $(K, F - q')$, respectively,

$$\begin{aligned}(x(q') - x(q))^T (q' - q) \geq \\ -\sigma \sum_{\nu \in I_+(x(q'),x(q))} (x_\nu(q') - x_\nu(q))^T (q'_\nu - q_\nu),\end{aligned}$$

where

$$I_+(x(q'), x(q)) \equiv \{\, \nu : (x_\nu(q') - x_\nu(q))^T (q'_\nu - q_\nu) > 0 \,\}.$$

(d) if F is a P function on K, then the solution function $x(q)$ is a P_0 plus function on its domain; that is, for any two vectors q and q' for which the VI $(K, F - q)$ and the VI $(K, F - q')$ have solutions, the respective solutions $x(q)$ and $x(q')$, which must necessarily be unique, satisfy

$$\begin{aligned}&[\, (x_\nu(q') - x_\nu(q))^T (q'_\nu - q_\nu) \leq 0 \quad \forall \, \nu \,] \\ &\Rightarrow \; [\, (x_\nu(q') - x_\nu(q))^T (q'_\nu - q_\nu) = 0 \quad \forall \, \nu \,].\end{aligned}$$

3.7.35 By definition, a matrix $M \in \mathbb{R}^{n \times n}$ is P if and only if the scalar:

$$\rho(M) \equiv \min_{\|x\|_2=1} \max_{1 \leq i \leq n} x_i(\,Mx\,)_i$$

is a positive constant. Let M be a P matrix. For every $q \in \mathbb{R}^n$, let $x(q)$ denote the unique solution of the LCP (q, M). Show that

$$\|\, x(q^1) - x(q^2)\,\| \leq \rho(M)^{-1} \,\|\, q^1 - q^2 \,\|.$$

Suppose that M is a symmetric positive definite matrix. Explore as much as possible the relationship between $\rho(M)$ and the spectrum of M. The significance of this exercise in comparison to part (b) of Exercise **3.7.34** is the emphasis on the special scalar $\rho(M)$ as a Lipschitz constant of the solution function $x(q)$.

3.8 Notes and Comments

Rademacher's Theorem **3.1.1** [706] provides the foundation for the differential theory of locally Lipschitz continuous functions. Our presentation of this theorem is based on Section 9J in [752]; see also the recent paper [63]. The B-derivative terminology was introduced in Robinson [737]. In an unpublished report [736], Robinson showed that under strong regularity a solution to a parametric generalized equation is implicitly B-differentiable. The strong property of the B-derivative was defined in [739], which is the source for Exercise **3.7.4**. Proposition **3.1.3** is due to Shapiro [774], who showed that in many situations the uniform limit (3.1.2) in this proposition is equivalent to several other definitions of the B-derivative. Exercise **3.7.6** is proved in [762]. Differentiability properties of the function in Example **3.1.5** were analyzed in [126].

Throughout the book we focus only on locally Lipschitz continuous functions. Thus Definition **3.1.2** of B-differentiability is sufficient for our purpose. For non-locally Lipschitz continuous functions, there is a notion of "semidifferenitability" that is based on the limit

$$\lim_{\substack{\tau \downarrow 0 \\ h' \to h}} \frac{F(x + \tau h') - F(x)}{\tau},$$

which, if exists, is called the *semiderivative* of F at x for h. See [752] for details.

Extending earlier work by Arrow, Hurwicz, and Uzawa [19] that dealt with the case of inequalities only, Mangasarian and Fromovitz [574] introduced their renowned CQ. Robinson [725] generalized the MFCQ to an abstract constraint system of the form

$$-F(x) \in S, \quad x \in C, \tag{3.8.1}$$

where C is a closed convex set in a real Banach space X, S is a closed convex cone in another real Banach space Y, and F is a Fréchet differentiable function from X into Y. Specifically, Robinson [733] called such a system *regular* at a feasible vector $\bar{x}$ if

$$0 \in \text{int}(\, F(\bar{x}) + JF(\bar{x})(C - \bar{x}) + S\,) \tag{3.8.2}$$

and showed that in the case where $C = \mathbb{R}^n$ and $S = \{0\}^\ell \times \mathbb{R}^m_+$, the regularity condition reduces to the MFCQ; see part (a) of Exercise **3.7.7**. More importantly, the system (3.8.1) is "stable" at $\bar{x}$ if (3.8.2) holds. We refer to [82] for an extensive treatment of stability and regularity of inequality systems in metric spaces.

Robinson [735] called condition (a) in Exercise **3.7.8** "nondegeneracy" of the feasible solution x^0 to the abstract system (3.8.1); this terminology originates from the concept of a nondegenerate solution in linear programming and should not be confused with that of a nondegenerate solution of a VI or CP in Definition **3.4.1**. The concept of an S-convex function appeared in [724].

In the context of a nonlinear program, Gauvin [287] observed that at a local minimizer the MFCQ is equivalent to the nonemptiness and boundedness of the KKT multipliers. Extending this result, Robinson [732] showed that under his regularity condition, the set-valued map of KKT multipliers of an abstract optimization problem is upper semicontinuous; see Proposition **3.2.2**. Kuntz and Scholtes [472] introduced a nonsmooth variant of the MFCQ. The SMFCQ was introduced in a paper by Fujiwara, Han, and Mangasarian [270]. The fact that the SMFCQ is equivalent to the uniqueness of the multiplier involved was observed by Kyparasis [477]. A similar uniqueness result in the context of linear programming was obtained independently by Mangasarian [565]. Shapiro [775] introduced a "strict constraint qualification" and established its sufficiency for the uniqueness of multipliers for optimization problems with abstract constraints in Banach cases; see also the subsequent paper [778]. Shapiro's CQ is stated in part (b) of Exercise **3.7.7** for a finitely representable set in an Euclidean space. The WCRCQ in Exercise **3.7.10** is due to Liu [515]; the two convex programs in this exercise and Exercise **3.7.9** are modifications of an example in the latter thesis.

Hoffman established his celebrated Lemma **3.2.3** in [349]. This classic paper is the foundation of the modern theory of error bounds of inequality systems. The latter theory has broad implications outside the VI/CP. Here we give a brief summary of the vast literature on error bounds for linear inequality systems and refer the reader to Section **6.10** for the bibliographic notes pertaining to extensions and to the VI/CP. Without using Hoffman's result, Walkup and Wets [865] established a Lipschitz characterization of convex polyhedra, which we phrased as the inclusion (3.2.12) and proved as a corollary of Lemma **3.2.3**. The multiplicative constant in Hoffman's error bound has been the subject of numerous studies. For the polyhedron $P(A,b)$, the best such constant can easily be seen to be

$$\sup_{x \notin P(A,b)} \frac{\operatorname{dist}(x, P(A,b))}{\|\,(\,Ax-b\,)_+\,\|},$$

which is clearly very difficult, if not impossible, to compute exactly. Robinson [724] suggested a computable error constant based on a simple construction. Mangasarian [567] derived computable estimates of the multiplicative constant based on linear programming. Other related estimates of this constant were obtained by Bergthaller and Singer [56], Güler, Hoffman, and Rothblum [317], Klatte and Thiere [436], Li [502], and Ng and Zheng [640]. Luo and Tseng [542] investigated the boundedness of the error constant as the system data undergo small changes. Deng [176, 177] extended the analysis of Luo and Tseng to convex inequality systems in Banach spaces. Lotov [524] gave an estimate of solution set perturbations for linear inequality systems. Li and Singer [507] obtained results pertaining to the boundedness of the error constant in a Banach space setting. Hu [357] established a characterization for the existence of a uniform global error bound of a system with possibly infinitely many linear inequalities under local perturbations of data parameters.

The Slater CQ appeared in an unpublished Cowles discussion paper [789]. Janin [374] introduced the CRCQ and proved that this CQ implied the KTCQ using the constant rank theorem of Malliavain in differential geometry. The characterization of CRCQ in Proposition **3.2.9** was proved in Pang and Ralph [676]. The SBCQ was introduced in [533]. Exercise **4.8.14**, where the latter CQ is used to establish the existence of a solution to a QVI, appears in [274]; this is in contrast to the epigraphical convergence approach employed by Robinson [744] for the same purpose. The SBCQ is closely related to a property defined by Klatte [430] for a set-valued map. Specifically, a multifunction $\Phi : \mathbb{R}^n \to \mathbb{R}^m$ is said to have a *non-trivial closed and locally bounded selection* near q^0 if there exist a neighborhood Q of q^0 and a multifunction $\Gamma : \mathbb{R}^n \to \mathbb{R}^m$ such that $\emptyset \neq \Gamma(q) \subseteq \Phi(q)$ for all $q \in Q \cap \operatorname{dom} \Phi$ and Γ is closed and locally bounded at q^0. Klatte then shows that the multiplier map associated with the KKT system of a nonlinear program has this selection property if either the MFCQ or the CRCQ (which includes the case of linear constraints) holds. As the reader can easily infer, the closedness and local boundedness of the multiplier map is the essence of the SBCQ.

McCormick [597] defined the second-order sufficiency condition for nonlinear programs using the critical cone in multiplier form; see Lemma **3.3.2**. Robinson [737] apparently was the first person to use the terminology "critical cone". Han and Mangasarian [324] observed that for a standard NLP the multiplier form of the critical cone is equivalent to the form using the gradient of the objective function.

The local uniqueness of solutions to LCPs and NCPs, Corollary **3.3.9**, was established by Mangasarian [566]. Before the latter paper, Kojima [442], based on the normal map formulation and the local nonsingularity theory of PC^1 maps, obtained sufficient conditions, which are more restrictive than b-regularity, for the isolatedness of a solution to an NLP. The equivalence of parts (b) and (c) in Proposition **3.3.7** is due to Reinoza [714], who called the implication (3.3.7) the total positivity condition. The terminology "b-regularity" in Definition **3.3.10** was coined by Pang and Gabriel [672] in the study of a nonsmooth equation based sequential quadratic programming (NE/SQP) method for solving the NCP. The letter "b" stood for "bounded": The b-regularity was used in the reference to demonstrate that the iterates produced by the NE/SQP method was bounded. Jiang [376] and Jiang and Qi [378] studied the local properties of solutions to VIs defined by nonsmooth functions. Most recently, Tawhid [828] extended the local uniqueness results in [378] by assuming that the defining function of the VI is H-differentiable; see Exercise **3.7.17**. The class of H-differentiable functions was introduced by Tawhid and Gowda [829] as a unification and generalization of several subclasses of nonsmooth functions.

Qiu and Magnanti [704] used the term "general second-order condition" (GSOC) to mean that for every $(\mu, \lambda) \in \mathcal{M}(x)$, where $x \in \mathrm{SOL}(K, F)$, $J_x L(x, \mu, \lambda)$ is strictly copositive on $\mathcal{C}(x; K, F)$. Extending a result in [732] for an NLP, Qiu and Magnanti showed that the MFCQ and the GSOC at a solution x of the VI (K, F) together imply that x is locally unique. Theorem **3.3.12** extends this result by relaxing the strict copositivity condition to the R_0 property. In spite of the improvement, the theorem left open an unanswered question. Namely, does part (a) of this theorem remain valid if the MFCQ holds at $x \in \mathrm{SOL}(K, F)$ and the homogeneous linear complementarity system (3.3.11) has a unique solution for some (instead of for all) $(\mu, \lambda) \in \mathcal{M}(x)$? The second NLP in Example **3.3.18**, which is due to

Robinson [732], shows that without the MFCQ the question cannot have an affirmative answer. Yet we are not aware of an answer to the question when the MFCQ is in place. Corollary **3.3.20** is proved in [732].

In [699], Qi called condition (3.3.20) "BD regularity" at x, where BD stands for B-derivative; Qi also proved Proposition **3.3.21** for a semismooth function. As we see in this proposition, B-differentiability is sufficient for the result to hold. The implication (a) $\Rightarrow$ (b) in Lemma **5.2.1** is also proved in Qi's paper.

Definition **3.4.1** of a nondegenerate solution to a VI is a natural generalization of the corresponding concept of a nondegenerate stationary point of an optimization problem due to Dunn [200], which in turn is a geometric characterization of the standard strict complementarity. The latter nondegeneracy condition plays a central role in the approach to the identification of active constraints in constrained optimization problems taken by Burke and Moré [100] and also in their subsequent study of exposing constraints [101]. See Section **6.10** for more notes and comments. Although he did not give a formal definition, Robinson [737] independently also studied the special case of the critical cone being a linear subspace, which is equivalent to the nondegeneracy condition. He noted in particular that in this case the constrained optimization problem behaves essentially like an unconstrained problem in a neighborhood of the stationary point in question. Somewhat unfortunately, Dunn's terminology of nondegeneracy, which we follow in this book, conflicts with Robinson's terminology of nondegeneracy for an abstract inequality system [735], which refers to an extension of the LICQ.

Partitioned VIs on Cartesian products received a treatment on its own for the first time in Pang [661]; see also [154]. Bertsekas and Tsitsiklis [62] studied parallel and distributed methods for solving these problems by exploiting the Cartesian structure. Semicopositive matrices on the nonnegative orthant were introduced by Eaves [204], who called them $\mathcal{L}_1$ matrices, where the letter "$\mathcal{L}$" was a tribute to Lemke. Attempting to unify the notation in LCP theory, Cottle, Pang, and Stone [142] used the symbol E_0 to denote such a matrix as a tribute to Eaves. Proposition **3.5.6** was proved in Pang and Huang [674].

The classes of P and P_0 matrices have a long history in the mathematical sciences. See [142] for a historical account of the role of these matrices in the study of the LCP. Extending these matrices, Moré and Rheinboldt [625] defined the P and P_0 functions and several related function classes. Traditionally, these functions were defined and their theory was developed relative to the natural Cartesian structure of a rectangle in $\mathbb{R}^n$. Facchinei and Pang [226] introduced Definition **3.5.8**, in which the univariate dimensional restriction of the component sets that constitute K was broadened. Proposition **3.5.9** was proved in [625]. The 2-dimensional example following this proposition, which shows that the Jacobian of a differentiable P function with a nonsingular Jacobian matrix is not necessarily a P matrix, also appeared in the latter reference. The recent article [802] studies the characterization of P and P_0-properties in nonsmooth functions using the broad class of H-subdifferentials.

Uniformly P functions on rectangles were defined in Moré's 1970 Ph.D. thesis. Proposition **3.5.10** generalizes the classic (existence and) uniqueness result for the NCP with a (uniformly) P function obtained by Moré [623, 624]. The 2-dimensional function in Exercise **3.7.24** was constructed by Megiddo and Kojima [601] to show that Cottle's condition of positively bounded Jacobians is different from Moré's uniformly P property, and thus for the GUS property to hold for an NCP (F) without F being a uniformly P function. This reference also gave several other examples to illustrate the distinctiveness of the GUS property; in particular, one such example shows that an NCP (F) of the non-P_0 type can still possess the GUS property.

It is somewhat surprising that the question of whether the function $F + \varepsilon \mathbf{I}$ is uniformly P for a P function F has evaded attention in the NCP literature for more than twenty-five years. It was only recently that Facchinei and Kanzow [224] constructed Example **3.5.14** to settle this question in the negative; more importantly, they established Theorem **3.5.15** despite the lack of the uniform P property in the perturbed functions.

In their unified study of interior point methods for solving the LCP, Kojima, Megiddo, Noma, and Yoshise [445] introduced the class of $P_*(\sigma)$ matrices and showed that such a matrix must be column sufficient. A remarkable result due to Väliaho [857] showed that the union of $P_*(\sigma)$ matrices for all $\sigma > 0$ is equal to the class of column sufficient matrices. It appears that the class of $P_*(\sigma)$ functions was defined independently in Lesaja's Ph.D. thesis [492] and also in the paper by Jansen, Roos, Terlaky, and Yoshise [375]. Lesaja's definition uses the function directly (see Definition **3.5.8**); in contrast, the definition by the other group of authors is based on the P_* property of the Jacobian matrices. The equivalence of these two definitions is shown in the subsequent papers [493, 689], where the authors study interior point methods for solving NCPs of the $P_*(\sigma)$ type. Zhao and Han [899] established a restricted version of Theorem **3.5.12** using their concept of an exceptional family of elements, which they defined assuming a finite representation of the defining set of the VI.

An important application of P matrices occurs in the theory of global univalence. Inspired by a conjecture of Samuelson [758], Gale and Nikaido

[283] showed that if the Jacobian matrix of a C^1 function is a P matrix everywhere in a rectangle, then the function must be globally univalent in the rectangle. Moré and Rheinboldt [625] used Exercise **2.9.6** to obtain a generalization of the Gale-Nikaido result, namely, if F is a continuously differentiable mapping from the open rectangle K into $\mathbb{R}^n$ with $JF(x)$ being a nonsingular P_0 matrix for each $x \in K$, then F is surjective in K. Partly due to its importance in mathematical economics, the Gale-Nikaido result has received a great deal of interest subsequently. Parthasarathy [684] presents in a single source many global univalence theorems. The class of weakly univalent functions was introduced by Gowda and Sznajder in [313]; the latter article is the source for the materials in Subsection **3.6.1**; see also [308].

Cao and Ferris [106] are the first to identify a class of real square matrices M for which the solution set of the LCP (q, M) is connected for all q. Using the theory of weakly univalent functions, Jones and Gowda [382] established Proposition **3.6.9**. By studying the connectedness of the solution set of an LCP of the P_0 kind, the latter study provides an elegant extension of a previous result (first proved by Cottle and Guu), which asserts that the cardinality of the solution set of such an LCP is either 0, 1, or infinity.

The concept of an off-diagonally antitone mapping, i.e., a Z function, was introduced by Rheinboldt [715]. Such a function is a nonlinear generalization of a Z matrix, which is a real square matrix with nonpositive off-diagonal entries. A real square matrix that is both Z and P is called an M (for Minkowski) matrix. There is a long history of Z and M matrices; for discussion of the role of these matrices in LCP theory, see [142]. In the pioneering paper [144], Cottle and Veinott established that a feasible LCP with a Z matrix must have a least-element solution. Tamir [827] generalized this classic result to the NCP (see Exercise **3.7.21**) and obtained a characterization of a Z matrix in terms of the existence of a least-element solution to the LCP; see also Moré [624]. The Ph.D. thesis of Pang [659] presents a comprehensive least-element complementarity theory. Exercise **3.7.22** is drawn from this thesis. For a connection of a Z function in a nonlinear Leontief input-output model, see [368]. Extending the M matrices, Moré and Rheiboldt [625] introduced the M functions, which are functions that are both Z and P. Exercise **3.7.23** gives one of several characterizations of an M function; see [625] for details.

Based on results in [142], Exercise **3.7.29** shows that there is a finite test for a matrix to be copositive or semicopositive on the nonnegative orthant. Nevertheless, the test derived from the exercise cannot be practically efficient, because it involves checking the S_0 property of *every* principal submatrix of the given matrix. In general, many matrix-theoretic properties are like these; namely, they can be tested in finite time albeit not necessarily efficiently. In the theory of computational complexity, which is a subject that this book has omitted, the classes of P, NP-complete, and co-NP-complete problems are well known [286]. For the problem of deciding whether a matrix with integer/rational entries has a certain property $\mathcal{P}$, various complexity results are available; these are summarized below:

(a) for $\mathcal{P}$ = S or S_0, problem is P via linear programming [142];

(b) for $\mathcal{P}$ = P [145], copositive [630], nondegenerate [114], column sufficient, R_0, strictly semicopositive, semicopositive, degenerate [851], all problems are co-NP-complete.

Mangasarian and Shiau [579] were the first to show that the unique solution of the LCP (q, M), where M is a P matrix, is a Lipschitz continuous function of q. Their proof does not exhibit the P matrix constant $\rho(M)$ as in Exercise **3.7.35**; see Section **6.10** for bibliographical remarks about this constant. Exercise **3.7.34** generalizes the Mangasarian-Shiau result in several directions.

Chapter 4

The Euclidean Projector and Piecewise Functions

The Euclidean projector is undoubtedly one of the most fundamental and useful mathematical tools for the study of the VI/CP. We know from Theorem **1.5.5** that the projector Π_K onto a closed convex set K is a globally Lipschitz continuous function. In this chapter, we undertake an in-depth study of the differentiability properties of this projector. Exercise **4.8.1** shows that the Euclidean projector is directionally differentiable at every point belonging to the set and gives an explicit formula for the directional derivative. Unfortunately, this exercise does not cover the situation where the base vector lies outside the given set; the latter situation is needed for the sensitivity analysis of the VI, which we develop in the next chapter. A 3-dimensional example in Exercise **4.8.5** shows that in general the projector Π_K does not necessarily have one-sided derivatives at a point outside K. A 2-dimensional set illustrating the same non-directional differentiability also exists; see Section **4.9** for a reference. Thus to obtain any differentiability property of Π_K at a vector $x \notin K$, the set K must be properly restricted.

Through the natural/normal map reformulation of the VI/CP, in which the Euclidean projector plays a central role, the results obtained in this chapter provide the cornerstone for the sensitivity and stability analysis of the VI/CP, which is the main topic of the next chapter. Since the projection $\Pi_K(x)$ is itself the solution of a parametric VI $(K, \mathbf{I} - x)$ with x as the parameter, many differentiability properties of $\Pi_K(x)$ persist in more general VIs. Thus the results obtained in this chapter provide a glimpse of the kind of results that we can obtain in the next chapter.

The organization of this chapter follows the complexity of the set K. Specifically, the chapter consists of three major parts. The first part begins in the next section with the case where K is polyhedral. Besides the important differentiability property of Π_K established in Theorem **4.1.1**, this part also includes two important topics: the normal manifold (Subsection **4.1.1**) and the theory of piecewise affine functions (Section **4.2**). A key result in the latter theory is a characterization of the global homeomorphism property of a piecewise affine map; see Theorem **4.2.11**. As an application of this theorem, we obtain a major result, Theorem **4.3.2** in Section **4.3**, that gives numerous equivalent conditions for a given affine pair (K, M), where K is a polyhedron and M is a matrix, such that the AVI (K, q, M) has a unique solution for all vectors q.

The second part of the chapter deals with a finitely representable convex set K. Sections **4.4** and **4.5** establish two important properties of Π_K under the SBCQ and the CRCQ, respectively; see Theorems **4.4.1** and **4.5.2**. In particular, the latter theorem is a direct extension of the polyhedral counterpart and shows that the projector Π_K is piecewise C^1 (PC^1) under the CRCQ (see Definition **4.5.1** for the formal definition of a PC^1 function). Due to the importance of this class of nonsmooth functions, we present in Section **4.6** an extensive theory of PC^1 functions in general. Of particular emphasis therein is the local homeomorphism property, which provides the key to the strong stability theory of VIs. The third and last part of the chapter covers the case where the set K is dependent on a varying parameter; see Section **4.7**. The main result obtained therein, Theorem **4.7.5**, shows that under the joint MFCQ and CRCQ the parametric projector remains a PC^1 function.

4.1 Polyhedral Projection

We begin with a quick summary of some basic facts pertaining to the Euclidean projector onto a finitely representable, convex set:

$$K \equiv \{ x \in \mathbb{R}^n : h(x) = 0, \ g(x) \leq 0 \}, \tag{4.1.1}$$

where $g : \mathbb{R}^n \to \mathbb{R}^m$ is a continuously differentiable function with each g_i being convex, and $h : \mathbb{R}^n \to \mathbb{R}^\ell$ is an affine function. For each vector $x \in \mathbb{R}^n$, the projected vector $\bar{x} \equiv \Pi_K(x)$ is the unique solution of the convex program:

$$\begin{array}{ll} \text{minimize} & \frac{1}{2}(y - x)^T(y - x) \\ \text{subject to} & h(y) = 0, \\ \text{and} & g(y) \leq 0. \end{array} \tag{4.1.2}$$

The first-order optimality condition of this projection problem is the VI $(K, \mathbf{I} - x)$. As such, we may consider the critical cone of this VI at the projected vector $\bar{x} \in K$. We denote this special cone by $\mathcal{C}_\pi(x; K)$ and call it the critical cone of K at x. By definition, we have

$$\mathcal{C}_\pi(x; K) \equiv \mathcal{T}(\bar{x}; K) \cap (\bar{x} - x)^\perp.$$

The KKT system of the projection problem (4.1.2) is:

$$\bar{x} - x + \sum_{j=1}^{\ell} \mu_j \nabla h_j(\bar{x}) + \sum_{i=1}^{m} \lambda_i \nabla g_i(\bar{x}) = 0,$$

$$h(\bar{x}) = 0,$$

$$0 \leq \lambda \perp g(\bar{x}) \leq 0.$$

Let $\mathcal{M}_\pi(x)$ denote the set of multiplier pairs (μ, λ) satisfying the above KKT system.

When K is a polyhedron, the above KKT system provides a necessary and sufficient condition for $\bar{x} = \Pi_K(x)$. Based on this equivalent system, the following result shows that the projector Π_K has an unusual property, implying in particular that it has a strong B-derivative everywhere on $\mathbb{R}^n$. See Figure 4.1 for an illustation of the theorem.

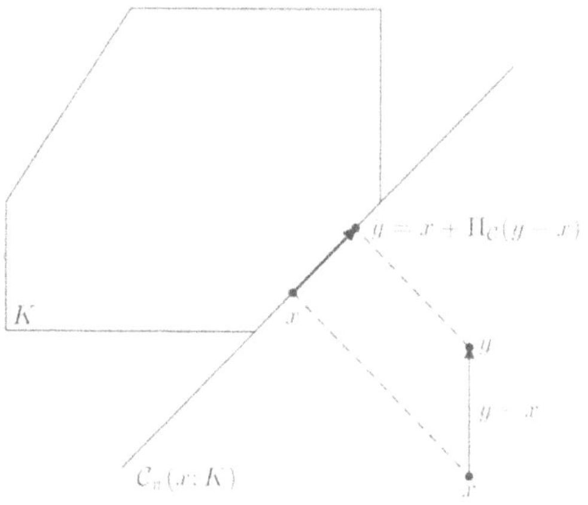

Figure 4.1: Illustration of Theorem **4.1.1**.

4.1.1 Theorem. Let K be a polyhedron in $\mathbb{R}^n$. For every $x \in \mathbb{R}^n$, there exists a neighborhood $\mathcal{N}$ of x such that

$$\Pi_K(y) = \Pi_K(x) + \Pi_{\mathcal{C}}(y - x), \quad \forall\, y \in \mathcal{N}, \tag{4.1.3}$$

where $\mathcal{C} \equiv \mathcal{C}_\pi(x; K)$. Therefore,

$$\Pi_K'(x; d) = \Pi_{\mathcal{C}}(d), \quad \forall\, d \in \mathbb{R}^n,$$

and Π_K is strongly B-differentiable everywhere on $\mathbb{R}^n$.

Proof. Assume for the sake of contradiction that no such neighborhood exists. There exists a sequence of vectors $\{x^k\}$ converging to x such that for all k,

$$\Pi_K(x^k) \neq \Pi_K(x) + \Pi_{\mathcal{C}}(x^k - x).$$

We write $\bar{x} \equiv \Pi_K(x)$ and $\bar{x}^k \equiv \Pi_K(x^k)$ for all k. For simplicity of notation, we let

$$K \equiv \{ y \in \mathbb{R}^n : Ay \leq b \} \tag{4.1.4}$$

for some matrix $A \in \mathbb{R}^{m \times n}$ and vector $b \in \mathbb{R}^m$. For each k, there exists $\lambda^k \in \mathbb{R}^m$ such that

$$\bar{x}^k - x^k + \sum_{i=1}^m \lambda_i^k \,(A_{i\cdot})^T = 0,$$

$$0 \leq \lambda^k \perp b - A\bar{x}^k \geq 0.$$

We may choose λ^k such that the row vectors

$$\{ A_{i\cdot} : i \in \operatorname{supp}(\lambda^k) \}$$

are linearly independent. Let B be the matrix with these vectors as its rows. By restricting to an appropriate subsequence of $\{x^k\}$ if necessary, we may further assume that the index sets $\operatorname{supp}(\lambda^k)$ are equal to a common index set $\mathcal{J}$ for all k. Clearly, $\mathcal{J} \subseteq \mathcal{I}(\bar{x})$. From the linear equations:

$$\bar{x}^k - x^k + \sum_{i \in \mathcal{J}} \lambda_i^k \,(A_{i\cdot})^T = 0,$$

$$A_{i\cdot}\bar{x}^k = b_i, \quad \forall\, i \in \mathcal{J},$$

we may solve for $(\lambda_i^k : i \in \mathcal{J})$ and $\bar{x}^k$ uniquely in terms of x^k, obtaining:

$$\lambda_{\mathcal{J}}^k = (BB^T)^{-1}(Bx^k - b_{\mathcal{J}}),$$

$$\bar{x}^k = [\mathbf{I} - B^T(BB^T)^{-1}B]\,x^k + B^T(BB^T)^{-1}b_{\mathcal{J}}.$$

Passing to the limit $k \to \infty$, we deduce

$$\bar{x} = [\mathbf{I} - B^T(BB^T)^{-1}B]\,x + B^T(BB^T)^{-1}b_{\mathcal{J}};$$

moreover, the sequence $\{\lambda^k_{\mathcal{J}}\}$ converges to

$$\bar{\lambda}_{\mathcal{J}} \equiv (BB^T)^{-1}(Bx - b_{\mathcal{J}}).$$

Defining $\bar{\lambda}_i \equiv 0$ for all $i \notin \mathcal{J}$, we see that $\bar{\lambda} \in \mathcal{M}_\pi(x)$. In terms of this multiplier, we can write, by Lemma **3.3.2**,

$$\mathcal{C} = \{ v \in \mathbb{R}^n : A_{i\cdot}v \le 0 \;\forall\, i \in \mathcal{I}(\bar{x}) \text{ with equality if } i \in \operatorname{supp}(\bar{\lambda}) \}.$$

Notice that $\operatorname{supp}(\bar{\lambda}) \subseteq \mathcal{J}$. We have

$$d^k \equiv \bar{x}^k - \bar{x} = [\mathbf{I} - B^T(BB^T)^{-1}B](x^k - x).$$

We claim that $d^k = \Pi_{\mathcal{C}}(x^k - x)$. Defining

$$\tilde{\lambda}^k_{\mathcal{J}} \equiv (BB^T)^{-1}B(x^k - x),$$

we see that

$$d^k - (x^k - x) + \sum_{i\in\mathcal{J}} \tilde{\lambda}^k_i (A_{i\cdot})^T = 0$$

and

$$A_{i\cdot}d^k = 0, \quad \forall\, i \in \mathcal{J}.$$

For $i \in \mathcal{I}(\bar{x}) \setminus \mathcal{J}$, we have

$$A_{i\cdot}d^k = A_{i\cdot}\bar{x}^k - A_{i\cdot}\bar{x} \le 0.$$

The last three equations are enough to establish that $d^k = \Pi_{\mathcal{C}}(x^k - x)$. Thus we obtain a contradiction. Consequently, we must have

$$\Pi_K(y) = \Pi_K(x) + \Pi_{\mathcal{C}}(y - x)$$

for all y sufficiently close to x. This identity immediately yields the formula for the directional derivative $\Pi'_K(x; d)$ and establishes that this derivative is strong. □

Based on the above result, we can characterize the F-differentiable points of the projector Π_K onto a polyhedron.

4.1.2 Corollary. Let K be a polyhedron in $\mathbb{R}^n$ and $x \in \mathbb{R}^n$. The following statements are equivalent.

(a) The projector Π_K is F-differentiable at x.

(b) The critical cone $\mathcal{C}_\pi(x; K)$ is a linear subspace.

(c) $\Pi_K(x)$ is a nondegenerate solution of the VI $(K, \mathbf{I} - x)$.

If $x \in K$, then (a) is further equivalent to

(b') the tangent cone $\mathcal{T}(x; K)$ is a linear subspace;

(d) the origin is a relative interior point of the normal cone $\mathcal{N}(x; K)$.

Proof. The equivalence of (b) and (c) follows from Proposition **3.4.2**. In what follows, we establish the equivalence of (a) and (b).

If $\mathcal{C} \equiv \mathcal{C}_\pi(x; K)$ is a linear subspace, then $\Pi_{\mathcal{C}}$ is easily shown to be a linear transformation. Thus $\Pi'_K(x; \cdot)$ is linear in the second argument. This is enough to establish the F-differentiability of Π_K at x. Conversely, if Π_K is F-differentiable at x, then for every vector d in $\mathcal{C}_\pi(x; K)$, we have

$$\Pi_{\mathcal{C}}(-d) = \Pi'_K(x; -d) = -\Pi'_K(x; d) = -\Pi_{\mathcal{C}}(d) = -d.$$

Thus $-d \in \mathcal{C}$. Hence $-\mathcal{C} \subseteq \mathcal{C}$. Since $\mathcal{C}$ is a polyhedral cone, it follows that $\mathcal{C}$ must be a linear subspace.

If x belongs to K then $\Pi_K(x) = x$, which implies $\mathcal{C}_\pi(x; K) = \mathcal{T}(x; K)$. Hence (b) becomes (b'). Finally, the equivalence of (b) and (d) follows from the definition of a nondegenerate solution. □

The proof of Theorem **4.1.1** reveals that Π_K has a piecewise affine structure. We formally define the class of piecewise affine functions in the following definition.

4.1.3 Definition. A continuous function $f : \mathbb{R}^n \to \mathbb{R}^m$ is said to be *piecewise affine (linear)* if there exists a finite family of affine (linear) functions $\{f^1, \cdots, f^k\}$ for some positive integer k, where each f^i maps $\mathbb{R}^n$ into $\mathbb{R}^m$, such that for all $x \in \mathbb{R}^n$, $f(x) \in \{f^1(x), \cdots, f^k(x)\}$. Each function f^i is called a *piece* of f. We use the acronym "PA (PL)" to mean "piecewise affine (linear)". If each $f^i(x) \equiv A^i x$ for some matrix $A \in \mathbb{R}^{m \times n}$, we also say that the family of matrices $\{A^1, \cdots, A^k\}$ are the linear pieces of the PL function f. □

The affine pieces of a PA function are clearly not unique, because we can add as many (spurious) affine functions as we like to an existing family of pieces. Subsequently, when we refer to a certain property of a PA function defined by its pieces, it is understood that this means the existence of a finite family of pieces, (the *effective pieces*), for which this property is valid.

There are many PA functions of interest, the projector Π_K, where K is polyhedral, being one; see Proposition **4.1.4** for a formal proof of this fact. More simply, if f and g are two affine functions defined on the same Euclidean space and with values in the same Euclidean space, then $\min(f, g)$ is a PA function. In general, the composition of two PA functions is PA. The following result provides the prime motivation for studying PA functions.

4.1.4 Proposition. Let K be a polyhedral set (cone) in $\mathbb{R}^n$. The Euclidean projector Π_K is PA (PL) on $\mathbb{R}^n$. Moreover, for any matrix M in $\mathbb{R}^{n\times n}$, both the natural map and the normal map of the pair (K, M):

$$x \mapsto x - \Pi_K(x - Mx) \quad \text{and} \quad z \mapsto M\Pi_K(z) + z - \Pi_K(z)$$

are PA maps.

Proof. For simplicity, we represent K by (4.1.4). Let $\mathcal{B}(A)$ be the family of matrices B whose rows are given by a subset of linearly independent rows of A. By the proof of Theorem **4.1.1**, for every $x \in \mathbb{R}^n$, there exists a matrix $B \in \mathcal{B}$ and a subset $\mathcal{J} \subseteq \mathcal{I}(\Pi_K(x))$ such that $B \equiv A_{\mathcal{J}\cdot}$ and

$$\Pi_K(x) = [\mathbf{I} - B^T(BB^T)^{-1}B]\,x + B^T(BB^T)^{-1}b_{\mathcal{J}}.$$

The right-hand side defines an affine function in x, one for each member of the finite family $\mathcal{B}$. Consequently, Π_K is a PA map. If K is a polyhedral cone, then $b = 0$ and hence Π_K is PL. For a matrix M and a polyhedron K, both the normal map and the natural map are the composites of a linear map with a PA map; thus they are piecewise affine. □

4.1.1 The normal manifold

The projector Π_K onto a polyhedron K induces a subdivision of $\mathbb{R}^n$ into a family of non-overlapping polyhedra; see Theorem **4.1.8**. The formal introduction of this subdivision and its geometric properties are the main topics of discussion in this subsection. Given an $m \times n$ matrix A and an m-vector b, let $P(A, b)$ denote the polyhedron (4.1.4); this notation was first introduced in the discussion preceding Corollary **3.2.5**. Define a collection of index sets:

$$\begin{aligned} \mathcal{I}(A,b) \equiv \{ \quad & I \subseteq \{1,\ldots,m\} \,:\, \exists\, x \in \mathbb{R}^n \text{ satisfying} \\ & A_{i\cdot}x = b_i \,\forall\, i \in I \text{ and } A_{i\cdot}x < b_i \,\forall\, i \notin I \}. \end{aligned} \tag{4.1.5}$$

Each element I of $\mathcal{I}(A, b)$ defines a nonempty face $\mathcal{F}_I$ of the polyhedron $P(A, b)$:

$$\mathcal{F}_I \equiv \{ x \in P(A,b) \,:\, A_{i\cdot}x = b_i \,\forall\, i \in I \}.$$

Conversely, every nonempty face of $P(A, b)$ gives rise to an index set in the collection $\mathcal{I}(A, b)$. Indeed, if the face is given by

$$\mathcal{F} \equiv \{ x \in P(A,b) \,:\, A_{i\cdot}x = b_i, \,\forall\, i \in J \}$$

for some index subset J of $\{1, \ldots, m\}$, it suffices to define I to be J union J', where J' is the set of indices corresponding to the "singular" inequalities that define the face $\mathcal{F}$; that is $j \in J'$ if $j \notin J$ and

$$x \in \mathcal{F} \Rightarrow A_{j \cdot} x = b_j.$$

For each index $i \in \{1, \ldots, m\} \setminus I$, there exists a vector $x^i \in \mathcal{F}$ such that $A_{i \cdot} x^i < b_i$; any convex combination of these vectors x^i is a vector $\bar{x}$ belonging to $\mathcal{F}$ that satisfies $A_{i \cdot} \bar{x} < b_i$ for all i in $\{1, \ldots, m\} \setminus I$. Thus I is a member of $\mathcal{I}(A, b)$ and $\mathcal{F} = \mathcal{F}_I$. Consequently,

$$\{ \mathcal{F}_I : I \in \mathcal{I}(A, b) \} \tag{4.1.6}$$

constitutes the collection of all nonempty faces of $P(A, b)$. It is easy to see that any two faces $\mathcal{F}_I$ and $\mathcal{F}_J$ corresponding to distinct index sets I and J in $\mathcal{I}(A, b)$ are distinct. Moreover, $I \cap J \in \mathcal{I}(A, b)$ if I and J both belong to $\mathcal{I}(A, b)$. The following result summarizes several important properties of each face in the collection (4.1.6); the result also identifies an alternative description of the piecewise affine structure of $\Pi_{P(A,b)}$.

4.1.5 Proposition. Let I be an arbitrary element of the family $\mathcal{I}(A, b)$.

(a) The relative interior of $\mathcal{F}_I$ is given by

$$\operatorname{ri} \mathcal{F}_I = \{ x \in \mathcal{F}_I : A_{i \cdot} x < b_i, \forall i \notin I \}.$$

(b) For each $x \in \operatorname{ri} \mathcal{F}_I$, the tangent cone of $P(A, b)$ at x, the normal cone of $P(A, b)$ at x, and the linear subspace spanned by $\mathcal{F}_I - x$ are all independent of $x \in \operatorname{ri} \mathcal{F}_I$ and depend only on the face $\mathcal{F}_I$; these sets are given, respectively, by

$$\begin{aligned} \mathcal{T}(x; P(A, b)) &= \{ v \in \mathbb{R}^n : A_{i \cdot} v \leq 0, \forall i \in I \}, \\ \mathcal{N}(x; P(A, b)) &= \operatorname{pos}\{ (A_{i \cdot})^T : i \in I \} \equiv \mathcal{N}_I, \\ \operatorname{lin}(\mathcal{F}_I - x) &= \{ v \in \mathbb{R}^n : A_{i \cdot} v = 0, \forall i \in I \}. \end{aligned}$$

(c) The set $\mathcal{F}_I + \mathcal{N}_I$ is full dimensional; that is, $\dim \operatorname{lin}(\mathcal{F}_I + \mathcal{N}_I) = n$.

(d) For each $z \in \mathcal{F}_I + \mathcal{N}_I$, $\Pi_{P(A,b)}(z) = \Pi_{S_I}(z)$, where

$$S_I \equiv \operatorname{aff} \mathcal{F}_I = \{ x \in \mathbb{R}^n : A_{i \cdot} x = b_i, \forall i \in I \}.$$

Thus $\Pi_{P(A,b)}$ is an affine map on $\mathcal{F}_I + \mathcal{N}_I$; in particular, $\Pi_{P(A,b)}$ is F-differentiable at every interior point of $\mathcal{F}_I + \mathcal{N}_I$.

(e) For each $x \in \mathcal{F}_I$,

$$\mathcal{N}_I = \mathcal{N}(x; P(A,b)) \cap \operatorname{aff} \mathcal{N}_I; \tag{4.1.7}$$

hence $\mathcal{N}_I$ is a face of $\mathcal{N}(x; P(A,b))$.

Proof. Part (a) is a well-known result for the relative interior of a polyhedron. The representation of $\operatorname{ri} \mathcal{F}_I$ from part (a) easily yields the displayed expression for the tangent cone $\mathcal{T}(x; P(A,b))$ for $x \in \operatorname{ri} \mathcal{F}_I$. This expression in turns yields the expression for the normal cone $\mathcal{N}(x; P(A,b))$. Clearly, we have

$$\operatorname{lin}(\mathcal{F}_I - x) \subseteq \{ v \in \mathbb{R}^n : A_{i\cdot} v = 0, \forall i \in I \}.$$

To show the reverse inclusion, let v be an arbitrary element of the right-hand set. For all $\tau > 0$ sufficiently small, the vector $x(\tau) \equiv x + \tau v$ belongs to $\mathcal{F}_I$. We have $v = \tau^{-1}(x(\tau) - x)$, which shows that v belongs to $\operatorname{lin}(\mathcal{F}_I - x)$. This establishes the desired expression for $\operatorname{lin}(\mathcal{F}_I - x)$.

To show that $\mathcal{F}_I + \mathcal{N}_I = x + (\mathcal{F}_I - x) + \mathcal{N}_I$ is full dimensional, it suffices to observe that the linear hull of $\mathcal{F}_I - x$ and the linear hull of $\mathcal{N}_I$ are orthogonal complements of each other, by part (b). Thus (c) holds.

Let $z \in \mathcal{N}_I + \operatorname{ri} \mathcal{F}_I$. There exist a vector $x \in \operatorname{ri} \mathcal{F}_I$ and nonnegative multipliers λ_i for all $i \in I$ such that

$$0 = x - z + \sum_{i \in I} \lambda_i (A_{i\cdot})^T.$$

Since x clearly belongs to $\mathcal{S}_I$, the above expression shows that $x = \Pi_{\mathcal{S}_I}(z)$. Furthermore, by the representation of $\operatorname{ri} \mathcal{F}_I$ from part (a), it follows that $x = \Pi_{P(A,b)}(z)$. More generally, if $z \in \mathcal{N}_I + \mathcal{F}_I$, then there exist $v \in \mathcal{N}_I$ and a sequence $\{x^k\} \subset \operatorname{ri} \mathcal{F}_I$ such that

$$z = v + \lim_{k \to \infty} x^k.$$

We have

$$\begin{aligned} \Pi_{P(A,b)}(z) &= \lim_{k \to \infty} \Pi_{P(A,b)}(v + x^k) \\ &= \lim_{k \to \infty} \Pi_{\mathcal{S}_I}(v + x^k) = \Pi_{\mathcal{S}_I}(z). \end{aligned}$$

This establishes (d).

Let $x \in \mathcal{F}_I$. We have

$$\mathcal{N}(x; P(A,b)) = \operatorname{pos}\{ (A_{i\cdot})^T : i \in \mathcal{I}(x) \}.$$

Therefore, since $I \subseteq \mathcal{I}(x)$ and

$$\mathcal{N}_I = \operatorname{pos}\{ (A_{i\cdot})^T : i \in I \},$$

it follows that

$$\mathcal{N}_I \subseteq \mathcal{N}(x; P(A,b)) \cap \operatorname{aff} \mathcal{N}_I.$$

Since $\mathcal{N}_I$ is a cone, we have

$$\operatorname{aff} \mathcal{N}_I = \operatorname{lin} \mathcal{N}_I = \operatorname{lin}\{ (A_{i\cdot})^T : i \in I \}.$$

Let $v \in \mathcal{N}(x; P(A,b)) \cap \operatorname{aff} \mathcal{N}_I$. There exist $\lambda_i \geq 0$ for $i \in \mathcal{I}(x)$ such that

$$v = \sum_{i \in \mathcal{I}(x)} \lambda_i (A_{i\cdot})^T.$$

There also exist λ_i' for $i \in I$ such that

$$v = \sum_{i \in I} \lambda_i' (A_{i\cdot})^T.$$

Let $\bar{x}$ be an arbitrary vector in $\operatorname{ri} \mathcal{F}_I$. Since $A_{i\cdot}\bar{x} = b_i = A_{i\cdot}x$ for all $i \in I$, we have

$$0 = v^T (\bar{x} - x).$$

On the other hand,

$$v^T (\bar{x} - x) = \sum_{i \in \mathcal{I}(x) \setminus I} \lambda_i A_{i\cdot} (\bar{x} - x),$$

and each summand in the right-hand sum is nonpositive; consequently, we must have $\lambda_i = 0$ for all $i \in \mathcal{I}(x) \setminus I$. Hence $v \in \mathcal{N}_I$. This establishes (4.1.7). Using a well-known fact from polyhedral theory that says a subset S of a polyhedral P is a face of P if and only if $S = P \cap \operatorname{aff} S$, (4.1.7) implies that $\mathcal{N}_I$ is a face of $\mathcal{N}(x; P(A,b))$. □

The collection of polyhedra

$$\{ \mathcal{F}_I + \mathcal{N}_I : I \in \mathcal{I}(A,b) \} \tag{4.1.8}$$

is called the *normal manifold* induced by the polyhedron $P(A,b)$. See Figure 4.2. Although we have defined this manifold using an algebraic representation of the polyhedron in terms of a system of linear inequalities, the normal manifold induced by a polyhedron P is independent of such a representation because every member $\mathcal{F}_I + \mathcal{N}_I$ is the sum of a nonempty face $\mathcal{F}_I$ of P and the normal cone $\mathcal{N}_I$ of P at a relative interior point of the face $\mathcal{F}_I$. Each such sum is a geometric set that is independent of the algebraic representation of the polyhedron.

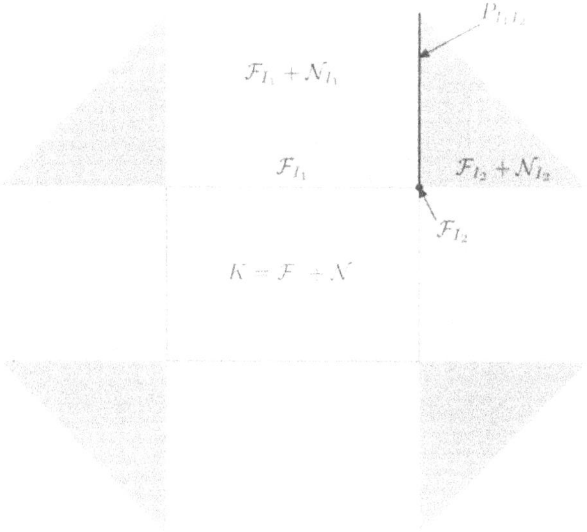

Figure 4.2: The normal manifold.

4.1.6 Example. We can easily illustrate the normal manifold induced by the nonnegative orthant of $\mathbb{R}^n$. For each subset I of $\{1, \ldots, n\}$, we have

$$\mathcal{F}_I = \{ x \in \mathbb{R}^n_+ : x_i = 0, \quad \forall i \in I \}$$

and

$$\mathcal{N}_I = \left\{ v \in \mathbb{R}^n : v_i \left\{ \begin{array}{ll} \leq 0 & \text{if } i \in I \\ = 0 & \text{if } i \notin I \end{array} \right\} \right\}.$$

Thus,

$$\mathcal{F}_I + \mathcal{N}_I \equiv \left\{ x \in \mathbb{R}^n : x_i \left\{ \begin{array}{ll} \leq 0 & \text{if } i \in I \\ \geq 0 & \text{if } i \notin I \end{array} \right\} \right\}.$$

Hence the normal manifold induced by $\mathbb{R}^n_+$ is simply the collection of 2^n orthants of $\mathbb{R}^n$. □

We illustrate the normal manifold with another example that pertains to a box constrained VI.

4.1.7 Example. As a contrast to Example **4.1.6**, we consider a compact rectangle K given by (1.1.7); that is,

$$K \equiv \{ x \in \mathbb{R}^n : a_i \leq x_i \leq b_i, i = 1, \ldots n \},$$

where for each i, $-\infty < a_i < b_i < \infty$. Omitting the derivations, we can deduce that the normal manifold induced by K is the collection of 3^n rectangles each of which is of the form

$$\prod_{i=1}^{n} S_i,$$

where each S_i is a (one-dimensional) interval of one of the following three types:

$$(-\infty, a_i\,], \quad [\,a_i, b_i\,], \quad \text{or} \quad [\,b_i, \infty\,). \tag{4.1.9}$$

It is easy to extend this example to the case where some of the bounds are infinite. □

Generalizing the above examples, we show below that the normal manifold induced by a polyhedron in $\mathbb{R}^n$ subdivides the entire space $\mathbb{R}^n$ into the family of polyhedra $\mathcal{F}_I + \mathcal{N}_I$, any two of which are either disjoint or intersect at a common face. We formally state and prove the latter statement in the following theorem. We continue to rely on the algebraic representation of the polyhedron $P(A, b)$ and present an algebraic proof of the theorem.

4.1.8 Theorem. Let $A \in \mathbb{R}^{m\times n}$ and $b \in \mathbb{R}^m$ be given. It holds that

$$\mathbb{R}^n = \bigcup_{I\in\mathcal{I}(A,b)} (\,\mathcal{F}_I + \mathcal{N}_I\,). \tag{4.1.10}$$

Moreover, if I and J are distinct index sets in $\mathcal{I}(A, b)$ such that

$$P_{IJ} \equiv (\,\mathcal{F}_I + \mathcal{N}_I\,) \cap (\,\mathcal{F}_J + \mathcal{N}_J\,) \neq \emptyset,$$

then

(a) $P_{IJ} = (\mathcal{F}_I \cap \mathcal{F}_J) + (\mathcal{N}_I + \mathcal{N}_J)$; and

(b) P_{IJ} is a common face of $\mathcal{F}_I + \mathcal{N}_I$ and $\mathcal{F}_J + \mathcal{N}_J$.

Proof. For any vector $z \in \mathbb{R}^n$, we have

$$z = \Pi_{P(A,b)}(z) + (\,z - \Pi_{P(A,b)}(z)\,).$$

There exists a unique face $\mathcal{F}_I$ of $P(A, b)$ containing the projected point $\Pi_{P(A,b)}(z)$ in its relative interior; namely, the one with $I \equiv \mathcal{I}(\Pi_{P(A,b)}(z))$. It is easy to see that $z - \Pi_{P(A,b)}(z)$ must belong to $\mathcal{N}_I$. Thus (4.1.10) holds.

To prove the two statements (a) and (b), write

$$P_I \equiv \mathcal{F}_I + \mathcal{N}_I, \quad \forall\, I \in \mathcal{I}(A, b).$$

Let I and J be two members of $\mathcal{I}(A,b)$ such that P_{IJ} is nonempty. Clearly, we have

$$(\mathcal{F}_I \cap \mathcal{F}_J) + (\mathcal{N}_I \cap \mathcal{N}_J) \subseteq P_{IJ}.$$

To prove the converse inclusion, let z be an arbitrary element of P_{IJ}. There exist vectors $u \in \mathcal{F}_I$, $u' \in \mathcal{F}_J$, $v \in \mathcal{N}_I$ and $v' \in \mathcal{N}_J$ such that

$$z = u + v = u' + v'.$$

By the proof of part (d) of Proposition **4.1.5**, we know that

$$u = \Pi_{P(A,b)}(z) = u'.$$

Thus $u = u' \in \mathcal{F}_I \cap \mathcal{F}_J$ and hence $v = v' \in \mathcal{N}_I \cap \mathcal{N}_J$. Consequently, (a) holds. To show (b), we show that P_{IJ} is a face of P_I; the proof for P_J is similar. Let z^1 and z^2 be two vectors in P_I such that for some scalar $\tau \in (0,1)$, $z \equiv \tau z^1 + (1-\tau)z^2$ belongs to P_{IJ}. Write $z^i \equiv u^i + v^i$ where $u^i \in \mathcal{F}_I$ and $v^i \in \mathcal{N}_I$. It follows that $u \equiv \tau u^1 + (1-\tau)u^2 \in \mathcal{F}_I$ and $v \equiv \tau v^1 + (1-\tau)v^2 \in \mathcal{N}_I$. Since $z = u + v \in P_{IJ}$, we have

$$u = \Pi_{P(A,b)}(z) \in \mathcal{F}_I \cap \mathcal{F}_J.$$

Thus $v = z - u \in \mathcal{N}_I \cap \mathcal{N}_J$. Since $\mathcal{F}_J$ is a face of $P(A,b)$, it follows that $u^i \in \mathcal{F}_J$ for $i = 1,2$. By part (e) of Proposition **4.1.5**, $\mathcal{N}_J$ is a face of $\mathcal{N}(u, P(A,b))$. Thus $v^i \in \mathcal{N}_J$ because the vector v belongs to $\mathcal{N}_J$ and both v^1 and v^2 are vectors in $\mathcal{N}_I \subset \mathcal{N}(u; P(A,b))$. Hence $z^i = u^i + v^i \in P_{IJ}$, establishing that P_{IJ} is a face of P_I. □

Associated with the polyhedron $K \equiv P(A,b)$, we compare the family of affine maps

$$\begin{aligned}\{\ & x \mapsto [\,\mathbf{I} - B^T(BB^T)^{-1}B\,]x + B^T(BB^T)^{-1}b_{\mathcal{J}} : \\ & \text{with } B = A_{\mathcal{J}\cdot} \text{ having linearly independent rows}\,\}\end{aligned} \tag{4.1.11}$$

with the family of affine maps derived from the normal manifold. Specifically, for each index set $I \in \mathcal{I}(A,b)$, let I^{LI} denote a subset of I such that the rows in the family

$$\{\, A_{i\cdot} : i \in I^{\text{LI}} \,\}$$

form a basis for the rows in $A_{I\cdot}$. Clearly, we have

$$\mathcal{S}_I = \operatorname{aff} \mathcal{F}_I = \{\, x \in \mathbb{R}^n : A_{i\cdot}x = b_i, \, \forall\, i \in I^{\text{LI}} \,\}.$$

Moreover, any such index set I^{LI} is a member of $\mathcal{B}(A)$. Thus the projection $\Pi_{\mathcal{S}_I}$ onto the affine hull of $\mathcal{F}_I$ is one of the affine maps in the family (4.1.11). By (4.1.10) and part (d) of Proposition **4.1.5**, the family

$$\{\, \Pi_{\mathcal{S}_I} : I \in \mathcal{I}(A,b) \,\} \tag{4.1.12}$$

also constitutes a family of affine pieces of the projector $\Pi_{P(A,b)}$. We call (4.1.12) the *normal pieces* of the PA projector Π_K. Interestingly, the normal pieces constitute the "smallest" family among all families of affine pieces of the Euclidean projector onto a polyhedron. We formally state this conclusion in the following proposition.

4.1.9 Proposition. If $\{f^1, \cdots, f^k\}$ is any family of affine pieces of $\Pi_{P(A,b)}$, then this family must contain the family (4.1.12); that is, for every I in $\mathcal{I}(A,b)$, there exists an index $i \in \{1,\ldots,k\}$ such that $\Pi_{S_I} = f^i$ on $\mathbb{R}^n$.

Proof. It suffices to show that for each $I \in \mathcal{I}(A,b)$, Π_{S_I} is equal to at least one member in the family $\{f^1, \cdots, f^k\}$. This follows from two simple facts: (i) Π_{S_I} coincides with Π_K on an open set $\mathcal{U}$ (namely, the interior of $\mathcal{F}_I + \mathcal{N}_I$), and (ii) for each $i = 1,\ldots,k$, the set $\{x : \Pi_K(x) \neq f^i(x)\}$ is open. In particular, the latter fact implies that if $\Pi_{S_I}(x) \neq f^i(x)$ for some $x \in \mathcal{U}$, then there exists a smaller open set $\mathcal{U}'$ contained in $\mathcal{U}$ such that $\Pi_K(x')$ is not equal to $f^i(x')$ for all $x' \in \mathcal{U}'$. Thus when restricted to $\mathcal{U}'$, the function f^i can be removed from the family of affine pieces of Π_K. By repeating this argument, we may deduce that Π_{S_I} must coincide with one function f^j on an open set. Since both of these functions are affine, their equality on an open set is enough to imply that they are equal on $\mathbb{R}^n$. □

4.2 Piecewise Affine Maps

If $f^i(x)$, $i = 1,\ldots,k$ is a finite family of affine functions each mapping $\mathbb{R}^n$ into $\mathbb{R}^m$ and if α_i is a subset of $\{1,\ldots,k\}$ for every $i = 1,\ldots,k$, then the function

$$f(x) \equiv \max_{1\leq i\leq k} \min_{j\in\alpha_i} f^j(x) \tag{4.2.1}$$

is piecewise affine. It turns out that the converse of this statement is valid in the case when $m = 1$. Specifically, every scalar-valued piecewise affine map has a max-min representation in terms of a finite family of affine functions $\{(a^i)^T x + b_i : i = 1,\ldots,k\}$ for some vectors a^i and scalars b_i. Instead of proving this representation, which we leave as an exercise for the reader, we present a "polyhedral representation" of a vector PA map. Let Ξ be a finite collection of polyhedra in $\mathbb{R}^n$. We say that Ξ is a *polyhedral subdivision* of $\mathbb{R}^n$ if

1. the union of all polyhedra in Ξ is equal to $\mathbb{R}^n$;

2. each polyhedron in Ξ is of dimension n; and

3. the intersection of any two polyhedra in Ξ is either empty or a common proper face of both polyhedra.

Roughly speaking, a polyhedral subdivision of $\mathbb{R}^n$ is a covering of $\mathbb{R}^n$ (point 1) by polyhedra with nonempty interiors (point 2) that are "glued" together by their faces; moreover, there is no duplication of these polyhedra (point 3). A simple polyhedral subdivision of $\mathbb{R}^n$ is the 2^n orthants. More generally, the normal manifold induced by a polyhedral set is also a polyhedral subdivision of $\mathbb{R}^n$, by Theorem **4.1.8** and part (c) of Proposition **4.1.5**. Among other things, the following result implies that every PA map from $\mathbb{R}^n$ into $\mathbb{R}^m$ induces a "natural" polyhedral subdivision of $\mathbb{R}^n$.

4.2.1 Proposition. A continuous function $f : \mathbb{R}^n \to \mathbb{R}^m$ is PA if and only if there exists a polyhedral subdivision Ξ of $\mathbb{R}^n$ and a finite family of affine functions $\{G^i\}$ such that f coincides with one of the functions G^i on each polyhedron in Ξ.

Proof. The sufficiency is obvious. To prove the necessity, let f be PA and let $\{f^1, \ldots, f^k\}$ be the affine pieces of f, where $f^i(x) \equiv A^i x + b^i$ for some matrix $A^i \in \mathbb{R}^{m\times n}$ and $b^i \in \mathbb{R}^m$. We may assume without loss of generality that for each $j = 1, \ldots, m$,

$$(A^i_{j\cdot}, b^i_j) \neq (A^{i'}_{j\cdot}, b^{i'}_j), \quad \forall\, i \neq i'.$$

For a tuple $\pi \equiv (\pi_1, \ldots, \pi_m)$ of permutations π_j of the numbers $\{1, \ldots, k\}$, define the set

$$P(\pi) \equiv \{ x \in \mathbb{R}^n : \\ A^{\pi_j(1)}_{j\cdot} x + b^{\pi_j(1)}_j \leq \ldots \leq A^{\pi_j(k)}_{j\cdot} x + b^{\pi_j(k)}_j ,\ 1 \leq j \leq m \}.$$

Each of these sets $P(\pi)$ is obviously a (possibly empty) polyhedron. Since each pair $(A^i_{j\cdot}, b^i_j)$ is distinct, it follows that

$$\operatorname{int} P(\pi) = \{ x \in \mathbb{R}^n : \\ A^{\pi_j(1)}_{j\cdot} x + b^{\pi_j(1)}_j < \ldots < A^{\pi_j(k)}_{j\cdot} x + b^{\pi_j(k)}_j ,\ j = 1, \ldots, m \}.$$

Thus for each set $P(\pi)$ with a nonempty interior, f coincides with a single affine function on $P(\pi)$.

We claim that the union of those sets $P(\pi)$ with a nonempty interior forms a polyhedral subdivision of $\mathbb{R}^n$. Obviously, the union of all sets $P(\pi)$ covers $\mathbb{R}^n$ and this covering property is not affected if we remove the polyhedra with empty interior. Since each set $P(\pi)$ with a nonempty interior must have dimension n, it remains to show the third property of a polyhedral subdivision; that is we need to show that if $P(\pi) \cap P(\pi')$ is nonempty for some $\pi \neq \pi'$ with $\operatorname{int} P(\pi)$ and $\operatorname{int} P(\pi')$ both nonempty, then the intersection $P(\pi) \cap P(\pi')$ is a proper face of both $P(\pi)$ and $P(\pi')$. For this purpose, we write $P(\pi)$ equivalently as:

$$P(\pi) \equiv \{ \; x \in \mathbb{R}^n \; : \; \sigma_{ii'j}(\pi) \, [\, (\, A^i_{j\cdot} - A^{i'}_{j\cdot} \,) x + (\, b^i_j - b^{i'}_j \,) \,] \leq 0, \\ i, i' = 1, \ldots, k, \; i \neq i', \; j = 1, \ldots, m \},$$

where

$$\sigma_{ii'j}(\pi) \equiv \begin{cases} 1 & \text{if } \pi_j^{-1}(i) < \pi_j^{-1}(i') \\ -1 & \text{if } \pi_j^{-1}(i) > \pi_j^{-1}(i'). \end{cases}$$

Hence $P(\pi) \cap P(\pi')$ is the subset of $P(\pi)$ obtained by turning inequalities into equalities whenever the relation $\sigma_{ii'j}(\pi) = -\sigma_{ii'j}(\pi')$ holds. Thus $P(\pi) \cap P(\pi')$ is a face of $P(\pi)$. Moreover, since $\pi \neq \pi'$, there exists at least one triple (i, i', j) satisfying $\sigma_{ii'j}(\pi) = -\sigma_{ii'j}(\pi')$ and since $\operatorname{int} P(\pi)$, which is nonempty, is the solution set of the strict inequality system, it follows that $P(\pi) \cap P(\pi')$ is a proper face of $P(\pi)$, and therefore also of $P(\pi')$ by interchanging the role of π and π'. □

Obviously, the affine functions G^i identified in the above proposition also constitute a family of affine pieces of f. These pieces can be used to establish some useful properties of f that extend some solution properties of an AVI. We first present the former properties and then discuss how they are related to the latter properties.

4.2.2 Proposition. Let $f : \mathbb{R}^n \to \mathbb{R}^m$ be a PA map. The following three properties hold:

(a) the range of f is the union of finitely many polyhedra and is thus a closed set;

(b) for every $q \in \mathbb{R}^n$, $f^{-1}(q)$ is the union of finitely many polyhedra and is thus a closed set; and

(c) f is globally Lipschitz continuous.

Proof. Let Ξ be a polyhedral subdivision of $\mathbb{R}^n$ induced by f and let $\{G^i\}$ be a finite family of affine functions such that f coincides with one of these functions on each polyhedron in Ξ. For each P in Ξ, let G_P be one of the affine members in $\{G^i\}$ that coincide with f on P. We then have

$$f(\mathbb{R}^n) = \bigcup_{P \in \Xi} G_P(P).$$

Since the image of a polyhedron under an affine transformation remains polyhedral, the above representation readily establishes part (a). Part (b) follows from a similar expression: for every $q \in \mathbb{R}^n$,

$$f^{-1}(q) = \bigcup_{P \in \Xi} [(G_P)^{-1}(q) \cap P].$$

Since G_P is affine, the set $(G_P)^{-1}(q)$ is an affine subspace; thus the intersection $(G_P)^{-1}(q) \cap P$ is polyhedral. Hence (b) holds.

To show part (c), let $\{f^1, \cdots, f^k\}$ be any finite family of affine pieces of f. For each $i = 1, \cdots, k$, write

$$f^i(x) \equiv A^i x - b^i, \quad \forall x \in \mathbb{R}^n$$

for some $m \times n$ matrix A^i and m-vector b^i. For each $x \in \mathbb{R}^n$, we let $\mathcal{P}(x)$ denote the set of indices $i \in \{1, \ldots, k\}$ such that $f(x) = f^i(x)$. It is easy to see that for each $x \in \mathbb{R}^n$, there exists an open neighborhood $\mathcal{N}(x)$ such that for every $y \in \mathcal{N}(x)$, $\mathcal{P}(y) \subseteq \mathcal{P}(x)$. To establish the globally Lipschitz continuity of f, let x and y be two arbitrary vectors in $\mathbb{R}^n$. Consider the line segment joining them:

$$[x, y] \equiv \{x + \tau(y - x) : \tau \in [0, 1]\}.$$

The family of open neighborhoods $\mathcal{N}(z)$, for all $z \in [x, y]$ is an open covering of this compact line segment. Thus there exits a finite family of these neighborhoods that covers $[x, y]$. This implies that there exists a partition of the interval $[0, 1]$,

$$0 = \tau_0 < \tau_1 < \cdots < \tau_p = 1$$

for some positive integer p, such that all $r = 0, \ldots p - 1$,

$$\mathcal{P}(x^r) \cap \mathcal{P}(x^{r+1}) \neq \emptyset,$$

where $x^r \equiv x + \tau_r(y - x)$. For each r, let i_r be a common element in the sets $\mathcal{P}(x^r)$ and $\mathcal{P}(x^{r+1})$. We have

$$\begin{aligned} f(x) - f(y) &= \sum_{r=0}^{p-1} (f(x^r) - f(x^{r+1})) = \sum_{r=0}^{p-1} A^{i_r} (x^r - x^{r+1}) \\ &= \sum_{r=0}^{p-1} A^{i_r} (\tau_r - \tau_{r+1})(y - x). \end{aligned}$$

Thus, with

$$L \equiv \max_{1 \leq i \leq k} \| A^i \|,$$

we have

$$\| f(x) - f(y) \| \leq L \| x - y \|,$$

establishing the globally Lipschitz continuity of f. □

Let (K, M) be an affine pair in $\mathbb{R}^n$; i.e., K is a polyhedron and M is an $n \times n$ matrix. The normal map $\mathbf{M}_K^{\text{nor}}$ is a PA map, by Proposition **4.1.4**. Thus $\mathbf{M}_K^{\text{nor}}(\mathbb{R}^n)$ is the union of finitely many polyhedra. It is easy to see that $\mathbf{M}_K^{\text{nor}}(\mathbb{R}^n)$ is equal to the negative of the AVI-range of the pair (K, M); that is

$$\mathbf{M}_K^{\text{nor}}(\mathbb{R}^n) = -\mathcal{R}(K, M).$$

Hence we recover Theorem **2.5.15** (b) from Proposition **4.2.2** (a). Similarly, we also recover part (a) of the former theorem from part (b) of the latter proposition.

4.2.1 Coherent orientation

A linear map from $\mathbb{R}^n$ into itself is a global homeomorphism if and only if it is represented by a nonsingular matrix. Central to the global homeomorphism property of a PA map from $\mathbb{R}^n$ into itself is the property defined below.

4.2.3 Definition. A PA map $f : \mathbb{R}^n \to \mathbb{R}^n$ is said to be *coherently oriented* if there exists a finite family of affine pieces $\{f^1, \cdots, f^k\}$ of f such that $\det Jf^i$ has the same nonzero sign for all i. □

Clearly, the composition of two coherently oriented PA maps is coherently oriented. Coherently oriented PA maps have many special properties. For one thing, there are several equivalent conditions for a PA map to be coherently oriented. The above definition is algebraic. An equivalent geometric characterization of coherent orientation is given by Proposition **4.2.5**; while an equivalent topological condition is given by Proposition **4.2.15**. To prepare for the geometric description, we prove a technical lemma below.

4.2.4 Lemma. Let P_1 and P_2 be two n-dimensional polyhedra intersecting in a common $(n-1)$-dimensional face. Let A^1 and A^2 be two nonsingular $n \times n$ matrices, and b^1 and b^2 be two n-vectors such that

$$A^1 x + b^1 = A^2 x + b^2, \quad \forall x \in P_1 \cap P_2. \tag{4.2.2}$$

The two polyhedra $A^1(P^1) + a^1$ and $A^2(P^2) + b^2$ intersect in a common $(n-1)$-dimensional face if and only if $\det A^1 \det A^2 > 0$.

Proof. It is fairly obvious that we may assume without loss of generality that the intersection $P_1 \cap P_2$ contains the origin, A^2 is the identity matrix, and $b^1 = b^2 = 0$. (Simply take a common point x^0 in P^1 and P^2, define $A \equiv (A^2)^{-1}A^1$, and consider the two polyhedra $A(P^1 - x^0)$ and $P^2 - x^0$.) Under this simplification, equation (4.2.2) becomes

$$A^1 x = x, \quad \forall x \in P_1 \cap P_2.$$

This implies in particular that $P^1 \cap P^2 \subseteq A^1(P^1) \cap P^2$, which implies in turn that $\dim(A^1(P^1) \cap P^2) \geq n - 1$, while $\dim(A^1(P^1) \cap P^2) = n - 1$ if and only if $A^1(P^1)$ and P^2 can be separated by a hyperplane.

Since $P^1 \cap P^2$ is contained in an $(n-1)$-dimensional linear subspace, we may further assume that this subspace is $\{x \in \mathbb{R}^n : x_n = 0\}$, which we denote E. Indeed E must be the linear hull of $P_1 \cap P_2$. Consequently, we have

$$A^1 x = x, \quad \forall x \in E.$$

It follows that A^1 must be an upper triangular matrix of the form

$$A^1 \equiv \begin{bmatrix} \mathbf{I}_{n-1} & a \\ 0 & s \end{bmatrix},$$

where $a \in \mathbb{R}^{n-1}$ and s is a certain nonzero scalar (because A^1 is nonsingular). Since $P^1 \cap P^2$ is a common face of P^1 and P^2, and P^1 and P^2 are full dimensional, it follows that P^1 and P^2 must be contained in opposite sides of the hyperplane E. Let's assume that

$$P^1 \subseteq E_+ \equiv \{x \in \mathbb{R}^n : x_n \geq 0\}$$
$$P^2 \subseteq E_- \equiv \{x \in \mathbb{R}^n : x_n \leq 0\}.$$

We need to show that $A^1(P^1) \cap P^2$ is a common $(n-1)$-dimensional face of $A^1(P^1)$ and P^2 if and only if $s > 0$. Suppose $s > 0$. It then follows that $A^1(P^1)$ is contained in E_+. It follows that E is a separating hyperplane of $A^1(P^1)$ and P^2. Hence $A^1(P^1) \cap P^2$ has dimension $n - 1$. We next show that $A^1(P^1) \cap P^2$ is a common face of $A^1(P^1)$ and P^2. We do this only for P^2. For this, it suffices to show that if $x \in A^1(P^1) \cap P^2$ is such that $x = \tau u + (1 - \tau)v$ for some u and v in P^2 and $\tau \in [0, 1]$, then both u and v must be in $A^1(P^1)$. Since $A^1(P^1)$ is contained in E_+ and P^2 is contained in E_-, it follows that x, u and v all belong to E. Thus $u = A^1 u$ and $v = A^1 v$. Let $y \in P^1$ be such that $x = A^1 y$. Since A^1 is nonsingular, we have $y = \tau u + (1 - \tau)v$. Hence $y \in P^2$ because both u and v are in P^2. Thus y belongs to $P^1 \cap P^2$. Since this intersection is a face of P^2, it follows that u and v both belong to P^1; thus u and v both belong to $A^1(P^1)$.

Conversely, suppose that $A^1(P^1) \cap P^2$ is a common $(n-1)$-dimensional face of $A^1(P^1)$ and P^2. We need to show that $s > 0$. By the above argument, it follows that E is a separating hyperplane of $A^1(P^1)$ and P^2. This implies that $A^1(P^1)$ is contained in E_+. Hence the scalar s must be positive because there must be at least one element in P^1 with $x_n > 0$. □

With the above lemma, we can prove the following equivalent condition of coherent orientation, which gives a geometric description of this concept.

4.2.5 Proposition. A PA map $f : \mathbb{R}^n \to \mathbb{R}^n$ is coherently oriented if and only if there exists a polyhedral subdivision Ξ of $\mathbb{R}^n$ with the following properties:

(a) f coincides with an affine mapping on each polyhedron $P \in \Xi$;

(b) for every $P \in \Xi$, $f(P)$ is of dimension n;

(c) for any two polyhedra P^1 and P^2 in Ξ, $f(P^1) \cap f(P^2)$ is a common $(n-1)$-dimensional face of $f(P^1)$ and $f(P^2)$.

Proof. The "only if" part is an immediate consequence of Lemma **4.2.4** and Definition **4.2.3**. To establish the "if" part, let conditions (a), (b), and (c) hold. Let $\{f^1, \cdots, f^k\}$ be the affine pieces of f. Clearly, Jf^i is nonsingular for all i by condition (b). Let Ξ_+ and Ξ_- be the subcollection of polyhedra in Ξ such that the determinants of the matrices corresponding to the polyhedra in Ξ_+ (Ξ_-) are positive (negative, respectively). It remains to show that either Ξ_+ or Ξ_- is empty. Assume for contradiction that both subcollections are nonempty. The two unions

$$\bigcup_{P_+ \in \Xi_+} P_+ \quad \text{and} \quad \bigcup_{P_- \in \Xi_-} P_-$$

have the same boundary, which is a subset of the union of the $(n-1)$-dimensional faces of the polyhedra in Ξ. If $\mathcal{F}$ is an $(n-1)$-dimensional boundary face of Ξ_+ and Ξ_-, then the subdivision property of Ξ implies that there are two polyhedra $P^1 \in \Xi_+$ and $P^2 \in \Xi_-$ whose intersection is $\mathcal{F}$. By property (c), $f(P^1)$ and $f(P^2)$ is a common $(n-1)$-dimensional face of $f(P^1)$ and $f(P^2)$. By Lemma **4.2.4**, the matrices of the corresponding affine functions have the same determinantal sign. This is a contradiction. □

We next state and prove two important properties of a PA map.

4.2.6 Proposition. Let $f : \mathbb{R}^n \to \mathbb{R}^n$ be a coherently oriented PA map. The following properties hold.

(a) f is norm-coercive; that is

$$\lim_{\|x\| \to \infty} \| f(x) \| = \infty.$$

(b) For each $y \in \mathbb{R}^n$, $f^{-1}(y)$ is a finite set.

Proof. Suppose for the sake of contradiction that f is not norm-coercive as stated. There exist a constant $c > 0$ and a sequence $\{x^\nu\}$ such that

$$\lim_{\nu \to \infty} \| x^\nu \| = \infty \quad \text{and} \quad \| f(x^\nu) \| \leq c, \quad \forall \nu.$$

For each ν, $f(x^\nu)$ is equal to $f^i(x^\nu)$ for some i in $\{1, \ldots, k\}$. Since there are only finitely many affine pieces of f, it follows that there exist an index i and an infinite index set $\kappa \in \{1, 2, \cdots\}$ such that $f(x^\nu) = f^i(x^\nu)$ for all $\nu \in \kappa$. Since f^i is an affine function with a constant nonsingular Jacobian matrix, and $\{\|x^\nu\| : \nu \in \kappa\}$ tends to infinity, it follows that $\{\|f(x^\nu)\| : \nu \in \kappa\}$ also tends to infinity. This is a contradiction. Consequently, (a) holds.

To prove (b), note that the equation $f(x) = y$ implies one of the k equations

$$f^i(x) = y, \quad i = 1, \ldots, k.$$

Since each f^i is an affine map with a nonsingular Jacobian matrix, each of the above equations has either zero or a unique solution. Thus $f^{-1}(y)$ is finite. □

Using degree theory, we show subsequently that every coherently oriented PA map must be surjective; cf. Proposition **4.2.12**. In what follows, we illustrate the coherent orientation concept with an important PA map.

When f is the normal map $\mathbf{M}_K^{\text{nor}}$ or the natural map $\mathbf{M}_K^{\text{nat}}$ of the affine pair (K, M), where K is a polyhedron in $\mathbb{R}^n$ and M is an $n \times n$ matrix, the coherent orientation of f can be characterized with the aid of the normal manifold induced by K. Specifically, for a matrix $A \in \mathbb{R}^{m \times n}$ and vector $b \in \mathbb{R}^m$, recall the normal pieces (4.1.12) of the projector Π_K and the associated family $\{I^{\text{LI}} : I \in \mathcal{I}(A, b)\}$ of index sets such that the rows in the family

$$\{ A_{i\cdot} : i \in I^{\text{LI}} \}$$

form a basis for the rows of $A_{I\cdot}$. Let $\mathcal{B}_{\text{bas}}(A, b)$ be the family of matrices:

$$\mathcal{B}_{\text{bas}}(A, b) \equiv \{ A_{I_{\text{LI}}\cdot} : I \in \mathcal{I}(A, b) \}.$$

In general, there are multiple bases for the rows of $A_{I\cdot}$. For each I in $\mathcal{I}(A, b)$, we pick one such basis. Thus $|\mathcal{B}_{\text{bas}}(A, b)| = |\mathcal{I}(A, b)|$. We call $\mathcal{B}_{\text{bas}}(A, b)$ a *normal family of basis matrices* of the polyhedron $P(A, b)$.

4.2.7 Proposition. Let $K \equiv P(A,b)$ be a polyhedron in $\mathbb{R}^n$ and M be a matrix in $\mathbb{R}^{n\times n}$. Let $\mathcal{B}_{\text{bas}}(A,b)$ be any normal family of basis matrices of $P(A,b)$. The following statements are equivalent.

(a) The normal map $\mathbf{M}_K^{\text{nor}}$ is coherently oriented.

(b) All matrices of the form:

$$M\,[\,\mathbf{I} - B^T(\,BB^T\,)^{-1}B\,] + B^T(\,BB^T\,)^{-1}B, \tag{4.2.3}$$

where $B \in \mathcal{B}_{\text{bas}}(A,b)$, have the same nonzero determinantal sign.

(c) All matrices of the form:

$$\begin{bmatrix} M & B^T \\ -B & 0 \end{bmatrix}, \tag{4.2.4}$$

where $B \in \mathcal{B}_{\text{bas}}(A,b)$, have the same nonzero determinantal sign.

(d) All matrices of the form:

$$[\,\mathbf{I} - B^T(\,BB^T\,)^{-1}B\,]M + B^T(\,BB^T\,)^{-1}B, \tag{4.2.5}$$

where $B \in \mathcal{B}_{\text{bas}}(A,b)$, have the same nonzero determinantal sign.

(e) The natural map $\mathbf{M}_K^{\text{nat}}$ is coherently oriented.

Proof. We prove that (a) $\Leftrightarrow$ (b) $\Leftrightarrow$ (c). The proof of (c) $\Leftrightarrow$ (d) $\Leftrightarrow$ (e) is similar. We have

$$\mathbf{M}_K^{\text{nor}}(z) \equiv M\Pi_K(z) + z - \Pi_K(z).$$

The affine pieces of $\mathbf{M}_K^{\text{nor}}$ induced by the normal pieces of the projector Π_K are given by

$$(\,M - \mathbf{I}\,)\,\{\,[\,\mathbf{I} - B^T(\,BB^T\,)^{-1}B\,]\,z + B^T(\,BB^T\,)^{-1}b_{\mathcal{J}}\,\} + z,$$

for $B \in \mathcal{B}_{\text{bas}}(A,b)$. Since this family of affine pieces is the smallest, it follows that $\mathbf{M}_K^{\text{nor}}$ is coherently oriented if and only if the Jacobian matrices of these affine pieces have the same nonvanishing determinantal sign. Consequently, (a) $\Leftrightarrow$ (b).

We claim that for any matrix B with full row rank, the determinant of the matrix (4.2.3) has the same sign as the determinant of the matrix (4.2.4). First of all, the former matrix is the Schur complement of BB^T in the matrix

$$\mathbf{Q} \equiv \begin{bmatrix} M\,[\,\mathbf{I} - B^T(\,BB^T\,)^{-1}B\,] & B^T \\ -B & BB^T \end{bmatrix}.$$

Therefore, by the Schur determinantal formula, the determinant of $\mathbf{Q}$ is the product of $\det BB^T$ and the determinant of (4.2.3). Since BB^T is a positive definite matrix, it has a positive determinant. Therefore, the sign of $\det \mathbf{Q}$ is the same as the sign of the determinant of (4.2.3). It is easy to verify

$$\mathbf{Q} = \begin{bmatrix} \mathbf{I} & MB^T(BB^T)^{-1} \\ 0 & \mathbf{I} \end{bmatrix} \begin{bmatrix} M & B^T \\ -B & 0 \end{bmatrix} \begin{bmatrix} \mathbf{I} & -B^T \\ 0 & \mathbf{I} \end{bmatrix},$$

from which it follows that $\det \mathbf{Q}$ is equal to the determinant of (4.2.4). □

Proposition **4.2.7** provides several equivalent matrix-theoretic descriptions of the coherent orientation of the normal or natural map associated with an affine pair (K, M). We illustrate these properties in the simple case where K is the nonnegative orthant.

4.2.8 Example. The nonnegative orthant $\mathbb{R}^n_+$ is equal to $P(-\mathbf{I}_n, 0)$. The normal map $\mathbf{M}^{\text{nor}}_{\mathbb{R}^n_+}$ and the natural map $\mathbf{M}^{\text{nat}}_{\mathbb{R}^n_+}$ are the familiar maps:

$$\mathbf{M}^{\text{nor}}_{\mathbb{R}^n_+}(z) \equiv Mz^+ - z^- \quad \text{and} \quad \mathbf{M}^{\text{nat}}_{\mathbb{R}^n_+}(x) \equiv \min(x, Mx).$$

Members of the family $\mathcal{B}_{\text{bas}}(-\mathbf{I}_n, , 0)$ are matrices whose rows are given by a subset of the rows of the negative identity matrix of order n. Thus if $B \equiv -\mathbf{I}_{J\cdot}$, it is easy to see that the columns of the matrix (4.2.3) are equal to $M_{\cdot i}$ if $i \notin J$ and $\mathbf{I}_{\cdot i}$ if $i \in J$. Therefore all these matrices have the same nonvanishing determinantal signs if and only if all principal submatrices of M have the same determinantal sign as the identity matrix (which corresponds to J being the entire set $\{1, \cdots, n\}$); in turn, the latter is equivalent to M being a P matrix. Similarly, we can show, by a direct argument as above, that $\mathbf{M}^{\text{nat}}_{\mathbb{R}^n_+}$ is coherent oriented if and only if M is a P matrix. □

We further illustrate the coherent orientation concept using the normal map of a box constrained VI. For simplicity, we consider only the case where the box is bounded.

4.2.9 Example. Let K be the compact rectangle in $\mathbb{R}^n$:

$$K \equiv \{ x \in \mathbb{R}^n : a_i \leq x_i \leq b_i, \, i = 1, \ldots n \},$$

where for each i, $-\infty < a_i < b_i < \infty$. From Example **4.1.7**, we know that the normal manifold induced by K consists of 3^n rectangles each of which is equal to the Cartesian product of 1-dimensional intervals of the form (4.1.9). Based on this structure of the normal manifold, it is easy to deduce that for any given matrix $M \in \mathbb{R}^{n \times n}$, the family of matrices (4.2.3) coincides with the family of matrices in the previous Example **4.2.8**. Consequently, it follows that $\mathbf{M}_K^{\rm nor}$ (also $\mathbf{M}_K^{\rm nat}$) is coherent oriented if and only if M is a P-matrix. □

The two PL maps $\mathbf{M}_{\mathbb{R}^n_+}^{\rm nor}$ and $\mathbf{M}_{\mathbb{R}^n_+}^{\rm nat}$ correspond to LCPs defined by the matrix M. More generally, we have the following result that pertains to PL maps of MLCPs.

4.2.10 Proposition. Let

$$M \equiv \begin{bmatrix} A & B \\ C & D \end{bmatrix}$$

be a square matrix with A and D both square. The PL map:

$$\Phi : \begin{pmatrix} u \\ v \end{pmatrix} \mapsto \begin{pmatrix} Au + Bv \\ \min(\, v, Cu + Dv\,) \end{pmatrix}$$

is coherently oriented if and only if A is nonsingular and the Schur complement $M/A \equiv D - CA^{-1}B$ is a P matrix.

Proof. This proposition is actually a consequence of Theorem **4.3.2** below. Here, we give an elementary matrix-theoretic proof. Let A and D be of order n_1 and n_2, respectively. The members of the smallest family of linear pieces of Φ are given by

$$\begin{pmatrix} u \\ v \end{pmatrix} \mapsto M(I) \begin{pmatrix} u \\ v \end{pmatrix}$$

where

$$M(I) \equiv \begin{bmatrix} A & B \\ C_{I\cdot} & D_{I\cdot} \\ \mathbf{0}_{J\cdot} & \mathbf{I}_{J\cdot} \end{bmatrix}$$

for arbitrary index subsets I of $\{1, \cdots, n_2\}$ with J being the complement of I. Suppose Φ is coherently oriented. Then

$$M(\emptyset) == \begin{bmatrix} A & B \\ 0 & \mathbf{I}_{n_2} \end{bmatrix}$$

is nonsingular; this implies that A is nonsingular. For an arbitrary index subset I of $\{1, \cdots, n_2\}$ with complement J, the Schur determinantal formula and an easy manipulation yield

$$\det M(I) \, = \, (\det A\,) \det(\, D_{II} - C_{I\cdot}A^{-1}B_{\cdot I}\,).$$

Since $\det M(I)$ must have the same nonzero sign as $\det A$, it follows that

$$D_{II} - C_{I\cdot}A^{-1}B_{\cdot I}$$

has a positive determinant. The latter displayed matrix is the principal submatrix of the Schur complement M/A indexed by I; since I is an arbitrary subset of $\{1, \cdots, n_2\}$, we conclude that M/A is a P matrix. The converse statement can be similarly proved by reversing the argument. □

We can extend Proposition **4.2.10** to the normal map $\mathbf{M}_K^{\text{nor}}$ or the natural map $\mathbf{M}_K^{\text{nat}}$ corresponding to a box constrained AVI defined on a bounded or unbounded rectangle K. The conclusion is very similar to that of the proposition; namely, the map $\mathbf{M}_K^{\text{nor}}$ (or $\mathbf{M}_K^{\text{nat}}$) is coherently oriented if and only if the principal submatrix of M corresponding to those variables that have no finite upper or lower bounds is nonsingular and the Schur complement of this principal submatrix in M is a P matrix.

It is known that P matrices are closely tied to the globally unique solvability of LCPs; more precisely, a matrix M is P if and only if the LCP (q, M) has a unique solution for all vectors q. In terms of the two PL maps $\mathbf{M}_{\mathbb{R}^n_+}^{\text{nor}}$ and $\mathbf{M}_{\mathbb{R}^n_+}^{\text{nat}}$, this result says that the LCP (q, M) has a unique solution for all q if and only if either one (or both) of these maps is (are) coherently oriented; cf. Example **4.2.8**. Inspired by this simple case, we are led to think that, in general, coherent orientation should have an important role to play in certain global properties of a PA map. This consideration is particularly important when the PA map in question is the normal (or natural) map of an affine pair (K, M) where K is a polyhedron and M is a square matrix. Before discussing this special case, we establish the following result, which summarizes some further properties of a (general) PA function and their interconnections. Among other things, the theorem below shows that PA maps inherit many fundamental properties of affine maps. The proof of this theorem makes use of Rademacher's Theorem **3.1.1** and the global homeomorphism Theorem **2.1.10**.

4.2.11 Theorem. Let $f : \mathbb{R}^n \to \mathbb{R}^n$ be a PA map. The following statements are equivalent:

(a) f is a homeomorphism on $\mathbb{R}^n$;

(b) f is a Lipschitz homeomorphism on $\mathbb{R}^n$;

(c) f is injective;

(d) f is bijective;

(e) f is coherently oriented and $\text{ind}(f, x)$ is well defined and equal to the same constant for all $x \in \mathbb{R}^n$; moreover, this constant is ± 1.

If any one of these statements (a)–(e) holds, then f^{-1} is a PA map and has the same properties as f.

Proof. We use the same notation as in the proof of Proposition **4.2.2**(b). Specifically, let $\{f^1, \cdots, f^k\}$ be a finite family of affine pieces of f. For each $i = 1, \cdots, k$, write

$$f^i(x) \equiv A^i x - b^i, \quad \forall x \in \mathbb{R}^n$$

for some $n \times n$ matrix A^i and n-vector b^i. For each $x \in \mathbb{R}^n$, we let $\mathcal{P}(x)$ denote the set of indices $i \in \{1, \ldots, k\}$ such that $f(x) = f^i(x)$. It is easy to see that for each $x \in \mathbb{R}^n$, there exists an open neighborhood $\mathcal{N}(x)$ such that for every $y \in \mathcal{N}(x)$, $\mathcal{P}(y) \subseteq \mathcal{P}(x)$.

(a) $\Rightarrow$ (b). If f is a homeomorphism on $\mathbb{R}^n$, then f^{-1} exists. It suffices that to show that f^{-1} is Lipschitz continuous. Let $x \in \mathbb{R}^n$ be an arbitrary vector. We claim that there exists an index $i \in \mathcal{P}(x)$ such that f^i is a homeomorphism. Let $\mathcal{N}(x)$ be a neighborhood of x such that $\mathcal{P}(y) \subseteq \mathcal{P}(x)$ for all $y \in \mathcal{N}(x)$. Suppose that the claim fails to hold. By the definition of the affine pieces, there exists an index i in $\mathcal{P}(x)$ such that A^i is singular. Let d be a nonzero vector such that $A^i d = 0$. Since f is a homeomorphism, thus injective, it follows that for all $\tau > 0$ sufficiently small, $i \in \mathcal{P}(x) \setminus \mathcal{P}(x + \tau d)$. Let $x' \equiv x + \tau d$ for one such τ. Then $\mathcal{P}(x')$ is a proper subset of $\mathcal{P}(x)$. Repeat the argument but apply it to $\mathcal{P}(x')$. Either one of two statements is valid: every index $i' \in \mathcal{P}(x')$ is such that $f^{i'}$ is a homeomorphism; or there exists a vector x'' such that

$$\mathcal{P}(x'') \subset \mathcal{P}(x') \subset \mathcal{P}(x).$$

By continuing this argument, since there are only finitely many indices in $\mathcal{P}(x)$, we obtain either a vector x^* with $\mathcal{P}(x^*)$ being a singleton contained in $\mathcal{P}(x)$ and that single piece of f at x^* must be a homeomorphism or an element $i \in \mathcal{P}(x)$ for which f^i is a nonsingular affine map. In either case, our claim is proved.

For every $x \in \mathcal{P}(x)$, let $\mathcal{P}'(x)$ be the largest subset of $\mathcal{P}(x)$ such that f^i is a homeomorphism for every $i \in \mathcal{P}'(x)$. Let $\mathcal{I}$ be the union of these index sets $\mathcal{P}'(x)$ over all $x \in \mathbb{R}^n$. The family $\{f^i : i \in \mathcal{I}\}$ defines an alternative family of affine pieces of f and has the additional property that each of its members is a homeomorphism. Clearly, the family $\{(f^i)^{-1} : i \in \mathcal{I}\}$ defines a family of affine pieces for f^{-1}. Therefore, f^{-1} is PA too. By property (A), f^{-1} is globally Lipschitz continuous. This completes the proof that (a) implies (b). Moreover, we have established at the same time that if (a) holds, then f^{-1} is a PA homeomorphism. With the equivalence of the five statements (a)–(e), whose proof will be completed below, the last assertion of the theorem about f^{-1} therefore holds.

(b) $\Rightarrow$ (c). This is obvious.

(c) $\Rightarrow$ (d). Let f be an injective PA map. Let u be an arbitrary vector in $\mathbb{R}^n$; also let $v = f(x)$ for some x and v. Consider the line segment $[v, u]$. For the sake of contradiction, assume that u does not belong to the range of f. We claim that there exists a vector $u' \in [v, u)$ (this notation means the segment $[v, u]$ without the end-point u) such that the entire subsegment $[u', u]$ lies outside the range of f. Indeed, if no such vector u' exists, then there exists a sequence of vectors $\{u^k\}$ converging to u such that each u^k belongs to the range of f. Since the latter range is closed by property (B), we obtain a contradiction. Thus, a vector u' with the desired property exists. Define the scalar τ^* to be the supremum of all scalars $\tau \in (0, 1)$ such that

$$[\, u(\tau), u\,] \cap \operatorname{ran} f \;=\; \emptyset,$$

where $u(\tau) = v + \tau(u - v)$. Thus, $u(\tau^*)$ must belong to the range of f; because if does not, then the above argument can be applied to $u(\tau^*)$, resulting in a contradiction to the definition of τ^*. We now use the injectivity of f to obtain our final contradiction. Let x^* be such that $f(x^*) = u(\tau^*)$. Since f is injective, x^* is unique. Moreover, with $\tilde{f} \equiv f - u(\tau^*)$, it follows that $\operatorname{ind}(\tilde{f}, x^*)$ is well defined and equal to ± 1; furthermore, for every open neighborhood $\mathcal{N}$ of x^* and every continuous function g satisfying

$$\sup_{y \in \operatorname{cl} \mathcal{N}} \| \tilde{f}(y) - g(y) \| \;<\; \operatorname{dist}_\infty(0, \tilde{f}(\operatorname{bd} \mathcal{N})),$$

g has a zero in $\mathcal{N}$. In particular, with $g(y) \equiv f(y) - \tilde{u}$, where $\tilde{u}$ satisfies

$$\| u^* - \tilde{u} \| \;<\; \operatorname{dist}_\infty(0, \tilde{f}(\partial \mathcal{N})),$$

we deduce that such a vector $\tilde{u}$ must belong to the range of f. This is the desired contradiction. Consequently, the range of f is equal to the entire $\mathbb{R}^n$; thus, f is surjective and (d) follows.

(d) $\Rightarrow$ (a). This is obvious because f is continuous to begin with.

(a) $\Rightarrow$ (e). Let f be a homeomorphism, thus an injective PA map. By the proof of (a) $\Rightarrow$ (b), we may assume without loss of generality that there exists a finite family of affine homeomorphisms

$$\mathcal{F} \equiv \{\, f^1, \ldots, f^k \,\}$$

such that $f(x) \in \{f^1(x), \ldots, f^k(x)\}$ for all $x \in \mathbb{R}^n$. The proof that (a) implies (e) will be complete if we can show the existence of a subfamily of $\mathcal{F}$ such that for all f^i belonging to this subfamily, Jf^i has the same nonzero determinantal sign. We use several facts.

(i) Rademacher's Theorem **3.1.1** implies that f is F-differentiable almost everywhere; therefore for every $x \in \mathbb{R}^n$, there exists a sequence $\{x^k\}$ of F-differentiable points of f converging to x.

(ii) Since f is injective, $\text{ind}(f, x)$ is well defined and equal to ± 1 for all $x \in \mathbb{R}^n$; moreover, if x is a F-differentiable point of f, then $\text{ind}(f, x)$ is equal to the sign of $\det Jf(x)$. See Proposition **2.1.4** and the discussion following it.

(iii) For every $x \in \mathbb{R}^n$, there exists a neighborhood $\mathcal{N}$ of x such that for all $y \in \mathcal{N}$, $\mathcal{P}(y) \subseteq \mathcal{P}(x)$ and $\text{ind}(f, y) = \text{ind}(f, x)$. To see why this is true, we note that $\text{ind}(f, x)$ is equal to the degree of f at $f(x)$ relative to any open neighborhood $\mathcal{N}$ of x. Fix an open neighborhood $\mathcal{N}'$ of x for which $\mathcal{P}(y) \subseteq \mathcal{P}(x)$ for all $y \in \mathcal{N}'$. By the continuity of f, there exists a neighborhood $\mathcal{N} \subseteq \mathcal{N}'$ such that

$$\sup_{y \in \mathcal{N}} \| f(y) - f(x) \| < \text{dist}_\infty(0, f_x(\partial \mathcal{N}')),$$

where f_x is the function f translated by $f(x)$; that is, $f_x(v) \equiv f(v) - f(x)$. Thus for $y \in \mathcal{N}$, we have

$$\sup_{v \in \text{cl}\, \mathcal{N}'} \| f_x(v) - f_y(v) \| < \text{dist}_\infty(0, f_x(\partial \mathcal{N}')).$$

Thus it follows that for all $y \in \mathcal{N}$,

$$\text{ind}(f, y) = \deg(f_y, \mathcal{N}') = \deg(f_x, \mathcal{N}') = \text{ind}(f, x).$$

(iv) If x is a F-differentiable point of f, then there exists $i \in \mathcal{P}(x)$ such that $Jf(x) = A^i$.

Combining facts (ii) and (iv), we deduce that for every F-differentiable point x of f, there exists a nonempty subset $\mathcal{P}'(x)$ of $\mathcal{P}(x)$ such that the determinantal signs of all matrices A^i with $i \in \mathcal{P}'(x)$ are the same and equal to $\text{ind}(f, x)$. This statement remains valid at a nondifferentiable point x by facts (i) and (iii) and also because there is only a finite number of such subfamilies. For each $x \in \mathbb{R}^n$, let $\mathcal{P}'(x)$ be the largest (in terms of inclusion) subset of $\mathcal{P}(x)$ that has this property. It remains to show that $\text{ind}(f, x)$ is a constant independent of x. We have shown above that for every $x \in \mathbb{R}^n$, there exists an open neighborhood $\mathcal{N}(x)$ such that for that $\text{ind}(f, y)$ is a constant for all $y \in \mathcal{N}(x)$. If x' is another vector distinct from x, then the closed line segment $[x, x']$ is covered by a finite number of such open neighborhoods, say

$$[x, x'] \subseteq \bigcup_{i=0}^{N} \mathcal{N}(x^i),$$

for some positive integer N, where $x^0 = x$ and x^1, x^2, ..., $x^N = x'$ are consecutive points on this closed line segment. Since $\operatorname{ind}(f,\cdot)$ is a constant in each neighborhood $\mathcal{N}(x^i)$, it follows that $\operatorname{ind}(f,x) = \operatorname{ind}(f,x')$ as desired. This completes the proof that (a) implies (e).

(e) $\Rightarrow$ (a). Suppose that f is coherently oriented and $\operatorname{ind}(f,x)$ is the same constant for all $x \in \mathbb{R}^n$ with this constant being ± 1. The coherent orientation of f implies that f is norm coercive. Thus, by Theorem **2.1.10**, it suffices to show that f is everywhere a local homeomorphism; in turn, it suffices to show that f is locally injective; that is, for every $x \in \mathbb{R}^n$, there exists an open neighborhood $\mathcal{N}$ of x such that f is one-to-one in $\mathcal{N}$. We note that since f is PA, for every $y \in \mathbb{R}^n$, $f^{-1}(y)$ is a finite set by the coherent orientation of f. Let $x \in \mathbb{R}^n$ be an arbitrary vector in $\mathbb{R}^n$ and let $y = f(x)$. Since $f^{-1}(y)$ is finite, there exists an open neighborhood $\mathcal{N}$ of x such that $y \notin f(\operatorname{bd}\mathcal{N})$; thus the degree of f at y relative to $\mathcal{N}$ is well defined. Moreover, similar to the proof of fact (iii) above, for all vectors x' sufficiently near x, the degree of f at $f(x')$ relative to $\mathcal{N}$ is well defined and equal to a constant. In particular, with x' being a F-differentiable point of f (such a point must exist by fact (i)), it follows that

$$\deg(f, \mathcal{N}, y) = \operatorname{sgn}\det Jf(x') = \pm 1,$$

because $Jf(x')$ must equal to the Jacobian matrix of one of the pieces of f. Moreover, we have

$$\deg(f, \mathcal{N}, y) = \sum_{x' \in f^{-1}(y) \cap \mathcal{N}} \operatorname{ind}(f, x').$$

Since $\operatorname{ind}(f,x')$ is a constant, it follows that $f^{-1}(y) \cap \mathcal{N}$ is a singleton. $\square$

The above proof reveals a surjectivity property of a coherently oriented PA map, which we formally state in the following result.

4.2.12 Proposition. Let $f : \mathbb{R}^n \to \mathbb{R}^n$ be a PA map. If f is coherently oriented, then f is surjective.

Proof. For every $y \in f(\mathbb{R}^n)$, $f^{-1}(y)$ is a finite set. Therefore, for every open set Ω containing $f^{-1}(y)$, we have by Proposition **2.1.6**,

$$\deg(f, \Omega, y) = \sum_{x \in f^{-1}(y)} \operatorname{ind}(f, x). \qquad (4.2.6)$$

In turn, by fact (iii) in the proof of Theorem **4.2.11**, it follows that whenever $\operatorname{ind}(f,x)$ is well defined, then $\operatorname{ind}(f,x')$ is well defined and equal to $\operatorname{ind}(f,x)$ for all x' sufficiently close to x. By Rademacher's Theorem, there are F-differentiable points of f that are sufficiently close to x and for all such points x', $\operatorname{ind}(f,x')$ is equal to $\operatorname{sgn}\det Jf(x')$. Since f is coherently oriented, the latter sign is equal to a constant (± 1). Applying this conclusion to the expression (4.2.6), we deduce that $\deg(f,\Omega,y)$ is a nonzero integer for all $y \in f(\mathbb{R}^n)$.

Let $y' \in \mathbb{R}^n$ be arbitrary. Fix an arbitrary $y \in f(\mathbb{R}^n)$. Consider the homotopy

$$H(x,t) \equiv f(x) - (ty + (1-t)y'), \quad \forall (x,t) \in \mathbb{R}^n \times [0,1].$$

Since f is coherently oriented, it is norm coercive. Hence, the union

$$\bigcup_{t\in[0,1]} H^{-1}(\cdot,t)$$

is bounded and thus contained in a bounded open set Ω. By the homotopy invariance property of the degree, we have

$$\deg(f,\Omega,y') = \deg(H(\cdot,0),\Omega) = \deg(H(\cdot,1),\Omega) = \deg(f,\Omega,y)$$

and the latter degree is a nonzero integer. Consequently, $f(x) = y'$ has a solution x in Ω. This establishes the surjectivity of f. □

As illustrated by the following example, not every coherently oriented PA map is injective. This shows that the index condition in part (e) of Theorem **4.2.11** can not be removed.

4.2.13 Example. Consider the vectors in $\mathbb{R}^2$:

$$x^1 \equiv \begin{pmatrix} 1 \\ 0 \end{pmatrix}, \quad x^2 \equiv \begin{pmatrix} 1 \\ 1 \end{pmatrix}, \quad x^3 \equiv \begin{pmatrix} 1 \\ 2 \end{pmatrix},$$

$$x^4 \equiv \begin{pmatrix} 1 \\ 3 \end{pmatrix}, \quad x^5 \equiv \begin{pmatrix} 1 \\ 4 \end{pmatrix}, \quad x^6 \equiv \begin{pmatrix} 0 \\ 1 \end{pmatrix}$$

and

$$y^1 \equiv \begin{pmatrix} 1 \\ 0 \end{pmatrix}, \quad y^2 \equiv \begin{pmatrix} 0 \\ 1 \end{pmatrix}, \quad y^3 \equiv \begin{pmatrix} -1 \\ 0 \end{pmatrix},$$

$$y^4 \equiv \begin{pmatrix} 0 \\ -1 \end{pmatrix}, \quad y^5 \equiv \begin{pmatrix} 1 \\ 0 \end{pmatrix}, \quad y^6 \equiv \begin{pmatrix} 0 \\ 1 \end{pmatrix}.$$

Define the PL function $f : \mathbb{R}^2 \to \mathbb{R}^2$, which coincides with the linear mapping that carries x^i onto y^i on the cone $\text{pos}(x^i, x^{i+1})$, for $i = 1, \ldots, 5$, and which is the identity outside the union of these 5 cones. The linear pieces of f are defined by the following six matrices:

$$A^1 \equiv \begin{pmatrix} 1 & -1 \\ 0 & 1 \end{pmatrix} \quad A^2 \equiv \begin{pmatrix} 1 & -1 \\ 2 & -1 \end{pmatrix} \quad A^3 \equiv \begin{pmatrix} -3 & 1 \\ 2 & -1 \end{pmatrix}$$

$$A^4 \equiv \begin{pmatrix} -3 & 1 \\ -4 & 1 \end{pmatrix} \quad A^5 \equiv \begin{pmatrix} 1 & 0 \\ -4 & 1 \end{pmatrix} \quad A^6 \equiv \begin{pmatrix} 1 & 0 \\ 0 & 1 \end{pmatrix}.$$

It is trivial to verify that all these matrices have determinants equal to one. Thus f is coherently oriented. Yet f is not injective because every nonzero vector in $\mathbb{R}^2$ has exactly two preimages. □

Proposition **4.2.12** can be used to establish yet another important characterization of a coherent oriented PA map, which is topological in nature. Namely, such a map must be open. This result complements the open mapping Theorem **2.1.11** because according to the latter theorem, every continuous, injective map must be open; yet as we see from the above example, a coherent oriented PA map is not necessarily injective. Before stating the result, we recall Theorem **4.1.1**, which presents a special differentiability property of the Euclidean projector onto a polyhedral set, the latter being a particular PA map. Extending this proposition, we remark that in general a PA map $f : \mathbb{R}^n \to \mathbb{R}^m$ must be everywhere B-differentiable; moreover, for every vector $x \in \mathbb{R}^n$, there exists a neighborhood $\mathcal{N}$ of x such that

$$f(y) = f(x) + f'(x; y - x), \quad \forall\, y \in \mathcal{N}. \tag{4.2.7}$$

A direct way to establish the directional differentiability and the above formula is to invoke the max-min representation of the (scalar-valued) component functions of f and to extend the simple argument in Example **3.1.4** to a multivariate setting. See Exercise **4.8.10** for details. As a by-product of this max-min representation, we can deduce that the B-derivative $f'(x; \cdot)$ of a coherently oriented PA map must be a coherently oriented PL map because the pieces of the B-derivative are a subset of the pieces of f. (The B-differentiability of a PA map follows immediately from Lemma **4.6.1** that pertains to the more general class of PC^1 maps. Nevertheless, the formula (4.2.7) is particular to a PA map.)

It is convenient to introduce a definition. We say that a set-valued map $\Phi : \mathbb{R}^n \to \mathbb{R}^m$ is lower semicontinuous at a pair $(u, v) \in \text{gph}\, \Phi$ if for every sequence $\{u^k\}$ converging to u, there exists a sequence of vectors $\{v^k\}$ converging to v such that $v^k \in \Phi(u^k)$ for all k. This definition is a refinement of part (c) of Definition **2.1.16** that defines the lower semicontinuity of a set-valued map at a point in its domain. Both concepts imply in particular that $\Phi(u')$ is nonempty-valued for all u' sufficiently close to u.

4.2.14 Lemma. Let $f : \mathcal{D} \to \mathbb{R}^m$ be a given function defined on the open set $\mathcal{D}$. The following statements are equivalent.

(a) f maps open subsets of $\mathcal{D}$ onto open subsets of $\mathbb{R}^m$.

(b) For every vector $x^\infty \in \mathcal{D}$ and every sequence $\{y^k\} \subset \mathbb{R}^m$ converging to $y^\infty = f(x^\infty)$, there exists a sequence $\{x^k\} \subset \mathcal{D}$ converging to x^∞ such that $f(x^k) = y^k$ for all k sufficiently large.

(c) f^{-1} is lower semicontinuous at every pair (y, x), where $x \in \mathcal{D}$ and $y = f(x)$.

Proof. The reader is asked to supply the proof in Exercise **4.8.11**. □

With the above preparations, we formally state and prove the openness characterization of the coherent orientation of a PA map. For a further equivalent condition of coherent orientation, see Theorem **6.3.19**.

4.2.15 Proposition. A PA map $f : \mathbb{R}^n \to \mathbb{R}^n$ is coherently oriented if and only if it maps open sets in $\mathbb{R}^n$ onto open sets in $\mathbb{R}^n$.

Proof. Suppose that f is coherently oriented. To show that f is open, let $\{y^k\}$ be a sequence converging to $y^\infty \equiv f(x^\infty)$. By (4.2.7) and part (b) in Lemma **4.2.14**, it suffices to find a sequence $\{x^k\}$ converging to x^∞ such that, for all k sufficiently large, $f'(x^\infty; x^k - x^\infty) = y^k - y^\infty$. In turn, it suffices to show that for a coherently oriented PL function from $\mathbb{R}^n$ into itself, the origin in the image space is an interior point of the image of any neighborhood of the origin in the domain space. To establish the latter property of a coherently oriented PL map, let f be such a map and let $\{y^k\}$ be any sequence converging to zero. By Proposition **4.2.12**, there exists x^k such that $f(x^k) = y^k$. It suffices to show that $\{x^k\}$ converges to zero. But this is easy because if $\{A^1, \cdots, A^m\}$ are the linear pieces of f, where each A^i is a nonsingular matrix, then $x^k \in \{(A^1)^{-1}(y^k), \cdots, (A^m)^{-1}(y^k)\}$, which clearly implies that $\{x^k\}$ converges to zero.

Conversely, suppose that f is open. Let Ξ be a polyhedral subdivision of $\mathbb{R}^n$ corresponding to f and such that each polyhedron in Ξ has a nonempty interior. Since f is open, $f(\operatorname{int} P)$ is an open set for every $P \in \Xi$. Hence $f(P)$ is of dimension n. It remains to show condition (c) in

Proposition **4.2.5**. Let P^1 and P^2 be two polyhedra in the subdivision Ξ intersecting in a common $(n-1)$-dimensional face $\mathcal{F}$ and let x be a relative interior point of F. Since the active affine pieces on P^1 and P^2 coincide on $\mathcal{F}$, the set $f(\mathcal{F})$ is a common face of $f(P^1)$ and $f(P^2)$. Thus $f(P^1)$ and $f(P^2)$ must be contained in one of the halfspaces induced by the affine hull of $f(\mathcal{F})$. If the intersection of $f(P^1)$ and $f(P^2)$ contains a point $y \notin f(\mathcal{F})$, then $f(P^1)$ and $f(P^2)$ lie on the same halfspace induced by the latter hull; in this case, for every sufficiently small open neighborhood U of x, $f(x)$ must be a boundary point of the image $f(U)$. This contradicts the openness of f. Hence $f(P^1)$ and $f(P^2)$ must be contained in opposite sides of the affine hull of $f(\mathcal{F})$, establishing that $f(P^1)$ and $f(P^2)$ intersect in the common $(n-1)$-dimensional face $f(\mathcal{F})$. □

4.3 Unique Solvability of AVIs

When specialized to the normal map $\mathbf{M}_K^{\text{nor}}$ or the natural map $\mathbf{M}_K^{\text{nat}}$ associated with a given affine pair (K, M), Theorem **4.2.11** results in various necessary and sufficient conditions for the AVI (K, q, M) to have a unique solution for all vectors q. These conditions generalize the well-known connection between a P matrix and the globally unique solvability of LCPs. Before presenting the grand result of this section, Theorem **4.3.2**, we state a special property of the normal map of an affine pair due to Robinson; see Section **4.9** for the source of the result.

4.3.1 Theorem. Let K be a polyhedral subset of $\mathbb{R}^n$ and M an $n \times n$ matrix. If the normal map $\mathbf{M}_K^{\text{nor}}$ is coherently oriented, then it is injective.

Sketch of the proof. The first important observation is that we can limit ourselves to studying the injectivity of $\mathbf{M}_K^{\text{nor}}$ when K is a polyhedral cone. In fact, $\mathbf{M}_K^{\text{nor}}$ is coercive and therefore, by Theorem **2.1.10**, it is sufficient to show that $\mathbf{M}_K^{\text{nor}}$ is locally one-to-one. However, by Theorem **4.1.1**, it is easy to see that for every $x \in \mathbb{R}^n$ there exists a neighborhood $\mathcal{N}$ of x such that

$$\mathbf{M}_K^{\text{nor}}(y) = \mathbf{M}_K^{\text{nor}}(x) + \mathbf{M}_C^{\text{nor}}(y - x), \quad \forall\, y \in \mathcal{N},$$

where $\mathcal{C} \equiv \mathcal{C}_\pi(x; K)$. From this fact it follows that $\mathbf{M}_K^{\text{nor}}$ is one-to-one near x if and only if $\mathbf{M}_C^{\text{nor}}$ is one-to-one near zero. Furthermore, the determinants of the pieces of $\mathbf{M}_C^{\text{nor}}$ are the determinants of the pieces of $\mathbf{M}_K^{\text{nor}}$ near x. So it is sufficient to prove that every coherently oriented normal map $\mathbf{M}_K^{\text{nor}}$ is one-to-one when K is a polyhedral cone. In order to analyze the latter case, we can proceed by induction on the dimension n of the space. For $n = 1$

the result is trivial. The case $n = 2$ can be studied in the following way. Observe that in this case the normal map has at most four pieces. Based on this observation, it is then easy to show that there is at least one point in the interior of one of the polyhedra of the normal manifold, which has exactly one preimage. By this fact and employing some degree-theoretic reasonings it is possible to show in a relatively direct way that $\mathbf{M}_K^{\text{nor}}$ is one-to-one. The general case $n \geq 3$ is much more technical and difficult. It involves a judicious use of the induction hypothesis, some nontrivial nonlinear analysis results about homeomorphisms and some conceptually simple but not straightforward arguments in analyzing the structure of not pointed polyhedral cones. We omit the details. □

We are ready to state and prove a summary of the unique solvability of the AVI (K, q, M) for all vectors q corresponding to a fixed affine pair (K, M), i.e., the GUS property of such a pair.

4.3.2 Theorem. Let K be a polyhedral subset of $\mathbb{R}^n$ and M an $n \times n$ matrix. Let f be either the normal map $\mathbf{M}_K^{\text{nor}}$ or natural map $\mathbf{M}_K^{\text{nat}}$ of the pair (K, M). The following statements are equivalent:

(a) f is bijective;

(b) for all vectors $q \in \mathbb{R}^n$, the AVI (K, q, M) has a unique solution;

(c) f is coherently oriented;

(d) for every q^∞ in the AVI range $\mathcal{R}(K, M)$ and every sequence $\{q^k\}$ converging to q^∞, q^k belongs to $\mathcal{R}(K, M)$ for all k sufficiently large;

(e) any one of the five statements (a)–(e) in Theorem **4.2.11** holds for f.

If any one of the above statements holds, then the unique solution of the AVI (K, q, M) is a PA, and thus B-differentiable, function of q.

Proof. Since both the normal map $\mathbf{M}_K^{\text{nor}}$ and the natural map $\mathbf{M}_K^{\text{nat}}$ are PA maps, either map is bijective if and only if any one of the five statements (a)–(e) in Theorem **4.2.11** is valid.

We first consider the case when f is the normal map $\mathbf{M}_K^{\text{nor}}$. By Theorems **4.2.11**, statement (a) implies (c). Conversely, by Theorem **4.3.1** and Proposition **4.2.12**, statement (c) implies (a). The equivalence of (b) and (a) follows from Proposition **1.5.11**. By Proposition **4.2.15**, the coherent orientation of $\mathbf{M}_K^{\text{nor}}$ is equivalent to the openness of $\mathbf{M}_K^{\text{nor}}$. By Lemma **4.2.14**, the openness of $\mathbf{M}_K^{\text{nor}}$ is equivalent to the following property: [for every vector $z^\infty \in \mathbb{R}^n$ and for every sequence $\{y^k\}$ converging to $y^\infty = \mathbf{M}_K^{\text{nor}}(z^\infty)$, there exists a sequence $\{z^k\}$ converging to z^∞ such that $\mathbf{M}_K^{\text{nor}}(z^k) = y^k$ for all k sufficiently large]. In turn the latter property is easily seen to be equivalent to statement (d).

The above completes the proof for the normal map. We next consider the natural map. By Theorem **4.2.11**, statement (a) implies (c) for the natural map $\mathbf{M}_K^{\text{nat}}$. By Proposition **4.2.7**, the natural map $\mathbf{M}_K^{\text{nat}}$ is coherently oriented if and only if the normal map $\mathbf{M}_K^{\text{nor}}$ is coherently oriented. Thus if (c) holds for $\mathbf{M}_K^{\text{nat}}$, then (b) holds. To complete the proof for the natural map, it remains to show that if (b) holds, then $\mathbf{M}_K^{\text{nat}}$ is injective. But this follows easily from Exercise **1.8.23**.

By Theorem **4.2.11**, if $\mathbf{M}_K^{\text{nor}}$ is bijective, then $(\mathbf{M}_K^{\text{nor}})^{-1}$ is PA. It is easy to see that if $x(q)$ denotes the unique solution of the AVI (K, q, M), then $x(q) = \Pi_K \circ (\mathbf{M}_K^{\text{nor}})^{-1}(-q)$. Since Π_K and $(\mathbf{M}_K^{\text{nor}})^{-1}$ are both PA, it follows that so is the solution function $x(\cdot)$. □

For a polyhedron K, the coherent orientation of the normal map $\mathbf{M}_K^{\text{nor}}$ and the natural map $\mathbf{M}_K^{\text{nat}}$ is characterized by the matrix-theoretic conditions given in Proposition **4.2.7**. These conditions can be added to the equivalences in Theorem **4.3.2** to yield further necessary and sufficient conditions for the affine pair (K, M) to have the GUS property. One such condition states that with $K = P(A, b)$, all matrices of the form (4.2.4):

$$\begin{bmatrix} M & B^T \\ -B & 0 \end{bmatrix},$$

where $B \in \mathcal{B}_{\text{bas}}(A, b)$, have the same nonzero determinantal sign. By this theorem, it follows that if the AVI (K, q, M) has a unique solution for all vectors $q \in \mathbb{R}^n$, then the unique solution is a globally Lipschitz continuous function of q. This conclusion raises the question of whether the converse is true. More specifically, we say that the pair (K, M) is *Lipschitzian* if there exists a constant $L > 0$ such that for all vectors q and q' belonging to the range of the pair (K, M), (i.e., $\text{SOL}(K, q, M)$ and $\text{SOL}(K, q', M)$ are both nonempty),

$$\text{SOL}(K, q', M) \subseteq \text{SOL}(K, q, M) + L \, \| \, q - q' \, \| \, \text{cl}\,\mathbb{B}(0, 1).$$

The question is: if K is polyhedral and (K, M) is a Lipschitzian pair, does it follow that $\text{SOL}(K, q, M)$ is a singleton for all q? In the context of the LCP (which has $K = \mathbb{R}^n_+$), we say that M is a *Lipschitzian matrix* if $(\mathbb{R}^n_+, M)$ is a Lipschitzian pair. Since M is a P matrix if and only if the LCP (q, M) has a unique solution for all q, the above question becomes: is a Lipschitzian matrix necessarily P? A complete answer to this question is not yet available. By Exercise **3.7.34**, it follows that a P matrix must necessarily be Lipschitzian. In general, the Lipschitzian property of the pair (K, M) is equivalent to a kind of "global error bound" with a common multiplicative constant for the solution sets of the VIs (K, q, M) for all vectors q. In the broader context of a PA map, such a global error bound along with the surjectivity of the map provides a necessary and sufficient condition for the map to be coherently oriented. See Subsection **6.3.2**, particularly Theorem **6.3.18**, for further discussion on these issues.

4.3.1 Inverse of $\mathbf{M}_K^{\rm nor}$

For a symmetric positive definite matrix M and a closed convex set K (not necessarily polyhedral), Proposition **1.5.11** implies that the normal map $\mathbf{M}_K^{\rm nor}$ is a global homeomorphism on $\mathbb{R}^n$. Moreover, $\mathbf{M}_K^{\rm nor}$ is clearly Lipschitz continuous. The inverse $(\mathbf{M}_K^{\rm nor})^{-1}$ is certainly a global homeomorphism. It turns out that $(\mathbf{M}_K^{\rm nor})^{-1}$ has a precise representation, which shows that it is also Lipschitz continuous. Thus $\mathbf{M}_K^{\rm nor}$ and $\mathbf{M}_K^{\rm nor}$ are both globally Lipschitz homeomorphisms. A formal statement of the latter assertion is given in Proposition **4.3.3**.

To derive the desired expression for $(\mathbf{M}_K^{\rm nor})^{-1}$, we recall the skewed projector $\Pi_{K,A}$ defined by a symmetric positive definite matrix A and a closed convex set K; see (1.5.9). Specifically, for every vector $x \in \mathbb{R}^n$, $\Pi_{K,A}(x)$ is the unique vector y that solves the convex program:

$$\begin{aligned} &\text{minimize} \quad \tfrac{1}{2}(\,y - x\,)^T A(\,y - x\,) \\ &\text{subject to} \quad y \in K. \end{aligned}$$

Associated with the pair (K, A), let us define the operator

$$\Pi_K^A \equiv \Pi_{K,A} \circ A^{-1}.$$

Since the skewed projector $\Pi_{K,A}$ is globally Lipschitz continuous on $\mathbb{R}^n$ (cf. the discussion preceding Subsection **1.5.3**), it follows that so is the operator Π_K^A. Moreover, it is easy to see that

$$\Pi_K^A(Ad) = d \quad \forall\, d \in K. \tag{4.3.1}$$

For every $x \in \mathbb{R}^n$, $\Pi_K^A(x)$ is the unique vector y that solves the convex program:

$$\begin{aligned} &\text{minimize} \quad \tfrac{1}{2}\,y^T A y - y^T x \\ &\text{subject to} \quad y \in K. \end{aligned}$$

By the variational inequality associated with the latter optimization problem, we deduce that $\Pi_K^A(x)$ is the unique vector in K that satisfies

$$(\,y - \Pi_K^A(x)\,)^T(\,A\Pi_K^A(x) - x\,) \geq 0, \quad \forall\, y \in K.$$

Let $B \equiv \sqrt{A}$ be the square root of the matrix A; that is B is the unique symmetric positive definite matrix such that $B^2 = A$. The above inequality is equivalent to

$$(By - B\Pi_K^A(x))^T (B\Pi_K^A(x) - B^{-1}x) \geq 0, \quad \forall y \in K;$$

which in turn is equivalent to

$$B\Pi_K^A(x) = \Pi_{BK}(B^{-1}x).$$

Since this holds for all $x \in \mathbb{R}^n$, we deduce that

$$\Pi_K^A = B^{-1} \circ \Pi_{BK} \circ B^{-1}.$$

Since the projector Π_{BK} is a co-coercive operator, by part (c) of Theorem **1.5.5**, it follows that

$$\begin{aligned}
&(x - y)^T (\Pi_K^A(x) - \Pi_K^A(y)) \\
&= (B^{-1}x - B^{-1}y)^T (\Pi_{BK}(B^{-1}x) - \Pi_{BK}(B^{-1}y)) \\
&\geq \| \Pi_{BK}(B^{-1}x) - \Pi_{BK}(B^{-1}y) \|_2^2.
\end{aligned}$$

We also have

$$\begin{aligned}
&\| \Pi_K^A(x) - \Pi_K^A(y) \|_2^2 \\
&= (\Pi_{BK}(B^{-1}x) - \Pi_{BK}(B^{-1}y))^T A^{-1} (\Pi_{BK}(B^{-1}x) - \Pi_{BK}(B^{-1}y)) \\
&\leq \frac{1}{\lambda_{\min}(A)} \| \| \Pi_{BK}(B^{-1}x) - \Pi_{BK}(B^{-1}y) \|_2^2,
\end{aligned}$$

because by the symmetry and positive definiteness of A, the largest eigenvalue of A^{-1} is equal to the inverse of smallest eigenvalue of A. Combining the above two inequalities, we deduce

$$(x - y)^T (\Pi_K^A(x) - \Pi_K^A(y)) \geq \lambda_{\min}(A) \| \Pi_K^A(x) - \Pi_K^A(y) \|_2^2, \qquad (4.3.2)$$

which shows that Π_K^A is a co-coercive, thus monotone, operator. If K is a closed convex cone, then $\Pi_K^A(x)$ is the unique vector $\hat{x}$ satisfying the complementarity system:

$$K \ni \hat{x} \perp A\hat{x} - x \in K^*.$$

In the next result, K is not assumed to be polyhedral but M is required to be symmetric; hence Π_K^M is well defined.

4.3.3 Proposition. Let M be a symmetric positive definite matrix of order n and K be a closed convex set in $\mathbb{R}^n$. The normal map $\mathbf{M}_K^{\text{nor}}$ of the pair (K, M) is a globally Lipschitz homeomorphism from $\mathbb{R}^n$ onto itself with inverse given by the mapping

$$H(d) \equiv (\mathbf{I} - M)\Pi_K^M(d) + d, \quad d \in \mathbb{R}^n.$$

Moreover, $\Pi_K \circ H = \Pi_K^M$ and $\Pi_K^M \circ \mathbf{M}_K^{\text{nor}} = \Pi_K$.

Proof. We have noted that $\mathbf{M}_K^{\text{nor}}$ maps $\mathbb{R}^n$ bijectively onto itself. We claim that $H \circ \mathbf{M}_K^{\text{nor}}$ is equal to the identity map. This is enough to show that H is the inverse of $\mathbf{M}_K^{\text{nor}}$. Consider $H \circ \mathbf{M}_K^{\text{nor}}(y)$. Let $d \equiv \mathbf{M}_K^{\text{nor}}(y)$. We claim that $\Pi_K^M(d) = \Pi_K(y)$. By the variational characterization of $\Pi_K^M(d)$, it suffices to show that for every vector $u \in K$,

$$(u - \Pi_K(y))^T(M\Pi_K(y) - d) \geq 0.$$

By the definition of d, we have $M\Pi_K(y) - d = \Pi_K(y) - y$. Thus the above inequality is obvious by the variational characterization of $\Pi_K(y)$. We have

$$\begin{aligned} H \circ \mathbf{M}_K^{\text{nor}}(y) &= H(d) = (\mathbf{I} - M)\Pi_K^M(d) + d \\ &= (\mathbf{I} - M)\Pi_K(y) + \mathbf{M}_K^{\text{nor}}(y) = y. \end{aligned}$$

Thus $H \circ \mathbf{M}_K^{\text{nor}}$ is the identity map. Moreover, the proof also establishes $\Pi_K^M \circ \mathbf{M}_K^{\text{nor}} = \Pi_K$. Since $H = (\mathbf{M}_K^{\text{nor}})^{-1}$, it follows that $\Pi_K^M = \Pi_K \circ H$. Since Π_K^M is globally Lipschitz continuous, it follows that so is H. Thus, both $\mathbf{M}_K^{\text{nor}}$ and its inverse are globally Lipschitz homeomorphisms from $\mathbb{R}^n$ onto itself. □

4.4 B-Differentiability under SBCQ

In this section, we establish an important differentiability property of the Euclidean projector Π_K for a finitely representable, convex set K. Since K is non-polyhedral, we need a suitable CQ at a projected vector $\bar{x}$ to make the analysis possible. Here we assume the broad SBCQ applied to $\bar{x}$ as a solution of the VI $(K, \mathbf{I} - x)$.

The main result of this section is Theorem **4.4.1** below, whose proof is nontrivial. Although there is a partial generalization of the result to the case where the set K is dependent on a parameter, by giving a separate treatment to parameter-free case, we can illustrate the key ideas more succinctly and facilitate the subsequent generalization.

4.4.1 Theorem. Let each $g_i : \mathbb{R}^n \to \mathbb{R}$ be convex and $h : \mathbb{R}^n \to \mathbb{R}^\ell$ be affine. Let K be given by (4.1.1). Let $x \in \mathbb{R}^n$ and $\bar{x} \equiv \Pi_K(x)$. Suppose that each g_i is twice continuously differentiable in a neighborhood of $\bar{x}$ and that the SBCQ holds at $\bar{x} \in \mathrm{SOL}(K, \mathbf{I} - x)$. For every vector $d \in \mathbb{R}^n$, $\Pi'_K(x; d)$ exists and is equal to the unique vector $u(x; d)$ that solves the following minimization problem in the variable v:

$$\text{minimize} \quad \tfrac{1}{2} \left[\| v - d \|_2^2 + \max_{(\mu,\lambda) \in \mathcal{M}_\pi(x)} \sum_{i=1}^m \lambda_i \, v^T \nabla^2 g_i(\bar{x}) v \right] \tag{4.4.1}$$

$$\text{subject to} \quad v \in \mathcal{C}_\pi(x; K).$$

Moreover, it holds that

$$\Pi_K(y) = \Pi_K(x) + \Pi'_K(x; y - x) + o(\|y - x\|). \tag{4.4.2}$$

When each g_i is also affine, Theorem **4.4.1** is subsumed by Theorem **4.1.1**. Notice that for y sufficiently close to x, the $o(\|y - x\|)$ term in (4.4.2) can be dropped when K is a polyhedral. Unlike the polyhedral case, the proof of Theorem **4.4.1** is quite involved and contains several main steps. To simplify the notation, we write

$$\mathcal{C} \equiv \mathcal{C}_\pi(x; K) \quad \text{and} \quad G(\lambda) \equiv \mathbf{I} + \sum_{i=1}^m \lambda_i \nabla^2 g_i(\bar{x}).$$

For every pair $(\mu, \lambda) \in \mathcal{M}_\pi(x)$, $G(\lambda)$ is a symmetric positive definite matrix due to the convexity of the functions g_i and the nonnegativity of λ; moreover, we have

$$\sum_{i=1}^m \lambda_i \nabla^2 g_i(\bar{x}) = \sum_{i \in \mathcal{I}(\bar{x})} \lambda_i \nabla^2 g_i(\bar{x}),$$

because $\lambda_i = 0$ for all $i \notin \mathcal{I}(\bar{x})$. Note that the function h does not appear in the matrix $G(\lambda)$ because h is assumed affine.

Let $\{\tau_k\}$ be an arbitrary sequence of positive scalars converging to zero; and let $d \in \mathbb{R}^n$ be an arbitrary vector. The first step in the proof of Theorem **4.4.1** is to establish a basic property of an accumulation point of the sequence of vectors given by

$$\left\{ \frac{\Pi_K(x + \tau_k d) - \Pi_K(x)}{\tau_k} \right\} \tag{4.4.3}$$

as $k \to \infty$. Since the projector Π_K is globally Lipschitz continuous, it follows that there exists a constant $c > 0$ such that for all vectors $d \in \mathbb{R}^n$ and scalars $\tau \geq 0$,

$$\| \Pi_K(x + \tau d) - \Pi_K(x) \| \leq c\tau \| d \|;$$

thus the sequence of vectors (4.4.3) is bounded. Hence the sequence has at least one accumulation point. To establish the desired property of such an accumulation point, we first prove a preliminary lemma, which is an easy consequence of the closedness of a polyhedron. Specifically, the following lemma states that if the polyhedron $P(A,b)$ is empty, then so is every $P(A,b')$, provided that b' is sufficiently close to b.

4.4.2 Lemma. Let $A \in \mathbb{R}^{m \times n}$ be a given matrix. Let $\{b^k\}$ be a sequence of m-vectors converging to $b^\infty \in \mathbb{R}^m$. If $P(A,b^k)$ is nonempty for each k, then so is $P(A,b^\infty)$.

Proof. The fact that the system $Ax \leq b^k$ is consistent means that b^k belongs to the polyhedral cone $A\mathbb{R}^n + \mathbb{R}^m_+$. Since the latter cone is closed, it follows that b^∞ is also an element of this cone; in turn, this means that the limiting system $Ax \leq b^\infty$ is consistent. □

The next lemma is central to the proof of Theorem **4.4.1**. It gives a desired characterization of a limit point of the sequence of vectors (4.4.3).

4.4.3 Lemma. Assume the setting of Theorem **4.4.1**. For every accumulation point u of the sequence of vectors (4.4.3) as τ_k tends to zero, there exists a $(\mu,\lambda) \in \mathcal{M}_\pi(x)$ such that

$$u = \Pi_C^{G(\lambda)}(d). \tag{4.4.4}$$

Moreover, such a pair (μ,λ) can be chosen to maximize the linear functional

$$(\mu',\lambda') \mapsto \tfrac{1}{2}\sum_{i=1}^{m} \lambda_i' \, u^T \nabla^2 g_i(\bar{x}) u \tag{4.4.5}$$

on the polyhedron $\mathcal{M}_\pi(x)$.

Proof. To simplify the notation, we assume that u is the limit of the vectors (4.4.3) as $\tau_k \to 0$. For each k, write

$$x^k \equiv x + \tau_k \, d \quad \text{and} \quad y^k \equiv \Pi_K(x + \tau_k d).$$

Then the sequence $\{y^k\}$ converges to $\bar{x}$; moreover, for all k sufficiently large, $\mathcal{I}(y^k) \subseteq \mathcal{I}(\bar{x})$. By the SBCQ, we may assume without loss of generality that for each k, there exists a pair $(\mu^k,\lambda^k) \in \mathcal{M}_\pi(x^k)$ and the sequence $\{(\mu^k,\lambda^k)\}$ converges to a pair $(\mu^\infty,\lambda^\infty)$, which must be an element in $\mathcal{M}_\pi(x)$. For each k, we have

$$y^k - x - \tau_k \, d + \sum_{j=1}^{\ell} \mu_j^k \, \nabla h_j(y^k) + \sum_{i \in \mathcal{I}(\bar{x})} \lambda_i^k \, \nabla g_i(y^k) = 0.$$

We also have

$$\bar{x} - x + \sum_{j=1}^{\ell} \mu_j^\infty \nabla h_j(\bar{x}) + \sum_{i \in \mathcal{I}(\bar{x})} \lambda_i^\infty \nabla g_i(\bar{x}) = 0.$$

To establish (4.4.4), it suffices to show:

$$\mathcal{C} \ni u \perp G(\lambda^\infty)u - d \in \mathcal{C}^*.$$

Noting the affinity of each h_j and the approximation:

$$\nabla g_i(y^k) = \nabla g_i(\bar{x}) + \nabla^2 g_i(\bar{x}) (y^k - \bar{x}) + e_i^g(y^k - \bar{x}), \quad \forall i = 1, \ldots, m,$$

where the vector error functions e_i^g satisfy

$$\lim_{v \to 0} \frac{e_i^g(v)}{\| v \|} = 0,$$

we can readily establish the desired properties of the vector u; the omitted arguments are very similar to the proof of Theorem **3.3.12**.

To establish that the pair $(\mu^\infty, \lambda^\infty)$ has the desired maximizing property, consider the linear program of maximizing the linear function (4.4.5) on the set $\mathcal{M}_\pi(x)$:

$$\begin{array}{ll} \text{maximize} & \frac{1}{2} \sum_{i=1}^{m} \lambda_i' \, u^T \nabla^2 g_i(\bar{x}) u \\[2ex] \text{subject to} & \bar{x} - x + \sum_{j=1}^{\ell} \mu_j' \nabla h_j(\bar{x}) + \sum_{i \in \mathcal{I}(\bar{x})} \lambda_i' \nabla g_i(\bar{x}) = 0 \\[2ex] & \lambda_i' \geq 0, \quad \forall i \in \mathcal{I}(\bar{x}) \\[1ex] & \lambda_i' = 0, \quad \forall i \notin \mathcal{I}(\bar{x}). \end{array} \tag{4.4.6}$$

The dual of this linear program is:

$$\begin{array}{ll} \text{maximize} & v^T (\bar{x} - x) \\[1ex] & v^T \nabla h_j(\bar{x}) = 0, \quad \forall j = 1, \ldots, \ell \\[1ex] & v^T \nabla g_i(\bar{x}) + \frac{1}{2} u^T \nabla^2 g_i(\bar{x}) u \leq 0, \quad \forall i \in \mathcal{I}(\bar{x}). \end{array} \tag{4.4.7}$$

The pair $(\mu^\infty, \lambda^\infty)$ is clearly feasible to (4.4.6); thus, it is optimal if and only if there exists a v^∞ such that v^∞ is feasible to (4.4.7) and complementary slackness holds between $(\mu^\infty, \lambda^\infty)$ and v^∞. Writing out the conditions for v^∞, we therefore need to show that there exists a vector v^∞ satisfying

(A) $(v^\infty)^T \nabla g_i(\bar{x}) + \frac{1}{2} u^T \nabla^2 g_i(\bar{x}) u \leq 0$ for all $i \in \mathcal{I}(\bar{x})$;

(B) $(v^\infty)^T \nabla g_i(\bar{x}) + \frac{1}{2} u^T \nabla^2 g_i(\bar{x}) u = 0$ for all $i \in \operatorname{supp}(\lambda^\infty)$; and

(C) $(v^\infty)^T \nabla h_j(\bar{x}) = 0$ for all j.

For each $i \in \mathcal{I}(\bar{x})$, we have $g_i(\bar{x}) = 0$. Thus, we can write

$$g_i(y^k) = (y^k - \bar{x})^T \nabla g_i(\bar{x}) + \tfrac{1}{2} (y^k - \bar{x})^T \nabla^2 g_i(\bar{x})(y^k - \bar{x}) + e_i(y^k - \bar{x}),$$

where the scalar error function e_i satisfies

$$\lim_{v \to 0} \frac{e_i(v)}{\| v \|^2} = 0.$$

Since each h_j is affine, we have

$$0 = h_j(y^k) = (y^k - \bar{x})^T \nabla h_j(\bar{x}), \quad \forall j = 1, \ldots, \ell.$$

Since $g_i(y^k) \le 0$ for all $i \in \mathcal{I}(\bar{x})$ with equality holding for all $i \in \operatorname{supp}(\lambda^\infty)$ and all k sufficiently large, it follows that the system of linear inequalities:

$(\mathrm{A})_k$ for all $i \in \mathcal{I}(\bar{x}) \setminus \operatorname{supp}(\lambda^\infty)$

$$v^T \nabla g_i(\bar{x}) + \tfrac{1}{2} \left(\frac{y^k - \bar{x}}{\| y^k - \bar{x} \|} \right)^T \nabla^2 g_i(\bar{x}) \left(\frac{y^k - \bar{x}}{\| y^k - \bar{x} \|} \right) + \frac{e_i(y^k - \bar{x})}{\| y^k - \bar{x} \|^2} \le 0,$$

$(\mathrm{B})_k$ for all $i \in \operatorname{supp}(\lambda^\infty)$,

$$v^T \nabla g_i(\bar{x}) + \tfrac{1}{2} \left(\frac{y^k - \bar{x}}{\| y^k - \bar{x} \|} \right)^T \nabla^2 g_i(\bar{x}) \left(\frac{y^k - \bar{x}}{\| y^k - \bar{x} \|} \right) + \frac{e_i(y^k - \bar{x})}{\| y^k - \bar{x} \|^2} = 0,$$

$(\mathrm{C})_k$, for all $j = 1, \ldots, \ell$, $v^T \nabla h_j(\bar{x}) = 0$

has a solution v^k for all k sufficiently large. By Lemma **4.4.2**, the existence of a vector v^∞ satisfying (A), (B), and (C) follows readily because the system defined by these three conditions is the limit of the system defined by $(\mathrm{A})_k$, $(\mathrm{B})_k$, and $(\mathrm{C})_k$ as $k \to \infty$. □

Lemma **4.4.3** does not directly establish the directional differentiability of Π_K at x. The reason is that although $\Pi_C^{G(\lambda)}(d)$ is uniquely defined for every vector d, the operator $\Pi_C^{G(\lambda)}$ depends on λ, which in turn depends on the accumulation point u. In order to establish the directional differentiability of Π_K at x, one approach is to show that the operator $\Pi_C^{G(\lambda)}$ is independent of the multiplier λ. There are two important situations where this independence is trivially valid. One is the polyhedral case that is treated thoroughly in Section **4.1**. The other case is when $\mathcal{M}_\pi(x)$ is a singleton; i.e., when the SMFCQ holds at a triple $(\bar{x}, \mu, \lambda)$, where $(\mu, \lambda) \in \mathcal{M}_\pi(x)$. In this case, the multiplier λ is unique and the directional differentiability of Π_K at x follows readily.

In the general situation, we need to show that the sequence of vectors (4.4.3) has a unique accumulation point. The following proof is based on the characterization of such a point; cf. Lemma **4.4.3**. Let $\theta(x)$ denote the objective function of (4.4.1).

Proof of Theorem 4.4.1. The objective function $\theta(v)$ of the program (4.4.1) is strongly convex, but possibly extended-valued. The strong convexity is due to the following two reasons: (i) the first summand $\|v - d\|^2$ is strongly convex in v, and (ii) the second summand

$$v \mapsto \max_{(\mu,\lambda)\in\mathcal{M}_\pi(x)} \sum_{i=1}^m \lambda_i \, v^T \nabla^2 g_i(\bar{x}) v,$$

being the pointwise maximum of a family of convex functions, is convex. By Lemma **4.4.3**, $\theta(v)$ is finite at some v. Consequently, θ attains its unique global minimum on the cone $\mathcal{C}_\pi(x; K)$. Let u be an arbitrary accumulation point of the sequence of vectors (4.4.3) as τ_k tends to zero, and let (μ, λ^∞) be in $\mathcal{M}_\pi(x)$ such that $u = \Pi_{\mathcal{C}}^{G(\lambda^\infty)}(d)$. By the definition of $\Pi_{\mathcal{C}}^{G(\lambda^\infty)}$, we have, for every $v \in \mathcal{C}$,

$$\begin{aligned}
\theta(v) &\geq \tfrac{1}{2} \| v - d \|_2^2 + \tfrac{1}{2} \sum_{i=1}^m \lambda_i^\infty \, v^T \nabla^2 g_i(\bar{x}) v \\
&= \tfrac{1}{2} v^T G(\lambda^\infty) v - v^T d + \tfrac{1}{2} d^T d \\
&\geq \tfrac{1}{2} u^T G(\lambda^\infty) u - u^T d + \tfrac{1}{2} d^T d \\
&= \tfrac{1}{2} \| u - d \|_2^2 + \sum_{i=1}^m \lambda_i^\infty \, u^T \nabla^2 g_i(\bar{x}) u \;=\; \theta(u)
\end{aligned}$$

Since u belongs to $\mathcal{C}_\pi(x; K)$, it follows that every accumulation point of the sequence (4.4.3) is equal to the global minimizer of θ on $\mathcal{C}_\pi(x; K)$. Since the latter minimizer is unique, the sequence (4.4.3) therefore converges to this unique minimizer. Since the sequence of scalars $\{\tau_k\}$ is arbitrary, it follows that $\Pi_K'(x; d)$ exists and is as described. Finally, the equation (4.4.2) is an immediate consequence of the B-differentiability of the projector. □

By Exercise **3.7.5**, it follows that for any closed convex set K, if Π_K is B-differentiable at a vector $x \in \mathbb{R}^n$, then $\Pi_K'(x; \cdot)$ is a co-coercive function. In particular, this holds under the assumptions of Theorem **4.4.1**.

It is useful to note that under the SBCQ, the set of multipliers $\mathcal{M}_\pi(x)$ is not necessarily bounded. We have not demonstrated that the linear function (4.4.5) attains its maximum on the set $\mathcal{M}_\pi(x)$ for all vectors $u \in \mathbb{R}^n$, but rather only for a special set of vectors u, namely, those u

belonging to the range of the directional derivative $\Pi'_K(x;\cdot)$. Specifically, for all vectors d, the linear program:

$$\begin{array}{ll} \text{maximize} & \displaystyle\sum_{i=1}^m \lambda_i \, (\, \Pi'_K(x;d)\,)^T \nabla^2 g_i(\bar{x}) \, \Pi'_K(x;d) \\ \text{subject to} & (\mu,\lambda) \in \mathcal{M}_\pi(x) \end{array} \tag{4.4.8}$$

must attain its finite maximum. Furthermore, if we let $\mathcal{M}^{\rm e}_\pi(x)$ be the subset of $\mathcal{M}_\pi(x)$ consisting of all pairs (μ,λ) such that the gradient vectors

$$\{\, \nabla h_j(\bar{x}) \,:\, j = 1,\ldots,\ell \,\} \cup \{\, \nabla g_i(\bar{x}) \,:\, i \in \operatorname{supp}(\lambda)\,\}$$

are linearly independent, then (i) $\mathcal{M}^{\rm e}_\pi(x)$ is a finite set, and (ii) the above linear program attains its maximum at a member of the set $\mathcal{M}^{\rm e}_\pi(x)$. The reason why $\mathcal{M}^{\rm e}_\pi(x)$ is finite is because for every subset $\mathcal{J} \subseteq \mathcal{I}(\bar{x})$ such that the gradients

$$\{\, \nabla h_j(\bar{x}) \,:\, j = 1,\ldots,\ell \,\} \cup \{\, \nabla g_i(\bar{x}) \,:\, i \in \mathcal{J}\,\}$$

are linearly independent, there is at most one member (μ,λ) belonging to $\mathcal{M}_\pi(x)$ satisfying $\operatorname{supp}(\lambda) = \mathcal{J}$, and also because there are only finitely many subsets of $\mathcal{I}(\bar{x})$. The fact that the directional linear program (4.4.8) must attain its maximum at a member in $\mathcal{M}^{\rm e}_\pi(x)$ is well known from linear programming. Needless to say, we may choose such a member to represent the directional derivative $\Pi'_K(x;d)$, as justified by Lemma **4.4.3**. The upshot of this discussion is that there exists a finite subset of KKT pairs (μ,λ) that can be used to describe the directional derivatives $\Pi'_K(x;d)$ for all direction vectors $d \in \mathbb{R}^n$. This observation is very important because subsequently we will impose conditions involving the multipliers in $\mathcal{M}^{\rm e}_\pi(z^*)$ for a certain vector z^*. Since there are only finitely many multipliers of this kind, the imposed conditions are much more easily verified than they would have been were they defined in terms of the entire continuum set of multipliers.

The problem (4.4.1) is a convex-concave minimax problem; as such, it can be equivalently cast as a saddle problem. Indeed, define the saddle function: for $(v,\mu,\lambda) \in \mathbb{R}^{n+\ell+m}$,

$$L(v,\mu,\lambda) \equiv \tfrac{1}{2}\left[\, \|\, v - d\,\|_2^2 + \sum_{i=1}^m \lambda_i \, v^T \nabla^2 g_i(\bar{x}) v \,\right], \tag{4.4.9}$$

which is clearly convex-concave in (v,y) on the set

$$C \equiv \mathcal{C}_\pi(x;K) \times \mathcal{M}_\pi(x),$$

where $y \equiv (\mu, \lambda)$. If $(\bar{v}, \bar{y})$ is a saddle point of L on C, where $\bar{y} \equiv (\bar{\mu}, \bar{\lambda})$, then, by definition, we have

$$L(\bar{v}, y) \leq L(\bar{v}, \bar{y}) \leq L(v, \bar{y}), \quad \forall\, (v, y) \in \mathcal{C}_\pi(x; K) \times \mathcal{M}_\pi(x).$$

Thus with $\theta(v)$ defined in (4.4.1), it follows that for all $v \in \mathcal{C}_\pi(x; K)$,

$$\theta(v) \geq L(v, \bar{y}) \geq \max_{y \in \mathcal{M}_\pi(x)} L(\bar{v}, y) = \theta(\bar{v}).$$

Conversely, if $\bar{v}$ is a minimizer of $\theta(v)$ on $\mathcal{C}_\pi(x; K)$, let $\bar{y} \equiv (\bar{\mu}, \bar{\lambda})$ maximize the linear function:

$$(\mu, \lambda) \mapsto \tfrac{1}{2} \sum_{i=1}^{m} \lambda_i\, \bar{v}^T \nabla^2 g_i(\bar{x}) \bar{v}$$

on the cone $\mathcal{C}_\pi(x; K)$; such a maximizer must exist because $\theta(\bar{v})$ is finite, by the definition of $\bar{v}$. It is then easy to show that $(\bar{v}, \bar{y})$ is a saddle point of the function $L(v, y)$ on $\mathcal{C}_\pi(x; K) \times \mathcal{M}_\pi(x)$.

Define the vector function

$$T : (v, \mu, \lambda) \in \mathbb{R}^{n+\ell+m} \mapsto \begin{pmatrix} G(\lambda) v - d \\ \mathbf{0}_\ell \\ -\frac{1}{2} \left(v^T \nabla^2 g_i(\bar{x}) v \right)_{i=1}^m \end{pmatrix} \in \mathbb{R}^{n+\ell+m},$$

Note that

$$T(v, \mu, \lambda) = \begin{pmatrix} \nabla_v L(v, \mu, \lambda) \\ -\nabla_\mu L(v, \mu, \lambda) \\ -\nabla_\lambda L(v, \mu, \lambda) \end{pmatrix}$$

The following result summarizes the above discussion and provides a variational characterization of the directional derivative $\Pi_K'(x; d)$ in terms of the VI (C, T). The proof of the result is an immediate consequence of the connection between the solutions of this VI and the saddle points of L on the set C, as per the discussion in Subsection **1.4.1**.

4.4.4 Proposition. Assume the setting of Theorem **4.4.1**. The vector v is equal to $\Pi_K'(x; d)$ if and only if there exists a pair $(\mu, \lambda) \in \mathcal{M}_\pi^e(x)$ such that (v, μ, λ) solves the VI (C, T).

Proof. If (v, μ, λ) solves the VI (C, T), then (v, y), where $y \equiv (\mu, \lambda)$, is a saddle point of the saddle function L defined by (4.4.9). As proved above, v minimizes the function θ on $\mathcal{C}_\pi(x; K)$. By Theorem **4.4.1**, v is therefore equal to $\Pi_K'(x; d)$. The converse can be proved by reversing the argument. The details are omitted. □

4.5 Piecewise Smoothness under CRCQ

The SBCQ is the main assumption in Theorem **4.4.1**. By assuming the CRCQ, we can show that the projector Π_K is a piecewise smooth function. The formal definition of this class of nonsmooth functions is given below.

4.5.1 Definition. A continuous function $G : \mathcal{D} \subseteq \mathbb{R}^n \to \mathbb{R}^m$ is said to be a *PC*^{1} function near the vector $x \in \mathcal{D}$ if there exist an open neighborhood $\mathcal{N} \subseteq \mathcal{D}$ of x and a finite family of C^1 functions defined on $\mathcal{N}$, $\{G^1, G^2, \cdots, G^k\}$, for some positive integer k, such that $G(y)$ is an element of $\{G^1(y), \cdots, G^k(y)\}$ for all $y \in \mathcal{N}$. Each function G^i is called a C^1 *piece* of G at x. Let $\mathcal{P}(y)$ denote the set of indices $i \in \{1, \cdots, k\}$ such that $G(y) = G^i(y)$. □

Every PA function is clearly PC^1. The componentwise minimum of two C^1 functions F and H each mapping $\mathbb{R}^n$ into $\mathbb{R}^m$ is a PC^1 function whose 2^m pieces G^i are obtained by letting $G^i_J \equiv F_J$ and $G^i_{\bar{J}} \equiv H_{\bar{J}}$, where J and $\bar{J}$ are any pair of complementary subsets of $\{1, \ldots, m\}$. The class of PC^1 functions is important in its own right. For our purpose here, we proceed to show that Π_K is a PC^1 function under the CRCQ.

For simplicity, we assume throughout this section that linear equality constraints are absent from K; thus K is of the form:

$$K \equiv \{ x \in \mathbb{R}^n : g(x) \leq 0 \} \tag{4.5.1}$$

where $g : \mathcal{D} \to \mathbb{R}^m$ is twice continuously differentiable on the open set $\mathcal{D}$ containing K, and each component function g_i is convex. Fix a vector $x \in \mathbb{R}^n$ and let $\bar{x} \equiv \Pi_K(x)$. The key assumption herein is that the CRCQ holds at $\bar{x}$. Let $\mathcal{B}(x)$ be the collection of index subsets $\mathcal{J} \subseteq \mathcal{I}(\bar{x})$ such that the family of gradient vectors

$$\{ \nabla g_i(\bar{x}) : i \in \mathcal{J} \}$$

are linearly independent. A certain collection of index sets similar to $\mathcal{B}(x)$ has played an important role in establishing the PA property of the Euclidean projector onto a polyhedron; see Proposition **4.1.4**. There, the affinity of the constraint functions greatly simplifies the proof of the property. Here, the nonlinearity of the functions g_i complicates matters substantially. The collection $\mathcal{B}(x)$ plays a similarly important role in the analysis to follow. The main tool in proving the next result is the classical implicit function theorem for smooth functions.

4.5.2 Theorem. Let K be given by (4.5.1) where each g_i is twice continuously differentiable and convex. Let $x \in \mathbb{R}^n$ be such that the CRCQ holds at $\bar{x} \equiv \Pi_K(x)$. The projector Π_K is a PC^1 function near x.

Proof. Let $\mathcal{B}(x)$ be as defined above. There exists at least one multiplier vector $\lambda \in \mathcal{M}_\pi(x)$ such that $\operatorname{supp}(\lambda) \in \mathcal{B}(x)$. Let $\mathcal{B}'(x)$ be the subcollection of $\mathcal{B}(x)$ consisting of index sets $\mathcal{J} \in \mathcal{B}(x)$ for which there exists a vector $\lambda \in \mathcal{M}_\pi(x)$ such that $\operatorname{supp}(\lambda) \subseteq \mathcal{J}$. Corresponding to each such index set $\mathcal{J} \in \mathcal{B}'(x)$, let $\Lambda(\mathcal{J})$ denote the subset of $\mathcal{M}_\pi(x)$ consisting of multipliers λ in $\mathcal{M}_\pi(x)$ such that $\operatorname{supp}(\lambda) \subseteq \mathcal{J}$. Notice that $\Lambda(\mathcal{J})$ is a finite set because corresponding to each subset $\mathcal{J}'$ of $\mathcal{J}$, there is at most one multiplier λ with $\operatorname{supp}(\lambda) = \mathcal{J}'$. For each $\mathcal{J} \in \mathcal{B}'(x)$, consider the function

$$\Phi_{\mathcal{J}} : (v, u, \eta_{\mathcal{J}}) \in \mathbb{R}^{2n+|\mathcal{J}|} \mapsto \begin{pmatrix} v - u + \sum_{j \in \mathcal{J}} \eta_j \nabla g_j(v) \\ \\ -g_{\mathcal{J}}(v) \end{pmatrix} \in \mathbb{R}^{n+|\mathcal{J}|},$$

which vanishes at the triple $(\bar{x}, x, \lambda_{\mathcal{J}})$ for all $\lambda \in \Lambda(\mathcal{J})$. The partial Jacobian matrix of $\Phi_{\mathcal{J}}$ with respect to the pair $(v, \eta_{\mathcal{J}})$ at any such triple is equal to

$$\begin{bmatrix} \mathbf{I}_n + \sum_{j \in \mathcal{J}} \lambda_j \nabla^2 g_j(\bar{x}) & Jg_{\mathcal{J}}(\bar{x})^T \\ \\ -Jg_{\mathcal{J}}(\bar{x}) & 0 \end{bmatrix},$$

which has a nonzero determinant, by the convexity of each g_j and the nonnegativity of each λ_j. By the classical implicit function theorem applied to $\Phi_{\mathcal{J}}$ and each triple $(\bar{x}, x, \lambda_{\mathcal{J}})$, where $\lambda \in \Lambda(\mathcal{J})$, with $(v, \eta_{\mathcal{J}})$ as the primary variable and u as the parameter, there exist open neighborhoods $\mathcal{V}(\mathcal{J}, \lambda)$ of $\bar{x}$, $\mathcal{N}(\mathcal{J}, \lambda)$ of x, and $\mathcal{U}(\mathcal{J}, \lambda)$ of $\lambda_{\mathcal{J}}$ and a continuously differentiable function

$$z : \mathcal{N}(\mathcal{J}, \lambda) \to \mathcal{V}(\mathcal{J}, \lambda) \times \mathcal{U}(\mathcal{J}, \lambda)$$

such that for every vector u in the neighborhood $\mathcal{N}(\mathcal{J}, \lambda)$, $z(u)$ is the unique pair $(y, \eta_{\mathcal{J}})$ in $\mathcal{V}(\mathcal{J}, \lambda) \times \mathcal{U}(\mathcal{J}, \lambda)$ satisfying

$$\Phi_{\mathcal{J}}(y, u, \eta_{\mathcal{J}}) = 0.$$

Let $z_{\mathcal{J},\lambda}$ be the y-part of the function z. By shrinking the neighborhood $\mathcal{N}(\mathcal{J}, \lambda)$ if necessary, we may assume that for all vectors $u \in \mathcal{N}(\mathcal{J}, \lambda)$, $\Pi_K(u)$ belongs to the neighborhood $\mathcal{V}(\mathcal{J}, \lambda)$. Let

$$\mathcal{N} \equiv \bigcap_{\mathcal{J} \in \mathcal{B}'(x)} \bigcap_{\lambda \in \Lambda(\mathcal{J})} \mathcal{N}(\mathcal{J}, \lambda).$$

Also let $\mathcal{U}(\mathcal{J})$ denote the finite family of neighborhoods

$$\{\mathcal{U}(\mathcal{J},\lambda) : \lambda \in \Lambda(\mathcal{J})\}.$$

It suffices to show that for all x' sufficiently close to x, $\Pi_K(x')$ belongs to the finite family

$$\{z_{\mathcal{J},\lambda}(x') : \lambda \in \Lambda(\mathcal{J}), \mathcal{J} \in \mathcal{B}'(x)\}.$$

Since the CRCQ holds at $\bar{x}$, it continues to hold at $\Pi_K(x')$ for all x' sufficiently close to x. Moreover, for such a vector x', $\mathcal{I}(\Pi_K(x'))$ is a subset of $\mathcal{I}(\bar{x})$. Thus, there exists a neighborhood $\mathcal{N}'$ of x such that for all $x' \in \mathcal{N}'$, a vector $\lambda' \in \mathbb{R}^m$ exists satisfying

$$\begin{gathered}\Pi_K(x') - x' - \sum_{j \in \mathcal{I}(\bar{x})} \lambda_j' \nabla g_j(\Pi_K(x')) = 0 \\ 0 \le \lambda_j \perp g_j(\Pi_K(x')) \le 0, \quad \forall j \in \mathcal{I}(\bar{x}).\end{gathered} \tag{4.5.2}$$

Without loss of generality, we may assume that the two neighborhoods $\mathcal{N}$ and $\mathcal{N}'$ coincide. We retain the former notation for this common neighborhood. We claim that by shrinking the neighborhood $\mathcal{N}$ if necessarily, we may choose for every $x' \in \mathcal{N}$ a multiplier λ' satisfying the above KKT system (4.5.2) for $\Pi_K(x')$ such that there exists an index set $\mathcal{J} \in \mathcal{B}'(x)$, which contains $\operatorname{supp}(\lambda')$ and $\lambda'_{\mathcal{J}}$ belongs to a member of the neighborhood family $\mathcal{U}(\mathcal{J})$. Assume the contrary. Then there exists a sequence of vectors $\{x^k\}$ converging to x such that for each k, every vector $\lambda' \in \mathcal{M}_\pi(x^k)$ is such that for any index set $\mathcal{J} \in \mathcal{B}'(x)$ containing $\operatorname{supp}(\lambda')$, $\lambda'_{\mathcal{J}}$ lies outside the union of the neighborhoods in the family $\mathcal{U}(\mathcal{J})$. For every k, choose a multiplier $\lambda^k \in \mathcal{M}_\pi(x^k)$ such that with $\mathcal{J}^k \equiv \operatorname{supp}(\lambda^k)$, the gradient vectors

$$\{\nabla g_j(\Pi_K(x^k)) : j \in \mathcal{J}^k\}$$

are linearly independent. Without loss of generality, by working with a subsequence of $\{x^k\}$ if necessary, we may assume that these index sets $\mathcal{J}^k$ are the same for all k; let $\mathcal{J}$ denote this common index set. By the CRCQ, the limiting gradients

$$\{\nabla g_j(\Pi_K(\bar{x})) : j \in \mathcal{J}\}$$

remain linearly independent. Hence $\mathcal{J} \in \mathcal{B}(x)$; moreover, the sequence of multipliers $\{\lambda^k\}$ must converge to a multiplier $\lambda^\infty \in \mathcal{M}_\pi(x)$ with

$$\operatorname{supp}(\lambda^\infty) \subseteq \mathcal{J}.$$

Hence $\mathcal{J} \in \mathcal{B}'(x)$ and $\lambda^\infty \in \Lambda(\mathcal{J})$. This implies that $\lambda^k_{\mathcal{J}}$ must belong to the neighborhood $\mathcal{U}(\mathcal{J}, \lambda^\infty)$ for all k sufficiently large. This is a contradiction and our claim is established.

Consequently, for every $x' \in \mathcal{N}$, there exist $\lambda' \in \mathcal{M}_\pi(\Pi_K(x'))$ and an index set $\mathcal{J} \in \mathcal{B}'(x)$ containing $\text{supp}(\lambda')$ such that the pair $(\Pi_K(x'), \lambda'_{\mathcal{J}})$ belongs to $\mathcal{V}(\mathcal{J}) \times \mathcal{U}(\mathcal{J}, \lambda)$ for some $\lambda \in \Lambda(\mathcal{J})$ and

$$\Phi_{\mathcal{J}}(\Pi_K(x'), x', \lambda'_{\mathcal{J}}) = 0.$$

By the uniqueness of $z(x')$ associated with the pair $(\mathcal{J}, \lambda)$, we must have $\Pi_K(x') = z_{\mathcal{J},\lambda}(x')$ as desired. □

By Theorem **4.4.1**, it follows that for every vector $d \in \mathbb{R}^n$, there exists $\lambda \in \mathcal{M}_\pi(x)$ such that

$$\Pi'_K(x; d) = \Pi^{G(\lambda)}_{\mathcal{C}}(d),$$

where $\mathcal{C} \equiv \mathcal{C}_\pi(x; K)$. In the next result, we show that under the CRCQ, *any* multiplier λ in the set $\mathcal{M}_\pi(x)$ can be used to represent the directional derivative $\Pi'_K(x; d)$. This result is computationally significant because it is no longer necessary to solve a linear program to obtain the desired multiplier (as required by Theorem **4.4.1** that assumes the SBCQ).

4.5.3 Theorem. Let K be given by (4.5.1) where each g_i is twice continuously differentiable and convex. Let $x \in \mathbb{R}^n$ be such that the CRCQ holds at $\bar{x} \equiv \Pi_K(x)$. For any multiplier $\lambda \in \mathcal{M}_\pi(x)$ and every vector $d \in \mathbb{R}^n$,

$$\Pi'_K(x; d) = \Pi^{G(\lambda)}_{\mathcal{C}}(d),$$

where $\mathcal{C} \equiv \mathcal{C}_\pi(x; K)$.

The proof of the above theorem is again nontrivial. We begin by defining an alternate multiplier set by perturbing the right-hand side of the constraints in K. Let

$$\mathcal{I}^+(x) \equiv \{ i : \exists \lambda \in \mathcal{M}_\pi(x) \text{ with } \lambda_i > 0 \}$$

and $\mathcal{I}^0(x) \equiv \mathcal{I}(\bar{x}) \setminus \mathcal{I}^+(x)$. Clearly, for any $\lambda \in \mathcal{M}_\pi(x)$, $\text{supp}(\lambda) \subseteq \mathcal{I}^+(x)$. The set $\mathcal{M}_\pi(x)$ is a polyhedron; thus its relative interior, denoted $\text{ri}\,\mathcal{M}_\pi(x)$, is nonempty. We observe that for every $\lambda \in \text{ri}\,\mathcal{M}_\pi(x)$, $\text{supp}(\lambda) = \mathcal{I}^+(x)$. Indeed, let λ be a relative interior point of $\mathcal{M}_\pi(x)$ and let i be an index in $\mathcal{I}^+(x)$. Let $\bar{\lambda} \in \mathcal{M}_\pi(x)$ be such that $\bar{\lambda}_i > 0$. For $\tau > 0$ sufficiently small $\lambda - \tau(\bar{\lambda} - \lambda)$ remains in the set $\mathcal{M}_\pi(x)$ because λ is a relative interior point of $\mathcal{M}_\pi(x)$. Consequently,

$$(1 + \tau)\lambda_i \geq \tau\bar{\lambda}_i > 0.$$

The indices in the sets $\mathcal{I}^+(x)$ and $\mathcal{I}^0(x)$ are called the *strongly active* and *strongly degenerate* indices at x, respectively. Associated with $\mathcal{I}^+(x)$, define the perturbation function $p(y)$ by:

$$p_i(y) \equiv \begin{cases} g_i(y) & \text{if } i \in \mathcal{I}^+(x) \\ 0 & \text{otherwise.} \end{cases}$$

Clearly, the function p majorizes the function g on the set K.

For any vector $y \in \mathbb{R}^n$ with $\bar{y} \equiv \Pi_K(y)$, consider the following nonlinear program in the variable z:

$$\begin{array}{ll} \text{minimize} & \frac{1}{2}(z-y)^T(z-y) \\ \text{subject to} & g(z) \le p(\bar{y}). \end{array} \tag{4.5.3}$$

It is easy to check that $\bar{y}$ is the unique solution of this problem. Moreover, the set of active indices at $\bar{y}$ with respect to the constraints in the feasible set in (4.5.3) must contain $\mathcal{I}^+(x)$. Let $\bar{\mathcal{M}}_\pi(y)$ denote the set of multipliers λ that satisfy the KKT conditions of (4.5.3); that is, λ belongs to $\bar{\mathcal{M}}_\pi(y)$ if and only if

$$\bar{y} - y + \sum_{i=1}^m \lambda_i \nabla g_i(\bar{y}) = 0$$

$$0 \le \lambda \perp g(\bar{y}) - p(\bar{y}) \le 0.$$

Since $p(\bar{x}) = 0$, $\bar{\mathcal{M}}_\pi(x) = \mathcal{M}_\pi(x)$. In general, we have $\mathcal{M}_\pi(y) \subseteq \bar{\mathcal{M}}_p(y)$ for all $y \in \mathbb{R}^n$. Moreover, for each $\lambda \in \bar{\mathcal{M}}_\pi(y)$,

$$\operatorname{supp}(\lambda) \subseteq \mathcal{I}(\bar{x}),$$

provided that y is sufficiently close to x.

The motivation to consider the modified projection problem (4.5.3) is that we hope to keep the same structure of the multiplier set for all points y near x; this will yield the important lower semicontinuity of the alternative multiplier map $\bar{\mathcal{M}}_\pi$ at x under CRCQ. The precise statement of the latter property of $\bar{\mathcal{M}}_\pi$ is contained in the lemma below.

4.5.4 Lemma. Under the assumptions of Theorem **4.5.3**, it holds that $\bar{\mathcal{M}}_\pi(y) \neq \emptyset$ for all y sufficiently close to x; moreover for every $\lambda \in \mathcal{M}_\pi(x)$,

$$\lim_{y \to x} \operatorname{dist}(\lambda, \bar{\mathcal{M}}_\pi(y)) = 0. \tag{4.5.4}$$

Proof. The CRCQ remains valid under a slight perturbation of the base vector. Consequently, $\mathcal{M}_\pi(y)$, and thus $\bar{\mathcal{M}}_\pi(y)$, is nonempty for all y sufficiently close to x. The relative interior of $\mathcal{M}_\pi(x)$ is a dense subset of $\mathcal{M}_\pi(x)$. Consequently, it suffices to show the desired limit (4.5.4) for a multiplier $\lambda \in \operatorname{ri} \mathcal{M}_\pi(x)$. In turn, it suffices to show that for every sequence $\{y^k\}$ converging to x, there exists for each sufficiently large k a $\lambda^k \in \bar{\mathcal{M}}_\pi(y^k)$ such that the sequence $\{\lambda^k\}$ converges to λ.

For each sufficiently large k, there exist $\lambda^k \in \bar{\mathcal{M}}_\pi(y^k)$ such that the gradients

$$\{ \nabla g_i(\bar{y}^k) \, : \, i \in \operatorname{supp}(\lambda^k) \}$$

are linearly independent, where $\bar{y}^k \equiv \Pi_K(y^k)$. Without loss of generality, we may assume that there exists a subset $\mathcal{K}$ of $\mathcal{I}(\bar{x})$ such that $\operatorname{supp}(\lambda^k) = \mathcal{K}$ for all k. By the CRCQ at x, it follows that the gradients

$$\{ \nabla g_i(\bar{x}) \, : \, i \in \mathcal{K} \}$$

remain linearly independent. It is easy to see that the sequence $\{\lambda^k\}$ is bounded. Without loss of generality, we may assume that this sequence converges to some λ^∞, which must belong to $\bar{\mathcal{M}}_\pi(x) = \mathcal{M}_\pi(x)$.

Let $\mathcal{J} \equiv \{i \in \mathcal{I}^+(x) : \lambda_i^\infty = 0\}$ and $\triangle\lambda \equiv \lambda - \lambda^\infty$. Since $\lambda \in \operatorname{ri} \mathcal{M}_\pi(x)$, we have $\operatorname{supp}(\lambda) = \mathcal{I}^+(x)$. Thus, $\triangle\lambda_i$ is positive for all $i \in \mathcal{J}$ and zero for all $i \notin \mathcal{I}^+(x)$. Since both λ and λ^∞ belong to $\mathcal{M}_\pi(x)$, we have

$$\sum_{i \in \mathcal{I}^+(x)} \triangle\lambda_i \, \nabla g_i(\bar{x}) \, = \, 0.$$

By Proposition **3.2.9**, there exists a sequence $\{d\lambda^k\}$ converging to $\triangle\lambda$ such that for every k,

$$\sum_{i \in \mathcal{I}^+(x)} d\lambda_i^k \, \nabla g_i(\bar{y}^k) \, = \, 0. \tag{4.5.5}$$

and $d\lambda_i^k = 0$ for every $i \notin \mathcal{I}^+(x)$. To complete the proof, we are going to employ the sequences $\{\lambda^k\}$ and $\{d\lambda^k\}$ to construct an alternate sequence $\{\bar{\lambda}^k\}$ such that $\bar{\lambda}^k \in \mathcal{M}_\pi(y^k)$ for all k and $\bar{\lambda}^k \to \lambda$.

For each $i \in \mathcal{I}^+(x) \setminus \mathcal{J}$, define

$$\tau_i^k \, \equiv \, \max\{ \tau \, \in \, [0,1] \, : \, \lambda_i^k + \tau d\lambda_i^k \, \geq \, 0 \},$$

and set $\bar{\tau}_k \equiv \min\{\tau_i^k : i \in \mathcal{I}^+(x) \setminus \mathcal{J}\}$. For each index $i \in \mathcal{I}^+(x) \setminus \mathcal{J}$, we have $\lambda_i^\infty > 0$; thus $\lambda_i^k > 0$ for all k sufficiently large. Since $\{\lambda_i^k + d\lambda_i^k\}$ converges to $\lambda_i^\infty + \triangle\lambda_i = \lambda_i \geq 0$, it follows that $\tau_i^k \to 1$. Consequently $\bar{\tau}_k \to 1$. Define

$$\bar{\lambda}^k \, \equiv \, \lambda^k + \bar{\tau}_k \, d\lambda^k.$$

We have $\bar{\lambda}^k \to \lambda^\infty + \triangle\lambda = \lambda$. It remains to show that $\bar{\lambda}^k$ belongs to $\mathcal{M}_\pi(\bar{y}^k)$ for all k sufficiently large. For this, we need to verify:

$$\bar{y}^k - y^k + \sum_{i=1}^m \bar{\lambda}^k \, \nabla g_i(\bar{y}^k) \, = \, 0 \tag{4.5.6}$$

$$0 \leq \bar{\lambda}^k \perp g(\bar{y}^k) - p(\bar{y}^k) \leq 0. \tag{4.5.7}$$

The first condition (4.5.6) is obvious in view of the definition of λ^k, $\bar{\lambda}^k$, and (4.5.5). For condition (4.5.7), we consider each component and divide the proof into several cases.

- $i \notin \mathcal{I}(\bar{x})$. We have $d\lambda_i^k = 0$ by definition of $d\lambda_i^k$. Moreover, by the definition of $p_i(\bar{y}^k)$, we further have $g_i(\bar{y}^k) - p_i(\bar{y}^k) = g_i(\bar{y}^k) < 0$; hence by complementarity $\lambda_i^k = 0$. Thus $\bar{\lambda}_i^k = 0$.

- $i \in \mathcal{I}^0(x)$. Then $i \notin \mathcal{I}^+(x)$ and $d\lambda_i^k = 0$. Hence $\bar{\lambda}_i^k = \lambda_i^k \geq 0$ and since $\lambda^k \in \mathcal{M}_\pi(y^k)$, we have $d\lambda_i^k(g_i(\bar{y}^k) - p_i(\bar{y}^k)) = \lambda_i^k(g_i(\bar{y}^k) - p_i(\bar{y}^k)) = 0$.

- $i \in \mathcal{I}^+(x) \setminus \mathcal{J}$. We have $\bar{\lambda}_i^k \geq 0$ by definition of $\bar{\lambda}_i^k$. Moreover, by definition of $p_i(\bar{y}^k)$, we have $g_i(\bar{y}^k) = p_i(\bar{y}^k)$.

- $i \in \mathcal{J}$. We have $d\lambda^k \to \Delta\lambda_i > 0$; consequently, $\bar{\lambda}_i^k$ must be nonnegative for all k sufficiently large. Since $\mathcal{J} \subseteq \mathcal{I}^+(x)$, we have $g_i(\bar{y}^k) = p_i(\bar{y}_k)$ as in the last case. □

Employing the above lemma, we can give the proof of Theorem **4.5.3**.

Proof of Theorem 4.5.3. Let $\lambda \in \mathcal{M}_\pi(x)$ and $d \in \mathbb{R}^n$ be arbitrary. By Theorem **4.4.1**, we know that $\Pi_K'(x; d)$ exists. Let $\{\tau_k\}$ be an arbitrary sequence of positive scalars converging to zero. Let $\{\lambda^k\}$ be a sequence converging λ, where for each k, λ^k belongs to $\bar{\mathcal{M}}_\pi(x + \tau_k d)$. For each k, let $y^k \equiv \Pi_K(x + \tau_k d)$. We have

$$\lim_{k\to\infty} \frac{y^k - \bar{x}}{\tau_k} = \Pi_K'(x; d).$$

The remainder of the proof is very similar to the proof of Lemma **4.4.3** and the proof of Theorem **3.3.12**. Due to the modification of the constraints, we give a complete proof for the sake of clarity.

As proved in Lemma **4.5.4**, we have

$$y^k - x - \tau_k d + \sum_{i=1}^m \lambda_i^k \nabla g_i(y^k) = 0 \tag{4.5.8}$$

$$0 \leq \lambda^k \perp g(y^k) - p(y^k) \leq 0.$$

We also have

$$\bar{x} - x + \sum_{i=1}^m \lambda_i \nabla g_i(\bar{x}) = 0. \tag{4.5.9}$$

Substituting the following expression into the equation (4.5.8):

$$\nabla g_i(y^k) = \nabla g_i(\bar{x}) + \nabla^2 g_i(\bar{x})(y^k - \bar{x}) + o(\|y^k - \bar{x}\|),$$

subtracting the equation (4.5.9), and simplifying, we obtain

$$G(\lambda^k)\frac{y^k - \bar{x}}{\tau_k} - d + \sum_{i:\lambda_i > 0} \frac{\lambda_i^k - \lambda_i}{\tau_k} \nabla g_i(\bar{x}) +$$

$$\sum_{i \in \mathcal{I}(\bar{x}):\lambda_i = 0} \frac{\lambda_i^k}{\tau_k} \nabla g_i(\bar{x}) + \frac{o(\| y^k - \bar{x} \|)}{\tau_k} = 0. \qquad (4.5.10)$$

In order to show that $\Pi'_K(x; d) = \Pi_C^{G(\lambda)}(d)$, it suffices to verify that

$$\mathcal{C} \ni \Pi'_K(x; d) \perp G(\lambda)\Pi'_K(x; d) - d \in \mathcal{C}^*.$$

In terms of any multiplier $\lambda' \in \mathcal{M}_\pi(x)$, Lemma **3.3.2** implies

$$\begin{aligned} \mathcal{C} = \{ v \in \mathbb{R}^n : \; & v^T \nabla g_i(x) = 0, \; \forall i \text{ such that } \lambda'_i > 0 \\ & v^T \nabla g_i(x) \leq 0, \; \forall i \text{ such that } \lambda'_i = 0 \}. \end{aligned}$$

In particular, $v^T \nabla g_i(\bar{x}) = 0$ for all $v \in \mathcal{C}$ and all $i \in \mathcal{I}^+(x)$. Let $v \in \mathcal{C}$ be arbitrary. From (4.5.10), it is easy to see that

$$v^T \left[G(\lambda^k)\frac{y^k - \bar{x}}{\tau_k} - d + \frac{o(\| y^k - \bar{x} \|)}{\tau_k} \right] \geq 0.$$

Letting $k \to \infty$, we deduce

$$v^T [G(\lambda)\Pi'_K(x; d) - d] \geq 0, \quad \forall v \in \mathcal{C}.$$

Thus, it remains to show that $\Pi'_K(x; d) \perp G(\lambda)\Pi'_K(x; d) - d$. In turn, it suffices to show that for all k sufficiently large,

$$\lambda_i^k \nabla g_i(\bar{x})^T \Pi'_K(x; d) = 0, \quad \forall i \in \mathcal{I}(\bar{x}) \text{ such that } \lambda_i = 0.$$

Suppose $\nabla g_i(\bar{x})^T \Pi'_K(x; d) < 0$ for an index $i \in \mathcal{I}(x)$ such that $\lambda_i = 0$. Such an index i must belong to $\mathcal{I}^0(x)$; thus $p_i(y^k) = 0$ Moreover, for all k sufficiently large, we have $p_i(y^k) = 0 > g_i(y^k)$. Hence $\lambda_i^k = 0$ by complementarity. □

An immediate consequence of Theorem **4.5.3** is that we can characterize the F-differentiable points of the projector Π_K under the CRCQ, which generalizes Corollary **4.1.2** that pertains to a polyhedral K.

4.5.5 Corollary. Assume the setting of Theorem **4.5.3**. The statements (a), (b), and (c) in Corollary **4.1.2** remain equivalent, and each is further equivalent to (b') and (d) if $x \in K$.

Proof. Fix any multiplier $\lambda \in \mathcal{M}_\pi(x)$ and apply the same proof of Corollary **4.1.2** to the operator $\Pi_C^{G(\lambda)}$. □

4.6 Local Properties of PC^1 Functions

Throughout this section, let $G : U \subseteq \mathbb{R}^n \to \mathbb{R}^n$ be a PC^1 function near a given vector x in the open set U. We establish the following lemma that describes some basic properties of G. The second property (b) is particularly important because it allows us to employ the results of PL maps established in Theorem **4.2.11**.

4.6.1 Lemma. Let G be a PC^1 map near x with C^1 pieces $\{G^1, \cdots, G^k\}$. The following statements are valid.

(a) G is B-differentiable at all points near x.

(b) $G'(x;\cdot)$ is piecewise linear with linear pieces $\{JG^1(x), \cdots, JG^k(x)\}$.

Proof. We first show that G is Lipschitz continuous in a neighborhood of x. The proof is quite similar to that of property (A) in Theorem **4.2.11**, except that the argument is localized to a neighborhood of x. To begin, since each function G^i is C^1, there exist a neighborhood $\mathcal{N}$ and a constant $L > 0$ such that

$$\| G^i(u) - G^i(v) \| \leq L \| u - v \|, \quad \forall u, v \in \mathcal{N}.$$

Without loss of generality, we may assume that this neighborhood $\mathcal{N}$ is such that for all $y \in \mathcal{N}$, $\mathcal{P}(y) \subseteq \mathcal{P}(x)$. Take any two points u and v in this neighborhood and consider the line segment joining them. Following the proof of Theorem **4.2.11**(B), we can establish that

$$\| G(u) - G(v) \| \leq L \| u - v \|,$$

as desired.

Next, we show that G is directionally differentiable at x. The same proof can then be applied to all points sufficiently close to x; this will establish statement (a) of the lemma. Let d be an arbitrary vector. Let $\mathcal{P}'(x;d)$ be the subset of $\mathcal{P}(x)$ consisting of indices i for which there exists a sequence of positive scalars $\{\tau_\nu\}$ converging to zero such that $G^i(x+\tau_\nu d) = G(x+\tau_\nu d)$ for all ν. We claim that

$$\lim_{\tau \to 0+} \frac{G(x+\tau d) - G(x)}{\tau}$$

exists and is equal to $JG^i(x)d$ for any $i \in \mathcal{P}'(x;d)$. For this purpose, we may assume without loss of generality that $\mathcal{P}'(x;d) = \mathcal{P}(x)$; this is because any element $i \in \mathcal{P}(x)$ that does not belong to $\mathcal{P}'(x;d)$ will not be used by G at all points $x + \tau d$ for $\tau > 0$ sufficiently small.

Since

$$\lim_{\tau \to 0+} \frac{G^i(x+\tau d) - G^i(x)}{\tau} = JG^i(x)d,$$

it follows that if for all indices i in $\mathcal{P}'(x;d)$, $JG^i(x)d$ are equal, then the claim holds and the proof of statement (a) is complete. For the sake of contradiction, we may therefore assume that there exist i and j belonging to $\mathcal{P}'(x;d)$ such that $JG^i(x)d \neq JG^j(x)d$. Since G^i and G^j are C^1 functions, it follows that there exists a scalar $\varepsilon_{ij} > 0$ such that $G^i(x+\tau d) \neq G^j(x+\tau d)$ for all $\tau \in (0, \varepsilon_{ij}]$. Applying this observation to all indices in $\mathcal{P}'(x;d)$, which is equal to $\mathcal{P}(x)$, we obtain a partition of $\mathcal{P}(x)$ into mutually disjoint subsets: for some integer $p \geq 2$,

$$\mathcal{P}(x) = \mathcal{P}'(x;d) = \bigcup_{r=1}^{p} \mathcal{P}_r,$$

such that $JG^i(x)d$ are all equal for all indices i belonging the same subset $\mathcal{P}_r$, whereas $JG^i(x)d \neq JG^j(x)d$ if i belongs $\mathcal{P}_r$ and j belongs to $\mathcal{P}_s$, where $r \neq s$. Furthermore, there exists a scalar $\varepsilon > 0$ such that for any such pair $r \neq s$, $G^i(x+\tau d) \neq G^j(x+\tau d)$ for all τ belonging to the interval $(0, \varepsilon]$ and all indices i and j belonging to $\mathcal{P}_r$ and $\mathcal{P}_s$, respectively. We may restrict ε such that $x + \tau d$ belongs to the neighborhood $\mathcal{N}$ of x for all such τ; thus $\mathcal{P}(x+\tau d) \subseteq \mathcal{P}(x)$. For each τ, there exists an open neighborhood $\mathcal{N}_\tau$ of the vector $x+\tau d$ such that for all $y \in \mathcal{N}_\tau$, $\mathcal{P}(y) \subseteq \mathcal{P}(x+\tau d)$. The family of neighborhoods $\mathcal{N}_\tau$ covers the line segment $[x, x+\varepsilon d]$ as τ ranges over the interval $[0, \varepsilon]$. Since a line segment is a compact set, there exists a partition of the interval $[0, \varepsilon]$: for some positive integer t,

$$0 \equiv \tau_0 < \tau_1 < \cdots, < \tau_t = \varepsilon,$$

such that the union

$$\bigcup_{r=0}^{t} \mathcal{N}_{\tau_r}$$

also covers $[x, x+\varepsilon d]$. For any two consecutive integers r and $r+1$, the two line subsegments:

$$\mathcal{N}_{\tau_r} \cap [\,x, x+\varepsilon\, d\,] \quad \text{and} \quad \mathcal{N}_{\tau_{r+1}} \cap [\,x, x+\varepsilon\, d\,] \tag{4.6.1}$$

must overlap, and thus contain a point in common, say y. Since $\mathcal{P}(y)$ is a subset of both $\mathcal{P}'(x+\tau_r d)$ and $\mathcal{P}'(x+\tau_{r+1} d)$, it follow that there exists a unique member in the family

$$\{\,\mathcal{P}_1, \cdots, \mathcal{P}_p\,\}, \tag{4.6.2}$$

say $\mathcal{P}'$, such that for all points z belonging to the union of the two subsegments (4.6.1), $\mathcal{P}(z) \subseteq \mathcal{P}'$. Applying this argument to all the subsegments, we deduce that $\mathcal{P}(x) = \mathcal{P}'$, contradicting (4.6.2), which shows that at least two members from the family (4.6.2) are needed to cover $\mathcal{P}(x)$. This contradiction establishes (a). Statement (b) follows easily from this proof because for all vectors d, $G'(x;d)$ must be an element of $\{JG^1(x)d, \cdots, JG^k(x)d\}$. □

By Proposition **2.1.14**, we know that if a function G is continuously differentiable in a neighborhood of x, then G is a locally Lipschitz homeomorphism at x if and only if the Jacobian matrix $JG(x)$ is invertible. It turns out that necessary and sufficient conditions for a PC^1 function to be a Lipschitz homeomorphism near a point x can be obtained in terms of an object that generalizes the Jacobian matrix of a smooth function. The cornerstone of this object is Rademacher's theorem, which implies that every locally Lipschitz continuous function is F-differentiable almost everywhere.

4.6.2 Definition. Let $G : \mathcal{U} \to \mathbb{R}^n$, where $\mathcal{U}$ is an open subset of $\mathbb{R}^n$, be a given function that is locally Lipschitz continuous in a neighborhood $\mathcal{N} \subseteq \mathcal{U}$ of a vector x. Define the *limiting Jacobian* $\mathrm{Jac}(G,x)$ to be the (nonempty) set of limit points of sequences $\{JG(x^\nu)\}$, where each $x^\nu \in \mathcal{U}$ is a F-differentiable point of G and the sequence $\{x^\nu\}$ converges to x. Another term for $\mathrm{Jac}(G,x)$ is the *B-subdifferential* of G at x, denoted $\partial_B G(x)$. □

By Rademacher's theorem, there is at least one sequence $\{x^\nu\}$ of F-differentiable points that converges to x. Since the Jacobian matrices of G at F-differentiable points near x are bounded in norm by the locally Lipschitz modulus of G at x, it follows that all sequences $\{JG(x^\nu)\}$ as specified in Definition **4.6.2** must have at least one accumulation point; thus $\mathrm{Jac}(G,x)$ is indeed a nonempty set. In general, $\mathrm{Jac}(G,x)$ contains multiple matrices for a nondifferentiable function G; see Example **4.6.4**. In addition to playing an important role in the theory herein, the set $\mathrm{Jac}(G,x)$ is the building block of the "Clarke generalized Jacobian" of G at x, which is a fundamental object in nonsmooth analysis. See Section **7.1** for details of the Clarke calculus and Proposition **7.4.11** for some general properties of the B-subdifferential $\partial_B G$ as a multifuction. In what follows, we restrict our discussion of $\mathrm{Jac}(G,x)$ to a PC^1 function G. For such a G, the result below identifies the elements of $\mathrm{Jac}(G,x)$ as the Jacobian matrices $JG^i(x)$ for all indices i such that G^i is a *locally effective* piece of G near x.

4.6.3 Lemma. Let G be a PC^1 function in a neighborhood $\mathcal{N}$ of x with C^1 pieces given by $\{G^i : i \in \mathcal{K}\}$, where $\mathcal{K} \equiv \{1, \cdots, k\}$. Then

$$\mathrm{Jac}(G'(x;\cdot), 0) \subseteq \mathrm{Jac}(G, x) = \{\, JG^i(x) \,:\, i \in \tilde{\mathcal{P}}(x)\,\}, \tag{4.6.3}$$

where

$$\tilde{\mathcal{P}}(x) \equiv \{\, i \,:\, x \in \mathrm{cl\,int}\{\, z \in \mathcal{N} \,:\, i \in \mathcal{P}(z)\,\}\,\}.$$

Proof. Certainly $\mathrm{Jac}(G, x)$ contains $\{\, JG^i(x) \,:\, i \in \tilde{\mathcal{P}}(x)\,\}$. To show the reverse inclusion, we examine the structure of G further. For $i \in \mathcal{K}$, G is continuously differentiable in the interior of the set

$$D_i \equiv \{\, z \in \mathcal{N} \,:\, i \in \mathcal{P}(z)\,\}.$$

Since

$$\bigcup_{i \in \mathcal{K}} D_i = \mathcal{N}$$

and $\mathcal{K}$ is finite, the union of $\mathrm{int}\, D_i$ for $i \in \mathcal{K}$ is a dense, open subset of $\mathcal{N}$. Let $\tilde{D}_i \equiv \mathrm{cl\,int}\, D_i$ for each $i \in \mathcal{K}$. Note that

$$\bigcup_{i \in \mathcal{K}} \tilde{D}_i = \mathrm{cl}\, \mathcal{N}$$

and $\tilde{\mathcal{P}}(x) = \{i : x \in \tilde{D}_i\}$.

Since $\mathcal{K} \backslash \tilde{\mathcal{P}}(x)$ is finite, there is a scalar $\varepsilon > 0$ such that for each $i \notin \tilde{\mathcal{P}}(x)$, $\mathrm{dist}(x, \tilde{D}_i) > \varepsilon$. To put it differently, the union over $i \in \tilde{\mathcal{P}}(x)$ of $\tilde{D}_i$ contains a neighborhood of x. Hence G is PC^1 near x with pieces $\{G^i : i \in \tilde{\mathcal{P}}(x)\}$. Let $\{x^k\}$ be a sequence converging to x such that $JG(x^k)$ exists for every k and such that $\{JG(x^k)\}$ converges. The directional derivative of G at x^k is therefore a linear function. By Lemma **4.6.1**(b), we know that $G'(x^k, \cdot)$ is a continuous selection of the directional derivatives of the locally active pieces at x. We therefore conclude that $JG(x^k)$ coincides with $JG^i(x^k)$ for some $i \in \tilde{\mathcal{A}}(x)$. Simple arguments then yield that

$$\mathrm{Jac}(G, x) \subseteq \{\, JG^i(x) \,:\, i \in \tilde{\mathcal{P}}(x)\,\};$$

thus equality holds. By Lemma **4.6.1**, we have

$$G'(x; d) \in \{JG^i(x)d \,:\, i \in \tilde{\mathcal{P}}(x)\,\}$$

for all $d \in \mathbb{R}^n$. Consequently, it follows that

$$\mathrm{Jac}(G'(x;\cdot), 0) \subseteq \{\, JG^i(x) \,:\, i \in \tilde{\mathcal{P}}(x)\,\}$$

as desired. □

The family of C^1 functions

$$\{ G^i : i \in \tilde{\mathcal{P}}(x) \}$$

are the locally effective pieces of G near x. By the proof of part (a) of the above proposition, it follows that for every vector $d \in \mathbb{R}^n$, we have

$$G'(x,d) = JG^i(x)d$$

for every index i belonging to the subset $\tilde{\mathcal{P}}'(x;d)$ of $\tilde{\mathcal{P}}(x)$ consisting of indices j for which there exists a sequence of positive scalars $\{\tau_\nu\}$ converging to zero such that $G^j(x+\tau_\nu d) = G(x+\tau_\nu d)$. The example below illustrates such a family of locally effective pieces and Lemma **4.6.3**.

4.6.4 Example. Let the function $G : \mathbb{R}^2 \to \mathbb{R}^2$ be given by

$$G(x,y) \equiv \begin{pmatrix} \min(x,y) \\ |x|^3 - y \end{pmatrix}.$$

We want to evaluate $\operatorname{Jac} G(0,0)$. In a neighborhood of the origin, it is easy to see that there are four effective pieces of G:

$$G^1(x,y) \equiv \begin{pmatrix} x \\ x^3 - y \end{pmatrix}, \quad G^2(x,y) \equiv \begin{pmatrix} y \\ x^3 - y \end{pmatrix},$$

$$G^3(x,y) \equiv \begin{pmatrix} x \\ -x^3 - y \end{pmatrix}, \quad G^4(x,y) \equiv \begin{pmatrix} y \\ -x^3 - y \end{pmatrix}.$$

Thus

$$\operatorname{Jac} G(0,0) = \left\{ \begin{bmatrix} 1 & 0 \\ 0 & -1 \end{bmatrix}, \begin{bmatrix} 0 & 1 \\ 0 & -1 \end{bmatrix} \right\}.$$

For this example, it is not hard to verify that the inclusion in (4.6.3) holds as an equality at the origin. □

For a PC^1 function G, Lemma **4.6.3** implies that $\operatorname{Jac}(G,x)$ is a finite set. This is distinguished feature of a PC^1 function. In general, the limiting Jacobian of a locally Lipschitz continuous function must be a compact set, albeit not necessarily finite; see Example **7.1.3**.

The family of the locally effective pieces of a PC^1 function plays an important role in the next theorem, which is the main result of this section. This result gives several necessary and sufficient conditions for a PC^1 map G to be a locally Lipschitz homeomorphism near a vector x in its domain in terms of some global properties of the directional derivative function

$G'(x;\cdot)$, which is a PL map by Lemma **4.6.1**. The result can be considered a nonlinear version of Theorem **4.2.11** that pertains to a PA map. The sign condition in statement (b) of the following result can be described as coherent orientation of G at x.

4.6.5 Theorem. Let G be a PC^1 map near x. Consider the following statements:

(a) G is a locally Lipschitz homeomorphism at x.

(b) The matrices in $\text{Jac}(G,x)$ have the same nonzero determinantal sign, and $\text{ind}(G,x)$ is well defined and equal to ± 1.

(c) $G'(x;\cdot)$ is a globally Lipschitz homeomorphism on $\mathbb{R}^n$.

(d) $G'(x;\cdot)$ is bijective on $\mathbb{R}^n$.

(e) $G'(x;\cdot)$ is injective on $\mathbb{R}^n$.

It holds that (a) $\Leftrightarrow$ (b) $\Rightarrow$ (c) $\Leftrightarrow$ (d) $\Leftrightarrow$ (e). Moreover, if (a) holds, then the local inverse of G near x is also PC^1. Finally, all five statements (a)–(e) are equivalent if

$$\text{Jac}(G'(x;\cdot),0) \supseteq \text{Jac}(G,x). \tag{4.6.4}$$

Proof. (a) $\Rightarrow$ (b). Let $\mathcal{N}$ be an open neighborhood of x such that G is a Lipschitz homeomorphism on $\text{cl}\,\mathcal{N}$. Let $\tilde{G}$ denote the restriction of G to $\mathcal{N}$. The index of G at each vector v near x is well defined and equal to ± 1. Next, we show that all matrices in $\text{Jac}(G,x)$ are nonsingular. Suppose that v is a F-differentiable point of G that lies in $\mathcal{N}$ and that $JG(v)$ is singular. Let d be a nonzero vector such that $JG(v)d = 0$. For $\tau > 0$ sufficiently small, the vector $y(\tau) \equiv G(v+\tau d)$ belongs to the neighborhood $G(\mathcal{N})$. We have

$$\|\,\tilde{G}^{-1}(y(\tau)) - \tilde{G}^{-1}(y(0))\,\| \;=\; \tau\,\|\,d\,\|;$$

whereas $y(\tau) - y(0) = o(\tau)$, contradicting the Lipschitz continuity of $\tilde{G}$. Thus $JG(v)^{-1}$ exists. Moreover, we have

$$JG(v)^{-1} \;=\; J\tilde{G}^{-1}(v).$$

Since $J\tilde{G}^{-1}(v)$ is bounded in norm by the Lipschitz modulus of $\tilde{G}^{-1}$, it follows that for any sequence of vectors $\{x^\nu\}$ converging to x with each x^ν being a F-differentiable point of G, the sequence $\{\|JG(x^\nu)^{-1}\|\}$ must be bounded. Consequently, every matrix in $\text{Jac}(G,x)$ must be nonsingular. This establishes statement (b).

(b) $\Rightarrow$ (a). We show that G is injective in a neighborhood of x. Let $\mathcal{N}$ be an open neighborhood of x such that $\mathcal{P}(x') \subseteq \mathcal{P}(x)$ for all $x' \in \mathcal{N}$ and

$JG(y)$ has constant nonzero determinantal sign for all F-differentiability points $y \in \mathcal{N}$ of G. Since each $JG^i(x)$ for $i \in \tilde{\mathcal{P}}(x)$ is nonsingular, it follows that G^i is a local homeomorphism near x. For all y sufficiently close to $G(x)$, it follows that $G^{-1}(y) \cap \mathcal{N}$ is a finite set. We may now apply the same proof for the implication (c) $\Rightarrow$ (a) in Theorem **4.2.11** to complete the proof that G is injective in a neighborhood of x. The details are not repeated.

The above proof shows that if (a) or (b) holds, then each G^i is a C^1 local homeomorphism near x for all $i \in \tilde{\mathcal{P}}(x)$. Thus each G^i has a local inverse near x that is C^1. Clearly, for all y sufficiently near $G(x)$,

$$G^{-1}(y) \in \{ (G^i)^{-1}(y) : i \in \tilde{\mathcal{P}}(x) \},$$

Therefore, the local inverse of G near x is also PC^1.

(b) $\Rightarrow$ (c). As mentioned above, $G'(x;\cdot)$ is a PL map. Moreover, Lemma **4.6.3** implies that $\text{Jac}(G'(x;\cdot),0)$ consists of matrices with the same nonzero determinantal sign. Thus $G'(x;d) \neq 0$ for all $d \neq 0$; thus $\text{ind}(G'(x;\cdot),0)$ is well defined. We claim that this index is equal to ± 1. Since G is B-differentiable, we have

$$G(x+d) - G(x) - G'(x;d) = o(\|d\|),$$

Therefore, $\text{ind}(G'(x;\cdot),0)$ is equal to $\text{ind}(G,x)$ and the claim is valid. Consequently, $G'(x;\cdot)$ is a local homeomorphism near the origin. Since the directional derivative is a positively homogeneous function in the second argument, it follows that $G'(x;\cdot)$ is a global homeomorphism, and thus a globally Lipschitz homeomorphism, on $\mathbb{R}^n$.

(c) $\Leftrightarrow$ (d) $\Leftrightarrow$ (e). This follows from Theorem **4.2.11** and the PL property of $G'(x;\cdot)$.

Finally, if $\text{Jac}(G'(x;\cdot),0) \supseteq \text{Jac}(G,x)$, then equality must hold. Therefore, the equivalence of (c) and (a) follows from the above proof by simply exchanging the roles of G at x with $G'(x;\cdot)$ at the origin. □

A key requirement of Theorem **4.6.5** is the inclusion (4.6.4). When G is the normal map $\mathbf{F}_K^{\text{nor}}$ associated with a VI (K,F), the following technical result provides the key to establish this requirement.

4.6.6 Lemma. Let K be given by (4.5.1), where $g : \mathcal{D} \supset K \to \mathbb{R}^m$ is twice continuously differentiable on the open set $\mathcal{D}$ with each component function g_i being convex. If the CRCQ holds at $\bar{x} \equiv \Pi_K(x)$, then

$$\text{Jac}(\Pi_K'(x;\cdot),0) \supseteq \text{Jac}(\Pi_K,x).$$

Proof. Let $\{x^k\}$ be a sequence of F-differentiable points of Π_K that converges to x and such that the sequence of Jacobian matrices $\{J\Pi_K(x^k)\}$ converges to a matrix P^∞. Write

$$\bar{x} \equiv \Pi_K(x), \quad \bar{x}^k \equiv \Pi_K(x^k) \quad \text{and} \quad P^k \equiv J\Pi_K(x^k).$$

Since the CRCQ continues to hold at $\bar{x}^k$ for all k sufficiently large, there exists $\lambda^k \in \mathcal{M}_\pi(x^k)$ such that the gradients

$$\{\, \nabla g_i(\bar{x}^k) \,:\, i \in \operatorname{supp}(\lambda^k) \,\}$$

are linearly independent. By taking subsequences, we may assume without loss of generality that the CRCQ holds at every $\bar{x}^k$ and $\operatorname{supp}(\lambda^k) = \mathcal{K}$, a constant index set. We may further assume that $\mathcal{I}(\bar{x}^k)$ is equal to a constant index set $\mathcal{J}$ for all k. Obviously, $\mathcal{K} \subseteq \mathcal{J} \subseteq \mathcal{I}(\bar{x})$. By the CRCQ at $\bar{x}$, the limiting gradients $\{\nabla g_i(\bar{x}) \,:\, i \in \mathcal{K}\}$ are linearly independent; moreover, the sequence $\{\lambda^k\}$ converges to some multiplier λ^∞ that belongs to $\mathcal{M}_\pi(x)$.

By Theorem **4.5.3**, for every vector $d \in \mathbb{R}^n$,

$$J\Pi_K(x^k)d = \Pi_{\mathcal{C}^k}^{G(\lambda^k)}(d),$$

where $\mathcal{C}^k \equiv \mathcal{C}_\pi(x^k, K)$ is the critical cone of K at x^k. In terms of the multiplier λ^k, we have

$$\mathcal{C}^k = \{d \,:\, \forall\, i \in \mathcal{J},\ \nabla g_i(\bar{x}^k)^T d \le 0 \text{ with equality if } \lambda_i^k > 0\,\}.$$

Since Π_K is F-differentiable at x^k, it follows from Corollary **4.5.5** that $\mathcal{C}^k$ is a linear subspace; as such we must have

$$\mathcal{C}^k = \{d \,:\, \forall\, i \in \mathcal{J},\ \nabla g_i(\bar{x}^k)^T d = 0\,\};$$

thus, $\mathcal{C}^k = \mathcal{N}(\bar{x}^k, \mathcal{J})$. By Proposition **3.2.9**, it follows that

$$\lim_{k \to \infty} \mathcal{C}^k = \mathcal{N}(\bar{x}, \mathcal{J}),$$

which we denote $\bar{\mathcal{L}}$. By Lemma **2.8.2**, we deduce

$$P^\infty = \lim_{k \to \infty} J\Pi_K(x^k) = \Pi_{\bar{\mathcal{L}}}^{G(\lambda^\infty)}.$$

Let $\mathcal{F} \equiv \bar{\mathcal{L}} \cap \mathcal{C}_\pi(x; K)$. We claim that $\mathcal{F}$ is a nonempty face of $\mathcal{C}_\pi(x; K)$ whose linear span is $\bar{\mathcal{L}}$; that is $\mathcal{F} - \mathcal{F} = \bar{\mathcal{L}}$. To see this, let $\mathcal{J}'$ be the set of indices $j \in \mathcal{I}(\bar{x})$ such that $\nabla g_j(\bar{x})$ is linearly dependent on the family of gradients $\{\nabla g_i(\bar{x}) : i \in \mathcal{J}\}$. We then have

$$\bar{\mathcal{L}} = \{d \,:\, \forall\, i \in \mathcal{J} \cup \mathcal{J}',\ \nabla g_i(\bar{x})^T d = 0\,\}.$$

Moreover, for each $i \in \mathcal{I}(\bar{x}) \setminus (\mathcal{J} \cup \mathcal{J}')$, there exists a vector $d^i \in \bar{\mathcal{L}}$ such that $\nabla g_i(\bar{x})^T d^i < 0$; hence there exists $\bar{d} \in \bar{\mathcal{L}}$ such that $\nabla g_i(\bar{x})^T \bar{d} < 0$ for all $i \in \mathcal{I}(\bar{x}) \setminus (\mathcal{J} \cup \mathcal{J}')$. By the representation of $\mathcal{C}_\pi(x; K)$ in terms of the multiplier λ^∞, we have

$$\mathcal{F} = \{ d : \forall i \in \mathcal{I}(\bar{x}), \quad \nabla g_i(\bar{x})^T d \leq 0 \quad \text{with equality if } \lambda_i^\infty > 0 \text{ or } i \in \mathcal{J} \cup \mathcal{J}' \}.$$

Since $\lambda_i^\infty > 0$ implies $\lambda_i^k > 0$ for all k sufficiently large, which means that $i \in \mathcal{J}$, it follows that

$$\mathcal{F} = \{ d : \forall i \in \mathcal{I}(\bar{x}), \, \nabla g_i(\bar{x}^k)^T d \leq 0 \text{ with equality if } i \in \mathcal{J} \cup \mathcal{J}' \}.$$

Thus $\mathcal{F}$ is a nonempty face of $\mathcal{C}_\pi(x; K)$. Since $\mathcal{F}$ is a polyhedral cone, its linear span is equal to $\mathcal{F} - \mathcal{F}$. Moreover, the existence of $\bar{d}$ implies that this span is equal to $\bar{\mathcal{L}}$.

By Theorem **4.5.3**, we have

$$\Pi'_K(x; \cdot) = \Pi_{\mathcal{C}}^{G(\lambda^\infty)},$$

where $\mathcal{C} \equiv \mathcal{C}_\pi(x; K)$. To show that $P^\infty = \Pi_{\bar{\mathcal{L}}}^{G(\lambda^\infty)}$ is an element of $\text{Jac}(\Pi'_K(x; \cdot), 0)$, we need to identify a sequence of F-differentiable points $\{y^k\}$ of $\Pi_{\mathcal{C}}^{G(\lambda^\infty)}$ that converges to zero such that

$$\lim_{k \to \infty} J\Pi_{\mathcal{C}}^{G(\lambda^\infty)}(y^k) = P^\infty.$$

Let $B \equiv \sqrt{G(\lambda^\infty)}$ be the square root of $G(\lambda^\infty)$. We have

$$\Pi_{\mathcal{C}}^{G(\lambda^\infty)} = B^{-1} \circ \Pi_{\mathcal{C}'} \circ B^{-1},$$

where $\mathcal{C}' \equiv B\mathcal{C}$. Since $B\mathcal{F}$ is a nonempty face of $\mathcal{C}'$, Proposition **4.1.5** implies the existence of an F-differentiable point y of $\Pi_{\mathcal{C}'}$ such that in a neighborhood of y, $\Pi_{\mathcal{C}'}$ is the Euclidean projection onto

$$B\mathcal{F} - B\mathcal{F} = B(\mathcal{F} - \mathcal{F}) = B\bar{\mathcal{L}}.$$

Consequently, $\Pi_{\mathcal{C}}^{G(\lambda^\infty)}$ is F-differentiable at the point By and its Jacobian matrix there is the mapping

$$B^{-1} \circ \Pi_{B\bar{\mathcal{L}}} \circ B^{-1} = \Pi_{\bar{\mathcal{L}}}^{G(\lambda^\infty)} = P^\infty.$$

Thus we have

$$\frac{\partial \Pi'_K(x; d)}{\partial d} \Big|_{d=By} = P^\infty.$$

It is only left to observe, since $\Pi'_K(x;\cdot)$ is positively homogenous, that P^∞ remains the Jacobian matrix of $\Pi'_K(x;\cdot)$ at tBy for each $t > 0$. Consequently, we have

$$P^\infty = \lim_{t\downarrow 0} J_d\Pi'_K(x;d)|_{d=tBy}.$$

This is what we need to show. □

Combining Theorem **4.6.5** and Lemma **4.6.6**, we deduce that under the assumptions of the lemma, $\mathbf{F}^{\text{nor}}_K$ is a locally Lipschitz homeomorphism near a vector z if and only if $(\mathbf{F}^{\text{nor}}_K)'(z;\cdot)$ is a bijection. The full implication of this conclusion is explored in the next chapter; see in particular Theorem **5.3.24**.

4.7 Projection onto a Parametric Set

In this section, we consider the Euclidean projector onto a parameter-dependent set. Specifically, assume that for each $\mathbf{p} \in \mathcal{P}$,

$$K(\mathbf{p}) \equiv \{ x \in \mathbb{R}^n : g(x,\mathbf{p}) \leq 0 \}, \tag{4.7.1}$$

where $g : \mathbb{R}^n \times \mathcal{P} \to \mathbb{R}^m$ is such that for each $\mathbf{p} \in \mathcal{P} \subseteq \mathbb{R}^p$, each $g_i(\cdot,\mathbf{p})$ is continuously differentiable and convex. We wish to establish some properties of the parametric projector $\Pi_{K(\mathbf{p})}(x)$ as a function of the pair $(x,\mathbf{p})$. From Lemma **2.8.2**, we know that if the limit holds:

$$\lim_{\mathbf{p}\to\mathbf{p}^*} K(\mathbf{p}) = K(\mathbf{p}^*), \tag{4.7.2}$$

then $\Pi_{K(\mathbf{p})}(x)$ is continuous at the pair $(x,\mathbf{p}^*)$ for every $x \in \mathbb{R}^n$. A natural question is then: with $K(\mathbf{p})$ finitely representable as given above, when does the limit (4.7.2) hold? The following result, which applies to a nonconvex set $K(\mathbf{p})$ and does not involve the projector, shows that the MFCQ is sufficient for this purpose. This result and its proof are both classical from nonlinear programming theory.

4.7.1 Proposition. Let $g : \mathbb{R}^n \times \mathcal{P} \to \mathbb{R}^m$ and $h : \mathbb{R}^n \times \mathcal{P} \to \mathbb{R}^\ell$ be continuously differentiable, where $\mathcal{P}$ is an open subset of $\mathbb{R}^p$. Let

$$K(\mathbf{p}) \equiv \{ x \in \mathbb{R}^n : g(x,\mathbf{p}) \leq 0,\ h(x,\mathbf{p}) = 0 \}.$$

If the MFCQ holds at a vector $x^* \in K(\mathbf{p}^*)$, then for every sequence $\{\mathbf{p}^k\}$ converging to $\mathbf{p}^*$, there exists a sequence $\{x^k\}$ converging to x^* such that $x^k \in K(\mathbf{p}^k)$ for all k. Consequently, if the MFCQ holds at every vector in $K(\mathbf{p}^*)$, then (4.7.2) holds.

Proof. The MFCQ at $x^* \in K(\mathbf{p}^*)$ means that the gradients

$$\{ \nabla_x h_j(x^*, \mathbf{p}^*) \, : \, j = 1, \ldots, \ell \} \tag{4.7.3}$$

are linearly independent and there exists a vector v satisfying

$$\begin{aligned} v^T \nabla_x h_j(x^*, \mathbf{p}^*) &= 0, \quad \forall j = 1, \ldots, \ell \\ v^T \nabla_x g_i(x^*, \mathbf{p}^*) &< 0, \quad \forall i \in \mathcal{I}(x^*, \mathbf{p}^*), \end{aligned} \tag{4.7.4}$$

where

$$\mathcal{I}(x^*, \mathbf{p}^*) \equiv \{ i \, : \, g_i(x^*, \mathbf{p}^*) = 0 \}.$$

Since the gradients (4.7.3) are linearly independent, we may assume without loss of generality that the square matrix

$$\mathbf{J} \equiv \left(\frac{\partial h_j(x^*, \mathbf{p}^*)}{\partial x_i} \right)_{i,j=1}^{\ell}$$

is nonsingular. We denote the first ℓ components of x by y and the last $n-\ell$ components by u. By the implicit function theorem for smooth functions, there exist an open neighborhood $\mathcal{V} \subset \mathbb{R}^\ell$ of the vector $y^* \equiv (x_1^*, \cdots, x_\ell^*)$, an open neighborhood $\mathcal{U} \subset \mathbb{R}^{n-\ell}$ of the vector $u^* \equiv (x_{\ell+1}^*, \cdots, x_n^*)$, an open neighborhood $\mathcal{W} \subset \mathbb{R}^p$ of $\mathbf{p}^*$, and a continuous function $y : \mathcal{U} \times \mathcal{W} \to \mathcal{V}$ such that (i) $y(u^*, \mathbf{p}^*) = y^*$, (ii) $h(y(u, \mathbf{p}), (u, \mathbf{p})) = 0$ for all $(u, \mathbf{p}) \in \mathcal{U} \times \mathcal{W}$, and (iii) $y(u, \mathbf{p})$ is F-differentiable at $(u^*, \mathbf{p}^*)$ with

$$J_u y(u^*, \mathbf{p}^*) = -\mathbf{J}^{-1} J_u h(x^*, \mathbf{p}^*)$$

and

$$J_{\mathbf{p}} y(u^*, \mathbf{p}^*) = -\mathbf{J}^{-1} J_{\mathbf{p}} h(x^*, \mathbf{p}^*).$$

Comparing the first identity with the first set of equations in (4.7.4) yields

$$J_u y(u^*, \mathbf{p}^*) v_u = v_y, \tag{4.7.5}$$

where v_y and v_u denote, respectively, the first ℓ and last $n - \ell$ components of the vector v. For each $\mathbf{p} \in \mathcal{W}$ and each $\tau > 0$ sufficiently small, define $x(\tau, \mathbf{p})$ to be the vector whose last $n - \ell$ components are given by the components of the vector $u^* + \tau v_u$ and the first ℓ components are given by the components of the vector $y(u^* + \tau v_u, \mathbf{p})$. Clearly $h(x(\tau, \mathbf{p}), \mathbf{p}) = 0$ for all such $(\tau, \mathbf{p})$ and

$$\lim_{(\tau, \mathbf{p}) \to (0, \mathbf{p}^*)} x(\tau, \mathbf{p}) = x^*.$$

Consequently, by restricting the neighborhood $\mathcal{W}$ and the scalar τ we have, for all $p \in \mathcal{W}$ and all $\tau > 0$ sufficiently small, $g_i(x(\tau, \mathbf{p}), \mathbf{p}) < 0$ for all $i \notin \mathcal{I}(x^*, \mathbf{p}^*)$. For each $i \in \mathcal{I}(x^*, \mathbf{p}^*)$, we have

$$\begin{aligned} g_i(x(\tau,\mathbf{p}),\mathbf{p}) &= g_i(x^*,\mathbf{p}^*) + \nabla_x g_i(x^*,\mathbf{p}^*)^T(\,x(\tau,\mathbf{p}) - x^*\,) + \\ &\quad \nabla_{\mathbf{p}} g_i(x^*,\mathbf{p}^*)^T(\,\mathbf{p}-\mathbf{p}^*\,) + o(\tau) + o(\|\,\mathbf{p}-\mathbf{p}^*\,\|) \\ &= \sum_{j=1}^{\ell} \frac{\partial g_i(x^*,\mathbf{p}^*)}{\partial x_j}\,(\,y_j(u^*+\tau v_u,\mathbf{p}) - y_j^*\,) + \tau \sum_{j=\ell+1}^{n} \frac{\partial g_i(x^*,\mathbf{p}^*)}{\partial x_j}\, v_j \\ &\quad + \nabla_{\mathbf{p}} g_i(x^*,\mathbf{p}^*)^T(\,\mathbf{p}-\mathbf{p}^*\,) + o(\tau) + o(\|\,\mathbf{p}-\mathbf{p}^*\,\|) \\ &= \tau\,\nabla_x g_i(x^*,\mathbf{p}^*)^T v + O(\,\|\,\mathbf{p}-\mathbf{p}^*\,\|\,) + o(\tau) + o(\|\,\mathbf{p}-\mathbf{p}^*\,\|), \end{aligned}$$

where the last equality follows from (4.7.5). Consequently, with $\mathbf{p}$ sufficiently close to $\mathbf{p}^*$, we may choose $\tau > 0$ such that $g_i(x(\tau, \mathbf{p}), \mathbf{p}) < 0$ for all $i \in \mathcal{I}(x^*, \mathbf{p}^*)$, due to the fact that $\nabla_x g_i(x^*, \mathbf{p}^*)^T v < 0$ for all such i. Consequently, for a sequence of $\{\mathbf{p}^k\}$ converging to $\mathbf{p}^*$, there exists for each k a vector $x^k \in K(\mathbf{p}^k)$ and the sequence $\{x^k\}$ converges to x^*.

To show the second assertion of the proposition, we note that the set-valued map $K(\mathbf{p})$ is clearly closed at $\mathbf{p}^*$, by continuity of g and h. Hence

$$\limsup_{\mathbf{p}\to\mathbf{p}^*} K(\mathbf{p}) \subseteq K(\mathbf{p}^*).$$

Let $\bar{x} \in K(\mathbf{p}^*)$ be arbitrary. Since the MFCQ holds at $\bar{x}$, the first part of the proposition implies

$$\lim_{\mathbf{p}\to\mathbf{p}^*} \operatorname{dist}(\bar{x}, K(\mathbf{p})) = 0.$$

This means that $\bar{x}$ belongs to $\liminf_{\mathbf{p}\to\mathbf{p}^*} K(\mathbf{p})$. Consequently, (4.7.2) follows. □

4.7.2 Remark. It is a known result from the stability theory of differentiable inequality systems that if the MFCQ holds at $x^* \in K(\mathbf{p}^*)$, then for all $(x, \mathbf{p})$ in a neighborhood of $(x^*, \mathbf{p}^*)$,

$$\operatorname{dist}(x, K(\mathbf{p})) = O(\,\|\,h(x,\mathbf{p})\,\| + \|\,\max(\,g(x,\mathbf{p}),\,0\,)\,\|\,).$$

Proposition **4.7.1** can be readily established based on this result. □

The assumption that the MFCQ holds at all vectors in $K(\mathbf{p}^*)$ is quite restrictive in general. Nevertheless, if for all $\mathbf{p}$ near $\mathbf{p}^*$, $K(\mathbf{p})$ is the convex set (4.7.1) defined by convex inequalities only, we can invoke Proposition **3.2.7** by assuming the MFCQ at the single vector $x^* \equiv \Pi_{K(\mathbf{p}^*)}(x)$.

Combining this observation with Proposition **4.7.1** and Lemma **2.8.2**, we obtain the following result, which shows that the MFCQ at x^* is sufficient for the parametric projector $\Pi_{K(\mathbf{p})}(x)$ to be continuous at every pair $(x, \mathbf{p})$, where $\mathbf{p}$ is sufficiently near $\mathbf{p}^*$.

4.7.3 Corollary. Let $K(\mathbf{p})$ be given by (4.7.1) where each $g_i(\cdot, \mathbf{p})$ is convex for each $\mathbf{p} \in \mathcal{P}$. Suppose that g is continuously differentiable in a neighborhood of the pair $(x^*, \mathbf{p}^*)$, where $\mathbf{p}^* \in \mathcal{P}$ and $x^* \equiv \Pi_{K(\mathbf{p}^*)}(x)$. If the MFCQ holds at $x^* \in K(\mathbf{p}^*)$, then there exists a neighborhood $\mathcal{W}$ of $\mathbf{p}^*$ such that the mapping

$$(x, \mathbf{p}) \mapsto \Pi_{K(\mathbf{p})}(x)$$

is continuous at every pair $(x, \mathbf{p}) \in \mathbb{R}^n \times \mathcal{W}$.

Proof. By the definition of $K(\mathbf{p}^*)$ and by Proposition **3.2.7**, it follows that there exists a vector u^* satisfying $g(u^*, \mathbf{p}^*) < 0$. Since g is continuous, it follows that there exists a neighborhood $\mathcal{W}$ of $\mathbf{p}^*$ such that for each $\mathbf{p}$ in $\mathcal{W}$, a vector u exists satisfying $g(u, \mathbf{p}) < 0$. In turn, this implies that the MFCQ holds at all feasible vectors belonging to $K(\mathbf{p})$ for all $\mathbf{p} \in \mathcal{W}$. By Proposition **4.7.1**, we have for all such $\mathbf{p}$,

$$\lim_{\mathbf{p}' \to \mathbf{p}} K(\mathbf{p}') = K(\mathbf{p}).$$

In turn, by Lemma **2.8.2**, it follows that $\Pi_{K(\cdot)}(\cdot)$ is continuous at every $(x, \mathbf{p}) \in \mathbb{R}^n \times \mathcal{W}$. □

Without further assumptions, it is possible for the parametric projector $\Pi_{K(\mathbf{p})}(x)$, with x fixed, to be directionally differentiable at a given $\mathbf{p}^*$ without the directional derivative being continuous. This then implies that the function $\mathbf{p} \mapsto \Pi_{K(\mathbf{p})}(x)$ is not Lipschitz continuous near $\mathbf{p}^*$. The following example illustrates this situation.

4.7.4 Example. Let

$$K(\mathbf{p}) \equiv \{ (x_1, x_2) \in \mathbb{R}^2 : x_1 \leq 0, \text{ and } x_1 + p_1 x_2 + p_2 \leq 0 \}, \quad \mathbf{p} \in \mathbb{R}^2.$$

Fix the point $x \equiv (1, 0)$ and consider $\Pi_{K(\mathbf{p})}(x)$ for all $\mathbf{p}$. By an easy calculation, we can show that

$$\Pi_{K(\mathbf{p})}(1,0) = \begin{cases} (0,0) & \text{if } p_2 \leq 0 \\ (0, -p_1^{-1} p_2) & \text{if } 0 < p_2 \leq p_1^2 \\ \dfrac{1}{1+p_1^2} (p_1^2 - p_2, -p_1 - p_1 p_2) & \text{if } p_1^2 \leq p_2. \end{cases}$$

We compute the directional derivative of the function $f(\mathbf{p}) \equiv \Pi_{K(\mathbf{p})}(1,0)$ at $\mathbf{p}^* \equiv (0,0)$ along the direction $d\mathbf{p} \equiv (1,\delta)$ for all $\delta > 0$. For $\tau > 0$ sufficiently small, we have $\tau^2 < \tau\delta$; hence

$$f(\tau d\mathbf{p}) = \frac{1}{1+\tau^2}\,(\,\tau^2 - \tau\,\delta,\; -\tau - \tau^2\delta\,).$$

Consequently,

$$f'(0, d\mathbf{p}) = \lim_{\tau \to 0+} \frac{f(\tau d\mathbf{p})}{\tau} = (\,-\delta,\; -1\,),$$

which yields

$$\lim_{\delta \to 0+} f'((0,0);(1,\delta)) = (\,0,\; -1\,).$$

This is not equal to $f'((0,0);(1,0)) = (0,0)$. Thus $f'(0;\cdot)$ is not continuous at $(1,0)$. □

It turns out that if the vector $\Pi_{K(\mathbf{p}^*)}(x)$ satisfies both the MFCQ and the CRCQ, then the parametric projector $\Pi_{K(\mathbf{p})}(z)$ is jointly Lipschitz continuous in $(z,\mathbf{p})$ near $(x,\mathbf{p}^*)$; in fact, it is a PC^1 function there. We formally state this assertion in the following result.

4.7.5 Theorem. Let $K(\mathbf{p})$ be given by (4.7.1) where each $g_i(\cdot,\mathbf{p})$ is continuous and convex for each $\mathbf{p} \in \mathcal{P}$. Suppose that g is twice continuously differentiable in a neighborhood of the pair $(x^*,\mathbf{p}^*)$, where $\mathbf{p}^* \in \mathcal{P}$ and $x^* \equiv \Pi_{K(\mathbf{p}^*)}(x)$. If the MFCQ and the CRCQ hold at $x^* \in K(\mathbf{p}^*)$, then the function

$$(\,z,\mathbf{p}\,) \mapsto \Pi_{K(\mathbf{p})}(z)$$

is PC^1 near $(x,\mathbf{p}^*)$; thus, it is Lipschitz continuous there.

Proof. The proof is very similar to that of Theorem **4.5.2**. Let $\mathcal{B}(x,\mathbf{p}^*)$ be the collection of index subsets $\mathcal{J} \subseteq \mathcal{I}(x^*,\mathbf{p}^*)$ such that the family of gradient vectors

$$\{\,\nabla_x g_i(x^*,\mathbf{p}^*)\,:\, i \in \mathcal{J}\,\}$$

are linearly independent. Let $\mathcal{M}_\pi(x,\mathbf{p}^*)$ denote the set of multipliers for the projection problem:

$$\begin{array}{ll} \text{minimize} & \tfrac{1}{2}\,(\,y-x\,)^T(\,y-x\,) \\[4pt] \text{subject to} & y \in K(\mathbf{p}^*); \end{array}$$

that is, λ belongs to $\mathcal{M}_\pi(x,\mathbf{p}^*)$ if and only if

$$\begin{gathered} x^* - x + \sum_{i \in \mathcal{I}(x^*,\mathbf{p}^*)} \lambda_i\,\nabla_x g_i(x^*,\mathbf{p}^*) = 0 \\ \lambda_i \geq 0 \quad \forall\, i \in \mathcal{I}(x^*,\mathbf{p}^*). \end{gathered}$$

There exists at least one multiplier $\lambda \in \mathcal{M}_\pi(x, \mathbf{p}^*)$ such that $\text{supp}(\lambda)$ belongs to $\mathcal{B}(x, \mathbf{p}^*)$. Let $\mathcal{B}'(x, \mathbf{p}^*)$ be the subcollection of $\mathcal{B}(x, \mathbf{p}^*)$ consisting of index sets $\mathcal{J} \in \mathcal{B}(x, \mathbf{p}^*)$ for which there exists a vector λ in $\mathcal{M}_\pi(x, \mathbf{p}^*)$ such that $\text{supp}(\lambda) \subseteq \mathcal{J}$. Corresponding to each $\mathcal{J}$ in $\mathcal{B}'(x, \mathbf{p}^*)$, let $\Lambda(\mathcal{J})$ denote the (finite) subset of $\mathcal{M}_\pi(x, \mathbf{p}^*)$ consisting of multipliers λ in $\mathcal{M}_\pi(x, \mathbf{p}^*)$ such that $\text{supp}(\lambda) \subseteq \mathcal{J}$. For each $\mathcal{J}$ in $\mathcal{B}'(x, \mathbf{p}^*)$, consider the function

$$\Phi_{\mathcal{J}} : \mathbb{R}^{2n+|\mathcal{J}|+p} \to \mathbb{R}^{n+|\mathcal{J}|}$$

$$\begin{pmatrix} y \\ z \\ \eta_{\mathcal{J}} \\ \mathbf{p} \end{pmatrix} \mapsto \begin{pmatrix} y - z + \sum_{j \in \mathcal{J}} \eta_j \nabla_x g_j(y, \mathbf{p}) \\ \\ -g_{\mathcal{J}}(y, \mathbf{p}) \end{pmatrix},$$

which vanishes at the tuple $(x^*, x, \lambda_{\mathcal{J}}, \mathbf{p}^*)$ for all λ in $\Lambda(\mathcal{J})$. The partial Jacobian matrix of $\Phi_{\mathcal{J}}$ with respect to the pair $(y, \eta_{\mathcal{J}})$ at this tuple is equal to

$$\begin{bmatrix} \mathbf{I}_n + \sum_{j \in \mathcal{J}} \lambda_j \nabla^2_{xx} g_j(x^*, \mathbf{p}^*) & J_x g_{\mathcal{J}}(x^*, \mathbf{p}^*)^T \\ \\ -J_x g_{\mathcal{J}}(x^*, \mathbf{p}^*) & 0 \end{bmatrix},$$

which has a nonzero determinant because the matrix

$$\mathbf{I}_n + \sum_{j \in \mathcal{J}} \lambda_j \nabla^2_{xx} g_j(x^*, \mathbf{p}^*)$$

is nonsingular by the convexity of each $g_j(\cdot, \mathbf{p}^*)$ and the nonnegativity of each λ_j and the matrix $J_x g_{\mathcal{J}}(x^*, \mathbf{p}^*)$ has full row rank (see Exercise **1.8.9**). By the classical implicit function theorem applied to $\Phi_{\mathcal{J}}$ and each tuple $(x^*, x, \lambda_{\mathcal{J}}, \mathbf{p}^*)$, where $\lambda \in \Lambda(\mathcal{J})$, with $(y, \eta_{\mathcal{J}})$ as the primary variable and $(z, \mathbf{p})$ as the parameter, there exist open neighborhoods $\mathcal{V}(\mathcal{J}, \lambda)$ of x^*, $\mathcal{Z}(\mathcal{J}, \lambda)$ of x, $\mathcal{W}(\mathcal{J}, \lambda)$ of $\mathbf{p}^*$, and $\mathcal{U}(\mathcal{J}, \lambda)$ of $\lambda_{\mathcal{J}}$ and a continuously differentiable function

$$\mathbf{x} : \mathcal{Z}(\mathcal{J}, \lambda) \times \mathcal{W}(\mathcal{J}, \lambda) \to \mathcal{V}(\mathcal{J}, \lambda) \times \mathcal{U}(\mathcal{J}, \lambda)$$

such that for every pair $(z, \mathbf{p})$ in the neighborhood $\mathcal{Z}(\mathcal{J}, \lambda) \times \mathcal{W}(\mathcal{J}, \lambda)$, $\mathbf{x}(z, \mathbf{p})$ is the unique pair $(y, \eta_{\mathcal{J}})$ in $\mathcal{V}(\mathcal{J}, \lambda) \times \mathcal{U}(\mathcal{J}, \lambda)$ satisfying

$$\Phi_{\mathcal{J}}(y, z, \eta_{\mathcal{J}}, \mathbf{p}) = 0.$$

Let $\mathbf{x}_{\mathcal{J},\lambda}$ be the y-part of the function $\mathbf{x}$. By shrinking the neighborhoods $\mathcal{Z}(\mathcal{J}, \lambda)$ and $\mathcal{W}(\mathcal{J}, \lambda)$ if necessary, we may assume that for all pairs $(z, \mathbf{p})$

in $\mathcal{Z}(\mathcal{J},\lambda) \times \mathcal{W}(\mathcal{J},\lambda)$, $\Pi_{K(\mathbf{p})}(z)$ belongs to the neighborhood $\mathcal{V}(\mathcal{J},\lambda)$, by the continuity of this parametric projector at $(x,\mathbf{p}^*)$. Let

$$\mathcal{Z} \equiv \bigcap_{\mathcal{J}\in\mathcal{B}'(x,\mathbf{p}^*)} \bigcap_{\lambda\in\Lambda(\mathcal{J})} \mathcal{Z}(\mathcal{J},\lambda)$$

and

$$\mathcal{W} \equiv \bigcap_{\mathcal{J}\in\mathcal{B}'(x,\mathbf{p}^*)} \bigcap_{\lambda\in\Lambda(\mathcal{J})} \mathcal{W}(\mathcal{J},\lambda).$$

It suffices to show that for all $(z,\mathbf{p})$ sufficiently close to $(x,\mathbf{p}^*)$, $\Pi_{K(\mathbf{p})}(z)$ belongs to the family

$$\{\,\mathbf{x}_{\mathcal{J},\lambda}(z,\mathbf{p}) \,:\, \lambda \in \Lambda(\mathcal{J}),\ \mathcal{J} \in \mathcal{B}'(x,\mathbf{p}^*)\,\}.$$

Utilizing the joint continuity of $\Pi_{K(\mathbf{p})}(z)$ at $(x,\mathbf{p}^*)$, we can complete the proof as in Theorem **4.5.2**. □

4.8 Exercises

4.8.1 Let K be a closed convex subset of $\mathbb{R}^n$ and $x \in K$ be arbitrary. Let $\mathcal{T}$ and $\mathcal{N}$ denote the tangent cone $\mathcal{T}(x;K)$ and normal cone $\mathcal{N}(x;K)$ of K at x, respectively.

(a) Show that

$$\lim_{\tau\downarrow 0} \frac{\Pi_K(x+\tau u) - x - \tau\, u}{\tau} = 0, \quad \forall\, u \in \mathcal{T}.$$

[Hint: let $\{x^k\} \subset K$ be a sequence converging to x and $\{\tau_k\}$ be a sequence of positive scalars converging to 0 such that

$$\lim_{k\to\infty} \frac{x^k - x}{\tau_k} = u.$$

Show that for $\tau \in (0,\tau_k]$,

$$\frac{\|\,\Pi_K(x+\tau u) - x - \tau\, u\,\|}{\tau} \leq \frac{\|\,x^k - x - \tau_k\, u\,\|}{\tau_k}.$$

Complete the proof by taking limits.] Deduce that

$$\lim_{\tau\downarrow 0} \frac{\operatorname{dist}(x+\tau u, K)}{\tau} = 0 \Leftrightarrow u \in \mathcal{T}.$$

(b) Similar to the proof of Proposition **3.1.3**, use (a) to show that

$$\lim_{u(\in\mathcal{T})\to 0} \frac{\Pi_K(x+u) - x - u}{\|\,u\,\|} = 0.$$

(c) For any $h \in \mathbb{R}^n$, write $h = u + v$, where $u \equiv \Pi_{\mathcal{T}}(h)$ and $v \equiv \Pi_{\mathcal{N}}(h)$. We may further write

$$\Pi_K(x+h) - x - h = \Pi_K(x+h) - \Pi_{x+\mathcal{T}}(x+h) + \Pi_{x+\mathcal{T}}(x+h) - x - h.$$

Taking the squared norm on both sides, expanding the right side, and noting $\Pi_{x+\mathcal{T}}(x+h) = x + \Pi_{\mathcal{T}}(h)$, show that

$$\| v \|^2 + \| \Pi_K(x+h) - x - u \|^2 \leq \| x + h - \Pi_K(x+u) \|^2.$$

Consequently,

$$\| \Pi_K(x+h) - x - \Pi_{\mathcal{T}}(h) \|^2 \leq \| x + u + v - \Pi_K(x+u) \|^2 - \| v \|^2.$$

Expand the right side and use part (b) and the fact that $\|v\| \leq \|h\|$ and $\|u\| \leq \|h\|$ to deduce

$$\lim_{h \to 0} \frac{\Pi_K(x+h) - x - \Pi_{\mathcal{T}}(h)}{\| h \|} = 0.$$

(d) Conclude from part (c) that Π_K is directionally differentiable at $x \in K$ and

$$\Pi_K'(x; h) = \Pi_{\mathcal{T}}(h), \quad \forall\, h \in \mathbb{R}^n.$$

(e) Deduce from (e) that $\psi(x) \equiv \operatorname{dist}(x, K)$ is directionally differentiable at $x \in K$ and

$$\psi'(x; d) = \operatorname{dist}(d, \mathcal{T}), \quad \forall\, d \in \mathbb{R}^n.$$

4.8.2 Let K be a closed convex subset of $\mathbb{R}^n$. Deduce from part (e) of the last exercise that $\operatorname{dist}(\cdot, K)$ is F-differentiable at $x \in K$ if and only if $\mathcal{T}(x; K)$ is a linear subspace. Use this observation and Rademacher's theorem to show that the set

$$\{ x \in K : \mathcal{T}(x; K) \text{ is not a linear subspace } \}$$

is negligible. Deduce from the latter result that the set of degenerate solutions of a VI must be negligible.

4.8.3 Recalling Exercise **1.8.28**, show that the Euclidean projector onto the Lorentz cone $K \subset \mathbb{R}^{n+1}$ is everywhere B-differentiable. Let (x, t) be an arbitrary vector in $\mathbb{R}^{n+1}$; write $\Pi_K(x, t) \equiv (\bar{x}, \bar{t})$. Define the symmetric positive definite matrix

$$A(x,t) \equiv \left(1 + \frac{\lambda}{\| \bar{x} \|} \right) I - \frac{\lambda}{\| \bar{x} \|} \frac{\bar{x}\bar{x}^T}{\bar{x}^T\bar{x}} \in \mathbb{R}^{(n+1)\times(n+1)},$$

where $\lambda \equiv \max(0, \frac{1}{2}(\|x\|_2 - t))$ and $0/0$ is defined to be 1. Show that $\Pi'_K((x,t);(dx,dt))$ is the unique minimizer of the convex program in the variable (y,τ):

$$\text{minimize} \quad \frac{1}{2}\begin{pmatrix} y \\ \tau \end{pmatrix}^T A(x,t)\begin{pmatrix} y \\ \tau \end{pmatrix} - \begin{pmatrix} y \\ \tau \end{pmatrix}^T \begin{pmatrix} dx \\ dt \end{pmatrix}$$

$$\text{subject to} \quad (y,\tau) \in \mathcal{C}$$

where $\mathcal{C} \equiv \mathcal{C}_\pi((x,t);K)$. (Hint: if (x,t) belongs to K, use Exercise **4.8.1**. If $(x,t) \notin K$ and $\bar{x} \neq 0$, use Theorem **4.4.1**. Complete the proof for the remaining case where $(x,t) \notin K$ and $\bar{x} = 0$.)

4.8.4 Let K be a closed convex subset of $\mathbb{R}^n$. Let $x \in K$ and $d \in \mathbb{R}^n$ be arbitrary. Show that

$$\psi_1(\tau) \equiv \frac{\| \Pi_K(x+\tau d) - x \|}{\tau}$$

is monotonically nonincreasing in $\tau > 0$. Incidentally, Exercise **4.8.1** shows that the above quotient has a limit as $\tau \downarrow 0$. Show further that

$$\psi_2(\tau) \equiv \| \Pi_K(x+\tau d) - x \|$$

is monotonically nondecreasing in $\tau > 0$. (Hint: for ψ_1, it suffices to show that

$$\| \Pi_K(x+\tau d) - x \| \leq \tau \| \Pi_K(x+d) - x \|, \quad \forall \tau > 1. \tag{4.8.1}$$

For ψ_2, it suffices to show that

$$\| \Pi_K(x+\tau d) - x \| \geq \| \Pi_K(x+d) - x \|, \quad \forall \tau > 1.$$

Write $a \equiv x + d$, $\bar{a} \equiv \Pi_K(a)$, $b \equiv x + \tau d$, and $\bar{b} \equiv \Pi_K(b)$. Without loss of generality, we may assume that $a \notin K$, $b \notin K$, $\bar{a} \neq \bar{b}$, $\bar{a} \neq x$, and $\bar{b} \neq x$. (Verify that the desired inequality (4.8.1) is trivially valid if any of these conditions holds.) Show that

$$\begin{aligned} \tau d^T(\bar{b} - \bar{a}) &\geq (\bar{b} - x)^T(\bar{b} - \bar{a}) \\ &\geq (\bar{a} - x)^T(\bar{b} - \bar{a}) \geq d^T(\bar{b} - \bar{a}). \end{aligned}$$

Therefore, $d^T(\bar{b} - \bar{a}) > 0$ and

$$\tau \geq \frac{(\bar{b} - x)^T(\bar{b} - \bar{a})}{(\bar{a} - x)^T(\bar{b} - \bar{a})}.$$

Verify the elementary inequality

$$\frac{y^T(y-z)}{\|y\|} \geq \frac{z^T(y-z)}{\|z\|}$$

and use it to complete the proof of (4.8.1). Use the two inequalities

$$(\bar{b}-x)^T(\bar{b}-\bar{a}) \geq 0 \quad \text{and} \quad (\bar{a}-x)^T(\bar{b}-\bar{a}) \geq 0$$

to show the desired monotonicity of ψ_2.)

Deduce from the above monotonic properties of $\psi_1(\tau)$ and $\psi_2(\tau)$ that for all $\tau > 0$,

$$\min(1,\tau)\,\|\mathbf{F}_K^{\mathrm{nat}}(x)\| \leq \|x - \Pi_K(x - \tau F(x))\| \leq \max(1,\tau)\,\|\mathbf{F}_K^{\mathrm{nat}}(x)\|.$$

In the language of error bounds (see Chapter **6**), the above derivation shows that the residual functions $\|x - \Pi_K(x - \tau F(x))\|$ are equivalent for all $\tau > 0$. Intuitively, it is natural to expect such an equivalence to hold; this exercise formalizes the intuition and the equivalence.

4.8.5 Let $\eta > 0$, $\delta \in (0,1)$ and $\lambda > 1$ be given scalars with η sufficiently large and δ sufficiently small. Let K be the convex hull of the following countable collection of points in $\mathbb{R}^3$:

$$q \equiv \begin{pmatrix} -\eta \\ 0 \\ 0 \end{pmatrix}, \quad p^\infty \equiv \begin{pmatrix} 0 \\ 0 \\ 0 \end{pmatrix} \quad \text{and} \quad p^i \equiv \begin{pmatrix} \delta^i \\ (-1)^i \delta^i \\ -\delta^{2i} \end{pmatrix}, \quad i \geq 0.$$

Define

$$x^\infty \equiv \begin{pmatrix} 0 \\ 0 \\ 1 \end{pmatrix}, \quad x^i \equiv \begin{pmatrix} \lambda\,\delta^i \\ 0 \\ 1 \end{pmatrix}, \quad i = 1, 2, \ldots.$$

(a) Show that K is compact.

(b) For any y, show that $\bar{y} = \Pi_K(y)$ if and only if

$$(p^i - \bar{y})^T(\bar{y} - y) \geq 0, \quad \forall\, i = 1, 2, \ldots, \infty$$

and

$$(q - \bar{y})^T(\bar{y} - y) \geq 0.$$

(c) Use (b) to show that $\Pi_K(x^i) = p^i$ for $i = 1, 2, \ldots, \infty$.

(d) Show that the sequence

$$\left\{ \frac{p^i - p^\infty}{\lambda\,\delta^i} \right\}$$

has two accumulation points $\lambda^{-1}(1, \pm 1, 0)$.

Together parts (c) and (d) show that $\Pi'_K(x^\infty; d)$ does not exist, where d is the direction $(1, 0, 0)$.

4.8.6 Let K be a closed convex subset of $\mathbb{R}^n$. Suppose that Π_K is directionally differentiable at $x \in \mathbb{R}^n$. Let $d \in \mathbb{R}^n$ be arbitrary. Show that

(a) $\Pi'_K(x; d)$ belongs to the critical cone $\mathcal{C}_\pi(x; K)$;

(b) for all $v \in (K - \Pi_K(x)) \cap (\Pi_K(x) - x)^\perp$,

$$(v - \Pi'_K(x; d))^T (\Pi'_K(x; d) - d) \geq 0;$$

(c) if additionally C is a self-dual cone, then $\Pi'_K(x; d) - d$ belongs to $\mathcal{T}(\Pi_K(x) - x; K) \cap \Pi_K(x)^\perp$. (Hint: recall $x - \Pi_K(x) = \Pi_{-K}(x)$.)

4.8.7 Let E be an $m \times n$ matrix with full row rank. Let $C \subseteq \mathbb{R}^n$ be the null space of E. Let A be a symmetric positive definite matrix. Show that

$$\Pi_C^A = A^{-1} - A^{-1}E^T(EA^{-1}E^T)^{-1}EA^{-1}$$

and the matrix on the right-hand side is symmetric positive semidefinite. Use (4.3.2) to show that

$$\| \Pi_C^A \|_2 \leq \frac{1}{\lambda_{\min}(A)}.$$

4.8.8 Let S be a convex subset of $\mathbb{R}^n$ and $L \subseteq \mathbb{R}^n$ an affine subspace. We can very naturally define the projection of S on L by

$$\Pi_L(S) \equiv \{ x \in L : x = \Pi_L(y) \text{ for some } y \in S \}.$$

(a) Show that $\Pi_L(S)$ is always a convex set, bounded if S is bounded.

(b) Show that if S is compact then so is $\Pi_L(S)$.

(c) Give an example of an unbounded, closed, convex set S for which $\Pi_L(S)$ is not closed.

4.8.9 For each vector $x \in \mathbb{R}^n$, arrange the components of x in nonincreasing order:

$$x_{[1]} \geq x_{[2]} \geq \cdots x_{[n]}$$

Define the mapping $f : \mathbb{R}^n \to \mathbb{R}^n$ by $f_i(x) \equiv x_{[i]}$ for $i = 1, \ldots, n$. Show that f is a PL map; identify a polyhedral subdivision of $\mathbb{R}^n$ and the linear pieces of f on such a subdivision.

4.8.10 Let $f^i : \mathbb{R}^n \to \mathbb{R}^m$, $i = 1, \ldots, k$, be a finite family of locally Lipschitz continuous functions that are B-differentiable at a point $\bar{x}$. Show that

$$f(x) \equiv \max_{1 \leq i \leq k} f^i(x)$$

is B-differentiable at $\bar{x}$; moreover,

$$f'(\bar{x}; d) \equiv \max_{i \in \mathcal{I}} (f^i)'(\bar{x}; d), \quad \forall\, d \in \mathbb{R}^n,$$

where $\mathcal{I}$ consists of the indices i for which $f(\bar{x}) = f^i(\bar{x})$. Use this fact to show that if f is a PA map with the max-min representation (4.2.1), where each f^i is an affine function, then for every $x \in \mathbb{R}^n$, there exists a neighborhood $\mathcal{N}$ of x such that (4.2.7) holds. Finally, show that the latter conclusion holds for all PA functions.

4.8.11 Prove Lemma **4.2.14**.

4.8.12 Let $f : \mathbb{R}^n \to \mathbb{R}^m$ be a PA map. Suppose $f(x) = q$. Show that

(a) there exists a neighborhood $\mathcal{W}$ of the pair (q, x) such that for every (q', x') in $\mathcal{W}$,

$$f(x') = q' \Leftrightarrow f'(x; x' - x) = q' - q.$$

(b) the multifunction f^{-1} is lower semicontinuous at every (q', x') that is sufficiently close to (q, x) if and only if $f'(x; \cdot)$ is an open map on $\mathbb{R}^n$, or equivalently, if and only if $f'(x; \cdot)$ is coherently oriented.

4.8.13 Let K be a polyhedron in $\mathbb{R}^n$. Let F be continuously differentiable in a neighborhood of a solution $x^* \in \text{SOL}(K, F)$.

(a) Show that if x^* is nondegenerate, then $\mathbf{F}_K^{\text{nat}}$ is F-differentiable at x^* and $\mathbf{F}_K^{\text{nor}}$ is F-differentiable at $z^* \equiv x^* - F(x^*)$.

(b) Show that x^* is nondegenerate if and only if

$$\begin{pmatrix} x \\ y \end{pmatrix} \mapsto \begin{pmatrix} y - F(x) \\ x - \Pi_K(x - y) \end{pmatrix}$$

is F-differentiable at $(x^*, F(x^*))$.

4.8.14 Consider a QVI (K, F), where

$$K(x) \equiv \{ x' \in X : G(x, x') \leq 0 \},$$

where

$$X \equiv \{ x' \in \mathbb{R}^n : H(x) \leq 0 \}.$$

Assume the following conditions, the last of which is a SBCQ:

(a) the function $H : \mathbb{R}^n \to \mathbb{R}^\ell$ is continuously differentiable and each component function H_i is convex;

(b) the function $F : \mathbb{R}^n \to \mathbb{R}^n$ is continuous and the set X is nonempty and bounded;

(c) for each $x \in X$, each component function $G_j(x, \cdot) : \mathbb{R}^n \to \mathbb{R}$, for $j = 1, \dots, m$, is continuously differentiable and convex;

(d) for every bounded sequence $\{x^k\} \subset X$, the VI $(K(x^k), F)$ has a solution along with a multiplier pair (μ^k, λ^k) for the constraints in the set $K(x^k)$ such that the sequence $\{(\mu^k, \lambda^k)\}$ is bounded.

Show that the QVI (K, F) has a nonempty bounded solution set.

4.8.15 Let (A, B) be a pair of real square matrices of order n. Consider the horizontal LCP:

$$0 = q + Ax - By$$

$$0 \leq x \perp y \geq 0.$$

Show that the following statements are equivalent.

(a) The PA map $z \mapsto Az^+ - Bz^-$ is a globally Lipschitz homeomorphism.

(b) For every $q \in \mathbb{R}^n$, there exists a unique pair $(x(q), y(q))$ that is a solution of the above horizontal LCP.

(c) The matrix A is nonsingular and $A^{-1}B$ is a P matrix.

(d) Every column representative matrix of the pair (A, B) has the same nonzero determinantal sign.

Show further that if any one of the above condition holds, then the unique pair $(x(q), y(q))$ in (b) is a PA function of q.

Historically, a pair of real square matrices (A, B) satisfying property (d) is said to have the *column $\mathcal{W}$ property*. See Exercise **11.9.4** for a related property. (Note: the proof of this exercise is made easy by the fundamental LCP result concerning a P matrix.)

4.8.16 Show that the pair of matrices

$$A \equiv \begin{bmatrix} 2 & 1 \\ -1 & -1 \end{bmatrix} \quad \text{and} \quad B \equiv \begin{bmatrix} 2 & 1 \\ 1 & 0 \end{bmatrix}$$

has the column $\mathcal{W}$ property; but the pair (A^T, B^T) does not; i.e., (A, B) does not have the *row $\mathcal{W}$ property*.

4.8.17 Let $f : \mathbb{R}^n \to \mathbb{R}^n$ be a coherently oriented PA map with affine pieces $\{f^1, \cdots, f^k\}$ such that $\operatorname{sgn} \det Jf^i$ is a nonzero constant.

(a) Suppose a matrix B exists such that $\tau B + (1-\tau) J f^i$ is nonsingular for all $\tau \in [0,1]$. Show that f is a homeomorphism from $\mathbb{R}^n$ onto itself. (Hint: verify the index condition in part (e) of Theorem **4.2.11**.)

(b) Suppose a matrix B exists such that $J f^i B^{-1}$ has all leading principal minors positive. Show that f is a homeomorphism from $\mathbb{R}^n$ onto itself. (Hint: use the matrix-theoretic properties in parts (c) and (d) to show that the assumption in part (b) implies the assumption in part (a).)

(c) Let A and B be two nonsingular matrices of the same order. Show that $\tau B + (1-\tau) A$ is nonsingular for all $\tau \in [0,1]$ if and only if AB^{-1} has no negative eigenvalues.

(d) Let A be a matrix with all leading principal minors positive. Show that there exists $\bar{\varepsilon} > 0$ such that for every $\varepsilon \in (0, \bar{\varepsilon})$,

$$\det(AE(\varepsilon) - \tau \mathbf{I}) > 0, \quad \forall \tau \leq 0,$$

where

$$E(\varepsilon) \equiv \operatorname{diag}(\varepsilon, \varepsilon^2, \varepsilon^{2^2}, \cdot, \varepsilon^{2^{n-1}}).$$

4.8.18 Use Proposition **4.3.3** to show that if M is a symmetric positive definite matrix of order n and K is a closed convex set in $\mathbb{R}^n$, then

$$(\mathbf{M}_K^{\text{nat}})^{-1} = \Pi_K^M \circ (\mathbf{I} - M) + \mathbf{I}.$$

4.9 Notes and Comments

Zarantonello [894] provides a comprehensive study of the projection operator in a Hilbert space. In particular, Exercise **4.8.1** is drawn from part I of this reference. Exercise **4.8.2** is inspired by an observation of Adrian Lewis in a private communication in June 2002. Exercise **4.8.4** is a result from Gafni and Bertsekas [281], which they use to study two-metric projection for constrained optimization. The 3-dimensional example in Exercise **4.8.5** is due to Kruskal [462]. Shapiro [776] gave a 2-dimensional example showing the non-directional differentiability of the Euclidean projector. In [326], Haraux introduced a class of closed convex sets in infinite-dimensional spaces called *polyhedric sets* that includes all finite-dimensional polyhedra; Haraux showed that the Euclidean projection onto such a set must be directionally differentiable. Thus Theorem **4.1.1** follows from Haraux' result. Nevertheless, the identity (4.1.3) in this theorem first appeared in Pang [663], who was inspired by an unpublished report of Robinson [736]. See Section **5.7** for comments on Corollary **4.1.2**.

Katzenelson appears to be the first person to use piecewise linear analysis to study nonlinear resistive networks [408]. Fujisawa and Kuh [269] obtained several fundamental properties of PA maps. In particular, they established the Lipschitz continuity of these maps, part (c) of Proposition **4.2.2**, and gave a determinantal condition that is sufficient for a PA map to be a global homeomorphism. Inspired by the Fujisawa-Kuh study, several authors obtained sufficient conditions for a PA map to be surjective in terms of a constant determinantal sign condition on the Jacobian matrices of a set of affine selection functions; see [121, 649, 716]. Extending the results in these papers, Kojima and Saigal [448, 449] obtained sufficient conditions ensuring that a PA map is a homeomorphism. Exercise **4.8.17** is based on the work of Kojima and Saigal on this topic. In the references, these authors also showed that if the polyhedral subdivision generated by the pieces of linearity of a PA map F are "regular", then F is a homeomorphism if and only if "the determinants of the Jacobians of the pieces of linearity of F have the same sign". The latter condition is, of course, just the coherent orientation in Definition **4.2.3**. This terminology was introduced several years later by Kuhn and Löwen [463], who also showed that every injective PA map must be coherently oriented. Proposition **4.2.12** was proved by Schramm [768].

The interest of Kojima and Saigal in piecewise affine functions was motivated by the central role of these functions in the family of fixed-point homotopy methods; see Eaves [206]. The article [212] investigates the relationships of properties of PA maps over ordered fields. Proposition **4.2.15** is proved in this article. Kojima [442] studied the approximation of PC^1 functions by PA maps. Several topics were investigated under nonsingularity conditions on piecewise approximations of PC^1 maps, with application to the NCP.

The normal manifold associated with a polyhedron was recognized by Robinson [740] as a special polyhedral subdivision of an Euclidean space. Robinson used this manifold to establish his remarkable characterization of the coherent orientation of the normal map of an affine pair in terms of the bijectivity of the map. More details are given below.

The systematic study of piecewise differentiable optimization problems originates from the theory of nondifferentiable penalty functions for constrained smooth optimization problems. Womersley [871] obtained optimality conditions for such nonsmooth optimization problems. Chaney [115] studied real-valued piecewise smooth functions from the point of view of nonsmooth analysis and optimization. The Habilitation thesis of Scholtes [762] studied these functions with an eye toward applications in VI and optimization. A significant portion of Section **4.2** is based on this thesis, which is by far the best comprehensive work on piecewise differentiable equations. The reader can find a proof of the max-min representation of a scalar PA map in Scholtes' thesis. Moreover, the proofs of Propositions **4.2.1**, Lemma **4.2.4**, and Proposition **4.2.5** as well as Example **4.2.13** are also from this thesis. The fact that a PC^1 function is locally Lipschitz continuous is an immediate consequence of an abstract and elegant result of Hager [321] having to do with Lipschitz continuous selections. The paper by Kuntz and Scholtes [471] studies structural properties of piecewise differentiable functions, showing in particular that Robinson's strong B-differentiability is a rather restrictive requirement for these functions. Kuntz and Scholtes further show that the Euclidean projection onto a finitely representable, convex set is a PC^1 function that is "PC^1-equivalent to a strongly B-differentiable PC^1 function", assuming that "every collection of at most n of the active gradients is linearly independent". Clearly the latter assumption implies the CRCQ. The related paper [473] studies qualitative aspects of the local approximation of a PC^2 function.

Although not phrased in terms of piecewise affine bijections, the seminal paper of Samelson, Thrall, and Wesler [757] is among the first that deals with this topic. The paper shows that for a matrix $M \in \mathbb{R}^{n \times n}$, the PL map $z \mapsto Mz^+ - z^-$ is a bijection, thus a global Lipschitz homeomorphism, from $\mathbb{R}^n$ onto itself if and only if M is a P matrix. This result was the root of substantial LCP research whose aim was to obtain matrix-theoretic characterizations of LCP properties. Examples of these include the column sufficiency that characterizes the convexity of solutions to LCPs and the P_0 property that characterizes the connectedness of such solutions, under the R_0 assumption (Proposition **3.6.9**).

Extending the result of Samelson, Thrall, and Wesler and building on the earlier work of Schramm [768], Kuhn and Löwen [463] investigated the unique solvability of the equation $Sx^+ - Tx^- = y$; see Exercise **4.8.15**. The notes and comments on this exercise and on the next one, Exercise **4.8.16**, can be found in Section **11.10**. Kuhn and Löwen further proved that if the polyhedral subdivision induced by a PL map has a "branching number" not exceeding 4, then the PL map is bijective if and only if it is coherently oriented. Bypassing the verification of the branching number of the normal manifold, Robsinon [740] showed that the normal map $\mathbf{M}_K^{\text{nor}}$ of an affine pair (K, M) is bijective if and only if it is coherently oriented. In particular, the original proof of Theorem **4.3.1** is given in this reference. Alternative proofs of Robinson's deep result can be found in [708, 763]. Our sketch of the proof in the text is based on [708]. Subsequently, Ralph

[709] and Scholtes [764] verified that the branching number condition is indeed satisfied by the normal manifold. The index condition in part (e) of Theorem **4.2.11** is due to Pang and Ralph [676]. The inverse of the normal map $\mathbf{M}_K^{\text{nor}}$ in Proposition **4.3.3** was also obtained in the latter article.

Differentiability properties of metric projections onto moving, closed convex sets were studied extensively by Shapiro, beginning with [773]. In particular, the directional derivative for such a projection was first obtained in the latter paper; the formula in Theorem **4.4.1** is a special case of Shapiro's result for a fixed set. Example **4.7.4**, which is a simplified example of Robinson [725], appeared in Shapiro's paper. The directional differentiability of the Euclidean projector onto a fixed, closed convex set under the SBCQ is established in this book for the first time. Sections **4.5** and **4.7**, which assume the CRCQ throughout, are based on the paper [676].

Using the theory of second-order regular sets [76], Bonnans, Cominetti, and Shapiro [75] established the directional differentiability of the projection onto the cone of symmetric positive semidefinite matrices under the Frobenius norm and showed that the directional derivative can be computed by solving a convex program similar to (4.4.1) for the case of a finitely representable set. Most recently, Sun and Sun [819] derived an explicit formula for the directional derivative of the "absolute-value" function

$$| A |_{\mathcal{M}^n_+} \equiv \Pi_{\mathcal{M}^n_+}(A) + \Pi_{\mathcal{M}^n_+}(-A),$$

from which one can obtain a corresponding explicit formula for the directional derivative of the projection $\Pi_{\mathcal{M}^n_+}$; more importantly, Sun and Sun also establish the semismoothness of this projector. A connection between the formula of Sun and Sun and the convex program of Bonnans, Cominetti, and Shapiro is noted in [678].

Although the B-subdifferential is the fundamental building block of the Clarke subdifferential, whose commentary is given in Section **7.7**, the former subdifferential had not received special attention until Qi popularized it in the paper [699], where he defined the important concept of (strong) BD regularity for solving nonsmooth equations. The significance of the limiting Jacobian of a PC^1 function was recognized by Pang and Ralph [676], who established Theorem **4.6.5**. Previously, using his theory of approximation by PA maps, Kojima [442] has studied inverse function theorems for PC^1 maps and used the results to establish the strong stability of an equilibrium point of an n-person noncooperative game [447]. Theorem **4.6.5** had inspired several subsequent papers, which were aimed at extending the result to broader class of functions. A first extension was obtained by Ralph and Scholtes [712] who considered a composite piecewise smooth function. Recently, Gowda [302] obtained inverse and implicit function theorems for semismooth functions. While the class of H-differentiable functions (see Exercise **3.7.17**) to which Gowda's results apply includes all locally Lipschitz continuous functions, the inverse and implicit function theorems in [302] do not address the directional differentiability of the resulting inverse and implicit functions. An important reason for the lack of treatment of the latter property is due to the fact that a H-differentiable function is itself not necessarily directionally differentiable; for a related study, see Sun [818]. A complete extension of Theorem **4.6.5** to semismooth functions was established in [678]; this extended result is the content of Exercise **7.6.18**, where the inclusion (4.6.4) continues to play an essential role. Exercise **5.6.22** is proved in the latter reference.

The Fréchet differentiability of the Euclidean projector onto a polyhedron was investigated by Pang [663], who established the equivalence of (a) and (b) in Corollary **4.1.2**. Based on a fundamental concept in classical differential topology, Sznajder and Gowda [826] called a zero of a PC^1 function from $\mathbb{R}^n$ into itself nondegenerate if the function is Fréchet differentiable there. They established part (b) of Exercise **4.8.13** and the equivalence of (b), (c), and (d) in Exercise **5.6.2**.

Chapter 5

Sensitivity and Stability

This chapter is concerned with a comprehensive analysis of the sensitivity of the solutions to the VI (K, F) when the pair (K, F) undergoes small changes. There are many important issues covered by the analysis. Needless to say, in order to develop the theory rigorously, we have to define "small changes" of the pair (K, F) formally. Such a formal definition requires the introduction of a metric on the family of closed convex sets and the space of continuous functions. This will be done in due time. In what follows, we give an overview of the subject of this chapter by assuming an intuitive understanding of what is meant by "change of (K, F)".

A common sensitivity issue concerns a given isolated solution x^* of the VI (K, F). We call this type of analysis an *isolated sensitivity analysis.* In this analysis, we are interested in the change of x^* as (K, F) is perturbed. The following are sample questions that we wish to address. Suppose that (L, G) is a slight perturbation of (K, F). Will the perturbed VI (L, G) have a solution that is close to x^*? Will such a (perturbed) solution be isolated? As (L, G) approaches (K, F), will the (possibly non-unique) solutions of the perturbed VI (L, G) that are close to x^* eventually tend to x^*? Answers to these questions lead to the concept of "solution stability" that we will formally define later. In essence, this concept has to do with the "continuity" of the perturbed solutions as a (most likely, multi)function of the pair (L, G).

Included in the general area of sensitivity analysis is the important topic of "parametric analysis" which we have introduced in Subsection **1.4.10**. Here, a family of parameter-dependent VIs is given:

$$\{ \text{VI } (K(\mathbf{p}), F(\cdot, \mathbf{p})) \, : \, \mathbf{p} \in \mathcal{P} \}$$

where $F : \mathcal{D} \times \mathcal{P} \subseteq \mathbb{R}^n \times \mathbb{R}^p \to \mathbb{R}^n$ is a function of two arguments $(x, \mathbf{p})$ and $K : \mathbb{R}^p \to \mathcal{D}$ is a multifunction with values in $\mathcal{D}$. We are given a solution x^* of the VI $(K(\mathbf{p}^*), F(\cdot, \mathbf{p}^*))$; the perturbed pair (L, G) is then $(K(\mathbf{p}), F(\cdot, \mathbf{p}))$ with $\mathbf{p}$ being close to $\mathbf{p}^*$. We refer to this kind of analysis as "isolated parametric analysis". In this analysis, in addition to the stability questions raised above, which have to do with the continuity of the perturbed solution trajectory $x(\mathbf{p})$ at $\mathbf{p}^*$, we can further ask about the "differentiability" of this trajectory at $\mathbf{p}^*$.

Much of the development of the isolated sensitivity theory is concerned with the case where only the function F is changed while K is held fixed. The analysis applies in particular to the sensitivity study of the MiCP (F) where we consider only "nearby" MiCP (G) with G being close to F; this yields an extensive set of results for the NCP.

In contrast to isolated sensitivity analysis, "total sensitivity analysis" refers to the investigation of the change of the entire solution set SOL(K, F) when the pair (K, F) undergoes small perturbation. Needless to say, the latter analysis is more complicated than the former; indeed, the state of the art of isolated sensitivity analysis is a lot more advanced than that of total sensitivity analysis. The presentation in this chapter reflects this unbalanced development of the two subareas.

5.1 Sensitivity of an Isolated Solution

Beginning in this section, we study the sensitivity of an isolated solution to a VI. Consider the VI (K, F) with a given solution x^* that we assume is locally unique. The set K is fixed throughout the section. Initially, we assume no special structure on K other than the fact that it is closed and convex. We want to study the change of x^* as F is perturbed. A central question is of course whether the perturbed VI (K, G), where G is a small perturbation of F, will have a solution that is near x^*. Using Example **3.3.11**, we demonstrate that the local uniqueness of x^* is not sufficient for this question to have an affirmative answer.

5.1.1 Example. Consider the LCP (q, M) from Example **3.3.11**, where

$$q = \begin{pmatrix} 0 \\ -1 \\ 1 \end{pmatrix} \quad \text{and} \quad M = \begin{bmatrix} 0 & -1 & -1 \\ 1 & 0 & 0 \\ -1 & 0 & 0 \end{bmatrix}.$$

This LCP has a unique solution $x = (1, 0, 0)$. Since the first row of M is nonpositive, the perturbed LCP $(q(\varepsilon), M)$ clearly has no solution for all $\varepsilon > 0$, where $q(\varepsilon) \equiv (-\varepsilon, -1, 1)$. The matrix M is positive semidefinite, but not positive definite. □

Although not sufficient to establish the solvability of the VIs (K, G) for G near F, the local uniqueness is nevertheless an important property in the sensitivity analysis of a solution to the VI (K, F). Indeed, every locally unique solution of a given VI is an "attractor" of all solutions of "nearby VIs". In order to make this statement precise, we introduce the concept of an ε-neighborhood of a function on a subset of $\mathbb{R}^n$. Given a function $H : \mathcal{D} \to \mathbb{R}^m$, where $\mathcal{D}$ is an open subset of $\mathbb{R}^n$, a positive scalar $\varepsilon > 0$, and a subset S of $\mathcal{D}$, let $\mathbb{B}(H; \varepsilon, S)$ be the ε-neighborhood of H restricted to S; that is $G \in \mathbb{B}(H; \varepsilon, S)$ if $G : \mathcal{D} \to \mathbb{R}^m$ is continuous and

$$\| G - H \|_S \equiv \sup_{y \in S} \| G(y) - H(y) \| < \varepsilon.$$

This neighborhood concept induces a functional convergence concept. We say that a sequence of continuous functions $\{G_k\}$ converges to H on S if

$$\lim_{k \to \infty} \| G_k - H \|_S = 0;$$

in other words, for every $\varepsilon > 0$, there exists $\bar{k}$ such that, for all $k \geq \bar{k}$, G_k belongs to $\mathbb{B}(H; \varepsilon, S)$.

5.1.2 Proposition. Let $F : \mathcal{D} \to \mathbb{R}^n$ be continuous, where $\mathcal{D}$ is an open subset of $\mathbb{R}^n$ containing K. If x^* is an isolated solution of the VI (K, F), then for every neighborhood $\mathcal{N}$ of x^* satisfying

$$\operatorname{cl} \mathcal{N} \subset \mathcal{D} \quad \text{and} \quad \operatorname{SOL}(K, F) \cap \operatorname{cl} \mathcal{N} = \{ x^* \}, \tag{5.1.1}$$

and for every sequence of continuous functions $\{G_k\}$ converging to F on $\mathcal{N}$, every sequence of vectors $\{x^k\}$, where $x^k \in \operatorname{SOL}(K, G_k) \cap \mathcal{N}$ for every k, converges to x^*.

Proof. Indeed, any such sequence $\{x^k\}$ must be bounded. Let x^∞ be any accumulation point of this sequence. It follows that $F(x^\infty)$ is an accumulation point of $\{F(x^k)\}$; in turn, by the convergence of $\{G_k\}$ to F, $F(x^\infty)$ is an accumulation point of the sequence $\{G_k(x^k)\}$. By a simple limiting argument, it is easy to show that x^∞ must be a solution of the VI (K, F) that lies in $\operatorname{cl} \mathcal{N}$. By (5.1.1), $x^\infty = x^*$. Hence $\{x^k\}$ has a unique accumulation point, namely x^*, to which it converges. □

From the discussion so far, it becomes obvious that a major challenge in studying the sensitivity of the isolated solution x^* is to identify situations when the perturbed VIs (K, G) will have solutions that are near x^*. The analytical tool that we employ to deal with this "local solvability" issue is degree theory. In principle, we can apply degree theory to either the natural map $\mathbf{F}_K^{\text{nat}}$ or the normal map $\mathbf{F}_K^{\text{nor}}$ of the pair (K, F). In this regard, we should caution the reader about the use of norms in the development to follow. The nearness property of the degree plays a major role throughout the analysis herein. On the one hand, as stated in part (c) of Proposition **2.1.3**, this property involves the max-norm. On the other hand, the definition of the projector Π_K is in terms of the Euclidean norm; in particular, the nonexpansiveness of this projector is proved under the Euclidean norm; cf. part (c) of Theorem **1.5.5**. There are occasions where these two properties are being used simultaneously in proving certain results. Consistent with the convention followed throughout the book, all norms are taken to be the Euclidean norm unless otherwise noted.

In what follows, we first present a sensitivity result based on $\mathbf{F}_K^{\text{nat}}$. Since x^* is isolated, it follows that for every open neighborhood $\mathcal{N}$ of x^* satisfying (5.1.1), $\deg(\mathbf{F}_K^{\text{nat}}, \mathcal{N})$ is well defined; moreover this degree is independent of the neighborhood $\mathcal{N}$ as long as $\mathcal{N}$ has the stated property. This common degree is the index of $\mathbf{F}_K^{\text{nat}}$ at x^*, which we denote $\text{ind}(\mathbf{F}_K^{\text{nat}}, x^*)$. The following result identifies a broad quantitative condition on the perturbed function G in order for the VI (K, G) to have a solution near x^*.

5.1.3 Proposition. Let $K \subseteq \mathbb{R}^n$ be closed convex and $F : \mathcal{D} \to \mathbb{R}^n$ be continuous, where $\mathcal{D}$ is an open set containing K. Let x^* be an isolated solution of the VI (K, F). Suppose that $\text{ind}(\mathbf{F}_K^{\text{nat}}, x^*)$ is nonzero. For every open neighborhood $\mathcal{N}$ of x^* satisfying (5.1.1) and every $G \in \mathbb{B}(F; \varepsilon, \text{cl}\,\mathcal{N})$, where $\varepsilon \equiv \text{dist}_\infty(0, \mathbf{F}_K^{\text{nat}}(\partial\mathcal{N}))$, the VI (K, G) has a solution in $\mathcal{N}$.

Proof. Since x^* is the only solution of the VI (K, F) in $\text{cl}\,\mathcal{N}$, we have $0 \notin \mathbf{F}_K^{\text{nat}}(\partial\mathcal{N})$; thus ε is positive. By the nonexpansiveness of the Euclidean projector, we have

$$\| \mathbf{F}_K^{\text{nat}}(x) - \mathbf{G}_K^{\text{nat}}(x) \|_2 \leq \| G(x) - F(x) \|_2.$$

Hence,

$$\begin{aligned} \sup_{x \in \text{cl}\,\mathcal{N}} \| \mathbf{F}_K^{\text{nat}}(x) - \mathbf{G}_K^{\text{nat}}(x) \|_\infty &\leq \sup_{x \in \text{cl}\,\mathcal{N}} \| \mathbf{F}_K^{\text{nat}}(x) - \mathbf{G}_K^{\text{nat}}(x) \|_2 \\ &< \text{dist}_\infty(0, \mathbf{F}_K^{\text{nat}}(\partial\mathcal{N})). \end{aligned}$$

Thus by the nearness property of the degree, it follows that $\deg(\mathbf{G}_K^{\text{nat}}, \mathcal{N})$ is well defined and equal to $\text{ind}(\mathbf{F}_K^{\text{nat}}, x^*)$. Since the latter index is nonzero, it follows that $\mathbf{G}_K^{\text{nat}}$ has a zero in $\mathcal{N}$; or equivalently, the VI (K, G) has a solution in $\mathcal{N}$. □.

An analogous result can be established by applying a similar degree-theoretic argument to the normal map $\mathbf{F}_K^{\text{nor}}$. By using the latter map, the condition on the perturbed function G can be slightly relaxed. It becomes only necessary to restrict the error function $G(x) - F(x)$ on the smaller set $K \cap \operatorname{cl} \mathcal{N}$. For this purpose, we write

$$K_{\mathcal{N}} \equiv K \cap \operatorname{cl} \mathcal{N}.$$

Throughout this section, we let $z^* \equiv x^* - F(x^*)$. We note that since $x^* = \Pi_K(z^*)$, for every open neighborhood $\mathcal{N}$ of x^* satisfying (5.1.1), there exists a neighborhood $\mathcal{Z}$ of z^* such that (i) $\Pi_K(\mathcal{Z}) \subseteq \mathcal{N}$, and hence, by continuity of Π_K, $\Pi_K(\operatorname{cl} \mathcal{Z}) \subseteq \operatorname{cl} \mathcal{N}$; and (ii) z^* is the unique zero of $\mathbf{F}_K^{\text{nor}}$ in $\operatorname{cl} \mathcal{Z}$, by (5.1.1). Thus $\operatorname{ind}(\mathbf{F}_K^{\text{nor}}, z^*)$ is well defined and the scalar

$$\varepsilon_{\mathcal{N}} \equiv \operatorname{dist}_\infty(0, \mathbf{F}_K^{\text{nor}}(\operatorname{bd} \mathcal{Z}))$$

is positive.

5.1.4 Proposition. Let $K \subseteq \mathbb{R}^n$ be closed convex and $F : \mathcal{D} \to \mathbb{R}^n$ be continuous, where $\mathcal{D}$ is open set containing K. Let x^* be an isolated solution of the VI (K, F). Suppose that $\operatorname{ind}(\mathbf{F}_K^{\text{nor}}, z^*)$ is nonzero. For every open neighborhood $\mathcal{N}$ of x^* satisfying (5.1.1), there exists a scalar $\varepsilon > 0$ such that, for every $G \in \mathbb{B}(H; \varepsilon, K_{\mathcal{N}})$, the VI (K, G) has a solution in $\mathcal{N}$.

Proof. Let $\mathcal{Z}$ and $\varepsilon \equiv \varepsilon_{\mathcal{N}}$ be as specified above. If G is any function belonging to $\mathbb{B}(H; \varepsilon, K_{\mathcal{N}})$, then for any $z \in \operatorname{cl} \mathcal{Z}$, we have $\Pi_K(z) \in K \cap \operatorname{cl} \mathcal{N}$. Hence,

$$\sup_{z \in \operatorname{cl} \mathcal{Z}} \| \mathbf{G}_K^{\text{nor}}(z) - \mathbf{F}_K^{\text{nor}}(z) \|_\infty \leq \sup_{x \in K \cap \operatorname{cl} \mathcal{N}} \| G(x) - F(x) \|_2 < \varepsilon.$$

As before, we therefore deduce that $\mathbf{G}_K^{\text{nor}}$ has a zero z in the neighborhood $\mathcal{Z}$. This implies that the VI (K, G) has a solution $\Pi_K(z)$ in the neighborhood $\mathcal{N}$ as desired. □

As an application of the above result, we present a corollary that pertains to a parametric VI with a fixed set. Subsequently, we extend this corollary to the case where the defining set of the VI also depends on the parameter; cf. Proposition **5.4.1**

5.1.5 Corollary. Let $F : \mathcal{D} \times \mathcal{P} \subseteq \mathbb{R}^{n+p} \to \mathbb{R}^n$ be a continuous mapping, where $\mathcal{D}$ is an open subset of $\mathbb{R}^n$ containing the closed convex set K. Let $\mathbf{p}^* \in \mathcal{P}$ be given and let x^* be an isolated solution of the VI $(K, F(\cdot, \mathbf{p}^*))$. Suppose that $\operatorname{ind}(\mathbf{F}_K^{\text{nor}}(\cdot, \mathbf{p}^*), z^*)$ is nonzero, where $z^* \equiv x^* - F(x^*, \mathbf{p}^*)$. For every pair of open neighborhoods $\mathcal{N}$ of x^* and $\mathcal{U}$ of $\mathbf{p}^*$ satisfying the following two conditions:

(a) there is a constant $L > 0$ such that, for all $\mathbf{p} \in \mathcal{U} \cap \mathcal{P}$,

$$\sup_{x \in K \cap \mathcal{N}} \| F(x,\mathbf{p}) - F(x,\mathbf{p}^*) \|_2 \leq L \| \mathbf{p} - \mathbf{p}^* \|_2, \tag{5.1.2}$$

(b) $\mathrm{SOL}(K, F(\cdot,\mathbf{p}^*)) \cap \mathrm{cl}\,\mathcal{N} = \{x^*\}$,

there exists a neighborhood $\mathcal{U}' \subseteq \mathcal{U}$ such that, for every $\mathbf{p} \in \mathcal{U}' \cap \mathcal{P}$,

$$S_{\mathcal{N}}(\mathbf{p}) \equiv \mathrm{SOL}(K, F(\cdot,\mathbf{p})) \cap \mathcal{N} \neq \emptyset;$$

moreover,

$$\lim_{\mathbf{p} \to \mathbf{p}^*} \sup \{ \| x - x^* \| \, : \, x \in S_{\mathcal{N}}(\mathbf{p}) \} = 0. \tag{5.1.3}$$

Proof. For simplicity, we take $\mathcal{P}$ to be the entire space $\mathbb{R}^p$. Choose an open neighborhood $\mathcal{Z}$ of z^* such that (i) $\Pi_K(\mathcal{Z}) \subseteq \mathcal{N}$ and (ii) z^* is the unique zero of $\mathbf{F}_K^{\mathrm{nor}}$ in $\mathrm{cl}\,\mathcal{Z}$; cf. the discussion before Proposition **5.1.4**. Furthermore, by choosing the neighborhood $\mathcal{U}'$ to be a subset of $\mathcal{U}$ such that

$$L \sup_{\mathbf{p} \in \mathcal{U}'} \| \mathbf{p} - \mathbf{p}^* \|_2 < \mathrm{dist}_\infty(0, \mathbf{F}_K^{\mathrm{nor}}(\partial \mathcal{Z}, \mathbf{p}^*)) \equiv \varepsilon,$$

we may apply the cited proposition to the function $G \equiv F(\cdot,\mathbf{p})$, which belongs to $\mathbb{B}(F(\cdot,\mathbf{p}^*); \varepsilon, K_{\mathcal{N}})$, thereby deducing that $S_{\mathcal{N}}(\mathbf{p})$ is nonempty for every $\mathbf{p} \in \mathcal{U}'$. The limit (5.1.3) follows easily from Proposition **5.1.2** because if $\{\mathbf{p}^k\}$ is an arbitrary sequence of perturbation vectors converging to $\mathbf{p}^*$, then the sequence of functions $F(\cdot,\mathbf{p}^k)$ converges to $F(\cdot,\mathbf{p}^*)$ on $K_{\mathcal{N}}$, by (5.1.2). □

Proposition **5.1.4** is the key to the local solvability issue that is part of the isolated sensitivity of the solution $x^* \in \mathrm{SOL}(K,F)$. Much of the subsequent analysis aims at obtaining sufficient conditions for the index assumption $\mathrm{ind}(\mathbf{F}_K^{\mathrm{nor}}, x^*) \neq 0$ in this proposition to hold. To motivate these conditions, we recall that in Proposition **3.3.4**, we have identified a sufficient condition for a solution x^* of a VI (K,F) to be locally unique. It turns out that this condition is also sufficient for $\mathrm{ind}(\mathbf{F}_K^{\mathrm{nat}}, x^*)$ to be nonzero, thus for the conclusion of Proposition **5.1.4** to hold. We recall from Section **3.3** that the critical cone $\mathcal{C}(x^*; K, F)$ is defined as the intersection of the tangent cone $\mathcal{T}(x^*; K)$ with the orthogonal complement of the vector $F(x^*)$.

5.1.6 Proposition. Let $K \subseteq \mathbb{R}^n$ be closed convex and $F : \mathcal{D} \to \mathbb{R}^n$ be B-differentiable at a solution x^* of the VI (K,F), where $\mathcal{D}$ is an open set containing K. If

$$v^T F'(x^*; v) > 0, \quad \forall v \in \mathcal{C}(x^*; K, F) \setminus \{0\}, \tag{5.1.4}$$

then both $\text{ind}(\mathbf{F}_K^{\text{nat}}, x^*)$ and $\text{ind}(\mathbf{F}_K^{\text{nor}}, z^*)$ are well defined and nonzero. Consequently, the conclusions of Proposition **5.1.3** and Proposition **5.1.4** are valid.

Proof. We prove that $\text{ind}(\mathbf{F}_K^{\text{nor}}, z^*)$ is nonzero; the proof for $\text{ind}(\mathbf{F}_K^{\text{nat}}, x^*)$ is similar. By Proposition **3.3.4**, x^* is an isolated solution of the VI (K, F). To show that $\text{ind}(\mathbf{F}_K^{\text{nor}}, z^*)$ is nonzero, let $F_t : x \mapsto tF(x) + (1-t)(x - z^*)$ be the homotopy between the function F and the identity map translated by z^*. Let $H(\cdot, t)$ be the normal map associated with the pair (K, F_t); i.e., for all $(z, t) \in \mathbb{R}^n \times [0, 1]$,

$$H(z,t) \equiv tF(\Pi_K(z)) + (1-t)(\Pi_K(z) - z^*) + z - \Pi_K(z).$$

We claim that there exists a scalar $\varepsilon > 0$ such that z^* is the only zero of $H(\cdot, 1)$ in the open ball $\mathcal{Z} \equiv \mathbb{B}(z^*, \varepsilon)$; and more importantly, for all $t \in [0,1]$, $H(\cdot, t) = 0$ has no solution lying on $\text{bd}\,\mathcal{Z}$. Assume for the sake of contradiction that the claim is false. Since x^* is a locally unique solution of the VI (K, F), and thus z^* is a locally unique zero of $H(\cdot, 1)$, this assumption implies that there exists a sequence of positive scalars $\{\varepsilon_k\}$ converging to zero such that, for each k, a pair

$$(z^k, t_k) \in \text{bd}\,\mathbb{B}(z^*, \varepsilon_k) \times [0, 1]$$

satisfying $H(z^k, t_k) = 0$ exists. With $x^k \equiv \Pi_K(z^k)$, it follows that x^k belongs to $\text{SOL}(K, F_{t_k})$. Thus, for all k,

$$(y - x^k)^T [t_k F(x^k) + (1 - t_k)(x^k - x^*)] \geq 0, \quad \forall y \in K.$$

Moreover, since $\|z^k - z^*\|_2 = \varepsilon_k > 0$, we have $x^k \neq x^*$; thus $t_k > 0$ for all k. Without loss of generality, we may assume that the normalized sequence of vectors $\{(x^k - x^*)/\|x^k - x^*\|_2\}$ converges to a limit v. By an argument similar to that used in the proof of Proposition **3.3.4**, we can establish that v is a critical vector of the pair (K, F) at x^*; moreover, this vector contradicts the condition (5.1.4). This contradiction therefore establishes the desired claim. Hence $\deg(H(\cdot, t), \mathcal{Z})$ is well defined for all $t \in [0, 1]$. By the homotopy invariance property of the degree, it follows that

$$\text{ind}(\mathbf{F}_K^{\text{nor}}, z^*) = \deg(\mathbf{F}_K^{\text{nor}}, \mathcal{Z}) = \deg(H(\cdot, 1), \mathcal{Z}) = \deg(H(\cdot, 0), \mathcal{Z}) = 1$$

because $H(z, 0) = z - z^*$ and $\mathcal{Z}$ contains z^*. □

In addition to establishing the local solvability of the perturbed VIs, the condition (5.1.4) also yields an upper bound on the distance between the perturbed solutions $x \in \mathcal{N} \cap \text{SOL}(K, G)$ and x^* in terms of maximum deviation of G from F in $\mathcal{N}$, where $\mathcal{N}$ is a suitable neighborhood of the solution $x^* \in \mathrm{SOL}(K, F)$. Such a bound is very important because it gives quantitative information on the change of x^* as F is being perturbed. We establish a preliminary lemma, which is an extension of Exercise **3.7.12**.

5.1.7 Lemma. Let $K \subseteq \mathbb{R}^n$ be closed convex and $F : \mathcal{D} \to \mathbb{R}^n$ be B-differentiable at a solution x^* of the VI (K, F), where $\mathcal{D}$ is an open set containing K. If (5.1.4) holds, then for every open neighborhood $\mathcal{N}$ of x^* satisfying (5.1.1), there exists a scalar $c > 0$ such that, for all vectors $q \in \mathbb{R}^n$,

$$\sup \{ \| x - x^* \| : x \in \mathrm{SOL}(K, q + F) \cap \mathrm{cl}\, \mathcal{N} \} \leq c \| q \|.$$

Proof. Let $\mathcal{N}$ be given. Assume for the sake of contradiction that no such scalar c exists. Then there exist a sequence of vectors $\{x^k\} \subset \mathrm{cl}\, \mathcal{N}$ and a sequence of vectors $\{q^k\}$ such that, for each k, $x^k \in \mathrm{SOL}(K, q^k + F)$ and

$$\| x^k - x^* \| > k \| q^k \|.$$

Clearly each x^k is distinct from x^*. Moreover, the sequence $\{x^k\}$ must be bounded and thus have an accumulation point x^∞, which must belong to $\mathrm{cl}\, \mathcal{N}$. The above inequality implies that the sequence $\{q^k\}$ converges to zero. For each k, we have

$$(y - x^k)^T (q^k + F(x^k)) \geq 0, \quad \forall\, y \in K.$$

Hence x^∞ solves the VI (K, F). But since x^* is the unique solution of this VI in the closed neighborhood $\mathrm{cl}\, \mathcal{N}$, it follows that $x^\infty = x^*$. Consequently, the sequence $\{x^k - x^*\}$ converges to x^*. Since

$$\lim_{k \to \infty} \frac{\| q^k \|}{\| x^k - x^* \|} = 0,$$

by the same argument as in the proof of Proposition **5.1.6**, we may show that every accumulation point of the normalized sequence

$$\left\{ \frac{x^k - x^*}{\| x^k - x^* \|} \right\}$$

yields a vector v that contradicts the condition (5.1.4). □

Combining Proposition **5.1.6** and Lemma **5.1.7**, we immediately obtain the following result.

5.1.8 Corollary. Let $K \subseteq \mathbb{R}^n$ be closed convex and $F : \mathcal{D} \to \mathbb{R}^n$ be B-differentiable at a solution x^* of the VI (K, F), where $\mathcal{D}$ is an open set containing K. If (5.1.4) holds, then for every open neighborhood $\mathcal{N}$ of x^* satisfying (5.1.1), there exist two positive scalars c and ε such that, for every continuous function $G : \mathcal{D} \to \mathbb{R}^n$ satisfying

$$\gamma \equiv \sup_{x \in K \cap \operatorname{cl} \mathcal{N}} \| G(x) - F(x) \| < \varepsilon,$$

the following two statements hold: $\mathrm{SOL}(K, G) \cap \mathcal{N} \neq \emptyset$, and

$$\sup \{ \| x - x^* \| : x \in \mathrm{SOL}(K, G) \cap \operatorname{cl} \mathcal{N} \} \leq c\gamma. \tag{5.1.5}$$

Proof. It suffices to observe that if x belongs to $\mathrm{SOL}(K, G)$, then x belongs to $\mathrm{SOL}(K, q + F)$ where $q \equiv G(x) - F(x)$. □

The condition (5.1.5) can be stated equivalently as the following set inclusion:

$$\mathrm{SOL}(K, G) \cap \operatorname{cl} \mathcal{N} \subseteq \operatorname{cl} \mathbb{B}(x^*, c\gamma).$$

We may specialize Corollary **5.1.8** to the parametric VI $(K, F(\cdot, \mathbf{p}^*))$ under the setting of Corollary **5.1.5**, with the nonzero index condition

$$\operatorname{ind}(\mathbf{F}_K^{\mathrm{nor}}(\cdot, \mathbf{p}^*), z^*) \neq 0$$

replaced by a condition analogous to (5.1.4). In this case, the continuity assertion (5.1.3) can be strengthened to become:

$$\sup \{ \| x - x^* \| : x \in S_{\operatorname{cl} \mathcal{N}}(\mathbf{p}) \} \leq c L \| \mathbf{p} - \mathbf{p}^* \|,$$

or equivalently,

$$S_{\operatorname{cl} \mathcal{N}}(\mathbf{p}) \subseteq S_{\operatorname{cl} \mathcal{N}}(\mathbf{p}^*) + c L \| \mathbf{p} - \mathbf{p}^* \| \mathbb{B}(0, 1);$$

the latter inclusion is referred to as an *upper Lipschitz* property of the localized solution map $S_{\operatorname{cl} \mathcal{N}}(\mathbf{p})$ at $\mathbf{p}^*$. Note that $S_{\operatorname{cl} \mathcal{N}}(\mathbf{p}^*) = \{x^*\}$. Since a compact-valued upper Lipschitz multifunction must be upper semicontinuous, it follows that $S_{\operatorname{cl} \mathcal{N}}$ is upper semicontinuous at $\mathbf{p}^*$.

5.2 Solution Stability of B-Differentiable Equations

The results obtained in the last section assume no special structure on K. In order to sharpen these results, we proceed as in Section **3.3** where we considered several special cases of the VI (K, F) and established the local uniqueness of a solution under conditions that are weaker than (5.1.4); see Proposition **3.3.7** for the case of the AVI, Corollary **3.3.9** for the NCP, and Theorem **3.3.12** for a VI with a finitely representable set. Parallel to these results, we consider the same special cases of the VI and derive results similar to Corollary **5.1.8** but under conditions weaker than (5.1.4). The approach is based on the theory of B-differentiable equations whose development began in Subsection **3.3.5**.

The results of the last section immediately yield some useful conclusions for a system of equations $H(z) = 0$, where H is a continuous mapping from $\mathbb{R}^n$ into itself. Firstly, by specializing Proposition **5.1.2** to the case where $K = \mathbb{R}^n$, we deduce that every isolated zero x of H attracts all zeros of nearby functions in a neighborhood of x whose closure contains no other zero of H. Secondly, if x is an isolated zero of H and $\operatorname{ind}(H, x)$ is nonzero, then for all continuous functions G sufficiently near H, the perturbed system $G(z) = 0$ has a solution that is near x. From Proposition **3.3.21**, we know that if $H(x) = 0$ and $H'(x; \cdot)$ has a unique zero, then x is an isolated zero of H. The following lemma is a refinement of this preliminary fact.

5.2.1 Lemma. Let $H : \mathcal{D} \to \mathbb{R}^n$ be continuous, where $\mathcal{D}$ is an open subset of $\mathbb{R}^n$. Suppose that $H(x) = 0$ and H is B-differentiable at x. The following three statements are equivalent.

(a) The implication below holds:

$$H'(x; v) = 0 \Rightarrow v = 0. \tag{5.2.1}$$

(b) There exists a scalar $c' > 0$ such that

$$\| v \| \leq c' \| H'(x; v) \|, \quad \forall v \in \mathbb{R}^n;$$

(c) There exist an open neighborhood $\mathcal{N} \subseteq \mathcal{D}$ of x and a scalar $c > 0$ such that

$$\| x' - x \| \leq c \| H(x') \| \quad \forall x' \in \mathcal{N}. \tag{5.2.2}$$

Proof. Since $H'(x; \cdot)$ is a positively homogeneous, continuous function of the second argument, it is easy to see that (a) $\Rightarrow$ (b).

Suppose that (b) holds. By the B-differentiability of H at x, we may choose an open neighborhood $\mathcal{N}$ of x contained in $\mathcal{D}$ such that

$$\| H(y) - H(x) - H'(x; y - x) \| \leq \varepsilon \| y - x \|, \quad \forall y \in \mathcal{N},$$

where $\varepsilon \in (0, 1/c')$ is an arbitrary given scalar. Since

$$H'(x; x' - x) = -[\, H(x') - H(x) - H'(x; x' - x) \,] + H(x'),$$

by (b), we obtain

$$\| x' - x \| \leq c' [\, \| H(x') - H(x) - H'(x; x' - x) \| + \| H(x') \| \,]$$

which yields

$$\| x' - x \| \leq c \| H(x') \|$$

where

$$c \equiv \frac{c'}{1 - c'\varepsilon}.$$

Thus, (c) holds.

Finally, assume that (c) holds. Suppose that $H'(x; v) = 0$. For all $\tau > 0$ sufficiently small, the vector $x' \equiv x + \tau v$ belongs to $\mathcal{N}$. Thus

$$\tau \| v \| \leq c \| H(x + \tau v) \|.$$

Since $H(x) = H'(x; \tau v) = 0$, we have

$$\| v \| \leq c \frac{\| H(x + \tau v) - H(x) - H'(x; \tau v) \|}{\tau}.$$

The left-hand side approaches zero as $\tau \to 0$. Therefore, $v = 0$. Thus (a) holds. □

Condition (c) implies that x is an isolated zero of H. Thus we recover Proposition **3.3.21** from Lemma **5.2.1**. By imposing an additional nonzero-index condition on the map $\Gamma(y) \equiv H'(x; y - x)$ at its zero $y = x$, we can establish the existence of a solution to the perturbed system $G(y) = 0$, $y \in \mathcal{N}$, where G is a sufficiently small perturbation of H near x as specified in the lemma below. In essence, the gist of the proof of this lemma lies in showing that $\operatorname{ind}(\Gamma, x) = \operatorname{ind}(H, x)$; thus if the former index is nonzero, then so is the latter, from which the rest of the proof of the lemma is then easy. The equality between these two indices is actually a consequence of Proposition **2.1.5**, due to the first-order approximation property of the B-derivative. In fact, the proof below is almost identical to the proof of this early proposition.

5.2.2 Lemma. Let $H : \mathcal{D} \to \mathbb{R}^n$ be continuous, where $\mathcal{D}$ is an open subset of $\mathbb{R}^n$. Suppose that $H(x) = 0$ and H is B-differentiable at x. If (5.2.1) holds and $\operatorname{ind}(\Gamma, x) \neq 0$, where $\Gamma(y) \equiv H'(x; y - x)$, then for every open neighborhood $\mathcal{N}$ of x such that

$$\operatorname{cl}\mathcal{N} \subset \mathcal{D} \quad \text{and} \quad H^{-1}(0) \cap \operatorname{cl}\mathcal{N} = \{ x \}, \tag{5.2.3}$$

there exists a positive scalar ε such that, for every $G \in \mathbb{B}(H; \varepsilon, \operatorname{cl}\mathcal{N})$, the set $G^{-1}(0) \cap \mathcal{N}$ is nonempty.

Proof. As noted above, (5.2.1) implies that x is isolated zero of H. Thus $\operatorname{ind}(H, x)$ is well defined. Moreover, there exists $\bar{\delta} > 0$ such that x is the only zero of H in the closed Euclidean ball with center at x and radius $\bar{\delta}$. We claim that there exists a constant $\gamma > 0$ such that, for all $\delta \in (0, \bar{\delta})$,

$$\gamma\,\delta \leq \mathrm{dist}_\infty(0, H(\partial\mathbb{B}(x,\delta))).$$

Suppose that the claim is not valid. Then there are sequences $\{\delta_k\} \subset (0, \bar{\delta})$ and $\{u^k\} \subset \mathrm{cl}\,\mathbb{B}(0,1)$ such that, for all k,

$$\|\, H(x + \delta_k u^k) \,\|_\infty < \frac{1}{k}\,\delta_k. \tag{5.2.4}$$

Without loss of generality, we may assume that the sequence $\{\delta_k\}$ converges to some limit $\delta_\infty \in [0, \bar{\delta}]$. Clearly, $\delta_\infty = 0$ because of the uniqueness of x as a zero of H in the closure of the ball $\mathbb{B}(x, \bar{\delta})$. Since

$$h(x + \delta_k u^k) \equiv H'(x; \delta_k u^k) = H(x + \delta_k u^k) + \mathbf{e}(\delta_k u^k),$$

where the error function $\mathbf{e}$ satisfies:

$$\lim_{v \to 0} \frac{\mathbf{e}(v)}{\|\, v \,\|} = 0,$$

we deduce, by the proof of Lemma **5.2.1**, that for some constant $c' > 0$,

$$\delta_k = \delta_k \,\|\, u^k \,\|_2 \leq c' \,\|\, H(x + \delta_k u^k) + \mathbf{e}(\delta_k u^k) \,\|_\infty, \quad \forall\, k.$$

Dividing by $\delta_k > 0$ and letting $k \to \infty$, we see that the left-most term is equal to unity while the right-most term approaches zero, by (5.2.4) and the property of the error function $\mathbf{e}$. This is a contradiction; hence the existence of the constant $\gamma > 0$ with the desired property follows.

By the uniform approximating property of the directional derivative $H'(x; \cdot)$ (cf. Proposition **3.1.3**), it follows that there exists a $\delta \in (0, \bar{\delta})$ such that, for all y in the closure of the ball $\mathbb{B}(x, \delta)$,

$$\|\, \Gamma(y) - H(y) \,\|_\infty \leq \tfrac{1}{2}\,\gamma\,\delta < \mathrm{dist}_\infty(0, H(\partial\mathbb{B}(x,\delta))).$$

This implies by the nearness property of the degree that Γ and H have the same index at x. If $\mathcal{N}$ is any open neighborhood of x such that (5.2.3) holds, then $\deg(H, \mathcal{N}) = \mathrm{ind}(H, x) \neq 0$. Moreover, for any function G in $\mathbb{B}(H; \varepsilon, \mathcal{N})$, where $\varepsilon \equiv \mathrm{dist}_\infty(0, H(\partial\mathcal{N})) > 0$, it follows that $\deg(G, \mathcal{N})$ is equal to $\deg(H, \mathcal{N})$ hence is nonzero. Thus such a function G must have a zero in the neighborhood $\mathcal{N}$. □

Lemmas **5.2.1** and **5.2.2** motivate the following definition, which will be generalized to the VI in the next section. The definition does not require the function H to be B-differentiable.

5.2.3 Definition. Let $H : \mathcal{D} \to \mathbb{R}^n$ be continuous, where $\mathcal{D}$ is an open subset of $\mathbb{R}^n$. An isolated zero $x \in \mathcal{D}$ of H is said to be

(a) *semistable* if for every open neighborhood $\mathcal{N}$ of x satisfying (5.2.3), there exist positive scalars ε and c such that, for all functions G in $\mathbb{B}(H; \varepsilon, \operatorname{cl} \mathcal{N})$,

$$\| x' - x \| \leq c \| H(x') \| \quad \forall x' \in G^{-1}(0) \cap \mathcal{N}; \tag{5.2.5}$$

(b) *stable* if for every open neighborhood $\mathcal{N}$ of x satisfying (5.2.3), there exist positive scalars ε and c such that, for all G in $\mathbb{B}(H; \varepsilon, \operatorname{cl} \mathcal{N})$, $G^{-1}(0) \cap \mathcal{N}$ is nonempty and (5.2.5) holds. □

Having given the above definition, we should immediately point out that stability is really a local concept that depends only on the local structure of the function H near the zero x. In fact, it is easy to see that we can restrict the neighborhood $\mathcal{N}$ in the above definition to be *sufficiently small.* This remark applies to several related concepts defined subsequently, such as regularity and strong stability for equations and for VIs.

Lemma **5.2.1** implies that if H is B-differentiable at $x \in H^{-1}(0)$ and $H'(x; \cdot)$ has a unique zero, then x is semistable. In practice, semistability is a fairly weak property because it does not deal with the solvability of the perturbed system of equations $G(y) = 0$ with G being a small perturbation of H. The next result is a restatement of Lemma **5.2.2** as providing a sufficient condition for stability.

5.2.4 Theorem. Let $H : \mathcal{D} \to \mathbb{R}^n$ be continuous, where $\mathcal{D}$ is an open set in $\mathbb{R}^n$. Suppose that H is B-differentiable at a zero $x \in \mathcal{D}$. If (5.2.1) holds and $\operatorname{ind}(\Gamma, x) \neq 0$, where $\Gamma(y) \equiv H'(x; y - x)$, then x is a stable zero of H.

Proof. By Lemma **5.2.1**, x is isolated. Let $\mathcal{N}_0$ be an open neighborhood of x and $c > 0$ be a positive scalar such that (5.2.2) holds. Let $\mathcal{N}$ be an arbitrary open neighborhood of x satisfying (5.2.3). Let $\varepsilon > 0$ be such that, for every $G \in \mathbb{B}(H; \varepsilon, \operatorname{cl} \mathcal{N})$, $G^{-1}(0) \cap \mathcal{N}$ is nonempty. Let $\delta > 0$ be such that $\operatorname{cl} \mathbb{B}(x, \delta)$ is contained in $\mathcal{N}_0$. By the isolatedness of x, it follows that there exists $\varepsilon' \in (0, \varepsilon)$ such that, for every $G \in \mathbb{B}(H; \varepsilon', \operatorname{cl} \mathcal{N})$,

$$G^{-1}(0) \cap \mathcal{N} \subseteq \operatorname{cl} \mathbb{B}(x, \delta).$$

(The boundedness of $\mathcal{N}$ is needed here.) Hence for every vector x' in $G^{-1}(0) \cap \mathcal{N}$, we have $\|x' - x\| \leq c \|H(x')\|$, establishing the stability of x. □

Ideally, it would be desirable to require a stable zero x of the map H to have the additional property such that the set $G^{-1}(0) \cap \mathcal{N}$ is a singleton, i.e., such that the perturbed equation $G(y) = 0$ has a unique solution near x. Realistically, to demand the latter property to hold for all continuous functions G that are slight perturbations of H turns out to be quite restrictive, even for a smooth and simple map such as the identity map. We explain this in detail in the following example.

5.2.5 Example. Consider the scalar function $H(t) \equiv t$ for $t \in \mathbb{R}$. This function has a unique zero at $t = 0$ and is clearly the simplest of all functions; moreover, H has a constant nonvanishing derivative and is a global homeomorphism. Nevertheless, for every neighborhood $\mathcal{N}$ of 0 and every scalar $\varepsilon > 0$, we can define the perturbed function:

$$G(t) \equiv t - \varepsilon' \sqrt{|t|} \sin(1/\sqrt{|t|}), \quad t \in \mathbb{R},$$

where the scalar ε' is to be properly chosen. Clearly, G is continuous everywhere. For an arbitrary positive integer k, let

$$\varepsilon' \equiv \frac{1}{2k\pi + \pi/2} \quad \text{and} \quad t_k \equiv \frac{1}{(2k\pi + \pi/2)^2}.$$

It is easy to see that $G(t_k) = 0 = G(0)$. By choosing k sufficiently large, we have

$$\sup_{t \in \mathrm{cl}\,\mathcal{N}} |H(t) - G(t)| \leq \varepsilon,$$

and t_k belongs to the neighborhood $\mathcal{N}$. Thus G is an ε-perturbation of H in $\mathrm{cl}\,\mathcal{N}$; yet G has multiple zeroes in this neighborhood. □

From the above simple example, we elicit that it is impractical to introduce a stability concept that requires the perturbed system $G(y) = 0$, $y \in \mathcal{N}$ to have a unique solution *for all* continuous functions G. This consideration leads to the following realistic definition of strong stability which differs from the concept of stability only in the continuity requirement of the perturbed solutions.

5.2.6 Definition. Let $H : \mathcal{D} \to \mathbb{R}^n$ be a given map, where $\mathcal{D}$ is an open subset of $\mathbb{R}^n$. An isolated zero $x \in \mathcal{D}$ of H is said to be *strongly stable* if for every open neighborhood $\mathcal{N}$ of x satisfying (5.2.3), there exist two positive scalars c and ε such that

(a) for every $G \in \mathbb{B}(H; \varepsilon, \mathrm{cl}\,\mathcal{N})$, the set $G^{-1}(0) \cap \mathcal{N}$ is nonempty; and

(b) for any two functions G and $\tilde{G}$ in $\mathbb{B}(H;\varepsilon,\mathrm{cl}\,\mathcal{N})$, and for all vectors $v \in G^{-1}(0) \cap \mathcal{N}$ and $\tilde{v} \in \tilde{G}^{-1}(0) \cap \mathcal{N}$,

$$\| v - \tilde{v} \| \leq c \| H(v) - H(\tilde{v}) \|.$$

With $\tilde{G} = H$ and $\tilde{v} = x$, condition (b) reduces to the semistability of x. Thus, every strongly stable zero must be stable. Consider a function G that is a translation of H by a vector u; that is, $G(y) \equiv H(y) - u$, where u is a fixed but arbitrary vector in $\mathbb{R}^n$. By the above definition, we see that if x is a strongly stable zero of H, then for every neighborhood $\mathcal{N}$ of x satisfying (5.1.1), there exists a scalar $\varepsilon > 0$ such that with $\|u\| \leq \varepsilon$, by condition (a) in Definition **5.2.6**, the system

$$H(x') = u, \quad x' \in \mathcal{N} \tag{5.2.6}$$

must have a solution, say $x(u)$; moreover, with $\tilde{G} = G$, (b) implies that this solution is unique. With $\tilde{G}(y) = H(y) - u'$ for some other vector u' with $\|u'\| \leq \varepsilon$, (b) further implies that

$$\| x(u) - x(u') \| \leq c \| u - u' \|.$$

The latter inequality says that the perturbed solution function $x(u)$ is Lipschitz continuous in the right-hand perturbation vector u. Consequently, if x is a strongly stable zero of H, then for every neighborhood $\mathcal{N}$ of x satisfying (5.2.3), there exists a neighborhood $\mathcal{U}$ of the origin such that $H^{-1}(u) \cap \mathcal{N}$ is a single-valued, Lipschitz continuous function of $u \in \mathcal{U}$. This property of a strongly stable zero of H is quite different from the situation in Example **5.2.5** because although u is an arbitrary vector with sufficiently small norm, the perturbed function $G \equiv H - u$ is a simple translation of H; therefore it is possible for $G(y) = 0$, $y \in \mathcal{N}$ to be uniquely solvable for this particular kind of perturbation function G.

If x is merely stable, then the mapping

$$H_{\mathcal{N}} : \mathcal{U} \ni u \mapsto H^{-1}(u) \cap \mathcal{N} \subseteq \mathcal{N}$$

is generally multivalued; it has the property that

$$H_{\mathcal{N}}(u) \subseteq \mathbb{B}(x, c\|u\|), \quad \forall\, u \in \mathcal{U}.$$

Note that $H_{\mathcal{N}}(0) = \{x\}$. In the set-valued literature, a multifunction $\Phi : \mathbb{R}^n \to \mathbb{R}^m$ is said to be *locally upper Lipschitz continuous* at a pair (u^0, v^0) if there exists a neighborhood V of v^0 and positive scalars δ and ε such that

$$\Phi(u) \cap V \subseteq \mathbb{B}(v^0, \varepsilon), \quad \forall\, u \in \mathbb{B}(u^0, \delta).$$

Thus if x is a semistable zero of the function H, then the set-valued inverse mapping H^{-1} is locally upper Lipschitz continuous at $(0, x)$.

Since the perturbed system (5.2.6) is of considerable importance, the above discussion motivates us to introduce the following simplified stability concepts, which we call "regularity".

5.2.7 Definition. Let $H : \mathcal{D} \to \mathbb{R}^n$ be a given map, where $\mathcal{D}$ is an open subset of $\mathbb{R}^n$. An isolated zero $x \in \mathcal{D}$ of H is said to be

(a) *regular* if for every open neighborhood $\mathcal{N}$ of x satisfying (5.2.3), there exist an open neighborhood $\mathcal{U}$ of the origin and a positive scalar c such that

$$\emptyset \neq H^{-1}(u) \cap \mathcal{N} \subseteq \mathbb{B}(x, c \|u\|), \quad \forall u \in \mathcal{U};$$

(b) *strongly regular* if in addition $H^{-1}(u) \cap \mathcal{N}$ is a singleton, say $\{x_{\mathcal{N}}(u)\}$, and furthermore,

$$\| x_{\mathcal{N}}(u) - x_{\mathcal{N}}(u') \| \leq c \| u - u' \|, \quad \forall u, u' \in \mathcal{U}.$$

In essence, the two regularity concepts differ from the corresponding stability concepts in that the perturbation of the base equation $H(y) = 0$ is restricted to the right-hand side in Definition **5.2.7** whereas an arbitrary small perturbation of H is allowed in Definition **5.2.3**. Thus it is clear that a (strongly) stable zero of H must be (strongly) regular.

Summarizing the observations made so far, we present the following diagram that clarifies the connection between the various concepts introduced: for a given $x \in H^{-1}(0)$ and with $\Gamma(y) \equiv H'(x; y - x)$,

[locally Lipschitz homeomorphism]		[$\Gamma^{-1}(0) = \{x\}$, $\operatorname{ind}(\Gamma, x) \neq 0$]	$\Rightarrow$	$\Gamma^{-1}(0) = \{x\}$
$\Downarrow$		$\Downarrow$		$\Downarrow$
strong stability	$\Rightarrow$	stability	$\Rightarrow$	semistability
$\Updownarrow$		$\Downarrow$		$\Downarrow$
strong regularity	$\Rightarrow$	regularity	$\Rightarrow$	isolatedness.

Contained in the above diagram is the assertion that strong stability and strong regularity are actually equivalent. This equivalence is formally stated and proved in Theorem **5.2.8** below, which shows that for a locally Lipschitz continuous function both stability properties are additionally equivalent to the function being a locally Lipschitz homeomorphism. No differentiability of any kind is required of the function in the result. For a B-differentiable function, we can invoke Exercise **3.7.6** to obtain further characterizations of strong stability in terms of the B-derivative; see Theorem **5.2.14**.

5.2.8 Theorem. Let $H : \mathcal{D} \to \mathbb{R}^n$ be continuous, where $\mathcal{D}$ is an open set in $\mathbb{R}^n$. Let $x \in \mathcal{D}$ be a zero of H. The following three statements are valid.

(a) x is strongly stable if and only if x is strongly regular.

(b) If H is a locally Lipschitz homeomorphism at x, then x is strongly stable.

(c) Suppose that H is locally Lipschitz continuous at x. If x is strongly stable, then H is a locally Lipschitz homeomorphism at x.

Proof. For simplicity, we assume that $\mathcal{D} = \mathbb{R}^n$. For (a), it suffices to show that if x is strongly regular, then it is strongly stable. Let $\mathcal{N} \equiv \mathbb{B}(x, \delta)$ be an open neighborhood of x satisfying (5.2.3), where δ is a positive scalar. Associated with the strong regularity of x, let ε and c be positive constants and let $x_{\mathcal{N}} : \mathbb{B}(0, \varepsilon) \to \mathcal{N}$ be a Lipschitz continuous function with modulus c such that, for every $u \in \mathbb{B}(0, \varepsilon)$, $x_{\mathcal{N}}(u)$ is the unique vector in $\mathcal{N}$ that satisfies $H(x_{\mathcal{N}}(u)) = u$; in particular, $x_{\mathcal{N}}(0) = x$. Let

$$\varepsilon' \equiv \min(\varepsilon, (2c)^{-1}\delta).$$

Let G be an arbitrary function in $\mathbb{B}(H; \varepsilon', \operatorname{cl} \mathcal{N})$ and let $e \equiv H - G$ be the difference function. For every $y \in \operatorname{cl} \mathcal{N}$, we have $e(y) \in \mathbb{B}(0, \varepsilon)$ and

$$\| x_{\mathcal{N}}(e(y)) - x \| \leq c \| e(y) \| \leq \delta/2.$$

Thus $x_{\mathcal{N}}(e(y))$ belongs to $\mathcal{N}$ for all $y \in \operatorname{cl} \mathcal{N}$. Define a function

$$\begin{aligned} \Phi : \operatorname{cl} \mathcal{N} &\to \mathcal{N} \\ y &\mapsto x_{\mathcal{N}}(e(y)). \end{aligned}$$

This is a continuous mapping from the compact convex set $\operatorname{cl} \mathcal{N}$ into itself. By Brouwer fixed-point theorem, Φ has a fixed point. Hence there exists a vector $y \in \operatorname{cl} \mathcal{N}$ such that $y = \Phi(y) = x_{\mathcal{N}}(e(y))$. We have

$$H(y) - G(y) = e(y) = H(x_{\mathcal{N}}(e(y))) = H(y).$$

Thus $G(y) = 0$. Hence, G has a zero in $\mathcal{N}$. If $\tilde{G}$ is another continuous function satisfying the same condition as G, and if $\tilde{y}$ is a zero of $\tilde{G}$ that lies in $\mathcal{N}$, then $\tilde{y} = x_{\mathcal{N}}(e(\tilde{y}))$ and hence

$$\| y - \tilde{y} \| \leq c \| e(y) - e(\tilde{y}) \| = c \| H(y) - H(\tilde{y}) \|.$$

Consequently, x is a strongly stable zero of H. Therefore, (a) follows.

Suppose that H is a locally Lipschitz homeomorphism at x. There exists an open neighborhood $\mathcal{N}_0$ of x such that H is a homeomorphism mapping $\mathcal{N}_0$ onto $H(\mathcal{N}_0)$ and the inverse of $H|_{\mathcal{N}_0}$, which maps $H(\mathcal{N}_0)$ onto $\mathcal{N}_0$, is Lipschitz continuous with modulus c. Without loss of generality, we may assume that $H^{-1}(0) \cap \operatorname{cl} \mathcal{N}_0 = \{x\}$. Thus $\operatorname{ind}(H, x)$ is well defined and equal to ± 1. Since the inverse of $H|_{\mathcal{N}_0}$ is Lipschitz continuous with modulus c, we have for any two vectors v^1 and v^2 in $\mathcal{N}_0$,

$$\| v^1 - v^2 \| \leq c \| H(v^1) - H(v^2) \|.$$

Let $\mathcal{N}$ be an arbitrary open neighborhood of x with $H^{-1}(0) \cap \operatorname{cl} \mathcal{N} = \{x\}$. Define

$$\varepsilon \equiv \operatorname{dist}_\infty(0, H(\partial \mathcal{N})).$$

By a familiar degree-theoretic argument we deduce that every function $G \in \mathbb{B}(H; \varepsilon, \operatorname{cl} \mathcal{N})$ must have a zero in $\mathcal{N}$. Moreover, by taking ε to be even smaller if necessary, the last assertion of Lemma **5.2.2**, which requires only the isolatedness of x in $\operatorname{cl} \mathcal{N}$, implies that any such zero must belong to the neighborhood $\mathcal{N}_0$. Consequently, condition (b) in Definition **5.2.6** follows. Therefore, x is a strongly stable zero of H. This establishes assertion (b).

Suppose that H is Lipschitz continuous in the open neighborhood $\mathcal{N}$ of x. Suppose further that x is a strongly regular zero of H. There exists an open neighborhood $\mathcal{U}$ of the origin and a Lipschitz continuous mapping $x_{\mathcal{N}} : \mathcal{U} \to \mathcal{N}$ with modulus $c > 0$ such that, for every $u \in \mathcal{U}$, $x_{\mathcal{N}}(u)$ is the only vector in $\mathcal{N}$ that satisfies $H(x_{\mathcal{N}}(u)) = u$. There exists an open subneighborhood $\mathcal{N}' \subseteq \mathcal{N}$ such that $H(\mathcal{N}') \subseteq \mathcal{U}$. We claim that $H|_{\mathcal{N}'}$ is injective. Assume $H(v^1) = H(v^2)$ for some v^1 and v^2 belonging to $\mathcal{N}'$. Writing $u \equiv H(v^1) = H(v^2)$, we have $v^1 = x_{\mathcal{N}}(u) = v^2$. Thus

$$H|_{\mathcal{N}'} : \mathcal{N}' \to H(\mathcal{N}')$$

is injective and Lipschitz continuous. To show that $(H|_{\mathcal{N}'})^{-1}$ is Lipschitz continuous on its domain, let v and v' be two vectors in $\mathcal{N}'$ and write $u \equiv H(v)$ and $u' \equiv H(v')$. We have $v = x_{\mathcal{N}}(u)$ and $v' = x_{\mathcal{N}}(u')$. Hence, by the Lipschitz continuity of $x_{\mathcal{N}}$, it follows that $\|v - v'\| \leq c\|u - u'\|$. Therefore H is a locally Lipschitz homeomorphism at x. □

It is natural to ask whether there is an analogous result for stability. The answer is no in general; subsequently, we show that in the context of the VI, stability is equivalent to regularity if monotonicity is present. For now, we show that if H is strongly F-differentiable, then all the concepts discussed thus far are equivalent.

5.2.9 Proposition. Let $H : \mathcal{D} \to \mathbb{R}^n$ be continuous, where $\mathcal{D}$ is an open set in $\mathbb{R}^n$. Suppose that $H(x) = 0$ and H is strongly F-differentiable at x. The following statements are equivalent.

(a) x is strongly stable;

(b) x is stable;

(c) x is regular;

(d) $JH(x)$ is nonsingular;

(e) H is a locally Lipschitz homeomorphism at x.

Proof. It is clear that (a) $\Rightarrow$ (b) $\Rightarrow$ (c); (e) $\Rightarrow$ (a) by Theorem **5.2.8**. The equivalence of (d) and (e) is the assertion in Proposition **2.1.14**. Therefore it remains to show that (c) implies (d). Suppose that x is a regular zero of H but there exists a nonzero vector v such that $JH(x)v = 0$. On the one hand, the regularity of x implies the existence of a constant $c > 0$ such that, for all $\tau > 0$ sufficiently small,

$$\| (x + \tau v) - x \| = \tau \| v \| \leq c \| H(x + \tau v) \|;$$

on the other hand, the F-differentiability of H at x implies

$$\| H(x + \tau v) \| = o(\| \tau \|),$$

because $H(x) = JH(x)v = 0$. Combining the two equalities, we obtain

$$\| v \| \leq c \frac{o(\| \tau \|)}{\tau}$$

which is a contradiction because v is nonzero. □

We illustrate Proposition **5.2.9** with a vertical CP at a nondegenerate solution. Let F and G be two continuously differentiable functions mapping an open subset $\mathcal{D}$ of $\mathbb{R}^n$ into $\mathbb{R}^n$. Consider the function

$$H(x) \equiv \min(F(x), G(x)), \quad x \in \mathcal{D},$$

The equation $H(x) = 0$ is equivalent to the CP (F, G):

$$0 \leq F(x) \perp G(x) \geq 0.$$

Let x^* be a solution of the latter CP. Define the index sets:

$$\begin{aligned} \alpha_F &\equiv \{ i : F_i(x^*) = 0 < G_i(x^*) \} \\ \beta &\equiv \{ i : F_i(x^*) = 0 = G_i(x^*) \} \\ \alpha_G &\equiv \{ i : F_i(x^*) > 0 = G_i(x^*) \}. \end{aligned}$$

Suppose that x^* is nondegenerate; that is, β is empty. It is then easy to show that H is F-differentiable at x^* and the i-th row of the Jacobian matrix $JH(x^*)$ is either $\nabla F_i(x^*)^T$ or $\nabla G_i(x^*)^T$ depending on whether i belongs to α_F or α_G. In this case, in a suitable neighborhood of x^*, H is essentially a smooth function with components uniquely determined by either F or G but not both. Specializing Proposition **5.2.9** to this application, we obtain the following simplified stability result for the vertical CP (F, G). Further specialization to the NCP is easy and omitted.

5.2.10 Corollary. Let F and G be two continuously differentiable functions mapping an open subset $\mathcal{D}$ in $\mathbb{R}^n$ into $\mathbb{R}^n$. Suppose that x^* is a nondegenerate solution of the CP (F, G). If the $n \times n$ matrix

$$\begin{bmatrix} \nabla F_i(x^*)^T & : i \in \alpha_F \\ \nabla G_i(x^*)^T & : i \in \alpha_G \end{bmatrix}$$

is nonsingular, then for every open neighborhood $\mathcal{N}$ of x^* such that x^* is the unique solution of the CP (F, G) in $\operatorname{cl}\mathcal{N}$, there exist two positive scalars ε and c such that, for any two pairs of continuous functions (F^i, G^i) for $i = 1, 2$ satisfying

$$\sup_{x \in \operatorname{cl}\mathcal{N}} \max(\| F(x) - F^i(x) \|, \| G(x) - G^i(x) \|) < \varepsilon, \tag{5.2.7}$$

the perturbed CP (F^i, G^i) has a unique solution x^i belonging to the neighborhood $\mathcal{N}$; moreover,

$$\| x^1 - x^2 \| \leq c \max(\| F(x^1) - F(x^2) \|, \| G(x^1) - G(x^2) \|)$$

Proof. Since the min function in Lipschitz continuous in its arguments, the function $H^i \equiv \min(F^i, G^i)$ is therefore an ε-perturbation of $H \equiv \min(F, G)$ on $\operatorname{cl}\mathcal{N}$; that is,

$$\sup_{x \in \operatorname{cl}\mathcal{N}} \| H(x) - H^i(x) \| < \varepsilon,$$

provided that (5.2.7) holds. The conclusion follows easily from the above discussion and Proposition **5.2.9**. □

Proposition **5.2.9** fails to apply if x is a non-F-differentiable point of H; in particular, the stability of a nonsmooth zero of H does not imply the strong stability of the zero. We illustrate this failure using the following example that is derived from an LCP.

5.2.11 Example. Let

$$H(x_1, x_2) \equiv \begin{pmatrix} \min(x_1, x_1 + x_2) \\ \min(x_2, x_1 + x_2). \end{pmatrix}, \quad \forall (x_1, x_2) \in \mathbb{R}^2.$$

The equation $H(x_1, x_2) = 0$ is equivalent to the homogeneous LCP:

$$\begin{pmatrix} 0 \\ 0 \end{pmatrix} \leq \begin{pmatrix} x_1 \\ x_2 \end{pmatrix} \perp \begin{bmatrix} 1 & 1 \\ 1 & 1 \end{bmatrix} \begin{pmatrix} x_1 \\ x_2 \end{pmatrix} \geq \begin{pmatrix} 0 \\ 0 \end{pmatrix},$$

which has a unique solution $(0, 0)$. This solution is a stable zero of H but it is not strongly stable. The stability can be proved by the following steps. Observe that the above 2×2 matrix of all ones is strictly copositive on $\mathbb{R}^2_+$; furthermore, the map H is the natural map of the displayed LCP; apply Theorem **2.5.10** to deduce that $\operatorname{ind}(H, (0,0))$ is equal to one; invoke Theorem **5.2.4**. Nevertheless, $(0, 0)$ is not a strongly stable zero of H because the equation $H(x_1, x_2) + (\varepsilon, \varepsilon) = 0$ has at least two solutions $(-\varepsilon, 0)$ and $(0, -\varepsilon)$ for every $\varepsilon > 0$. □

An important contribution of the sensitivity and stability theory developed in this chapter is the ability to deal with "degenerate" VIs and CPs that correspond to genuinely nonsmooth equations to which the classical smooth analysis fails to be applicable.

5.2.1 Characterizations in terms of the B-derivative

In what follows, we derive necessary and sufficient conditions for the stability and the strong stability of a zero of a B-differentiable function in terms of properties of the directional derivative. For the purpose of obtaining the characterization of stability, we introduce a technical condition that is used in Theorem **5.2.12**. Specifically, we say that a mapping $\Gamma : \mathcal{D} \subseteq \mathbb{R}^n \to \mathbb{R}^n$ has the *nonvanishing property* at a zero $x \in \Gamma^{-1}(0)$ if there exists a continuous function $\Phi : \mathcal{D} \to \mathbb{R}^n$ such that (a) for all scalars $\delta \geq 0$, x is the only zero of $\Phi + \delta\Gamma$ (thus $\operatorname{ind}(\Phi, x)$ is well defined), and (b) $\operatorname{ind}(\Phi, x)$ is nonzero. Classes of mappings Γ that have this nonvanishing property include (i) affine mappings $\Gamma(y) \equiv a + Ay$ for some nonsingular matrix $A \in \mathbb{R}^{n \times n}$, and (ii) P_0 functions on $\mathbb{R}^n$. Indeed, if Γ is a mapping of type (i) and $x \equiv A^{-1}a$, then with $\Phi \equiv \Gamma$, condition (a) clearly holds; moreover, we have $\operatorname{ind}(\Phi, x) = \det A \neq 0$. If Γ is a P_0 function on $\mathbb{R}^n$ and $x \in \Gamma^{-1}(0)$, then with $\Phi(y) \equiv y - x$, we clearly have $\operatorname{ind}(\Phi, x) = 1$; moreover, for every $\delta \geq 0$, $\Phi + \delta\Gamma$ is a P function and thus is injective. Hence $\Phi + \delta\Gamma$ has a unique zero.

The following result provides necessary and sufficient conditions for a zero of a B-differentiable function to be stable, assuming the nonvanishing property at the zero. In particular, part (e) of the result shows that the requirement "for every open neighborhood $\mathcal{N}$ of x" in Defintion **5.2.3** can be weakened to "there exists an open neighborhood".

5.2.12 Theorem. Let x be a zero of the continuous map $H : \mathcal{D} \to \mathbb{R}^n$, where $\mathcal{D}$ is an open subset of $\mathbb{R}^n$. Suppose that H is B-differentiable at x and $\Gamma(y) \equiv H'(x; y-x)$ has the nonvanishing property at x. The following statements are equivalent:

(a) x is a stable zero of Γ;

(b) Γ is surjective and $\Gamma^{-1}(\mathbb{B}(0,1))$ is bounded;

(c) x is the unique zero of Γ;

(d) x is a stable zero of H;

(e) there exist an open neighborhood $\mathcal{N}$ of x satisfying (5.2.3) and two positive scalars ε and c such that, for all $G \in \mathbb{B}(H; \varepsilon, \operatorname{cl}\mathcal{N})$, $G^{-1}(0) \cap \mathcal{N}$ is nonempty and (5.2.5) hold.

Proof. (a) $\Rightarrow$ (b). If x is a stable zero of Γ, then there exist neighborhoods $\mathcal{N}$ of x and $\mathcal{U}$ of the origin, and a constant $c > 0$ such that the system

$$\Gamma(y) = u, \quad y \in \mathcal{N}$$

has a solution for all $u \in \mathcal{U}$; moreover,

$$\| y - x \| \leq c \| u \|.$$

To show the surjectivity of Γ, let $u \in \mathbb{R}^n$ be arbitrary. Then $\tau u \in \mathcal{U}$ for all $\tau > 0$ sufficiently small. Hence there exists a vector $y \in \mathcal{N}$ such that $\tau u = \Gamma(y) = H'(x; y - x)$. Since $H'(x; \cdot)$ is positively homogeneous in the second argument, we have $u = H'(x; \tau y - \tau x) = \Gamma(\tau y - \tau x + x)$. Thus Γ is surjective. To show the boundedness of $\Gamma^{-1}(\mathbb{B}(0,1))$, let $u \in \mathbb{B}(0,1)$ and y be such that $u = \Gamma(y) = H'(x; y - x)$. Thus for all $\tau > 0$ sufficiently small, we have $\tau u \in \mathcal{U}$ and $\tau u = H'(x; \tau y - \tau x) = \Gamma(\tau y - \tau x + x)$. Hence, it follows that

$$\tau \| y - x \| \leq c\tau \| u \|,$$

which implies that $\|y\| \leq \|x\| + c$. Consequently, (b) holds.

(b) ⇒ (c). If y is a zero of Γ, then so is $\tau y - \tau x + x$ for all $\tau > 0$. Since (b) implies that $G^{-1}(0)$ is bounded, it follows that y must equal to x.

(c) ⇒ (d). If (c) holds, then $\operatorname{ind}(\Gamma, x)$ is well defined. By considering the homotopy:

$$H(y,t) \equiv (1-t)\Phi(y) + t\Gamma(y), \quad (y,t) \in \mathbb{R}^n \times [0,1],$$

and using the nonvanishing property of Γ and the invariance property of the degree, we easily deduce $\operatorname{ind}(\Gamma, x) \neq 0$. Thus by Theorem **5.2.4**, (d) follows.

(d) ⇒ (e). This is trivial.

(e) ⇒ (c). Let $\mathcal{N}$, ε, and c be as given in condition (e). Suppose $\Gamma(y) = H'(x; y - x) = 0$. For all $\tau > 0$, we have

$$H(\tau(y-x)+x) = H(x) + \tau H'(x; y-x) + \mathbf{e}(\tau) = \mathbf{e}(\tau),$$

where the error function $\mathbf{e}(\tau)$ satisfies

$$\lim_{\tau \to 0} \frac{\mathbf{e}(\tau)}{\tau} = 0.$$

Consequently, for all $\tau > 0$ sufficiently small, the function

$$G(y') \equiv H(y') - H(\tau(y-x)+x), \quad y' \in \mathcal{D},$$

is a member of $\mathbb{B}(H; \varepsilon, \operatorname{cl}\mathcal{N})$ and $\tau(y-x)+x$ belongs to $G^{-1}(0) \cap \mathcal{N}$. Consequently, it follows that

$$\tau \| y - x \| \leq c \| H(\tau(y-x)+x) \| = \mathbf{e}(\tau).$$

Dividing by $\tau > 0$ and letting $\tau \to 0$, we deduce that $y = x$.

(c) ⇒ (a). As above, we have $\operatorname{ind}(\Gamma, x) \neq 0$. Moreover, Γ is easily seen to be B-differentiable at x with $\Gamma'(x; y-x) = \Gamma(y)$ for all y. Consequently, the stability of x as a zero of Γ follows from Theorem **5.2.4**. □

5.2.13 Remark. The implications (a) ⇒ (b) ⇒ (c) ⇐ (d) do not require the nonvanishing property of Γ. □

To derive the next result which pertains to the strong stability of x, we recall that the B-derivative $H'(x; \cdot)$ is strong if the error function

$$\mathbf{e}(y) \equiv H(y) - H(x) - H'(x; y-x)$$

satisfies

$$\lim_{\substack{y^1 \neq y^2 \\ (y^1, y^2) \to (x,x)}} \frac{\mathbf{e}(y^1) - \mathbf{e}(y^2)}{\| y^1 - y^2 \|} = 0.$$

An important difference between the two results, Theorems **5.2.12** and **5.2.14**, is that the nonvanishing property of the map Γ is a precondition for several of the implications in the previous theorem to hold; this property is no longer needed in the following theorem. Another difference is that the former theorem does not require the B-derivative to be strong, whereas the latter theorem requires this derivative to be strong. The significance of part (d) in Theorem **5.2.14** is similar to that of part (e) in Theorem **5.2.12**.

5.2.14 Theorem. Let $H : \mathcal{D} \to \mathbb{R}^n$ be continuous, where $\mathcal{D}$ is an open subset of $\mathbb{R}^n$. Suppose that H has a strong B-derivative at a zero x of H. The following statements are equivalent:

(a) the origin is a strongly stable zero of the B-derivative $H'(x;\cdot)$;

(b) $H'(x;\cdot)$ is a globally Lipschitz homeomorphism on $\mathbb{R}^n$;

(c) x is a strongly stable zero of H;

(d) there exist an open neighborhood $\mathcal{N}$ of x satisfying (5.2.3) and two positive scalars c and ε such that conditions (a) and (b) in Definition **5.2.6** holds.

Proof. (a) $\Rightarrow$ (b). Since the B-derivative $H'(x;\cdot)$ is Lipschitz continuous in the second argument, part (c) of Theorem **5.2.8** implies that $H'(x;\cdot)$ is a locally Lipschitz homeomorphism at the origin. By part (a) of Exercise **3.7.6**, this is equivalent to (b).

(b) $\Rightarrow$ (c). Since $H'(x;\cdot)$ is a globally Lipschitz homeomorphism on $\mathbb{R}^n$, it follows from part (c) of Exercise **3.7.6** that H is a locally Lipschitz homeomorphism at x. In turn, part (b) of Theorem **5.2.8** then implies that x is a strongly stable zero of H.

(c) $\Rightarrow$ (d). This is trivial.

(d) $\Rightarrow$ (a). Let $\mathcal{N}$, ε and c be as given in condition (d). Choose a neighborhood $\mathcal{N}' \subseteq \mathcal{N}$ of x such that $\|H(x')\| < \varepsilon$ for all $x' \in \mathcal{N}'$. For any two vectors x^1 and x^2 in $\mathcal{N}'$, define

$$G^i(y) \equiv H(y) - H(x^i), \quad \forall\, y \in \mathcal{D}.$$

Then $G^i \in \mathbb{B}(H;\varepsilon,\operatorname{cl}\mathcal{N})$, $G^i(x^i) = 0$, and $x^i \in \mathcal{N}$. Hence, we have

$$\| x^1 - x^2 \| \leq c \| H(x^1) - H(x^2) \|.$$

Therefore, by Corollary **2.1.13**, it follows that H is a locally Lipschitz homeomorphism at x. Part (b) of Exercise **3.7.6** implies $H'(x;\cdot)$ is a globally Lipschitz homeomorphism on $\mathbb{R}^n$. Part (b) of Theorem **5.2.8** applied to $H'(x;\cdot)$ then shows that the origin is a strongly stable zero of the B-derivative. □

If the B-derivative $H'(x;\cdot)$ is piecewise linear in the second argument, the equivalent conditions in Theorem **4.2.11** for $H'(x;\cdot)$ to be a global Lipschitz homeomorphism on $\mathbb{R}^n$ can be added to Theorem **5.2.14** to yield more necessary and sufficient conditions for the strong stability of x.

5.2.2 Extensions to locally Lipschitz functions

The development in the last subsection has focused on the class of B-differentiable functions, where the directional derivative was used as the principal tool in the two main results, Theorems **5.2.12** and **5.2.14**. For a locally Lipschitz continuous function that is not necessarily B-differentiable, the above theory can be extended via the concept of a "first-order approximation" (FOA). Throughout this subsection, all functions are assumed to be locally Lipschitz continuous.

Formally, a function h is said to be a FOA of H at a vector x if

$$\lim_{x \neq y \to x} \frac{H(y) - h(y)}{\| y - x \|} = 0.$$

With h and H both continuous at x, the above definition implies in particular that $h(x) = H(x)$. A FOA h of H is said to be *strong* if the error function $\mathbf{e}(y) \equiv H(y) - h(y)$ satisfies

$$\lim_{\substack{y^1 \neq y^2 \\ (y^1,y^2) \to (x,x)}} \frac{\mathbf{e}(y^1) - \mathbf{e}(y^2)}{\| y^1 - y^2 \|} = 0.$$

Notice that h is a (strong) FOA of H at x if and only if H is a (strong) FOA of h at x; thus "being a FOA" is a symmetric relation. If H is (strongly) B-differentiable at x, then

$$h(y) \equiv H(x) + H'(x; y - x)$$

is a (strong) FOA of H at x. More generally, consider the composite map: $H \equiv S \circ N$, where S is (strongly) B-differentiable at $N(x)$ and N is an arbitrary locally Lipschitz continuous function. Then the function:

$$h(y) \equiv S(N(x)) + S'(N(x); N(y) - N(x))$$

is a (strong) FOA of H at x. An example of a composite map of the latter type is the normal map $\mathbf{F}_K^{\text{nor}}$ of the VI (K, F), where F is a F-differentiable function. We have $\mathbf{F}_K^{\text{nor}} = S \circ N$, where $S : \mathbb{R}^{2n} \to \mathbb{R}^n$ is given by $S(a,b) \equiv F(a) + b$, which is clearly continuously differentiable, while $N : \mathbb{R}^n \to \mathbb{R}^{2n}$ is given by

$$N(z) \equiv \begin{pmatrix} \Pi_K(z) \\ z - \Pi_K(z) \end{pmatrix}$$

and is only Lipschitz continuous. The corresponding FOA of $\mathbf{F}_K^{\text{nor}}$ at a vector $\bar{z}$ is given by

$$z \mapsto \mathbf{F}_K^{\text{nor}}(\bar{z}) + JF(\Pi_K(\bar{z}))(\,\Pi_K(z) - \Pi_K(\bar{z})\,) + z - \Pi_K(z), \tag{5.2.8}$$

which we easily recognize as the normal map of the semi-linearized VI $(K, L_{\bar{z}})$, where $L_{\bar{z}}(x)$ is the linearization of the function $F(x)$ at the vector $\bar{x} \equiv \Pi_K(\bar{z})$; that is,

$$L_{\bar{z}}(x) \equiv F(\bar{z}) + JF(\bar{z})(\,x - \bar{z}\,).$$

The term "semi-linearized" refers to the fact that in the VI (K, F), only the function F is approximated by its linearization at $\bar{x}$, resulting in the VI $(K, L_{\bar{x}})$ where the set K is not modified. This kind of approximation is different from the B-derivative approximation of $\mathbf{F}_K^{\text{nor}}$ that also requires an approximation of the projector Π_K; see Chapter **4** for details on the latter approximation.

The following result relates the (strong) stability of a zero of a nonsmooth function in terms of the (strong) stability of the same zero of a FOA of the function. For an application of part (b) of the result, see Exercise **5.6.10**.

5.2.15 Proposition. Let H and h be locally Lipschitz continuous functions at x. Suppose that h is a FOA of H at x. The following two statements are valid.

(a) If x is a stable zero of H and $\text{ind}(H, x)$ is nonzero, then x is a stable zero of h.

(b) If h is a strong FOA of H of x, then x a strongly stable zero of H if and only if x is a strongly stable zero of h.

Proof. Assume the conditions in (a). The stability of x implies the existence of a constant $c > 0$ such that, for all y sufficiently close to x,

$$\|\,H(y)\,\| \geq c\,\|\,y - x\,\|.$$

Since $h(x) = 0$ and

$$h(y) = h(y) - H(y) + H(y),$$

it follows that

$$\liminf_{x \neq y \to x} \frac{h(y) - h(x)}{\| y - x \|} \geq c.$$

Hence by Proposition **2.1.5**, we deduce that x is an isolated zero of h and $\operatorname{ind}(h, x) = \operatorname{ind}(H, x)$. The rest of the proof is very similar to that of Theorem **5.2.4**. The details are not repeated.

Since being a strong FOA is a symmetric relation, if suffices to show that if x is a strongly stable zero of H, then it is a strongly stable zero of h. We have $H(x) = h(x) = 0$. By Theorem **5.2.8**, it follows that H is a locally Lipschitz homeomorphism near x. Thus there exist a neighborhood $\mathcal{N}_0$ of x and a constant $c > 0$ such that

$$\| H(y) - H(y') \| \geq c \| y - y' \|, \quad \forall y, y' \in \mathcal{N}_0.$$

Let η be any positive scalar in the interval $(0, c)$. By restricting the neighborhood $\mathcal{N}_0$ if necessary, we may assume that for any vectors y and y' in $\mathcal{N}_0$,

$$\| \mathbf{e}(y) - \mathbf{e}(y') \| \leq \eta \| y - y' \|.$$

We have, for any y and y' in $\mathcal{N}_0$,

$$h(y) - h(y') = [\, h(y) - H(y) - (h(y') - H(y')) \,] + H(y) - H(y'),$$

which implies

$$\| h(y) - h(y') \| \geq (c - \eta) \| y - y' \|.$$

By Corollary **2.1.13**, it follows that h is a locally Lipschitz homeomorphism at x. Part (b) of Theorem **5.2.8** then implies that x is a strongly stable zero of h. □

5.2.16 Remark. It is not clear whether the index condition in statement (a) of Proposition **5.2.15** can be dropped. □

5.3 Solution Stability: The Case of a Fixed Set

We formally define two important concepts of stability of an isolated solution to a VI with an unperturbed set. In principle, we could define these concepts by directly specializing Definitions **5.2.3** and **5.2.6** to the natural or normal map of the VI. Nevertheless, instead of employing this somewhat indirect approach, we introduce the stability concepts for a solution to the VI by modifying these previous definitions in order to properly reflect the inherent structure of the VI. Such a consideration leads to Definition **5.3.1** that generalizes Definitions **5.2.3** and **5.2.6**. With $K = \mathbb{R}^n$, the definition below reduces to the previous definitions for the respective concepts of a stable zero and a strongly stable zero of an equation. Subsequently, we show in Proposition **5.3.6** that the strong stability (as defined below) of a solution to a VI is equivalent to the strong stability (as defined in Definition **5.2.6**) of a corresponding zero of the normal map of the VI.

5.3.1 Definition. A solution x^* of the VI (K, F) is said to be *semistable* if for every open neighborhood $\mathcal{N}$ of x^* satisfying (5.1.1), there exist two positive scalars c and ε such that, for every G in $\mathbb{B}(F; \varepsilon, K_{\mathcal{N}})$ and every $x \in \mathrm{SOL}(K, G) \cap \mathcal{N}$,

$$\| x - x^* \| \leq c \| e(x) \|,$$

where $e(v) \equiv F(v) - G(v)$ is the difference function. The solution x^* is *stable* if in addition $\mathrm{SOL}(K, G) \cap \mathcal{N} \neq \emptyset$. This solution x^* is said to be *strongly stable* if x^* is stable and for every neighborhood $\mathcal{N}$ with corresponding scalars c and ε as above, and for any two continuous functions G and $\tilde{G}$ belonging to $\mathbb{B}(F; \varepsilon, K_{\mathcal{N}})$,

$$\| x - x' \| \leq c \| e(x) - \tilde{e}(x') \|,$$

for every $x \in \mathrm{SOL}(K, G) \cap \mathcal{N}$ and $x' \in \mathrm{SOL}(K, G') \cap \mathcal{N}$, where $\tilde{e}$ is the difference function between F and $\tilde{G}$; that is, $\tilde{e}(v) \equiv F(v) - \tilde{G}(v)$. □

By definition, every stable solution of a VI is semistable; every strongly stable solution is stable. In turn, every semistable solution must be isolated. As in Definition **5.2.7**, we introduce a simplified case of stability that pertains to perturbing the function F by constant vectors with small norms.

5.3.2 Definition. A solution x^* of the VI (K, F) is said to be

(a) *semiregular* if for every open neighborhood $\mathcal{N}$ of x^* satisfying (5.1.1), there exist two positive scalars c and ε such that, for every vector $q \in \mathbb{R}^n$ satisfying $\|q\| \leq \varepsilon$, it holds that

$$\sup \{ \| x - x^* \| \, : \, x \in \mathrm{SOL}(K, F + q) \cap \mathcal{N} \} \leq c \| q \|;$$

(b) *regular* if in addition $\mathrm{SOL}(K, F + q) \cap \mathcal{N} \neq \emptyset$ for all q satisfying $\|q\| \leq \varepsilon$;

(c) *strongly regular* if x^* is regular and

$$\max(\|q\|, \|q'\|) \le \varepsilon \Rightarrow \|x - x'\| \le c\|q - q'\| \tag{5.3.1}$$

for all $x \in \text{SOL}(K, F + q) \cap \mathcal{N}$ and $x' \in \text{SOL}(K, F + q') \cap \mathcal{N}$. □

Suppose that x^* is strongly regular, by (5.3.1) with $q' = q$, it follows that $\text{SOL}(K, F+q) \cap \mathcal{N}$ is a singleton, which we denote $\{x_{\mathcal{N}}(q)\}$; thus $x_{\mathcal{N}}(q)$ is a single-valued function of q. Moreover, $x_{\mathcal{N}}(q)$ is Lipschitz continuous on its domain. This observation is similar to the case of a nonsmooth equation; see the discussion following Definition **5.2.6**. Furthermore, as in this previous case, we can summarize the relation between stability and regularity in the following diagram:

$$\begin{array}{ccccc} \text{strong stability} & \Rightarrow & \text{stability} & \Rightarrow & \text{semistability} \\ \Updownarrow & & \Downarrow & & \Updownarrow \\ \text{strong regularity} & \Rightarrow & \text{regularity} & \Rightarrow & \text{semiregularity} \\ & & & & \Downarrow \\ & & & & \text{isolatedness.} \end{array}$$

All relations are fairly obvious, the only two exceptions are (i) semiregularity implies semistability and (ii) strong regularity implies strong stability. The former implication (i) is not difficult because $x \in \text{SOL}(K, G)$ if and only if $x \in \text{SOL}(K, F + q)$, where $q \equiv G(x) - F(x)$. The latter implication (ii) is formally stated in the following result.

5.3.3 Proposition. Let K be a closed convex set in $\mathbb{R}^n$ and $F : \mathcal{D} \to \mathbb{R}^n$ be a continuous mapping, where $\mathcal{D}$ is an open set containing K. If x^* is a strongly regular solution of the VI (K, F), then x^* is strongly stable.

Proof. The proof is similar to that of part (a) of Theorem **5.2.8**. □

Since semiregularity is equivalent to semistability, the only reason why regularity is not necessarily equivalent to stability is due to the possible unsolvability of the perturbed VI (K, G), for G sufficiently close to F. Using Kakutani's fixed-point theorem instead of Brouwer's fixed-point theorem, we can establish an analogous result showing that regularity implies stability for a solution to a pseudo monotone VI.

5.3.4 Proposition. Let K be a closed convex set in $\mathbb{R}^n$ and $F : \mathcal{D} \to \mathbb{R}^n$ be a continuous, pseudo monotone mapping on K, where $\mathcal{D}$ is an open set containing K. If x^* is a regular solution of the VI (K, F), then x^* is unique and stable.

Proof. Since a regular solution must be isolated and since the solution set of a pseudo monotone VI is convex, it follows that x^* must be the unique solution of the VI (K, F). To show that x^* is stable, let $\mathcal{N} \equiv \mathbb{B}(x, \delta)$ be an open neighborhood of x^* satisfying (5.1.1). Let c and ε be two positive scalars associated with the regularity of x^*. Let ε' be the positive scalar:

$$\varepsilon' \equiv \min(\varepsilon,\ \delta/2c).$$

Let $G \in \mathbb{B}(H; \varepsilon', K_{\mathcal{N}})$ and $e \equiv G - F$ be the difference function. We define a set-valued mapping

$$\begin{array}{rcll} \Phi : & K \cap \operatorname{cl}\mathcal{N} & \to & K \cap \operatorname{cl}\mathcal{N} \\ & v & \mapsto & \operatorname{SOL}(K, F + e(v)) \cap \operatorname{cl}\mathcal{N}. \end{array}$$

It is easy to show that Φ is a closed multifunction from the convex compact set $K \cap \operatorname{cl}\mathcal{N}$ into itself; moreover, for each $v \in K \cap \operatorname{cl}\mathcal{N}$, $\Phi(v)$ is a nonempty, closed, and convex set; convex because of the pseudo monotonicity of F on K. Therefore, by Kakutani's fixed point theorem, Φ has a fixed point; that is, there exists a vector $v \in \operatorname{SOL}(K, F + e(v)) \cap \operatorname{cl}\mathcal{N}$. Such a fixed point must be a solution of the VI (K, G). Moreover, we have

$$\| v - x^* \| \leq c \| e(v) \| \leq c\varepsilon' < \delta.$$

Thus the vector v belongs to the neighborhood $\mathcal{N}$. This shows that the set $\operatorname{SOL}(K, G) \cap \mathcal{N}$ is nonempty. Moreover, if x is an arbitrary vector in the latter set, then $x \in \operatorname{SOL}(K, F + e(x)) \cap \mathcal{N}$ and we have

$$\| x - x^* \| \leq c \| e(x) \|.$$

This establishes the stability of x^*. □

Employing Example **5.2.11**, we illustrate below that a monotone VI can have a stable (thus unique) solution which is not strongly stable.

5.3.5 Example. Consider the LCP:

$$\begin{pmatrix} 0 \\ 0 \end{pmatrix} \leq \begin{pmatrix} x_1 \\ x_2 \end{pmatrix} \perp \begin{bmatrix} 1 & 1 \\ 1 & 1 \end{bmatrix} \begin{pmatrix} x_1 \\ x_2 \end{pmatrix} \geq \begin{pmatrix} 0 \\ 0 \end{pmatrix},$$

which has a unique solution $(0, 0)$. By either direct verification or Corollary **5.3.19**, we can show that this solution is stable. Yet it is not strongly stable, as can be seen by either direct verification or Corollary **5.3.20**. □

Proposition **5.3.3** has an interesting consequence. Indeed, we use this proposition to show that a solution x^* of the VI (K, F) is strongly stable if and only if the vector $z^* \equiv x^* - F(x^*)$ is a strongly stable zero of the normal map $\mathbf{F}_K^{\text{nor}}$. Although this assertion seems intuitive, the proof is not trivial for two reasons. One, the normal map $\mathbf{F}_K^{\text{nor}}$ involves a change of variables; two, Definition **5.2.6** involves perturbing the map $\mathbf{F}_K^{\text{nor}}$ by an arbitrary continuous function G that is sufficiently close to $\mathbf{F}_K^{\text{nor}}$, whereas the strong stability in Definition **5.3.1** restricts the perturbation to be of the form $\mathbf{G}_K^{\text{nor}}$ with G being close to F. Thus, it is interesting to deduce that the latter kind of strong stability, which seemingly is more restrictive, is nevertheless equivalent to the former kind of strong stability. The following result requires that F be Lipschitz continuous in a neighborhood of the solution $x^* \in \text{SOL}(K, F)$.

5.3.6 Proposition. Let $K \subseteq \mathbb{R}^n$ be closed convex and $F : \mathcal{D} \to \mathbb{R}^n$ be Lipschitz continuous in a neighborhood of a solution x^* of the VI (K, F), where $\mathcal{D}$ is an open set containing K. The vector x^* is a strongly stable solution of the VI (K, F) in the sense of Definition **5.3.1** if and only if the vector $z^* \equiv x^* - F(x^*)$ is a strongly stable zero of the normal map $\mathbf{F}_K^{\text{nor}}$.

Proof. Suppose that x^* is a strongly stable solution of the VI (K, F). It then follows that x^* is a strongly regular solution of the same VI; moreover, z^* is an isolated zero of $\mathbf{F}_K^{\text{nor}}$. Let $\mathcal{Z}$ be an open neighborhood of z^* satisfying

$$(\mathbf{F}_K^{\text{nor}})^{-1}(0) \cap \text{cl}\,\mathcal{Z} = \{z^*\}.$$

Let $\mathcal{N}$ be an open neighborhood of x^* containing $\Pi_K(\mathcal{Z})$. Associated with $\mathcal{N}$, there exist two positive scalars ε and c such that, for every vector q satisfying $\|q\| \leq \varepsilon$, there exists a unique solution $x_{\mathcal{N}}(q)$ of the VI $(K, F+q)$ in the neighborhood $\mathcal{N}$, moreover, for another q' satisfying the same norm condition as q,

$$\| x_{\mathcal{N}}(q) - x_{\mathcal{N}}(q') \| \leq c \| q - q' \|.$$

We may restrict ε such that, for all q satisfying $\|q\| \leq \varepsilon$, $x_{\mathcal{N}}(q)$ lies in the neighborhood of x^* in which F is Lipschitz continuous. Choose a positive scalar $\varepsilon' < \varepsilon$ such that

$$\| q \| \leq \varepsilon' \Rightarrow x_{\mathcal{N}}(q) - F(x_{\mathcal{N}}(q)) \in \mathcal{Z}.$$

For an arbitrary vector q satisfying $\|q\| \leq \varepsilon'$, the unique solution $x_{\mathcal{N}}(q)$ of the VI $(K, F+q)$ in the neighborhood $\mathcal{N}$ induces a zero

$$z_{\mathcal{N}}(q) \equiv x_{\mathcal{N}}(q) - F(x_{\mathcal{N}}(q))$$

of the normal map $\mathbf{F}_K^{\rm nor}$ that belongs to the neighborhood $\mathcal{Z}$. Moreover, for another vector q' satisfying $\|q'\| \leq \varepsilon'$, we have

$$\begin{aligned} \| z_{\mathcal{N}}(q) - z_{\mathcal{N}}(q') \| &\leq \| x_{\mathcal{N}}(q) - x_{\mathcal{N}}(q') \| + \| F(x_{\mathcal{N}}(q)) - F(x_{\mathcal{N}}(q')) \| \\ &\leq (1+L) \| x_{\mathcal{N}}(q) - x_{\mathcal{N}}(q') \| \\ &\leq c(1+L) \| q - q' \|, \end{aligned}$$

where $L > 0$ is the local Lipschitz modulus of F near x^*. Consequently, we have shown that z^* is a strongly regular zero of the normal map $\mathbf{F}_K^{\rm nor}$. By part (a) of Theorem **5.2.8**, it follows that z^* is a strongly stable zero of $\mathbf{F}_K^{\rm nor}$.

Conversely, suppose that z^* is a strongly stable zero of $\mathbf{F}_K^{\rm nor}$. Then x^* is an isolated solution of the VI (K, F). Let $\mathcal{N} \equiv \mathbb{B}(x^*, \delta)$ be an open neighborhood of x^* satisfying (5.1.1). Let $\mathcal{Z}$ be an open neighborhood of z^* such that

$$x \in \mathcal{N} \Rightarrow x - F(x) \in \mathcal{Z}.$$

There exist positive scalars ε and c such that, for every vector q satisfying $\|q\| \leq \varepsilon$, the system:

$$\mathbf{F}_K^{\rm nor}(z) + q = 0, \quad z \in \mathcal{Z}$$

has a unique solution, which we denote $z(q)$; moreover, for another q' satisfying the same norm condition as q,

$$\| z(q) - z(q') \| \leq c \| q - q' \|.$$

Since $\mathbf{F}_K^{\rm nor} + q$ is the normal map of the VI $(K, F + q)$, it follows that for every q satisfying $\|q\| \leq \varepsilon$, the perturbed VI $(K, F+q)$ has a solution $x(q)$ with $x(q) = \Pi_K(z(q))$. By the nonexpansiveness of the projector, we have

$$\| x(q) - x(q') \| \leq \| z(q) - z(q') \| \leq c \| q - q' \|.$$

Choose a positive scalar $\varepsilon' < \min(\varepsilon, \delta/c)$. For every q satisfying $\|q\| \leq \varepsilon'$, the perturbed VI $(K, F + q)$ has a solution $x(q)$ satisfying

$$\begin{aligned} \| x(q) - x^* \| &= \| x(q) - x(0) \| \\ &\leq c \| q \| \leq c\varepsilon' < \delta. \end{aligned}$$

Thus $x(q) \in \mathcal{N}$. We have therefore established that x^* is a strongly regular solution of the VI (K, F). By Proposition **5.3.3**, it follows that x^* is a strongly stable solution of the VI (K, F). □

Before we proceed to derive conditions that ensure the stability and strong stability of a solution to a VI, we first state an immediate consequence of the stability of such a solution. The following result gives a necessary and sufficient for a solution of a VI to be semistable in terms of an error bound near the solution. A globalization of this result is established in Proposition **6.2.1**.

5.3.7 Proposition. Let $K \subseteq \mathbb{R}^n$ be closed convex and $F : \mathcal{D} \to \mathbb{R}^n$ be Lipschitz continuous in a neighborhood of a solution x^* of the VI (K, F), where $\mathcal{D}$ is an open set containing K. The following two statements are equivalent.

(a) x^* is semistable.

(b) There exist a constant $\eta > 0$ and a neighborhood $\mathcal{N}' \subseteq \mathcal{D}$ of x^* such that, for every $x \in \mathcal{N}'$,

$$\| x - x^* \| \leq \eta \| \mathbf{F}_K^{\text{nat}}(x) \|.$$

Proof. (a) $\Rightarrow$ (b). Suppose that x^* is a solution of the VI (K, F). Let c, ε and $\mathcal{N}$ be, respectively, the two positive constants and the neighborhood of x^* associated with the semistability of x^*. Without loss of generality, by shrinking $\mathcal{N}$ if necessary, we may assume that F is Lipschitz continuous in $\operatorname{cl}\mathcal{N}$. Let $L > 0$ be such that

$$\| F(z) - F(z') \| \leq L \| z - z' \| \quad \forall z, z' \in \mathcal{N}.$$

For an arbitrary vector $x \in \mathcal{D}$, let $r \equiv \mathbf{F}_K^{\text{nat}}(x)$. We have

$$0 = x - r - \Pi_K(x - F(x)).$$

Hence the vector $y \equiv x - r$ is a solution of the VI (K, G_r), where $G_r(z) \equiv F(z + r) - r$. We may choose a neighborhood $\mathcal{N}' \subseteq \mathcal{D}$ such that, for every $x \in \mathcal{N}'$, we have $x - r \in \mathcal{N}$ and

$$\sup_{z \in K_{\mathcal{N}}} \| G_r(z) - F(z) \| \leq (L + 1) \| r \| \leq \varepsilon.$$

Therefore, by semistability, it follows that

$$\| x - r - x^* \| \leq c(L + 1) \| r \|,$$

which implies

$$\| x - x^* \| \leq [1 + c(L + 1)] \| r \|.$$

Therefore, (b) holds with $\eta \equiv 1 + c(L + 1)$.

(b) $\Rightarrow$ (a). Suppose that $x \in \text{SOL}(K, G)$. Since $x = \Pi_K(x - G(x))$, we have

$$\| \mathbf{F}_K^{\text{nat}}(x) \| = \| \Pi_K(x - G(x)) - \Pi_K(x - F(x)) \| \leq \| F(x) - G(x) \|.$$

If $x \in \mathcal{N}'$, then

$$\| x - x^* \| \leq \eta \| \mathbf{F}_K^{\text{nat}}(x) \| \leq \eta \| F(x) - G(x) \|.$$

From this inequality, the semistability of x^* follows readily. □

5.3.1 The case of a finitely representable set

In this subsection, we consider the case of a finitely representable, convex set K given by

$$K \equiv \{ x \in \mathbb{R}^n : h(x) = 0, \ g(x) \leq 0 \},$$

where each g_i is convex and h is affine. We further assume that F is continuously differentiable and every g_i is twice continuously differentiable in a neighborhood of a given solution $x^* \in \text{SOL}(K, F)$. From Corollary **5.1.8**, we know that if the Jacobian matrix $JF(x^*)$ is strictly copositive on the critical cone $\mathcal{C}(x^*; K, F)$, then x^* is a stable solution of the VI (K, F). By exploiting the finite representation of K and by employing Lemmas **5.2.1** and **5.2.2** and the directional differentiability of the Euclidean projector Π_K, we show that x^* is a stable solution of the VI (K, F) under conditions weaker than the strict copositivity of $JF(x^*)$.

We assume that the SBCQ holds at x^*. The KKT system of the VI (K, F) at x^* is:

$$\begin{aligned} F(x^*) + \sum_{j=1}^{\ell} \mu_j \nabla h_j(x^*) + \sum_{i=1}^{m} \lambda_i \nabla g_i(x^*) &= 0 \\ h(x^*) &= 0 \\ 0 \leq \lambda \perp g(x^*) &\leq 0. \end{aligned}$$

We let $\mathcal{M}(x^*)$ be the set of pairs (μ, λ) satisfying this KKT system. Let $z^* \equiv x^* - F(x^*)$. Note that $x^* = \Pi_K(z^*)$; moreover, we have

$$\mathcal{M}_\pi(z^*) = \mathcal{M}(x^*) \quad \text{and} \quad \mathcal{C}_\pi(z^*; K) = \mathcal{C}(x^*; K, F).$$

We write $\mathcal{M}^e(x^*)$ for $\mathcal{M}_\pi^e(z^*)$. As noted before, $\mathcal{M}^e(x^*)$ is a finite set.

The vector Lagrangian function of the VI (K, F) is:

$$\mathbf{L}(x, \mu, \lambda) = F(x) + \sum_{j=1}^{\ell} \mu_j \nabla h_j(x) + \sum_{i=1}^{m} \lambda_i \nabla g_i(x).$$

Since h is affine, we write

$$J_x\mathbf{L}(x,\lambda) \equiv J_x\mathbf{L}(x,\mu,\lambda) = JF(x) + \sum_{i=1}^{m} \lambda_i \nabla^2 g_i(x).$$

The normal map associated with the VI (K, F) is given by

$$\mathbf{F}_K^{\mathrm{nor}}(z) \equiv F(\Pi_K(z)) + z - \Pi_K(z), \quad z \in \mathbb{R}^n.$$

It is easy to show, by the chain rule and Theorem **4.4.1**, that $\mathbf{F}_K^{\mathrm{nor}}$ is B-differentiable at z^*; moreover,

$$(\mathbf{F}_K^{\mathrm{nor}})'(z^*; dz) \equiv (JF(x^*) - \mathbf{I}_n)\Pi_{\mathcal{C}}^{G(\lambda)}(dz) + dz, \quad \forall\, dz \in \mathbb{R}^n,$$

where $\mathcal{C} \equiv \mathcal{C}(x^*; K, F)$ and (μ, λ) is a pair in the finite set $\mathcal{M}^e(x^*)$ that maximizes the following linear function on $\mathcal{M}(x^*)$:

$$(\mu', \lambda') \mapsto \tfrac{1}{2}\sum_{i=1}^{m} \lambda_i' u^T \nabla^2 g_i(x^*) u,$$

where $u \equiv \Pi_K'(z^*; dz)$. Consistent with the notation in Lemma **5.2.2**, we let

$$\Gamma(z) \equiv (\mathbf{F}_K^{\mathrm{nor}})'(z^*; z - z^*), \quad z \in \mathbb{R}^n.$$

In order to apply this lemma and Lemma **5.2.1**, we need to seek conditions to ensure that $\Gamma(z)$ has z^* as the unique zero and $\mathrm{ind}(\Gamma, z^*)$ is nonzero. The approach is the familiar homotopy argument. Notice that Γ is itself not the normal map of an AVI; nevertheless, it bears close relation to one, see Lemma **5.3.8** below.

For each $(\mu, \lambda) \in \mathcal{M}^e(x^*)$, let

$$H_\lambda(d) \equiv (\mathbf{I}_n - G)\Pi_{\mathcal{C}}^{G}(d) + d, \quad d \in \mathbb{R}^n,$$

where $G \equiv G(\lambda)$. By Proposition **4.3.3**, the map H_λ is a globally Lipschitz homeomorphism on $\mathbb{R}^n$ with inverse $\mathbf{G}_{\mathcal{C}}^{\mathrm{nor}}$, which is the normal map of the pair $(\mathcal{C}, G)$. We further write the normal map of the pair $(\mathcal{C}, J_x\mathbf{L}(x,\lambda))$ as h_λ; that is,

$$h_\lambda(z) \equiv J_x\mathbf{L}(x,\lambda)\Pi_{\mathcal{C}}(z) + z - \Pi_{\mathcal{C}}(z), \quad z \in \mathbb{R}^n.$$

As described in the following lemma, the directional derivative $(\mathbf{F}_K^{\mathrm{nor}})'(z^*; \cdot)$ is closely related to the composition $h_\lambda \circ H_\lambda$ for various λ.

5.3.8 Lemma. Assume the setting of this subsection. For each vector dz in $\mathbb{R}^n$, there exists a pair $(\mu, \lambda) \in \mathcal{M}^e(x^*)$ such that, for $G \equiv G(\lambda)$, we have $\Pi_K'(z^*; dz) = \Pi_{\mathcal{C}}^{G}(dz)$ and

$$(\mathbf{F}_K^{\mathrm{nor}})'(z^*; dz) = h_\lambda \circ H_\lambda(dz).$$

Proof. We have by Proposition **4.3.3**,

$$
\begin{aligned}
(\mathbf{F}_K^{\text{nor}})'(z^*; dz) &= (\,JF(x^*) - \mathbf{I}_n\,)\Pi_{\mathcal{C}}^G(dz) + dz \\
&= (\,JF(x^*) - \mathbf{I}_n\,)\Pi_{\mathcal{C}} \circ H_\lambda(dz) + dz \\
&= h_\lambda \circ H_\lambda(dz) - (G - \mathbf{I}_n)\Pi_{\mathcal{C}} \circ H_\lambda(dz) - H_\lambda(dz) + dz \\
&= h_\lambda \circ H_\lambda(dz),
\end{aligned}
$$

as desired. □

Using the above lemma, we are ready to present a sufficient condition for $\text{ind}(\Gamma, z^*)$ to be well defined and nonzero.

5.3.9 Lemma. Assume the setting of this subsection. Suppose that for every scalar $\tau \geq 0$ and every $(\mu, \lambda) \in \mathcal{M}^e(x^*)$, $(\mathcal{C}, J_x\mathbf{L}(x^*, \lambda) + \tau G(\lambda))$ is a R_0-pair; that is, the following homogeneous linear complementarity system:

$$
\mathcal{C} \ni v \perp (\,J_x\mathbf{L}(x^*, \lambda) + \tau\, G(\lambda)\,)\, v \in \mathcal{C}^* \tag{5.3.2}
$$

has $v = 0$ as the unique solution. Then $(\mathbf{F}_K^{\text{nor}})'(z^*; dz)$ has $dz = 0$ as the unique zero; moreover $\text{ind}(\Gamma, z^*)$ is well defined and equal to unity.

Proof. Define the homotopy

$$
J(z,t) \equiv t\,(\mathbf{F}_K^{\text{nor}})'(z^*; z - z^*) + (\,1 - t\,)\,(\,z - z^*\,). \quad (\,z, t\,) \in \mathbb{R}^n \times [0,1].
$$

This is a continuous function in (z,t) because $(\mathbf{F}_K^{\text{nor}})'(z^*; \cdot)$ is (Lipschitz) continuous. Suppose $J(z,t) = 0$ for some (z,t) with $t \in [0,1]$. If $t = 0$, then $z = z^*$. Assume that $t > 0$. Let $dz \equiv z - z^*$. Let $(\mu, \lambda) \in \mathcal{M}^e(x^*)$ be such that $\Pi_K'(z^*; dz) = \Pi_K^G(dz)$ and

$$
(\mathbf{F}_K^{\text{nor}})'(z^*, dz) = h_\lambda \circ H_\lambda(dz),
$$

where $G \equiv G(\lambda)$. Letting $y \equiv H_\lambda(dz)$ so that $dz = \mathbf{G}_{\mathcal{C}}^{\text{nor}}(y)$, we deduce

$$
\begin{aligned}
0 &= J(z,t) = t\,h_\lambda(y) + (\,1 - t\,)\,\mathbf{G}_{\mathcal{C}}^{\text{nor}}(y) \\
&= [\,t\,J_x\mathbf{L}(x^*, \lambda) + (\,1 - t\,)\,G(\lambda)\,]\,\Pi_{\mathcal{C}}(y) + y - \Pi_{\mathcal{C}}(y).
\end{aligned}
$$

Hence the vector $v \equiv \Pi_{\mathcal{C}}(y)$ is a solution of the CP:

$$
\mathcal{C} \ni v \perp (\,J_x\mathbf{L}(x^*, \lambda) + \tau\, G(\lambda)\,)\, v \in \mathcal{C}^*,
$$

where $\tau \equiv (1 - t)/t$. By assumption, we have $0 = v = \Pi_{\mathcal{C}}(y)$. Thus, $y = 0$ and $dz = 0$. This shows for all $t \in [0,1]$, $J(\cdot, t)$ has $z = z^*$ as the unique solution. A standard homotopy argument completes the proof. □

5.3.10 Remark. It is possible to extend Lemma **5.3.9**, thus the stability theorem below, by considering a more general homotopy of the type:

$$J(z,t) \equiv t\,(\mathbf{F}_K^{\rm nor})'(z^*; z - z^*) + (1-t)\,(\mathbf{E}_K^{\rm nor})'(z^*; z - z^*)$$

where $\mathbf{E}_K^{\rm nor}$ is the normal map of the pair (K, E) with the matrix E satisfying an appropriate condition. We do not treat this extension because it does not add much value to the analysis. □

It is important to note that if the CRCQ holds at x^*, then instead of requiring that for every scalar $\tau \geq 0$ and *every* multiplier $(\mu, \lambda) \in \mathcal{M}^e(x^*)$, $(\mathcal{C}, J_x\mathbf{L}(x^*, \lambda) + \tau G(\lambda))$ is a R_0 pair, Lemma **5.3.9** remains valid under the assumption that for every $\tau \geq 0$, *there exists* $(\mu, \lambda) \in \mathcal{M}(x^*)$ such that $(\mathcal{C}, J_x\mathbf{L}(x^*, \lambda) + \tau G(\lambda))$ is a R_0 pair. This relaxation of the assumption is possible because by Theorem **4.5.3**, *any* member $(\mu, \lambda) \in \mathcal{M}(x^*)$ can be used to represent the directional derivative $\Pi_K'(z^*; dz)$ for any vector dz.

Combining Theorems **5.2.4** and **5.3.9**, we obtain readily the following result which provides a set of sufficient conditions for x^* to be a stable solution of the VI (K, F). Implicit in the proof of this result is the fact that if z^* is a stable zero of the normal map $\mathbf{F}_K^{\rm nor}$, then x^* is a stable solution of the VI (K, F).

5.3.11 Theorem. Assume the setting of this subsection. The solution x^* is stable under either one of the following two conditions:

(a) for every $\tau \geq 0$ and every $(\mu, \lambda) \in \mathcal{M}^e(x^*)$, $(\mathcal{C}, J_x\mathbf{L}(x^*, \lambda) + \tau G(\lambda))$ is a R_0 pair;

(b) the CRCQ holds at x^* and for every $\tau \geq 0$, there exists $(\mu, \lambda) \in \mathcal{M}(x^*)$ such that $(\mathcal{C}, J_x\mathbf{L}(x^*, \lambda) + \tau G(\lambda))$ is a R_0 pair.

Proof. The assumptions imply that x^* is an isolated solution of the VI (K, F) and z^* is a stable zero of $\mathbf{F}_K^{\rm nor}$. Hence, for every neighborhood $\mathcal{Z}$ of z^* such that

$$(\mathbf{F}_K^{\rm nor})^{-1}(0) \cap \operatorname{cl}\mathcal{Z} = \{ z^* \}, \tag{5.3.3}$$

there exists $\varepsilon > 0$ such that, for every function $J \in \mathbb{B}(\mathbf{F}_K^{\rm nor}; \varepsilon, \operatorname{cl}\mathcal{Z})$, the set $J^{-1}(0) \cap \mathcal{Z}$ is nonempty; moreover, there exist a neighborhood $\mathcal{Z}_0$ of z^* and a scalar $c > 0$ such that, for every $z' \in J^{-1}(0) \cap \mathcal{Z}_0$, we have

$$\| z' - z^* \| \leq c \| \mathbf{F}_K^{\rm nor}(z') \|.$$

Let $\mathcal{N}$ be an open neighborhood of x^* satisfying (5.1.1). Since x^* is an attractor of all solutions of nearby VIs, there exist a scalar $\varepsilon' \in (0, \varepsilon)$ and a subneighborhood $\mathcal{N}' \subseteq \mathcal{N}$ such that

$$x \in \mathcal{N}' \Rightarrow x - F(x) \in \mathcal{Z}_0.$$

and for every $G \in \mathrm{SOL}(F; \varepsilon', K_{\mathcal{N}})$, every solution x of the VI (K, G) belongs to $\mathcal{N}'$. Let $\mathcal{Z}$ be a neighborhood of z^* satisfying (5.3.3) and such that

$$z \in \mathcal{Z} \Rightarrow \Pi_K(x) \in \mathcal{N}'.$$

Let $G \in \mathbb{B}(F; \varepsilon', K_{\mathcal{N}})$. With $\mathbf{G}_K^{\mathrm{nor}}$ denoting the normal map associated with the pair (K, G), we have

$$\sup_{z \in \mathrm{cl}\, \mathcal{Z}} \| \mathbf{F}_K^{\mathrm{nor}}(z) - \mathbf{G}_K^{\mathrm{nor}}(z) \| \leq \sup_{x \in K \cap \mathrm{cl}\, \mathcal{N}} \| F(x) - G(x) \| < \varepsilon' < \varepsilon.$$

Thus, $\mathbf{G}_K^{\mathrm{nor}}$ has a zero in $\mathcal{Z}$; hence, $\mathrm{SOL}(K, G) \cap \mathcal{N}$ is nonempty by the choice of $\mathcal{Z}$. If x' is an element of $\mathrm{SOL}(K, G) \cap \mathcal{N}$, then $z' \equiv x' - F(x')$ belongs to the neighborhood $\mathcal{Z}_0$. Hence,

$$\begin{aligned} \| x' - x^* \| &= \| \Pi_K(z') - \Pi_K(z^*) \| \\ &\leq \| z' - z^* \| \leq c \| \mathbf{F}_K^{\mathrm{nor}}(z') \| \\ &= c \| F(x') - G(x') \|, \end{aligned}$$

where the last inequality holds because $\mathbf{G}_K^{\mathrm{nor}}(z') = 0$. This establishes the stability of the solution x^*. □

We next proceed to derive some necessary and sufficient conditions for the stability of x^*. Although not being directly used in the proof of the main result, Theorem **5.3.14**, the following lemma provides an important motivation for a key assumption in this theorem, in light of Theorem **5.2.12**. Specifically, we show that the copositivity of $J_x\mathbf{L}(x^*, \lambda)$ on $\mathcal{C}$ for all (μ, λ) in $\mathcal{M}^e(x^*)$ is a sufficient condition for the map Γ to have the nonvanishing property at x^*.

5.3.12 Lemma. Assume the setting of this subsection. If $J_x\mathbf{L}(x^*, \lambda)$ is copositive on $\mathcal{C}$ for all (μ, λ) in $\mathcal{M}^e(x^*)$, then the map Γ has the nonvanishing property at x^*.

Proof. We show that for every $\delta > 0$, the map

$$\Psi(z) \equiv z - z^* + \delta\, \Gamma(z), \quad z \in \mathbb{R}^n$$

has z^* as the only zero. Suppose $\Psi(z) = 0$. Let $dz \equiv z - z^*$. By the proof of Lemma **5.3.9**, there exists $(\mu, \lambda) \in \mathcal{M}^e(x^*)$ such that the vector $v \equiv \Pi_{\mathcal{C}} \circ H_\lambda(dz)$ satisfies:

$$\mathcal{C} \ni v \perp (\, \delta\, J_x\mathbf{L}(x^*, \lambda) + G(\lambda)\,)\, v \in \mathcal{C}^*.$$

Since $J_x\mathbf{L}(x^*, \lambda)$ is copositive on $\mathcal{C}$ and $G(\lambda)$ is positive definite, it follows that $v = 0$. As before, this implies $dz = 0$. Consequently, Γ has the nonvanishing property at z^*. □

In Theorem **5.2.12**, we have shown that if x is a stable zero of a B-differentiable equation $H(v) = 0$, then the origin is the unique zero of the directional derivative $H'(x; \cdot)$. This result does not apply directly to the VI because we have not demonstrated the complete equivalence between the stability of a solution x^* to the VI (K, F) and the stability of the corresponding zero $z^* \equiv x^* - F(x^*)$ of the normal map $\mathbf{F}_K^{\text{nor}}$. Nevertheless, we can establish the following analogous lemma.

5.3.13 Lemma. Assume the setting of this subsection. Assume further that either the CRCQ holds at x^* or $\mathcal{M}^e(x^*)$ is a singleton. Consider the following three statements:

(a) x^* is a stable solution of the VI (K, F);

(b) the origin is the unique zero of $(\mathbf{F}_K^{\text{nor}})'(z^*; \cdot)$;

(c) for every $(\mu, \lambda) \in \mathcal{M}^e(x^*)$, the pair $(\mathcal{C}, J_x\mathbf{L}(x^*, \lambda))$ has the R_0 property.

It holds that (a) $\Rightarrow$ (b) $\Leftrightarrow$ (c).

Proof. Under the assumed CQs, it follows that for all (μ, λ) in $\mathcal{M}^e(x^*)$ and for all vectors $d \in \mathbb{R}^n$, $\Pi'_K(z^*; d) = \Pi_{\mathcal{C}}^G(d)$, where $G \equiv G(\lambda)$.

(a) $\Rightarrow$ (b). Let x^* be a stable solution of the VI (K, F). Assume that dz is a nonzero vector satisfying $(\mathbf{F}_K^{\text{nor}})'(z^*; dz) = 0$. Let $G \equiv G(\lambda)$ be as described above. As in the proof of Lemma **5.3.9**, there exists a vector y such that $dz \equiv \mathbf{G}_{\mathcal{C}}^{\text{nor}}(y)$, and the vector $v \equiv \Pi_{\mathcal{C}}^G(dz) = \Pi_{\mathcal{C}}(y)$ satisfies:

$$(\mathbf{F}_K^{\text{nor}})'(z^*; dz) = JF(x^*)v + dz - v = 0.$$

Since $v \in \mathcal{C}$, thus $v \in \mathcal{T}(x^*; K)$, it follows that there exist a sequence of vectors $\{x^k\}$ in K converging to x^* and a sequence of positive scalars $\{\tau_k\}$ converging to zero such that

$$v = \lim_{k \to \infty} \frac{x^k - x^*}{\tau_k}.$$

Let $z^k \equiv x^k - F(x^k)$ and $q^k \equiv \mathbf{F}_K^{\text{nor}}(z^k)$. Clearly, the sequences $\{z^k\}$ and $\{\mathbf{F}_K^{\text{nor}}(z^k)\}$ converge to $x^* - F(x^*) = z^*$ and $\mathbf{F}_K^{\text{nor}}(z^*) = 0$, respectively. The vector $\Pi_K(z^k)$ is clearly a solution of the VI $(K, F - q^k)$. Consequently, there exists a constant $c > 0$ such that, for all k sufficiently large,

$$\| \Pi_K(z^k) - x^* \| \leq c \, \| q^k \|.$$

We claim that

$$\lim_{k\to\infty} \frac{\Pi_K(z^k) - x^*}{\tau_k} = v \quad \text{and} \quad \lim_{k\to\infty} \frac{q^k}{\tau_k} = 0.$$

This yields a contradiction because v is a nonzero vector. We have

$$\begin{aligned} \Pi_K(z^k) - x^* &= \Pi_K(x^k - F(x^k)) - \Pi_K(x^* - F(x^*)) \\ &= \Pi^G_{\mathcal{C}}(x^k - x^* - JF(x^*)(x^k - x^*)) + o(\|x^k - x^*\|). \end{aligned}$$

Consequently,

$$\lim_{k\to\infty} \frac{\Pi_K(z^k) - x^*}{\tau_k} = \Pi^G_{\mathcal{C}}(v - JF(x^*)v) = \Pi^G_{\mathcal{C}}(dz) = v.$$

Furthermore, we have

$$q^k = \mathbf{F}^{\text{nor}}_K(z^k) - \mathbf{F}^{\text{nor}}_K(z^*) = (\mathbf{F}^{\text{nor}}_K)'(z^*; z^k - z^*) + o(\|z^k - z^*\|)$$

which implies

$$\lim_{k\to\infty} \frac{q^k}{\tau_k} = (\mathbf{F}^{\text{nor}}_K)'(z^*; v - JF(x^*)v) = (\mathbf{F}^{\text{nor}}_K)'(z^*; dz) = 0.$$

Thus we have obtained the desired contradiction; (b) therefore holds.

(b) $\Leftrightarrow$ (c). The proof of this equivalence is more or less contained in the above proof. For completeness, we repeat some details. For each vector dz in $\mathbb{R}^n$, we have

$$(\mathbf{F}^{\text{nor}}_K)'(z^*; dz) = h_\lambda \circ H_\lambda(dz),$$

by Lemma **5.3.8**. Since h_λ is the normal map of the pair $(\mathcal{C}, J_x\mathbf{L}(x^*, \lambda))$ and H_λ is a global homeomorphism, the equivalence of (b) and (c) follows readily. □

We are ready to establish the desired necessary and sufficient conditions for x^* to be a stable solution of the VI (K, F).

5.3.14 Theorem. Assume the setting of this subsection. Suppose that for all $(\mu, \lambda) \in \mathcal{M}^e(x^*)$, $J_x\mathbf{L}(x^*, \lambda)$ is copositive on $\mathcal{C}$. Consider the following five statements.

(a) For all $(\mu, \lambda) \in \mathcal{M}^e(x^*)$, $J_x\mathbf{L}(x, \lambda)$ is strictly copositive on $\mathcal{C}$.

(b) For all $(\mu, \lambda) \in \mathcal{M}^e(x^*)$, there exists a constant $c > 0$ such that, for all $q \in \mathbb{R}^n$, the affine complementarity system:

$$\mathcal{C} \ni v \perp q + J_x\mathbf{L}(x^*, \lambda)v \in \mathcal{C}^* \tag{5.3.4}$$

has a nonempty solution set; moreover, $\| x \| \le c \| q \|$ for all solutions x of (5.3.4).

(c) For all $(\mu, \lambda) \in \mathcal{M}^e(x^*)$ and all $q \in \mathbb{R}^n$, the affine complementarity system (5.3.4) has a nonempty bounded solution set.

(d) For all $(\mu, \lambda) \in \mathcal{M}^e(x^*)$, the pair $(\mathcal{C}, J_x\mathbf{L}(x^*, \lambda))$ has the R_0 property.

(e) The solution x^* is stable.

The following implications hold:

$$(\mathrm{a}) \Rightarrow (\mathrm{b}) \Leftrightarrow (\mathrm{c}) \Leftrightarrow (\mathrm{d}) \Rightarrow (\mathrm{e}).$$

Furthermore, the first four statements are equivalent if $JF(x^*)$ is symmetric. Finally, the last four statements are equivalent if either the CRCQ holds at x^* or $\mathcal{M}^e(x^*)$ is a singleton.

Proof. Clearly, (a) implies (b) which in turn implies (c); moreover, (c) implies (d) by taking q to be the zero vector. The implication (d) $\Rightarrow$ (b) follows from Theorem **2.5.10** and Proposition **2.5.6**.

It is easy to see that under condition (d) and the copositivity assumption of $J_x\mathbf{L}(x^*, \lambda)$, the pair $(\mathcal{C}, J_x\mathbf{L}(x^*, \lambda) + \tau G(\lambda))$ has the R_0 property for all $(\mu, \lambda) \in \mathcal{M}^e(x^*)$ and all $\tau \geq 0$. Thus, (e) follows from Theorem **5.3.11**. If $JF(x^*)$ is symmetric, the equivalence of (c) and (a) follows from Proposition **2.5.7**. Finally, if either the CRCQ holds at x^* or $\mathcal{M}^e(x^*)$ is a singleton, (e) implies (d) by Lemma **5.3.13**. □

If the set K is a Cartesian product of finitely representable sets of lower dimensions, then the copositivity assumption in Theorem **5.3.14** can be weakened to semicopositivity. In this case, Corollary **3.5.3** can be used to establish a stability result analogous to the theorem. To illustrate a result of this kind, we present a simplified situation where K is the Cartesian product of finitely many polyhedra of lower dimensions. Further specialization to the NCP is given in the next subsection.

5.3.15 Proposition. Suppose that K is the Cartesian product of N polyhedra $K_\nu \subseteq \mathbb{R}^{n_\nu}$. Let x^* be a solution of the (linearly constrained) VI (K, F) and assume that F is continuously differentiable in a neighborhood of x^*. If $JF(x^*)$ is semicopositive on $\mathcal{C}(x^*; K, F)$, then x^* is a stable solution of the VI (K, F) if and only if $(\mathcal{C}(x^*; K, F), JF(x^*))$ is a R_0 pair.

Proof. Since K is polyhedral, the problem (5.3.2) reduces to

$$\mathcal{C} \ni v \perp JF(x^*)v + \tau v \in \mathcal{C}^*. \tag{5.3.5}$$

Since K is the Cartesian product of the N polyhedra K_ν, it follows that $\mathcal{C}$ is the Cartesian product of the N polyhedral cones

$$\mathcal{C}_\nu \equiv \mathcal{T}(x^*_\nu; K_\nu) \cap F_\nu(x^*)^\perp.$$

If $JF(x^*)$ is semicopositive on $\mathcal{C}$ and $(\mathcal{C}(x^*; K, F), JF(x^*))$ is a R_0 pair, then the CP (5.3.5) has a unique solution for all $\tau \geq 0$. Therefore, by Theorem **5.3.11**, x^* is a stable solution of the VI (K, F). Conversely, if x^* is a stable solution of the latter VI, then since K is polyhedral, Lemma **5.3.13** applies and it follows that $(\mathcal{C}(x^*; K, F), JF(x^*))$ is a R_0 pair. □

It is natural to ask whether it is possible to establish necessary and sufficient conditions for a solution of a VI to be stable, without assuming *a priori* any special property of the Jacobian matrix of the VI Lagrangian function, such as the copositivity of $J_x\mathbf{L}(x^*, \lambda)$ in Theorem **5.3.14** or the semicopositivity of $JF(x^*)$ in Proposition **5.3.15**. The answer to this question is open to date. This situation is quite different from the strongly stable case where a complete characterization is available, as we see from the subsequent development.

We illustrate Proposition **5.3.15** using an LCP.

5.3.16 Example. Consider the LCP (q, M) with data:

$$q = \begin{pmatrix} 0 \\ -1 \\ 1 \end{pmatrix} \quad \text{and} \quad M = \begin{bmatrix} 0 & -2 & 2 \\ 1 & 1 & 2 \\ -1 & -1 & 0 \end{bmatrix}.$$

This LCP has a unique solution $x^* = (1, 0, 0)$. The matrix M is copositive on $\mathbb{R}^3_+$ but is not positive semidefinite. The critical cone of the LCP at this solution is given by:

$$\mathcal{C} = \{ (x_1, x_2, x_3) \in \mathbb{R}^3 : (x_2, x_3) \geq 0 \}.$$

It is not difficult to verify that M is semicopositive on $\mathcal{C}$, that is for every nonzero (x_1, x_2, x_3) belonging to $\mathcal{C}$, there exists an index $i \in \{1, 2, 3\}$ such that

$$x_i \neq 0 \quad \text{and} \quad x_i(Mx)_i \geq 0.$$

It is also not difficult to verify that the homogeneous system:

$$\begin{aligned} 0 &= -2dx_2 + 2dx_3 \\ 0 &\leq dx_1 + dx_2 + 2x_3 \quad \perp \quad dx_2 \geq 0 \\ 0 &\leq -dx_1 - dx_2 \qquad\quad \perp \quad dx_3 \geq 0 \end{aligned}$$

has $(dx_1, dx_2, dx_3) = (0, 0, 0)$ as the unique solution. Consequently, it follows that x^* is a stable solution of the LCP (q, M). By Corollary **5.3.20**, this solution is not strongly stable because the first diagonal entry of M is zero. □

We next turn our attention to strong stability. We can apply the results in Section **4.1** and Theorem **5.2.14** to obtain some analogous results pertaining to the strong stability of a solution to a VI. Since this theorem requires a strong B-differentiability condition on the underlying nonsmooth-equation map representing the VI, the following preliminary strong stability result is restricted to a linearly constrained VI. Although restricted in scope, the result can be specialized to the NCP and the KKT system of a nonlinearly constrained VI; the specialized results are presented in Corollary **5.3.20** and Corollary **5.3.22**. In Subsection **5.3.3**, we extend the theorem below to a non-polyhedral set K.

5.3.17 Theorem. Let K be a polyhedron in $\mathbb{R}^n$ and $F : \mathcal{D} \to \mathbb{R}^n$ be given, where $\mathcal{D}$ is an open set containing K. Let x^* be a solution of the VI (K, F). Suppose that F is continuously differentiable in a neighborhood of x^*. Let $\mathcal{C} \equiv \mathcal{C}(x^*; K, F)$ and $J\mathbf{F}^{\text{nor}}_{\mathcal{C}}$ be the normal map of the pair $(\mathcal{C}, JF(x^*))$. The following statements are equivalent.

(a) x^* is a strongly stable solution of the VI (K, F).

(b) $J\mathbf{F}^{\text{nor}}_{\mathcal{C}}$ is a Lipschitz homeomorphism on $\mathbb{R}^n$.

(c) $J\mathbf{F}^{\text{nor}}_{\mathcal{C}}$ is coherently oriented.

(d) $J\mathbf{F}^{\text{nor}}_{\mathcal{C}}$ is an open map.

(e) For every vector $q \in \mathbb{R}^n$, the AVI $(\mathcal{C}, q, JF(x^*))$ has a unique solution.

(f) The zero vector is a strongly stable solution of the homogeneous AVI $(\mathcal{C}, 0, JF(x^*))$.

Proof. Using the same notation as above, we have, by the polyhedrality of K,

$$(\mathbf{F}^{\text{nor}}_K)'(z^*; d) \;=\; JF(x^*)\Pi_{\mathcal{C}}(d) + d - \Pi_{\mathcal{C}}(d) \;=\; J\mathbf{F}^{\text{nor}}_{\mathcal{C}}(d), \quad \forall\, d \in \mathbb{R}^n.$$

Thus, $\Gamma(z) = J\mathbf{F}^{\text{nor}}_{\mathcal{C}}(z - z^*)$. We claim that $\mathbf{F}^{\text{nor}}_K$ has a strong B-derivative at z^*. We have

$$\begin{aligned}
\mathbf{e}(z) &\equiv \mathbf{F}^{\text{nor}}_K(z) - \mathbf{F}^{\text{nor}}_K(z^*) - (\mathbf{F}^{\text{nor}}_K)'(z^*; z - z^*) \\
&= F(\Pi_K(z)) - JF(x^*)\Pi_{\mathcal{C}}(z - z^*) + z^* - \Pi_K(z) + \Pi_{\mathcal{C}}(z - z^*) \\
&= F(\Pi_K(z)) - JF(x^*)[\,\Pi_K(z) - \Pi_K(z^*)\,] + z^* - \Pi_K(z^*),
\end{aligned}$$

where the last equality holds for all z sufficiently close to z^*, by Theorem **4.1.1**. Consequently,

$$\begin{aligned}
\mathbf{e}(z) - \mathbf{e}(z') &= F(\Pi_K(z)) - F(\Pi_K(z')) - JF(x^*)[\,\Pi_K(z) - \Pi_K(z')\,] \\
&= o(\,\|\, z - z'\,\|),
\end{aligned}$$

by the continuous differentiability of F near x^* and the nonexpansiveness of the Euclidean projector. This establishes our claim.

The equivalence of (b)–(e) follows from Theorem **4.3.2** and Proposition **4.2.15**. The equivalence of all statements (a)–(f) follows from Proposition **5.3.6** and Theorem **5.2.14**. □

We can invoke Corollary **4.3.2** and Proposition **4.2.7** to obtain many more equivalent conditions in Theorem **5.3.17**. A complete set of equivalent conditions for strong stability is presented in Theorem **5.3.24**.

5.3.2 The NCP and the KKT system

In this subsection, we specialize the results in the last subsection to the NCP and the KKT system. To begin, we recall the following three index sets associated with a solution x^* of the NCP (F):

$$\alpha = \{ i : x_i^* = 0 < F_i(x^*) \}$$
$$\beta = \{ i : x_i^* = 0 = F_i(x^*) \}$$
$$\gamma = \{ i : x_i^* > 0 = F_i(x^*) \} = \operatorname{supp}(x^*);$$

see Section **3.3**. The critical cone $\mathcal{C}$ of the pair $(\mathbb{R}^n_+, F)$ at x^* comprises of those vectors $v \in \mathbb{R}^n$ for which

$$[\, v_i \geq 0 \quad \forall i \in \beta \,] \quad \text{and} \quad [\, v_i = 0 \quad \forall i \in \alpha \,];$$

see the discussion following Remark **3.3.3**. Hence, for every $q \in \mathbb{R}^n$, the AVI $(\mathcal{C}, q, JF(x^*))$ is equivalent to the following MLCP in the variable dx:

$$(q + JF(x^*)dx)_\gamma = 0$$
$$0 \leq dx_\beta \perp (q + JF(x^*)dx)_\beta \geq 0$$
$$dx_\alpha = 0.$$

Specializing Theorem **5.3.11**, we immediately obtain the following result which gives a sufficient condition for a solution of the NCP to be stable.

5.3.18 Corollary. Let $F : \mathbb{R}^n \to \mathbb{R}^n$ be continuously differentiable in a neighborhood of a solution x^* of the NCP (F). If the homogeneous MLCP:

$$0 = (\tau \mathbf{I}_{|\gamma|} + J_\gamma F_\gamma(x^*)) dx_\gamma + J_\beta F_\gamma(x^*) dx_\beta$$
$$0 \leq J_\gamma F_\beta(x^*) dx_\gamma + (\tau \mathbf{I}_{|\beta|} + J_\beta F_\beta(x^*)) dx_\beta \perp dx_\beta \geq 0$$

has $(dx_\gamma, dx_\beta) = (0,0)$ as the unique solution for all $\tau \geq 0$, where γ and β are the support and degenerate set of x^* respectively, then x^* is a stable solution of the NCP (F). □

Specializing Proposition **5.3.15** to the NCP (F), we obtain the following result which gives a necessary and sufficient condition for a solution of the NCP to be stable, under a P_0 assumption on the principle submatrix of $JF(x^*)$ corresponding to the active constraints at $F(x^*)$.

5.3.19 Corollary. Let $F : \mathbb{R}^n \to \mathbb{R}^n$ be continuously differentiable in a neighborhood of a solution x^* of the NCP (F). If the principal submatrix

$$\begin{bmatrix} J_\gamma F_\gamma(x^*) & J_\beta F_\gamma(x^*) \\ J_\gamma F_\beta(x^*) & J_\beta F_\beta(x^*) \end{bmatrix}$$

is a P_0 matrix, then x^* is stable if and only if the homogeneous MLCP:

$$0 = J_\gamma F_\gamma(x^*)dx_\gamma + J_\beta F_\gamma(x^*)dx_\beta$$

$$0 \leq J_\gamma F_\beta(x^*)dx_\gamma + J_\beta F_\beta(x^*)dx_\beta \perp dx_\beta \geq 0$$

has $(dx_\gamma, dx_\beta) = (0,0)$ as the unique solution. □

We next discuss the specialization of Theorem **5.3.17** to the NCP (F). With the critical cone $\mathcal{C}$ identified above, combining Proposition **4.2.10** and Theorem **5.3.17**, we immediately obtain Corollary **5.3.20** below, which provides two necessary and sufficient conditions for the strong stability of a solution to the NCP (F). Both of these conditions are matrix-theoretic. The first of them is in terms of a certain "reduced P property" of the Jacobian matrix $JF(x^*)$; the second condition is an equivalent way of describing this P property that exposes its combinatorial nature.

5.3.20 Corollary. Let F be continuously differentiable in a neighborhood of a solution x^* of the NCP (F). The following three statements are equivalent.

(a) x^* is a strongly stable solution of the NCP (F).

(b) The principal submatrix

$$J_\gamma F_\gamma(x^*) \tag{5.3.6}$$

is nonsingular and the Schur complement:

$$J_\beta F_\beta(x^*) - J_\gamma F_\beta(x^*)(\,J_\gamma F_\gamma(x^*)\,)^{-1} J_\beta F_\gamma(x^*) \tag{5.3.7}$$

is a P matrix.

(c) There exists a constant $\sigma = \pm 1$ such that, for every index subset γ' of $\{1, \ldots, n\}$ satisfying

$$\gamma \subseteq \gamma' \subseteq \gamma \cup \beta,$$

it holds that

$$\det JF(x)_{\gamma'\gamma'} = \sigma.$$

Proof. The equivalence of (a) and (b) does not require further proof. A direct proof of the equivalence of (b) and (c) is similar to the proof of the equivalence of (a) and (b) in Corollary **3.3.9**, using the Schur determinantal formula. □

We easily recognize the principal submatrix (5.3.6) as the basic matrix of the solution x^*; see Definition **3.3.10**. Due to the close connection between condition (c) in the above corollary and the b-regularity property of x^* introduced in this definition, we call a solution x^* satisfying condition (c), or equivalently (b), a *strongly b-regular* solution of the NCP. Thus, Corollary **5.3.20** has shown that a solution of the NCP is strongly stable if and only if it is strongly b-regular. If in addition $JF(x^*)_{\bar{\alpha}\bar{\alpha}}$ is a P_0 matrix, where $\bar{\alpha}$ is the complement of α in $\{1, \ldots, n\}$, the strong stability of x is further equivalent to the b-regularity of x. This result, which is formally stated in the proposition below, is an easy consequence of Corollary **5.3.20.**

5.3.21 Proposition. Suppose that $F : \mathbb{R}^n \to \mathbb{R}^n$ is continuously differentiable in a neighborhood of a given solution x^* of the NCP (F). Consider the following three statements:

(a) x^* is strongly stable;

(b) x^* is strongly b-regular;

(c) x^* is b-regular.

Conditions (a) and (b) are equivalent. If $JF(x^*)_{\bar{\alpha}\bar{\alpha}}$ is a P_0 matrix, then all three statements (a), (b), and (c) are equivalent.

Proof. If suffices to show the equivalence of (b) and (c). The implication (b) $\Rightarrow$ (c) follows from Corollary **5.3.20** and the Definition **3.3.10** of b-regularity. Conversely, assume that x^* is a b-regular solution of NCP (F). By the definition of b-regularity and Corollary **3.3.9**, it follows that for every index subset γ' of $\{1, \ldots, n\}$ satisfying

$$\gamma \subseteq \gamma' \subseteq \gamma \cup \beta,$$

the principal submatrix

$$JF(x^*)_{\gamma'\gamma'}$$

is nonsingular; by the P_0 property, the latter principal submatrix has positive determinant. Consequently, x^* is a strongly b-regular solution of the NCP (F). □

The reader may wonder why it is useful to introduce the concept of a strongly b-regular solution of an NCP; after all this concept is equivalent to a strongly stable solution. One reason is that like b-regularity, strong b-regularity is a matrix-theoretic condition, which happens to be equivalent to strong stability if the vector under consideration is a solution of an NCP to start with. As matrix-theoretic conditions, b-regularity and strong b-regularity can be extended to an arbitrary vector x in $\mathbb{R}^n$, without regard to whether x is a solution of an NCP. Such an extended definition is highly relevant to the analysis of the FB C-function reformulation of the NCP. We do not give the extended definition explicitly but will make use of the key idea in Subsection **9.1.1**.

A result similar to Corollary **5.3.20** can be obtained for the strong stability of a KKT triple of a VI. Specifically, consider the following KKT system:

$$\begin{aligned} 0 = \mathbf{L}(x,\mu,\lambda) &\equiv F(x) + \sum_{j=1}^{\ell} \mu_j \nabla h_j(x) + \sum_{i=1}^{m} \lambda_i \nabla g_i(x), \\ 0 &= h(x), \\ 0 \leq \lambda &\perp g(x) \leq 0, \end{aligned} \tag{5.3.8}$$

which is an MiCP equivalent to the VI $(\mathbb{R}^{n+\ell} \times \mathbb{R}^m_+, \mathbf{F})$ where

$$\mathbf{F}(x,\mu,\lambda) \equiv \begin{pmatrix} \mathbf{L}(x,\mu,\lambda) \\ -h(x) \\ -g(x) \end{pmatrix}.$$

Let $z^* \equiv (x^*, \mu^*, \lambda^*)$ be a given KKT triple satisfying (5.3.8). Associated with this triple, we define the index sets:

$$\begin{aligned} \alpha &\equiv \{ i : \lambda_i^* > 0 = g_i(x^*) \} = \operatorname{supp}(\lambda^*) \\ \beta &\equiv \{ i : \lambda_i^* = 0 = g_i(x^*) \} \\ \gamma &\equiv \{ i : \lambda_i^* = 0 > g_i(x^*) \}. \end{aligned}$$

Let

$$\mathcal{I} \equiv \alpha \cup \beta$$

be the index set of active (inequality) constraints at x^*. Corollary **5.3.22** addresses the strong stability of z^* as a solution of the KKT system (5.3.8) considered as an MiCP. We present the corollary without assuming the convexity of g_i or the affinity of h. Needless to say, the special form of the function $\mathbf{F}(x,\mu,\lambda)$ plays an important role. We refer the reader to Exercises **5.6.12** and **5.6.13** for the treatment of a stable KKT triple; in particular, the latter exercise is the analog of Corollary **5.3.23** in the stable case.

5.3.22 Corollary. Let $z^* \equiv (x^*, \mu^*, \lambda^*)$ be a solution of (5.3.8). Let F be continuously differentiable and let each g_i and h_j be twice continuously differentiable in a neighborhood of x^*. The four statements (a)–(d) below are equivalent. Furthermore, if any one of these statements holds, then the LICQ holds at x^*.

(a) z^* is a strongly stable solution of (5.3.8).

(b) The matrix

$$\mathbf{B} \equiv \begin{bmatrix} J_x\mathbf{L}(z^*) & Jh(x^*)^T & Jg_\alpha(x^*)^T \\ -Jh(x^*) & 0 & 0 \\ -Jg_\alpha(x^*) & 0 & 0 \end{bmatrix}$$

is nonsingular, and the Schur complement

$$\mathbf{C} \equiv \begin{bmatrix} Jg_\beta(x^*) & 0 & 0 \end{bmatrix} \mathbf{B}^{-1} \begin{bmatrix} Jg_\beta(x^*)^T \\ 0 \\ 0 \end{bmatrix}$$

is a P matrix.

(c) The following implication holds

$$\left.\begin{array}{r} J_x\mathbf{L}(z^*)v + \displaystyle\sum_{j=1}^{\ell} \mu_j \nabla h_j(x^*) + \sum_{i \in \mathcal{I}} \lambda_i \nabla g_i(x^*) = 0 \\ \nabla h_j(x^*)^T v = 0, \quad \forall j = 1, \ldots, \ell \\ \nabla g_i(x^*)^T v = 0, \quad \forall i \in \alpha \\ \lambda_i \nabla g_i(x^*)^T v \leq 0, \quad \forall i \in \beta \end{array}\right\} \Rightarrow (v, \mu, \lambda_{\mathcal{I}}) = 0.$$

(d) For all triples $(q, r, s) \in \mathbb{R}^{n+\ell+|\mathcal{I}|}$, the affine KKT system

$$\begin{aligned} q + J_x\mathbf{L}(z^*)v + \sum_{j=1}^{\ell} \mu_j \nabla h_j(x^*) + \sum_{i \in \mathcal{I}} \lambda_i \nabla g_i(x^*) &= 0 \\ r_j + \nabla h_j(x^*)^T v = 0, \quad \forall j = 1, \ldots, \ell \\ s_i + \nabla g_i(x^*)^T v = 0, \forall i \in \alpha \\ 0 \leq \lambda_i \perp s_i + \nabla g_i(x^*)^T v \leq 0, \forall i \in \beta, \end{aligned}$$

has a unique solution $(v, \mu, \lambda_{\mathcal{I}})$.

Proof. The equivalence of (a), (b), and (d) is similar to Corollary **5.3.20**. In what follows, we show the equivalence of (b) and (c). Suppose (b) holds. Let $(v, \mu, \lambda_{\mathcal{I}})$ satisfy the left-hand conditions in the displayed implication in (c). We have

$$\begin{pmatrix} v \\ \mu \\ \lambda_\alpha \end{pmatrix} = \mathbf{B}^{-1} \begin{bmatrix} Jg_\beta(x^*)^T \\ 0 \\ 0 \end{bmatrix} \lambda_\beta.$$

Thus,

$$0 \geq \lambda_\beta \circ Jg_\beta(x^*)v = \lambda_\beta \circ \mathbf{C}\lambda_\beta.$$

Since $\mathbf{C}$ is a P matrix, by the sign reversal property of such a matrix, it follows that $\lambda_\beta = 0$. This implies that (v, μ, λ_α) is equal to zero also. Thus (c) holds. The reverse implication (c) $\Rightarrow$ (b) can be proved similarly by first showing the nonsingularity of the matrix $\mathbf{B}$ and then showing the P property of Schur complement, using the same sign reversal property of a P matrix. The details are omitted. Finally, with $v = 0$ in the displayed implication in (c), it is easy to see that the LICQ must hold at x^* if any one of the statements (a)–(d) holds. □

A further special case of Corollary **5.3.22** is worth mentioning. Let

$$\mathcal{C} = \{ v \in \mathbb{R}^n : Jh(x^*)v = 0, Jg_\alpha(x^*)v = 0, Jg_\beta(x^*)v \leq 0 \}$$

be the critical cone of the pair (K, F) at the solution x^*, expressed in terms of the index sets α and β associated with the KKT vector z^*; see Lemma **3.3.2**.

5.3.23 Corollary. Assume the setting of Corollary **5.3.22**. The following condition is sufficient for z^* to be a strongly stable KKT triple.

(e) The LICQ holds at x^* and $J_x\mathbf{L}(z^*)$ is positive definite on the null space of the vectors:

$$\{ \nabla g_i(x^*) : i \in \alpha \} \cup \{ \nabla h_j(x^*) : j = 1, \ldots, \ell \}. \tag{5.3.9}$$

Conversely, if z^* is a strongly stable KKT triple, $JF(x^*)$ is symmetric, and $J_x\mathbf{L}(z^*)$ is copositive on $\mathcal{C}$, then (e) holds.

Proof. It is clear that (e) implies (c). It remains to show that (c) implies the desired positive definiteness of $J_x\mathbf{L}(z^*)$ under the symmetry assumption on $JF(x^*)$ and the copositivity of $J_x\mathbf{L}(z^*)$ on $\mathcal{C}$. We first prove several consequences of these two assumptions.

(A) The matrix $\mathbf{C}$ is symmetric; thus it is positive definite. Indeed, if we partition $\mathbf{B}^{-1}$ in the same form as $\mathbf{B}$, say

$$\mathbf{B}^{-1} = \begin{bmatrix} A & * & * \\ * & * & * \\ * & * & * \end{bmatrix},$$

then it is easy to deduce $AJ_x\mathbf{L}(z^*)A^T = A^T$. Since $JF(x^*)$ is symmetric, thus so is $J_x\mathbf{L}(z^*)$; therefore, A^T, and hence A, is symmetric. Clearly, we have $\mathbf{C} = J_x g_\beta(x^*) A J_x g_\beta(x^*)^T$, establishing the symmetry of $\mathbf{C}$.

(B) For every nonzero vector v belonging to the null space of the vectors (5.3.9) such that $J_x\mathbf{L}(z^*)v$ belongs the linear span of the gradients of the active (inequality and equality) constraints at x^*, we have $v^T J_x\mathbf{L}(z^*)v > 0$. This follows by using the positive definiteness of $\mathbf{C}$ and the same argument to prove (b) $\Rightarrow$ (c) in Corollary **5.3.22**.

(C) The matrix $J_x\mathbf{L}(z^*)$ is strictly copositive on $\mathcal{C}$. Indeed, let u be a nonzero vector in $\mathcal{C}$. If $u^T J_x\mathbf{L}(z^*)u = 0$, then u minimizes the quadratic form $v^T J_x\mathbf{L}(z^*)v$ on the cone $\mathcal{C}$. Thus, there exist multipliers μ and $\lambda_{\mathcal{I}}$ such that

$$J_x\mathbf{L}(x^*)u + \sum_{j=1}^{\ell} \mu_j \nabla h_j(x^*) + \sum_{i\in\mathcal{I}} \lambda_i \nabla g_i(x^*)v = 0$$
$$0 \le \lambda_i \perp \nabla g_i(x^*)^T u = 0, \quad \forall\, i \in \beta.$$

This contradicts (c). Therefore, $J_x\mathbf{L}(z^*)$ is strictly copositive on $\mathcal{C}$.

To prove the positive definiteness of $J_x\mathbf{L}(z^*)$ on the null space of the gradient vectors (5.3.9), let u be an arbitrary nonzero vector in this space. Consider the quadratic program in the variable v:

$$\begin{aligned} \text{minimize} \quad & v^T J_x\mathbf{L}(z^*)v \\ \text{subject to} \quad & \nabla h_j(x^*)^T v = 0, \quad j = 1,\ldots,\ell \\ & \nabla g_i(x^*)^T v = 0, \quad i \in \alpha \\ & \nabla g_i(x^*)^T v = \nabla g_i(x^*)^T u, \quad i \in \beta. \end{aligned} \tag{5.3.10}$$

The vector u is feasible to this program. We claim that the objective function of this program is bounded below on its feasible set. Otherwise, there exists a vector v satisfying:

$$v^T J_x\mathbf{L}(z^*)v < 0$$
$$\nabla h_j(x^*)^T v = 0, \quad j = 1,\ldots,\ell$$
$$\nabla g_i(x^*)^T v = 0, \quad i \in \mathcal{I}.$$

The last two conditions show that the vector v must be an element of the critical cone $\mathcal{C}$; the first condition then contradicts the copositivity of $J_x\mathbf{L}(z^*)$ on this cone. Thus our claim holds. By Frank-Wolfe's theorem, it follows that the quadratic program (5.3.10) has an optimal solution $\bar{v}$. If $\bar{v} = 0$, then $u \in \mathcal{C}$ and $u^T J_x\mathbf{L}(z^*)u > 0$ because u is nonzero and $J_x\mathbf{L}(z^*)$ is strictly copositive on $\mathcal{C}$. If $\bar{v} \neq 0$, then we have

$$u^T J_x\mathbf{L}(z^*)u \geq \bar{v}^T J_x\mathbf{L}(z^*)\bar{v} > 0$$

where the last inequality follows from (B) because $J_x\mathbf{L}(z^*)\bar{v}$ must belong to the linear span of the gradients of the active (inequality and equality) constraints at x^*, by the optimality of $\bar{v}$. □

5.3.3 Strong stability under CRCQ

We return to the setting of Subsection **5.3.1**. For simplicity, we again omit the linear equality constraint $h(x) = 0$ from the set K; thus

$$K \equiv \{ x \in \mathbb{R}^n : g(x) \leq 0 \} \tag{5.3.11}$$

where $g : \mathcal{D} \to \mathbb{R}^m$ is twice continuously differentiable on the open set $\mathcal{D}$ containing K, and each component function g_i is convex.

To obtain conditions for the solution x^* to be strongly stable, we rely on the theory of PC^1 local homeomorphisms; see Section **4.6**. For this purpose, we assume that the CRCQ holds at x^*. By Theorem **4.5.2**, the Euclidean projection Π_K is a PC^1 map near $z^* \equiv x^* - F(x^*)$. Thus provided that F is continuously differentiable near x^*, the normal map $\mathbf{F}_K^{\text{nor}}$ is also PC^1 near z^*. Hence the results in Section **4.6** are applicable. To apply these results, we first note that by Theorem **5.2.8** if $\mathbf{F}_K^{\text{nor}}$ is a locally Lipschitz homeomorphism at z^*, then z^* is a strongly stable zero of $\mathbf{F}_K^{\text{nor}}$. In turn, by Proposition **5.3.6**, this implies that x^* is a strongly stable solution of the VI (K, F). Thus, the missing piece is to know when $\mathbf{F}_K^{\text{nor}}$ is a locally Lipschitz homeomorphism at z^*. Theorem **4.6.5** provides the needed connection because by Lemma **4.6.6**, we have

$$\text{Jac}((\mathbf{F}_K^{\text{nor}})'(z^*;\cdot), 0) \supseteq \text{Jac}(\mathbf{F}_K^{\text{nor}}, z^*),$$

verifying a key assumption of the theorem.

Collecting all the relevant results, namely, Theorem **5.2.8**, Proposition **5.3.6**, Theorem **4.6.5**, Lemma **4.6.6**, Proposition **4.2.7**, and Theorem **4.3.2**, we can state a unifying result, Theorem **5.3.24**, that gives numerous equivalent conditions for a solution to the VI (K, F) to be strongly stable under the CRCQ. This result generalizes Theorem **5.3.17** which pertains to a linearly constrained VI. Before formally presenting this result, we need one last piece of preparation.

The critical cone $\mathcal{C} \equiv \mathcal{C}(x^*; K, F)$ is a polyhedron of the type $P(A, 0)$ for some suitable matrix A. We intend to apply Proposition **4.2.7**. For this purpose, we need to determine a normal family of basis matrices of the cone $\mathcal{C}$, which we denote $\mathcal{B}_{\text{bas}}(\mathcal{C})$. More specifically, we wish to express the elements of this family in terms of a fixed but arbitrary multiplier $\lambda \in \mathcal{M}(x^*)$. We have

$$\mathcal{C} = \{ v \in \mathbb{R}^n : \nabla g_i(x^*)^T v \leq 0 \ \forall i \in \mathcal{I}(x^*) \text{ with equality if } i \in \alpha \}$$

where

$$\alpha \equiv \{ i : \lambda_i > 0 = g_i(x^*) \} = \text{supp}(\lambda)$$

$$\beta \equiv \{ i : \lambda_i = 0 = g_i(x^*) \}$$

$$\gamma \equiv \{ i : \lambda_i = 0 > g_i(x^*) \},$$

are the three fundamental index set associated with the pair (x^*, λ). Define the family of index sets $\mathcal{J}(\lambda)$ as follows. An index set J belongs to $\mathcal{J}(\lambda)$ if

$$J \subseteq \alpha \cup \beta'$$

for some $\beta' \subseteq \beta$ such that there exists a vector $v \in \mathbb{R}^n$ satisfying

$$\nabla g_i(x^*)^T v = 0, \quad \forall i \in \alpha \cup \beta'$$

$$\nabla g_i(x^*)^T v < 0, \quad \forall i \in \beta \setminus \beta'$$

and the subfamily of gradients

$$\{ \nabla g_i(x^*) : i \in J \} \tag{5.3.12}$$

forms a basis of the family of gradients

$$\{ \nabla g_i(x^*) : i \in \alpha \cup \beta' \}.$$

By definition, a matrix B belongs to the family $\mathcal{B}_{\text{bas}}(\mathcal{C})$ if the rows of B are given by the family of gradients (5.3.12) transposed. To denote the dependence of these matrices on λ, we write this particular family of basic matrices of $\mathcal{C}$ as $\mathcal{B}^{\lambda}_{\text{bas}}(\mathcal{C})$.

We now have all the necessary preparations to state and prove the following major strong stability result.

5.3.24 Theorem. Let K be given by (5.3.11) where $g : \mathcal{D} \to \mathbb{R}^m$ is twice continuously differentiable with each component function g_i being convex, where $\mathcal{D}$ is an open subset of $\mathbb{R}^n$. Let $F : \mathcal{D} \to \mathbb{R}^n$ be continuously differentiable. Let $x^* \in \text{SOL}(K, F)$ and let $z^* \equiv x^* - F(x^*)$. Let $\mathcal{C} \equiv \mathcal{C}(x^*; K, F)$. Let (P) be any of one of the following four terms: "injective", "bijective", "a globally Lipschitz homeomorphism", or "coherently oriented". Suppose that the CRCQ holds at x^*. The following statements are equivalent.

(a) x^* is a strongly stable solution of the VI (K, F).

(b) z^* is a strongly stable zero of the normal map $\mathbf{F}_K^{\text{nor}}$.

(c) $\mathbf{F}_K^{\text{nor}}$ is a locally Lipschitz homeomorphism at z^*.

(d) $(\mathbf{F}_K^{\text{nor}})'(z^*; \cdot)$ is (P).

(e) For each $\lambda \in \mathcal{M}(x^*)$, the map f_λ is (P), where

$$f_\lambda \equiv JF(x^*) \circ \Pi_{\mathcal{C}}^{G(\lambda)} + \mathbf{I} - \Pi_{\mathcal{C}}^{G(\lambda)}.$$

(f) There exists $\lambda \in \mathcal{M}(x^*)$ such that f_λ is (P).

(g) For each $\lambda \in \mathcal{M}(x^*)$, the linear normal map h_λ^{nor} is (P), where

$$h_\lambda^{\text{nor}} \equiv J_x\mathbf{L}(x^*, \lambda) \circ \Pi_{\mathcal{C}} + \mathbf{I} - \Pi_{\mathcal{C}}.$$

(h) There exists $\lambda \in \mathcal{M}(x^*)$ such that the h_λ^{nor} is (P).

(i) There exists $\lambda \in \mathcal{M}(x^*)$ such that all matrices of the form

$$\begin{bmatrix} J_x\mathbf{L}(x^*, \lambda) & B^T \\ -B & 0 \end{bmatrix},$$

where $B \in \mathcal{B}_{\text{bas}}^{\lambda}(\mathcal{C})$, have the same nonzero determinantal sign.

Proof. The equivalence of (a)–(d) follows from the aforementioned results. For the remaining equivalences, it suffices to note that for any $\lambda \in \mathcal{M}(x^*)$,

$$(\mathbf{F}_K^{\text{nor}})' = f_\lambda = h_\lambda^{\text{nor}} \circ (\mathbf{G}(\lambda)_{\mathcal{C}}^{\text{nor}})^{-1}$$

where $\mathbf{G}(\lambda)_{\mathcal{C}}^{\text{nor}}$ is the normal map associated with the affine pair $(\mathcal{C}, G(\lambda))$. The last equality in the above expression is due to Proposition **4.3.3** and Lemma **5.3.8**. Since $\mathbf{G}(\lambda)_{\mathcal{C}}^{\text{nor}}$, and thus its inverse, is a globally Lipschitz homeomorphism, it follows that f_λ is injective, bijective, or a globally Lipschitz homeomorphism if and only if h_λ^{nor} is injective, bijective, or a globally Lipschitz homeomorphism, respectively. By Theorem **4.3.2**, h_λ^{nor}, being the normal map of the affine pair $(\mathcal{C}, J_x\mathbf{L}(x^*, \lambda))$, is coherently oriented if and only if it is (P). Moreover, both $\mathbf{G}(\lambda)_{\mathcal{C}}^{\text{nor}}$ and its inverse are coherently oriented. Since the composition of two coherently oriented PA maps are coherently oriented, it follows that f_λ is coherently oriented if and only if h_λ^{nor} is so. Therefore, all statements (a)–(h) are equivalent. Finally, the equivalence of (h) and (i) follows from Proposition **4.2.7**. □

5.4 Parametric Problems

We extend the previous results to the family of parametric VIs:

$$\{ \text{VI } (K(\mathbf{p}), F(\cdot, \mathbf{p})) \, : \, \mathbf{p} \in \mathcal{P} \}. \tag{5.4.1}$$

Let x^* be a given solution of the VI $(K(\mathbf{p}^*), F(\cdot, \mathbf{p}^*))$ corresponding to a given parameter $\mathbf{p}^* \in \mathcal{P}$. Let $z^* \equiv x^* - F(x^*, \mathbf{p}^*)$. For each $\mathbf{p} \in \mathcal{P}$, the normal map associated with the VI $(K(\mathbf{p}), F(\cdot, \mathbf{p}))$ is

$$\mathbf{F}^{\text{nor}}_{K(\mathbf{p})}(z, \mathbf{p}) \equiv F(\Pi_{K(\mathbf{p})}(z), \mathbf{p}) + z - \Pi_{K(\mathbf{p})}(z), \quad \forall \, z \in \mathbb{R}^n.$$

We are interested in analyzing the behavior of the solution x^* as the parameter $\mathbf{p}$ is being perturbed near the base vector $\mathbf{p}^*$. Part of the complication of this parametric analysis is due to the complex properties of the projection $\Pi_{K(\mathbf{p})}$ onto a varying set $K(\mathbf{p})$. From Lemma **2.8.2**, we know that if the limit holds:

$$\lim_{\mathbf{p} \to \mathbf{p}^*} K(\mathbf{p}) = K(\mathbf{p}^*), \tag{5.4.2}$$

then $\Pi_{K(\mathbf{p})}(z)$ is continuous at the pair $(z, \mathbf{p}^*)$ for every $z \in \mathbb{R}^n$. Hence for every compact subset S of $\mathbb{R}^n$ and for every scalar $\varepsilon > 0$, there exists a neighborhood $\mathcal{W}$ of $\mathbf{p}^*$ such that, for every pair $(z, \mathbf{p}) \in S \times (\mathcal{W} \cap \mathcal{P})$, we have

$$\| \, \Pi_{K(\mathbf{p})}(z) - \Pi_{K(\mathbf{p}^*)}(z) \, \| \, \leq \, \varepsilon.$$

Indeed, if for some compact S and positive ε, no such neighborhood $\mathcal{W}$ exists, then there exist a sequence of vectors $\{\mathbf{p}^k\}$ converging to $\mathbf{p}^*$ and a sequence $\{z^k\} \subset S$ such that, for every k,

$$\| \, \Pi_{K(\mathbf{p}^k)}(z^k) - \Pi_{K(\mathbf{p}^*)}(z^k) \, \| \, > \, \varepsilon.$$

Without loss of generality, we may assume that the sequence $\{z^k\}$ converges to some vector z^∞. Since $\Pi_{K(\mathbf{p})}(z)$ is continuous at $(z^\infty, \mathbf{p}^*)$, the left-hand norm in the above expression converges to zero, contradicting the fact that ε is positive.

Based on the above observation, we establish a result that extends Corollary **5.1.5** without assuming any particular structure on $K(\mathbf{p})$. We need to assume that F is continuous and satisfies two pointwise Lipschitz conditions at $(x^*, \mathbf{p}^*)$; see conditions (5.4.3) and (5.4.4) below. The latter conditions are easily satisfied if F is continuously differentiable near $(x^*, \mathbf{p}^*)$.

5.4.1 Proposition. Let $F : \mathbb{R}^{n+p} \to \mathbb{R}^n$ be a continuous mapping and $K : \mathcal{P} \to \mathbb{R}^n$ be a closed-valued and convex-valued point-to-set map. Let

$\mathbf{p}^* \in \mathcal{P}$ be such that (5.4.2) holds. Suppose that x^* is an isolated solution of the VI $(K(\mathbf{p}^*), F(\cdot, \mathbf{p}^*))$ and $\operatorname{ind}(\mathbf{F}^{\text{nor}}_{K(\mathbf{p}^*)}(\cdot, \mathbf{p}^*), z^*)$ is nonzero. Assume that there exist an open neighborhood $\mathcal{V}$ of x^*, an open neighborhood $\mathcal{U}$ of $\mathbf{p}^*$, and positive constants L and L' such that

$$\sup_{x \in K(\mathbf{p}) \cap \mathcal{V}} \| F(x, \mathbf{p}) - F(x, \mathbf{p}^*) \| \leq L \| \mathbf{p} - \mathbf{p}^* \|, \quad \forall \mathbf{p} \in \mathcal{U} \cap \mathcal{P}, \tag{5.4.3}$$

and

$$\| F(x, \mathbf{p}^*) - F(x', \mathbf{p}^*) \| \leq L' \| x - x' \|, \quad \forall x, x' \in \mathcal{V}. \tag{5.4.4}$$

For every neighborhood $\mathcal{N}$ of x^*, there exists a neighborhood $\mathcal{U}'$ of $\mathbf{p}^*$ such that, for every $\mathbf{p} \in \mathcal{U}' \cap \mathcal{P}$,

$$S_{\mathcal{N}}(\mathbf{p}) \equiv \text{SOL}(K(\mathbf{p}), F(\cdot, \mathbf{p})) \cap \mathcal{N} \neq \emptyset;$$

moreover, if $\text{SOL}(K(\mathbf{p}^*), F(\cdot, \mathbf{p}^*)) \cap \operatorname{cl} \mathcal{N} = \{x^*\}$, then

$$\lim_{\mathbf{p} \to \mathbf{p}^*} \sup \{ \| x - x^* \| \, : \, x \in S_{\mathcal{N}}(\mathbf{p}) \} = 0. \tag{5.4.5}$$

Proof. For simplicity, we assume that $\mathcal{P} = \mathbb{R}^p$. By restricting the neighborhood $\mathcal{V}$ if necessary, we may assume without loss of generality that

$$\text{SOL}(K(\mathbf{p}^*), F(\cdot, \mathbf{p}^*)) \cap \operatorname{cl} \mathcal{V} = \{ x^* \},$$

Let $\mathcal{N}$ be an arbitrary neighborhood of x^*. Let

$$\mathcal{N}' \equiv \mathcal{N} \cap \mathcal{V}.$$

Since $\operatorname{cl} \mathcal{N}'$ is clearly a subset of $\operatorname{cl} \mathcal{V}$, it follows that x^* is the unique solution of the VI $(K(\mathbf{p}^*), F(\cdot, \mathbf{p}^*))$ in $\operatorname{cl} \mathcal{N}'$. By Lemma **2.8.2**, the parametric projector $\Pi_{K(\mathbf{p})}(z)$ is continuous at $(z^*, \mathbf{p}^*)$. Thus there exist an open neighborhood $\mathcal{Z}$ of z^* and an open neighborhood $\mathcal{W}_0$ of $\mathbf{p}^*$ such that (i) $\Pi_{K(\mathbf{p})}(\operatorname{cl} \mathcal{Z}) \subseteq \mathcal{N}'$ for all $\mathbf{p} \in \mathcal{W}_0$, and (ii) z^* is the unique zero of $\mathbf{F}^{\text{nor}}_{K(\mathbf{p}^*)}(\cdot, \mathbf{p}^*)$ in $\operatorname{cl} \mathcal{Z}$. Let

$$\varepsilon \equiv \operatorname{dist}_\infty(0, \mathbf{F}^{\text{nor}}_{K(\mathbf{p}^*)}(\partial \mathcal{Z}, \mathbf{p}^*)),$$

which is a positive scalar. There exists a neighborhood $\mathcal{W}_1$ of $\mathbf{p}^*$ such that

$$(L' + 1) \sup_{z \in \operatorname{cl} \mathcal{Z}} \| \Pi_{K(\mathbf{p})}(z) - \Pi_{K(\mathbf{p}^*)}(z) \| + L \| \mathbf{p} - \mathbf{p}^* \| < \varepsilon.$$

Let $\mathcal{W} \equiv \mathcal{W}_0 \cap \mathcal{W}_1$. We may write

$$\begin{aligned}
&\mathbf{F}^{\text{nor}}_{K(\mathbf{p}^*)}(z, \mathbf{p}^*) - \mathbf{F}^{\text{nor}}_{K(\mathbf{p})}(z, \mathbf{p}) \\
&= F(\Pi_{K(\mathbf{p}^*)}(z), \mathbf{p}^*) - F(\Pi_{K(\mathbf{p})}(z), \mathbf{p}) - \Pi_{K(\mathbf{p}^*)}(z) + \Pi_{K(\mathbf{p})}(z) \\
&= F(\Pi_{K(\mathbf{p}^*)}(z), \mathbf{p}^*) - F(\Pi_{K(\mathbf{p})}(z), \mathbf{p}^*) + \\
&\quad F(\Pi_{K(\mathbf{p})}(z), \mathbf{p}^*) - F(\Pi_{K(\mathbf{p})}(z), \mathbf{p}) - \Pi_{K(\mathbf{p}^*)}(z) + \Pi_{K(\mathbf{p})}(z).
\end{aligned}$$

For $z \in \operatorname{cl} \mathcal{Z}$ and $\mathbf{p} \in \mathcal{W}$, we have $\Pi_{K(\mathbf{p}^*)}(z)$ and $\Pi_{K(\mathbf{p})}(z)$ both belong to $\mathcal{N}'$, which is a subset of $\mathcal{V}$. Thus, we have by (5.4.4),

$$\| F(\Pi_{K(\mathbf{p}^*)}(z), \mathbf{p}^*) - F(\Pi_{K(\mathbf{p})}(z), \mathbf{p}^*) \| \\ \leq L' \| \Pi_{K(\mathbf{p}^*)}(z) - \Pi_{K(\mathbf{p})}(z) \|,$$

and by (5.4.3),

$$\| F(\Pi_{K(\mathbf{p})}(z), \mathbf{p}^*) - F(\Pi_{K(\mathbf{p})}(z), \mathbf{p}) \| \leq L \| \mathbf{p} - \mathbf{p}^* \|.$$

Hence,

$$\sup_{z \in \operatorname{cl} \mathcal{Z}} \| \mathbf{F}^{\text{nor}}_{K(\mathbf{p}^*)}(z, \mathbf{p}^*) - \mathbf{F}^{\text{nor}}_{K(\mathbf{p})}(z, \mathbf{p}) \| \\ \leq (L' + 1) \sup_{z \in \operatorname{cl} \mathcal{Z}} \| \Pi_{K(\mathbf{p})}(z) - \Pi_{K(\mathbf{p}^*)}(z) \| + L \| \mathbf{p} - \mathbf{p}^* \| < \varepsilon.$$

Consequently, by the homotopy invariance property of the degree, we have

$$\deg(\mathbf{F}^{\text{nor}}_{K(\mathbf{p})}(\cdot, \mathbf{p}), \mathcal{Z}) = \deg(\mathbf{F}^{\text{nor}}_{K(\mathbf{p}^*)}(\cdot, \mathbf{p}^*), \mathcal{Z}) = \operatorname{ind}(\mathbf{F}^{\text{nor}}_{K(\mathbf{p}^*)}(\cdot, \mathbf{p}^*), z^*).$$

Since the last index is nonzero, it follows that $\mathbf{F}^{\text{nor}}_{K(\mathbf{p})}(\cdot, \mathbf{p})$ has a zero in $\mathcal{Z}$. Thus the VI $(K(\mathbf{p}), F(\cdot, \mathbf{p}))$ has a solution in $\mathcal{N}$. The proof of (5.4.5) is similar to that of (5.1.3) in Corollary **5.1.5**. □

5.4.2 Remark. The assumptions of Proposition **5.4.1** imply that x^* is a stable solution of the VI $(K(\mathbf{p}^*), F(\cdot, \mathbf{p}^*))$. Although (5.4.3) implies that $F(\cdot, \mathbf{p})$ is close to $F(\cdot, \mathbf{p}^*)$ for $\mathbf{p}$ close to $\mathbf{p}^*$, the stability of x^* is not sufficient to deduce that $S_{\mathcal{N}}(\mathbf{p})$ is nonempty because the perturbed VI $(K(\mathbf{p}), F(\cdot, \mathbf{p}))$ does not have the same defining set as the base VI $(K(\mathbf{p}^*), F(\cdot, \mathbf{p}^*))$. This explains why it is necessary to resort to the degree-theoretic proof that is the basis of all existential results. □

We next proceed to derive some sufficient conditions for the assumptions of Proposition **5.4.1** to hold and to analyze the "parametric stability" of x^* under the assumption that $K(\mathbf{p})$ is finitely representable:

$$K(\mathbf{p}) \equiv \{ x \in \mathbb{R}^n : g(x, \mathbf{p}) \leq 0 \}; \tag{5.4.6}$$

where $g : \mathbb{R}^n \times \mathcal{P} \to \mathbb{R}^m$ is such that, for each $\mathbf{p} \in \mathcal{P} \subset \mathbb{R}^p$, $g_i(\cdot, \mathbf{p})$ is continuously differentiable for each i. For each pair $(x, \mathbf{p})$ with $x \in \operatorname{SOL}(K(\mathbf{p}), F(\cdot, \mathbf{p}))$, let $\mathcal{M}(x, \mathbf{p})$ denote the set of multipliers λ satisfying

$$F(x, \mathbf{p}) + \sum_{i=1}^{m} \lambda_i \nabla_x g_i(x, \mathbf{p}) = 0 \\ 0 \leq \lambda \perp g(x, \mathbf{p}) \leq 0. \tag{5.4.7}$$

Let $\mathcal{M}^e(x, \mathbf{p})$ be the subset of $\mathcal{M}(x, \mathbf{p})$ consisting of $\lambda \in \mathcal{M}(x, \mathbf{p})$ for which the gradients

$$\{ \nabla_x g_i(x, \mathbf{p}) \, : \, i \in \mathcal{I}(x, \mathbf{p}) \}$$

are linearly independent, where

$$\mathcal{I}(x, \mathbf{p}) \equiv \{ i \, : \, g_i(x, \mathbf{p}) = 0 \}$$

is the index set of active constraints at x. As a set consisting of vectors λ satisfying the KKT system (5.4.7), $\mathcal{M}(x, \mathbf{p})$ is well defined without $g_i(\cdot, \mathbf{p})$ being convex. This set of multipliers is nonempty if x belongs to $\mathrm{SOL}(K(\mathbf{p}), F(\cdot, \mathbf{p}))$ and the MFCQ holds at x. In turn, if the MFCQ holds at $x^* \in \mathrm{SOL}(K(\mathbf{p}^*), F(\cdot, \mathbf{p}^*))$ and provided that the pair $(x, \mathbf{p})$ is sufficiently close to $(x^*, \mathbf{p}^*)$, then the MFCQ continues to hold at $x \in K(\mathbf{p})$ and hence $\mathcal{M}(x, \mathbf{p})$ is nonempty. The latter conclusion was the content of part (d) of Proposition **3.2.1** in the parameter-free case; the same proof applies to the parametric case provided that each g_i is continuously differentiable in a neighborhood of $(x^*, \mathbf{p}^*)$.

The following result, which does not require the convexity of each $g_i(\cdot, \mathbf{p})$, extends Proposition **3.2.2** to the parametric context. The proof follows immediately from Corollary **3.2.5**, which is in turn a consequence of Hoffman's inequality for polyhedra. Since the proof is the same as Proposition **3.2.2**, it is not repeated here.

5.4.3 Proposition. Let g be continuously differentiable and F be continuous near the pair $(x^*, \mathbf{p}^*)$, where $x^* \in \mathrm{SOL}(K(\mathbf{p}^*), F(\cdot, \mathbf{p}^*))$. If the MFCQ holds at $x^* \in K(\mathbf{p}^*)$, then there exist a scalar $c > 0$, a neighborhood $\mathcal{N}$ of x^* and a neighborhood $\mathcal{W}$ of $\mathbf{p}^*$ such that, for all $(x, \mathbf{p}) \in \mathcal{N} \times \mathcal{W}$ with $x \in \mathrm{SOL}(K(\mathbf{p}), F(\cdot, \mathbf{p}))$,

$$\emptyset \neq \mathcal{M}(x, \mathbf{p}) \subseteq \mathcal{M}(x^*, \mathbf{p}^*) + c\, \gamma(x, \mathbf{p})\, \mathbb{B}(0, 1),$$

where

$$\gamma(x, \mathbf{p}) \equiv \| F(x, \mathbf{p}) - F(x^*, \mathbf{p}^*) \| + \sum_{i \in \mathcal{I}(x^*, \mathbf{p}^*)} \| \nabla_x g_i(x, \mathbf{p}) - \nabla_x g_i(x^*, \mathbf{p}^*) \|. \tag{5.4.8}$$

Based on the joint continuity of the parametric projector $\Pi_{K(\mathbf{p})}(z)$ at $(z^*, \mathbf{p}^*)$, we are ready to present the promised sufficient conditions for the parametric stability of a solution x^* of the VI $(K(\mathbf{p}^*), F(\cdot, \mathbf{p}^*))$ as we perturb the parameter $\mathbf{p}$ around $\mathbf{p}^*$. In addition to proving the same conclusions as those in Proposition **5.4.1**, we are also interested in establishing a "pointwise Lipschitz continuity" of the perturbed solution multifunction

$$\mathbf{p} \in \mathcal{W} \mapsto x(\mathbf{p}) \in S_{\mathcal{N}}(\mathbf{p});$$

see (5.4.10). It turns out that there are two ways to guarantee the latter property; one is by assuming a strict copositivity condition (see Theorem **5.4.4**); and the other is by assuming the Lipschitz continuity of $\Pi_{K(\mathbf{p})}(z)$ near $(z^*, \mathbf{p}^*)$ (see Theorem **5.4.5**).

The specific setting is as follows. Let $K(\mathbf{p})$ be given by (5.4.6), where each $g_i(\cdot, \mathbf{p})$ is continuous and convex for each $\mathbf{p} \in \mathcal{P}$. Let x^* be a solution of the VI $(K(\mathbf{p}^*), F(\cdot, \mathbf{p}^*))$. We assume that there exist an open neighborhood $\mathcal{V}$ of x^*, an open neighborhood $\mathcal{U}$ of $\mathbf{p}^*$, and positive constants L and L' such that $F(\cdot, \mathbf{p}^*)$ is continuously differentiable in $\mathcal{V}$, g is twice continuously differentiable in $\mathcal{V} \times \mathcal{U}$, and (5.4.3) and (5.4.4) hold. We further assume that the MFCQ holds at $x^* \in K(\mathbf{p}^*)$. The Lagrangian function for the VI $(K(\mathbf{p}), F(\cdot, \mathbf{p}))$ is denoted:

$$\mathbf{L}(x, \lambda, \mathbf{p}) \equiv F(x, \mathbf{p}) + \sum_{i=1}^{m} \lambda_i \nabla_x g_i(x, \mathbf{p});$$

the critical cone of the VI $(K(\mathbf{p}^*), F(\cdot, \mathbf{p}^*))$ at the solution x^* is:

$$\mathcal{C} \equiv \mathcal{C}(x^*; K(\mathbf{p}^*), F(\cdot, \mathbf{p}^*));$$

and we write the matrices:

$$G(\lambda) \equiv \mathbf{I} + \sum_{i=1}^{m} \lambda_i \, \nabla^2_{xx} g_i(x^*, \mathbf{p}^*).$$

The next result pertains to the parametric stability of a solution x^* in $\mathrm{SOL}(K(\mathbf{p}^*), F(\cdot, \mathbf{p}^*))$; this is a restricted kind of stability of x^* that is the result of the perturbation of the parameter $\mathbf{p}$. The last assertion of the theorem below is closely related to the bound (5.1.5) in Corollary **5.1.8**, which deals with a parameter-free set K.

5.4.4 Theorem. In the above setting, if for all $\lambda \in \mathcal{M}^e(x^*, \mathbf{p}^*)$ and all scalars $\tau \geq 0$, $(\mathcal{C}, J_x\mathbf{L}(x^*, \lambda, \mathbf{p}^*) + \tau G(\lambda))$ is a R_0 pair, then there exist a neighborhood $\mathcal{N}$ of x^* and a neighborhood $\mathcal{W}$ of $\mathbf{p}^*$ such that, for all $\mathbf{p} \in \mathcal{W}$,

$$S_{\mathcal{N}}(\mathbf{p}) \equiv \mathrm{SOL}(K(\mathbf{p}), F(\cdot, \mathbf{p})) \cap \mathcal{N} \neq \emptyset;$$

moreover,

$$\lim_{\mathbf{p} \to \mathbf{p}^*} \sup \{ \, \| x(\mathbf{p}) - x^* \| \, : \, x(\mathbf{p}) \in S_{\mathcal{N}}(\mathbf{p}) \, \} = 0. \tag{5.4.9}$$

If in addition $J_x\mathbf{L}(x^*, \lambda, \mathbf{p}^*)$ is strictly copositive on $\mathcal{C}(x^*; K(\mathbf{p}^*), F(\cdot, \mathbf{p}^*))$ for all $\lambda \in \mathcal{M}(x^*, \mathbf{p}^*)$, then there exists a constant $c > 0$ such that, for all $\mathbf{p}$ sufficiently close to $\mathbf{p}^*$,

$$\sup \{ \, \| x(\mathbf{p}) - x^* \| \, : \, x(\mathbf{p}) \in S_{\mathcal{N}}(\mathbf{p}) \, \} \leq c \, \| \mathbf{p} - \mathbf{p}^* \|. \tag{5.4.10}$$

Proof. The assumptions imply that z^* is a stable zero of $\mathbf{F}^{\text{nor}}_{K(\mathbf{p}^*)}(\cdot,\mathbf{p}^*)$ and $\text{ind}(\mathbf{F}^{\text{nor}}_{K(\mathbf{p}^*)}(\cdot,\mathbf{p}^*), z^*)$ is nonzero, by Lemma **5.3.9** and Theorem **5.2.4**. Let $\mathcal{Z}$ be a neighborhood of z^* such that $\Pi_{K(\mathbf{p})}(z)$ is continuous at $(z,\mathbf{p}^*)$ for every $z \in \text{cl}\,\mathcal{Z}$. For every scalar $\varepsilon > 0$ a neighborhood $\mathcal{W}_0$ of $\mathbf{p}^*$ exists such that, for all $\mathbf{p} \in \mathcal{W}_0$,

$$\sup_{z \in \text{cl}\,\mathcal{Z}} \| \Pi_{K(\mathbf{p})}(z) - \Pi_{K(\mathbf{p}^*)}(z) \| \leq \varepsilon.$$

By choosing ε appropriately as in the proof of Proposition **5.4.1**, we can establish the existence of the desired neighborhoods $\mathcal{N}$ of x^* and $\mathcal{W}$ of $\mathbf{p}^*$ as well as the limit (5.4.9) of the perturbed solutions. The details are not repeated.

The proof of (5.4.10) is more involved and utilizes the solution continuity property (5.4.9). This proof does not require the convexity of each $g_i(\cdot,\mathbf{p})$. For each $\lambda \in \mathcal{M}(x,\mathbf{p})$ with x sufficiently close to x^* such that $\mathcal{I}(x,\mathbf{p})$ is contained in $\mathcal{I}(x^*,\mathbf{p}^*)$, we have

$$F(x,\mathbf{p}) + \sum_{i \in \mathcal{I}(x^*,\mathbf{p}^*)} \lambda_i \nabla_x g_i(x,\mathbf{p}) = 0.$$

We can write

$$\begin{aligned}
&F(x,\mathbf{p}) \\
&= F(x,\mathbf{p}) - F(x,\mathbf{p}^*) + F(x,\mathbf{p}^*) - F(x^*,\mathbf{p}^*) + F(x^*,\mathbf{p}^*) \\
&= F(x,\mathbf{p}) - F(x,\mathbf{p}^*) + J_x F(x^*,\mathbf{p}^*)(x - x^*) + F(x^*,\mathbf{p}^*) + o(\|x - x^*\|);
\end{aligned}$$

similarly,

$$\begin{aligned}
\nabla_x g_i(x,\mathbf{p}) = \; & \nabla_x g_i(x,\mathbf{p}) - \nabla_x g_i(x,\mathbf{p}^*) \\
& + \nabla^2_{xx} g_i(x^*,\mathbf{p}^*)(\, x - x^* \,) + \nabla_x g_i(x^*,\mathbf{p}^*) + o(\|\, x - x^* \,\|).
\end{aligned}$$

Let $\lambda^*(\mathbf{p}) \in \mathcal{M}(x^*,\mathbf{p}^*)$ be such that

$$\|\, \lambda - \lambda^*(\mathbf{p}) \,\| \leq c\,\gamma(x,\mathbf{p}),$$

where $\gamma(x,\mathbf{p})$ is given by (5.4.8) and c is a constant independent of λ. Consequently, we have

$$\begin{aligned}
0 = \; & J_x \mathbf{L}(x^*,\lambda,\mathbf{p}^*)(\, x - x^* \,) + F(x,\mathbf{p}) - F(x,\mathbf{p}^*) + \\
& \sum_{i \in \mathcal{I}(x^*,\mathbf{p}^*)} \lambda_i (\nabla_x g_i(x,\mathbf{p}) - \nabla_x g_i(x,\mathbf{p}^*)) + \\
& \sum_{i \in \mathcal{I}(x^*,\mathbf{p}^*)} (\lambda_i - \lambda_i^*(\mathbf{p})) \nabla_x g_i(x^*,\mathbf{p}^*) + o(\|\, x - x^* \,\|).
\end{aligned} \tag{5.4.11}$$

Consider the product $(\lambda_i - \lambda_i^*(\mathbf{p}))(x - x^*)^T \nabla_x g_i(x^*, \mathbf{p}^*)$ for each i in $\mathcal{I}(x^*, \mathbf{p}^*)$. We have, for all i,

$$\begin{aligned} g_i(x, \mathbf{p}) &= g_i(x^*, \mathbf{p}^*) + \nabla_x g_i(x^*, \mathbf{p}^*)^T (x - x^*) \\ &+ \nabla_{\mathbf{p}} g_i(x^*, \mathbf{p}^*)^T (\mathbf{p} - \mathbf{p}^*) + O(\|x - x^*\|^2 + \|\mathbf{p} - \mathbf{p}^*\|^2). \end{aligned} \tag{5.4.12}$$

For $i \in \mathcal{I}(x, \mathbf{p})$, which is a subset of $\mathcal{I}(x^*, \mathbf{p}^*)$, we have

$$g_i(x, \mathbf{p}) = g_i(x^*, \mathbf{p}^*) = 0;$$

thus

$$\begin{aligned} &(\lambda_i - \lambda_i^*(\mathbf{p}))(x - x^*)^T \nabla_x g_i(x^*, \mathbf{p}) \\ &\quad = (\lambda_i - \lambda_i^*(\mathbf{p}))[O(\|\mathbf{p} - \mathbf{p}^*\|) + O(\|x - x^*\|^2 + \|\mathbf{p} - \mathbf{p}^*\|^2)]. \end{aligned}$$

For $i \in \mathcal{I}(x^*, \mathbf{p}^*) \setminus \mathcal{I}(x, \mathbf{p})$, we have

$$\lambda_i = 0 \quad \text{and} \quad 0 = g_i(x^*, \mathbf{p}^*) \geq g_i(x, \mathbf{p});$$

thus

$$\begin{aligned} &(\lambda_i - \lambda_i^*(\mathbf{p}))(x - x^*)^T \nabla_x g_i(x^*, \mathbf{p}^*) \\ &\quad \geq (\lambda_i - \lambda_i^*(\mathbf{p}))[O(\|\mathbf{p} - \mathbf{p}^*\|) + O(\|x - x^*\|^2 + \|\mathbf{p} - \mathbf{p}^*\|^2)]. \end{aligned}$$

Premultiplying (5.4.11) by $(x - x^*)^T$, we deduce

$$\begin{aligned} &(x - x^*)^T J_x \mathbf{L}(x^*, \lambda, \mathbf{p}^*)(x - x^*) \\ &\leq -(x - x^*)^T [F(x, \mathbf{p}) - F(x, \mathbf{p}^*)] \\ &-(x - x^*)^T \sum_{i \in \mathcal{I}(x^*, \mathbf{p}^*)} \lambda_i (\nabla_x g_i(x, \mathbf{p}) - \nabla_x g_i(x, \mathbf{p}^*)) \\ &+o(\| x - x^* \|^2) \\ &- \sum_{i \in \mathcal{I}(x^*, \mathbf{p}^*)} (\lambda_i - \lambda_i^*(\mathbf{p}))[O(\|\mathbf{p} - \mathbf{p}^*\|) + O(\|x - x^*\|^2 + \|\mathbf{p} - \mathbf{p}^*\|^2)]. \end{aligned} \tag{5.4.13}$$

Suppose for the sake of contradiction that no constant $c > 0$ exists such that (5.4.10) holds for all $\mathbf{p}$ near $\mathbf{p}^*$. A sequence $\{\mathbf{p}^k\}$ converging to $\mathbf{p}^*$ exists such that, for each k, a solution $x^k \in S_{\mathcal{N}}(\mathbf{p}^k)$ exists satisfying

$$\| x^k - x^* \| > k \| \mathbf{p}^k - \mathbf{p}^* \|.$$

Thus

$$\lim_{k \to \infty} \frac{\| \mathbf{p}^k - \mathbf{p}^* \|}{\| x^k - x^* \|} = 0. \tag{5.4.14}$$

Clearly $x^k \neq x^*$ for all k and the sequence $\{x^k\}$ converges to x^* as $k \to \infty$ by (5.4.9). Without loss of generality, we may assume that the normalized sequence

$$\left\{ \frac{x^k - x^*}{\| x^k - x^* \|} \right\}$$

converges to a vector $\bar{v}$, which must be nonzero. We claim that $\bar{v}$ belongs to the critical cone $\mathcal{C}$. By (5.4.12) and a standard normalization argument, and also by (5.4.14), it is easy to show that $\bar{v}$ must belong to $\mathcal{T}(x^*; K(\mathbf{p}^*))$. For each k, let $\lambda^k \in \mathcal{M}(x^k, \mathbf{p}^k)$ and let $\tilde{\lambda}^k \in \mathcal{M}(x^*, \mathbf{p}^*)$ be such that

$$\| \lambda^k - \tilde{\lambda}^k \| \leq c\,\gamma(x^k, \mathbf{p}^k).$$

Since $\mathcal{M}(x^*, \mathbf{p}^*)$ is compact, it follows that both sequences $\{\tilde{\lambda}^k\}$ and $\{\lambda^k\}$ are bounded. Moreover, since $\{\gamma(x^k, \mathbf{p}^k)\}$ converges to zero, we may assume without loss of generality that $\{\tilde{\lambda}^k\}$ and $\{\lambda^k\}$ both converge to the same vector $\lambda^\infty \in \mathcal{M}(x^*, \mathbf{p}^*)$. For $i \in \mathrm{supp}(\lambda^\infty)$, we have $\lambda_i^k > 0$ for all k sufficiently large; thus $g_i(x^k, \mathbf{p}^k) = 0$. We can then show that $\bar{v}^T \nabla_x g_i(x^*, \mathbf{p}^*) = 0$. This is enough to deduce that $\bar{v}$ is a critical vector of the VI $(K(\mathbf{p}^*), F(\cdot, \mathbf{p}^*))$ at x^*.

We next show that $\bar{v}$ provides a vector which contradicts the strict copositivity of $J_x\mathbf{L}(x^*, \lambda^\infty; \mathbf{p}^*)$ on $\mathcal{C}$. We have

$$\frac{\gamma(x^k, \mathbf{p}^k)}{\| x^k - x^* \|} = \frac{\| F(x^k, \mathbf{p}^k) - F(x^*, \mathbf{p}^*) \|}{\| x^k - x^* \|} +$$

$$\sum_{i \in \mathcal{I}(x^*, \mathbf{p}^*)} \frac{\| \nabla_x g_i(x, \mathbf{p}) - \nabla_x g_i(x^*, \mathbf{p}^*) \|}{\| x^k - x^* \|}$$

which implies, in view of (5.4.14), that the sequence

$$\left\{ \frac{\| \lambda^k - \tilde{\lambda}^k \|}{\| x^k - x^* \|} \right\}$$

is bounded. Dividing (5.4.13) by $\|x^k - x^*\|^2$, we obtain

$$\left(\frac{x^k - x^*}{\| x^k - x^* \|} \right)^T J_x\mathbf{L}(x^*, \lambda^k, \mathbf{p}^*) \left(\frac{x^k - x^*}{\| x^k - x^* \|} \right) \leq T_1 + T_2 + T_3,$$

where

$$T_1 \equiv \left(\frac{x^k - x^*}{\| x^k - x^* \|} \right)^T \left(\frac{F(x^k, \mathbf{p}^k) - F(x^k, \mathbf{p}^*)}{\| x^k - x^* \|} \right),$$

$$T_2 \equiv \left(\frac{x^k - x^*}{\| x^k - x^* \|} \right)^T \sum_{i \in \mathcal{I}(x^*, \mathbf{p}^*)} \lambda_i^k \left(\frac{\nabla_x g_i(x^k, \mathbf{p}^k) - \nabla_x g_i(x^k, \mathbf{p}^*)}{\| x^k - x^* \|} \right),$$

and

$$T_3 \leq \left(\frac{\| \lambda^k - \tilde{\lambda}^k \|}{\| x^k - x^* \|} \right) \left(\frac{O(\|\mathbf{p}^k - \mathbf{p}^*\|)}{\| x^k - x^* \|} + \frac{O(\|x^k - x^*\|^2 + \|\mathbf{p}^k - \mathbf{p}^*\|^2))}{\| x^k - x^* \|} \right).$$

By (5.4.14), it follows that all three terms T_1, T_2, and T_3 converge to zero. Consequently, we deduce $\bar{v}^T J_x \mathbf{L}(x^*, \lambda^\infty, \mathbf{p}^*) \bar{v} \leq 0$, contradicting the strict copositivity of $J_x \mathbf{L}(x^*, \lambda^\infty, \mathbf{p}^*)$ on $\mathcal{C}$. □

In the next result, we assume that the CRCQ holds at x^* in addition to the MFCQ. The CRCQ results in two major changes in Theorem **5.4.4**. First, the assumption that for all $\lambda \in \mathcal{M}^e(x^*, \mathbf{p}^*)$ and $\tau \geq 0$, $(\mathcal{C}, J_x \mathbf{L}(x^*, \lambda, \mathbf{p}^*) + \tau G(\lambda))$ is a R_0 pair is replaced by the weaker requirement that there is some $\lambda \in \mathcal{M}(x^*, \mathbf{p}^*)$ such that, for all $\tau \geq 0$ the pair $(\mathcal{C}, J_x \mathbf{L}(x^*, \lambda, \mathbf{p}^*) + \tau G(\lambda))$ has the R_0-property. This is possible because Theorem **4.5.3** allows the use of any multiplier in $\mathcal{M}(x^*, \mathbf{p}^*)$ to represent the directional derivative $(\mathbf{F}^{\text{nor}}_{K(\mathbf{p}^*)}(\cdot, \mathbf{p}^*))'(z^*; dz)$ for all vectors dz in $\mathbb{R}^n$. Second, it is no longer necessary to assume the strict copositivity of $J_x \mathbf{L}(x^*, \lambda, \mathbf{p}^*)$ in order to establish the pointwise Lipschitz condition (5.4.10). This is due to the Lipschitz continuity of the parametric projector $\Pi_{K(\mathbf{p})}(x)$ near $(x^*, \mathbf{p}^*)$.

5.4.5 Theorem. Assume in addition to the setting stated before Theorem **5.4.4** that the CRCQ hold at $x^* \in K(\mathbf{p}^*)$. If there exists λ in $\mathcal{M}(x^*, \mathbf{p}^*)$ such that, for all scalars $\tau \geq 0$, $(\mathcal{C}, J_x \mathbf{L}(x^*, \lambda, \mathbf{p}^*) + \tau G(\lambda))$ is a R_0 pair, then there exist a neighborhood $\mathcal{N}$ of x^*, a neighborhood $\mathcal{W}$ of $\mathbf{p}^*$, and a constant $c > 0$ such that, for all $\mathbf{p} \in \mathcal{W}$,

$$\emptyset \neq S_{\mathcal{N}}(\mathbf{p}) \subset \{ x^* \} + c \| \mathbf{p} - \mathbf{p}^* \| \operatorname{cl} \mathbb{B}(0,1).$$

Proof. The proof is a refinement of the proof of Proposition **5.4.1** and Theorem **5.4.4**, using the Lipschitz continuity of the parametric projector $\Pi_{K(\mathbf{p})}(z)$ near $(z^*, \mathbf{p}^*)$. As in these previous results, we can deduce the existence of a neighborhood $\mathcal{N}$ of x^*, a neighborhood $\mathcal{W}$ of $\mathbf{p}^*$ in which $\Pi_{K(\cdot)}(z^*)$ is Lipschitz continuous with modulus $L_0 > 0$ and a constant $c' > 0$ such that, for all $\mathbf{p} \in \mathcal{W}$, $S_{\mathcal{N}}(\mathbf{p})$ is nonempty and

$$\| z' - z^* \| \leq c' \| \mathbf{F}^{\text{nor}}_{K(\mathbf{p}^*)}(z', \mathbf{p}^*) \|,$$

where $z' \equiv x' - F(x', \mathbf{p})$ with x' being an arbitrary vector in $S_{\mathcal{N}}(\mathbf{p})$. The above inequality is the result of z^* being a stable zero of $\mathbf{F}^{\text{nor}}_{K(\mathbf{p}^*)}(\cdot, \mathbf{p}^*)$, by Lemma **5.3.9**. It remains to show that for some constant $c > 0$, $S_{\mathcal{N}}(\mathbf{p})$ is contained in $\mathbb{B}(x^*, c\|\mathbf{p} - \mathbf{p}^*\|)$. By the Cauchy-Schwarz inequality, the nonexpansiveness of the projector $\Pi_{K(\mathbf{p})}$, the Lipschit continuity of $\Pi_{K(\cdot)}(z^*)$, and the last displayed inequality, we have

$$\begin{aligned}\|x' - x^*\| &= \|\Pi_{K(\mathbf{p})}(z') - \Pi_{K(\mathbf{p}^*)}(z^*)\| \\ &\leq \|\Pi_{K(\mathbf{p})}(z') - \Pi_{K(\mathbf{p})}(z^*)\| + \|\Pi_{K(\mathbf{p})}(z^*) - \Pi_{K(\mathbf{p}^*)}(z^*)\| \\ &\leq \|z' - z^*\| + L_0 \|\mathbf{p} - \mathbf{p}^*\| \\ &\leq c' \|\mathbf{F}^{\text{nor}}_{K(\mathbf{p}^*)}(z', \mathbf{p}^*)\| + L_0 \|\mathbf{p} - \mathbf{p}^*\|.\end{aligned}$$

Since $\mathbf{F}^{\text{nor}}_{K(\mathbf{p})}(z', \mathbf{p}) = 0$, we have

$$\begin{aligned}&\mathbf{F}^{\text{nor}}_{K(\mathbf{p}^*)}(z', \mathbf{p}^*) = \mathbf{F}^{\text{nor}}_{K(\mathbf{p}^*)}(z', \mathbf{p}^*) - \mathbf{F}^{\text{nor}}_{K(\mathbf{p})}(z', \mathbf{p}) \\ &= F(\Pi_{K(\mathbf{p}^*)}(z'), \mathbf{p}^*) - F(\Pi_{K(\mathbf{p})}(z'), \mathbf{p}) - \Pi_{K(\mathbf{p}^*)}(z') + \Pi_{K(\mathbf{p})}(z') \\ &= F(\Pi_{K(\mathbf{p}^*)}(z'), \mathbf{p}^*) - F(\Pi_{K(\mathbf{p})}(z'), \mathbf{p}^*) - \Pi_{K(\mathbf{p}^*)}(z') + \Pi_{K(\mathbf{p})}(z') + \\ &\quad F(\Pi_{K(\mathbf{p})}(z'), \mathbf{p}^*) - F(\Pi_{K(\mathbf{p})}(z'), \mathbf{p}).\end{aligned}$$

Consequently, by (5.4.3) and (5.4.4), we deduce

$$\|\mathbf{F}^{\text{nor}}_{K(\mathbf{p}^*)}(z')\| \leq (L' L_0 + L + L_0) \|\mathbf{p} - \mathbf{p}^*\|,$$

which yields

$$\|x' - x^*\| \leq [c'(L' L_0 + L + L_0) + L_0] \|\mathbf{p} - \mathbf{p}^*\| \equiv c \|\mathbf{p} - \mathbf{p}^*\|,$$

where $c \equiv c'(L'L_0 + L + L_0) + L_0$. □

The next result pertains to the "parametric strong stability" of the solution x^* in $\text{SOL}(K(\mathbf{p}^*), F(\cdot, \mathbf{p}^*))$. This result requires the function F to be C^1 in a neighborhood of $(x^*, \mathbf{p}^*)$; this differentiability assumption and the PC^1 property of the parametric projector $\Pi_{K(\mathbf{p})}(z)$ allow us to use the theory of PC^1 local homeomorphisms presented in Section **4.6**.

5.4.6 Theorem. Assume in addition to the setting of Theorem **5.4.5** that F is continuously differentiable in a neighborhood of $(x^*, \mathbf{p}^*)$. If x^* is a strongly stable solution of the VI $(K(\mathbf{p}^*), F(\cdot, \mathbf{p}^*))$, there exist a neighborhood $\mathcal{N}$ of x^*, a neighborhood $\mathcal{W}$ of $\mathbf{p}^*$ and a PC^1 function $x_{\mathcal{N}} : \mathcal{W} \to \mathcal{N}$ such that $x_{\mathcal{N}}(\mathbf{p}^*) = x^*$ and for every $\mathbf{p} \in \mathcal{W}$, $x_{\mathcal{N}}(\mathbf{p})$ is the only solution of the VI $(K(\mathbf{p}), F(\cdot, \mathbf{p}))$ that lies in $\mathcal{N}$.

Proof. Consider the mapping

$$\Xi : \begin{pmatrix} z \\ \mathbf{p} \end{pmatrix} \mapsto \begin{pmatrix} \mathbf{F}^{\text{nor}}_{K(\mathbf{p})}(z, \mathbf{p}) \\ \mathbf{p} - \mathbf{p}^* \end{pmatrix},$$

which is PC^1 near $(z^*, \mathbf{p}^*)$. Since x^* is a strongly stable solution of the VI $(K(\mathbf{p}^*), F(\cdot, \mathbf{p}^*))$, $(\mathbf{F}^{\rm nor}_{K(\mathbf{p}^*)}(\cdot, \mathbf{p}^*))'(z^*; dz)$ is a globally Lipschitz homeomorphism in dz. Thus $\Xi'((z^*, \mathbf{p}^*); (dz, d\mathbf{p}))$ is a globally Lipschitz homeomorphism in $(dz, d\mathbf{p})$. Hence Ξ is a locally Lipschitz homeomorphism near $(z^*, \mathbf{p}^*)$. Therefore there exist a neighborhood $\mathcal{Z}$ of z^* and a neighborhood $\mathcal{W}$ of $\mathbf{p}^*$ such that the restricted map

$$\Psi \equiv \Xi|_{\mathcal{Z}\times\mathcal{W}} : \mathcal{Z} \times \mathcal{W} \to \Xi(\mathcal{Z} \times \mathcal{W})$$

is a Lipschitz homeomorphism. Let $z_{\mathcal{Z}}(\mathbf{p})$ denote the z-part of the vector $\Psi^{-1}(0, \mathbf{p} - \mathbf{p}^*)$ for all $\mathbf{p} \in \mathcal{W}$. Being the inverse of a PC^1 function, Ψ^{-1} is PC^1. Thus $z_{\mathcal{Z}}$ defines a PC^1 function from $\mathcal{W}$ into $\mathcal{Z}$ with the property that $z_{\mathcal{Z}}(\mathbf{p}^*) = z^*$ and for each $\mathbf{p} \in \mathcal{W}$, $z_{\mathcal{Z}}(\mathbf{p})$ is the unique vector in $\mathcal{Z}$ satisfying

$$\mathbf{F}^{\rm nor}_{K(\mathbf{p})}(z_{\mathcal{Z}}(\mathbf{p}), \mathbf{p}) = 0, \quad \forall\, \mathbf{p} \in \mathcal{W}.$$

Hence $\Pi_{K(\mathbf{p})}(z_{\mathcal{Z}}(\mathbf{p}))$ is a solution of the VI $(K(\mathbf{p}), F(\cdot, \mathbf{p}))$. Choose a neighborhood $\mathcal{N}$ of x^* such that

$$x \in \mathcal{N} \;\Rightarrow\; x - F(x) \in \mathcal{Z}.$$

Letting $\gamma > 0$ be the Lipschitz modulus of Ψ^{-1}, we have for any $\mathbf{p}$ and $\mathbf{p}'$ in $\mathcal{W}$,

$$\begin{aligned} &\| \Pi_{K(\mathbf{p})}(z_{\mathcal{Z}}(\mathbf{p})) - \Pi_{K(\mathbf{p})}(z_{\mathcal{Z}}(\mathbf{p}')) \| \\ &\leq \| z_{\mathcal{Z}}(\mathbf{p}) - z_{\mathcal{Z}}(\mathbf{p}') \| \leq \gamma \| \mathbf{p} - \mathbf{p}' \|. \end{aligned}$$

Thus by restricting $\mathcal{W}$ if necessary, we can ensure that $\Pi_{K(\mathbf{p})}(z_{\mathcal{Z}}(\mathbf{p}))$ belongs to the neighborhood $\mathcal{N}$ for all $\mathbf{p} \in \mathcal{W}$. This defines the PC^1 function $x_{\mathcal{N}}$ mapping $\mathcal{W}$ into $\mathcal{N}$ with the property that for each $\mathbf{p}$ in $\mathcal{W}$, $x_{\mathcal{N}}(\mathbf{p})$ is a solution of the VI $(K(\mathbf{p}), F(\cdot, \mathbf{p}))$ that lies in $\mathcal{N}$. It remains to establish that $x_{\mathcal{N}}(\mathbf{p})$ is the only solution of the latter VI in $\mathcal{N}$. Let x' be another such solution. Then $z' \equiv x' - F(x', \mathbf{p})$ is a zero of the normal map $\mathbf{F}^{\rm nor}_{K(\mathbf{p})}(\cdot, \mathbf{p})$ that belongs to $\mathcal{Z}$. Thus we have $z' = z_{\mathcal{Z}}(\mathbf{p})$, which implies that $x' = \Pi_{K(\mathbf{p})}(z') = x_{\mathcal{N}}(\mathbf{p})$, establishing the uniqueness of $x_{\mathcal{N}}(\mathbf{p})$. □

Theorem **5.4.6** is an implicit function theorem for a parametric VI. The distinguished feature of the implicit solution function $x_{\mathcal{N}}$ is its piecewise smooth property. In the next subsection, we discuss how to compute the directional derivative of this function; see Theorem **5.4.10**.

5.4.1 Directional differentiability

The concepts of stability and strong stability provide useful information pertaining to two important sensitivity properties of an isolated solution

x^* of the VI $(K(\mathbf{p}^*), F(\cdot, \mathbf{p}^*))$; namely, the solvability of the perturbed problems VI $(K(\mathbf{p}), F(\cdot, \mathbf{p}))$ and the quantitative change of the perturbed solutions with reference to x^*. Information of the latter kind is of the zeroth order because it concerns the simple difference $\| x(\mathbf{p}) - x^* \|$, where $x(\mathbf{p})$ is a solution of the perturbed VI $(K(\mathbf{p}), F(\cdot, \mathbf{p}))$. Parametric change of the first order is of interest also; thus we are led to consider questions pertaining to the following limit of divided quotients:

$$\lim_{\mathbf{p} \to \mathbf{p}^*} \frac{x(\mathbf{p}) - x^*}{\| \mathbf{p} - \mathbf{p}^* \|}.$$

One such question is whether the solution function $x(\mathbf{p})$ is directionally differentiable at $\mathbf{p}^*$, if such a (single-valued) function indeed exists locally for $\mathbf{p}$ near $\mathbf{p}^*$. The rest of this subsection is concerned with the investigation of this first-order parametric analysis.

Throughout this subsection, $K(\mathbf{p})$ is assumed to be given by (5.4.6), where for each $\mathbf{p}$ in $\mathcal{P}$, the function $g_i(\cdot, \mathbf{p})$ is convex. Let x^* be a given solution of the VI $(K(\mathbf{p}^*), F(\cdot, \mathbf{p}^*))$. Assume further that F is continuously differentiable and each g_i is twice continuously differentiable near $(x^*, \mathbf{p}^*)$. We further assume that the MFCQ holds at x^*. (The CRCQ is not assumed until later. Incidentally, when K is a fixed polyhedral set, such as in the case of a parametric NCP, no CQ is needed in the following analysis.) Our goal is to establish results that extend those in Section **4.4**, which pertain to a very special instance of parametric VIs, namely the Euclidean projector.

As a first step in this analysis, we define for each vector $d\mathbf{p} \in \mathbb{R}^p$ and each multiplier $\lambda \in \mathcal{M}(x^*, \mathbf{p}^*)$ the *directional critical set*:

$$\begin{aligned} \mathcal{C}(x^*, \lambda; \mathbf{p}^*, d\mathbf{p}) \equiv \{ v \in \mathbb{R}^n : \\ \nabla_x g_i(x^*, \mathbf{p}^*)^T v + \nabla_{\mathbf{p}} g_i(x^*, \mathbf{p}^*)^T d\mathbf{p} = 0, \ \forall i \in \operatorname{supp}(\lambda) \\ \nabla_x g_i(x^*, \mathbf{p}^*)^T v + \nabla_{\mathbf{p}} g_i(x^*, \mathbf{p}^*)^T d\mathbf{p} \le 0, \ \forall i \in \beta \}, \end{aligned}$$

where β is the degenerate index set associated with λ; i.e.,

$$\beta = \{ i \in \mathcal{I}(x^*, \mathbf{p}^*) : \lambda_i = 0 \}.$$

Clearly when $d\mathbf{p} = 0$, the critical set $\mathcal{C}(x^*, \lambda; \mathbf{p}^*, 0)$ reduces to the critical cone $\mathcal{C}(x^*; K(\mathbf{p}^*), F(\cdot, \mathbf{p}^*))$. In general, $\mathcal{C}(x^*, \lambda; \mathbf{p}^*, d\mathbf{p})$ is a polyhedral set that is possibly empty for some nonzero $d\mathbf{p}$. The following proposition identifies a necessary and sufficient condition for $\mathcal{C}(x^*, \lambda; \mathbf{p}^*, d\mathbf{p})$ to be nonempty.

5.4.7 Proposition. For each $\bar{\lambda} \in \mathcal{M}(x^*, \mathbf{p}^*)$ and $d\mathbf{p} \in \mathbb{R}^n$, the critical set $\mathcal{C}(x^*, \bar{\lambda}; \mathbf{p}^*, d\mathbf{p})$ is nonempty if and only if $\bar{\lambda}$ solves the linear program:

$$\begin{array}{ll} \text{maximize} & \displaystyle\sum_{i=1}^{m} \lambda_i \nabla_{\mathbf{p}} g_i(x^*, \mathbf{p}^*)^T d\mathbf{p} \\ \text{subject to} & \lambda \in \mathcal{M}(x^*, \mathbf{p}^*). \end{array} \tag{5.4.15}$$

Proof. Writing out the inequalities defining the multiplier set $\mathcal{M}(x^*, \mathbf{p}^*)$:

$$F(x^*, \mathbf{p}^*) + \sum_{i \in \mathcal{I}(x^*, \mathbf{p}^*)} \lambda_i \nabla_x g_i(x^*, \mathbf{p}^*) = 0$$
$$\lambda_i \geq 0, \quad \forall i \in \mathcal{I}(x^*, \mathbf{p}^*)$$
$$\lambda_i = 0, \quad \forall i \notin \mathcal{I}(x^*, \mathbf{p}^*),$$

we can state the dual of the linear program (5.4.15) as:

$$\begin{array}{l} \text{minimize} \quad dx^T F(x^*, \mathbf{p}^*) \\ \text{subject to} \\ \nabla_{\mathbf{p}} g_i(x^*, \mathbf{p}^*)^T d\mathbf{p} + \nabla_x g_i(x^*, \mathbf{p}^*)^T dx \leq 0, \quad \forall i \in \mathcal{I}(x^*, \mathbf{p}^*). \end{array} \tag{5.4.16}$$

It follows that $\bar{\lambda}$ is a maximizer of (5.4.15) if and only if there exists a vector dx that is feasible to the above dual linear program and $(\bar{\lambda}, dx)$ satisfies the complementary slackness condition. Such an optimal dual solution dx is easily seen to be an element of the critical set $\mathcal{C}(x^*, \bar{\lambda}; \mathbf{p}^*, d\mathbf{p})$. □

Let $\mathcal{M}^c(x^*, \mathbf{p}^*; d\mathbf{p})$ and $\mathcal{D}^c(x^*, \mathbf{p}^*; d\mathbf{p})$ denote, respectively, the set of optimal solutions of (5.4.15) and (5.4.16). Since the MFCQ holds at x^*, it is easy to see that the dual linear program (5.4.16) is always feasible for all $d\mathbf{p} \in \mathbb{R}^p$. Hence both $\mathcal{M}^c(x^*, \mathbf{p}^*; d\mathbf{p})$ and $\mathcal{D}^c(x^*, \mathbf{p}^*; d\mathbf{p})$ are nonempty for all $d\mathbf{p}$ in $\mathbb{R}^p$. Moreover,

$$\mathcal{C}(x^*, \lambda, \mathbf{p}^*, d\mathbf{p}) = \begin{cases} \mathcal{D}^c(x^*, \mathbf{p}^*; d\mathbf{p}) & \text{if } \lambda \in \mathcal{M}^c(x^*, \mathbf{p}^*; d\mathbf{p}) \\ \emptyset & \text{if } \lambda \in \mathcal{M}(x^*, \mathbf{p}^*) \setminus \mathcal{M}^c(x^*, \mathbf{p}^*; d\mathbf{p}). \end{cases}$$

The latter identity follows easily from Proposition **5.4.7**. Consequently, for all $\lambda \in \mathcal{M}(x^*, \mathbf{p}^*)$, the directional critical set $\mathcal{C}(x^*, \lambda; \mathbf{p}^*, d\mathbf{p})$ is either empty or a constant polyhedral set dependent only on the triple $(x^*, \mathbf{p}^*, d\mathbf{p})$. For computational purposes, it is useful to point out that the set $\mathcal{D}^c(x^*, \mathbf{p}^*, d\mathbf{p})$ has the following linear inequality representation: namely dx belongs to $\mathcal{D}^c(x^*, \mathbf{p}^*, d\mathbf{p})$ if and only if

$$\nabla_{\mathbf{p}} g_i(x^*, \mathbf{p}^*) d\mathbf{p} + \nabla_x g_i(x^*, \mathbf{p}^*)^T dx \leq 0, \quad \forall i \in \mathcal{I}(x^*, \mathbf{p}^*)$$
$$dx^T F(x^*, \mathbf{p}^*) \leq \theta_{\text{opt}}(x^*, \mathbf{p}^*, d\mathbf{p}),$$

where $\theta_{\rm opt}(x^*, \mathbf{p}^*, d\mathbf{p})$ is the optimal objective value of the LP (5.4.16) or its dual (5.4.15). This representation requires the actual solution of either one of these two LPs, and is more convenient to manipulate in practical calculations than the representation in terms of the complementary slackness between the two LPs, which is a nonlinear relation.

Under the assumptions of Theorems **5.4.4**–**5.4.6**, there exists a neighborhood $\mathcal{N}$ of x^* such that, for every sequence $\{\mathbf{p}^k\}$ converging to $\mathbf{p}^*$, any sequence

$$\left\{ \frac{x^k - x^*}{\| \mathbf{p}^k - \mathbf{p}^* \|} \right\},$$

where $x^k \in \mathrm{SOL}(K(\mathbf{p}^k), F(\cdot, \mathbf{p}^k)) \cap \mathcal{N}$ for all k, must be bounded and thus have at least one accumulation point. The following result identifies such a point as a solution to an appropriate AVI. In presenting this result, we bypass some technical assumptions of the cited theorems (mainly the copositivity and the CRCQ) that guarantee the existence of an accumulation point of the above sequence of divided differences and focus on establishing the essential property of such a point. Such technical assumptions are not needed in the proof.

5.4.8 Proposition. Assume the setting in this subsection (including the MFCQ). Let $\{\mathbf{p}^k\}$ be a sequence such that

$$\lim_{k \to \infty} \mathbf{p}^k = \mathbf{p} \quad \text{and} \quad \lim_{k \to \infty} \frac{\mathbf{p}^k - \mathbf{p}^*}{\| \mathbf{p}^k - \mathbf{p}^* \|} = d\mathbf{p}.$$

Suppose for each k, x^k is a solution of the VI $(K(\mathbf{p}^k), F(\cdot, \mathbf{p}^k))$ such that

$$\lim_{k \to \infty} x^k = x^* \quad \text{and} \quad \lim_{k \to \infty} \frac{x^k - x^*}{\| \mathbf{p}^k - \mathbf{p}^* \|} = \bar{v}.$$

There exists $\bar{\lambda} \in \mathcal{M}^c(x^*, \mathbf{p}^*; d\mathbf{p})$ such that $\bar{v}$ is a solution of the

$$\text{AVI } (\mathcal{D}^c(x^*, \mathbf{p}^*; d\mathbf{p}), J_{\mathbf{p}}\mathbf{L}(x^*, \bar{\lambda}, \mathbf{p}^*)d\mathbf{p}, J_x\mathbf{L}(x^*, \bar{\lambda}, \mathbf{p}^*)). \tag{5.4.17}$$

Moreover, such a $\bar{\lambda}$ can be chosen to maximize the linear function

$$\lambda \mapsto \sum_{i=1}^m \lambda_i \begin{pmatrix} \bar{v} \\ d\mathbf{p} \end{pmatrix} \begin{bmatrix} \nabla^2_{xx} g_i(x^*, \mathbf{p}^*) & \nabla^2_{x\mathbf{p}} g_i(x^*, \mathbf{p}^*) \\ \nabla^2_{\mathbf{p}x} g_i(x^*, \mathbf{p}^*) & \nabla_{\mathbf{p}\mathbf{p}} g_i(x^*, \mathbf{p}^*) \end{bmatrix} \begin{pmatrix} \bar{v} \\ d\mathbf{p} \end{pmatrix}$$

on the directional critical set $\mathcal{M}^c(x^*, \mathbf{p}^*; d\mathbf{p})$.

The proof of this proposition is very similar to that of Lemma **4.4.3** except for several differences. One is that the present situation concerns a parametric VI instead of a parametric optimization problem (the projection problem); the other is that the constraint set is parameterized instead of being fixed. This last difference is the main reason for assuming the MFCQ instead of the SBCQ, which was assumed in the earlier lemma. In essence, the following proof remains valid under a parametric version of the SBCQ. The reason the SBCQ is not assumed here is that it is not known if this CQ is sufficient to ensure the existence of solutions to the perturbed VI $(K(\mathbf{p}^k), F(\cdot, \mathbf{p}^k))$; thus although we could establish the desired property of the limit vector $\bar{v}$ under the SBCQ, the existence of the sequence $\{x^k\}$ is in jeopardy. Therefore, unlike the treatment in Section **4.4**, we have chosen to assume a stronger CQ to avoid this uncertainty.

Proof of Proposition 5.4.8. For each k, let λ^k be a vector in $\mathcal{M}(x^k, \mathbf{p}^k)$. By working with appropriate subsequences, we may assume without loss of generality that the sequence $\{\lambda^k\}$ converges to a multiplier $\lambda^\infty \in \mathcal{M}(x^*, \mathbf{p}^*)$ and a subset $\mathcal{J}$ of $\mathcal{I}(x^*, \mathbf{p}^*)$ exists such that $\mathcal{I}(x^k, \mathbf{p}^k) = \mathcal{J} \supseteq \operatorname{supp}(\lambda^\infty)$ for all k. We have

$$F(x^k, \mathbf{p}^k) + \sum_{i \in \mathcal{J}} \lambda_i^k \nabla_x g_i(x^k, \mathbf{p}^k) = 0. \tag{5.4.18}$$

We can write

$$\begin{aligned} F(x^k, \mathbf{p}^k) &= F(x^*, \mathbf{p}^*) + J_x F(x^*, \mathbf{p}^*)(x^k - x^*) \\ &+ J_{\mathbf{p}} F(x^*, \mathbf{p}^*)(\mathbf{p} - \mathbf{p}^*) + o(\| x^k - x^* \|) + o(\| \mathbf{p} - \mathbf{p}^* \|) \end{aligned} \tag{5.4.19}$$

and for each $i \in \mathcal{I}(x^*, \mathbf{p}^*)$, since $g_i(x^*, \mathbf{p}^*) = 0$,

$$\begin{aligned} 0 \geq g_i(x^k, \mathbf{p}^k) &= \nabla_x g_i(x^*, \mathbf{p}^*)^T (x^k - x^*) + \nabla_{\mathbf{p}} g_i(x^*, \mathbf{p}^*)^T (\mathbf{p}^k - \mathbf{p}^*) \\ &+ \frac{1}{2} \begin{pmatrix} x^k - x^* \\ \mathbf{p}^k - \mathbf{p}^* \end{pmatrix} \begin{bmatrix} \nabla^2_{xx} g_i(x^*, \mathbf{p}^*) & \nabla^2_{x\mathbf{p}} g_i(x^*, \mathbf{p}^*) \\ \nabla^2_{\mathbf{p}x} g_i(x^*, \mathbf{p}^*) & \nabla_{\mathbf{p}\mathbf{p}} g_i(x^*, \mathbf{p}^*) \end{bmatrix} \begin{pmatrix} x^k - x^* \\ \mathbf{p}^k - \mathbf{p}^* \end{pmatrix} \\ &+ o(\| x^k - x^* \|^2) + o(\| \mathbf{p}^k - \mathbf{p}^* \|^2). \end{aligned}$$

Dividing the last inequality by $\|\mathbf{p}^k - \mathbf{p}^*\|$ and letting $k \to \infty$, we deduce

$$\nabla_x g_i(x^*, \mathbf{p}^*)^T \bar{v} + \nabla_{\mathbf{p}} g_i(x^*, \mathbf{p}^*)^T d\mathbf{p} \leq 0, \quad \forall\, i \in \mathcal{I}(x^*, \mathbf{p}^*).$$

Moreover, if $\lambda_i^\infty > 0$, then $\lambda_i^k > 0$ for all k sufficiently large, which implies $g_i(x^k, \mathbf{p}^k) = 0$. Hence,

$$\nabla_x g_i(x^*, \mathbf{p}^*)^T \bar{v} + \nabla_{\mathbf{p}} g_i(x^*, \mathbf{p}^*)^T d\mathbf{p} = 0, \quad \forall\, i \in \operatorname{supp}(\lambda^\infty).$$

Consequently, $\bar{v}$ belongs to the directional critical set $\mathcal{C}(x^*, \lambda^\infty, \mathbf{p}^*, d\mathbf{p})$. Moreover, since

$$\begin{aligned} \nabla_x g_i(x^k, \mathbf{p}^k) &= \nabla_x g_i(x^*, \mathbf{p}^*) + \nabla^2_{xx} g_i(x^*, \mathbf{p}^*)(x^k - x^*) + \\ \nabla^2_{x\mathbf{p}} g_i(x^*, \mathbf{p}^*)(\mathbf{p}^k - \mathbf{p}^*) &+ o(\| x^k - x^* \|) + o(\| \mathbf{p}^k - \mathbf{p}^* \|), \end{aligned}$$

substituting this expression and (5.4.19) into (5.4.18) and utilizing

$$F(x^*, \mathbf{p}) + \sum_{i \in \mathcal{J}} \lambda_i^\infty \nabla_x g_i(x^*, \mathbf{p}^*) = 0$$

we obtain

$$0 = J_x \mathbf{L}(x^*, \lambda^\infty, \mathbf{p}^*)(x^k - x^*) + + J_{\mathbf{p}} \mathbf{L}(x^*, \lambda^\infty, \mathbf{p}^*)(\mathbf{p}^k - \mathbf{p}^*) +$$
$$\sum_{i \in \mathcal{J}} (\lambda_i^k - \lambda_i^\infty) \nabla_x g_i(x^*, \mathbf{p}^*) + o(\| x^k - x^* \|) + o(\| \mathbf{p}^k - \mathbf{p}^* \|).$$

Dividing by $\|\mathbf{p}^k - \mathbf{p}^*\|$ and letting $k \to \infty$, we can readily establish that $\bar{v}$ solves the AVI (5.4.17).

By the same proof as that in Lemma **4.4.3**, we can show that λ^∞ can be chosen to maximize the linear function

$$\lambda \mapsto \sum_{i=1}^m \lambda_i \left\{ \nabla_{\mathbf{p}} g_i(x^*, \mathbf{p}^*)^T d\mathbf{p} + \right.$$
$$\left. \frac{1}{2} \begin{pmatrix} \bar{v} \\ d\mathbf{p} \end{pmatrix} \begin{bmatrix} \nabla^2_{xx} g_i(x^*, \mathbf{p}^*) & \nabla^2_{x\mathbf{p}} g_i(x^*, \mathbf{p}^*) \\ \nabla^2_{\mathbf{p}x} g_i(x^*, \mathbf{p}^*) & \nabla_{\mathbf{p}\mathbf{p}} g_i(x^*, \mathbf{p}^*) \end{bmatrix} \begin{pmatrix} \bar{v} \\ d\mathbf{p} \end{pmatrix} \right\}$$

on $\mathcal{M}(x^*, \mathbf{p}^*)$. Since $\mathcal{M}^c(x^*, \mathbf{p}^*; d\mathbf{p})$ is a subset of $\mathcal{M}(x^*, \mathbf{p}^*)$ and contains λ^∞, it follows that λ^∞ also maximizes the same linear function on $\mathcal{M}^c(x^*, \mathbf{p}^*; d\mathbf{p})$. Since

$$\sum_{i=1}^m \lambda_i \nabla_{\mathbf{p}} g_i(x^*, \mathbf{p}^*)^T d\mathbf{p}$$

is a constant for λ belonging to $\mathcal{M}^c(x^*, \mathbf{p}^*; d\mathbf{p})$, which is equal to the optimal solution of the dual linear program (5.4.16), it follows that such a λ^∞ maximizes

$$\lambda \mapsto \sum_{i=1}^m \lambda_i \begin{pmatrix} \bar{v} \\ d\mathbf{p} \end{pmatrix} \begin{bmatrix} \nabla^2_{xx} g_i(x^*, \mathbf{p}^*) & \nabla^2_{x\mathbf{p}} g_i(x^*, \mathbf{p}^*) \\ \nabla^2_{\mathbf{p}x} g_i(x^*, \mathbf{p}^*) & \nabla_{\mathbf{p}\mathbf{p}} g_i(x^*, \mathbf{p}^*) \end{bmatrix} \begin{pmatrix} \bar{v} \\ d\mathbf{p} \end{pmatrix}$$

on $\mathcal{M}^c(x^*, \mathbf{p}^*; d\mathbf{p})$ as claimed. □

Based on Proposition **5.4.8** and the results in the last section, we can provide several sufficient conditions for the following limit to hold for a given pair $(\bar{v}, d\mathbf{p}) \in \mathbb{R}^{n+p}$:

$$\lim_{\tau \downarrow 0} \sup \left\{ \frac{\| x(\tau) - x^* - \tau \bar{v} \|}{\tau} \right. \tag{5.4.20}$$
$$\left. x(\tau) \in \text{SOL}(K(\mathbf{p}^* + \tau d\mathbf{p}), F(\cdot, \mathbf{p}^* + \tau d\mathbf{p})) \cap \mathcal{N} \right\} = 0.$$

In the following result, we continue to assume the existence of the perturbed solutions $x(\tau)$ for $\tau > 0$ sufficiently small, without referring to the technical conditions that are needed for this purpose.

5.4.9 Corollary. Assume the setting of this subsection (including the MFCQ). Let $d\mathbf{p} \in \mathbb{R}^p$ be given. Suppose that there exists a constant $c > 0$ such that, for all $\tau > 0$ sufficiently small,

$$\| x(\tau) - x^* \| \leq c\tau,$$

for all $x(\tau) \in \text{SOL}(K(\mathbf{p}^* + \tau d\mathbf{p}), F(\cdot, \mathbf{p}^* + \tau d\mathbf{p})) \cap \mathcal{N}$. Under any one of the following conditions, the limit (5.4.20) holds:

(a) for all $\lambda \in \mathcal{M}^c(x^*, \mathbf{p}^*; d\mathbf{p})$, $\bar{v}$ is the unique solution of the

$$\text{AVI } (\mathcal{D}^c(x^*, \mathbf{p}^*; d\mathbf{p}), J_{\mathbf{p}}\mathbf{L}(x^*, \lambda, \mathbf{p}^*)d\mathbf{p}, J_x\mathbf{L}(x^*, \lambda, \mathbf{p}^*)); \quad (5.4.21)$$

(b) the SMFCQ holds at x^* and $\bar{v}$ is the unique solution of (5.4.21), where $\mathcal{M}(x^*, \mathbf{p}^*) = \{\lambda\}$;

(c) $K(\mathbf{p}) = K$ is a constant finitely representable set for all $\mathbf{p}$ near $\mathbf{p}^*$ and for all $\lambda \in \mathcal{M}(x^*, \mathbf{p}^*)$, $\bar{v}$ is unique solution of the homogeneous CP:

$$\mathcal{C} \ni v \perp J_{\mathbf{p}}F(x^*, \mathbf{p}^*)d\mathbf{p} + J_x\mathbf{L}(x^*, \lambda, \mathbf{p})v \in \mathcal{C}^*,$$

where $\mathcal{C}$ is the critical cone of the VI $(K, F(\cdot, \mathbf{p}^*))$ at x^*.

Proof. By Proposition **5.4.8**, condition (a) implies that for any sequence of positive scalars $\{\tau_k\}$ converging to zero, any sequence

$$\left\{ \frac{x(\tau_k) - x^*}{\tau_k} \right\}$$

has a unique accumulation point. Thus (5.4.20) holds readily. Condition (b) clearly implies (a) because in this case $\mathcal{M}^c(x^*, \mathbf{p}^*; d\mathbf{p})$ is the singleton $\{\lambda\}$. Condition (c) also implies (a) because under (c), we have

$$\mathcal{M}^c(x^*, \mathbf{p}^*; d\mathbf{p}) = \mathcal{M}(x^*, \mathbf{p}^*), \quad \mathcal{D}^c(x^*, \mathbf{p}^*; d\mathbf{p}) = \mathcal{C},$$

and

$$J_{\mathbf{p}}\mathbf{L}(x^*, \lambda, \mathbf{p}^*) = J_{\mathbf{p}}F(x^*, \mathbf{p}^*), \quad \forall \lambda \in \mathcal{M}(x^*, \mathbf{p}^*).$$

Consequently, the limit (5.4.20) holds under any of the three conditions (a), (b), or (c). □

We next consider the full setting of Theorem **5.4.6**. That is we assume both the MFCQ and CRCQ at x^* and that x^* is a strongly stable solution of the VI $(K(\mathbf{p}^*), F(\cdot, \mathbf{p}^*))$. Under these assumptions, there exist a neighborhood $\mathcal{N}$ of x^*, a neighborhood $\mathcal{W}$ of $\mathbf{p}^*$ and a PC^1 function $x_{\mathcal{N}} : \mathcal{W} \to \mathcal{N}$ such that $x_{\mathcal{N}}(\mathbf{p}^*) = x^*$ and for every $\mathbf{p} \in \mathcal{W}$, $x_{\mathcal{N}}(\mathbf{p})$ is the only solution of the VI $(K(\mathbf{p}), F(\cdot, \mathbf{p}))$ that lies in $\mathcal{N}$. Being a PC^1 function, $x_{\mathcal{N}}$ is directionally differentiable at $\mathbf{p}^*$; for every vector $d\mathbf{p} \in \mathbb{R}^p$, Proposition **5.4.8** gives the directional derivative $x'_{\mathcal{N}}(\mathbf{p}^*; d\mathbf{p})$ as a solution of an AVI. Furthermore, as a result of the CRCQ, we can use any multiplier $\lambda \in \mathcal{M}^c(x^*, \mathbf{p}^*, d\mathbf{p})$ to describe this directional derivative. The result below formalizes this observation.

5.4.10 Theorem. Assume the full setting of Theorem **5.4.6**. For each $d\mathbf{p} \in \mathbb{R}^p$ and each $\lambda \in \mathcal{M}^c(x^*, \mathbf{p}^*; d\mathbf{p})$, $x'_{\mathcal{N}}(\mathbf{p}^*, d\mathbf{p})$ is a solution of (5.4.21).

Proof. The proof is similar to the proof of Theorem **4.5.3**. It is based on redefining the multiplier set $\mathcal{M}(x^*, \mathbf{p}^*)$ in order to ensure the lower semicontinuity of the modified multipliers. The technique is the same as before and not repeated here. □

5.4.2 The strong coherent orientation condition

It is natural to ask when the AVI (5.4.21) has a unique solution. This question is important for computational reasons because we can obtain the directional derivative $x'_{\mathcal{N}}(\mathbf{p}^*, d\mathbf{p})$ by safely solving such an AVI, knowing for sure that its only solution will provide the desired derivative. For a given pair $(d\mathbf{p}, \lambda)$ with $\lambda \in \mathcal{M}^c(x^*, \mathbf{p}^*; d\mathbf{p})$ Corollary **4.3.2** implies that if the normal map of the pair $(\mathcal{D}^c, J_x\mathbf{L}(x^*, \lambda, \mathbf{p}^*))$, i.e., the map

$$z \mapsto J_x\mathbf{L}(x^*, \lambda, \mathbf{p}^*)\Pi_{\mathcal{D}^c}(z) + z - \Pi_{\mathcal{D}^c}(z),$$

where $\mathcal{D}^c$ is a shorthand for $\mathcal{D}^c(x^*, \mathbf{p}^*; d\mathbf{p})$, is coherently oriented, then the AVI $(\mathcal{D}^c, q, J_x\mathbf{L}_x(x^*, \lambda, \mathbf{p}^*))$ has a unique solution for all vectors q. In turn, by Proposition **4.2.7**, this normal map is coherently oriented if all matrices of the form:

$$\begin{bmatrix} J_x\mathbf{L}(x^*, \lambda, \mathbf{p}^*) & B^T \\ -B & 0 \end{bmatrix}, \tag{5.4.22}$$

where B is a member of the normal family of basis matrices of the polyhedron $\mathcal{D}^c$, have the same nonzero determinantal sign. In order to reveal the explicit form of such a matrix B, we represent $\mathcal{D}^c$ using the multiplier λ:

$$\begin{aligned} \mathcal{D}^c = \{ v \in \mathbb{R}^n : & \\ & \nabla_x g_i(x^*, \mathbf{p}^*)^T v + \nabla_{\mathbf{p}} g_i(x^*, \mathbf{p}^*)^T d\mathbf{p} = 0, \ \forall i \in \operatorname{supp}(\lambda) \\ & \nabla_x g_i(x^*, \mathbf{p}^*)^T v + \nabla_{\mathbf{p}} g_i(x^*, \mathbf{p}^*)^T d\mathbf{p} \leq 0, \ \forall i \in \beta \}, \end{aligned}$$

where β is the degenerate index set of the multiplier λ; i.e.,

$$\beta = \{ i \in \mathcal{I}(x^*, \mathbf{p}^*) : \lambda_i = 0 \}.$$

Let $\mathcal{J}(\lambda, d\mathbf{p})$ denote the family of index sets β' where β' is a subset of β for which there exists a vector $v \in \mathcal{D}^c$ such that

$$\nabla_x g_i(x^*, \mathbf{p}^*)^T v + \nabla_{\mathbf{p}} g_i(x^*, \mathbf{p}^*)^T d\mathbf{p} = 0, \ \forall i \in \beta'$$
$$\nabla_x g_i(x^*, \mathbf{p}^*)^T v + \nabla_{\mathbf{p}} g_i(x^*, \mathbf{p}^*)^T d\mathbf{p} < 0, \ \forall i \in \beta \setminus \beta'.$$

For each $\beta' \in \mathcal{J}(\lambda, d\mathbf{p})$, a basis matrix B of $\mathcal{D}^c$ consists of a maximal subset of linearly independent rows from the family of transposed gradients:

$$\{ \nabla_x g_i(x^*, \mathbf{p}^*)^T : i \in \operatorname{supp}(\lambda) \cup \beta' \}.$$

In general, the family $\mathcal{J}(\lambda, d\mathbf{p})$ varies with $d\mathbf{p}$. We have noted that when $d\mathbf{p} = 0$, the directional critical set $\mathcal{D}^c(x^*, \mathbf{p}^*; 0)$ reduces to the critical cone $\mathcal{C}$ of the VI $(K(\mathbf{p}^*), F(\cdot, \mathbf{p}^*))$ at x^*; moreover, it is easy to see that in this case the normal family of basis matrices B identified above (as defined by the index sets in the family $\mathcal{J}(\lambda, 0)$) is precisely the normal family $\mathcal{B}^{\lambda}_{\text{bas}}(\mathcal{C})$ of basis matrices of the critical cone $\mathcal{C}(x^*; K(\mathbf{p}^*), F(\cdot, \mathbf{p}^*))$ defined just before Theorem **5.3.24**.

We introduce a constraint qualification, known as the *strong coherent orientation condition* (SCOC), under which all matrices of the form (5.4.22) will have the same nonzero determinantal sign for all vectors $d\mathbf{p} \in \mathbb{R}^p$, provided that λ is an extreme multiplier; that is, if the gradients:

$$\{ \nabla_x g_i(x^*, \mathbf{p}^*) : i \in \operatorname{supp}(\lambda) \}$$

are linearly independent. See Theorem **5.4.12**. This CQ is most useful for dealing with the fully parametric VI $(K(\mathbf{p}), F(\cdot, \mathbf{p}))$, where both the defining set and function are parameter dependent. For the "semi-parametric" VI $(K, F(\cdot, \mathbf{p}))$, where the set K is parameter free, there is no need to use the SCOC; for such a VI, it suffices to use a normal family of basis matrices of the critical cone $\mathcal{C}(x^*; K, F(\cdot, \mathbf{p}^*))$ as described above.

Let $\mathcal{B}(x^*, \mathbf{p}^*)$ be the family of index sets $\mathcal{J} \subseteq \mathcal{I}(x^*, \mathbf{p}^*)$ for which there exists a multiplier $\lambda \in \mathcal{M}(x^*, \mathbf{p}^*)$ such that $\operatorname{supp}(\lambda) \subseteq \mathcal{J}$ and the gradients

$$\{ \nabla_x g_i(x^*, \mathbf{p}^*) : i \in \mathcal{J} \} \tag{5.4.23}$$

are linearly independent. We have seen families like this in the proof of Theorem **4.5.2**. We call $\mathcal{B}(x^*, \mathbf{p}^*)$ the *SCOC family of index sets*. The SCOC family of basis matrices is defined in an obvious manner. Specifically, for each $\mathcal{J}$ in $\mathcal{B}(x^*, \mathbf{p}^*)$, let B be the matrix whose rows are the transpose of the gradients (5.4.23). The family $\mathcal{B}(x^*, \mathbf{p}^*)$ is nonempty and finite: nonempty because any extreme multiplier in $\mathcal{M}^e(x^*, \mathbf{p}^*)$ induces an index set belonging to $\mathcal{B}(x^*, \mathbf{p}^*)$; finite because there are only many finitely constraints. Label the index sets in $\mathcal{B}(x^*, \mathbf{p}^*)$; specifically, let

$$\mathcal{B}(x^*, \mathbf{p}^*) = \{ \mathcal{J}^1, \cdots, \mathcal{J}^s \}.$$

For each $j = 1, \ldots, s$, the multiplier $\lambda^j \in \mathcal{B}(x^*, \mathbf{p}^*)$ satisfying the defining properties of $\mathcal{J}^j$ must necessarily be unique and belong to $\mathcal{M}^e(x^*, \mathbf{p}^*)$. Conversely, given a multiplier $\lambda \in \mathcal{M}^e(x^*, \mathbf{p}^*)$, there may be multiple index sets $\mathcal{J} \in \mathcal{B}(x^*, \mathbf{p}^*)$ that contain $\mathrm{supp}(\lambda)$. Hence $|\mathcal{M}^e(x^*, \mathbf{p}^*)| \leq |\mathcal{B}(x^*, \mathbf{p}^*)|$. Moreover, if the SMFCQ holds at the pair (x^*, λ^*) and λ^* is a nondegenerate multiplier (or equivalently, if x^* is a strongly nondegenerate solution of the VI $(K(\mathbf{p}^*), F(\cdot, \mathbf{p}^*)))$, then $|\mathcal{B}(x^*, \mathbf{p}^*)| = |\mathcal{M}(x^*, \mathbf{p}^*)| = 1$. For each $j = 1, \ldots, s$, define the matrix

$$\Lambda^j \equiv \begin{bmatrix} J_x \mathbf{L}(x^*, \lambda^j, \mathbf{p}^*) & \nabla_x g_{\mathcal{J}^j}(x^*, \mathbf{p}^*)^T \\ -\nabla_x g_{\mathcal{J}^j}(x^*, \mathbf{p}^*) & 0 \end{bmatrix}.$$

5.4.11 Definition. The vector $x^* \in K(\mathbf{p}^*)$ satisfies the *strong coherent orientation condition* (SCOC) if all s matrices

$$\{ \Lambda^j : j = 1, \ldots, s \}$$

have the same nonzero determinantal sign. □

We illustrate the SCOC with $K(\mathbf{p}^*) = \mathbb{R}^n_+$, i.e., for the parametric NCP:

$$0 \leq x \perp F(x, \mathbf{p}) \geq 0;$$

A subset $\mathcal{J}$ of $\{1, \ldots, n\}$ is a SCOC index set if $\mathcal{J}$ satisfies

$$\{ i : F_i(x^*, \mathbf{p}^*) > 0 = x_i^* \} \subseteq \mathcal{J} \subseteq \{ i : x_i^* = 0 \}.$$

For any such index set $\mathcal{J}^j$, the corresponding matrix Λ^j is equal to

$$\Lambda^j = \begin{bmatrix} J_x F(x^*, \mathbf{p}^*) & (\mathbf{I}_{\mathcal{J}^j \cdot})^T \\ -\mathbf{I}_{\mathcal{J}^j \cdot} & 0 \end{bmatrix}.$$

The determinant of this matrix is equal to that of the principal submatrix

$$(J_x F(x^*, \mathbf{p}^*))_{\mathcal{K}^j \mathcal{K}^j},$$

where $\mathcal{K}^j$ is the complement of $\mathcal{J}^j$ in $\{1, \ldots, n\}$. We have

$$\mathcal{K}^j = \mathrm{supp}(x^*) \cup \beta^j,$$

where β^j is some subset of the degenerate set

$$\beta \equiv \{ i : x_i^* = F_i(x^*, \mathbf{p}^*) = 0 \}.$$

As β^j ranges over the subsets of β, $\mathcal{J}^j$ ranges over all the SCOC index sets. In particular, with β^j equal to the empty set, we must have

$$\det(J_x F(x^*, \mathbf{p}^*))_{\gamma\gamma} \neq 0,$$

where $\gamma \equiv \operatorname{supp}(x^*)$. The above matrix is precisely the basic matrix of the solution x^*. Moreover, for each subset β^j of β, we may write in partitioned form:

$$(J_x F(x^*, \mathbf{p}^*))_{\mathcal{K}^j \mathcal{K}^j} = \begin{bmatrix} (J_x F(x^*, \mathbf{p}^*))_{\gamma\gamma} & (J_x F(x^*, \mathbf{p}^*))_{\gamma\beta^j} \\ (J_x F(x^*, \mathbf{p}^*))_{\beta^j\gamma} & (J_x F(x^*, \mathbf{p}^*))_{\beta^j\beta^j} \end{bmatrix}.$$

By the Schur determinantal formula, we have

$$\det(J_x F(x^*, \mathbf{p}^*))_{\mathcal{K}^j \mathcal{K}^j} = \det(J_x F(x^*, \mathbf{p}^*))_{\gamma\gamma} \det S^j,$$

where S^j is the Schur complement of $(J_x F(x^*, \mathbf{p}^*))_{\gamma\gamma}$ in $(J_x F(x^*, \mathbf{p}^*))_{\mathcal{K}^j \mathcal{K}^j}$:

$$J_x F(x^*, \mathbf{p}^*)_{\beta^j\beta^j} - J_x F(x^*, \mathbf{p}^*)_{\beta^j\gamma} [J_x F(x^*, \mathbf{p}^*)_{\gamma\gamma}]^{-1} J_x F(x^*, \mathbf{p}^*)_{\gamma\beta^j}.$$

Since $\det(J_x F(x^*, \mathbf{p}^*))_{\mathcal{K}^j \mathcal{K}^j}$ must have the same sign as the basic matrix of x^*, it follows that $\det S^j$ must be positive for all subsets β^j of β. We have therefore established that the SCOC is equivalent to two conditions:

(a) the basic matrix $(J_x F(x^*, \mathbf{p}^*))_{\gamma\gamma}$ is nonsingular,

(b) the following Schur complement is a P matrix:

$$J_x F(x^*, \mathbf{p}^*)_{\beta\beta} - J_x F(x^*, \mathbf{p}^*)_{\beta\gamma} [J_x F(x^*, \mathbf{p}^*)_{\gamma\gamma}]^{-1} J_x F(x^*, \mathbf{p}^*)_{\gamma\beta}.$$

In turn, these two conditions are precisely the defining conditions for the strong b-regularity of x^*, which is equivalent to the strong stability of x^* as a solution of the NCP $(F(\cdot, \mathbf{p}^*))$; see Corollary **5.3.20**. Consequently, for the parametric NCP, the SCOC reduces to a familiar condition. Incidentally, the above derivation is very similar to the proof of the equivalence of the first two statements in Corollary **3.3.9**. Subsequently, we generalize this discussion to the KKT system of a VI; see Proposition **5.4.17**.

The next result, Theorem **5.4.12**, gives two important consequences of the SCOC. In essence, the proof of this result consists of showing that the SCOC family of basis matrices contains a normal family of basis matrices of the directional critical set $\mathcal{D}^c(x^*, \mathbf{p}^*; d\mathbf{p})$ for every $d\mathbf{p} \in \mathbb{R}^p$ including $d\mathbf{p} = 0$; thus the SCOC family of basis matrices also contains a normal family of basis matrices of the critical cone $\mathcal{C}(x^*; K(\mathbf{p}^*), F(\cdot, \mathbf{p}^*))$.

5.4.12 Theorem. Let x^* be a solution of the VI $(K(\mathbf{p}^*), F(\cdot, \mathbf{p}^*))$ satisfying the MFCQ, CRCQ, and SCOC. The following two statements are valid.

(a) x^* is a strongly stable of the VI $(K(\mathbf{p}^*), F(\cdot, \mathbf{p}^*))$; thus the parametric solution function $\mathbf{x}_{\mathcal{N}}$ is PC^1 near $\mathbf{p}^*$.

(b) For all $d\mathbf{p} \in \mathbb{R}^p$ and $\lambda \in \mathcal{M}^e(x^*, \mathbf{p}^*) \cap \mathcal{M}^c(x^*, \mathbf{p}^*, d\mathbf{p})$, the AVI (5.4.21) has a unique solution, which is equal to $\mathbf{x}'_{\mathcal{N}}(\mathbf{p}^*; d\mathbf{p})$.

Proof. We prove (b) first. Let $(\lambda, d\mathbf{p})$ be as given. For each index set $\beta' \in \mathcal{J}(\lambda, d\mathbf{p})$, let $\mathcal{J}^j \equiv \text{supp}(\lambda) \cup \beta''$, where β'' is the subset of β' such that the gradients

$$\{ \nabla_x g_i(x^*, \mathbf{p}^*) \,:\, i \in \mathcal{J}^j \}$$

form a basis of

$$\{ \nabla_x g_i(x^*, \mathbf{p}^*) \,:\, i \in \text{supp}(\lambda) \cup \beta' \}.$$

Consequently, in view of the above discussion, it is easy to see that SCOC implies (b). In a similar fashion, it can be proved that SCOC implies that the normal map:

$$z \mapsto J_x \mathbf{L}(x^*, \lambda, \mathbf{p}^*) \Pi_{\mathcal{C}}(z) + z - \Pi_{\mathcal{C}}(z)$$

is coherently oriented for all $\lambda \in \mathcal{M}^e(x^*, \mathbf{p}^*)$. Thus, we have verified condition (h) in Theorem **5.3.24** for the VI $(K(\mathbf{p}^*), F(\cdot, \mathbf{p}^*))$. Hence x^* is a strongly stable solution of this VI. □

The next issue we address is the question of F-differentiability of the solution function $x_{\mathcal{N}}(\mathbf{p})$ at $\mathbf{p}^*$. Equivalently, the question is: when is the directional derivative $x'(\mathbf{p}^*, d\mathbf{p})$ linear in $d\mathbf{p}$? To deal with this question, we introduce some further notation. For $d\mathbf{p} \in \mathbb{R}^p$, define

$$\begin{aligned} \mathcal{E}(x^*, \mathbf{p}^*; d\mathbf{p}) \equiv \{ & v \in \mathbb{R}^n : \\ & \nabla_x g_i(x^*, \mathbf{p}^*)^T v + \nabla_{\mathbf{p}} g_i(x^*, \mathbf{p}^*)^T d\mathbf{p} = 0, \ \forall\, i \in \mathcal{I}(x^*, \mathbf{p}^*) \}, \end{aligned}$$

which is a (possibly empty) affine subspace contained in the directional critical set $\mathcal{D}^c(x^*, \mathbf{p}^*; d\mathbf{p})$. By using the maximal elements in the SCOC family $\mathcal{B}(x^*, \mathbf{p}^*)$ of index sets, we may obtain a reduced representation of $\mathcal{E}(x^*, \mathbf{p}^*; d\mathbf{p})$. Specifically, an index set $\mathcal{J} \in \mathcal{B}(x^*, \mathbf{p}^*)$ is *maximal* if there exists no index set in $\mathcal{B}(x^*, \mathbf{p}^*)$ that properly contains $\mathcal{J}$. Clearly, for each $\lambda \in \mathcal{M}^e(x^*, \mathbf{p}^*)$, there exists at least one maximal element in $\mathcal{B}(x^*, \mathbf{p}^*)$

containing supp(λ). In general, for each maximal element $\mathcal{J} \in \mathcal{B}(x^*, \mathbf{p}^*)$, and every $d\mathbf{p} \in \mathbb{R}^p$, we have

$$\begin{aligned}\mathcal{E}(x^*, \mathbf{p}^*; d\mathbf{p}) \equiv \{ v \in \mathbb{R}^n : \\ \nabla_x g_i(x^*, \mathbf{p}^*)^T v + \nabla_{\mathbf{p}} g_i(x^*, \mathbf{p}^*)^T d\mathbf{p} = 0, \ \forall i \in \mathcal{J} \},\end{aligned} \tag{5.4.24}$$

Based on this reduction, we may obtain several necessary and sufficient conditions for the implicit solution function $x_{\mathcal{N}}(\mathbf{p})$ to be F-differentiable at $\mathbf{p}^*$.

5.4.13 Theorem. Let x^* be a solution of the VI $(K(\mathbf{p}^*), F(\cdot, \mathbf{p}^*))$ satisfying the MFCQ, CRCQ, and SCOC. Let $x_{\mathcal{N}}(\mathbf{p})$ be the implicit PC^1 solution function of the parametric VI $(K(\mathbf{p}), F(\cdot, \mathbf{p}))$ guaranteed by Theorem **5.4.6**. The following statements are equivalent.

(a) $x_{\mathcal{N}}(\mathbf{p})$ is F-differentiable at $\mathbf{p}^*$.

(b) For all $d\mathbf{p} \in \mathbb{R}^n$, $x'_{\mathcal{N}}(\mathbf{p}^*; d\mathbf{p})$ belongs to $\mathcal{E}(x^*, \mathbf{p}^*; d\mathbf{p})$.

(c) For all $d\mathbf{p} \in \mathbb{R}^n$ and $\lambda \in \mathcal{M}^e(x^*, \mathbf{p}^*)$, $x'_{\mathcal{N}}(\mathbf{p}^*; d\mathbf{p})$ is the unique solution of the

$$\text{AVI } (\mathcal{D}^c(x^*, \mathbf{p}^*; d\mathbf{p}), J_{\mathbf{p}}\mathbf{L}(x^*, \lambda, \mathbf{p}^*)d\mathbf{p}, J_x\mathbf{L}(x^*, \lambda, \mathbf{p}^*)). \tag{5.4.25}$$

(d) For all $d\mathbf{p} \in \mathbb{R}^n$, $\lambda \in \mathcal{M}^e(x^*, \mathbf{p}^*)$, and maximal element $\mathcal{J} \in \mathcal{B}(x^*, \mathbf{p}^*)$ containing supp(λ), $x'_{\mathcal{N}}(\mathbf{p}^*; d\mathbf{p})$, along with some $d\lambda_{\mathcal{J}}(\mathbf{p}^*; d\mathbf{p})$, is the unique solution $(dx, d\lambda_{\mathcal{J}})$ of the system of linear equations:

$$\begin{aligned} J_{\mathbf{p}}\mathbf{L}(x^*, \lambda, \mathbf{p}^*)d\mathbf{p} + J_x\mathbf{L}(x^*, \lambda, \mathbf{p}^*)dx + \sum_{i \in \mathcal{J}} d\lambda_i \nabla_x g_i(x^*, \mathbf{p}^*) &= 0 \\ \nabla_{\mathbf{p}} g_i(x^*, \mathbf{p}^*)^T d\mathbf{p} + \nabla_x g_i(x^*, \mathbf{p}^*)^T dx = 0, \quad \forall i \in \mathcal{J}. & \end{aligned} \tag{5.4.26}$$

(e) There exist $\lambda \in \mathcal{M}^e(x^*, \mathbf{p}^*)$ and maximal element $\mathcal{J} \in \mathcal{B}(x^*, \mathbf{p}^*)$ containing supp(λ) such that, for all $d\mathbf{p} \in \mathbb{R}^p$, $x'_{\mathcal{N}}(\mathbf{p}^*; d\mathbf{p})$, along with some $d\lambda_{\mathcal{J}}(\mathbf{p}^*; d\mathbf{p})$, is the unique solution $(dx, d\lambda_{\mathcal{J}})$ of the system of linear equations (5.4.26).

If any one of the above conditions holds, then

$$\mathcal{M}^c(x^*, \mathbf{p}^*; d\mathbf{p}) = \mathcal{M}(x^*, \mathbf{p}^*), \quad \forall\, d\mathbf{p} \in \mathbb{R}^p. \tag{5.4.27}$$

Proof. As a PC^1 function, $x_{\mathcal{N}}(\mathbf{p})$ is B-differentiable at $\mathbf{p}^*$. Thus $x_{\mathcal{N}}(\mathbf{p})$ is F-differentiable at $\mathbf{p}^*$ if and only if the directional derivative $x'_{\mathcal{N}}(\mathbf{p}^*, d\mathbf{p})$ is linear in the second argument.

(a) ⇒ (b). If (a) holds, then

$$x'_{\mathcal{N}}(\mathbf{p}^*, -d\mathbf{p}) = -x'_{\mathcal{N}}(\mathbf{p}^*, d\mathbf{p}), \quad \forall\, d\mathbf{p} \in \mathbb{R}^p.$$

Since $x'(\mathbf{p}^*, d\mathbf{p}) \in \mathcal{D}^c(x^*, \mathbf{p}^*; d\mathbf{p})$, it follows easily from the above identity that $x'_{\mathcal{N}}(\mathbf{p}^*, d\mathbf{p}) \in \mathcal{E}(x^*, \mathbf{p}^*, d\mathbf{p})$. Hence (b) holds.

(b) ⇒ (c) ⇔ (d). Since $\mathcal{E}(x^*, \mathbf{p}^*; d\mathbf{p})$ is a subset of $\mathcal{D}^c(x^*, \mathbf{p}^*; d\mathbf{p})$, it follows that any solution of the AVI (5.4.21) that lies in $\mathcal{E}(x^*, \mathbf{p}^*; d\mathbf{p})$ must be a solution of the AVI (5.4.25). It remains to show that the latter AVI has a unique solution if $\lambda \in \mathcal{M}^e(x^*, \mathbf{p}^*)$ and the AVI is equivalent to the system of linear equations (5.4.26). This follows by combining several observations: one, for any such λ, $\text{supp}(\lambda)$ is contained in a maximal element $\mathcal{J}$ of $\mathcal{B}(x^*, \mathbf{p}^*)$; the AVI (5.4.25) is equivalent to its KKT system, which by (5.4.24) is precisely the system of linear equations (5.4.26) for this index set $\mathcal{J}$; and three, for any such λ and maximal element $\mathcal{J}$ of $\mathcal{B}(x^*, \mathbf{p}^*)$ that contains $\text{supp}(\lambda)$, the matrix

$$\begin{bmatrix} J_x\mathbf{L}(x^*, \lambda, \mathbf{p}^*) & \nabla_x g_{\mathcal{J}}(x^*, \mathbf{p}^*)^T \\ -\nabla_x g_{\mathcal{J}}(x^*, \mathbf{p}^*) & 0 \end{bmatrix}$$

is equal to Λ^j, for some $j \in \{1, \ldots, s\}$, thus nonsingular by SCOC.

(d) ⇒ (e). This is obvious.

(e) ⇒ (a). Since (5.4.26) is a nonsingular system of linear equations in $(dx, d\lambda_{\mathcal{J}})$ for λ and $\mathcal{J}$ as given, it follows that $x'_{\mathcal{N}}(\mathbf{p}^*; d\mathbf{p})$ is linear in $d\mathbf{p}$.

Finally, if any one of the conditions (a)–(e) holds, then $D^c(x^*, \mathbf{p}^*; d\mathbf{p})$ is nonempty for all $d\mathbf{p} \in \mathbb{R}^p$, or equivalently, the directional critical set $\mathcal{C}(x^*, \lambda; \mathbf{p}^*, d\mathbf{p})$ is nonempty for all λ in $\mathcal{M}(x^*, \mathbf{p}^*)$. This implies (5.4.27) because $\mathcal{M}^c(x^*, \mathbf{p}^*; d\mathbf{p})$ is precisely the set of λ in $\mathcal{M}(x^*, \mathbf{p}^*)$ for which the directional critical set $\mathcal{C}(x^*, \lambda; \mathbf{p}^*, d\mathbf{p})$ is nonempty. □

An important special case of Theorem **5.4.13** arises when $K(\mathbf{p})$ is a parameter-free set, which we denote K. In this case, the directional critical set $\mathcal{D}^c(x^*, \mathbf{p}^*; d\mathbf{p})$ coincides with the critical cone $\mathcal{C}(x^*; K, F(\cdot, \mathbf{p}^*))$, which we denote $\mathcal{C}(x^*, \mathbf{p}^*)$, for all vectors $d\mathbf{p}$. The lineality space of $\mathcal{C}(x^*, \mathbf{p}^*)$ is equal to

$$\mathcal{E}(x^*, \mathbf{p}^*) \equiv \{ v \in \mathbb{R}^n : \nabla g_i(x^*)^T v = 0, \ \forall\, i \in \mathcal{I}(x^*) \}.$$

Thus $\mathcal{C}(x^*, \mathbf{p}^*)$ is a linear subspace if and only if $\mathcal{C}(x^*, \mathbf{p}^*)$ coincides with $\mathcal{E}(x^*, \mathbf{p}^*)$. Based on this observation and Proposition **3.4.2**, we can establish the following corollary of Theorem **5.4.13**.

5.4.14 Corollary. Let $F : \mathbb{R}^{n+p} \to \mathbb{R}^n$ be continuous differentiable and

$$K \equiv \{ x \in \mathbb{R}^n : g(x) \leq 0 \}$$

where each $g_i : \mathbb{R}^n \to \mathbb{R}$ is convex and twice continuously differentiable. Suppose that $x^* \in \text{SOL}(K, F(\cdot, \mathbf{p}^*))$ satisfies the MFCQ, CRCQ, and SCOC. Let $x_{\mathcal{N}}(\mathbf{p})$ be the PC^1 function guaranteed by Theorem **5.4.6**. If $J_x F(x^*, \mathbf{p}^*)(\mathcal{C}(x^*, \mathbf{p}^*))$ is contained in the column space of $J_{\mathbf{p}} F(x^*, \mathbf{p}^*)$, then the following three statements are equivalent.

(a) $x_{\mathcal{N}}$ is F-differentiable at $\mathbf{p}^*$.

(b) $\mathcal{C}(x^*, \mathbf{p}^*)$ is a linear subspace.

(c) x^* is a nondegenerate solution of the VI $(K, F(\cdot, \mathbf{p}^*))$.

Proof. (a) $\Rightarrow$ (b). This follows easily from Theorem **5.4.13** and the above remarks.

(b) $\Rightarrow$ (a). Conversely, suppose that $x_{\mathcal{N}}$ is F-differentiable at $\mathbf{p}^*$. Let $v \in \mathcal{C}(x^*, \mathbf{p}^*)$ be arbitrary. By assumption, there exists a vector $d\mathbf{p}$ such that

$$J_x F(x^*, \mathbf{p}^*) v + J_{\mathbf{p}} F(x^*, \mathbf{p}^*) d\mathbf{p} = 0,$$

which implies that $x'_{\mathcal{N}}(\mathbf{p}^*; d\mathbf{p}) = v$. Hence by (b) of Theorem **5.4.13**, v belongs to $\mathcal{E}(x^*, \mathbf{p}^*)$. Consequently, $\mathcal{C}(x^*, \mathbf{p}^*) = \mathcal{E}(x^*, \mathbf{p}^*)$, and thus $\mathcal{C}(x^*, \mathbf{p}^*)$ is a linear subspace.

(b) $\Leftrightarrow$ (c). This follows from Proposition **3.4.2**. □

Informally, the assumption that $J_x F(x^*, \mathbf{p}^*)(\mathcal{C}(x^*, \mathbf{p}^*))$ is contained in the column space of $J_{\mathbf{p}} F(x^*, \mathbf{p}^*)$ can be thought of as requiring that the parameter dependence has to be sufficiently broad in the function F. An easy example where this assumption holds easily is when $p = n$ and

$$F(x, \mathbf{p}) \equiv \tilde{F}(x) + \mathbf{p}$$

for some function $\tilde{F}$ from $\mathbb{R}^n$ into itself. A specialization of the theory developed so far to this case can be easily obtained. The details are omitted.

5.4.3 PC^1 multipliers and more on SCOC

When the LICQ holds at x^*, the multiplier set $\mathcal{M}(x^*, \mathbf{p}^*)$ is a singleton. Moreover, for all $\mathbf{p}$ sufficiently near $\mathbf{p}^*$, the LICQ remains valid at $x_{\mathcal{N}}(\mathbf{p})$. Thus in addition to the implicit solution function $x_{\mathcal{N}}(\mathbf{p})$ to the VI $(K(\mathbf{p}), F(\cdot, \mathbf{p}))$, there exists a single-valued function of multipliers corresponding to the solution function. The following result shows that such a multiplier function is also PC^1 and its directional derivative is obtained as a by-product when the AVI (5.4.21) is solved.

5.4.15 Theorem. Suppose that the LICQ and the SCOC hold at the solution $x^* \in \mathrm{SOL}(K(\mathbf{p}^*), F(\cdot, \mathbf{p}^*))$. Let λ^* be the unique KKT multiplier corresponding to x^*. The following two statements are valid.

(a) There exist a neighborhood $\mathcal{N}$ of x^* and a neighborhood $\mathcal{W}$ of $\mathbf{p}^*$ and PC^1 functions $x_{\mathcal{N}} : \mathcal{W} \to \mathcal{N}$ and $\lambda : \mathcal{W} \to \mathbb{R}^m$ such that, for each $\mathbf{p} \in \mathcal{W}$, the pair $(x_{\mathcal{N}}(\mathbf{p}), \lambda(\mathbf{p}))$ is the unique solution in $\mathcal{N} \times \mathbb{R}^m$ of the KKT conditions (5.4.7).

(b) For each vector $d\mathbf{p} \in \mathbb{R}^p$, $(x'_{\mathcal{N}}(\mathbf{p}^*; d\mathbf{p}), \lambda'(\mathbf{p}^*; d\mathbf{p}))$ is the unique pair $(dx, d\lambda)$ that satisfies the KKT system of the AVI (5.4.21); that is,

$$J_{\mathbf{p}}\mathbf{L}(x^*, \lambda, \mathbf{p}^*)d\mathbf{p} + J_x\mathbf{L}(x^*, \lambda, \mathbf{p}^*)dx + \sum_{i \in \mathcal{I}(x^*, \mathbf{p}^*)} d\lambda_i \, \nabla_x g_i(x^*, \mathbf{p}^*) = 0$$

$$0 \le d\lambda_i \perp \nabla_{\mathbf{p}} g_i(x^*, \mathbf{p}^*)^T d\mathbf{p} + \nabla_x g_i(x^*, \mathbf{p}^*)^T dx \le 0, \quad \forall\, i \in \beta$$

$$\nabla_{\mathbf{p}} g_i(x^*, \mathbf{p}^*)^T d\mathbf{p} + \nabla_x g_i(x^*, \mathbf{p}^*)^T dx = 0, \quad \forall\, i \in \mathrm{supp}(\lambda^*)$$

$$d\lambda_i = 0, \quad \forall\, i \notin \mathcal{I}(x^*, \mathbf{p}^*).$$

(c) If in addition λ^* is nondegenerate, that is if $\lambda^* - g(x^*, \mathbf{p}^*) > 0$, then $(x_{\mathcal{N}}(\mathbf{p}), \lambda(\mathbf{p}))$ is a C^1 function near $\mathbf{p}^*$.

Proof. In view of the remarks made before the proposition, it suffices to show the PC^1 property of the multiplier function and the uniqueness of the pair $(x_{\mathcal{N}}(\mathbf{p}), \lambda(\mathbf{p}))$. The latter follows easily from the LICQ and part (b) of Theorem **5.4.12**. For the former property, note that the LICQ implies that for each $\mathbf{p}$ sufficiently near $\mathbf{p}^*$, we have $\lambda_i(\mathbf{p}) = 0$ for all $i \notin \mathcal{I}(x^*, \mathbf{p}^*)$ and

$$F(x_{\mathcal{N}}(\mathbf{p}), \mathbf{p}) + \sum_{i \in \mathcal{I}(x^*, \mathbf{p}^*)} \lambda_i(\mathbf{p}) \, \nabla_x g_i(x_{\mathcal{N}}(\mathbf{p}), \mathbf{p}) = 0, \tag{5.4.28}$$

from which we can solve for $\lambda_{\mathcal{I}}(\mathbf{p})$ uniquely, obtaining

$$\begin{aligned} &\lambda_{\mathcal{I}}(\mathbf{p}) = \\ &-\left(\nabla_x g_{\mathcal{I}}(x_{\mathcal{N}}(\mathbf{p}), \mathbf{p}) \nabla_x g_{\mathcal{I}}(x_{\mathcal{N}}(\mathbf{p}), \mathbf{p})^T \right)^{-1} F(x_{\mathcal{N}}(\mathbf{p}), \mathbf{p}), \end{aligned} \tag{5.4.29}$$

where $\mathcal{I}$ is a shorthand for $\mathcal{I}(x^*, \mathbf{p}^*)$. This expression easily establishes the desired PC^1 property of $\lambda(\mathbf{p})$.

For part (b), it suffices to show that the pair $(x'_{\mathcal{N}}(\mathbf{p}^*, d\mathbf{p}), \lambda'(\mathbf{p}^*, d\mathbf{p}))$ satisfies the KKT conditions of the AVI (5.4.21). Noting (5.4.28) and the definition

$$\lambda'(\mathbf{p}^*, d\mathbf{p}) = \lim_{\tau \downarrow 0} \frac{\lambda(\mathbf{p}^* + \tau d\mathbf{p}) - \lambda^*}{\tau},$$

we can easily show that the pair of directional derivatives satisfies:

$$J_x\mathbf{L}(x^*,\lambda^*\mathbf{p}^*)x'_{\mathcal{N}}(\mathbf{p}^*,d\mathbf{p}) + J_{\mathbf{p}}\mathbf{L}(x^*,\lambda^*\mathbf{p}^*)d\mathbf{p}+$$

$$\sum_{i\in\mathcal{I}(x^*,\mathbf{p}^*)} \lambda_i'(\mathbf{p}^*,d\mathbf{p})\,\nabla_x g_i(x^*,\mathbf{p}^*) = 0$$

$$\lambda_i'(\mathbf{p}^*,d\mathbf{p}) \geq 0, \quad \forall\, i \in \beta^*$$

$$\lambda_i'(\mathbf{p}^*,d\mathbf{p}) = 0, \quad \forall\, i \notin \mathcal{I}(x^*,\mathbf{p}^*)$$

$$\lambda_i'(\mathbf{p}^*,d\mathbf{p}) \geq 0, \quad \forall\, i \in \operatorname{supp}(\lambda^*),$$

where β^* is the degenerate index set of λ^*, i.e.,

$$\beta^* \equiv \{\, i : \in \mathcal{I}(x^*,\mathbf{p}^*) \,:\, \lambda_i^* = 0\,\}.$$

The only thing left to be shown is the complementarity condition:

$$d\lambda_i\left[\,\nabla_{\mathbf{p}} g_i(x^*,\mathbf{p}^*)^T d\mathbf{p} + \nabla_x g_i(x^*,\mathbf{p}^*)^T dx\,\right] = 0, \quad \forall \in \beta^*.$$

If $\nabla_{\mathbf{p}} g_i(x^*,\mathbf{p}^*)^T d\mathbf{p} + \nabla_x g_i(x^*,\mathbf{p}^*)^T dx < 0$, then

$$g_i(x_{\mathcal{N}}(\mathbf{p}^* + \tau d\mathbf{p}), \mathbf{p}^* + \tau d\mathbf{p}) < 0$$

for all $\tau > 0$ sufficiently small, hence $\lambda_i(\mathbf{p}^* + \tau d\mathbf{p}) = 0$ for all such τ. This clearly implies that $\lambda_i'(\mathbf{p}^*;d\mathbf{p}) = 0$ as desired.

Finally, if λ^* is nondegenerate, then so is $\lambda(\mathbf{p})$ for all $\mathbf{p}$ sufficiently close to $\mathbf{p}^*$. Theorem **5.4.13** implies that $x_{\mathcal{N}}(\mathbf{p})$ is C^1 for $\mathbf{p}$ sufficiently near $\mathbf{p}^*$ and (5.4.29) shows that so is $\lambda(\mathbf{p})$. □

In Corollary **5.3.22**, we have considered the strong stability of a KKT triple for a VI and shown that this property can be characterized by a matrix-theoretic condition (part (b) of the corollary), which implies the LICQ. It turns out that this matrix-theoretic condition is equivalent to the LICQ and SCOC combined. In order to establish this equivalence, which is presented in Proposition **5.4.17**, we state an equivalent way of expressing SCOC. Since this discussion applies to a general VI with a finitely representable set and not restricted to parametric VIs, we drop the parameter $\mathbf{p}$ in the rest of this subsection.

Specifically, we consider the following parameter-free KKT system

$$\begin{aligned} F(x) + \sum_{i=1}^{m} \lambda_j \nabla g_i(x) &= 0 \\ 0 \leq \lambda \perp g(x) &\leq 0, \end{aligned} \tag{5.4.30}$$

where we do not assume the convexity of the functions g_i. Nonlinear equalities can also be included in this treatment; for simplicity, we continue not to include them. For each $\lambda \in \mathcal{M}(x^*)$, define the matrix

$$\mathbf{B}(\lambda) \equiv \begin{bmatrix} J_x\mathbf{L}(x^*, \lambda) & \nabla g_\alpha(x^*)^T \\ -\nabla g_\alpha(x^*) & 0 \end{bmatrix}$$

where $\alpha \equiv \operatorname{supp}(\lambda)$. Under the SCOC, this matrix is nonsingular if λ is in $\mathcal{M}^e(x^*)$ because $\operatorname{supp}(\lambda)$ is a SCOC index set, by the extreme property of the multiplier λ. For each SCOC index set $\mathcal{J}^j \in \mathcal{B}(x^*)$, there exists a unique multiplier $\lambda^j \in \mathcal{M}^e(x^*)$ such that $\operatorname{supp}(\lambda^j) \subseteq \mathcal{J}^j$. Let

$$\mathcal{J}_0^j \equiv \mathcal{J}^j \setminus \operatorname{supp}(\lambda^j).$$

Let S^j denote the Schur complement of $B(\lambda^j)$ in Λ^j; that is,

$$S^j \equiv [\ \nabla g_{\mathcal{J}_0^j}(x^*) \quad 0\ ]\,(\mathbf{B}(\lambda^j))^{-1} \begin{bmatrix} \nabla g_{\mathcal{J}_0^j}(x^*)^T \\ 0 \end{bmatrix}.$$

This Schur complement is uniquely determined by the SCOC index set $\mathcal{J}^j$.

5.4.16 Lemma. Let x^* be a solution of (5.4.30) satisfying the MFCQ. The following two statements are equivalent.

(a) The SCOC holds at x^*.

(b) There is an integer $\sigma = \pm 1$ such that

$$\operatorname{sgn}\ \det \mathbf{B}(\lambda) = \sigma, \quad \forall\, \lambda \in \mathcal{M}^e(x^*);$$

moreover, for each index set $\mathcal{J}^j$ in $\mathcal{B}(x^*)$, the Schur complement S^j has positive determinant.

Proof. (a) $\Rightarrow$ (b). Since for each $\lambda \in \mathcal{M}^e(x^*)$, $\operatorname{supp}(\lambda)$ is a SCOC index set, it follows that $\mathbf{B}(\lambda)$ has the same nonzero determinantal sign for all such λ. Let $\mathcal{J}^j$ be an arbitrary SCOC index set. By the Schur determinantal formula, we have

$$\det \Lambda^j = \det \mathbf{B}(\lambda^j) \det S^j.$$

Since $\det \Lambda^j$ and $\det \mathbf{B}(\lambda^j)$ have the same nonzero sign, it follows that $\det S^j$ is positive.

(b) $\Rightarrow$ (a). The same Schur formula implies that for all SCOC index set $\mathcal{J}^j$, the sign of $\det \Lambda^j$ is equal to the constant σ. □

Based on the above lemma, we can establish the aforementioned equivalence of the LICQ/SCOC and condition (b) in Corollary **5.3.22**. The result below generalizes the special case of the NCP discussed previously pertaining to the equivalence of the SCOC and the strong b-regularity.

5.4.17 Proposition. Let (x^*, λ^*) be a KKT pair satisfying (5.4.30). The following two statements are equivalent.

(a) The matrix $\mathbf{B}(\lambda^*)$ is nonsingular, and the Schur complement

$$\mathbf{C} \equiv \left[\begin{array}{cc} Jg_\beta(x^*) & 0 \end{array} \right] \mathbf{B}(\lambda^*)^{-1} \left[\begin{array}{c} Jg_\beta(x^*)^T \\ 0 \end{array} \right],$$

where β is the degenerate index set of λ^*, is a P matrix.

(b) Both the LICQ and the SCOC hold at x^*.

Proof. (a) $\Rightarrow$ (b). By Corollary **5.3.22**, the LICQ holds. Hence for every $\mathcal{J}^j \in \mathcal{B}(x^*)$, $\lambda^j = \lambda^*$. It is easy to see that the Schur complement S^j is a principal submatrix of $\mathbf{C}$. Since the latter is a P matrix, it follows that $\det S^j$ is positive.

(b) $\Rightarrow$ (a). The LICQ implies that $\mathcal{I}(x^*)$ is a SCOC index set with λ^* as its unique multiplier. By SCOC, $B(\lambda^*)$ is nonsingular. If C is a principal submatrix of $\mathbf{C}$, there exists a subset β' of β such that

$$C = \left[\begin{array}{cc} Jg_{\beta'}(x^*) & 0 \end{array} \right] \mathbf{B}(\lambda^*)^{-1} \left[\begin{array}{c} Jg_{\beta'}(x^*)^T \\ 0 \end{array} \right].$$

Since $\alpha \cup \beta'$ is also a SCOC index set, it follows that C has positive determinant. Hence $\mathbf{C}$ is a P matrix. □

5.5 Solution Set Stability

Throughout this section, we are given a continuous function $F : \mathbb{R}^n \to \mathbb{R}^n$ and a closed convex set K. In order to avoid some unnecessary details, we have taken the domain of F to be the entire space $\mathbb{R}^n$; furthermore we assume that SOL(K, F) is nonempty. We are interested in the "stability" of the entire solution set of the VI (K, F) when the function F is being perturbed. We first define this concept formally.

5.5.1 Definition. The VI (K, F) is said to be *semistable* if for every open set $\mathcal{U}$ containing SOL(K, F), there exist two positive scalars c and

ε such that, for every function $G \in \mathbb{B}(F; \varepsilon, K_{\mathcal{U}})$, where $K_{\mathcal{U}} \equiv K \cap \mathrm{cl}\,\mathcal{U}$, it holds that

$$\mathrm{SOL}(K, G) \cap \mathcal{U} \subset \mathrm{SOL}(K, F) + c\,\omega\,\mathbb{B}(0, 1),$$

where

$$\omega \equiv \sup_{x \in K_{\mathcal{U}}} \| G(x) - F(x) \|;$$

the VI (K, F) is said to be *stable* if in addition $\mathrm{SOL}(K, G) \cap \mathcal{U} \neq \emptyset$. We also say that $\mathrm{SOL}(K, F)$, which we assume is nonempty, is (semi)stable if the VI is so. □

It is important to point out that the open set $\mathcal{U}$ in the above definition is not required to be bounded. On the one hand, allowing an unbounded $\mathcal{U}$ (e.g., $\mathcal{U} = \mathbb{R}^n$) is appropriate in view of the fact that we have not assumed the boundedness of $\mathrm{SOL}(K, F)$ in the above definition. Moreover, the semistability concept as defined is equivalent to the validity of a "local error bound" for $\mathrm{SOL}(K, F)$ (see Proposition **5.5.5**), which is an useful property by itself. On the other hand, allowing an unbounded $\mathcal{U}$ restricts the class of semistable VIs.

The case where $\mathrm{SOL}(K, F)$ is a singleton, say $\{x^*\}$, deserves special consideration. In this case, one would hope that the solution set semistability of the VI (K, F) as given in Definition **5.5.1** would coincide with the pointwise semistability of x^* as given in Definition **5.3.1**. Unfortunately, while the semistability of $\mathrm{SOL}(K, F) = \{x^*\}$ according to Definition **5.5.1** can easily be seen to imply the pointwise semistability of x^* according to Definition **5.3.1**, the converse is false as illustrated by the example below.

5.5.2 Example. Let $K \equiv \mathbb{R}$ and $F(t) \equiv \min(t, e^{-t})$ for $t \in \mathbb{R}$. It is easy to see that $t = 0$ is a strongly stable zero of F because in a sufficiently small neighborhood of this zero, F is the identity map. Nevertheless, by condition (b) in Proposition **5.5.5**, the zero set of F is not semistable, thus not stable, according to Definition **5.5.1**. □

In view of the above example, additional explanation of the distinction between the two kinds of semistability is in order. In essence, pointwise (semi)stability focuses on the behavior of a single solution of the VI (K, F) when F is perturbed and pays absolutely no attention to the global change of the solution set of the VI. In contrast, the latter change is the main concern of solution set (semi)stability. In particular, if the solution set of the base VI is bounded, we expect solution set (semi)stability to imply that the solution set of the perturbed problems to be at least bounded. In order to guarantee this desirable consequence, the open set $\mathcal{U}$ in Definition **5.5.1** cannot be too restrictive. Returning to Example **5.5.2**, we see that for any $\varepsilon > 0$, there exist perturbation functions $G(t)$ such that $|G(t) - F(t)| \leq \varepsilon$ for all t and yet G can have infinitely many zeros that tend to plus infinity. Consequently, the unique zero of this function F is pointwise stable but the zero set of F is not even semistable.

With $\mathcal{U}$ possibly unbounded, it is instructive to say a few more words about the restriction $G \in \mathbb{B}(F; \varepsilon, K_{\mathcal{U}})$ in Definition **5.5.1**. At first sight, this condition seems fairly restrictive. For instance, consider an affine map $F(x) \equiv q + Mx$ and suppose we are interested in perturbing q and M so that the perturbed function $G(x) \equiv q' + M'x$. If $K_{\mathcal{U}}$ is an unbounded set, then G cannot belong to the neighborhood $\mathbb{B}(F; \varepsilon, K_{\mathcal{U}})$ if $M' \neq M$. To handle this case, a more appropriate restriction would be: for some $\delta > 0$,

$$\| G(x) - F(x) \| < \delta (1 + \| x \|), \quad \forall x \in K_{\mathcal{U}}, \tag{5.5.1}$$

which automatically includes the affine perturbation. The reason we do not use the latter restriction in Definition **5.5.1** is because in the key stability results established in this section, we always assume that the solution set $\text{SOL}(K, F)$ is contained in a bounded open set Ω to begin with. Within this framework, we have the following clarifying result.

5.5.3 Proposition. Consider the following statements:

(a) For every open set $\mathcal{U}$ containing $\text{SOL}(K, F)$, there exists $\delta > 0$ such that, for every continuous function G satisfying (5.5.1), $\text{SOL}(K, G) \cap \mathcal{U}$ is nonempty.

(b) For every open set $\mathcal{U}$ containing $\text{SOL}(K, F)$, there exists $\varepsilon > 0$ such that, for every function $G \in \mathbb{B}(F; \varepsilon, K_{\mathcal{U}})$, $\text{SOL}(K, G) \cap \mathcal{U}$ is nonempty.

It holds that (a) $\Rightarrow$ (b). The reverse implication (b) $\Rightarrow$ (a) holds if $\text{SOL}(K, F)$ is bounded.

Proof. The implication (a) $\Rightarrow$ (b) is obvious. To prove the reverse implication, assume that (b) holds and $\text{SOL}(K, F)$ is bounded. Fix a bounded open set $\mathcal{U}_0$ containing $\text{SOL}(K, F)$. Let $\mathcal{U}$ be an arbitrary open set containing $\text{SOL}(K, F)$. Associated with the bounded open set

$$\mathcal{U}' \equiv \mathcal{U} \cap \mathcal{U}_0,$$

there exists a scalar $\varepsilon > 0$ such that, for every function $G \in \mathbb{B}(F; \varepsilon, K_{\mathcal{U}'})$, we have $\text{SOL}(K, G) \cap \mathcal{U}' \neq \emptyset$. Let $\delta > 0$ be such that

$$\delta (1 + \| x \|) < \varepsilon, \quad \forall x \in \mathcal{U}'.$$

If G is any continuous function satisfying (5.5.1), then $G \in \mathbb{B}(F; \varepsilon, K_{\mathcal{U}'})$. Hence $\mathrm{SOL}(K, G) \cap K_{\mathcal{U}'}$ is nonempty, which implies that $\mathrm{SOL}(K, G) \cap K_{\mathcal{U}}$ is nonempty. Thus (a) holds. □

Specializing Proposition **5.5.3** to the VI (K, q, M), we obtain the following corollary immediately, which pertains to the solvability of the perturbed VI (K, q', M'), when only the pair (q, M) is changed.

5.5.4 Corollary. Assume that the VI (K, q, M) is stable and its solution set is bounded. For every bounded open set $\mathcal{U}$ containing $\mathrm{SOL}(K, q, M)$, there exists $\varepsilon > 0$ such that the VI (K, q', M') has a solution in $\mathcal{U}$ for all pairs (q', M') satisfying

$$\| q - q' \| + \| M - M' \| < \varepsilon.$$

Proof. The assertion is a special case of statement (a) in Proposition **5.5.3**. □

5.5.1 Semistability

If the VI (K, F) is semistable, then with $\mathcal{U} = \mathbb{R}^n$, it follows that there exist two positive scalars c and ε such that, for every function G belonging to $\mathbb{B}(F; \varepsilon, K)$, we have

$$\mathrm{SOL}(K, G) \subset \mathrm{SOL}(K, F) + c \sup_{x \in K} \| F(x) - G(x) \| \, \mathbb{B}(0, 1). \tag{5.5.2}$$

It turns out that semistability is equivalent to an important "local error bound" property of the solution set of the VI, which in turn is related to the notion of an inexact solution of the VI. We know that

$$\mathbf{F}_K^{\mathrm{nor}}(z) = 0 \Leftrightarrow [\, x \equiv \Pi_K(z) \in \mathrm{SOL}(K, F) \text{ and } z = x - F(x) \,].$$

Thus if we are given a vector z with $\mathbf{F}_K^{\mathrm{nor}}(z) \neq 0$, then $\Pi_K(z)$ is not a solution of the VI (K, F). Nevertheless if $\|\mathbf{F}_K^{\mathrm{nor}}(z)\|$ is small, then a natural thing to expect is that $\Pi_K(z)$ will be close to some solution of the VI (K, F), or equivalently, $\mathrm{dist}(\Pi_K(z), \mathrm{SOL}(K, F))$ will be accordingly small. Conversely, if the latter distance is small but nonzero, then one would hope the former norm of the normal map to be small. Mathematically, these intuitive ideas are justified if there exist positive constants c_1 and c_2 such that

$$c_1 \| \mathbf{F}_K^{\mathrm{nor}}(z) \| \leq \mathrm{dist}(\Pi_K(z), \mathrm{SOL}(K, F)) \leq c_2 \| \mathbf{F}_K^{\mathrm{nor}}(z) \|,$$

for all vectors z of interest. An inequality such as the above is called an *error bound* for the solutions of the VI (K, F); the quantity $\|\mathbf{F}_K^{\mathrm{nor}}(z)\|$ is called the *residual* of the vector z with the normal map $\mathbf{F}_K^{\rm nor}$ being the *residual function*. (With an abuse of language, we often do not distinguish the residual from the residual function.)

The following result relates the semistability of the VI (K, F) to an error bound of the set SOL(K, F) and also to an upper Lipschitz property of this set. For a related result based on the natural map, see Proposition **6.2.1**.

5.5.5 Proposition. Let K be closed convex and $F : \mathbb{R}^n \to \mathbb{R}^n$ be continuous. The following three statements are equivalent.

(a) The VI (K, F) is semistable.

(b) Two positive scalars c and ε exist such that, for all vectors $q \in \mathbb{R}^n$,

$$\| q \| < \varepsilon \Rightarrow \text{SOL}(K, q + F) \subset \text{SOL}(K, F) + \mathbb{B}(0, c \| q \|). \tag{5.5.3}$$

(c) There exist two positive scalars c and ε such that, for all $z \in \mathbb{R}^n$,

$$\| \mathbf{F}_K^{\rm nor}(z) \| < \varepsilon \Rightarrow \text{dist}(\Pi_K(z), \text{SOL}(K, F)) \leq c \| \mathbf{F}_K^{\rm nor}(z) \|. \tag{5.5.4}$$

Proof. (a) $\Rightarrow$ (b). Suppose that the VI (K, F) is semistable. Let c and ε be positive scalars such that (5.5.2) holds for every continuous function G belonging to $\mathbb{B}(F; \varepsilon, K)$. Let q satisfy $\|q\| < \varepsilon$. The function $G \equiv q + F$ clearly belongs to $\mathbb{B}(F; \varepsilon, K)$. Consequently, (5.5.3) follows easily from (5.5.2).

(b) $\Rightarrow$ (c). Let z be such that $\|\mathbf{F}_K^{\rm nor}(z)\| < \varepsilon$. The vector $\Pi_K(z)$ is clearly a solution of the VI $(K, q + F)$, where $q \equiv -\mathbf{F}_K^{\rm nor}(z)$. Hence (5.5.4) follows easily from (5.5.3).

(c) $\Rightarrow$ (a). Let an open set $\mathcal{U}$ containing SOL(K, F) be given. Let c and ε be two positive scalars such that (5.5.4) holds. Let G be a function belonging to $\mathbb{B}(F; \varepsilon, K_{\mathcal{U}})$. Let $\bar{x} \in \text{SOL}(K, G) \cap \mathcal{U}$ be given and define $\bar{z} \equiv \bar{x} - G(\bar{x})$. We have

$$\mathbf{F}_K^{\rm nor}(\bar{z}) = F(\bar{x}) + \bar{z} - \bar{x} = F(\bar{x}) - G(\bar{x}).$$

Hence,

$$\| \mathbf{F}_K^{\rm nor}(\bar{z}) \| = \| F(\bar{x}) - G(\bar{x}) \| < \varepsilon.$$

Consequently,

$$\text{dist}(\bar{x}, \text{SOL}(K, F)) \leq c \| \mathbf{F}_K^{\rm nor}(\bar{z}) \| \leq c \sup_{x \in K_{\mathcal{U}}} \| F(x) - G(x) \|,$$

from which the semistability of the VI (K, F) follows. □

In what follows, we identify two situations in which the VI (K, F) is semistable. We say that the VI (K, F) is *pointwise semiregular* if every solution is semiregular. For a given scalar $\varepsilon > 0$, we say that this VI is *ε-solution bounded* if the set

$$S_\varepsilon(K,F) \equiv \bigcup_{q \in \mathbb{B}(0,\varepsilon)} \mathrm{SOL}(K, q+F) \tag{5.5.5}$$

is bounded. The latter condition implies in particular that SOL(K,F) is bounded. In turn, the above set $S_\varepsilon(K,F)$ is bounded if and only if the ε-level set of the normal map $\mathbf{F}_K^{\mathrm{nor}}$, i.e., the set

$$L_\varepsilon^{\mathrm{nor}}(K,F) \equiv \{\, z \in \mathbb{R}^n \,:\, \|\,\mathbf{F}_K^{\mathrm{nor}}(z)\,\| < \varepsilon\,\}$$

is bounded. To see this, let $x \in \mathrm{SOL}(K, q+F)$ be given, where q is an arbitrary element of $\mathbb{B}(0,\varepsilon)$. Let $z \equiv x - q - F(x)$. We have

$$\Pi_K(z) = x \quad \text{and} \quad \mathbf{F}_K^{\mathrm{nor}}(z) = -q.$$

Consequently, $z \in L_\varepsilon^{\mathrm{nor}}(K,F)$; hence

$$\bigcup_{q \in \mathbb{B}(0,\varepsilon)} \mathrm{SOL}(K, q+F) \subseteq \Pi_K(L_\varepsilon^{\mathrm{nor}}(K,F)).$$

So if the right-hand set is bounded, then so is the left. Conversely, let z be an arbitrary vector in $L_\varepsilon^{\mathrm{nor}}(K,F)$. Write $q \equiv -\mathbf{F}_K^{\mathrm{nor}}(z)$. Then $x \equiv \Pi_K(z)$ is a solution of the VI $(K, q+F)$ and $z = x - q - F(x)$. Consequently,

$$L_\varepsilon^{\mathrm{nor}}(K,F) \subseteq (\mathbf{I} - F)\left[\bigcup_{q \in \mathbb{B}(0,\varepsilon)} \mathrm{SOL}(K, q+F)\right] - \mathbb{B}(0,\varepsilon),$$

which shows that if the right-hand set is bounded, then so is the left.

If either the normal map $\mathbf{F}_K^{\mathrm{nor}}$ is norm-coercive on $\mathbb{R}^n$ or the natural map $\mathbf{F}_K^{\mathrm{nat}}$ is norm-coercive on K, then the level set $L_\varepsilon^{\mathrm{nor}}(K,F)$ is bounded, and thus the VI (K,F) is ε-solution bounded, for every $\varepsilon > 0$. This is clear in the former case and also in the latter case in view of Exercise **1.8.34**.

5.5.6 Theorem. Let K be closed convex and $F : \mathbb{R}^n \to \mathbb{R}^n$ be continuous. Assume that SOL(K,F) is nonempty. The following two statements are valid.

(a) If SOL(K,F) is bounded and the VI (K,F) is semistable, then the VI (K,F) is ε-solution bounded for some $\varepsilon > 0$.

(b) Conversely, if the VI (K,F) is ε-solution bounded for some $\varepsilon > 0$ and SOL(K,F) is pointwise semiregular, then SOL(K,F) is finite and the VI (K,F) is semistable.

Proof. By Proposition **5.5.5**, if SOL(K,F) is bounded and the VI (K,F) is semistable, then the set $\Pi_K(L_\varepsilon^{\rm nor}(K,F))$ is bounded for some $\varepsilon > 0$. Since

$$\mathbf{F}_K^{\rm nor}(z) = F(\Pi_K(z)) + z - \Pi_K(z),$$

it follows that $L_\varepsilon^{\rm nor}(K,F)$ is bounded. Hence (a) holds.

Assume that the VI (K,F) is ε_0-solution bounded for some $\varepsilon_0 > 0$ and SOL(K,F) is pointwise semiregular. If SOL(K,F) contains infinitely many elements, then it contains an infinite sequence $\{x^k\}$ of distinct vectors, which must be bounded and hence have at least one accumulation point. Such a point is a non-isolated thus non-semiregular solution of the VI (K,F). This contradicts the pointwise semiregularity of the solution set SOL(K,F). We claim that there exist positive scalars ε and c such that the local error bound (5.5.4) holds for all $z \in \mathbb{R}^n$. Assume for the sake of contradiction that no such scalars exist. It follows that a sequence of vectors $\{z^k\}$ and a sequence of positive scalars $\{\varepsilon_k\}$ converging to zero exist such that, for each k, z^k satisfies $\|\mathbf{F}_K^{\rm nor}(z^k)\| < \varepsilon_k$ and

$$\operatorname{dist}(\Pi_K(z^k), \text{SOL}(K,F)) > k \, \| \, \mathbf{F}_K^{\rm nor}(z^k) \, \|. \tag{5.5.6}$$

Except for possibly finitely many vectors, the sequence $\{z^k\}$ must be contained in $L_{\varepsilon_0}^{\rm nor}(K,F)$. Hence $\{z^k\}$ is bounded and thus has an accumulation point z^∞, which must necessarily be a zero of $\mathbf{F}_K^{\rm nor}$. Let $x^\infty \equiv \Pi_K(z^\infty)$; then x^∞ is a solution of the VI (K,F). By assumption, x^∞ is semiregular. Hence there exist a neighborhood $\mathcal{N}$ of x^∞ and two positive scalars ε' and c' such that, for all

$$\| \, q \, \| < \varepsilon' \;\Rightarrow\; \text{SOL}(K, q+F) \cap \mathcal{N} \subseteq \mathbb{B}(x^\infty, c\|q\|).$$

Let $q^k \equiv -\mathbf{F}_K^{\rm nor}(z^k)$. For k sufficiently large, we have $\|q^k\| < \varepsilon'$; moreover, $x^k \equiv \Pi_K(z^k)$ is a solution of the VI $(K, q^k + F)$ that belongs to $\mathcal{N}$. Consequently, we have

$$\| \, x^k - x^\infty \, \| \leq c \, \| \, \mathbf{F}_K^{\rm nor}(z^k) \, \|.$$

But this contradicts (5.5.6). Hence (b) holds. □

Another situation where the VI (K,F) is semistable is when (K,F) is an affine pair, i.e., when we have an AVI. To establish this fact, let K be a polyhedron in $\mathbb{R}^n$ and M be an $n \times n$ matrix and consider the normal map of the pair (K,M):

$$\mathbf{M}_K^{\rm nor}(z) \equiv M\Pi_K(z) + z - \Pi_K(z), \quad z \in \mathbb{R}^n.$$

For every $q \in \mathbb{R}^n$, we have

$$\begin{aligned} & z \in (\mathbf{M}_K^{\text{nor}})^{-1}(-q) \\ \Leftrightarrow \; & [x \equiv \Pi_K(z) \in \text{SOL}(K, q, M) \text{ and } z = x - (q + Mx)]. \end{aligned}$$

By Theorem **2.5.15**, SOL(K, q, M) is piecewise polyhedral; i.e., this set is the union of finitely many polyhedra. Thus so is $(\mathbf{M}_K^{\text{nor}})^{-1}(-q)$. The map

$$(\mathbf{M}_K^{\text{nor}})^{-1} : \mathbb{R}^n \to \mathbb{R}^n$$

is a multifunction with domain equal to the negative of the AVI range of the pair (K, M); that is the set $-\mathcal{R}(K, M)$. This multifunction has an important property stated in the result below.

5.5.7 Proposition. Let K be a polyhedron in $\mathbb{R}^n$ and M be an $n \times n$ matrix. The graph of the multifunction $(\mathbf{M}_K^{\text{nor}})^{-1}$ is piecewise polyhedral.

Proof. The graph of the multifunction $(\mathbf{M}_K^{\text{nor}})^{-1}$ consists of all pairs (z, q) such that $\mathbf{M}_K^{\text{nor}}(z) = q$. The proof of part (a) of Theorem **2.5.15** easily establishes that the set

$$\{ (x, q) \in \mathbb{R}^{2n} : x \in \text{SOL}(K, q, M) \}$$

is piecewise polyhedral. By the canonical relation between the equation $\mathbf{M}_K^{\text{nor}}(z) = q$ and the AVI $(K, -q, M)$, the piecewise polyhedrality of the graph of $(\mathbf{M}_K^{\text{nor}})^{-1}$ follows readily. □

In general, a set-valued map $\Phi : \mathbb{R}^n \to \mathbb{R}^m$ whose graph is piecewise polyhedral is called a *polyhedral multifunction*. A broad class of polyhedral multifunctions is obtained by considering the (set-valued) inverse of a PA map. Specifically, for every PA map $f : \mathbb{R}^n \to \mathbb{R}^m$, the inverse f^{-1} is a polyhedral multifunction from $\mathbb{R}^m$ into $\mathbb{R}^n$. The proof is based on the polyhedral representation of f; see Proposition **4.2.1**. Indeed if Ξ is a polyhedral subdivision of $\mathbb{R}^n$ induced by f and for each $P \in \Xi$, G_P is an affine function that coincides with f on P, then

$$\text{gph}\, f^{-1} = \bigcup_{P \in \Xi} \{ (q, x) \in \mathbb{R}^m \times P : G_P(x) = q \},$$

which shows that gph f^{-1} is the union of finitely many polyhedra in $\mathbb{R}^{m+n}$. This proof is very similar to that of Proposition **4.2.2**(b).

Theorem **5.5.8** below asserts that every polyhedral multifunction is everywhere pointwise upper Lipschitz continuous; that is, for every vector $q \in \mathbb{R}^n$, there exist positive scalars c and ε such that

$$\| q' - q \| < \varepsilon \Rightarrow \Phi(q') \subset \Phi(q) + c \, \| q' - q \| \, \mathbb{B}(0, 1).$$

In the case where $\Phi(q)$ is empty, this inequality implies that $\Phi(q')$ is also empty for all q' sufficiently close to q. This fact is merely an equivalent statement of the closedness of the range of Φ; in turn, the latter property is an easy consequence of the polyhedrality of Φ.

5.5.8 Theorem. Every polyhedral multifunction is everywhere pointwise upper Lipschitz continuous.

Proof. Let Φ be a polyhedral multifunction from $\mathbb{R}^n$ into subsets of $\mathbb{R}^n$. Write

$$\operatorname{gph} \Phi \;=\; \bigcup_{i=1}^{N} P_i$$

where N is positive integer and each P_i is a polyhedral set in $\mathbb{R}^{2n}$ given by

$$P_i \;\equiv\; \{\, (x,q) \in \mathbb{R}^{2n} \;:\; A^i x + B^i q \,\leq\, b^i \,\}$$

for some matrices A^i and B^i and vector b^i of appropriate dimensions. For each vector $q \in \mathbb{R}^n$, let

$$P_i(q) \equiv \{\, x \in \mathbb{R}^n \;:\; (x,q) \in P_i \,\}$$

and let $\mathcal{I}(q)$ be the set of indices i such that $P_i(q)$ is nonempty. We have

$$\Phi(q) \;=\; \bigcup_{i \in \mathcal{I}(q)} P_i(q).$$

By Lemma **4.4.2** and the finiteness of the family of polyhedra $\{P_i\}$, it follows that for every vector q, there exists a neighborhood $\mathcal{Q}$ such that, for every $q' \in \mathcal{Q}$, $\mathcal{I}(q')$ is contained in $\mathcal{I}(q)$. Furthermore, by part (a) of Corollary **3.2.5**, for each index $i \in \mathcal{I}(q)$, there exists a positive scalar L_i (that also depends on q, which is a fixed but arbitrary vector in this discussion) such that

$$P_i(q') \subset P_i(q) + L_i \,\|\, q - q' \,\|\, \mathbb{B}(0,1), \qquad \forall\, q' \in \mathcal{Q}.$$

Consequently, for all $q' \in \mathcal{Q}$, we have

$$\Phi(q') \;=\; \bigcup_{i \in \mathcal{I}(q')} P_i(q') \subset \Phi(q) + L \,\|\, q - q' \,\|\, \mathbb{B}(0,1),$$

where $L \equiv \max\{L_i : i \in \mathcal{I}(q)\}$. □

Specializing the above theorem to the multifunction $(\mathbf{M}_K^{\mathrm{nor}})^{-1}$, we immediately obtain the following result.

5.5.9 Corollary. Let K be a polyhedron in $\mathbb{R}^n$ and M be an $n \times n$ matrix. For every vector $q \in \mathbb{R}^n$, the AVI (K, q, M) is semistable.

Proof. By Theorem **5.5.8**, there exist positive constants c and ε such that

$$\| q' - q \| < \varepsilon \Rightarrow (\mathbf{M}_K^{\text{nor}})^{-1}(-q') \subset (\mathbf{M}_K^{\text{nor}})^{-1}(-q) + c \| q - q' \| \mathbb{B}(0,1).$$

Let x' be a solution of the VI (K, q', M). Then

$$z' \equiv x' - q' - Mx'$$

belongs to $(\mathbf{M}_K^{\text{nor}})^{-1}(-q')$ and $\Pi_K(z') = x'$. There exists a vector z in $(\mathbf{M}_K^{\text{nor}})^{-1}(-q)$ such that

$$\| z - z' \| \leq c \| q - q' \|.$$

The vector $x \equiv \Pi_K(z)$ solves the AVI (K, q, M); we have

$$\| x - x' \| \leq \| z - z' \|.$$

Therefore, we have established condition (b) in Proposition **5.5.5** for the AVI (K, q, M), which is equivalent to its semistability. □

5.5.2 Solvability of perturbed problems and stability

We next consider the issue of the existence of a solution to the perturbed VI (K, G), which is the other requirement for the stability of the VI (K, F). Relying on degree theory, we establish such an existence result that is analogous to Proposition **5.1.4** but pertains to the entire solution set rather than just a particular solution.

5.5.10 Theorem. Let $K \subseteq \mathbb{R}^n$ be closed convex and $F : \mathbb{R}^n \to \mathbb{R}^n$ be continuous. If SOL(K, F) is nonempty and there exists a bounded open set Ω containing $(\mathbf{F}_K^{\text{nor}})^{-1}(0)$ such that $\deg(\mathbf{F}_K^{\text{nor}}, \Omega)$ is nonzero, then for every open set $\mathcal{U}$ containing SOL(K, F), there exists a scalar $\delta > 0$ such that, for every continuous function G satisfying (5.5.1), the VI (K, G) has a solution in $\mathcal{U}$.

Proof. Let $\mathcal{U}$ be any open set containing SOL(K, F). It suffices to show that an $\varepsilon > 0$ exists such that, for every function $G \in \mathbb{B}(F; \varepsilon, K_{\mathcal{U}})$, the VI (K, F) has a solution in $\mathcal{U}$; see Proposition **5.5.3**. Since Π_K is continuous and $\mathcal{U}$ is open, it follows that $\Pi_K^{-1}(\mathcal{U})$ is open. The set $\mathcal{Z} \equiv \Pi_K^{-1}(\mathcal{U}) \cap \Omega$ is open and bounded and contains $(\mathbf{F}_K^{\text{nor}})^{-1}(0)$. Indeed if z belongs to $(\mathbf{F}_K^{\text{nor}})^{-1}(0)$, then $\Pi_K(z)$ solves the VI (K, F), thus belongs to $\mathcal{U}$; hence

z belongs to $\mathcal{Z}$. The openness and boundedness of $\mathcal{Z}$ are obvious. By the discussion following Proposition **2.1.6**, we deduce that $\deg(\mathbf{F}_K^{\text{nor}}, \mathcal{Z})$ is equal to $\deg(\mathbf{F}_K^{\text{nor}}, \Omega)$ and thus is nonzero. Let ε be a positive scalar less than $\text{dist}_\infty(0, \mathbf{F}_K^{\text{nor}}(\partial\mathcal{Z}))$. For every function G belonging to $\mathbb{B}(F; \varepsilon, K_{\mathcal{U}})$, we have

$$\max_{z \in \text{cl}\,\mathcal{Z}} \| \mathbf{F}_K^{\text{nor}}(z) - \mathbf{G}_K^{\text{nor}}(z) \|_\infty \leq \max_{x \in K \cap \text{cl}\,\mathcal{U}} \| F(x) - G(x) \|_2 < \varepsilon.$$

Therefore, by the nearness property of the degree, it follows that $\mathbf{G}_K^{\text{nor}}$ has a zero in $\mathcal{Z}$, which implies that the VI (K, G) has a solution in $\mathcal{U}$. □

Specializing the above theorem to an AVI and invoking Corollary **5.5.9**, we immediately obtain the following corollary, which gives a simple degree condition for the AVI to be stable. No proof is needed.

5.5.11 Corollary. Let K be a polyhedron in $\mathbb{R}^n$ and M be an $n \times n$ matrix. If $\text{SOL}(K, F)$ is nonempty and a bounded open set Ω containing $(\mathbf{F}_K^{\text{nor}})^{-1}(0)$ exists such that $\deg(\mathbf{F}_K^{\text{nor}}, \Omega)$ is nonzero, where $F(x) \equiv q + Mx$, then the AVI (K, q, M) is stable. □

In general, many conditions introduced in Chapter **2** are sufficient for $\deg(\mathbf{F}_K^{\text{nor}}, \Omega)$ to be well defined and nonzero for some bounded open set Ω containing $(\mathbf{F}_K^{\text{nor}})^{-1}(0)$. One broad condition of this kind is given in Proposition **2.2.3**.

5.5.12 Corollary. Let $K \subseteq \mathbb{R}^n$ be closed convex and $F : K \to \mathbb{R}^n$ be continuous. Suppose there exists a vector $x^{\text{ref}} \in K$ such that

$$L_{\leq} \equiv \{ x \in K : F(x)^T(x - x^{\text{ref}}) \leq 0 \},$$

is bounded. For every open set $\mathcal{U}$ containing $\text{SOL}(K, F)$, there exists a scalar $\delta > 0$ such that, for every continuous function G satisfying (5.5.1), the VI (K, G) has a solution in $\mathcal{U}$.

Proof. By Proposition **2.2.3**, $\text{SOL}(K, F)$ is nonempty; moreover this solution set is bounded because it is contained in $L_{\leq}$. Let $\mathcal{U}$ be a bounded open set containing the set $L_{\leq}$. Define $\Omega \equiv \Pi_K^{-1}(\mathcal{U})$. The set Ω is bounded and open and contains $(\mathbf{F}_K^{\text{nor}})^{-1}(0)$ and x^{ref}. It remains to show that $\deg(\mathbf{F}_K^{\text{nor}}, \Omega)$ is nonzero. For this purpose, let $H(z, t)$ be the normal map of the VI (K, F_t) for $t \in [0, 1]$, where

$$F_t(x) \equiv tF(x) + (1 - t)(x - x^{\text{ref}}).$$

By a direct verification (see the proof of the cited proposition), we can show that the set

$$\bigcup_{t\in[0,1]} H(\cdot,t)^{-1}(0)$$

is contained in Ω. By a familiar homotopy argument, we can easily establish that $\deg(\mathbf{F}_K^{\rm nor},\Omega)$ is equal to one. □

In what follows, we consider a broad coercivity condition that yields the stability of the VI (K,F).

5.5.13 Proposition. Let $K \subseteq \mathbb{R}^n$ be closed convex and $F : \mathbb{R}^n \to \mathbb{R}^n$ be continuous. The VI (K,F) is stable if the following two conditions hold:

(a) there exists a vector $x^{\rm ref} \in K$ such that

$$\liminf_{\substack{x\in K\\ \|x\|\to\infty}} \frac{F(x)^T(x - x^{\rm ref})}{\|x\|} > 0,$$

(b) every solution of the VI (K,F) is semiregular.

Proof. Condition (a) clearly implies that the set $L_{\leq}$ in Corollary **5.5.12** is bounded. We claim that under condition (a), there exists an $\varepsilon > 0$ such that the VI (K,F) is ε-solution bounded. Assume the contrary. Then there exist a sequence of vectors $\{q^k\}$ converging to zero and a sequence of vectors $\{x^k\}$ such that, for every k, x^k belongs to $\mathrm{SOL}(K,q^k+F)$; moreover,

$$\lim_{k\to\infty} \|x^k\| = \infty.$$

Since $x^{\rm ref} \in K$, we have for every k,

$$(x^{\rm ref} - x^k)^T(q^k + F(x^k)) \geq 0,$$

which implies

$$\frac{(x^{\rm ref} - x^k)^T F(x^k)}{\|x^k\|} \geq \frac{(x^{\rm ref} - x^k)^T q^k}{\|x^k\|}.$$

The right-hand side converges to zero as k tends to infinity. This contradicts the limit condition in (a). The desired stability of the VI (K,F) follows readily from Corollary **5.5.12** and part (b) of Theorem **5.5.6**. □

By imposing an index condition on the solutions of the VI in part (a) of Theorem **5.5.6**, we obtain another sufficient condition for the stability of the VI.

5.5.14 Corollary. Let $K \subseteq \mathbb{R}^n$ be closed convex and $F : \mathbb{R}^n \to \mathbb{R}^n$ be continuous. The VI (K, F) is stable if the following three conditions hold:

(a) the VI (K, F) is semistable;

(b) $\text{SOL}(K, F)$ is finite;

(c) $\operatorname{sgn} \operatorname{ind}(\mathbf{F}_K^{\text{nor}}, z)$ is a nonzero constant for every $z \in (\mathbf{F}_K^{\text{nor}})^{-1}(0)$.

Proof. Since

$$(\mathbf{F}_K^{\text{nor}})^{-1}(0) \subseteq \{ x - F(x) : x \in \text{SOL}(K, F) \},$$

it follows by (b) that the left-hand set is finite. Let Ω be any bounded open set containing $(\mathbf{F}_K^{\text{nor}})^{-1}(0)$. By Proposition **2.1.6**, we have

$$\deg(\mathbf{F}_K^{\text{nor}}, \Omega) = \sum_{z \in (\mathbf{F}_K^{\text{nor}})^{-1}(0)} \operatorname{ind}(\mathbf{F}_K^{\text{nor}}, z).$$

By condition (c), it follows that the left-hand degree is nonzero. The corollary follows easily from Theorem **5.5.10**. □

Unlike Corollary **5.5.11**, a major assumption in Corollary **5.5.14** is the finiteness of $\text{SOL}(K, F)$. Presently, there are not many known sufficient conditions that will imply the stability of the VI (K, F) with a non-affine pair (K, F) and with $\text{SOL}(K, F)$ being a continuum. The difficulty is with the semistability of the VI, which is a fairly demanding "upper Lipschitz continuity" requirement of the perturbed solutions. For a class of nonlinear VIs that are semistable, see Theorem **6.2.8**. Although the latter VIs all have nonlinear defining functions, their solution sets are nevertheless polyhedral.

5.5.3 Partitioned VIs with P_0 pairs

When (K, F) is a P_0 pair, the converse of Theorem **5.5.10** is valid. This is formally stated in the next result, which concerns only with the solvability of the perturbed VIs but does not deal with the semistability issue.

5.5.15 Theorem. Let K be the Cartesian product of N closed convex sets $K_\nu \subseteq \mathbb{R}^{n_\nu}$. Let F be a continuous P_0 function on K. Suppose that $\text{SOL}(K, F)$ is nonempty. The following three statements are equivalent.

(a) For every open set $\mathcal{U}$ containing $\text{SOL}(K, F)$, there exists a scalar $\delta > 0$ such that, for every continuous function G satisfying (5.5.1), the VI (K, G) has a solution in $\mathcal{U}$.

(b) The solution set SOL(K, F) is bounded.

(c) There exists a bounded open set Ω containing $(\mathbf{F}_K^{\mathrm{nor}})^{-1}(0)$ such that $\deg(\mathbf{F}_K^{\mathrm{nor}}, \Omega)$ is nonzero.

If in addition F is monotone, then any one of the above statements is further equivalent to:

(d) For every open set $\mathcal{U}$ containing SOL(K, F), there exists a scalar $\varepsilon > 0$ such that, for every vector q satisfying $\|q\| < \varepsilon$, the VI $(K, q+F)$ has a solution in $\mathcal{U}$.

Proof. (a) $\Rightarrow$ (b). We first show that SOL(K, F) is bounded. Assume by way of contradiction that a sequence $\{x^k\} \subset$ SOL(K, F) exists satisfying

$$\lim_{k\to\infty} \| x^k \| = \infty.$$

Without loss of generality we may assume, by working with an appropriate subsequence of $\{x^k\}$ if necessary, that there exists index sets α_∞, $\alpha_{-\infty}$, and α_c satisfying $\alpha_\infty \cup \alpha_{-\infty} \neq \emptyset$ and

$$\begin{aligned}
\alpha_\infty &= \left\{ i : \lim_{k\to\infty} x_i^k = \infty \right\} \\
\alpha_{-\infty} &= \left\{ i : \lim_{k\to\infty} x_i^k = -\infty \right\} \\
\alpha_c &= \left\{ i : \lim_{k\to\infty} x_i^k = x_i^\infty \right\}
\end{aligned}$$

for some scalars x_i^∞, $i \in \alpha_c$. Define the vector function $g : \mathbb{R}^n \to \mathbb{R}^n$ as follows. For $i = 1, \ldots, n$,

$$g_i(x) \equiv \begin{cases} \arctan(x_i) + \frac{\pi}{2}, & i \in \alpha_{-\infty} \\ \arctan(x_i) - \frac{\pi}{2}, & i \in \alpha_\infty \\ x_i - x_i^\infty & i \in \alpha_c. \end{cases}$$

Each function g_i is continuous and strictly increasing; moreover

$$\lim_{k\to\infty} g_i(x^k) = 0.$$

For every $\varepsilon > 0$, the function $F_\varepsilon \equiv F + \varepsilon g$ is a P function on K. Thus by Corollary **3.6.2** the natural map of the pair (K, F_ε), i.e., the map

$$\Phi_\varepsilon(x) \equiv x - \Pi_K(x - F_\varepsilon(x))$$

is weakly univalent; moreover, $\Phi_\varepsilon^{-1}(0)$ is either empty or a singleton. For $\varepsilon > 0$ sufficiently small, F_ε is a legitimate perturbation of F satisfying $\|F_\varepsilon(x) - F(x)\| < \delta(1 + \|x\|)$. By assumption, it follows that $\Phi_\varepsilon^{-1}(0)$ is a singleton for all $\varepsilon > 0$ sufficiently small. Fix one such ε. There exists $\eta > 0$ such that the level set

$$L_\varepsilon(\eta) \equiv \{ x \in \mathbb{R}^n : \| \Phi_\varepsilon(x) \| \leq \eta \}$$

is bounded, by part (c) of Theorem **3.6.6** applied to Φ_ε. By the nonexpansiveness of the projector, we have, for all k sufficiently large,

$$\begin{aligned} &\|\Phi_\varepsilon(x^k)\| \\ &= \| x^k - \Pi_K(x^k - F(x^k) - \varepsilon\, g(x^k)) - (x^k - \Pi_K(x^k - F(x^k))) \| \\ &\leq \varepsilon \| g(x^k) \| \leq \eta, \end{aligned}$$

which implies that x^k belongs to $L_\varepsilon(\eta)$. Consequently, $\{x^k\}$ is bounded. This contradiction shows that SOL(K, F) must be bounded.

(b) $\Rightarrow$ (c). The normal map $\mathbf{F}_K^{\rm nor}$ is weakly univalent because it is the uniform limit of $\mathbf{F}_{K,\varepsilon}^{\rm nor}$, where $\mathbf{F}_{K,\varepsilon}^{\rm nor}$ is the normal map of the perturbed VI $(K, F + \varepsilon\mathbf{I})$. Since the set $(\mathbf{F}_K^{\rm nor})^{-1}(0)$ is compact, by part (c) of Theorem **3.6.4**, for every bounded open set Ω containing $(\mathbf{F}_K^{\rm nor})^{-1}(0)$, $\deg(\mathbf{F}_K^{\rm nor}, \Omega)$ is equal to ± 1.

(c) $\Rightarrow$ (a). This is Theorem **5.5.10**.

(a) $\Rightarrow$ (d). This is obvious.

(d) $\Rightarrow$ (b) if F is monotone. Suppose that there exist a sequence $\{x^k\}$ of solutions of the VI (K, F) and a nonzero vector $d \in \mathbb{R}^n$ satisfying the following properties:

$$\lim_{k\to\infty} \| x^k \| = \infty \quad \text{and} \quad \lim_{k\to\infty} \frac{x^k}{\| x^k \|} = d.$$

Let $q^k \equiv -d/\|x^k\|$. For k sufficiently large, the VI $(K, q^k + F)$ has a solution, say $\tilde{x}^k$. For a fixed k_0 sufficiently large, we have for all $k > k_0$,

$$0 \leq F(\tilde{x}^{k_0})^T(x^k - \tilde{x}^{k_0}) - \frac{d^T(x^k - \tilde{x}^{k_0})}{\| x^{k_0} \|}.$$

By the monotonicity of F,

$$(F(x^k) - F(\tilde{x}^{k_0}))^T(x^k - \tilde{x}^{k_0}) \geq 0.$$

Thus we deduce

$$0 \leq F(x^k)^T(x^k - \tilde{x}^{k_0}) - \frac{d^T(x^k - \tilde{x}^{k_0})}{\| x^{k_0} \|} \leq -\frac{d^T(x^k - \tilde{x}^{k_0})}{\| x^{k_0} \|}.$$

Dividing by $\|x^k\|$ and passing to the limit $k \to \infty$, we obtain a contradiction because $d \neq 0$. □

The last part of the above theorem complements Theorem **2.3.16**, which gives a necessary and sufficient condition for a monotone VI to have a nonempty bounded solution set. Instead of the geometric condition (2.3.16):

$$K_\infty \cap [\,-(\,F(K)^*\,)\,] = \{0\},$$

condition (d) in Theorem **5.5.15** pertains to the solvability of all perturbed VIs $(K, q + F)$, where q has sufficiently small norm. Although the semistability of the VI cannot be established, we can deduce a limit condition of the perturbed solutions that is weaker than semistability.

5.5.16 Proposition. Let K be the Cartesian product of N closed convex sets $K_\nu \subseteq \mathbb{R}^{n_\nu}$. Let F be a continuous P_0 function on K. Suppose that SOL(K, F) is nonempty and bounded. For every open set $\mathcal{U}$ containing SOL(K, F), there exists a $\bar{\varepsilon} > 0$ such that, for every G in $\mathbb{B}(F, \bar{\varepsilon}, \mathcal{U})$, SOL$(K, G) \cap \mathcal{U}$ is nonempty and

$$\lim_{\varepsilon \downarrow 0} \left[\sup_{G \in \mathbb{B}(F,\varepsilon,K_{\mathcal{U}})} \{\, \mathrm{dist}(\tilde{x}, \mathrm{SOL}(K,F)) \,:\, \tilde{x} \,\in\, \mathrm{SOL}(K,G) \,\cap\, \mathcal{U} \,\} \right] = 0.$$

Proof. Theorem **3.6.6** implies the existence of $\eta > 0$ such that the level set

$$\{\, x \,\in\, \mathbb{R}^n \,:\, \|\, \mathbf{F}^{\mathrm{nat}}_K(x) \,\| \,\leq\, \eta \,\} \tag{5.5.7}$$

is bounded. We claim that

$$\bigcup_{\varepsilon \in (0,\bar{\varepsilon})} \; \bigcup_{G \in \mathbb{B}(F,\varepsilon,K_{\mathcal{U}})} \mathrm{SOL}(K, G) \cap \mathcal{U} \tag{5.5.8}$$

is bounded for all $\bar{\varepsilon} \in (0, \eta)$. Indeed, if $\tilde{x} \in \mathrm{SOL}(K, G) \cap \mathcal{U}$, we have

$$\begin{aligned} \|\, \mathbf{F}^{\mathrm{nat}}_K(\tilde{x}) \,\| &= \|\, \Pi_K(\tilde{x} - F(\tilde{x})) - \Pi_K(\tilde{x} - G(\tilde{x})) \,\| \\ &\leq \|\, F(\tilde{x}) - G(\tilde{x}) \,\| \,\leq\, \sup_{x \in K \cap \mathrm{cl}\,\mathcal{U}} \|\, F(x) - G(x) \,\|. \end{aligned}$$

Therefore, provided that $\bar{\varepsilon} < \eta$, the set (5.5.8) is contained in the set (5.5.7) and the boundedness of the former set is therefore immediate. From this boundedness property, the desired limit follows readily. □

We next give a necessary and sufficient condition for an AVI of the P_0 type to be stable. This result follows easily from Theorem **5.5.15** and Corollary **5.5.9**. No proof is required.

5.5.17 Proposition. Let K be the Cartesian product of N polyhedra $K_\nu \subseteq \mathbb{R}^{n_\nu}$. Let M be a P_0 matrix on $\mathbb{R}^n \equiv \prod_{\nu=1}^N \mathbb{R}^{n_\nu}$. The AVI (K, q, M) is stable if and only if $\mathrm{SOL}(K, q, M)$ is nonempty and bounded. □

5.6 Exercises

5.6.1 Let $f : \mathbb{R}^{n+m+1} \to \mathbb{R}^n$ and $F : \mathbb{R}^{n+m} \to \mathbb{R}^m$ be locally Lipschitz continuous functions. Let K a closed convex set in $\mathbb{R}^m$. Let x^0 and v^0 be two given vectors in $\mathbb{R}^n$ and t_0 be a given scalar. Consider an ordinary differential complementarity system of finding two functions $x(t)$ and $y(t)$ satisfying

$$\begin{aligned} \dot{x} &= f(x, y, t) \\ y &\in \mathrm{SOL}(K, F(x, \cdot)) \end{aligned}$$

for all $t \in [t_0, \bar{t}]$ and $x(t_0) = x^0$ and $\dot{x}(t_0) = v^0$, where $\dot{x} \equiv dx/dt$. Suppose that the VI $(K, F(x^0, \cdot))$ has a strongly stable solution. Show that there exist a scalar $\bar{t} > t_0$, a continuously differentiable function $x : [t_0, \bar{t}] \to \mathbb{R}^n$ and a Lipschitz continuous function $y : [t_0, \bar{t}] \to \mathbb{R}^m$ that solves the above differential complementarity system.

5.6.2 Let $f : \mathbb{R}^n \to \mathbb{R}^n$ be a PA map. Show that if f is F-differentiable at a zero x^*, then the following four statements are equivalent:

(a) x^* is strongly stable;

(b) x^* is stable;

(c) f is injective in a neighborhood of x^*;

(d) x^* is isolated.

5.6.3 Let x^* be a nondegenerate solution of the AVI (K, q, M) and let $\mathcal{C}$ be the critical cone of this AVI at x^*, which must be a linear subspace by the nondegeneracy of x^*. Show that the following five statements are equivalent.

(a) x^* is strongly stable;

(b) x^* is stable;

(c) x^* is isolated;

(d) $M\Pi_{\mathcal{C}} + \mathbf{I} - \Pi_{\mathcal{C}}$ is a nonsingular linear transformation;

(e) for each nonzero $v \in \mathcal{C}$, we have $Mv \notin \mathcal{C}^\perp$.

5.6.4 Consider the AVI (K, q, M) in $\mathbb{R}^4$, where

$$K \equiv \{ (x_1, x_2, x_3, x_4) \in \mathbb{R}^4_+ : x_1 + x_3 = 16, \, x_2 + x_4 = 4 \},$$

$$q \equiv \begin{pmatrix} 30 \\ 28 \\ 30 \\ 28 \end{pmatrix}, \quad \text{and} \quad M \equiv \begin{bmatrix} 1.5 & 5 & 0 & 0 \\ 1.3 & 2.6 & 0 & 0 \\ 0 & 0 & 1.5 & 5 \\ 0 & 0 & 1.3 & 2.6 \end{bmatrix}.$$

Show that

$$x^1 \equiv \begin{pmatrix} 4/3 \\ 4 \\ 44/3 \\ 0 \end{pmatrix}, \quad x^2 \equiv \begin{pmatrix} 8 \\ 2 \\ 8 \\ 2 \end{pmatrix}, \quad \text{and} \quad x^3 \equiv \begin{pmatrix} 44/3 \\ 0 \\ 4/3 \\ 4 \end{pmatrix}$$

are the only solutions of the AVI; show further that all three solutions are nondegenerate and strongly stable. Finally, show that M is strictly copositive on $\mathcal{C}(x^1; K, F)$ but not on $\mathcal{C}(x^2; K, F)$.

5.6.5 Let $H : \mathbb{R}^n \to \mathbb{R}^n$ be a B-differentiable function at $x \in \mathbb{R}^n$. Show that if x is a strongly stable zero of H, then $H'(x; \cdot)$ is a globally Lipschitz homeomorphism. Give an example to show that the converse does not always hold.

5.6.6 A set-valued map $\Phi : \mathbb{R}^m \to \mathbb{R}^n$ is said to have the *Aubin property* at a pair $(u^0, v^0) \in \operatorname{gph} \Phi$ if there exist a constant $L > 0$ and an open neighborhood $U \times V$ of (u^0, v^0) such that

$$\Phi(u) \cap V \subseteq \Phi(u') + L \, \| u - u' \|, \quad \forall \, u, u' \in U.$$

(a) Let $F : \mathbb{R}^n \to \mathbb{R}^m$ and let $\Phi \equiv F^{-1}$. Show that Φ has the Aubin property at a pair $(y^0, x^0) \in \operatorname{gph} \Phi$ if and only if there exist an open neighborhood $\mathcal{N}$ of x^0 and positive constants ε and c such that, for any y and y' in $\mathbb{B}(y^0, \varepsilon)$,

$$F^{-1}(y) \cap \mathcal{N} \subseteq F^{-1}(y') + c \, \| y - y' \|.$$

(b) Deduce from (a) that if G is a strong FOA of F at a zero x^0 of F, then F^{-1} has the Aubin property at $(0, x^0)$ if and only if G^{-1} has the Aubin property at the same pair.

5.6.7 Consider the affine pair (K, M), where K is a polyhedral set in $\mathbb{R}^n$ and M is an $n \times n$ matrix. For every $q \in \mathbb{R}^n$, let $\Phi(q) \equiv \mathrm{SOL}(K, q, M)$. Let (q^0, x^0) be a given pair in the graph of Φ. Show that the following three statements are equivalent.

(a) x^0 is a strongly stable/regular solution of the AVI (K, q^0, M).

(b) Φ has the Aubin property at (q^0, x^0).

(c) Φ is lower semicontinuous at every (q, x) that is sufficiently close to (q^0, x^0).

(Hint: see Exercise **4.8.12** for (c) $\Rightarrow$ (a).)

5.6.8 Let x^* be a solution of the linearly constrained VI (K, F), where K is a polyhedral set in $\mathbb{R}^n$ and F is continuously differentiable mapping in a neighborhood of x^*. Show that the following three statements are equivalent.

(a) x^* is a strongly stable/regular solution of the VI (K, F).

(b) The inverse of the normal map $\mathbf{F}_K^{\mathrm{nor}}$ has the Aubin property at the pair $(0, z^*)$, where $z^* \equiv x^* - F(x^*)$.

(c) The inverse of the affine normal map of the semi-linearized AVI (K, F^*), where

$$F^*(x) \equiv F(x^*) + JF(x^*)(x - x^*), \quad \forall x \in K,$$

is the linearization of F at x^*, has the Aubin property at the pair $(0, z^*)$.

(Hint: use part (b) of Exercise **5.6.6** to establish the equivalence of (b) and (c) herein.)

5.6.9 Let K be a polyhedron in $\mathbb{R}^n$ and $F : \mathbb{R}^n \to \mathbb{R}^n$ be continuously differentiable in a neighborhood of a solution $x^* \in \mathrm{SOL}(K, F)$. Suppose that $(\mathcal{C}(x^*; K, F), JF(x^*))$ is an R_0 pair and $\mathrm{ind}\,\mathbf{M}_{\mathcal{C}}^{\mathrm{nor}}$ is nonzero, where $M \equiv JF(x^*)$ and $\mathcal{C}$ is the shorthand for $\mathcal{C}(x^*; K, F)$. Show that x^* is a stable solution of the VI (K, F).

5.6.10 Let K be a closed convex set in $\mathbb{R}^n$ and $F : \mathbb{R}^n \to \mathbb{R}^n$ be continuously differentiable in a neighborhood of a solution $x^* \in \mathrm{SOL}(K, F)$. Suppose that $JF(x^*)$ is strictly copositive on $K - K$. Use Propositions **5.2.15** and **5.3.6** to show that x^* is strongly stable.

5.6.11 Let $A \in \mathbb{R}^{n\times n}$, $B \in \mathbb{R}^{n\times m}$, $C \in \mathbb{R}^{m\times n}$, and $D \in \mathbb{R}^{m\times m}$ be given matrices. Suppose that the homogenous MLCP:

$$0 = Ax + By$$
$$0 \leq y \perp Cx + Dy \geq 0$$

has the origin as the unique solution. Write

$$M \equiv \begin{bmatrix} A & B \\ C & D \end{bmatrix}$$

and let $K \equiv \mathbb{R}^n \times \mathbb{R}^m_+$. Show that there exists $\delta > 0$ such that, for all matrices

$$\tilde{M} \equiv \begin{bmatrix} \tilde{A} & \tilde{B} \\ \tilde{C} & \tilde{D} \end{bmatrix}$$

partitioned similarly to M and satisfying two conditions: (a) $\tilde{A}$ is nonsingular, and (b) $\|M - \tilde{M}\| < \delta$, it holds that

$$\operatorname{ind} \mathbf{M}^{\rm nor}_K = (\operatorname{sgn}\det \tilde{A}) \operatorname{ind} \tilde{\mathbf{E}}^{\rm nor}_{\mathbb{R}^m_+},$$

where $\tilde{E} \equiv \tilde{M}/\tilde{A}$ is the Schur complement of $\tilde{A}$ in $\tilde{M}$. (Hint consider the two homotopies:

$$H^1(x,y,t) \equiv \begin{pmatrix} \tilde{A}x + \tilde{B}y^+ \\ (1-t)(\tilde{C}x + \tilde{D}y^+) + t\,\tilde{E}y^+ - y^- \end{pmatrix}$$

and

$$H^2(x,y,t) \equiv \begin{pmatrix} \tilde{A}x + (1-t)\tilde{B}y^+ \\ \tilde{E}y^+ - y^- \end{pmatrix}.$$

Use the homotopy invariance property and the Cartesian product property of the degree to show that $\operatorname{ind} \tilde{\mathbf{M}}^{\rm nor}_K$ is equal to the product of $\operatorname{sgn}\det \tilde{A}$ and $\operatorname{ind} \tilde{\mathbf{E}}^{\rm nor}_{\mathbb{R}^m_+}$. Use the nearness property of the degree to complete the proof.)

5.6.12 This exercise concerns the KKT system (5.3.8) where the functions g_i are not assume convex and the functions h_j are not assumed affine. A KKT triple $z^* \equiv (x^*, \mu^*, \lambda^*)$ is said to be stable if for every open neighborhood

$$\mathcal{Z} \equiv \mathcal{N} \times \mathcal{U} \times \mathcal{V} \subset \mathbb{R}^{n+\ell+m}$$

of z^* such that z^* is the unique KKT triple in $\operatorname{cl}\mathcal{Z}$, positive scalars ε and c exist such for every function $\tilde{F}$ in $\mathbb{B}(F;\varepsilon,\operatorname{cl}\mathcal{N})$, every C^1 function

$\tilde{h}$ in $\mathbb{B}(h;\varepsilon,\mathrm{cl}\,\mathcal{N})$ with $J\tilde{h}$ in $\mathbb{B}(Jh;\varepsilon,\mathrm{cl}\,\mathcal{N})$, and every C^1 function $\tilde{g}$ in $\mathbb{B}(g;\varepsilon,\mathrm{cl}\,\mathcal{N})$ with $J\tilde{g}$ in $\mathbb{B}(Jg;\varepsilon,\mathrm{cl}\,\mathcal{N})$, the perturbed KKT system

$$\tilde{F}(x) + \sum_{j=1}^{\ell} \mu_j \nabla \tilde{h}_j(x) + \sum_{i=1}^{m} \lambda_i \nabla \tilde{g}_i(x) = 0$$

$$\tilde{h}(x) = 0$$

$$0 \leq \lambda \perp \tilde{g}(x) \leq 0.$$

has a solution in $\mathcal{Z}$; moreover, any such perturbed KKT triple (x,μ,λ) in $\mathcal{Z}$ satisfies

$$\|\,(x^*,\mu^*,\lambda^*) - (x,\mu,\lambda)\,\| \leq c\,\max\{\,\|\,F(x) - \tilde{F}(x)\,\|,\ \|\,g(x) - \tilde{g}(x)\,\|,$$
$$\|\,Jg(x) - J\tilde{g}(x)\,\|,\ \|\,h(x) - \tilde{h}(x)\,\|,\ \|\,Jh(x) - J\tilde{h}(x)\,\|\,\}.$$

(a) Show that if z^* is stable KKT triple, then (x^*,λ^*) satisfies the SMFCQ. Show further that an open neighborhood $\mathcal{N}'$ of x^* and positive scalars ε' and c' exist such that, for every function $\tilde{F}$ in $\mathbb{B}(F;\varepsilon',\mathrm{cl}\,\mathcal{N}')$, every C^1 function $\tilde{h}$ in $\mathbb{B}(h;\varepsilon',\mathrm{cl}\,\mathcal{N}')$ with $J\tilde{h}$ in $\mathbb{B}(Jh;\varepsilon',\mathrm{cl}\,\mathcal{N}')$, and every C^1 function $\tilde{g}$ in $\mathbb{B}(g;\varepsilon',\mathrm{cl}\,\mathcal{N}')$ with $J\tilde{g}$ in $\mathbb{B}(Jg;\varepsilon',\mathrm{cl}\,\mathcal{N}')$, every solution $\tilde{x}$ of the perturbed VI $(\tilde{K},\tilde{F})$ that lies in $\mathcal{N}'$, where

$$\tilde{K} \equiv \{\, x \in \mathbb{R}^n : \tilde{h}(x) = 0,\ \tilde{g}(x) \leq 0\,\},$$

must satisfy

$$\|\,\tilde{x} - x^*\,\| \leq c'\,\max\{\,\|\,F(\tilde{x}) - \tilde{F}(\tilde{x})\,\|,\ \|\,g(\tilde{x}) - \tilde{g}(\tilde{x})\,\|,$$
$$\|\,Jg(\tilde{x}) - J\tilde{g}(\tilde{x})\,\|,\ \|\,h(\tilde{x}) - \tilde{h}(\tilde{x})\,\|,\ \|\,Jh(\tilde{x}) - J\tilde{h}(\tilde{x})\,\|\,\}.$$

(b) Conversely, suppose that z^* is a KKT triple satisfying the SMFCQ. Let

$$\alpha \equiv \{\, i : \lambda_i^* > 0 = g_i(x^*)\,\}$$
$$\beta \equiv \{\, i : \lambda_i^* = 0 = g_i(x^*)\,\}$$
$$\gamma \equiv \{\, i : \lambda_i^* = 0 > g_i(x^*)\,\}$$

be the strongly active, degenerate, and inactive set of z^*, respectively. Suppose further that the matrix

$$\begin{bmatrix} J_x\mathbf{L}(z^*) & Jh(x^*)^T & Jg_\alpha(x^*)^T \\ -Jh(x^*) & 0 & 0 \\ -Jg_\alpha(x^*) & 0 & 0 \end{bmatrix}$$

is nonsingular and that the Schur complement of the above matrix in

$$\begin{bmatrix} J_x\mathbf{L}(z^*) & Jh(x^*)^T & Jg_\alpha(x^*)^T & Jg_\beta(x^*)^T \\ -Jh(x^*) & 0 & 0 & 0 \\ -Jg_\alpha(x^*) & 0 & 0 & 0 \\ -Jg_\beta(x^*) & 0 & 0 & 0 \end{bmatrix}$$

is R_0 and semicopositive on $\mathbb{R}^{|\beta|}_+$. Show that z^* is a stable KKT triple.

5.6.13 This exercise is a continuation of Exercise **5.6.12** with F being the gradient map of a twice continuously differentiable, real-valued function θ. Show that the following two statements are equivalent for a KKT triple $z^* \equiv (x^*, \mu^*, \lambda^*)$.

(a) z^* is stable and x^* is a local minimizer of θ on K.

(b) The SMFCQ holds at z^* and $\nabla_x^2 L(z^*)$ is strictly copositive on the critical cone $\mathcal{C}(x^*; K, F)$, where $L(x, \mu, \lambda) \equiv \theta(x) + \mu^T h(x) + \lambda^T g(x)$.

The strict copositivity property in (b) is precisely the classical second-order sufficiency condition in NLP.

5.6.14 Let $\Phi : \mathbb{R}^n \to \mathbb{R}^m$ be a single-valued polyhedral (multi)function; that is, the graph of Φ is piecewise polyhedral (it is the union of finitely many polyhedra in $\mathbb{R}^{n+m}$) and $\Phi(u)$ is a singleton for all $u \in \operatorname{dom}\Phi$.

(a) Show that Φ is a PA map on its domain; that is, (i) Φ is continuous on its domain, meaning that if $\{u^k\} \subset \operatorname{dom}\Phi$ is a sequence of vectors converging to u^∞, then $\{\Phi(u^k)\}$ converges to $\Phi(u^\infty)$, and (ii) there exist finitely many affine functions Φ^i for $i = 1, \ldots, k$ such that $\Phi(u) \in \{\Phi^1(u), \cdots, \Phi^k(u)\}$ for all $u \in \operatorname{dom}\Phi$. (Hint: First use Theorem **5.5.8** to show that Φ is continuous on its domain. Next let $\{P_i : i = 1, \ldots N\}$ be finitely many polyhedra in $\mathbb{R}^{n+m}$ whose union is $\operatorname{gph}\Phi$. Let $u \in \mathbb{R}^n$ be arbitrary and suppose $(u, \Phi(u))$ belongs to P_i for some i. Write

$$P_i \equiv \{ (x, y) \in \mathbb{R}^{n+m} : A^i x + B^i y \leq a^i \}$$

for some matrices A^i and B^i of appropriate dimensions and some vector a^i of corresponding dimension. Let $\mathcal{I}$ be the active index set at $(u, \Phi(u))$; thus

$$(A^i u + B^i \Phi(u))_j = a^i_j, \quad \forall j \in \mathcal{I}$$

and

$$(A^i u + B^i \Phi(u))_j < a^i_j, \quad \forall j \notin \mathcal{I}.$$

Show that $\mathcal{I}$ is nonempty and the columns of $(B^i)_{\mathcal{I}\cdot}$ must be linearly independent. It is now easy to define the affine pieces of Φ.)

(b) Show that if $\operatorname{dom}\Phi$ is convex, then Φ is Lipschitz continuous on its domain.

5.6.15 Consider the pair $(K, A^T EA)$, where K is a polyhedron and E is a symmetric positive semidefinite matrix. Define the single-valued map Φ, where for each vector q belonging to the AVI range of the pair $(K, A^T EA)$, $\Phi(q) \equiv EAx$ for any $x \in \text{SOL}(K, q, A^T EA)$, and $\Phi(q)$ is the empty set otherwise; cf. Exercise **2.9.20**. Use Exercise **5.6.14** to show that Φ is Lipschitz continuous on its domain. A related property of the solutions to a nonlinear generalization of this class of AVIs can be found in Corollary **6.2.11**.

5.6.16 Let

$$X \equiv \{ x \in \mathbb{R}^n : Ax \leq b \}$$

and

$$Y \equiv \{ y \in \mathbb{R}^m : Cy \leq d \}$$

be two given polyhedra. Let

$$L(x,y) = p^T x + q^T y + \tfrac{1}{2} x^T P x + x^T R y - \tfrac{1}{2} y^T Q y, \quad (x,y) \in \mathbb{R}^{n+m},$$

where the matrices P and Q are symmetric positive semidefinite. By Exercise **2.9.18**, if (x^1, y^1) and (x^2, y^2) are any two pairs of saddle points of the triple (L, X, Y), then

$$Px^1 = Px^2 \quad \text{and} \quad Qy^1 = Qy^2.$$

Show that, for fixed matrices A, C, P, R, and Q, the function

$$(b, d, p, q) \mapsto (Px, Qy),$$

where (x, y) is any saddle point of the triple (L, X, Y) if such a point exists, is Lipschitz continuous on its domain.

5.6.17 Let K be a polyhedral cone in $\mathbb{R}^n$ and M be an $n \times n$ copositive matrix on K. Show that for every $q \in \operatorname{int} \mathcal{K}(K, M)^*$, the CP (K, q, M) is stable.

5.6.18 Let K be a closed convex subset of $\mathbb{R}^n$ and F be a continuous mapping from K into $\mathbb{R}^n$. Suppose that $\text{SOL}(K, F) \neq \emptyset$ and that the solutions of the VI (K, F) are pointwise semiregular.

(a) Show that for every compact set S, there exist positive scalars ε and c such that

$$\| q \| < \varepsilon \Rightarrow \text{SOL}(K, q + F) \cap S \subseteq \text{SOL}(K, F) + \mathbb{B}(0, c \| q \|).$$

(b) Suppose that the natural map $\mathbf{F}_K^{\text{nat}}$ is norm-coercive on K. Show that the VI (K, F) is semistable. In particular, this must be true if K is compact.

(c) Suppose that $\text{SOL}(K, F)$ is a finite set. Show that the VI (K, F) is semistable if and only if for every $\varepsilon > 0$, there exists $\delta > 0$, such that

$$\| q \| < \delta \Rightarrow \text{SOL}(K, q + F) \subset \text{SOL}(K, F) + \mathbb{B}(0, \varepsilon).$$

5.6.19 Let $C \subseteq \mathbb{R}^m$ be a pointed, solid, closed convex cone and let F be a continuous monotone mapping from $C \times \mathbb{R}^\ell$ into $\mathbb{R}^{m+\ell}$. Write

$$F(u, v) \equiv \begin{pmatrix} H(u, v) \\ L(u, v) \end{pmatrix} \in \mathbb{R}^{m+\ell}, \quad (u, v) \in C \times \mathbb{R}^\ell,$$

where $H : C \times \mathbb{R}^\ell \to \mathbb{R}^m$ is continuous and $L : \mathbb{R}^{m+\ell} \to \mathbb{R}^\ell$ is a surjective affine map. Suppose that $\text{SOL}(K, F) \neq \emptyset$ and that there exists a vector $x^0 \equiv (u^0, v^0) \in \text{int}\, C \times \mathbb{R}^\ell$ satisfying $L(x^0) = 0$. Show that $\text{SOL}(K, F)$ is bounded if and only if a vector $\hat{x} \in C \times \mathbb{R}^\ell$ exists such that $H(\hat{x}) \in \text{int}\, C^*$ and $L(\hat{x}) = 0$.

5.6.20 This and the next exercise pertain to the implicit horizontal CP:

$$\begin{aligned} H(x, y) &= 0 \\ K \ni x \perp y &\in K^*, \end{aligned} \tag{5.6.1}$$

where $H : \mathbb{R}^{2n} \to \mathbb{R}^n$ is a continuous function and K is a closed convex cone in $\mathbb{R}^n$. This exercise concerns the local uniqueness of a solution and the next one concerns the stability issues. A solution (x^*, y^*) of (5.6.1) is said to be locally unique, or isolated, if there exists an open neighborhood $\mathcal{W} \subset \mathbb{R}^{2n}$ of the pair (x^*, y^*) such that (5.6.1) has no other solution in $\text{cl}\, \mathcal{W}$. Suppose that H is C^1 in a neighborhood of (x^*, y^*).

(a) Show that if the following implication holds:

$$\left.\begin{aligned} J_x H(x^*, y^*) u + J_y H(x^*, y^*) v &= 0 \\ u \in \mathcal{T}(x^*; K) \cap (y^*)^\perp & \\ v \in \mathcal{T}(y^*; K^*) \cap (x^*)^\perp & \\ (u, v) \neq 0 & \end{aligned}\right\} \Rightarrow u^T v > 0,$$

then (x^*, y^*) is isolated.

(b) Suppose that K is polyhedral. Show that if the following implication holds:

$$\left.\begin{array}{r} J_xH(x^*,y^*)u + J_yH(x^*,y^*)v = 0 \\ \mathcal{T}(x^*;K) \cap (y^*)^\perp \ni u \perp v \in \mathcal{T}(y^*;K^*) \cap (x^*)^\perp \end{array}\right\}$$
$$\Rightarrow (u,v) = 0,$$

then (x^*,y^*) is isolated.

(c) Finally, show that if K is polyhedral and H is affine, then the implication in (b) is also necessary for (x^*,y^*) to be isolated.

5.6.21 A solution pair (x^*,y^*) of (5.6.1) is said to be *stable* if for every open neighborhood W of (x^*,y^*) such that (5.6.1) has no other solution in $\operatorname{cl} W$, there exist two positive scalars c and ε such that, for every continuous function G satisfying

$$\sup_{(x,y)\in(K\times K^*)\cap \operatorname{cl} W} \| G(x,y) - H(x,y) \| \leq \varepsilon,$$

the problem

$$G(x,y) = 0$$
$$K \ni x \perp y \in K^*,$$

has a solution pair in W; moreover, for any such solution (x,y),

$$\| (x,y) - (x^*,y^*) \| \leq c \| G(x,y) - H(x,y) \|.$$

(a) Let $z^* \equiv x^* - y^*$ and

$$\Phi(z) \equiv H(\Pi_K(z), \Pi_K(z) - z), \quad \forall z \in \mathbb{R}^n.$$

Show that $\Phi(z^*) = 0$. Show further that z^* is an isolated zero of Φ if and only if (x^*,y^*) is an isolated solution of (5.6.1).

(b) Show that if z^* is a stable zero of Φ, then (x^*,y^*) is a stable solution of (5.6.1).

5.6.22 Let $\Phi : \Omega \subseteq \mathbb{R}^n \to \mathbb{R}^m$ be B-differentiable on the open set Ω. Suppose that for every $x \in \Omega$,

$$\partial_B\Phi(x) = \partial_B\Psi(0),$$

where $\Psi \equiv \Phi'(x;\cdot)$. Let $F : \mathbb{R}^m \to \mathbb{R}^\ell$ be continuously differentiable in an open neighborhood of $\Phi(\bar{x})$, where $\bar{x} \in \Omega$. Show that

$$\partial_B(F \circ \Phi)(\bar{x}) = JF(\Phi(\bar{x})) \circ \partial_B\Phi(\bar{x}) = \partial_B\Xi(0),$$

where $\Xi \equiv (F \circ \Phi)'(\bar{x};\cdot)$.

5.6.23 Let K be the Lorentz cone in $\mathbb{R}^{n+1}$; see Exercise **1.8.27**. Show that $\partial_B\Pi_K(x,t) = \partial_B\Pi'_K((x,t);\cdot)(0)$ for any $(x,t) \in \mathbb{R}^{n+1}$.

5.7 Notes and Comments

Sensitivity and stability analysis of nonlinear programs began with the pioneering work of Fiacco and McCormick [246]. Under the restrictive strict complementarity assumption and the LICQ, these authors were able to use the classical implicit function theorem for smooth equations to obtain sensitivity results for parametric optimization problems. Subsequent parametric analysis of optimization problems have focused on the relaxation of these two restrictions. Today this subject is very well developed, with a rich literature (see the survey [78]) and many excellent books such as [36, 79, 245, 316, 433], which contain numerous references.

In contrast, the systematic study of the sensitivity and stability of VIs and CPs has a shorter history. As explained in the paper [514], there are substantial differences between the sensitivity and stability analysis of an NLP and that of a VI. Most importantly, the lack of symmetry in the defining function of a VI invalidates a straightforward application of the results in optimization. In this regard, one can credit Robinson and Kojima for laying the foundation for the contemporary sensitivity and stability analysis of VI/CPs.

Building on his earlier work pertaining to normed convex processes and parametric differentiable inequality and related systems [722, 723, 725, 726], Robinson introduced the broad framework of generalized equations in several seminal papers [728, 730, 732, 734, 736], which contain many fundamental results that have inspired much subsequent research. Among Robinson's major contributions is the definition of a strongly regular solution to a generalized equation [730]. The consequence of this concept is far reaching, having a strong impact in both the parametric analysis of NLPs and VIs and the convergence of Newton's method for solving these problems. For the application of the theory of generalized equations to parameterized optimization problems with cone constraints, see Bonnans and Shapiro [77, 780] and the references in [78, 79].

At about the same time as Robinson introduced his theory of generalized equations, Kojima published his influential paper [443], in which he defined the concept of a strongly stable stationary solution to a nonlinear program. Like Robinson's theory, a major departure in Kojima's work from the classical sensitivity analysis of Fiacco and McCormick is that the restrictive strict complementarity assumption is removed. In the absence of this assumption, the implicit function theorem for smooth equations can no longer be used; thus new tools need to be introduced. While Robinson used Brouwer's fixed-point theorem in his work on strong regularity, Kojima developed his results using degree theory. Kojima and Hiabayashi [444] discuss continuous one-parameter families of nonlinear programs, paying particular attention to the question of how the Morse index of a stationary solution varies with the parameter. Relying on Kojima's theory [442] of locally nonsingular PC^1 functions, Kojima, Okada, and Shindoh [447] study strongly stable Nash equilibrium points of multi-person noncooperative games.

The paper [443] is arguably the earliest to use a degree approach for the sensitivity and stability analysis of a parametric KKT system using a nonsmooth-equation formulation. While Robinson's concept of a strongly regular solution is defined for a generalized equation, when specialized to the KKT system, this concept has much in common with Kojima's concept of a strongly stable stationary point for a nonlinear program. The equivalence of these two concepts is investigated by several authors. According to the bibliographic notes in [194, 431], Jongen, Möbert, Rückmann, and Tammer [387], assuming the LICQ, showed in 1987 that the respective matrix-theoretic characterizations of Robinson [730] and Kojima [443] are equivalent. So, except for the LICQ, the paper [387] is the first to establish the equivalence of these two important concepts for a KKT point of a smooth nonlinear program. In [435, 468], the LICQ was shown to be unnecessary for the equivalence to hold. The paper [431] gives a unified framework for the simultaneous treatment of strong stability and strong regularity. The article by Bonnans and Sulem [80] contains a proof of the converse statement in Corollary **5.3.23**; see also Dontchev [191]. The paper [386] establishes the equivalence of the strong stability of a KKT triple of an optimization problem and the nonsingularity of the Clarke Jacobian of Kojima's nonsmooth-equation formulation of the KKT system. The most recent book by Klatte and Kummer [433] establishes detailed links between regularity and derivative concepts of nonsmooth analysis as well as solution methods and stability for optimization, complementarity, and equilibrium problems. These topics are treated in various parts of our book.

Ha [318] introduced a stability concept for a solution to an LCP. In essence, Ha's definition replaces the semistability requirement by a continuity condition on the perturbed solutions, which is necessarily valid if the base solution is isolated. Ha obtained a number of useful results, including the characterization of a strongly stable solution to an LCP and the construction of Example **5.3.16** to demonstrate that the nonsingularity of the principal submatrix $M_{\gamma\gamma}$ is not a necessary condition for a solution $x^* \in \mathrm{SOL}(q, M)$ to be stable, where $\gamma \equiv \mathrm{supp}(x^*)$. Subsequently, employing degree theory, Ha [319] extended his stability study to the NCP. Gowda and Pang [303] give complete characterizations for x^* to be a solution of the LCP (q, M). Ha's work has further influenced subsequent research on the use of degree theory in complementarity. Stewart [808] obtained an index formula for a degenerate LCP and used it to explain the degenerate solutions of a particular problem due to Broyden [91]. Gowda [300] used Stewart's formula in two interesting applications of the LCP. These authors' work accumulates in [305], where degree theory was the main tool employed for an extensive study of the stability of the mixed NCP and VI. Exercise **5.6.11** contains a summary of some results in [305] pertaining to the mixed LCP.

In the late 1980s and early 1990s, many authors contributed to the sensitivity study of the VI; some of their results were derived under restrictive assumptions, however. The late Dafermos [152] obtained sensitivity results by assuming some strong monotonicity conditions; she and Nagurney [156] considered solution sensitivity of asymmetric network equilibrium problems; see also [840]. Qiu and Kyparisis [484] considered solution differentiability for oligopolisitic network equilibria. Kyparisis and Ip [483] studied sensitivity analysis of parametric implicit complementarity problems. In addition, Kyparisis [478, 479, 480, 482], Qiu and Magnanti [703, 704], and Tobin [838] have all made important contributions to this area.

The paper [667] employs degree theory to investigate the stability and proto-differentiability of an isolated solution to a B-differentiable equation. This paper is the source for much of the stability theory presented in Subsections **5.2.1** and **5.2.2**. The theory of proto-differentiability of set-valued maps originated from [748]. The identification of the semistability property as a separate component in stability theory originated from Bonnans [72, 73, 74], who recognized the importance of this property in the convergence analysis of Newton's method for solving VIs and NLPs; see also [666]. Robinson [739] introduced the concepts of pointwise FOAs and strong FOAs of locally Lipschitz continuous functions in normed linear spaces and obtained an implicit-function theorem for the class of locally Lipschitz equations admitting strong FOAs. For related implicit-function theorems, see [192, 193].

The applicability of Robinson's strong regularity theory to the VI (K, F) is restricted to two situations: one situation is that the set K is convex, and another situation pertains to the KKT system of the VI where K is finitely representable by differentiable inequalities but is not necessarily convex. In the latter situation, the uniqueness of a KKT triple must be in place for the theory to be applicable. Robinson's original treatment assumes the LICQ. In contrast, Kojima's theory of a strongly stable stationary point is applicable to a nonconvex NLP under the MFCQ, which allows nonunique KKT multipliers. In other words, the latter theory is applicable to a KKT point of the VI (K, F), where F is a gradient map. The latter assumption is employed in a nontrivial way in Kojima's analysis. Gowda and Pang [305] were able to obtain sensitivity results for VIs with convex perturbed feasible sets without uniqueness of the multipliers. Liu, in his Ph.D. thesis [515] and in two accompanying papers [513, 514], made significant advances in the sensitivity and stability theory of NLPs and VIs. Most importantly, he extended Kojima's strong stability result to the VI by removing the symmetry assumption on F. Specifically, Liu established that for a VI (K, F), where K is finitely generated but not necessarily convex, under the MFCQ, a KKT point of the VI is strongly stable if and only if the "general coorientedness condition" holds at the point. For details of Liu's deep result, see [514].

In an important paper [194], Dontchev and Rockafellar studied the strong regularity of a solution to a linearly constrained VI in terms of some Lipschitz properties of the solution (set-valued) map. Specifically, inspired by Aubin's paper [23] that deals with the Lipschitz behavior of solutions to convex minimization problems, Dontchev and Rockafellar defined the Aubin property of a set-valued map at an element in its graph; see Exercise **5.6.6**. Exercises **4.8.12**, **5.6.7**, and **5.6.8** are derived from [194]. In the subsequent paper [195], Dontchev and Rockafellar extended their study to investigate sensitivity properties of stationary points and KKT triples of parameterized nonlinear programs. Robinson [745] studies stability conditions for saddle problems on products of polyhedra.

The various characterizations of strong stability motivated Pang [669] to seek analogous characterizations for the stability of a solution to a parametric nonsmooth equation, and in particular, to an NCP and the KKT system of a VI. This reference is the source of the stability results in Subsection **5.3.2** and also Exercise **5.6.12**. Exercise **5.6.13**, which is the specialization of the previous exercise to the KKT system of an NLP, is inspired by [190]; see also the earlier paper [74] for a related result.

Part (b) in Proposition **5.2.15**, Part (b) in Exercise **5.6.6**, and their consequences such as the equivalence of (b) and (c) in Exercise **5.6.8**, as well as part (c) of Exercise **3.7.6** all share a common theme; namely, certain locally Lipschitz properties are invariant under strong first-order approximations. For details of these properties, we refer the reader to [189, 190].

Jittorntrum [379] published the first paper on the directional differentiability of the solution function to a parametric nonlinear program without strict complementarity. Inspired by Robinson's unpublished paper [736], Pang [664] established the directional differentiability of the solution function to a linearly constrained parametric VI under the strong regularity property; a continuation embedding method based on the directional differentiability of the parametric solution function was proposed in the cited reference for solving a VI. Subsequently, Robinson [738] developed an extended mathematical theory of such a method; Sellami and Robinson [770, 771] reported computational results with the implementation of an embedding method for solving nonlinear VIs. Kyparisis [481] obtained further results on the directional differentiability of parametric VIs and NLPs. The articles [171, 711] establish the directional differentiability of optimal solutions of NLPs under Slater's condition and the CRCQ, respectively. The Fréchet differentiability of the parametric solution function to VIs was discussed in [664, 481]. The SCOC was introduced in [533].

In much of the isolated sensitivity analysis, the defining function F in the VI (K, F) is assumed to be continuously differentiable in a neighborhood of the solution in question. When specialized to the NLP, these results require the objective function to be C^2. Kummer [468, 469] initiated the study of the locally Lipschitz behavior of solutions to perturbed $C^{1,1}$ NLPs via their KKT systems; $C^{1,1}$ means that the objective and constraint functions all have locally Lipschitz continuous derivatives. Subsequent papers on the sensitivity analysis of the $C^{1,1}$ programs include [427, 432, 470]. A main tool in the latter papers is the contingent derivative of a multifunction [24] in variational analysis. The use of such a derivative in characterizations of the locally upper Lipschitz stationary solutions of C^2 programs originates from King and Rockafellar [411].

The origin of the stability concept in Definition **5.5.1** for the solution set of a VI can be traced to a paper of Robinson [727], who showed that a primal-dual pair of linear programs is stable if and only if their solution sets are nonempty and bounded. This result was subsequently extended to a broad class of generalized equations [728] that includes the monotone LCP. In turn, this extension in the polyhedral case relies heavily on the upper Lipschitz continuity of a polyhedral multifunction (Theorem **5.5.8**), which was established by Robinson in [731]. Gowda and Pang [306] studied in detail the stability of the AVI, using degree theory and the latter result of Robinson. McLinden [598] presented a comprehensive study of the stability of the solution set of a VI defined by a maximal monotone set-valued map; see also [31].

In an unpublished manuscript [226], Facchinei and Pang studied the "total stability" of the VI (K, F), which allows the perturbation of both members of the pair (K, F). Unlike Definition **5.5.1**, total stability as defined in the reference concerns only the solvability of the perturbed VIs and does not stipulate the continuity of the perturbed solutions. To the best of our knowledge, this study of Facchinei and Pang is the only one that deals with the perturbation of both F and K simultaneously, where K is taken to be an arbitrary closed convex set and the perturbation of F can be arbitrary. Motivated by several papers [221, 314, 713] on the NCP of the P_0 kind and on the VI on a rectangle, a major focus of [226] was devoted to the partitioned VI (K, F) with F being a P_0 function on the product set K. Theorem **5.5.15** is derived from a result in the latter reference. As can be seen from Subsection **5.5.3**, the theory of weakly univalent functions of Gowda and Sznajder [313] plays a central role in the stability of this class of partitioned VIs.

The example in Exercise **5.6.4** appears in [688], which defines a concept of "robust solutions" to general VIs with applications to multi-class network equilibrium problems; for more discussion on the latter class of applied problems, see [587].

A major application of sensitivity and stability analysis of VIs is to the study of the class of MPECs. Indeed, the strong regularity property and the associated implicit function results for parametric VIs provided a fruitful avenue for deriving a set of first-order stationarity conditions of an MPEC [533, 654]. Scheel and Scholtes [761] obtained second-order optimality conditions of the Fiacco-McCormick type and initiated a stability analysis for this class of optimization problems.

Various CQs have played a major role throughout this chapter. One CQ that we have not utilized is Gollan's directional constraint qualification [295]. In essence, this CQ is a directional regularity condition that extends the MFCQ. Based on earlier developments that include [75], Bonnans and Shapiro [79, Section 4.5] present a theory of directional stability of optimal solutions based on Gollan's CQ instead of the MFCQ. To date, the application of Gollan's CQ is restricted to an optimization problem. The extent to which this CQ can be applied to a variational inequality with an asymmetric function deserves to be investigated.

There is a substantial literature on the sensitivity and stability analysis of optimization problems, variational inequalities, and generalized equations that utilizes the nonsmooth analytic tools of Mordukhovich [617, 618, 621]; see e.g. [496, 497, 498, 499, 500, 619, 620, 622]. The treatise by Rockafellar and Wets [752] on variational analysis provides a related set of tools that is most suitable for dealing with parameter-dependent mathematical programs with multi-valued solution sets.

Chapter 6

Theory of Error Bounds

An error bound for a subset of an Euclidean space is an inequality that bounds the distance from vectors in a *test set* to the given set in terms of a *residual function*. We have already encountered such error bounds for a polyhedron in Section **3.2** where we introduced the renowned Hoffman result, Lemma **3.2.3**. We have also very briefly discussed how a kind of local error bound for the solution set of a VI is related to its semistability; see Section **5.5**. In the present chapter, we give a comprehensive treatment of error bounds for solutions to CPs and VIs and discuss several major applications of these error bounds. The literature on the theory of error bounds for general inequality systems is vast. As always, the last section of the chapter gives references on the results presented herein.

6.1 General Discussion

We begin with a general introduction to the theory of error bounds. Let X be a given subset of $\mathbb{R}^n$. We are interested in obtaining inequalities that bound the distance $\operatorname{dist}(x, X)$ from points x in a given set S to X in terms of a computable residual function $r(x)$. The latter is a nonnegative-valued function $r : S \cup X \to \mathbb{R}_+$ whose zeros coincide with the elements of X; that is,

$$r(x) = 0 \Leftrightarrow x \in X.$$

When X is the solution set of a system of finitely many inequalities, say

$$X \equiv \{ x \in \mathbb{R}^n : g(x) \leq 0, \ h(x) = 0 \},$$

where $g : \mathbb{R}^n \to \mathbb{R}^m$ and $h : \mathbb{R}^n \to \mathbb{R}^\ell$, a natural residual function is:

$$r(x) \equiv \| \max(0, g(x)) \|^\gamma + \| h(x) \|^\eta,$$

where γ and η are suitable positive constants. When X is the solution set of a VI (K, F), where K is a closed convex subset of $\mathbb{R}^n$ and $F : \mathbb{R}^n \to \mathbb{R}^n$ is continuous, there are many residual functions of interest. In general, given an equivalent reformulation of this VI as a system of equations, i.e.,

$$x \in \text{SOL}(K, F) \Leftrightarrow G(x) = 0,$$

where $G : \mathbb{R}^n \to \mathbb{R}^n$ is continuous, a straightforward choice of a residual function is

$$r(x) \equiv \| G(x) \|^{\gamma},$$

for a suitable positive constant γ. When G is the normal (natural) map $\mathbf{F}_K^{\text{nor}}$ ($\mathbf{F}_K^{\text{nat}}$), we call the resulting residual $\|\mathbf{F}_K^{\text{nor}}\|$ ($\|\mathbf{F}_K^{\text{nat}}\|$) the normal (natural) residual. (Note: $\|\mathbf{F}_K^{\text{nor}}(z)\|$ does not provide a residual function for the solution set SOL(K, F) directly; instead, $\|\mathbf{F}_K^{\text{nor}}(z)\|$ provides a residual function for the set $(\mathbf{F}_K^{\text{nor}})^{-1}(0)$, which is equal to the image of SOL(K, F) under a transformation.) Often, we will also employ residual functions of a composite kind, such as

$$r(x) \equiv \| G(x) \|^{\gamma_1} + \| \tilde{G}(x) \|^{\gamma_2},$$

where G and $\tilde{G}$ correspond to two equation reformulations of the same VI and γ_1 and γ_2 are two possibly different positive scalars. Such composite residuals also apply to general inequality systems.

Many iterative algorithms for solving the VI (K, F) are based on the minimization of certain residual functions of this problem. Since these algorithms are only convergent in the limit, it is important to understand how these residual functions are related to the distance function to the solution set of the VI; ideally, one would want to be in the situation where the smaller the residual is the closer the iterate is to the solution set of the VI. To give a simple example to illustrate that this situation is generally not valid, consider a one-dimensional NCP (F), where $F(x) = e^{-x}$ for $x \in \mathbb{R}$. This NCP has $x = 0$ as the unique solution. By letting

$$r(x) \equiv | \min(x, F(x)) |$$

we see that $r(k) = e^{-k}$, which converges to zero as k tends to ∞. Thus, the fact that the residual is small does not necessarily imply that the point is close to the solution of the problem. As we see later, part of the reason why in this example the smallness of the residual cannot be used to predict the closeness of the point to the solution is that the test set is unbounded.

In general, the theory of error bounds provides a useful aid for understanding the connection between a residual function and the actual distance function to the set X. This theory has an important role to play in the study of unbounded asymptotics and provides valuable quantitative information about the iterates obtained at the termination of iterative algorithms for computing the elements of X.

Formally, given a residual function r of the set X and a subset (the test set) S of $\mathbb{R}^n$, we wish to establish the existence of positive scalars c_1, c_2, γ_1, and γ_2 such that

$$c_1\, r(x)^{\gamma_1} \leq \operatorname{dist}(x, X) \leq c_2\, r(x)^{\gamma_2}, \quad \forall\, x \in S. \tag{6.1.1}$$

Such an inequality is called an *error bound* for the set X with residual r and with respect to the (test) set S. Since the residual function is a computable quantity whereas the exact distance function is not (because we do not know the elements of the set X), the former function can therefore be used as a surrogate of the latter function for computational purposes, such as in the design of solution algorithms for computing an (approximate) element of X.

If r is Hölder continuous on $S \cup X$, that is if there exist positive constants L and η such that

$$|\, r(x) - r(y)\,| \leq L\, \|\, x - y\,\|^{\eta}, \quad \forall\, x, y \in S \cup X,$$

then the left-hand inequality in (6.1.1) clearly holds with $c_1 \equiv L^{-1}$ and $\gamma_1 \equiv \eta^{-1}$. In the case where $X = \operatorname{SOL}(K, F)$, an example of a Lipschitz continuous residual function is the Euclidean norm of the natural map; that is,

$$r(x) \equiv \|\, \mathbf{F}_K^{\text{nat}}(x)\,\| = \|\, x - \Pi_K(x - F(x))\,\|.$$

If F is Lipschitz continuous on $S \cup \operatorname{SOL}(K, F)$, then so is r. Another example of a Lipschitz continuous residual function is

$$r(x) \equiv \|\, (\, Ax - b\,)_+\,\|$$

that is the natural residual of the polyhedron

$$P(A, b) \equiv \{\, x \in \mathbb{R}^n \,:\, Ax \leq b\,\}.$$

In view of the above observation, the study of error bounds has focused largely on the right-hand inequality in (6.1.1) because it is the more restrictive part of the two inequalities. The expression (6.1.1) is sometimes called an *absolute error bound.* Together, these two inequalities can be used to derive a "relative error bound", which is formally described in the following simple result.

6.1.1 Proposition. Suppose that (6.1.1) holds. It holds that

$$\frac{c_1}{c_2}\frac{r(x)^{\gamma_1}}{r(a)^{\gamma_2}} \leq \frac{\operatorname{dist}(x,X)}{\operatorname{dist}(a,X)} \leq \frac{c_2}{c_1}\frac{r(x)^{\gamma_2}}{r(a)^{\gamma_1}} \qquad \text{for all } x \in S$$

for every vector a in $S \setminus X$. □

A special case of this proposition is when X consists of a singleton, say $X = \{x^*\}$ (for instance, when the VI (K, F) has a unique solution). In this case, a relative error bound for all vectors in S can be obtained as follows, provided that $x^* \neq 0$ and S contains the zero vector:

$$\frac{c_1}{c_2}\frac{r(x)^{\gamma_1}}{r(0)^{\gamma_2}} \leq \frac{\|x - x^*\|}{\|x^*\|} \leq \frac{c_2}{c_1}\frac{r(x)^{\gamma_2}}{r(0)^{\gamma_1}} \qquad \text{for all } x \in S.$$

The rest of this chapter focuses on the right-hand inequality of (6.1.1); we will also drop the subscript "2" in the constants c_2 and γ_2. Thus we are interested in the validity of the expression:

$$\operatorname{dist}(x,X) \leq c\, r(x)^{\gamma} \quad \text{for all } x \in S \tag{6.1.2}$$

for suitable positive constants c and γ. Corresponding to a given value of γ, the existence of a constant $c > 0$ such that (6.1.2) holds is clearly equivalent to the scalar

$$\hat{c} \equiv \sup_{x \in S \setminus X} \frac{\operatorname{dist}(x,X)}{r(x)^{\gamma}}$$

being finite. This scalar provides the best multiplicative constant c for the error bound (6.1.2) with γ fixed to hold. Since it is impractical, if not impossible, to compute $\hat{c}$ exactly, obtaining a (finite) upper bound for $\hat{c}$ therefore becomes a primary concern in the study of error bounds. Needless to say, it would be desirable to obtain as tight a bound for $\hat{c}$ as possible.

For a given residual function $r(x)$, we say that a *Lipschitzian error bound* holds for the set X with respect to S if there exists a constant $c > 0$ such that (6.1.2) holds with $\gamma = 1$; if this inequality holds for some $\gamma \neq 1$, we say that a *Hölderian error bound* holds for X with respect to S. For most applications, a Lipschitzian error bound is more desirable than a Hölderian error bound with an exponent $\gamma \neq 1$; however, the validity of the former typically requires more restrictive properties on the pair (X, S) and on the residual function $r(x)$ than those needed for the latter.

If $S = \mathbb{R}^n$, we say that (6.1.2) is a *global* error bound. If S is a set of the type

$$S = \{ x \in \mathbb{R}^n : r(x) \leq \varepsilon \}$$

for some $\varepsilon > 0$, we say that (6.1.2) is a *local* error bound. Another useful concept is a "pointwise" error bound. Specifically, we say that the set X admits an error bound (with residual r) near a vector $x^* \in X$ if there exist positive scalars c, γ, and ε such that (6.1.2) holds for $S \equiv \mathrm{cl}\,\mathbb{B}(x^*, \varepsilon)$. It is clear that the following two statements (a) and (b) are equivalent: for a given vector $x^* \in X$ and positive scalars c and γ,

(a) x^* is an isolated vector in X and there exists $\varepsilon > 0$ such that

$$\| x - x^* \| \leq \varepsilon \Rightarrow \mathrm{dist}(x, X) \leq c\, r(x)^\gamma;$$

(b) there exists $\varepsilon' > 0$ such that

$$\| x - x^* \| \leq \varepsilon' \Rightarrow \| x - x^* \| \leq c\, r(x)^\gamma.$$

That (b) implies (a) is obvious. If (a) holds, it suffices to restrict ε so that

$$\| x - x^* \| = \mathrm{dist}(x, X), \quad \forall\, x \in \mathrm{cl}\,\mathbb{B}(x^*, \varepsilon).$$

The following result clarifies the difference between a local error bound and a pointwise error bound.

6.1.2 Proposition. Let $r : \mathbb{R}^n \to \mathbb{R}_+$ be a continuous residual function of the nonempty closed set $X \subseteq \mathbb{R}^n$. Among the following four statements, it holds that (a) $\Rightarrow$ (b) $\Rightarrow$ (c) $\Rightarrow$ (d).

(a) There exist positive scalars c and ε such that

$$\mathrm{dist}(x, X) \leq c\, r(x), \quad \forall\, x \text{ satisfying } r(x) \leq \varepsilon.$$

(b) There exists a positive constant c' such that

$$\mathrm{dist}(x, X) \leq c' (1 + \| x \|)\, r(x), \quad \forall\, x \in \mathbb{R}^n.$$

(c) For every compact subset S of $\mathbb{R}^n$, a constant $c > 0$ exists such that

$$\mathrm{dist}(x, X) \leq c\, r(x), \quad \forall\, x \in S.$$

(d) X admits a Lipschitzian error bound with residual r near every $x \in X$.

All four statements (a)–(d) are equivalent if the level set

$$\{ x \in \mathbb{R}^n : r(x) \leq \varepsilon \}$$

is bounded for some $\varepsilon > 0$.

Proof. (a) $\Rightarrow$ (b). Without loss of generality, we may assume that the set

$$Y \equiv \{ x \in \operatorname{cl} \mathbb{B}(0,1) : r(x) \geq \varepsilon \}$$

is nonempty. Clearly, Y is compact. Thus the function $\operatorname{dist}(x, X)$ attains a finite maximum value on Y, say δ, which must be positive. Hence for every $x \in Y$,

$$\operatorname{dist}(x, X) \leq \delta \leq \frac{\delta}{\varepsilon} r(x).$$

By assumption, for every $x \in \mathbb{R}^n$ such that $r(x) \leq \varepsilon$, we have

$$\operatorname{dist}(x, X) \leq c\, r(x).$$

If $x \in \mathbb{R}^n$ is such that $\|x\| > 1$ and $r(x) \geq \varepsilon$, then for any $a \in X$,

$$\begin{aligned} \operatorname{dist}(x, X) &\leq \| x - a \| \leq \| x \| + \| a \| \\ &\leq \varepsilon^{-1} \max(\| a \|, 1) (1 + \| x \|) r(x). \end{aligned}$$

By combining the above three inequalities, (b) follows readily.

(b) $\Rightarrow$ (c) $\Rightarrow$ (d). These implications are obvious.

Suppose there exists $\bar{\varepsilon} > 0$ such that $T \equiv r^{-1}(\operatorname{cl} \mathbb{B}(0, \bar{\varepsilon}))$ is bounded. It suffices to show that (d) $\Rightarrow$ (a). Assume for contradiction that (d) holds but no positive scalars c and ε exist satisfying (a). There exist a sequence of positive scalars $\{\varepsilon_k\}$ converging to zero and a sequence of vectors $\{x^k\}$ such that for every k, $r(x^k) \leq \varepsilon_k$ and

$$\operatorname{dist}(x^k, X) > k\, r(x^k).$$

For all but finitely many members, the sequence $\{x^k\}$ belongs to the level set T. Thus without loss of generality, we may assume that this sequence converges to a vector x^∞, which must belong to X because $r(x^\infty) = 0$. By (d), there exists a neighborhood $\mathcal{N}$ of x^∞ such that an error bound holds on $\mathcal{N}$. This neighborhood must contain x^k for all k sufficiently large. Thus for some constant $c' > 0$,

$$\operatorname{dist}(x^k, X) \leq c'\, r(x^k)$$

for all k sufficiently large. This contradiction establishes (a). □

The next result shows that any local error bound must be a global error bound if the residual function $r(x)$ is convex.

6.1.3 Proposition. Let $r : \mathbb{R}^n \to \mathbb{R}_+$ be a residual function of the nonempty closed set $X \subseteq \mathbb{R}^n$. Suppose that r is convex. If there exist scalars $c > 0$, $\varepsilon > 0$, and $\gamma \geq 0$ such that

$$\operatorname{dist}(x, X) \leq c (r(x) + r(x)^\gamma), \quad \forall\, x \text{ satisfying } r(x) \leq \varepsilon,$$

then there exists $c' > 0$ such that

$$\operatorname{dist}(x, X) \leq c' \, (r(x) + r(x)^\gamma), \quad \forall \, x \in \mathbb{R}^n.$$

Proof. Let x be such that $r(x) > \varepsilon$ and $\bar{x} \equiv \Pi_X(x)$. Let

$$y \equiv \left(1 - \frac{\varepsilon}{2\,r(x)} \right) \bar{x} + \frac{\varepsilon}{2\,r(x)} \, x.$$

Clearly $\bar{x} = \Pi_X(y)$. Moreover, by the convexity of r, we have

$$r(y) \leq \left(1 - \frac{\varepsilon}{2\,r(x)} \right) r(\bar{x}) + \frac{\varepsilon}{2\,r(x)} \, r(x) = \frac{\varepsilon}{2}.$$

Thus

$$\begin{aligned} \| \, y - \bar{x} \, \| &= \operatorname{dist}(y, X) \\ &\leq c \, (r(y) + r(y)^\gamma) \leq c \left[\frac{\varepsilon}{2} + \left(\frac{\varepsilon}{2} \right)^\gamma \right]. \end{aligned}$$

Since

$$y - \bar{x} = \frac{\varepsilon}{2\,r(x)} \, (\, x - \bar{x} \,),$$

we deduce

$$\| \, x - \bar{x} \, \| = \frac{2 \, \| \, y - \bar{x} \, \|}{\varepsilon} \, r(x) \leq c' \, r(x)$$

where

$$c' \equiv c \left[1 + \left(\frac{\varepsilon}{2} \right)^{\gamma - 1} \right].$$

Consequently, $\operatorname{dist}(x, X) \leq c'(r(x) + r(x)^\gamma)$ for all $x \in \mathbb{R}^n$. □

Proposition **6.1.2** implies that if a set X admits a local error bound with an (arbitrary) residual function, then X has a global error bound with a multiplicative factor $1 + \|x\|$ that depends on the norm of the test vector x. The next result shows that if X is a convex set with a Slater point, then such a global error bound with a variable multiplicative constant always holds. It therefore follows that every solid, compact, convex set must admit a global error bound with a fixed multiplicative constant.

6.1.4 Proposition. Let $g : \mathbb{R}^n \to \mathbb{R}$ be convex and let $X \equiv g^{-1}(-\infty, 0]$. If there exists $\hat{x}$ such that $g(\hat{x}) < 0$, then statement (b) of Proposition **6.1.2** holds with $r(x) \equiv g(x)_+$. If in addition X is bounded, then X has a global Lipschitzian error bound with residual r.

Proof. Let $x \in \mathbb{R}^n$ be arbitrary. Without loss of generality, we may assume that $g(x) > 0$. Define

$$\delta \equiv \frac{g(x)}{g(x) - g(\hat{x})} \leq \frac{g(x)}{-g(\hat{x})};$$

clearly $\delta \in (0,1)$. By the convexity of g, it is easy to show that the vector $x_\delta \equiv (1-\delta)x + \delta\hat{x}$ belongs to X. Consequently, we have

$$\begin{aligned} \operatorname{dist}(x,X) &\leq \| x - x_\delta \| \\ &= \delta \| x - \hat{x} \| \leq \frac{\| x - \hat{x} \|}{-g(\hat{x})} g(x). \end{aligned}$$

From this inequality, statement (b) of Proposition **6.1.2** follows readily. If in addition X is bounded, then so is the level set

$$\{ x \in \mathbb{R}^n : r(x) \leq \varepsilon \}$$

for all $\varepsilon > 0$ because $r(x) = \max(0, g(x))$ is a convex function and the base set

$$X = \{ x \in \mathbb{R}^n : r(x) \leq 0 \}$$

is bounded by assumption. Hence X has a local Lipschitzian error bound with residual function r. That this is a global error bound follows from Proposition **6.1.3**. □

For an extension of the above proposition to an unbounded convex set satisfying a "strong Slater CQ", see Exercise **6.9.15**. See also Corollary **6.6.5** for a global Hölderian error bound for a bounded convex set defined by finitely many convex, "subanalytic" inequalities.

Extending the implication (a) ⇒ (b) in Proposition **6.1.2**, we show in the next result that it is possible to derive a global error bound for a set X using two residual functions, one of which provides a local error bound for X, and the other one satisfies a certain growth property. (To recover the said implication, take $r_2(x) \equiv \|x\| r_1(x)$.)

6.1.5 Proposition. Let $r_i : \mathbb{R}^n \to \mathbb{R}_+$ for $i = 1,2$ be two given functions and let X be a nonempty closed subset of $\mathbb{R}^n$. Suppose that there exist positive scalars c and ε such that

$$\operatorname{dist}(x,X) \leq c\,(\,r_1(x) + r_2(x)\,), \quad \forall\, x \text{ satisfying } r_1(x) \leq \varepsilon.$$

and for any unbounded sequence $\{x^k\}$ satisfying $r_1(x^k) > \varepsilon$ for every k,

$$\limsup_{k\to\infty} \frac{r_2(x^k)}{\| x^k \|} > 0.$$

There exists a constant $\eta > 0$ such that

$$\operatorname{dist}(x,X) \leq \eta\,(\,r_1(x) + r_2(x)\,) \quad \forall\, x \in \mathbb{R}^n.$$

Proof. Assume for contradiction that no such η exists. There exists then a sequence $\{x^k\}$ such that, for every k,

$$\text{dist}(x^k, X) > k(r_1(x^k) + r_2(x^k)).$$

We must have $r_1(x^k) > \varepsilon$ for all k sufficiently large. Moreover $\{x^k\}$ must be unbounded. For any fixed vector $\bar{x} \in X$, the above inequality yields, for all k sufficiently large,

$$\| x^k - \bar{x} \| > k(\varepsilon + r_2(x^k)),$$

which implies

$$\lim_{k \to \infty} \frac{r_2(x^k)}{\| x^k \|} = 0,$$

which is a contradiction. □

6.2 Pointwise and Local Error Bounds

Pointwise and local error bounds play a central role in the convergence analysis of many iterative algorithms, the former for the convergence to an isolated limit point and the latter for problems with special structures. In this section, we establish the close connection between these two kinds of error bounds and semistability and derive error bounds for some special VIs including the KKT systems.

6.2.1 Semistability and error bounds

Proposition **5.5.5** provides a kind of local error bound for the solution set of the VI (K, F) in terms of the semistability of this VI, using the normal map as the residual function. However, part (c) of this proposition is not exactly the kind of local error bound described in part (a) of Proposition **6.1.2** because the inequality (5.5.4) bounds $\text{dist}(\Pi_K(z), \text{SOL}(K, F))$ for all z with a small normal residual $\|\mathbf{F}_K^{\text{nor}}(z)\|$; to qualify for the kind of local error bound in Proposition **6.1.2**, we need to bound $\text{dist}(x, \text{SOL}(K, F))$ for all x in $\mathbb{R}^n$ with a small residual, including those vectors x that lie outside of the set K. For this purpose, we resort to the natural residual $\|\mathbf{F}_K^{\text{nat}}(x)\|$ and impose a uniformly, locally Lipschitz property of F on K.

6.2.1 Proposition. Let K be a closed convex set in $\mathbb{R}^n$ and F be a continuous mapping from $\mathbb{R}^n$ into itself. Assume that $\text{SOL}(K, F)$ is nonempty. The following two statements are valid.

(a) If $\text{SOL}(K, F)$ admits a local Lipschitzian error bound with the residual function $\|\mathbf{F}_K^{\text{nat}}(x)\|$, then the VI (K, F) is semistable.

(b) Conversely, if the VI (K, F) is semistable and there exist positive scalars L and δ such that for all $x \in K$,

$$\| y \| \le \delta \Rightarrow \| F(x+y) - F(x) \| \le L \| y \|,$$

then SOL(K, F) admits a local Lipschitzian error bound with the residual function $\|\mathbf{F}_K^{\mathrm{nat}}(x)\|$.

Proof. Suppose that there exist positive constants c and ε such that

$$\| \mathbf{F}_K^{\mathrm{nat}}(x) \| \le \varepsilon \Rightarrow \mathrm{dist}(x, \mathrm{SOL}(K, F)) \le c \| \mathbf{F}_K^{\mathrm{nat}}(x) \|.$$

Let z satisfy $\|\mathbf{F}_K^{\mathrm{nor}}(z)\| < \varepsilon$. Write $x \equiv \Pi_K(z)$. By Proposition **1.5.14**, we have

$$\| \mathbf{F}_K^{\mathrm{nat}}(x) \| \le \| \mathbf{F}_K^{\mathrm{nor}}(z) \| < \varepsilon.$$

Therefore it follows that

$$\mathrm{dist}(\Pi_K(z), \mathrm{SOL}(K, F)) \le c \| \mathbf{F}_K^{\mathrm{nat}}(x) \| \le c \| \mathbf{F}_K^{\mathrm{nor}}(z) \|,$$

which establishes the semistability of the VI (K, F), by Proposition **5.5.5**.

Conversely, let $c > 0$ and $\varepsilon > 0$ be such that

$$\| \mathbf{F}_K^{\mathrm{nor}}(z) \| \le \varepsilon \Rightarrow \mathrm{dist}(\Pi_K(z), \mathrm{SOL}(K, F)) \le c \| \mathbf{F}_K^{\mathrm{nor}}(z) \|.$$

Let

$$\varepsilon' \equiv \frac{\varepsilon}{1+L}$$

and let x satisfy $\|\mathbf{F}_K^{\mathrm{nat}}(x)\| \le \varepsilon'$. Define $z \equiv x - F(x)$. We have

$$\mathbf{F}_K^{\mathrm{nor}}(z) = F(\Pi_K(z)) + z - \Pi_K(z).$$

With $r \equiv \mathbf{F}_K^{\mathrm{nat}}(x) = x - \Pi_K(z)$, we obtain

$$\mathbf{F}_K^{\mathrm{nor}}(z) = F(\Pi_K(z)) - F(x) + r = F(\Pi_K(z)) - F(\Pi_K(z) + r) + r,$$

which implies

$$\| \mathbf{F}_K^{\mathrm{nor}}(z) \| \le (L+1) \| r \| \le \varepsilon.$$

Hence by assumption, we deduce

$$\mathrm{dist}(\Pi_K(z), \mathrm{SOL}(K, F)) \le c \| \mathbf{F}_K^{\mathrm{nor}}(z) \|,$$

which implies

$$\mathrm{dist}(x - r, \mathrm{SOL}(K, F)) \le c(L+1) \| \mathbf{F}_K^{\mathrm{nat}}(x) \|.$$

Hence $\text{dist}(x, \text{SOL}(K,F)) \leq [1 + c(L+1)] \| \mathbf{F}_K^{\text{nat}}(x) \|$. Thus SOL$(K,F)$ admits a local Lipschitzian error bound with the natural residual. □

Specializing the above proposition to an affine function F and invoking Corollary **5.5.9**, we obtain the following consequence readily. No proof is required.

6.2.2 Corollary. Let K be a closed convex set in $\mathbb{R}^n$ and F be an affine map from $\mathbb{R}^n$ into itself. Assume that SOL(K,F) is nonempty. The VI (K,F) is semistable if and only if SOL(K,F) admits a local Lipschitzian error bound with the natural residual $\|\mathbf{F}_K^{\text{nat}}(x)\|$. Hence the solution set of every solvable AVI admits a local, Lipschitzian error bound with the natural residual. □

When K is a polyhedral cone, the solution set of the homogeneous CP $(K, 0, M)$, which coincides with the CP kernel $\mathcal{K}(K,M)$ of the pair (K,M), is a cone; moreover, the natural residual $\mathbf{M}_K^{\text{nat}}(x) \equiv x - \Pi_K(x - Mx)$ is a piecewise linear function. Due to these two special properties, the CP kernel of the pair (K,M) admits a global, Lipschitzian error bound with $\|\mathbf{M}_K^{\text{nat}}(x)\|$ as the residual.

6.2.3 Corollary. Let K be a polyhedral cone in $\mathbb{R}^n$ and M be an $n \times n$ matrix. There exists a constant $c > 0$ such that

$$\text{dist}(x, \mathcal{K}(K,M)) \leq c \| \mathbf{M}_K^{\text{nat}}(x) \|, \quad \forall x \in \mathbb{R}^n.$$

Proof. By Corollary **6.2.2**, there exist positive constants c and ε such that

$$\text{dist}(x, \mathcal{K}(K,M)) \leq c \| \mathbf{M}_K^{\text{nat}}(x) \|, \quad \forall x \text{ such that } \| \mathbf{M}_K^{\text{nat}}(x) \| \leq \varepsilon.$$

Since the two functions $\text{dist}(x, \mathcal{K}(K,M))$ and $\|\mathbf{M}_K^{\text{nat}}(x)\|$ are both positively homogeneous of degree one, the above local error bound easily holds globally for all x in $\mathbb{R}^n$. □

Localizing Propositions **5.5.5** and **6.2.1**, we recover Proposition **5.3.7**, which we rephrase below using the language of error bounds of this chapter. Specifically, the following proposition characterizes the existence of a pointwise error bound for the solution set of a VI near an isolated solution in terms of the semistability of the solution.

6.2.4 Proposition. Let $K \subseteq \mathbb{R}^n$ be closed convex and $F : \mathbb{R}^n \to \mathbb{R}^n$ be Lipschitz continuous in a neighborhood of an isolated solution x^* in SOL(K,F). The solution x^* is semistable if and only if SOL(K,F) admits a Lipschitzian error bound near x^* with the residual function $\|\mathbf{F}_K^{\text{nat}}(x)\|$.

Proof. By Proposition **5.3.7**, x^* is semistable if and only if there exist positive scalars ε and η such that for every $x \in \mathrm{cl}\,\mathbb{B}(x^*, \varepsilon)$,

$$\| x - x^* \| \leq \eta \| \mathbf{F}_K^{\mathrm{nat}}(x) \|.$$

Since x^* is isolated, the above is equivalent to the desired Lipschitzian error bound of SOL(K, F) near x^*. □

In particular, if x^* is a (strongly) stable solution of the VI (K, F), then the solution set SOL(K, F) admits a Lipschitzian error bound near x^* with the residual function $\|\mathbf{F}_K^{\mathrm{nat}}(x)\|$. Thus the stability theory of Chapter **5** is applicable and yields many sufficient conditions for a VI to have a Lipschitzian error bound near a solution. In what follows, we use the results in the previous chapter to derive a broad condition for such an error bound to hold in the case where F is continuously differentiable and K is a finitely representable, convex set given by:

$$K \equiv \{ x \in \mathbb{R}^n : g(x) \leq 0, \ h(x) = 0 \}, \tag{6.2.1}$$

where $h : \mathbb{R}^n \to \mathbb{R}^\ell$ is affine and each $g_i : \mathbb{R}^n \to \mathbb{R}$ is convex and twice continuously differentiable for $i = 1, \ldots, m$. Let $x^* \in$ SOL(K, F) be given. We assume that the SBCQ holds at x^*. Thus the normal map $\mathbf{F}_K^{\mathrm{nor}}$ is B-differentiable at $z^* \equiv x^* - F(x^*)$. Referring to the setting and notation in Section **5.3**, we note that by the proof of Lemma **5.3.9**, the following holds. Namely, if for every $(\mu, \lambda) \in \mathcal{M}^e(x^*)$, $(\mathcal{C}, J_x\mathbf{L}(x^*, \lambda))$ is an R_0 pair, where $\mathcal{C} \equiv \mathcal{C}(x^*; K, F)$ is the critical cone of the VI (K, F) at x^*, then the equation $(\mathbf{F}_K^{\mathrm{nor}})'(z^*; dz) = 0$ has $dz = 0$ has the unique solution. In turn, by Lemma **5.2.1**, it follows that there exist a constant $c > 0$ and a neighborhood $\mathcal{Z}$ of z^* such that for all z in $\mathcal{Z}$,

$$\| z - z^* \| \leq c \| \mathbf{F}_K^{\mathrm{nor}}(z) \|,$$

which implies

$$\| \Pi_K(z) - x^* \| \leq c \| \mathbf{F}_K^{\mathrm{nor}}(z) \|.$$

With the above preparation, we can establish the following pointwise error bound for a VI with a finitely representable convex set.

6.2.5 Theorem. Let K be a finitely representable convex set given by (6.2.1), where h is affine and each g_i is convex and twice continuously differentiable. Let $F : \mathbb{R}^n \to \mathbb{R}^n$ be continuously differentiable. Let $x^* \in$ SOL(K, F) satisfy the SBCQ. Consider the following two statements.

(a) For every $(\mu, \lambda) \in \mathcal{M}^e(x^*)$, $(\mathcal{C}, J_x\mathbf{L}(x^*, \lambda))$ is an R_0 pair.

(b) The set SOL(K, F) has a Lipschitzian error bound with residual function $\|\mathbf{F}_K^{\text{nat}}(x)\|$ near x^*.

It holds that (a) $\Rightarrow$ (b). If x^* is isolated and either the CRCQ holds or $\mathcal{M}^e(x^*)$ is a singleton, then (b) $\Rightarrow$ (a).

Proof. (a) $\Rightarrow$ (b). If (a) holds, Theorem **3.3.12** and the subsequent remark imply that x^* is a locally unique solution of the VI (K, F). By the differentiability of F, there exist a neighborhood $\mathcal{N}_0$ of x^* and a scalar $L > 0$ such that

$$u, v \in \mathcal{N}_0 \;\Rightarrow\; \|\, F(u) - F(v)\,\| \leq L \,\|\, u - v\,\|.$$

Let $\mathcal{N}_1$ be a neighborhood of x^* such that

$$x \in \mathcal{N}_1 \;\Rightarrow\; x - F(x) \in \mathcal{Z}.$$

For any vector x, let $z \equiv x - F(x)$. Thus

$$\Pi_K(z) = \Pi_K(x - F(x)) = x - \mathbf{F}_K^{\text{nat}}(x).$$

Without loss of generality, we may take the neighborhood $\mathcal{N}_1$ to be contained in $\mathcal{N}_0$ and such that

$$x \in \mathcal{N}_1 \;\Rightarrow\; x - \mathbf{F}_K^{\text{nat}}(x) \in \mathcal{N}_0.$$

Thus for every $x \in \mathcal{N}_1$, we have

$$\|\, F(x) - F(x - r)\,\| \leq L \,\|\, r\,\|,$$

where $r \equiv \mathbf{F}_K^{\text{nat}}(x)$. Following the proof of part (b) Proposition **6.2.1**, we may deduce

$$\|\, x - x^*\,\| \leq [\, 1 + c\,(\, L + 1\,)\,]\, \|\, \mathbf{F}_K^{\text{nat}}(x)\,\|$$

for all $x \in \mathcal{N}_1$. This is equivalent to the Lipschitzian error bound of the set SOL(K, F) near the isolated solution x^*.

Suppose that the CRCQ holds. By Theorem **4.5.3** and the chain rule of differentiation, we have for any $(\mu, \lambda) \in \mathcal{M}^e(x^*)$ and any $dx \in \mathbb{R}^n$,

$$(\, \mathbf{F}_K^{\text{nat}}\,)'(x^*; dx) = dx - \Pi_{\mathcal{C}}^{G(\lambda)}(dx - JF(x^*)dx),$$

where $\mathcal{C} \equiv \mathcal{C}(x^*; K, F)$. The above expression remains valid if $\mathcal{M}^e(x^*)$ is a singleton. Suppose that (b) holds. By Lemma **5.2.1**, it follows that

$$(\, \mathbf{F}_K^{\text{nat}}\,)'(x^*; dx) = 0 \;\Rightarrow\; dx = 0.$$

Let dx belong to $\mathrm{SOL}(\mathcal{C}, 0, J_x\mathbf{L}(x^*, \lambda))$ for some $(\mu, \lambda) \in \mathcal{M}^e(x^*)$. It suffices to show that dx is a zero of $(\mathbf{F}_K^{\mathrm{nat}})'(x^*; \cdot)$. We have

$$\mathcal{C} \ni dx \perp J_x\mathbf{L}(x^*, \lambda)dx \in \mathcal{C}^*,$$

which is equivalent to

$$\mathcal{C} \ni dx \perp G(\lambda)dx + JF(x^*)dx - dx \in \mathcal{C}^*.$$

In turn, the latter yields

$$0 = dx - \Pi_{\mathcal{C}}^{G(\lambda)}(dx - JF(x^*)dx).$$

Consequently, $(\mathbf{F}_K^{\mathrm{nat}})'(x^*; dx) = 0$. □

6.2.6 Remark. The last part of the above proof is the counterpart of Lemma **5.3.13**, which pertains to the normal map. □

6.2.2 Local error bounds for KKT triples

Consider a VI (K, F) defined by a finitely representable, nonconvex set K given by (6.2.1), where $F : \mathbb{R}^n \to \mathbb{R}^n$ is continuously differentiable, $g : \mathbb{R}^n \to \mathbb{R}^m$ and $h : \mathbb{R}^n \to \mathbb{R}^\ell$ are both twice continuously differentiable. Throughout this subsection, g_i is not assumed convex and h_j is not assumed affine. The KKT system of such a VI is:

$$\begin{aligned} \mathbf{L}(x, \mu, \lambda) &\equiv F(x) + \sum_{j=1}^{\ell} \mu_j \nabla h_j(x) + \sum_{i=1}^{m} \lambda_i \nabla g_i(x) = 0 \\ h(x) &= 0 \\ 0 \leq \lambda &\perp g(x) \leq 0. \end{aligned} \tag{6.2.2}$$

Since K is not convex, the above KKT system is only a necessary, but not sufficient, condition of a solution to the VI under an appropriate CQ.

The focus of the discussion herein is on an isolated KKT point $\bar{x}$ satisfying the MFCQ. The associated set of multipliers $\mathcal{M}(\bar{x})$ is thus nonempty and bounded; however, we do not assume that $\mathcal{M}(\bar{x})$ is a singleton. In what follows, we show that under the R_0 assumption in Theorem **3.3.12**, we can obtain a "local" error bound of the set $\{\bar{x}\} \times \mathcal{M}(\bar{x})$ in terms of the natural residual of the KKT system (6.2.2). This result does not follow from Theorem **6.2.5** because the setting herein does not assume that K is convex. Furthermore, although the KKT system is a MiCP in the (x, μ, λ) variable, the error bound derived below does not pertain to the entire solution set of the KKT system, but rather to the special subset $\{\bar{x}\} \times \mathcal{M}(\bar{x})$.

Therefore, none of the error bound results obtained so far have addressed the issue considered herein. In fact, the result below is neither a local error bound in the sense of part (a) in Proposition **6.1.2**; nor is it a pointwise error bound near $\bar{x}$. With an abuse of terminology, we refer to the result below as a local error bound of the set $\{\bar{x}\} \times \mathcal{M}(\bar{x})$.

6.2.7 Proposition. Let F be continuously differentiable and all h_j and g_i be twice continuously differentiable in a neighborhood of a KKT point $\bar{x}$. Suppose that $\bar{x}$ satisfies the MFCQ and that for every $(\mu, \lambda) \in \mathcal{M}(\bar{x})$, the homogeneous linear complementarity system:

$$\mathcal{C}(\bar{x}; K, F) \ni v \perp J_x \mathbf{L}(\bar{x}, \mu, \lambda) v \in \mathcal{C}(\bar{x}; K, F)^*, \tag{6.2.3}$$

has a unique solution $v = 0$. There exist positive constants $c > 0$ and $\varepsilon > 0$ such that for all triples (x, μ, λ) satisfying

$$\| x - \bar{x} \| + \operatorname{dist}((\mu, \lambda), \mathcal{M}(\bar{x})) \leq \varepsilon,$$

it holds that

$$\| x - \bar{x} \| + \operatorname{dist}((\mu, \lambda), \mathcal{M}(\bar{x})) \leq c\, r(x, \mu, \lambda),$$

where

$$r(x, \mu, \lambda) \equiv \| \mathbf{L}(x, \mu, \lambda) \| + \| h(x) \| + \| \min(\lambda, -g(x)) \|$$

is the residual of the KKT system (6.2.2). Thus $\bar{x}$ is an isolated KKT point.

Proof. The proof is a refinement of that of Proposition **3.2.2** and Theorem **3.3.12**. A difference between the following proof and that of Theorem **3.3.12** is that we are dealing with the solutions of the KKT system, which are not necessarily solutions of the VI (K, F) due to the lack of convexity of the set K.

Assume for contradiction that no such scalars c and ε exist. There exist a sequence of positive scalars $\{\varepsilon_k\}$ converging to zero and a sequence of triples $\{(x^k, \mu^k, \lambda^k)\}$ such that for every k,

$$\| x^k - \bar{x} \| + \operatorname{dist}((\mu^k, \lambda^k), \mathcal{M}(\bar{x})) \leq \varepsilon_k$$

and

$$\| x^k - \bar{x} \| + \operatorname{dist}((\mu^k, \lambda^k), \mathcal{M}(\bar{x})) > k\, r(x^k, \mu^k, \lambda^k).$$

Since $\varepsilon_k \to 0$, the following four limits hold:

$$\lim_{k \to \infty} x^k = \bar{x}, \quad \lim_{k \to \infty} \operatorname{dist}((\mu^k, \lambda^k), \mathcal{M}(\bar{x})) = 0,$$

$$\lim_{k \to \infty} r(x^k, \mu^k, \lambda^k) = 0,$$

and

$$\lim_{k \to \infty} \frac{r(x^k, \mu^k, \lambda^k)}{\| x^k - \bar{x} \| + \operatorname{dist}((\mu^k, \lambda^k), \mathcal{M}(\bar{x}))} = 0.$$

Writing, for every k,

$$u^k \equiv \mathbf{L}(x^k, \mu^k, \lambda^k), \quad w^k \equiv h(x^k), \quad \text{and} \quad v^k \equiv \min(\lambda^k, -g(x^k)),$$

we have

$$F(x^k) - u^k + \sum_{i=1}^m v_i^k \nabla g_i(x^k) + \sum_{i=1}^m (\lambda_i^k - v_i^k) \nabla g_i(x^k) + \sum_{j=1}^\ell \mu_j \nabla h_j(x^k) = 0,$$

$$h(x^k) - w^k = 0,$$

and

$$0 \leq \lambda^k - v^k \perp g(x^k) + v^k \leq 0.$$

Moreover,

$$\lim_{k \to \infty} \frac{(u^k, w^k, v^k)}{\| x^k - \bar{x} \| + \operatorname{dist}((\mu^k, \lambda^k), \mathcal{M}(\bar{x}))} = 0.$$

We further let

$$s^k \equiv u^k - \sum_{i=1}^m v_i^k \nabla g_i(x^k) \quad \text{and} \quad \tilde{\lambda}^k \equiv \lambda^k - v^k.$$

We have

$$\lim_{k \to \infty} \frac{s^k}{\| x^k - \bar{x} \| + \operatorname{dist}((\mu^k, \lambda^k), \mathcal{M}(\bar{x}))} = 0 \tag{6.2.4}$$

and

$$\lim_{k \to \infty} \operatorname{dist}((\mu^k, \tilde{\lambda}^k), \mathcal{M}(\bar{x})) = 0.$$

Moreover,

$$F(x^k) - s^k + \sum_{i=1}^m \tilde{\lambda}_i^k \nabla g_i(x^k) + \sum_{j=1}^\ell \mu_j \nabla h_j(x^k) = 0.$$

Let

$$\mathcal{I}(\bar{x}) \equiv \{ i : g_i(\bar{x}) = 0 \}$$

be the index set of active constraints at $\bar{x}$. Without loss of generality, we may assume that for each $i \notin \mathcal{I}(\bar{x})$, we have, for every k,

$$g_i(x^k) + v_i^k < 0,$$

which implies $\tilde{\lambda}_i^k = 0$. Therefore by part (b) of Corollary **3.2.5**, we deduce the existence of a constant $L > 0$ such that for every k,

$$\operatorname{dist}((\mu^k, \tilde{\lambda}^k), \mathcal{M}(\bar{x})) \leq L \left[\| F(\bar{x}) - F(x^k) + s^k \| + \sum_{j=1}^{\ell} \| \nabla h_j(x^k) - \nabla h_j(\bar{x}) \| + \sum_{i \in \mathcal{I}(\bar{x})} \| \nabla g_i(x^k) - \nabla g_i(\bar{x}) \| \right].$$

By (6.2.4) and the differentiability assumption on F, g, and h, we can deduce that

$$\limsup_{k \to \infty} \frac{\operatorname{dist}((\mu^k, \tilde{\lambda}^k), \mathcal{M}(\bar{x}))}{\| x^k - \bar{x} \|} < \infty.$$

Since

$$\frac{(u^k, w^k, v^k)}{\| x^k - \bar{x} \|} = \frac{(u^k, w^k, v^k)}{\| x^k - \bar{x} \| + \operatorname{dist}((\mu^k, \lambda^k), \mathcal{M}(\bar{x}))} \left[1 + \frac{\operatorname{dist}((\mu^k, \tilde{\lambda}^k), \mathcal{M}(\bar{x}))}{\| x^k - \bar{x} \|} \right],$$

we deduce

$$\lim_{k \to \infty} \frac{(u^k, w^k, v^k)}{\| x^k - \bar{x} \|} = 0, \tag{6.2.5}$$

which implies

$$\lim_{k \to \infty} \frac{s^k}{\| x^k - \bar{x} \|} = 0.$$

Without loss of generality, we may assume that the normalized sequence $\{(x^k - \bar{x})/\|x^k - \bar{x}\|\}$ converges to a limit, which we denote d^∞. We claim that the latter vector belongs to the critical cone $\mathcal{C}(\bar{x}, K, F)$. We first show that d^∞ belongs to the tangent cone $\mathcal{T}(\bar{x}; K)$, which is equal to the linearization cone:

$$\{ d \in \mathbb{R}^n : d^T \nabla g_i(\bar{x}) \leq 0, \ \forall i \in \mathcal{I}(\bar{x}); \ d^T \nabla h_j(\bar{x}) = 0, \ \forall j = 1, \ldots, \ell \}.$$

For an index $i \in \mathcal{I}(\bar{x})$, we have

$$0 \geq v_i^k + g_i(x^k) = v_i^k + \nabla g_i(\bar{x})^T (x^k - \bar{x}) + o(\| x^k - \bar{x} \|).$$

Thus dividing by $\|x^k - \bar{x}\|$ and using (6.2.5), we deduce

$$\nabla g_i(\bar{x})^T d^\infty \leq 0, \quad \forall i \in \mathcal{I}(\bar{x}).$$

Similarly, we can deduce

$$\nabla h_j(\bar{x})^T d^\infty = 0, \quad \forall j = 1, \ldots, \ell.$$

Thus $d^\infty \in \mathcal{T}(\bar{x}; K)$. To show that $d^\infty \perp F(\bar{x})$, assume without loss of generality that the sequence $\{(\mu^k, \tilde{\lambda}^k)\}$ converges to a pair $(\mu^\infty, \lambda^\infty)$, which must necessarily belong to $\mathcal{M}(\bar{x})$. It suffices to show that $\nabla g_i(\bar{x})^T d^\infty = 0$ for all i such that $\lambda_i^\infty > 0$. For such an index i, we must have $\tilde{\lambda}_i^k > 0$ for all k sufficiently large, which implies

$$0 = v_i^k + g_i(x^k).$$

Proceeding as above, we can establish $\nabla g_i(\bar{x})^T d^\infty = 0$. Therefore d^∞ is a critical vector of the pair (K, F).

As in the proof of Theorem **3.3.12**, we can show that for all k sufficiently large,

$$\tilde{\lambda}_i^k \nabla g_i(\bar{x})^T d^\infty = 0, \quad \forall i.$$

Moreover, proceeding as in the previous proposition, we can deduce that d^∞ satisfies

$$d^\infty \perp J_x \mathbf{L}(\bar{x}, \mu^\infty, \lambda^\infty) d^\infty \in \mathcal{C}(\bar{x}; K, F)^*.$$

Since d^∞ is nonzero, we obtain a contradiction.

To establish that $\bar{x}$ is an isolated KKT point, assume that there exists a sequence $\{x^\nu\}$ converging to $\bar{x}$ such that for each ν, the triple $(x^\nu, \mu^\nu, \lambda^\nu)$ satisfies the KKT system (6.2.2) for some (μ^ν, λ^ν). For ν sufficiently large, the triple $(x^\nu, \mu^\nu, \lambda^\nu)$ is within ε distance from the set $\{\bar{x}\} \times \mathcal{M}(\bar{x})$. Since $r(x^\nu, \mu^\nu, \lambda^\nu) = 0$, it follows that $x^\nu = \bar{x}$ for all ν sufficiently large. □

6.2.3 Linearly constrained monotone composite VIs

In this section, we consider a linearly constrained, strongly monotone composite VI (K, F) where K is a polyhedron and F is of the form (2.3.10):

$$F(x) \equiv A^T G(Ax) + b, \quad x \in \mathbb{R}^n,$$

where $A \in \mathbb{R}^{m \times n}$, $b \in \mathbb{R}^n$, and $G : \mathbb{R}^m \to \mathbb{R}^m$ is strongly monotone on the range of A. We introduce a notation that we use throughout this subsection. For a given matrix $A \in \mathbb{R}^{m \times n}$, a mapping $G : \mathbb{R}^m \to \mathbb{R}^m$, and a set $K \subseteq \mathbb{R}^n$, let $\mathcal{R}(K, G, A)$ denote the set of all vectors $b \in \mathbb{R}^m$ for which the VI (K, F) has a solution, where F is as given above. The reader can surmise that the notation $\mathcal{R}(K, G, A)$ has the intended meaning as the "range" of the triple (K, G, A). We further write the VI (K, F) explicitly as VI (K, G, A, b) and write SOL(K, G, A, b) for its solution set.

Since G is strongly monotone on the range of A, by Corollary **2.3.13**, it follows that SOL(K, G, A, b) is a polyhedron for every $b \in \mathcal{R}(K, G, A)$.

Let $v(b) \in \mathbb{R}^m$ and $w(b) \in \mathbb{R}^n$ be the constant vectors associated with the solutions of the VI (K, G, A, b); that is, for any $x \in \text{SOL}(K, G, A, b)$,

$$v(b) = Ax \quad \text{and} \quad w(b) = F(x) = A^T G(v(b)) + b;$$

see the cited corollary. A vector x belongs to $\text{SOL}(K, G, A, b)$ if and only if $Ax = v(b)$ and x is a solution of the linear program

$$\begin{array}{ll} \text{minimize} & y^T w(b) \\ \text{subject to} & y \in K. \end{array}$$

Our goal is to establish the following local Lipschitzian error bound for this class of VIs.

6.2.8 Theorem. Let $G : \mathbb{R}^m \to \mathbb{R}^m$ be strongly monotone and Lipschitz continuous on the range of the matrix $A \in \mathbb{R}^{m \times n}$. Let K be a polyhedron in $\mathbb{R}^n$. For every $b \in \mathcal{R}(K, G, A)$, there exist positive constants η and δ, dependent on b, such that

$$\| \mathbf{F}_K^{\text{nat}}(x) \| \leq \delta \Rightarrow \text{dist}(x, \text{SOL}(K, F)) \leq \eta \| \mathbf{F}_K^{\text{nat}}(x) \|,$$

where $F(x) \equiv A^T G(Ax) + b$. Therefore the VI (K, G, A, b) is semistable.

Only the first statement of the above theorem requires a proof; the second statement follows from part (a) of Proposition **6.2.1**. We establish two lemmas. The first lemma, which does not require the Lipschitz continuity of G, asserts three properties associated with the triple (K, G, A). First, a certain boundedness property holds for the solutions of the VI (K, G, A, b); second, $\mathcal{R}(K, G, A)$ is a closed set; third, $v(b)$ and $w(b)$ are continuous functions of $b \in \mathcal{R}(K, G, A)$.

6.2.9 Lemma. Let $G : \mathbb{R}^m \to \mathbb{R}^m$ be continuous and strongly monotone on the range of the matrix $A \in \mathbb{R}^{m \times n}$. Let K be a polyhedron in $\mathbb{R}^n$.

(a) Positive constants η_1 and η_2 exist such that for every $b \in \mathcal{R}(K, G, A)$, a solution $x \in \text{SOL}(K, G, A)$ exists satisfying

$$\| x \| \leq \eta_1 (\| b \| + \| v(b) \|) + \eta_2.$$

(b) The set $\mathcal{R}(K, G, A)$ is closed.

(c) The following two limits hold:

$$\lim_{\substack{b' \in \mathcal{R}(K,G,A) \\ b' \to b}} v(b') = v(b) \quad \text{and} \quad \lim_{\substack{b' \in \mathcal{R}(K,G,A) \\ b' \to b}} w(b') = w(b).$$

Proof. Write

$$K \equiv \{ x \in \mathbb{R}^n : Cx \leq d \} \tag{6.2.6}$$

for some matrix C and vector d. By the KKT conditions of the VI (K, G, A, b), a vector x is a solution of the VI (K, G, A, b) if and only if $Ax = v(b)$ and there exists $\lambda \in \mathbb{R}^m$ satisfying

$$\begin{aligned} 0 &= b + A^T G(v(b)) + C^T \lambda \\ 0 &\leq \lambda \perp Cx - d \leq 0. \end{aligned}$$

Letting $\mathcal{I}(x)$ denote the index set of active constraints at x, we deduce that a vector $x \in K$ belongs to SOL(K, G, A, b) if and only if there exists $\lambda \in \mathbb{R}^m$ satisfying

$$\begin{aligned} 0 &= v(b) - Ax \\ 0 &= b + A^T G(v(b)) + \sum_{i \in \mathcal{I}(x)} \lambda_i \, (C_{i\cdot})^T \\ 0 &\geq (-d + Cx)_i, \quad i \notin \mathcal{I}(x) \\ 0 &= (-d + Cx)_i \text{ and } \lambda_i \geq 0, \quad i \in \mathcal{I}(x). \end{aligned}$$

For a fixed but arbitrary vector $b \in \mathcal{R}(K, G, A)$ and an index set $\mathcal{I}(x) = \mathcal{I}$, the latter can be considered a system of linear inequalities in the variables x and $\lambda_{\mathcal{I}}$. By Hoffman's error bound, it follows that there exists a positive constant $\eta_{\mathcal{I}}$ such that for all b for which the above linear inequality system is consistent, a solution $(x, \lambda_{\mathcal{I}})$ exists satisfying

$$\| (x, \lambda_{\mathcal{I}}) \| \leq \eta_{\mathcal{I}} \, [\, \| b \| + \| v(b) \| + \| d \| \,].$$

Since there are only finitely many index sets $\mathcal{I}$, part (a) follows.

To prove part (b), let $\{b^k\} \subset \mathcal{R}(K, G, A)$ be a sequence of vectors converging to some limit b^∞. For each k, let $x^k \in$ SOL(K, G, A, b^k) be such that

$$\| x^k \| \leq \eta_1 (\| b^k \| + \| v(b^k) \|) + \eta_2. \tag{6.2.7}$$

We have $v(b^k) = Ax^k$. Suppose for the sake of contradiction that

$$\lim_{k \to \infty} \| x^k \| = \infty.$$

Take any vector $y \in K$. For every k, we have,

$$(y - x^k)^T (b^k + A^T G(Ax^k)) \geq 0.$$

Letting $c > 0$ be a strong monotonicity constant of G, we deduce

$$\begin{aligned}(y - x^k)^T b^k &\geq (Ax^k - Ay)^T G(Ax^k) \\ &\geq c \| v(b^k) - Ay \|^2 - (Ax^k - Ay)^T G(y).\end{aligned}$$

Dividing by $\|x^k\|^2$ and letting $k \to \infty$, we obtain

$$\lim_{k \to \infty} \frac{v(b^k)}{\| x^k \|} = 0.$$

But this contradicts (6.2.7). Consequently $\{x^k\}$ is bounded. Any accumulation point of this sequence must be a solution of the VI (K, G, A, b^∞). Hence b^∞ is an element of $\mathcal{R}(K, G, A)$. This establishes part (b). Part (c) follows easily because of the uniqueness of the vector $v(b)$ associated with the VI (K, G, A, b). □

The next lemma is an intermediate step toward the desired error bound in Theorem **6.2.8**.

6.2.10 Lemma. Let $G : \mathbb{R}^m \to \mathbb{R}^m$ be strongly monotone and Lipschitz continuous on the range of the matrix $A \in \mathbb{R}^{m \times n}$. Let K be a polyhedron in $\mathbb{R}^n$. For every $b \in \mathcal{R}(K, G, A)$, there exist positive constants η_1 and δ_1, dependent on b, such that

$$\| \mathbf{F}_K^{\text{nat}}(x) \| \leq \delta_1 \Rightarrow \text{dist}(x, \text{SOL}(K, F)) \leq \eta_1 [\| \mathbf{F}_K^{\text{nat}}(x) \| + \| Ax - v(b) \|],$$

where $F(x) \equiv A^T G(Ax) + b$.

Proof. Let K be given by (6.2.6). Consider the KKT system of the VI (K, G, A, b):

$$\begin{aligned} 0 &= b + A^T G(v(b)) + C^T \lambda \\ 0 &\leq \lambda \perp Cx - d \leq 0. \end{aligned}$$

With b fixed, this is an MLCP in the (x, λ) variable. Equivalently, this MLCP is the primal-dual formulation of the linear program

$$\begin{array}{ll} \text{minimize} & v^T w(b) \\ \text{subject to} & v \in K. \end{array} \tag{6.2.8}$$

where $w(b) \equiv b + A^T G(v(b))$. Since an MLCP is semistable, it follows that there exist positive scalars c_0 and ε_0 such that for every vector q satisfying

$$\| q - w(b) \| \leq \varepsilon_0$$

and for every pair (x', λ') satisfying

$$\begin{aligned} 0 &= q + C^T \lambda' \\ 0 &\le \lambda' \perp Cx' - d \le 0, \end{aligned}$$

there exists a pair $(\hat{x}, \lambda)$ with $\hat{x} \in \mathrm{SOL}(K, G, A, b)$ such that

$$\| (x', \lambda') - (\hat{x}, \lambda) \| \le c_0 \| q - w(b) \|.$$

Let $x \in \mathbb{R}^n$ be arbitrary. Write $r \equiv \mathbf{F}_K^{\mathrm{nat}}(x) = x - \Pi_K(x - b - A^T G(Ax))$. With $y \equiv x - r$ and $s \equiv -r + A^T(G(Ax) - G(Ay))$, it follows that

$$0 = y - \Pi_K(y - b - A^T G(Ay) - s).$$

Thus $y \in \mathrm{SOL}(K, G, A, b + s)$. With $L > 0$ denoting the Lipschitz constant of G, we have

$$\| s \| \le (1 + L \| A^T \| \| A \|) \| r \|.$$

By part (c) of Lemma **6.2.9**, for every $\varepsilon > 0$, there exists $\delta > 0$, such that

$$\begin{aligned} \| r \| \le \delta \;&\Rightarrow\; \| Ay - v(b) \| \le \varepsilon \\ &\Rightarrow\; \| Ax - v(b) \| \le \varepsilon + \| A \| \| r \|. \end{aligned}$$

Consequently, we have

$$\| r \| \le \delta \Rightarrow \| G(Ax) - G(v(b)) \| \le L [\varepsilon + \| A \| \| r \|].$$

The vector y satisfies

$$\begin{aligned} 0 &= b - r + A^T G(Ax) + C^T \lambda' \\ 0 &\le \lambda' \perp Cy - d \le 0, \end{aligned} \tag{6.2.9}$$

for some λ'. We may choose $\delta_1 > 0$ such that

$$\| r \| \le \delta_1 \Rightarrow \| b - r + A^T G(Ax) - w(b) \| \le \varepsilon_0.$$

It follows that a vector $\hat{x} \in \mathrm{SOL}(K, G, A, b)$ exists satisfying

$$\| y - \hat{x} \| \le c_0 [\| r \| + \| A^T (G(Ax) - G(v(b))) \|].$$

Since $y = x - r$, we have

$$\| x - \hat{x} \| \le (1 + c_0) \| r \| + c_0 \| A^T \| L \| Ax - v(b) \|.$$

Letting $\eta_1 \equiv \max(1 + c_0, c_0 L \| A^T \|)$, we easily obtain the desired claim of the lemma. □

Proof of Theorem 6.2.8. We continue to use the notation in the previous lemma. Subtracting the first equation in (6.2.8) from the first equation in (6.2.9), we obtain

$$0 = -r + A^T G(Ax) - A^T G(v(b)) + C^T(\lambda' - \lambda).$$

Premultiplying the above equation by $(x - \hat{x})^T$, rearranging terms, and utilizing the KKT system for the VI (K, G, A, b), we obtain

$$\begin{aligned}
&(Ax - A\hat{x})^T(G(Ax) - G(A\hat{x})) \\
&= r^T(x - \hat{x}) + (\lambda' - \lambda)^T(C\hat{x} - Cx) \\
&\leq r^T(x - \hat{x}) - (Cr)^T(\lambda' - \lambda) \\
&\leq \| r \| (\|x - \hat{x}\| + \| C \| \| \lambda - \lambda' \|).
\end{aligned}$$

By the strong monotonicity of G, we have

$$c \| Ax - A\hat{x} \|^2 \leq (Ax - A\hat{x})^T(G(Ax) - G(A\hat{x}));$$

consequently, for some constant $c' > 0$ independent of x, it follows that

$$\| Ax - v(b) \|^2 \leq c' \| r \| (\|x - \hat{x}\| + \| \lambda - \lambda' \|).$$

By Lemma **6.2.10**, we have

$$\begin{aligned}
\| x - \hat{x} \|^2 &\leq \eta^2 [\| r \| + \| Ax - v(b) \|]^2 \\
&\leq 2\eta^2 [\| r \|^2 + \| Ax - v(b) \|^2] \\
&\leq \max(2\eta^2, c') [\| r \|^2 + \| r \| (\|x - \hat{x}\| + \| \lambda - \lambda' \|)].
\end{aligned}$$

From the proof of Lemma **6.2.10**, we have

$$\begin{aligned}
\| \lambda - \lambda' \| &\leq c_0 [\| r \| + \| A \| L \| Ax - v(b) \|] \\
&\leq c_0 [\| r \| + \| A \|^2 L \| x - \hat{x} \|].
\end{aligned}$$

Consequently, for some constant $\sigma > 0$ independent of x,

$$\| x - \hat{x} \|^2 \leq \sigma [\| r \|^2 + \| r \| \, \| x - \hat{x} \|],$$

which implies

$$\| x - \hat{x} \| \leq \tfrac{1}{2}(\sigma + \sqrt{\sigma^2 + 4\sigma}) \| r \|.$$

This easily yields the desired error bound of $\text{dist}(x, \text{SOL}(K, G, A, b))$ in terms of the natural residual $\|\mathbf{F}_K^{\text{nat}}(x)\|$. □

Part (c) of Lemma **6.2.9** shows that the functions $v(b)$ and $w(b)$ are continuous on their common domain. By using Theorem **6.2.8**, we can show that these functions are "pointwise Lipschitz continuous".

6.2.11 Corollary. Let $G : \mathbb{R}^m \to \mathbb{R}^m$ be strongly monotone and Lipschitz continuous on the range of the matrix $A \in \mathbb{R}^{m \times n}$. Let K be a polyhedron in $\mathbb{R}^n$. For every $b \in \mathcal{R}(K, G, A)$, there exist positive constants δ and L, dependent on b, such that

$$\| b' - b \| \leq \delta \Rightarrow \max(\| v(b' - v(b) \|, \| w(b') - w(b) \|) \leq L \| b' - b \|.$$

Proof. This follows easily from the semistability of the VI (K, G, A, b) and the uniqueness of the two vectors $v(b')$ and $w(b')$ for any b'. □

6.3 Global Error Bounds for VIs/CPs

In general, global Lipschitzian error bounds for a VI with the natural (or other) residual are rare and require fairly strong asymptotic conditions. In what follows, we derive a result of this type for a VI whose defining set is a Cartesian product. This result applies to both the strongly monotone VI and the NCP with a uniformly P function. In the case of a strongly monotone VI, the local error bound of Theorem **6.2.8** is hereby strengthened to a global bound. See also Theorem **2.3.3**. Specifically, let

$$K = \prod_{\nu=1}^{N} K_\nu, \tag{6.3.1}$$

where N is a positive integer and each K_ν is a closed convex subset of $\mathbb{R}^{n_\nu}$ with

$$\sum_{\nu=1}^{N} n_\nu = n.$$

We assume that $F : \mathbb{R}^n \to \mathbb{R}^n$ is a continuous uniformly P function on $\mathbb{R}^n$. Thus the VI (K, F) has a unique solution, say x^*. Under a kind of restricted globally Lipschitz property of F at x^* or a compactness assumption on K along with a locally Lipschitz property of F, we can establish the following result.

6.3.1 Proposition. In the above setting, assume either (a) there exists a constant $L > 0$ such that

$$\| F(x) - F(x^*) \| \leq L \| x - x^* \|, \quad \forall x \in \mathbb{R}^n,$$

or (b) K is compact and F is locally Lipschitz continuous on $\mathbb{R}^n$. A constant $c > 0$ exists such that

$$\| x - x^* \| \leq c \| \mathbf{F}_K^{\text{nat}}(x) \|, \quad \forall x \in \mathbb{R}^n.$$

Proof. Let $\eta > 0$ be the uniformly P modulus of F; that is, for all x and x' in $\mathbb{R}^n$,

$$\max_{1 \leq \nu \leq N} (x_\nu - x'_\nu)^T (F_\nu(x) - F_\nu(x')) \geq \eta \| x - x^\nu \|^2.$$

Assume condition (a). Let $x \in \mathbb{R}^n$ be given. Write $r \equiv \mathbf{F}_K^{\text{nat}}(x)$. Thus $x - r = \Pi_K(x - F(x))$. By the variational principle for the Euclidean projection, we have

$$(y - x + r)^T (F(x) - r) \geq 0, \quad \forall y \in K.$$

Let ν_0 be an index such that

$$(x_{\nu_0} - x^*_{\nu_0})^T (F_{\nu_0}(x) - F_{\nu_0}(x^*)) = \max_{1 \leq \nu \leq N} (x_\nu - x'_\nu)^T (F_\nu(x) - F_\nu(x')).$$

With $y \in K$ defined as

$$y_\nu \equiv \begin{cases} x_\nu - r_\nu & \text{if } \nu \neq \nu_0 \\ x^*_\nu & \text{if } \nu = \nu_0, \end{cases}$$

we deduce

$$(x^* - x + r)^T_{\nu_0} (F(x) - r)_{\nu_0} \geq 0.$$

Since $x^* \in \text{SOL}(K, F)$ and $x - r \in K$, we also have

$$(x - r - x^*)^T_{\nu_0} F_{\nu_0}(x^*) \geq 0.$$

Adding the last two inequalities and rearranging terms, we obtain

$$\begin{aligned} &(x - x^*)^T_{\nu_0} (F(x) - F(x^*))_{\nu_0} \\ &\quad \leq -(x - x^*)^T_{\nu_0} r_{\nu_0} - \| r_{\nu_0} \|^2 + (F(x) - F(x^*))^T_{\nu_0} r_{\nu_0}. \end{aligned}$$

Hence by the definition of ν_0 and by an easy manipulation, it follows that

$$\| x - x^* \| \leq \eta^{-1} (1 + L) \| \mathbf{F}_K^{\text{nat}}(x) \|,$$

which is the desired error bound.

Assume condition (b). Since K is compact, K is contained in a closed Euclidean ball $\text{cl}\, \mathbb{B}(0, \rho)$ for some scalar $\rho > 0$. Since F is locally Lipschitz continuous on $\mathbb{R}^n$, it must be Lipschitz continuous on compact sets. In particular, F is Lipschitz continuous on the ball $\text{cl}\, \mathbb{B}(0, 3\rho)$. By the above proof, it follows that there exists a constant $c' > 0$ such that

$$\| x - x^* \| \leq c' \| \mathbf{F}_K^{\text{nat}}(x) \|, \quad \forall x \in \text{cl}\, \mathbb{B}(0, 3\rho).$$

We now compare $\|x - x^*\|$ and $\|\mathbf{F}_K^{\text{nat}}(x)\|$ for $x \notin \text{cl}\,\mathbb{B}(0, 3\rho)$. For such an x, we have,

$$\begin{aligned} \| \mathbf{F}_K^{\text{nat}}(x) \| &= \| x - \Pi_K(x - F(x)) \| \\ &\geq \| x \| - \| \Pi_K(x - F(x)) \| > 2\rho \end{aligned}$$

where the last inequality follows because x lies outside the ball $\text{cl}\,\mathbb{B}(0, 3\rho)$ and K is contained in $\text{cl}\,\mathbb{B}(0, \rho)$. Thus,

$$\begin{aligned} \| x - x^* \| &\leq \| x - \Pi_K(x - F(x)) \| + \| \Pi_K(x - F(x)) \| + \| x^* \| \\ &\leq \| \mathbf{F}_K^{\text{nat}}(x) \| + 2\rho < 2 \| \mathbf{F}_K^{\text{nat}}(x) \|. \end{aligned}$$

Consequently,

$$\| x - x^* \| \leq \max(2, c') \| \mathbf{F}_K^{\text{nat}}(x) \|, \quad \forall x \in \mathbb{R}^n,$$

which is the desired global Lipschitzian error bound. □

6.3.2 Remark. If instead of being uniformly P, F is ξ-monotone on K (cf. part (d) of Definition **2.3.1**), that is, there exist constants $\eta > 0$ and $\xi > 1$ such that for all x and x' in $\mathbb{R}^n$,

$$\max_{1 \leq \nu \leq N} (x_\nu - x'_\nu)^T (F_\nu(x) - F_\nu(x')) \geq \eta \| x - x' \|^\xi,$$

then under the assumptions of Proposition **6.3.1**, there exists a constant $c > 0$ such that

$$\| x - x^* \| \leq c \max\left(\|\mathbf{F}_K^{\text{nat}}(x)\|, \|\mathbf{F}_K^{\text{nat}}(x)\|^{\frac{1}{\xi - 1}} \right), \quad \forall x \in \mathbb{R}^n.$$

This remark broadens somewhat the class of VIs for which a global error bound of the Hölderian type exists. □

Besides its own merit, Proposition **6.3.1** has two important implications. First, the proof under assumption (b) can be used to establish a global error bound from a local error bound for a VI with a compact defining set, without requiring the uniformly P assumption of the defining function. We formally state this error bound in the following result.

6.3.3 Proposition. Let K be a compact convex subset of $\mathbb{R}^n$ and let $F : \mathbb{R}^n \to \mathbb{R}^n$ be continuous. If SOL(K, F) admits a local Lipschitzian error bound with the natural residual, then SOL(K, F) admits a global Lipschitzian error bound with the same residual. Hence, the solution set of every solvable AVI defined by a compact polyhedron admits a global Lipschitzian error bound with the natural residual.

Proof. As in the proof of Proposition **6.3.1**, let $\rho > 0$ be such that K is contained in $\operatorname{cl}\mathbb{B}(0,\rho)$. By assumption and Proposition **6.1.2**, there exists a constant $c' > 0$ such that

$$\operatorname{dist}(x, \operatorname{SOL}(K,F)) \leq c' \, \| \, \mathbf{F}_K^{\mathrm{nat}}(x) \, \|, \quad \forall \, x \in \operatorname{cl}\mathbb{B}(0,3\rho).$$

By a similar argument as in the proof of the previous proposition, we can deduce

$$\operatorname{dist}(x, \operatorname{SOL}(K,F)) \leq 2 \, \| \, \mathbf{F}_K^{\mathrm{nat}}(x) \, \|, \quad \forall \, x \notin \operatorname{cl}\mathbb{B}(0,3\rho).$$

Consequently, combining the above two inequalities, we obtain

$$\operatorname{dist}(x, \operatorname{SOL}(K,F)) \leq \max(\, 2, c \,) \, \| \, \mathbf{F}_K^{\mathrm{nat}}(x) \, \|, \quad \forall \, x \in \mathbb{R}^n,$$

which is the desired global Lipschitzian error bound. The last assertion of the proposition follows readily from Corollary **6.2.2** and what has just been proved. □

The other important implication of Proposition **6.3.1** is that under the assumptions of this proposition, if $r(x)$ is any residual function of the VI that majorizes a positive multiple of the natural residual, then a global error bounds also holds with $r(x)$ as the residual function. See Proposition **10.3.11** for a result of this kind and Section **10.3** where an entire family of parameterized natural maps and an induced family of D-gap functions are shown to be equivalent to the natural residual; see Propositions **10.3.6** and **10.3.7**.

6.3.1 Without Lipschitz continuity

From Proposition **6.3.1**, it follows that for the NCP (F) with a (globally) Lipschitz continuous, uniformly P function F, a global Lipschitzian error bound holds with the min residual $\| \min(x, F(x)) \|$ and any other residual that majorizes or is equivalent to it. The following example shows that such an error bound cannot hold without the Lipschitz property; see also Exercise **6.9.5** for another example and Theorem **6.3.8** for a sufficient condition in lieu of Lipschitz continuity that guarantees such a global Lipschitzian bound.

6.3.4 Example. Consider the NCP (F) with F given by:

$$F(x_1, x_2) \equiv \begin{pmatrix} x_1 - x_2^2 \\ \\ x_2 + x_2^3 \end{pmatrix}, \quad (\, x_1, x_2 \,) \in \mathbb{R}^2.$$

We have

$$JF(x_1, x_2) = \begin{bmatrix} 1 & -2x_2 \\ 0 & 1 + 3x_2^2 \end{bmatrix}, \quad \forall (x_1, x_2) \in \mathbb{R}^2.$$

It is easy to see that the eigenvalues of the symmetric part of $JF(x_1, x_2)$ are bounded below by 1/2 for all (x_1, x_2). Thus F is a strongly monotone, thus uniformly P, function on $\mathbb{R}^2$. Moreover, the unique solution of the NCP (F) is $(0, 0)$. By considering the vector $x^k \equiv (k^2, k)$, where $k > 0$, we see that $\min(x^k, F(x^k)) = (0, k)$, which implies that

$$\lim_{k \to \infty} \frac{\| x^k \|}{\| \min(x^k, F(x^k)) \|} = \infty.$$

Thus a global Lipschitzian error bound cannot hold with the min residual.
□

With an alternative residual function, a global square-root error bound can be derived for an NCP with a uniformly P function that is not required to be Lipschitz continuous. Let

$$r_{\text{LTKYF}}(x) \equiv \sum_{i=1}^{n} [\, (x_i F_i(x))_+ + (x_i)_- + (F_i(x))_- \,],$$

which is the ℓ_1-residual of the equation formulation (1.5.5) of the NCP (F). Variants of this residual function can be used; the key feature of the residuals in this family is the presence of the products $x_i F_i(x)$, which are absent in both the min residual.

6.3.5 Proposition. Let $F : \mathbb{R}^n \to \mathbb{R}^n$ be a continuous uniformly P function on $\mathbb{R}^n$. There exists a constant $c > 0$ such that

$$\| x - x^* \| \le c \sqrt{r_{\text{LTKYF}}(x)}, \quad \forall x \in \mathbb{R}^n, \tag{6.3.2}$$

where x^* is the unique solution of the NCP (F).

Proof. Let x be an arbitrary vector in $\mathbb{R}^n$. We have

$$\begin{aligned} &(x - x^*) \circ (F(x) - F(x^*)) \\ &= x \circ F(x) - x^* \circ F(x) - x \circ F(x^*) \\ &\le (x \circ F(x))_+ + x^* \circ F(x)_- + F(x^*) \circ x_- \end{aligned}$$

because both x^* and $F(x^*)$ are nonnegative. Let j be an index such that

$$(x_j - x_j^*)(F_j(x) - F_j(x^*)) = \max_{1 \le i \le n} (x_i - x_i^*)(F_i(x) - F_i(x^*)).$$

With $\eta > 0$ being the uniformly P modulus of F, we easily obtain

$$\begin{aligned} \eta \, \| \, x - x^* \, \|^2 &\leq \; (\, x_j F_j(x) \,)_+ + x_j^* (\, F_j(x) \,)_- + F_j(x^*)(\, x_j \,)_- \\ &\leq \; \max_{1 \leq i \leq n} \, \max(\, 1, x_i^*, F_i(x^*) \,) \, r_{\text{LTKYF}}(x), \end{aligned}$$

from which the desired error bound (6.3.2) follows. □

In general, the residual $r_{\text{LTKYF}}(x)$ is not compatible with either the min or its equivalents in the sense that there exist no positive scalars c_1, c_2, η_1, and η_2 such that for all $x \in \mathbb{R}^n$,

$$c_1 \, \| \, \min(x, F(x)) \, \|^{\eta_1} \; \leq \; r_{\text{LTKYF}}(x) \; \leq \; c_2 \, \| \, \min(x, F(x)) \, \|^{\eta_2}; \tag{6.3.3}$$

this is illustrated in following example.

6.3.6 Example. Let

$$F(x_1, x_2) \equiv \begin{pmatrix} e^{x_1} \\ \\ x_2 \end{pmatrix}, \qquad (\, x_1, x_2 \,) \in \mathbb{R}^2,$$

which is clearly a strongly monotone, and hence uniformly P, function on $\mathbb{R}^2$. We claim that there exist no positive scalars c_1, c_2, η_1, and η_2 such that (6.3.3) holds for all $x \in \mathbb{R}^2$. Assume for contradiction that for some positive scalars c_1 and η_1, the left-hand inequality holds for all $x \in \mathbb{R}^2$. In what follows, let k be a sufficiently large positive integer. With $(x_1, x_2) = (-k, 0)$, we must have $\eta_1 \leq 1$. With $(x_1, x_2) = (1/k^{1/\eta_1}, 0)$, we see that $\eta_1 = 1$. With $(x_1, x_2) = (0, k)$, we arrive at a contradiction. With $(x_1, x_2) = (k, 0)$, we see that the right-hand inequality in (6.3.3) can not hold for any positive c_2 and η_2. □

It follows that if r is any residual of the NCP that majorizes a positive multiple of r_{LTKYF}, then (6.3.2) holds with r_{LTKYF} replaced by the majorizing residual r and with a different error bound constant. An example of such a residual is derived from the C-function ψ_{YYF} introduced in Exercise **1.8.21**; e.g.,

$$r_{\text{YYF}}(x) \equiv \sqrt{\sum_{i=1}^{n} \left[\, (\, (\, x_i F_i(x) \,)_+ \,)^2 + \left(\, \sqrt{x_i^2 + F_i(x)^2} - x_i - F_i(x) \, \right)^2 \, \right]}. \tag{6.3.4}$$

The reader is asked to show in Exercise **6.9.3** that the above residual also provides a global Hölderian error bound for the NCP with a uniformly P function that is not necessarily Lipschitz continuous.

Returning to the setting at the opening of this section, we present a global error bound for a VI (K, F) with a uniformly P function F. Unlike Proposition **6.3.1**, this error bound does not require F to be Lipschitz continuous but is applicable only to vectors belonging to K. The residual is:

$$r_{\mathcal{N}}(x) \equiv \operatorname{dist}(-F(x), \mathcal{N}(x; K)), \quad \forall x \in K.$$

Since the normal cone $\mathcal{N}(x; K)$ is needed in this definition, it follows that this residual is not meaningful for a test vector that does not belong to K. The quantity $r_{\mathcal{N}}(x)$ appeared for the first time in Proposition **1.5.14**, which shows that $r_{\mathcal{N}}$ majorizes the natural residual $\|\mathbf{F}_K^{\mathrm{nat}}\|$.

6.3.7 Proposition. Let K be given by (6.3.1) where each K_ν is closed convex. Let F be a continuous and uniformly P function on K. Let x^* be the unique solution of the VI (K, F). There exists a constant $c > 0$ such that

$$\| x - x^* \| \leq c\, r_{\mathcal{N}}(x), \quad \forall x \in K.$$

Proof. Let $v \equiv \Pi_{\mathcal{N}(x;K)}(-F(x))$. We have $r_{\mathcal{N}}(x) = \|v + F(x)\|$. Moreover, for each ν,

$$(\,(y - x)_\nu\,)^T v_\nu \leq 0, \quad \forall y_\nu \in K_\nu.$$

In particular, with $y_\nu = x^*_\nu$, we have

$$(\,(x^* - x)_\nu\,)^T [\,v + F(x) - F(x)\,]_\nu \leq 0.$$

Since $x \in$K, we also have

$$(\,(x^* - x)_\nu\,)^T F_\nu(x^*) \leq 0.$$

Adding the last two inequalities, using the uniformly P property of F, and following the argument used in the proof of Proposition **6.3.1**, we can easily deduce the desired error bound in terms of $r_{\mathcal{N}}(x)$. □

We can give a brief comparison of the computations involved in evaluating the two residuals $r_{\mathcal{N}}(x)$ and $\|\mathbf{F}_K^{\mathrm{nat}}(x)\|$ for a vector x in K. In general, to compute the former residual, we need to solve the following optimization problem:

$$\begin{array}{ll} \text{minimize} & \| F(x) + v \| \\ \text{subject to} & v \in \mathcal{N}(x; K), \end{array}$$

which requires an explicit expression of $\mathcal{N}(x; K)$. In contrast, to compute $\|\mathbf{F}_K^{\mathrm{nat}}(x)\|$, we need to compute the projection $\Pi_K(x - F(x))$, which amounts to solving the minimization problem:

$$\begin{array}{ll} \text{minimize} & \| F(x) + y - x \| \\ \text{subject to} & y \in K. \end{array}$$

For a non-polyhedral finitely representable (convex) set K, the latter minimization problem is a convex program; whereas in this case if $x \in K$ satisfies a suitable CQ and if the Euclidean norm is used, then the former minimization problem is a convex quadratic program. Thus computationally the residual $r_{\mathcal{N}}(x)$ is easier to obtain than the natural residual $\|\mathbf{F}_K^{\text{nat}}(x)\|$. Nevertheless, the main weakness of the former residual is that it is restricted to test vectors that belong to K whereas the latter residual is applicable to all vectors in the domain of definition of F.

When $K = \mathbb{R}^n_+$, the residual $r_{\mathcal{N}}(x)$ has a simple form; indeed we have

$$r_{\mathcal{N}}(x) = \sqrt{\sum_{i:x_i>0} F_i(x)^2 + \sum_{i:x_i=0} (\,(F_i(x))_-\,)^2}, \quad \forall\, x \in \mathbb{R}^n_+.$$

The reader is asked to extend this expression to the case where K is a general rectangle in Exercise **6.9.2**.

In spite of Example **6.3.4**, it is possible to combine Propositions **6.1.2**, **6.2.4**, and **6.1.5** to establish a global Lipschitzian error bound for the NCP (F) in terms of the min residual under several conditions. We state the result in a slightly more general form in the following proposition.

6.3.8 Theorem. Let $F : \mathbb{R}^n \to \mathbb{R}^n$ be continuous and $r : \mathbb{R}^n \to \mathbb{R}_+$ be given. Assume that

(a) the NCP (F) has a semistable solution;

(b) for some $\varepsilon > 0$, the level set

$$\{\, x \in \mathbb{R}^n \,:\, \| \min(x, F(x)) \| \leq \varepsilon \,\}$$

is bounded;

(c) for every sequence $\{x^k\}$ satisfying $\min(x^k, F(x^k)) \neq 0$ for every k,

$$\lim_{k\to\infty} \| x^k \| = \infty, \quad \lim_{k\to\infty} \frac{(-x^k)_+}{\| x^k \|} = 0 \quad \lim_{k\to\infty} \frac{(-F(x^k))_+}{\| x^k \|} = 0,$$

it holds that

$$\limsup_{k\to\infty} \frac{r(x^k)}{\| x^k \|} > 0.$$

There exists $\eta > 0$ such that

$$\mathrm{dist}(x, S) \leq \eta\,[\,\|\min(x, F(x)\,\| + r(x)\,], \quad \forall\, x \in \mathbb{R}^n,$$

where S is the solution set of the NCP (F).

Proof. By Propositions **6.1.2** and **6.2.4**, it follows that $\|\min(x, F(x))\|$ provides a local error bound for S. To complete the proof, it suffices to show that if $\{x^k\}$ is an arbitrary sequence satisfying

$$\lim_{k\to\infty} \|\,x^k\,\| = \infty,$$

then

$$\limsup_{k\to\infty} \frac{\|\min(x^k, F(x^k))\,\| + r(x^k)}{\|\,x^k\,\|} > 0. \tag{6.3.5}$$

This clearly holds if

$$\lim_{k\to\infty} \frac{(\,-x^k\,)_+}{\|\,x^k\,\|} = 0 \quad \text{and} \quad \lim_{k\to\infty} \frac{(\,-F(x^k)\,)_+}{\|\,x^k\,\|} = 0,$$

by assumption. Suppose that one of the latter two limits, say the second one, fails. There exists an index i such that

$$\limsup_{k\to\infty} \frac{-F_i(x^k)}{\|\,x^k\,\|} > 0.$$

Since for any two scalars a and b,

$$|\min(a, b)\,| \geq \max(a_-, b_-)$$

it follows that

$$\|\min(x^k, F(x^k))\,\| \geq -F_i(x^k),$$

which easily implies (6.3.5). □

As an application of the above proposition, we derive a global Lipschitzian error bound for an NCP of the $P_*(\sigma)$ type for some $\sigma > 0$ in terms of the sum of min residual and an additional term. The key assumption is that the NCP has a semistable solution, which must necessarily be unique.

6.3.9 Corollary. Let $F : \mathbb{R}^n \to \mathbb{R}^n$ be a continuous $\mathrm{P}_*(\sigma)$ function for some $\sigma > 0$. If the NCP (F) has a semistable (thus unique) solution x^*, then there exists $\eta > 0$ such that

$$\|\,x - x^*\,\| \leq \eta \left[\,\|\min(x, F(x)\,\| + \sum_{i=1}^n (\,x_i F_i(x)\,)_+\,\right], \quad \forall\, x \in \mathbb{R}^n.$$

Proof. Since a $\mathrm{P}_*(\sigma)$ function must be a P_0 function, by Theorem **3.6.6**, x^* must be unique; moreover, there exists $\varepsilon > 0$ such that the level set $\{x \in \mathbb{R}^n : \|\min(x, F(x))\| \leq \varepsilon\}$ is bounded. Thus it remains to show that if $\{x^k\}$ is a sequence satisfying the hypothesis in part (c) of Theorem **6.3.8**, then

$$\limsup_{k\to\infty} \sum_{i=1}^{n} \frac{(x_i^k F_i(x^k))_+}{\| x^k \|} > 0.$$

By Theorem **3.5.12**, the NCP (F) is strictly feasible. Let $\bar{x} > 0$ satisfy $F(\bar{x}) > 0$. By the $\mathrm{P}_*(\sigma)$ property of F, we have, for every k

$$(x^k - \bar{x})^T (F(x^k) - F(\bar{x})) \geq -\sigma \sum_{i\in I_k} (x_i^k - \bar{x}_i)(F_i(x^k) - F_i(\bar{x})),$$

where

$$I_k \equiv \{ i : (x_i^k - \bar{x}_i)(F_i(x^k) - F_i(\bar{x})) > 0 \}.$$

We may rewrite the above inequality as follows:

$$\begin{aligned}
&\sum_{i=1}^{n} x_i^k F_i(x^k) + \sigma \sum_{i\in I_k} x_i^k F_i(x^k) \\
&\quad \geq \sum_{i=1}^{n} [\bar{x}_i F_i(x^k) + F_i(\bar{x}) x_i^k] - \bar{x}^T F(\bar{x}) + \sigma \sum_{i\in I_k} [\bar{x}_i F_i(x^k) + F_i(\bar{x}) x_i^k] \\
&\qquad + \sigma \sum_{i\in I_k} \bar{x}_i F_i(\bar{x}).
\end{aligned}$$

Since

$$\lim_{k\to\infty} \frac{(-x^k)_+}{\| x^k \|} = 0,$$

it follows that every limit point of the normalized sequence $\{x^k/\|x^k\|\}$ must be a nonzero, nonnegative vector; thus since $F(\bar{x}) > 0$, we deduce

$$\liminf_{k\to\infty} \frac{x_i^k F_i(\bar{x})}{\| x^k \|} \geq 0, \quad \forall i$$

and

$$\liminf_{k\to\infty} F(\bar{x})^T \frac{x^k}{\| x^k \|} > 0.$$

Similarly, we have

$$\liminf_{k\to\infty} \frac{\bar{x}_i F_i(x^k)}{\| x^k \|} \geq 0, \quad \forall i.$$

Consequently,

$$(1+\sigma)\limsup_{k\to\infty}\sum_{i=1}^{n}\frac{(x_i^k F_i(x^k))_+}{\|x^k\|}$$

$$\geq \limsup_{k\to\infty}\frac{\sum_{i=1}^{n} x_i^k F_i(x^k) + \sigma \sum_{i\in I_k} x_i^k F_i(x^k)}{\|x^k\|} > 0.$$

By Theorem **6.3.8**, the desired error bound follows. □

6.3.10 Remark. When F is monotone instead and the NCP (F) has a semistable solution, it is clear from the above proof that the bound in Corollary **6.3.9** can be improved slightly as follows:

$$\|x - x^*\| \leq \eta[\|\min(x, F(x)\| + (x^T F(x))_+], \quad \forall x \in \mathbb{R}^n.$$

6.3.2 Affine problems

It turns out that for the AVI (K,q,M), where K is a polyhedron in $\mathbb{R}^n$, q is a vector in $\mathbb{R}^n$ and M is a matrix in $\mathbb{R}^{n\times n}$, the validity of a global Lipschitzian error bound for SOL(K,q,M) in terms of the natural residual can be completely characterized. Before presenting this characterization, we first give an example to show that such an error bound is generally not valid even for an LCP, for which the natural residual becomes the min residual.

6.3.11 Example. Consider the LCP (q,M) with

$$M \equiv \begin{bmatrix} 0 & 1 \\ 1 & 1 \end{bmatrix} \quad \text{and} \quad q \equiv \begin{pmatrix} -1 \\ -2 \end{pmatrix}.$$

It is easily seen that SOL$(q,M) = \{(1,1),(0,2)\}$. Consider the vector $x(t) \equiv (t,1)$, where $t \in [0,\infty)$ is a parameter. As $t \to \infty$, we have $\text{dist}(x(t), \text{SOL}(q,M)) \to \infty$. Since

$$\mathbf{F}_K^{\text{nat}}(x(t)) = \min(x(t), q + Mx(t)) = (0,1)$$

for all $t \geq 2$, there can not exist any positive scalars c and η such that

$$\text{dist}(x(t), \text{SOL}(q,M)) \leq c\,\|\min(x(t), q + Mx(t))\|^{\eta}$$

for all $t \geq 2$ sufficiently large. Thus a global error bound with the min residual, even raised to any positive power, can not hold in this case. □

By Proposition **2.5.3**, we have

$$\mathrm{SOL}(K, q, M)_\infty \subseteq \mathcal{K}(K, M)$$

for all affine pairs (K, M) and vectors q. The equality between the above two sets provides a necessary and sufficient condition for $\mathrm{SOL}(K, q, M)$ to possess a global Lipschitzian error bound in terms of the natural residual. The proof of this assertion is not trivial and relies heavily on the piecewise affine nature of the AVI. Indeed, a similar characterization holds for the zero set of an arbitrary piecewise affine equation. In turn, the proof of the latter more general result is based on the relation between the "recession equation" and the given PA equation. For our purpose here, we give a direct proof of the following result.

6.3.12 Theorem. Let K be a polyhedral set in $\mathbb{R}^n$ and let M be an $n \times n$ matrix. Let $q \in \mathcal{R}(K, M)$ be arbitrary. There exists a constant $c > 0$ such that

$$\mathrm{dist}(x, \mathrm{SOL}(K, q, M)) \leq c \, \| \, x - \Pi_K(x - q - Mx) \, \|, \quad \forall \, x \in \mathbb{R}^n \quad (6.3.6)$$

if and only if $\mathrm{SOL}(K, q, M)_\infty = \mathcal{K}(K, M)$.

We continue to write $\mathbf{F}_K^{\mathrm{nat}}(x)$ for $x - \Pi_K(x - q - Mx)$, which is the natural map of the VI (K, q, M). The proof of Theorem **6.3.12** is done in several steps. We first establish a limiting property of the natural map of the VI (K, q, M), where the set K is not necessarily polyhedral.

6.3.13 Lemma. Let K be a closed convex set in $\mathbb{R}^n$ and let M be an $n \times n$ matrix. Let $q \in \mathbb{R}^n$ be arbitrary. For any pair of vectors x and d in $\mathbb{R}^n$,

$$\lim_{\tau \to \infty} \frac{\mathbf{F}_K^{\mathrm{nat}}(x + \tau d)}{\tau} = \mathbf{M}_{K_\infty}^{\mathrm{nat}}(d), \quad (6.3.7)$$

where $\mathbf{F}_K^{\mathrm{nat}}(x)$ is the natural map of the VI (K, q, M).

Proof. By the respective definition of $\mathbf{F}_K^{\mathrm{nat}}$ and $\mathbf{M}_K^{\mathrm{nat}}$, it suffices to show that

$$\lim_{\tau \to \infty} \frac{\Pi_K(x + \tau d - q - M(x + \tau d))}{\tau} = \Pi_{K_\infty}(d - Md). \quad (6.3.8)$$

Let $\{\tau_k\}$ be any sequence of scalars converging to ∞. Write

$$r^k \equiv \Pi_K(x + \tau_k d - q - M(x + \tau_k d)).$$

By the nonexpansiveness of the Euclidean projection, it follows easily that the sequence $\{(\tau_k)^{-1} r^k\}$ is bounded. Without loss of generality, we may assume that the latter sequence converges to some vector v. To complete the proof, it suffices to verify that $v = \Pi_{K_\infty}(d - Md)$, or equivalently,

$$K_\infty \ni v \perp (v - d + Md) \in (K_\infty)^*.$$

We can write

$$v = \lim_{k\to\infty} \frac{\Pi_K(x + \tau_k d - q - M(x + \tau_k d)) - \Pi_K(x - q - Mx)}{\tau_k},$$

which implies that $v \in K_\infty$. Let u be an arbitrary vector in K_∞. For any vector y in K, we have

$$(y + \tau_k u - r^k)^T(r^k - x - \tau_k d + q + Mx + \tau Md) \geq 0.$$

Dividing by τ_k^2 and letting $k \to \infty$, we obtain

$$(u - v)^T(v - d + Md) \geq 0.$$

Since K_∞ is a cone and $u \in K_\infty$ is arbitrary, the above inequality is enough to yield $v \perp (v - d + Md) \in (K_\infty)^*$. □

Borrowing a terminology from convex analysis, we see that $\mathbf{M}^{\text{nat}}_{K_\infty}$ is the *recession function* of $\mathbf{F}^{\text{nat}}_K$, even though the latter function is not convex. (See Subsection **12.7.1** where the recession function of a convex function is formally reviewed.) Notice that in the two left-hand limits in (6.3.7) and (6.3.8), the step τ tends to infinity instead of zero. This is consistent with the concept of the recession cone. From (6.3.8), we deduce

$$\lim_{\tau\to\infty} \frac{\Pi_K(x + \tau y)}{\tau} = \Pi_{K_\infty}(y), \quad \forall\, x, y \in \mathbb{R}^n.$$

Let K be a polyhedron given by

$$K \equiv \{ x \in \mathbb{R}^n : Ax \leq b \}$$

for some $m \times n$ matrix A and m-vector b. Fix the vector x. For each τ, write $\bar{y}(\tau) \equiv \Pi_K(x + \tau y)$. There exists $\lambda^\tau \geq 0$ such that

$$\Pi_K(x + \tau y) - x - \tau y + \sum_{i \in \mathcal{I}(\bar{y}(\tau))} \lambda_i^\tau (A_{i\cdot})^T = 0,$$

where

$$\mathcal{I}(\bar{y}(\tau)) \equiv \{ i : A_{i\cdot}\bar{y}(\tau) = b_i \}$$

is the index set of active constraints at $\bar{y}(\tau)$. Since there are only finitely many such index sets, it follows that there exist a sequence $\{\tau_k\} \to \infty$ and an index set $\mathcal{J}$, such that $\mathcal{I}(\bar{y}(\tau_k)) = \mathcal{J}$ for all k; moreover, we must have $\mathcal{J} \subseteq \mathcal{I}(\Pi_{K_\infty}(y))$ and, for some nonnegative λ^∞,

$$\Pi_{K_\infty}(y) - y + \sum_{i \in \mathcal{J}} \lambda_i^\infty \, (A_{i\cdot})^T = 0.$$

It therefore follows that for any vector $w \in K$ satisfying $A_{i\cdot} w = b_i$ for all $i \in \mathcal{J}$, we must have

$$\Pi_K(y + w) = \Pi_{K_\infty}(y) + w, \tag{6.3.9}$$

because

$$\begin{aligned} \Pi_{K_\infty}(y) + w - (y + w) + \sum_{i \in \mathcal{J}} \lambda_i^\infty \, (A_{i\cdot})^T &= 0 \\ A_{i\cdot} (\Pi_{K_\infty}(y) + w) &= b_i, \quad \forall \, i \in \mathcal{J} \\ A_{i\cdot} (\Pi_{K_\infty}(y) + w) &\leq b_i, \quad \forall \, i \notin \mathcal{J}, \end{aligned}$$

which is exactly the set of KKT conditions for $\Pi_{K_\infty}(y) + w$ to be the Euclidean projector of $y + w$ onto K; that is for (6.3.9) to hold. Moreover, at least one such vector w must exist, which can be chosen to depend only on the index set $\mathcal{J}$ and not on the vector y. Therefore we have proved that there is a finite family $\mathcal{W}$ of vectors in K such that for every vector $y \in \mathbb{R}^n$, a vector $w \in \mathcal{W}$ exists such that (6.3.9) holds. Based on this conclusion, we can establish the following lemma, which asserts that for an affine pair (K, M), the difference between $\mathbf{F}_K^{\text{nat}}(x)$ and $\mathbf{M}_{K_\infty}^{\text{nat}}(x)$ is bounded on $\mathbb{R}^n$.

6.3.14 Lemma. Let K be a polyhedron in $\mathbb{R}^n$ and let M be an $n \times n$ matrix. Let $q \in \mathbb{R}^n$ be arbitrary. It holds that

$$\sup_{x \in \mathbb{R}^n} \| \mathbf{F}_K^{\text{nat}}(x) - \mathbf{M}_{K_\infty}^{\text{nat}}(x) \| < \infty. \tag{6.3.10}$$

Proof. Let $\mathcal{W}$ be a finite subset of K such that for every $x \in \mathbb{R}^n$, a $w \in \mathcal{W}$ exists satisfying

$$\Pi_{K_\infty}(x - Mx) + w = \Pi_K(x - Mx + w).$$

We have

$$\begin{aligned} \mathbf{F}_K^{\text{nat}}(x) - \mathbf{M}_{K_\infty}^{\text{nat}}(x) &= \Pi_{K_\infty}(x - Mx) - \Pi_K(x - Mx) \\ &= \Pi_K(x - Mx + w) - \Pi_K(x - Mx) - w. \end{aligned}$$

Since $\mathcal{W}$ is finite, the bound (6.3.10) follows easily from the nonexpansiveness of the Euclidean projector. □

Every polyhedron P is the (vector) sum of a compact polyhedron, denoted $C(P)$, and its recession cone P_∞. For any two closed sets X and Y in $\mathbb{R}^n$, the Hausdorff distance between them is by definition the quantity:

$$\mathcal{H}(X,Y) \equiv \max\{\, e(X,Y), e(Y,X)\,\},$$

where

$$e(X,Y) \equiv \sup_{x \in X} \operatorname{dist}(x,Y).$$

Some of these quantities may be equal to ∞. When X is the union of finitely many polyhedra, let $C(X)$ denote the union of $C(X_i)$, with X_i ranging over all the polyhedral pieces of X. Such a set X needs not be convex, but $C(X)$ must be compact. The following result pertains to two sets X and Y that are unions of finitely many polyhedra.

6.3.15 Lemma. Let X and Y be two nonempty sets in $\mathbb{R}^n$ that are unions of finitely many polyhedra. It holds that $e(X,Y) < \infty$ if and only if $X_\infty \subseteq Y_\infty$. Hence $\mathcal{H}(X,Y) < \infty$ if and only if $X_\infty = Y_\infty$.

Proof. Suppose $e(X,Y)$ is finite. Let d be an arbitrary recession direction in X_∞. There exists a vector x in X such that $x + kd$ is in X for all positive integers k. For each k, let y^k be in Y such that $\|x + kd - y^k\| \leq e(X,Y)$. Without loss of generality, we may assume that all y^k are contained in the same polyhedral piece, say Y_1, of Y. We may write $y^k = v^k + ke^k$, where v^k belongs to $C(Y_1)$ and e^k belongs to $(Y_1)_\infty$. Thus

$$\|\, x + kd - v^k - ke^k \,\| \leq e(X,Y).$$

Dividing by k and letting $k \to \infty$, we deduce that d is the limit of $\{e^k\}$. Since $(Y_1)_\infty$ is a polyhedral cone, it is closed; hence d is a recession direction of Y. This establishes the inclusion $X_\infty \subseteq Y_\infty$.

Conversely, suppose that $X_\infty \subseteq Y_\infty$. Let x be an arbitrary vector in a polyhedral piece, say X_1 of X. Write $x = \bar{v} + d$, where $\bar{v} \in C(X_1)$ and $d \in (X_1)_\infty$. For some $\bar{u} \in C(Y)$, we have $\bar{u} + d \in Y$. Thus

$$\operatorname{dist}(x,Y) \leq \|\, \bar{u} - \bar{v}\,\|.$$

Hence,

$$e(X,Y) = \sup_{x \in X} \operatorname{dist}(x,Y) \leq \sup\{\, \|\, u - v\,\| \,:\, (\,u,v\,) \in C(X) \times C(Y)\,\};$$

the latter supremum is finite because $C(X)$ and $C(Y)$ are compact sets. The assertion about the Hausdorff distance is obvious. □

We have all the necessary preparations to prove Theorem **6.3.12**.

Proof of Theorem 6.3.12. By Corollary **6.2.2** and Corollary **6.2.3**, there exist positive constants c and ε such that

$$\text{dist}(x, \text{SOL}(K,q,M)) \leq c \,\|\, \mathbf{F}^{\text{nat}}_K(x) \,\|, \quad \forall\, x \text{ such that } \|\, \mathbf{F}^{\text{nat}}_K(x) \,\| \leq \varepsilon,$$

and

$$\text{dist}(x, \mathcal{K}(K,M)) \leq c \,\|\, \mathbf{M}^{\text{nat}}_K(x) \,\|, \quad \forall\, x \in \mathbb{R}^n.$$

Moreover, by Theorem **2.5.15**, both $\text{SOL}(K,q,M)$ and $\mathcal{K}(K,M)$ are the unions of finitely many polyhedra, with the latter being a cone. Suppose $\text{SOL}(K,q,M)_\infty = \mathcal{K}(K,M)$. Let x be such that $\|\mathbf{F}^{\text{nat}}_K(x)\| > \varepsilon$. Clearly,

$$\text{dist}(x, \text{SOL}(K,q,M)) \leq \text{dist}(x, \mathcal{K}(K,M)) + e(\mathcal{K}(K,M), \text{SOL}(K,q,M)).$$

By Lemma **6.3.15**, $e(\mathcal{K}(K,M), \text{SOL}(K,q,M))$ is finite. We have

$$\begin{aligned}
\text{dist}(x, \mathcal{K}(K,M)) &\leq c \,\|\, \mathbf{M}^{\text{nat}}_K(x) \,\| \\
&\leq c \,\|\, \mathbf{F}^{\text{nat}}_K(x) \,\| \, \frac{\|\, \mathbf{M}^{\text{nat}}_K(x) \,\|}{\|\, \mathbf{F}^{\text{nat}}_K(x) \,\|} \\
&\leq c \,\|\, \mathbf{F}^{\text{nat}}_K(x) \,\| \left[1 + \frac{\|\, \mathbf{M}^{\text{nat}}_K(x) - \mathbf{F}^{\text{nat}}_K(x) \,\|}{\|\, \mathbf{F}^{\text{nat}}_K(x) \,\|} \right] \\
&\leq c' \,\|\, \mathbf{F}^{\text{nat}}_K(x) \,\|,
\end{aligned}$$

where

$$c' \equiv c \left(1 + \varepsilon^{-1} \sup_{x \in \mathbb{R}^n} \|\, \mathbf{M}^{\text{nat}}_K(x) - \mathbf{F}^{\text{nat}}_K(x) \,\| \right).$$

By Lemma **6.3.14**, c' is finite. Thus,

$$\text{dist}(x, \text{SOL}(K,q,M)) \leq \left(c' + \frac{e(\mathcal{K}(K,M), \text{SOL}(K,q,M))}{\varepsilon} \right) \|\, \mathbf{F}^{\text{nat}}_K(x) \,\|$$

for all x such that $\|\mathbf{F}^{\text{nat}}_K(x)\| > \varepsilon$. Combining this with the local error bound for x with natural residual not exceeding ε, we have therefore established the sufficiency of the theorem.

To prove the necessity, suppose that (6.3.6) holds. It suffices to show that $\mathcal{K}(K,M)$ is contained in $\text{SOL}(K,q,M)_\infty$; in turn, we need to show that $e(\mathcal{K}(K,M), \text{SOL}(K,q,M))$ is finite. Let x be an arbitrary vector in $\mathcal{K}(K,M)$. By the assumed error bound,

$$\text{dist}(x, \text{SOL}(K,q,M)) \leq c \,\|\, \mathbf{F}^{\text{nat}}_K(x) \,\| \leq c \,\|\, \mathbf{F}^{\text{nat}}_K(x) - \mathbf{M}^{\text{nat}}_K(x) \,\|$$

because $\mathbf{M}^{\text{nat}}_K(x) = 0$. The desired finiteness of $e(\mathcal{K}(K,M), \text{SOL}(K,q,M))$ follows from Lemma **6.3.14**. □

Next, we present a corollary of Theorem **6.3.12** that gives a sufficient condition for the AVI (K, q, M) to have a global Lipschitzian error bound in terms of the natural residual for all vectors q in the AVI range of the pair (K, M); this condition turns out to be necessary under a mild assumption that is valid in many interesting cases.

6.3.16 Corollary. Let K be a polyhedral set in $\mathbb{R}^n$ and let M be an $n \times n$ matrix. Consider the following two statements.

(a) (K, M) is an R_0 pair.

(b) For all vectors $q \in \mathcal{R}(K, M)$, a global Lipschitzian error bound holds for $\mathrm{SOL}(K, q, M)$ in terms of the natural residual of the AVI (K, q, M).

It holds that (a) $\Rightarrow$ (b). The reverse implication holds if there exists a vector $\bar{q} \in \mathcal{R}(K, M)$ such that $\mathrm{SOL}(K, \bar{q}, M)$ is bounded.

Proof. By Theorem **6.3.12**, (a) clearly implies (b). Conversely, if for some $\bar{q}$ in $\mathcal{R}(K, M)$, $\mathrm{SOL}(K, \bar{q}, M)$ is bounded, then $\mathrm{SOL}(K, \bar{q}, M)_\infty$ is the singleton $\{0\}$. Since a global Lipschitzian error bound holds for $\mathrm{SOL}(K, \bar{q}, M)$ in terms of the natural residual, (a) follows also from the previous theorem. □

Proposition **6.3.17** below identifies a broad class of affine pairs (K, M) for which a vector $\bar{q} \in \mathcal{R}(K, M)$ exists such that $\mathrm{SOL}(K, \bar{q}, M)$ is bounded, and thus the two statements (a) and (b) in Corollary **6.3.16** are equivalent. This class includes all affine pairs (K, M) with K having an extreme point; cf. Remark **2.5.21**. It also includes all affine pairs (K, M) satisfying the sharp property (2.5.6) and such that $\operatorname{int} \mathcal{K}(K, M)^* \subseteq \mathcal{R}(K, M)$; cf. Proposition **2.5.5**. In particular, it follows that for a given matrix M, a global Lipschitzian error bound holds for the solutions of the LCP (q, M) in terms of the min residual $\| \min(x, q + Mx) \|$ for all vectors q if and only if M is an R_0 matrix.

6.3.17 Proposition. Let K be a polyhedral set in $\mathbb{R}^n$ and let M be an $n \times n$ matrix. If $\mathcal{R}(K, M)$ contains an open set Ω, then there exists an open and dense subset Ω_0 of Ω such that, for all $q \in \Omega_0$, $\mathrm{SOL}(K, q, M)$ is nonempty and finite.

Proof. We prove a more general result for a piecewise affine map f from $\mathbb{R}^n$ into itself. Namely, if the range of such a map contains an open set Ω, then there exists an open and dense subset Ω_0 such that for all $q \in \Omega_0$, the inverse image $f^{-1}(q)$ is nonempty and finite. Let Ξ be a polyhedral subdivision of $\mathbb{R}^n$ and let $\{f^i\}$ be a corresponding finite family of affine functions such that each polyhedron P_i in Ξ has a nonempty interior and f coincides with one of the functions f^i on P_i; cf. Proposition **4.2.1**. Write $f^i(x) \equiv A^i x + a^i$ for some $n \times n$ matrix A^i and n-vector a^i. Since the range of f is open, at least one matrix A^i is nonsingular. Let $\{A^i : i \in \mathcal{S}\}$ and $\{A^i : i \in \mathcal{N}\}$ be the subfamily of singular and nonsingular matrices, respectively, within the family $\{A^i\}$. Since A^i is nonsingular for each $i \in \mathcal{N}$, we have $f^i(\operatorname{int} P_i) = \operatorname{int} f^i(P_i)$. At the same time,

$$\bigcup_{i \in \mathcal{S}} f^i(P_i)$$

is a finite union of lower-dimensional polyhedral sets and so its complement is open and dense in $\mathbb{R}^n$. Put

$$\Omega_0 \equiv \Omega \bigcap \left(\bigcup_{i \in \mathcal{S}} f^i(P_i) \right)^c \bigcap \left(\bigcup_{i \in \mathcal{N}} f^i(\operatorname{int} P_i) \right).$$

It is easy to see that for each $q \in \Omega_0$, $f^{-1}(q)$ is nonempty and contained in the union of $\operatorname{int} P_i$ for $i \in \mathcal{N}$. Since $\mathcal{N}$ is a finite index set, $f^{-1}(q)$ is nonempty and finite.

To apply the above result to the proposition, let f be the PL map $\mathbf{M}_K^{\text{nor}}$. The range of $\mathbf{M}_K^{\text{nor}}$ coincides with $-\mathcal{R}(K, M)$. Thus there exists an open dense subset Ω_0 of Ω such that $(\mathbf{M}_K^{\text{nor}})^{-1}(-q)$ is a nonempty finite subset for all $q \in \Omega_0$. This implies that the AVI (K, q, M) has a finite number of solutions. □

In general, the multiplicative factor in the error bound for SOL(K, q, M) is dependent on q. It is natural to ask when a common multiplicative constant exists that applies to all vectors q. A question related to this was first asked at the end of Section **4.3**. We present a partial answer to this question in Theorem **6.3.18** below. Specifically, it was asked previously whether for an affine Lipschitzian pair (K, M), the AVI (K, q, M) must necessarily have a unique solution for all $q \in \mathbb{R}^n$. The following theorem provides an affirmative answer to this question under the assumption that a solution to the AVI (K, q, M) exists for all $q \in \mathbb{R}^n$, that is, if the AVI range of the pair (K, M) is equal to $\mathbb{R}^n$ to begin with.

6.3.18 Theorem. Let K be a closed convex set in $\mathbb{R}^n$ and let M be an $n \times n$ matrix. The following two statements are equivalent.

(a) There exists a constant $c > 0$ such that, for all $q \in \mathcal{R}(K, M)$,

$$\operatorname{dist}(x, \operatorname{SOL}(K, q, M)) \leq c \, \| \, x - \Pi_K(x - q - Mx) \, \|, \quad \forall \, x \in \mathbb{R}^n.$$

(b) (K, M) is a Lipschitzian pair; that is, there exists a constant $c' > 0$ such that, for all q and q' in $\mathcal{R}(K, M)$,

$$\text{SOL}(K, q', M) \subseteq \text{SOL}(K, q, M) + c' \| q - q' \| \text{ cl } \mathbb{B}(0, 1).$$

If K is polyhedral and $\mathcal{R}(K, M) = \mathbb{R}^n$, the above two statements (a) and (b) are further equivalent to $\text{SOL}(K, q, M)$ being a singleton for all $q \in \mathbb{R}^n$.

Proof. The equivalence of (a) and (b) is not difficult to prove; in fact, similar equivalences in related contexts have been noted in previous chapters. Assume that (a) holds. Let x be a solution of the VI (K, q', M). We have $x = \Pi_K(x - q' - Mx)$. By (a), it follows that

$$\text{dist}(x, \text{SOL}(K, q, M)) \leq c \| x - \Pi_K(x - q - Mx) \| \leq c \| q - q' \|,$$

where the second inequality is due to the nonexpansiveness of the Euclidean projector. This establishes (b). Conversely, suppose (b) holds. Let $x \in \mathbb{R}^n$ be given. Write $r \equiv x - \Pi_K(x - q - Mx)$. The vector $y \equiv x - r$ is then a solution of the VI (K, q', M), where $q' \equiv q - r + Mr$. By (b), it follows that

$$\text{dist}(x, \text{SOL}(K, q, M)) \leq c' \| r - Mr \|,$$

from which (a) follows.

Suppose that K is polyhedral and $\mathcal{R}(K, M) = \mathbb{R}^n$. We claim that the normal map $\mathbf{M}_K^{\text{nor}}$, which is a PA map, is open if (b) holds. The openness of $\mathbf{M}_K^{\text{nor}}$ is equivalent to saying that for every pair (z', q') such that $q' \equiv \mathbf{M}_K^{\text{nor}}(z')$ and for every sequence $\{q^k\}$ converging to q^∞, there exists a sequence $\{z^k\}$ converging to z^∞ such that $q^k = \mathbf{M}_K^{\text{nor}}(z^k)$ for all k sufficiently large. Let $x^\infty \equiv \Pi_K(z^\infty)$. Since $\mathcal{R}(K, M) = \mathbb{R}^n$, part (b) implies that there exists a vector $x^k \in \text{SOL}(K, q^k, M)$ such that

$$\| x^\infty - x^k \| \leq c' \| q^\infty - q^k \|.$$

Hence the sequence $\{x^k\}$ converges to x^∞. Letting $z^k \equiv x^k - q^k - Mx^k$, we have $q^k = \mathbf{M}_K^{\text{nor}}(z^k)$ and $\{z^k\}$ clearly converges to $z^\infty = x^\infty - q^\infty - Mx^\infty$. Hence $\mathbf{M}_K^{\text{nor}}$ is an open PA map. By Proposition **4.2.15**, $\mathbf{M}_K^{\text{nor}}$ is coherently oriented. By Theorem **4.3.2**, the AVI (K, q, M) has a unique solution for all $q \in \mathbb{R}^n$.

Conversely, if the AVI (K, q, M) has a unique solution for all $q \in \mathbb{R}^n$, Theorem **4.3.2** implies that this unique solution is a Lipschitz continuous function of q. This is precisely the statement (b) when the solution sets of the AVIs with various q are singletons. □

The affine case of Theorem **6.3.18** can be extended to a general PA map. We remark that the inverse of every PA map is a polyhedral multifunction; cf. the discussion preceding Theorem **5.5.8**. We recall that, in general, a multifunction Φ is Lipschitz continuous on a set $\Omega \subseteq \operatorname{dom}\Phi$ if there exists a constant $L > 0$ such that for every x and y in Ω,

$$\Phi(x) \subseteq \Phi(y) + L \| x - y \| \operatorname{cl}\mathbb{B}(0,1).$$

6.3.19 Proposition. Let $f : \mathbb{R}^n \to \mathbb{R}^m$ be a PA map. The following two statements are equivalent.

(a) f is open.

(b) f is surjective and f^{-1} is Lipschitz continuous on $\mathbb{R}^n$.

If $n = m$, these are further equivalent to f being coherently oriented.

Proof. It suffices to show that (a) and (b) are equivalent.

(b) $\Rightarrow$ (a). The proof is the same as in Theorem **6.3.18**.

(a) $\Rightarrow$ (b). Being a PA map, the range of f is a closed set, by part (a) of Proposition **4.2.2**. Since f is open, the range of f, being the image of the open set $\mathbb{R}^n$, is open. Hence $f(\mathbb{R}^n) = \mathbb{R}^m$ and f is surjective. It remains to show that f^{-1} is Lipschitz continuous. We need to show the existence of a constant $L > 0$ such that for every q and q' in $\mathbb{R}^m$,

$$f^{-1}(q') \subseteq f^{-1}(q) + L \| q - q' \| \operatorname{cl}\mathbb{B}(0,1).$$

As in the proof of Proposition **4.2.2**, let Ξ be a polyhedral subdivision of $\mathbb{R}^n$ induced by f and let $\{G^i\}$ be a finite family of affine functions such that f coincides with one of these functions on each polyhedron in Ξ. For each P in Ξ, let G_P be one of the affine members in $\{G^i\}$ that coincide with f on P. We then have

$$\bigcup_{P \in \Xi} P = \mathbb{R}^n = f(\mathbb{R}^n) = \bigcup_{P \in \Xi} G_P(P).$$

and for every $q \in \mathbb{R}^n$,

$$f^{-1}(q) = \bigcup_{P \in \Xi} [\,(G_P)^{-1}(q) \cap P\,].$$

Expressed in linear inequalities, each set

$$\Omega_P \equiv (G_P)^{-1}(q) \cap P$$

is a polyhedron with q appearing as a constant vector in the linear equation $G_P(x) = q$. By Hoffman's error bound for polyhedra, there exists for each P in Ξ a constant $L_P > 0$ such that for all $q \in \mathbb{R}^n$,

$$\operatorname{dist}(x, \Omega_P) \le L_P\,[\,\| G_P(x) - q \| + \operatorname{dist}(x, P)\,], \quad \forall\, x \in \mathbb{R}^n.$$

Let

$$L \equiv \max_{P \in \Xi} L_P.$$

We claim that this is the desired constant.

Let q and q' be two given vectors in $\mathbb{R}^n$ and let $x \in f^{-1}(q')$ be arbitrary, so that $f(x) = q'$. Consider the line segment joining q and q'. We claim that there exist a partition of this segment:

$$\{ q' = q^0, q^1, \ldots, q^\ell = q \}$$

for some positive integer ℓ, and a corresponding set:

$$\{ x^0 = x, x^1, \ldots, x^\ell \}$$

such that $f(x^i) = q^i$ for all $i = 0, \ldots, \ell$, any two consecutive vectors x^i and x^{i+1} belong to the same affine piece of f for all $i = 0, \ldots, \ell - 1$, and

$$\| x^i - x^{i+1} \| \leq L \| q^i - q^{i+1} \|, \quad \forall i = 0, \ldots, \ell - 1.$$

Once this is established, the proof of the proposition will be complete. Let Ξ_0 be the subcollection of polyhedra P in Ξ that contains x^0. There exists an open neighborhood $\mathcal{N}_0$ of x^0 such that

$$x^0 \in \mathcal{N}_0 \subseteq \bigcup_{P \in \Xi_0} P$$

and

$$q^0 \in f(\mathcal{N}_0) \subseteq \bigcup_{P \in \Xi_0} G_P(P)$$

By the openness of f, $f(\mathcal{N}_0)$ is an open set. Let τ_0 denote the supremum of real numbers τ between 0 and 1 with the property that the closed line segment joining q^0 and $(1-\tau)q^0 + \tau q$ is contained in some $G_{P_0}(P_0)$ for some P_0 in Ξ_0. The scalar τ_0 is well defined and positive, due to the openness of $f(\mathcal{N}_0)$ and the fact that it is contained in at least one such $G_{P_0}(P_0)$. Furthermore, the vector

$$q^1 \equiv (1 - \tau_0) \, q^0 + \tau \, q$$

belongs to $G_{P_0}(P_0)$. Let $x^1 \equiv \Pi_{\Omega_0}(x^0)$. Both x^0 and x^1 belong to P_0; moreover,

$$\| x^0 - x^1 \| \leq L \, \| q^0 - q^1 \|.$$

If $q^1 = q$, we are done. Otherwise, repeat the above construction with (x^0, q^0) replaced by (x^1, q^1) to obtain (x^2, q^2) and a polyhedron P_1 in Ξ, which must necessarily be distinct from P_0 by the maximality of the scalar

τ_0 and because the step τ_1 obtained in the process is positive. Since the number of polyhedra in Ξ is finite, the process must eventually reach the vector q and our claim follows. □

6.3.20 Remark. There is an extended version of the equivalence between statements (a) and (b) in Proposition **6.3.19** when f is a polyhedral multifunction. In fact, the above proof needs only be slightly modified to establish such an extension. □

6.4 Monotone AVIs

We present a theory of (global) error bounds for the AVI (K, q, M) with the gap function as the residual, where K is a polyhedron in $\mathbb{R}^n$, q is a vector in $\mathbb{R}^n$, and M is a positive semidefinite matrix in $\mathbb{R}^{n \times n}$. The motivation for using the gap residual is derived from the LCP (q, M), where the gap function reduces to the complementarity gap $x^T(q + Mx)$ for all vectors x feasible to the problem. This is a reasonable residual function because feasible vectors to the LCP are not difficult to obtain (via linear programming). Extending this consideration to the AVI leads to the gap residual.

The cornerstone of the error bound theory developed in this section is Theorem **2.4.15**, which gives a polyhedral representation of SOL(K, q, M). For ease of reference, we repeat the key facts and definitions that lead to this theorem. First, we have the two invariants of the monotone AVI (K, q, M):

$$d \equiv (M + M^T)x \quad \text{and} \quad \sigma \equiv x^T M x, \quad \forall x \in \text{SOL}(K, q, M).$$

Next, we recall the gap function

$$\theta_{\text{gap}}(x) = x^T(q + Mx) - \omega(x),$$

where

$$\omega(x) \equiv \min_{y \in K} y^T(q + Mx), \tag{6.4.1}$$

associated with which are two polyhedral sets:

$$\begin{aligned} \Omega &\equiv \{ x \in \mathbb{R}^n : \omega(x) > -\infty \} \\ &= \{ x \in \mathbb{R}^n : q + Mx \in (K_\infty)^* \} \end{aligned}$$

and

$$\begin{aligned} \Omega' &\equiv \{ x \in \mathbb{R}^n : \omega(x) - (\sigma + q^T x) \geq 0 \} \\ &= \{ x \in \Omega : v^T(q + Mx) \geq \sigma + q^T x \text{ for all } v \in E \}, \end{aligned}$$

where $K = \mathcal{H}E + K_\infty$. Theorem **2.4.15** then yields

$$\mathrm{SOL}(K,q,M) = \{ x \in K \cap \Omega' : (M + M^T)x = d \}.$$

For every $x \in \mathrm{SOL}(K,q,M)$, we have

$$\begin{aligned} \omega(x) &\leq x^T(q + Mx) \\ &= (q + d - M^T x)^T x = (q + d)^T x - \sigma; \end{aligned}$$

simple algebra gives the alternative representations:

$$\begin{aligned} &\mathrm{SOL}(K,q,M) = \\ &\{ x \in K : \omega(x) - (q + d)^T x + \sigma \geq 0, \ (M + M^T)x = d \} = \\ &\{ x \in K : \omega(x) - (q + d)^T x + \sigma = 0, \ (M + M^T)x = d \} = \\ &\bigcap_{z \in E} \{ x \in K \cap \Omega : z^T(q + Mx) \geq (q + d)^T x - \sigma, \ (M + M^T)x = d \}. \end{aligned}$$

Applying Hoffman's error bound to the latter expression, we can easily derive a global Lipschitzian error bound for $\mathrm{SOL}(K,q,M)$ whose residual would involve the distance to the set Ω and the two constants σ and d. Alternatively, our goal is to derive an error bound that employs the gap function $\theta_{\mathrm{gap}}(x)$ as the residual function. In order for $\theta_{\mathrm{gap}}(x)$ to qualify as a legitimate residual, we need to have (a) $\theta_{\mathrm{gap}}(x) \geq 0$ and (b) $\theta_{\mathrm{gap}}(x) = 0$ if and only if $x \in \mathrm{SOL}(K,q,M)$. For this to hold, it is imperative for x to be an element of K to begin with. The following result, which is the promised error bound for a monotone AVI with the gap residual, therefore focuses on test vectors that belong to the set K.

6.4.1 Theorem. Let K be a polyhedron in $\mathbb{R}^n$ and let M be an $n \times n$ positive semidefinite matrix. Assume that $\mathrm{SOL}(K,q,M)$ is nonempty. Let d and σ be the two invariants associated with the solutions of the AVI (K,q,M) (see Lemma **2.4.14**). There exists a constant $L > 0$ such that

$$\mathrm{dist}(x, \mathrm{SOL}(K,q,M)) \leq L \left[\theta_{\mathrm{gap}}(x) + \sqrt{\theta_{\mathrm{gap}}(x)} \right], \quad \forall x \in K.$$

Proof. Since $\theta_{\mathrm{gap}}(x) = \infty$ for all x not in Ω, the above error bound holds trivially for all $x \notin \Omega$. Therefore, in the analysis below, we focus on vectors x that lie in the intersection $K \cap \Omega$.

By Hoffman's error bound, we deduce the existence of a constant $L > 0$ such that for all x in $K \cap \Omega$,

$$\begin{aligned} &\mathrm{dist}(x, \mathrm{SOL}(K,q,M)) \leq L \, [\, \| (M + M^T)x - d \| + \\ &\max_{z \in E} \{ (z^T(q + Mx) - (q + d)^T x + \sigma)_- \} \Big]. \end{aligned}$$

By the definition ω, we have for all $x \in \Omega$ and all $z \in E \subset K$,

$$z^T(q + Mx) - (q + d)^T x + \sigma \geq \omega(x) - (q + d)^T x + \sigma.$$

Let $\bar{x}$ be an arbitrary solution of the AVI (K, q, M), we have

$$\begin{aligned}
&(q + d)^T x - \sigma \\
&= q^T x + \bar{x}^T(M + M^T)^T x - \bar{x}^T M \bar{x} \\
&= (q + Mx)^T x - (x - \bar{x})^T M(x - \bar{x}) \leq (q + Mx)^T x.
\end{aligned}$$

Hence, for all $x \in \Omega$ and all $z \in E$,

$$z^T(q + Mx) - (q + d)^T x + \sigma \geq -\theta_{\text{gap}}(x),$$

which implies,

$$\max_{z \in E} \{ (z^T(q + Mx) - (q + d)^T x + \sigma)_- \} \leq \theta_{\text{gap}}(x) \tag{6.4.2}$$

because $\theta_{\text{gap}}(x)$ is nonnegative. It remains to bound $\|(M + M^T)x - d\|$. Since $M + M^T$ is symmetric positive semidefinite, we have

$$\|(M + M^T)v\|^2 \leq 2\lambda_{\max}(M + M^T)\, v^T M v, \quad \forall\, v \in \mathbb{R}^n,$$

where $\lambda_{\max}(M + M^T)$ is the largest eigenvalue of $M + M^T$. Let $\bar{x}$ be a solution of the AVI (K, q, M). Consequently, with $c \equiv 2\lambda_{\max}(M + M^T)$,

$$\begin{aligned}
\|(M + M^T)x - d\|^2 &= \|(M + M^T)(x - \bar{x})\|^2 \\
&\leq c(x - \bar{x})^T M(x - \bar{x}) \\
&= c[(x - \bar{x})^T(q + Mx) - (x - \bar{x})^T(q + M\bar{x})] \\
&\leq c[x^T(q + Mx) - \omega(x)] = c\,\theta_{\text{gap}}(x),
\end{aligned}$$

which yields,

$$\|(M + M^T)x - d\| \leq \sqrt{c}\sqrt{\theta_{\text{gap}}(x)}.$$

Combining this bound with (6.4.2), we obtain the desired error bound of $\text{dist}(x, \text{SOL}(K, q, M))$ in terms of $\theta_{\text{gap}}(x) + \sqrt{\theta_{\text{gap}}(x)}$ for all $x \in K \cap \Omega$. □

For the monotone affine CP (K, q, M), the above result gives the error bound: for all $x \in \text{FEA}(K, q, M)$,

$$\text{dist}(x, \text{SOL}(K, q, M)) \leq L\left[x^T(q + Mx) + \sqrt{x^T(q + Mx)} \right]. \tag{6.4.3}$$

This error bound, which requires the test vector x to be feasible to the CP (K, q, M), is generalized in Corollary **6.4.11**, where the feasibility requirement is removed. The following example shows that the square root term $\sqrt{x^T(q + Mx)}$ is essential and can not be dropped.

6.4.2 Example. Consider the LCP (q, M) with data

$$q \equiv \begin{pmatrix} 0 \\ 1 \end{pmatrix} \quad \text{and} \quad M \equiv \begin{bmatrix} 1 & -1 \\ 1 & 1 \end{bmatrix}.$$

Clearly $\text{SOL}(q, M) = \{(0,0)\}$ and $\text{FEA}(q, M) = \{x \in \mathbb{R}^2_+ : x_2 \leq x_1\}$. Let $x(\varepsilon) \equiv (\varepsilon, \varepsilon^2)$ for $\varepsilon \in [0,1]$. We have

$$\frac{\|x(\varepsilon)\|_\infty}{x(\varepsilon)^T(\,q + Mx(\varepsilon)\,)} = \frac{\varepsilon}{2\varepsilon^2 + \varepsilon^4},$$

which tends to ∞ as ε tends to zero. □

In general, it would be desirable to obtain an error bound for the monotone AVI (K, q, M) where the square-root term $\sqrt{\theta_{\text{gap}}(x)}$ is not needed. It turns out that besides being important in its own right, the existence of such a Lipschitzian error bound is equivalent to several interesting properties of the problem. Before formally stating the main result, Theorem **6.4.6**, we need to discuss some related properties of the KKT system of the AVI (K, q, M).

With $K \equiv P(A, b)$ for some $m \times n$ matrix A and m-vector b, the KKT system of the AVI (K, q, M) is:

$$q + Mx + A^T\lambda = 0$$

$$0 \leq \lambda \perp b - Ax \geq 0.$$

By Corollary **3.4.3**, a solution x of the AVI (K, q, M) is nondegenerate if and only if there exists a multiplier λ such that (x, λ) is a nondegenerate KKT pair, that is, (x, λ) satisfies the above KKT system and $\lambda + b - Ax > 0$. Associated with this KKT system is the natural quadratic program:

$$\begin{aligned} \text{minimize} \quad & x^T(\,q + Mx\,) + \lambda^T b \\ \text{subject to} \quad & q + Mx + A^T\lambda = 0 \\ & Ax \leq b, \quad \lambda \geq 0. \end{aligned} \tag{6.4.4}$$

In what follows, we let $y \equiv (x, \lambda)$; also let $f(y)$ denote the objective function of (6.4.4). Notice that $f(y) = y^T(p + Ny)$, where

$$p \equiv \begin{pmatrix} q \\ b \end{pmatrix} \quad \text{and} \quad N \equiv \begin{bmatrix} M & A^T \\ -A & 0 \end{bmatrix},$$

Moreover since M is positive semidefinite, so is N. Let $\text{FEA}(K, p, N)$ and $S_{\text{opt}}(K, p, N)$ denote the feasible region and optimal solution set of (6.4.4), respectively. It is easy to see that f is nonnegative on FEA(K, q, N); hence if FEA(K, p, M) is nonempty, then so is $S_{\text{opt}}(K, p, N)$, regardless of whether the AVI (K, q, M) is solvable or not. The AVI (K, q, M) has a solution if and only if $f(y) = 0$ for all $y \in S_{\text{opt}}(K, p, N)$; moreover, the x part of such a vector y must be a solution of the AVI (K, q, M). Note that FEA$(K, p, N) = K \cap \Omega$. By Corollary **2.3.7**, it follows that

$$S_{\text{opt}}(K, p, N) = \{ y \in \text{FEA}(K, p, N) : Ny = N\bar{y},\ p^T y = p^T \bar{y} \}$$

for any $\bar{y} \in S_{\text{opt}}(K, p, N)$. The following lemma shows that a certain error bound holds for $S_{\text{opt}}(K, p, N)$ if and only if this set has a simplified representation.

6.4.3 Lemma. Let M be a positive semidefinite matrix. Assume that FEA(K, p, N) is nonempty. There exists a constant $L > 0$ such that

$$\text{dist}(y, S_{\text{opt}}(K, p, N)) \leq L (f(y) - f(\bar{y})), \quad \forall y \in \text{FEA}(K, p, N),$$

if and only if

$$S_{\text{opt}}(K, p, N) = \{ y \in \text{FEA}(K, p, N) : \nabla f(\bar{y})^T (y - \bar{y}) \leq 0 \} \tag{6.4.5}$$

for any $\bar{y} \in S_{\text{opt}}(K, p, N)$.

Proof. Since $\bar{y} \in S_{\text{opt}}(K, p, N)$, the minimum principle implies that

$$\nabla f(\bar{y})^T (y - \bar{y}) \geq 0, \quad \forall y \in \text{FEA}(K, p, N).$$

Suppose that (6.4.5) holds. By Hoffman's error bound for polyhedra, there exists a constant $L > 0$ such that

$$\text{dist}(y, S_{\text{opt}}(K, p, N)) \leq L \nabla f(\bar{y})^T (y - \bar{y}), \quad \forall y \in \text{FEA}(K, p, N).$$

By the gradient inequality for a convex function, we have

$$f(y) - f(\bar{y}) \geq \nabla f(\bar{y})^T (y - \bar{y}).$$

Combining the above two inequalities easily establishes the "if" statement of the lemma.

Conversely, suppose the constant L exists with the prescribed property. Let $\bar{S}$ denote the right-hand set in (6.4.5). We obviously have

$$\nabla f(\bar{y})^T (y - \bar{y}) = 0, \quad \forall y \in \bar{S}.$$

Clearly, $S_{\text{opt}}(K, p, N)$ is a subset of $\bar{S}$. Conversely, let y be an arbitrary vector $\bar{S}$. For each $\tau > 0$, write

$$y(\tau) \equiv \bar{y} + \tau (y - \bar{y}) \quad \text{and} \quad \bar{y}(\tau) \equiv \Pi_{S_{\text{opt}}}(y(\tau)),$$

where S_{opt} is a shorthand for $S_{\text{opt}}(K, p, N)$. Since S_{opt} is a polyhedron, the Euclidean projector $\Pi_{S_{\text{opt}}}$ onto S_{opt} is B-differentiable. Moreover, by Theorem **4.1.1**,

$$\Pi'_{S_{\text{opt}}}(\bar{y}; v) = \Pi_{\mathcal{C}_{\text{opt}}}(v), \quad \forall v \in \mathbb{R}^{n+m},$$

where $\mathcal{C}_{\text{opt}}$ is the critical cone of S_{opt} at $\bar{y}$. This critical cone is contained in the linear subspace $\{v \in \mathbb{R}^{n+m} : Nv = 0, p^T v = 0\}$. We have

$$\lim_{\tau \downarrow 0} \frac{\bar{y}(\tau) - \bar{y}}{\tau} = \Pi'_{S_{\text{opt}}}(\bar{y}; y - \bar{y}) = \Pi_{\mathcal{C}_{\text{opt}}}(y - \bar{y}).$$

By assumption, for every $\tau > 0$,

$$f(y(\tau)) - f(\bar{y}) \geq L^{-1} \| \bar{y}(\tau) - y(\tau) \| = L^{-1} \| \bar{y}(\tau) - \bar{y} - \tau (y - \bar{y}) \|;$$

dividing by $\tau > 0$ and letting $\tau \to 0$, we deduce

$$\nabla f(\bar{y})^T (y - \bar{y}) \geq L^{-1} \| \Pi_{\mathcal{C}_{\text{opt}}}(y - \bar{y}) - (y - \bar{y}) \|.$$

Since the left-hand side is equal to zero, we deduce that $y - \bar{y}$ belongs to $\mathcal{C}_{\text{opt}}$; hence $Ny = N\bar{y}$ and $p^T y = N^T \bar{y}$. Therefore y belongs to S_{opt}. □

Although Lemma **6.4.3** is stated and proved for the special quadratic program (6.4.4), the lemma is valid for any convex quadratic program; this is clear because the above proof does not rely on the special form of (6.4.4). In the context of an optimization problem in the general form:

$$\begin{array}{ll} \text{minimize} & f(y) \\ \text{subject to} & y \in S, \end{array} \tag{6.4.6}$$

with a nonempty optimal solution set S_{opt} and a finite optimum objective value f_{inf}, if an error bound of the following type holds for some constant $L > 0$:

$$\text{dist}(y, S_{\text{opt}}) \leq L (f(y) - f_{\text{inf}}), \quad \forall y \in S,$$

we refer to S_{opt} as the set of *weak sharp minima* of the given optimization problem and say that the pair (f, S) has weak sharp minima. If S is convex, then any vector $\bar{y} \in S_{\text{opt}}$ must be an optimal solution of the linear program:

$$\begin{array}{ll} \text{minimize} & \nabla f(\bar{y})^T y \\ \text{subject to} & y \in S, \end{array}$$

by the well-known minimum principle for (6.4.6). The optimal solution set of this linear program is precisely the right-hand set $\bar{S}$ in (6.4.5) (with

$S \equiv \text{FEA}(K, p, N)$). The assertion that $\bar{S}$ is contained in S_{opt} can be thought of as a converse of the minimum principle, and this is referred to as the *minimum principle sufficiency* (MPS). Thus Lemma **6.4.3** states that a convex quadratic program possesses weak sharp minima if and only if the minimum principle sufficiency holds. Without repeating the proof, we formally state the latter conclusion in the following result.

6.4.4 Proposition. Let S be a polyhedron in $\mathbb{R}^n$ and f be a convex quadratic function bounded below on S. Let S_{opt} denote the set of minimizers of f on S. There exists a constant $L > 0$ such that

$$\text{dist}(x, S_{\text{opt}}) \leq L\,(\,f(x) - f(\bar{x})\,), \quad \forall\, x \in S$$

if and only if

$$S_{\text{opt}} = \{\, x \in S \,:\, \nabla f(\bar{x})^T(\,x - \bar{x}\,) \leq 0\,\}$$

for any $\bar{x} \in S_{\text{opt}}$. □

Since the gradient of the objective function of a linear program is a constant, it follows that the minimum principle sufficiency holds trivially for a linear problem; hence a solvable linear program must have weak sharp minima. Thus we recover Exercise **3.7.11**, which the reader was asked to prove by a direct argument.

In Section **12.7.2**, we will define an analogous minimum principle sufficiency and a weak minimum principle sufficiency for a general VI; see Definition **12.7.13**. The latter property turns out to be very important in the convergence analysis of a broad class of iterative methods for solving monotone VIs. For now, we show that the quadratic program (6.4.4) has weak sharp minima if and only if the AVI (K, q, M) has a Lipschitzian error bound with the gap residual for all test vectors in K. To prepare for this result, we consider the dual of the linear program corresponding to $\omega(x)$, which we denote $\Delta(x)$:

$$\begin{aligned} &\text{maximize} \quad -b^T\lambda \\ &\text{subject to} \quad q + Mx + A^T\lambda = 0, \quad \text{and} \quad \lambda \geq 0. \end{aligned}$$

Let $\Lambda(x)$ denote the optimal solution set of $\Delta(x)$. For every $x \in \Omega$, $\Lambda(x)$ is nonempty. Since solvable linear programs have weak sharp minima, it follows that there exists a constant $c > 0$ such that for all x in Ω and λ feasible to $\Delta(x)$,

$$b^T\lambda + \omega(x) \geq c\,\text{dist}(\lambda, \Lambda(x));$$

moreover, since the proof of this bound is an application of Hoffman's error bound to $\Lambda(x)$, the multiplicative constant c is independent of x (it depends on A and b). Furthermore, by the Lipschitz property of the solutions to a linear inequality system with a parametric right-hand side, (cf. Corollary **3.2.5**), it follows that there exists a constant $c' > 0$ such that for all x and x' belonging to Ω,

$$\Lambda(x) \subseteq \Lambda(x') + c' \| x - x' \| \operatorname{cl} \mathbb{B}(0,1).$$

These facts are used freely in the proof of the following lemma.

6.4.5 Lemma. Let K be a polyhedron in $\mathbb{R}^n$ and let M be an $n \times n$ positive semidefinite matrix. Assume that SOL(K, q, M) is nonempty. The QP (6.4.4) has weak sharp minima if and only if there exists a constant $L > 0$ such that

$$\operatorname{dist}(x, \operatorname{SOL}(K, q, M)) \leq L\, \theta_{\text{gap}}(x), \quad \forall\, x \in K.$$

Proof. Suppose that the AVI (K, q, M) has the Lipschitzian error bound. The optimal objective value of the QP (6.4.4) is zero. Let $y \equiv (x, \lambda)$ be a feasible solution to this QP . It follows that $x \in \Omega$ and λ is feasible to $\Delta(x)$. Thus $\Lambda(x)$ is nonempty and

$$b^T \lambda + \omega(x) \geq c \operatorname{dist}(\lambda, \Lambda(x)).$$

We have

$$\begin{aligned} f(y) &= \theta_{\text{gap}}(x) + \omega(x) + \lambda^T b \\ &\geq L^{-1} \operatorname{dist}(x, \operatorname{SOL}(K, q, M)) + c \operatorname{dist}(\lambda, \Lambda(x)). \end{aligned}$$

Pick $(x', \lambda') \in \operatorname{SOL}(K, q, M) \times \Lambda(x)$ such that

$$\operatorname{dist}(x, \operatorname{SOL}(K, q, M)) = \| x - x' \| \quad \text{and} \quad \operatorname{dist}(\lambda, \Lambda(x)) = \| \lambda - \lambda' \|.$$

Since $x' \in \operatorname{SOL}(K, q, M)$, it follows that $\omega(x')$ is finite, and thus $\Lambda(x')$ is nonempty. Hence there exists $\tilde{\lambda} \in \Lambda(x')$ such that

$$\| \lambda' - \tilde{\lambda} \| \leq c' \| x - x' \|.$$

The pair $y' \equiv (x', \tilde{\lambda})$ belongs to $S_{\text{opt}}(K, p, N)$. Consequently, we have

$$\begin{aligned} &\operatorname{dist}(y, S_{\text{opt}}(K, p, N)) \\ &\quad \leq \| x - x' \| + \| \lambda - \tilde{\lambda} \| \\ &\quad \leq \operatorname{dist}(x, \operatorname{SOL}(K, q, M)) + \| \lambda' - \tilde{\lambda} \| + \| \lambda - \lambda' \| \\ &\quad \leq (1 + c') \operatorname{dist}(x, \operatorname{SOL}(K, q, M)) + \operatorname{dist}(\lambda, \Lambda(x)). \end{aligned}$$

Consequently,

$$f(y) \geq \min\left(\frac{1}{L(1+c')}, c \right) \operatorname{dist}(y, S_{\text{opt}}(K, p, N)),$$

for all y feasible to (6.4.4). Thus this QP has weak sharp minima.

Conversely, suppose that there exists a constant $c > 0$ such that

$$f(y) \geq c \operatorname{dist}(y, S_{\text{opt}}(K, p, N))$$

for all y feasible to the QP (6.4.4). Let $x \in K$ be given. Since $\theta_{\text{gap}}(x)$ is equal to ∞ if $x \notin \Omega$, we may assume without loss of generality that $x \in \Omega$. Pick any λ in $\Lambda(x)$. The pair $y \equiv (x, \lambda)$ is feasible to (6.4.4). We have

$$\begin{aligned} \operatorname{dist}(x, \operatorname{SOL}(K, q, M)) &\leq \operatorname{dist}(y, S_{\text{opt}}(K, p, N)) \\ &\leq c^{-1} f(y) \\ &= c^{-1} [x^T(q + Mx) + \lambda^T b] = c^{-1} \theta_{\text{gap}}(x), \end{aligned}$$

as desired. □

We have all the necessary preparatory results to prove the following main theorem pertaining to a Lipschitzian error bound for the monotone AVI (K, q, M) in terms of the gap residual.

6.4.6 Theorem. Let K be a polyhedron in $\mathbb{R}^n$ and let M be an $n \times n$ positive semidefinite matrix. Assume that $\operatorname{SOL}(K, q, M)$ is nonempty. Let d and σ be the two invariants associated with the solutions of the AVI (K, q, M). The following three statements are equivalent.

(a) The AVI (K, q, M) has a nondegenerate solution.

(b) The following representation holds:

$$\operatorname{SOL}(K, q, M) = \{ x \in K : \omega(x) - (q + d)^T x + \sigma \geq 0 \}. \quad (6.4.7)$$

(c) There exists a constant $L > 0$ such that

$$\operatorname{dist}(x, \operatorname{SOL}(K, q, M)) \leq L\, \theta_{\text{gap}}(x), \quad \forall\, x \in K.$$

Proof. (a) $\Rightarrow$ (b). Let S denote the right-hand set in (6.4.7). It suffices to verify that

$$S \subseteq \operatorname{SOL}(K, q, M).$$

Let $x \in S$ and let $\hat{x}$ be a nondegenerate solution of the AVI (K, q, M). Since $\omega(x)$ is finite, there exists $\lambda \in \Lambda(x)$ satisfying $b^T \lambda + \omega(x) = 0$. Since

$\hat{x}$ is nondegenerate, there exists a $\hat{\lambda}$ such that

$$\begin{aligned} q + M\hat{x} + A^T\hat{\lambda} &= 0 \\ 0 \leq \hat{\lambda} \perp A\hat{x} - b &\leq 0 \\ \hat{\lambda} + b - A\hat{x} &> 0. \end{aligned}$$

We have

$$\begin{aligned} \omega(x) &\geq (q+d)^T x - \sigma \\ &= [q + (M + M^T)\hat{x}]^T x - \hat{x}^T M\hat{x} \\ &= (q + M\hat{x})^T x + \hat{x}^T M(x - \hat{x}) \\ &= -\hat{\lambda}^T Ax + (\hat{\lambda} - \lambda)A\hat{x} \\ &= \hat{\lambda}^T(b - Ax) + \lambda^T(b - A\hat{x}) - \lambda^T b \\ &\geq -\lambda^T b. \end{aligned}$$

Hence equalities hold throughout, and we obtain

$$\hat{\lambda}^T(b - Ax) = \lambda^T(b - A\hat{x}) = 0.$$

Since $\hat{\lambda} + b - A\hat{x} > 0$ and $\hat{\lambda} \perp (b - A\hat{x}) = 0$, it follows easily that λ is complementary to $b - Ax$. Hence x belongs to SOL(K, q, M).

(b) $\Rightarrow$ (c). This follows from the proof of Theorem **6.4.1**.

(c) $\Rightarrow$ (a). It suffices to show that the KKT system has a nondegenerate solution. Combining Lemmas **6.4.5** and **6.4.3** and expanding the expression $\nabla f(\bar{y})^T(y - \bar{y})$, such a solution exists if and only if the following linear program in the variables $(x, \lambda, \varepsilon)$ has a feasible solution with a negative objective value:

$$\begin{aligned} \text{minimize} \quad & -\varepsilon \\ \text{subject to} \quad & q + Mx + A^T\lambda = 0 \\ & Ax \leq b, \quad \lambda \geq 0 \\ & [q + (M + M^T)\bar{x}]^T(x - \bar{x}) + b^T(\lambda - \bar{\lambda}) \leq 0 \\ & \lambda + b - Ax \geq \varepsilon\, \mathbf{1}_m, \end{aligned}$$

where $(\bar{x}, \bar{\lambda})$ is an arbitrary KKT pair. Assume for contradiction that the KKT system of the AVI (K, q, M) has no nondegenerate solution. Since the above linear program is feasible, with $(x, \lambda, \varepsilon) \equiv (\bar{x}, \bar{\lambda}, 0)$ as a feasible solution, the assumption implies that the program has an optimal solution with zero objective value. By letting (u, v, ζ, w) be an optimal dual solution, we have

$$\begin{aligned} & M^T u - A^T(v+w) - \zeta\,[\,q + (M+M^T)\bar{x}\,] = 0 \\ & Au - \zeta\, b + w \leq 0 \\ & \mathbf{1}_m^T w = 1 \\ & (v, \zeta, w) \geq 0 \\ & -q^T u - b^T(v+w) + \zeta\,[\,-b^T\bar{\lambda} - \bar{x}^T(q + (M+M^T)\bar{x})\,] = 0. \end{aligned}$$

Premultiplying the first equation by u^T, the second constraint by $(v+w)^T$, and the last equation by $-\zeta$, adding the resulting constraints, using the fact that

$$\bar{x}^T(q + M\bar{x}) + b^T\bar{\lambda} = 0,$$

and simplifying, we deduce

$$(u - \zeta\bar{x})^T M (u - \zeta\bar{x}) + (v+w)^T w \leq 0.$$

Since M is positive semidefinite and both v and w are nonnegative, the last inequality implies that $w = 0$, which contradicts the equation $\mathbf{1}_m^T w = 1$. □

The following result is an immediate corollary of the above theorem; the result gives a sufficient condition for a monotone AVI to have a nondegenerate solution. A further specialization of the corollary yields the classical Goldman-Tucker theorem in linear programming, which asserts that every solvable linear program has a nondegenerate optimal solution.

6.4.7 Corollary. Let K be a polyhedron in $\mathbb{R}^n$ and let M be an $n \times n$ positive semidefinite matrix. Assume that SOL(K, q, M) is nonempty and $K \cap \Omega$ is contained in the null space of $M + M^T$, then the AVI (K, q, M) has a nondegenerate solution.

Proof. Since $K \cap \Omega$ is contained in the null space of $M + M^T$, the two invariants σ and d of the AVI are both equal to zero. Moreover, it is easy to verify that the right-hand set in (6.4.7) reduces to

$$\{\, x \in K : (z - x)^T(q + Mx) \geq 0 \;\forall\, z \in K \,\},$$

which is exactly SOL(K, q, M). Hence condition (b) in Theorem **6.4.6** holds, and the corollary follows. □

When K is a polyhedral cone, the expression (6.4.7) becomes:

$$\text{SOL}(K, q, M) = \{\, x \in K : q + Mx \in K^*,\ (q+d)^T x - \sigma \leq 0 \,\}.$$

Applying Hoffman's error bound to this representation, we readily obtain the following corollary of Theorem **6.4.6**, which gives a global error bound for the solution set of an affine CP with a nondegenerate solution, where the residual does not contain any square-root term.

6.4.8 Corollary. Let K be a polyhedral cone in $\mathbb{R}^n$ and M be a positive semidefinite matrix. Assume that $\mathrm{SOL}(K, q, M)$ has a nondegenerate solution. There exists a constant $c > 0$ such that for all $x \in \mathbb{R}^n$,

$$\begin{aligned}&\operatorname{dist}(x, \mathrm{SOL}(K, q, M)) \\ &\quad \leq c\,[\,\operatorname{dist}(x, K) + \operatorname{dist}(q + Mx, K^*) + (\,x^T(\,q + Mx\,)\,)_+\,].\end{aligned}$$

Proof. It suffices to note that as proved in Theorem **6.4.6**,

$$(\,q + d\,)^T x - \sigma \leq x^T(\,q + Mx\,),$$

which implies

$$[\,(\,q + d\,)^T x - \sigma\,]_+ \leq (\,x^T(\,q + Mx\,)\,)_+.$$

and the desired error bound follows easily as indicated above. □

6.4.1 Convex quadratic programs

Since a convex quadratic program is equivalent to a symmetric, monotone AVI, the error bounds obtained in the last section can be employed to derive error bounds for the optimal solution set of a convex QP. Nevertheless, such an approach has not taken into account the symmetry of the defining matrix of the AVI arising from a QP. In what follows, bypassing the equivalent AVI, we employ a more direct approach to obtain an alternative error bound for the solution set of a convex QP, with the residual being the deviation of the objective value at the test vector from the optimum objective value. From a practical point of view, this residual is computationally useful only if the latter value is known. In contrast to Proposition **6.4.4** the error bound obtained below holds for all vectors in $\mathbb{R}^n$, and not only for the feasible vectors; the latter is what the previous proposition deals with.

Consider the convex quadratic program:

$$\begin{aligned}&\text{minimize} \quad \theta(x) \equiv q^T x + \tfrac{1}{2} x^T M x \\ &\text{subject to} \quad x \in P(A, b),\end{aligned} \tag{6.4.8}$$

where q is an n-vector, M is an $n \times n$ symmetric positive semidefinite matrix, A is an $m \times n$ matrix and b is an m-vector. We recall that

$$P(A, b) \equiv \{\, x \in \mathbb{R}^n \,:\, Ax \leq b\,\}.$$

Let S_{opt} denote the optimal solution set of this QP, which we assume is nonempty. By Corollary **2.3.8**, we have

$$\begin{aligned} S_{\text{opt}} &= \\ \{ x \in P(A,b) \,:\, Mx &= M\bar{x},\ (q+M\bar{x})^T(x-\bar{x}) \leq 0 \}, \end{aligned} \tag{6.4.9}$$

where $\bar{x} \in S_{\text{opt}}$ is arbitrary. Let θ_{opt} denote the optimum objective value of (6.4.8). By the minimum principle, we have

$$x \in P(A,b) \;\Rightarrow\; \nabla\theta(\bar{x})^T(x-\bar{x}) \geq 0.$$

We establish a lemma that pertains to an implication of the above type; the lemma roughly says that if a linear inequality is implied by a system of linear inequalities, then the residual associated with the latter inequalities majorizes (up to a multiplicative constant) the residual associated with the former inequality.

6.4.9 Lemma. Let $\bar{x} \in P(A,b)$. If the inequality $a^T(x-\bar{x}) \geq 0$ holds for every x in $P(A,b)$, then there exists a constant $\eta > 0$ such that

$$[\, a^T(x-\bar{x}) \,]_- \leq \eta \,\|\, (Ax-b)_+ \,\|, \quad \forall\, x \in \mathbb{R}^n.$$

Proof. By Farkas' lemma, there exists a nonnegative vector λ such that

$$a + A^T\lambda = 0 \quad \text{and} \quad a^T\bar{x} + \lambda^T b = 0.$$

We have for all x in $\mathbb{R}^n$

$$\begin{aligned} [\, a^T(x-\bar{x}) \,]_- &= [\, \lambda^T(b-Ax) \,]_- \\ &\leq \lambda^T(Ax-b)_+ \leq \|\,\lambda\,\|\,\|\,(Ax-b)_+\,\|, \end{aligned}$$

where the first inequality follows from the fact that $\lambda \geq 0$ and the last step is simply the Cauchy-Schwarz inequality. □

Applying Hoffman's error bound to the representation (6.4.9) of S_{opt} and by some algebraic manipulation, we can establish the following global error bound.

6.4.10 Proposition. There exists a constant $L > 0$ such that for all x in $\mathbb{R}^n$,

$$\begin{aligned} &\operatorname{dist}(x, S_{\text{opt}}) \leq \\ &L\left[\operatorname{dist}(x,P) + \sqrt{\operatorname{dist}(x,P)} + (\,\theta(x)-\theta_{\text{opt}}\,)_+ + \sqrt{(\,\theta(x)-\theta_{\text{opt}}\,)_+} \right], \end{aligned}$$

where $P \equiv P(A,b)$.

Proof. There exists a constant $L' > 0$ such that for all x in $\mathbb{R}^n$,

$$\begin{array}{l}\operatorname{dist}(x, S_{\text{opt}}) \leq \\ \quad L'\{\operatorname{dist}(x,P) + \|M(x-\bar{x})\| + [(q+M\bar{x})^T(x-\bar{x})]_+\}.\end{array} \tag{6.4.10}$$

Since M is symmetric positive semidefinite definite, there exists a constant $c_1 > 0$ such that for all $x \in \mathbb{R}^n$,

$$\|M(x-\bar{x})\| \leq c_1\sqrt{(x-\bar{x})^T M(x-\bar{x})};$$

in turn, we have, for some constant $c_2 > 0$

$$\begin{aligned}(x-\bar{x})^T M(x-\bar{x}) &= 2[\theta(x) - \theta_{\text{opt}} - \nabla\theta(\bar{x})^T(x-\bar{x})] \\ &\leq 2(\theta(x) - \theta_{\text{opt}})_+ + 2[\nabla\theta(\bar{x})^T(x-\bar{x})]_- \\ &\leq 2(\theta(x) - \theta_{\text{opt}})_+ + 2c_2\|(Ax-b)_+\|.\end{aligned}$$

Hence there exists a constant $c' > 0$ such that for all x in $\mathbb{R}^n$,

$$\|M(x-\bar{x})\| \leq c'\left[\sqrt{(\theta(x)-\theta_{\text{opt}})_+} + \sqrt{\|(Ax-b)_+\|}\right]. \tag{6.4.11}$$

By the gradient inequality of a convex function, we have

$$\theta(x) - \theta_{\text{opt}} \geq \nabla\theta(\bar{x})^T(x-\bar{x}),$$

which implies

$$(\theta(x) - \theta_{\text{opt}})_+ \geq (\nabla\theta(\bar{x})^T(x-\bar{x}))_+.$$

Combining this last inequality with (6.4.10) and (6.4.11), we easily obtain the desired error bound asserted by the proposition. □

We can apply Proposition **6.4.10** to the natural QP associated with the affine CP (K, q, M). Specifically, let K be a polyhedral cone. Consider the QP:

$$\begin{array}{ll}\text{minimize} & x^T(q+Mx) \\ \text{subject to} & x \in K \quad \text{and} \quad q+Mx \in K^*.\end{array}$$

With a positive semidefinite matrix M (not assumed symmetric), this is a convex QP because the dual of a polyhedral cone is polyhedral. Moreover, if the affine CP (K, q, M) has a solution, then the above QP has an optimal solution with a zero objective value. The following corollary is immediate; it generalizes the error bound obtained from Theorem **6.4.1** by removing the feasibility requirement of the test vector x.

6.4.11 Corollary. Let K be a polyhedral cone in $\mathbb{R}^n$ and M be a positive semidefinite matrix. Assume that SOL(K, q, M) is nonempty. There exists a constant $c > 0$ such that for all $x \in \mathbb{R}^n$,

$$\operatorname{dist}(x, \operatorname{SOL}(K,q,M)) \leq c[\operatorname{dist}(x,K) + \sqrt{\operatorname{dist}(x,K)} + \operatorname{dist}(q+Mx, K^*)$$
$$+\sqrt{\operatorname{dist}(q+Mx,K^*)} + (x^T(q+Mx))_+ + \sqrt{(x^T(q+Mx))_+}].$$

We point out that by Corollary **6.4.8**, all three square-root terms are not needed when the affine CP (K, q, M) has a nondegenerate solution.

6.5 Global Bounds via a Variational Principle

In this section, we introduce an alternative approach to derive global error bounds for nonlinear VIs and CPs; namely, via a powerful variational principle due to Ekeland. The theory of global error bounds developed herein has a direct bearing on the important concepts of a "minimizing sequence" and a "stationary sequence" associated with a constrained minimization problem. A detailed discussion of these topics is beyond the scope of this section; see Exercises **6.9.7**, **6.9.8**, and **6.9.10** for a summary of some results and examples.

We begin by presenting Ekeland's variational principle. In essence, this principle, which can be stated in various equivalent forms, is an existence theorem about ε-solutions to an optimization problem for any positive scalar ε. The principle is valid in general metric spaces; for our purpose, we present it in the setting of a constrained optimization problem in an Euclidean space.

6.5.1 Theorem. Let S be a closed set in $\mathbb{R}^n$ and f be a continuous real-valued function defined on X and bounded below there. For every vector $u \in S$ satisfying

$$f(u) \leq \inf_{x \in S} f(x) + \varepsilon,$$

and every scalar $\delta > 0$, there exists a vector $\bar{x} \in S \cap \operatorname{cl} \mathbb{B}(u, \varepsilon/\delta)$ such that $f(\bar{x}) \leq f(u)$ and

$$f(y) + \delta \| \bar{x} - y \| > f(\bar{x}) \quad \forall y \in S \setminus \{\bar{x}\}. \tag{6.5.1}$$

Proof. Consider the function

$$f_u(x) \equiv f(x) + \delta \| x - u \|, \quad x \in S.$$

Since f is bounded below on S, f_u has bounded level sets on S; i.e., for every scalar τ, the set

$$\{ x \in S : f_u(x) \leq \tau \}$$

is bounded. Consequently, being continuous, f_u attains its minimum on S. Moreover, the argmin set of f_u on S, which we denote U, is nonempty and compact. Thus, f attains its minimum on U. Let $\bar{x}$ be a minimizer of f on U. We verify that $\bar{x}$ has the desired properties. We have

$$f(u) = f_u(u) \geq f_u(\bar{x}) = f(\bar{x}) + \delta \| \bar{x} - u \| \geq f(\bar{x}).$$

Furthermore, this implies

$$\varepsilon \geq \delta \| \bar{x} - u \|,$$

which yields $\bar{x} \in \operatorname{cl} \mathbb{B}(u, \varepsilon/\delta)$. For any vector y in S distinct from $\bar{x}$, if y belongs to U, then

$$f(\bar{x}) \leq f(y) < f(y) + \delta \| \bar{x} - y \|.$$

If y does not belong to U, then

$$f(y) + \delta \| y - u \| = f_u(y) > f_u(\bar{x}) = f(\bar{x}) + \delta \| \bar{x} - u \|,$$

which implies

$$f(y) + \delta \| y - \bar{x} \| \geq f(y) + \delta \| y - u \| - \delta \| \bar{x} - u \| > f(\bar{x}).$$

This completes the proof of the claimed properties of $\bar{x}$. □

The expression (6.5.1) says that the function $f + \delta\|\bar{x} - \cdot\|$ attains its (unique) minimum on S at the point $\bar{x}$ whose f-value is no worse than the f-value at the reference vector u. Roughly speaking, Theorem **6.5.1** says that every ε-solution u of the minimization problem:

$$\begin{array}{ll} \text{minimize} & f(x) \\ \text{subject to} & x \in S, \end{array} \tag{6.5.2}$$

is arbitrarily close to an exact optimal solution $\bar{x}$ of a certain perturbed minimization problem. In what follows, we employ Theorem **6.5.1** to establish the existence of weak sharp minima for a constrained optimization problem in terms of a condition known in the nonlinear analysis literature as the Takahashi condition; see statement (b) below. Contained in the theorem below is the assertion that the Takahashi condition implies the existence of a global minimizer to the optimization problem in question.

6.5.2 Theorem. Let S be a closed set in $\mathbb{R}^n$ and f be a continuous real-valued function defined on S and bounded below there. Let f_{inf} denote the infimum value of f on S. For a scalar $c > 0$, the following two statements are equivalent.

(a) The pair (f, S) has weak sharp minima with constant c; that is, the optimal set

$$S_{\text{opt}} \equiv \{ x \in S : f(x) = f_{\text{inf}} \}.$$

is nonempty and

$$f(x) - f_{\text{inf}} \geq c \,\text{dist}(x, S_{\text{opt}}), \quad \forall x \in S. \tag{6.5.3}$$

(b) For each $x \in S$ with $f(x) > f_{\text{inf}}$, there exists a vector y in S distinct from x such that

$$f(y) + c \| x - y \| \leq f(x). \tag{6.5.4}$$

Proof. (a) $\Rightarrow$ (b). This is the easy part of the equivalence. Indeed, let $x \in S$ be such that $f(x) > f_{\text{inf}}$. Let $y \in S_{\text{opt}}$ satisfy $\|x - y\| = \text{dist}(x, S_{\text{opt}})$. Such a vector y, which may not be unique, must exist because S_{opt} is a closed set. The inequality (6.5.4) then follows easily from (6.5.3).

(b) $\Rightarrow$ (a). Let $x \in S$ be arbitrary. Without loss of generality, we may assume that $f(x) > f_{\text{inf}}$. Define the closed set

$$S_c(x) \equiv \{ y \in S : f(y) + c \| x - y \| \leq f(x) \},$$

which must contain x. Assumption (b) implies that $S_c(x)$ contains at least a vector that is distinct from x. By Ekeland's Theorem **6.5.1** applied to the function f and the set $S_c(x)$, we deduce the existence of a vector $z \in S_c(x)$ such that

$$f(y) + c \| z - y \| > f(z) \quad \forall y \in S_c(x) \setminus \{ z \}.$$

We claim that the above inequality holds for all y in S distinct from z. If not, then for some y belonging to $S \setminus S_c(x)$ and distinct from z, we have

$$f(y) + c \| z - y \| \leq f(z).$$

Since $z \in S_c(x)$, we have

$$f(z) + c \| x - z \| \leq f(x).$$

Adding the last two inequalities and using the triangle inequality, we obtain

$$f(y) + c \| x - y \| \leq f(x),$$

which implies that $y \in S_c(x)$; a contradiction. Consequently, we have shown that there exists a vector $z \in S_c(x)$ such that

$$f(y) + c \| z - y \| > f(z) \quad \forall y \in S \setminus \{ z \}.$$

By condition (b), it follows that $f(z) = f_{\text{inf}}$; thus z belongs to S_{opt} and S_{opt} is nonempty. Since z belongs to $S_c(x)$, we have

$$f(x) - f(z) \geq c \| z - x \| \geq c \operatorname{dist}(x, S_{\text{opt}}),$$

which is the desired inequality in (a). □

Takahashi's condition has a simple geometric meaning; namely, for every non-minimizer x of the optimization problem (6.5.2), a feasible vector y distinct from x exists such that the objective value $f(y)$ is smaller than $f(x)$ by at least a constant multiple of the distance between x and y. It is possible to give various sufficient conditions for this postulate to hold, thereby deducing the existence of weak sharp minima. In what follows, we present one such condition for a B-differentiable function. The B-differentiability assumption is made here for convenience; indeed, one can obtain similar results for a locally Lipschitz continuous function f by employing a suitable generalized directional derivative of f. We avoid such a general setting in order to simplify the discussion. To motivate the sufficient condition, we recall that if $f : \mathcal{D} \supset S \to \mathbb{R}$ is B-differentiable on the open set $\mathcal{D} \subseteq \mathbb{R}^n$, and if x is a local minimizer of f on S, then

$$f'(x; d) \geq 0, \quad \forall d \in \mathcal{T}(x; S).$$

The key assumption of the following proposition is that there is a uniform negative upper bound on the latter directional derivative for all non-optimal vectors x in S and an associated tangent vector d.

6.5.3 Proposition. Let S be a closed set in $\mathbb{R}^n$ and $f : \mathcal{D} \supset S \to \mathbb{R}$ be a B-differentiable function defined on the open set $\mathcal{D}$ and bounded below on S. Let f_{inf} denote the infimum value of f on S. If there exists a scalar $\delta > 0$ such that for every $x \in S$ with $f(x) > f_{\text{inf}}$, a vector $d \in \mathcal{T}(x; S)$ with unit (Euclidean) length exists satisfying $f'(x; d) \leq -\delta$, then the pair (f, S) has weak sharp minima with constant δ.

Proof. Let $x \in S$ with $f(x) > f_{\text{inf}}$ be given. Let $d \in \mathcal{T}(x; S) \cap \operatorname{cl} \mathbb{B}(0,1)$ satisfy $f'(x; d) \leq -\delta$. There exists a sequence $\{x^k\} \subset S$ converging to x such that $x^k \neq x$ for all k and

$$\lim_{k \to \infty} \frac{x^k - x}{\| x^k - x \|} = d.$$

We have

$$\lim_{k \to \infty} \frac{f(x^k) - f(x)}{\| x^k - x \|} = f'(x; d) \leq -\delta.$$

Thus for every $\varepsilon \in (0, \delta)$ there exists $x^k \in S$ satisfying

$$f(x^k) - f(x) \leq -(\delta - \varepsilon)\,\| x^k - x \|,$$

or equivalently,

$$f(x^k) + (\delta - \varepsilon)\,\| x^k - x \| \leq f(x).$$

This shows that statement (b) in Theorem **6.5.2** holds with $\delta - \varepsilon$. Hence (f, S) has weak sharp minima with constant $\delta - \varepsilon$. Since ε can be made arbitrarily small, it follows that (f, S) has weak sharp minima with constant δ. □

We next use Proposition **6.5.3** to derive a Hölderian error bound for the feasible vectors of a constrained minimization problem.

6.5.4 Corollary. Let $S \subseteq \mathbb{R}^n$ be closed convex and $f : \mathcal{D} \supset S \to \mathbb{R}$ be a continuously differentiable function defined on the open set $\mathcal{D}$ and bounded below on S. Let $\gamma \in (0, 1]$ and $\delta > 0$ be given scalars. Suppose that for every $x \in S$ with $f(x) > f_{\text{inf}}$,

$$\text{dist}(-\nabla f(x), \mathcal{N}(x; S)) \geq \delta\,(f(x) - f_{\text{inf}})^{1-\gamma}. \tag{6.5.5}$$

The following three statements hold.

(a) For every sequence $\{x^k\} \subset S$,

$$\lim_{k\to\infty} \text{dist}(-\nabla f(x^k), \mathcal{N}(x^k; S)) = 0 \Rightarrow \lim_{k\to\infty} f(x^k) = f_{\text{inf}}. \tag{6.5.6}$$

(b) The optimal set

$$S_{\text{opt}} \equiv \{ x \in S : f(x) = f_{\text{inf}} \}$$

is nonempty.

c) The following error bound holds:

$$\text{dist}(x, S_{\text{opt}}) \leq \frac{1}{\gamma\,\delta}\,(f(x) - f_{\text{inf}})^{\gamma}, \quad \forall\, x \in S.$$

Proof. Let $\{x^k\} \subset S$ satisfy the left-hand limit in (6.5.6). If

$$\liminf_{k\to\infty} f(x^k) > f_{\text{inf}}.$$

it then follows that

$$\liminf_{k\to\infty} \text{dist}(-\nabla f(x^k), \mathcal{N}(x^k; S)) > 0,$$

which is a contradiction. Thus (a) holds. To prove (b) and (c), we apply Proposition **6.5.3** to the function $\tilde{f} \equiv (f - f_{\inf})^\gamma$ on the set S. We have $\tilde{f}_{\inf} = 0$. For a given vector $x \in S$ satisfying $\tilde{f}(x) > 0$, we have

$$\nabla \tilde{f}(x) = \gamma\,(f(x) - f_{\inf})^{\gamma-1} \nabla f(x).$$

Write $\mathcal{N} \equiv \mathcal{N}(x; S)$ and $\mathcal{T} \equiv \mathcal{T}(x; S)$. The vector $-\nabla f(x) - \Pi_{\mathcal{N}}(-\nabla f(x))$ is nonzero, belongs to $\mathcal{T}$, and is orthogonal to $\Pi_{\mathcal{N}}(-\nabla f(x))$. Let

$$d \equiv -\frac{\nabla f(x) + \Pi_{\mathcal{N}}(-\nabla f(x))}{\|\,\nabla f(x) + \Pi_{\mathcal{N}}(-\nabla f(x))\,\|}.$$

We have

$$\begin{aligned} \nabla \tilde{f}(x)^T d &= -\gamma\,(f(x) - f_{\inf})^{\gamma-1} \|\,\nabla f(x) + \Pi_{\mathcal{N}}(-\nabla f(x))\,\| \\ &= -\gamma\,(f(x) - f_{\inf})^{\gamma-1} \operatorname{dist}(-\nabla f(x), \mathcal{N}) \\ &\leq -\gamma\,\delta. \end{aligned}$$

Thus by Proposition **6.5.3**, the pair $(\tilde{f}, S)$ has weak sharp minima with constant $\gamma\delta$. This means that S_{opt} is nonempty and

$$\operatorname{dist}(x, S_{\mathrm{opt}}) \leq (\gamma\,\delta)^{-1}\,(f(x) - f_{\inf})^{\gamma}, \quad \forall\, x \in S,$$

completing the proof of (b) and (c). □

The implication (6.5.6) deserves further discussion. Recall that a stationary point of the minimization problem (6.5.2) is by definition a solution of the VI $(S, \nabla f)$; thus $x \in S$ is stationary if and only if

$$\operatorname{dist}(-\nabla f(x), \mathcal{N}(x; S)) = 0.$$

Hence a feasible sequence $\{x^k\} \subset S$ that satisfies the left-hand limit in (6.5.6) can be thought of as being asymptotically stationary. Similarly, a feasible sequence $\{x^k\} \subset S$ that satisfies the right-hand limit in (6.5.6) is asymptotically minimizing. Corollary **6.5.4** has given a sufficient (albeit not necessary) condition for every stationary sequence to be minimizing; and this condition is also sufficient for the existence of "Hölderian minima".

Borrowing a terminology from the theory of unbounded asymptotics in convex optimization, we say that the pair (f, S) is *well-behaved* if (6.5.6) holds; we say that the pair (f, S) is *strongly well-behaved* if there exist constants $\delta > 0$ and $\gamma \in (0, 1]$ such that (6.5.5) holds for all $x \in S$ with $f(x) > f_{\inf}$. Consequently, a strongly well-behaved pair has several desirable asymptotic properties.

To prepare for the application of the general results developed so far, we consider the problem of solving a system of semismooth equations with a C^1 squared norm function. The proof of the following result combines ideas from the proof of Corollary **6.5.4** and Proposition **6.5.3**.

6.5.5 Proposition. Let $G : \mathbb{R}^n \to \mathbb{R}^n$ be a semismooth function with $G^{-1}(0) \neq \emptyset$. Suppose that $G^T G$ is continuously differentiable. If there exist a subset W of $\mathbb{R}^n$ and a constant $c > 0$ such that (a) for every $x \in W \setminus G^{-1}(0)$, a matrix $H \in \partial G(x)$ exists satisfying

$$\| H^T G(x) \| \geq c \| G(x) \|, \tag{6.5.7}$$

and (b) for every $x \notin W$,

$$\operatorname{dist}(x, G^{-1}(0)) \leq c^{-1} \| G(x) \|,$$

then

$$\operatorname{dist}(x, G^{-1}(0)) \leq c^{-1} \| G(x) \|, \quad \forall x \in \mathbb{R}^n.$$

Proof. By Theorem **6.5.2** and a continuity argument, it suffices to show that for every $\varepsilon \in (0, c)$ and for every $x \in \mathbb{R}^n$ with $G(x) \neq 0$, a vector $y \neq x$ exists such that

$$\| G(y) \| + (c - \varepsilon) \| x - y \| \leq \| G(x) \|. \tag{6.5.8}$$

By assumption, the latter inequality clearly holds if $x \notin W$ and $G(x) \neq 0$; simply let y be any vector in $G^{-1}(0)$. Consider a vector $x \in W \setminus G^{-1}(0)$. For any vector $d \in \mathbb{R}^n$ and $\tau > 0$ sufficiently small, we can write, for any $\tilde{H} \in \partial G(x)$,

$$\| G(x + \tau d) \| - \| G(x) \| = \tau \frac{d^T \tilde{H}^T G(x)}{\| G(x) \|} + o(\tau).$$

With $d \equiv -H^T G(x)/\|H^T G(x)\|$, where H satisfies (6.5.7), we have, for every $\varepsilon \in (0, c)$ and for all $\tau > 0$ sufficiently small,

$$\| G(x + \tau d) \| - \| G(x) \| \leq -(c - \varepsilon) \tau,$$

or equivalently,

$$\| G(x + \tau d) \| + (c - \varepsilon) \| \tau d \| \leq \| G(x) \|.$$

Thus (6.5.8) holds with $y \equiv x + \tau d$. This is enough to establish the desired global error bound for the zero set $G^{-1}(0)$. □

Roughly speaking, Proposition **6.5.5** asserts that for a semismooth vector function G with a C^1 merit function $\theta \equiv G^T G$, the zero set $G^{-1}(0)$ has a global Lipschitzian error bound if it has such an error bound on a set W outside of which θ is "strongly well-behaved". In the case where G is itself continuously differentiable, the assumption of the proposition implies that $\| JG(x)^T \| \geq c$ for all $x \in W \setminus G^{-1}(0)$. This assumption does not involve any requirement on the Jacobian of G at its zeros.

Since the application of Proposition **6.5.5** to the NCP requires the introduction of an asymptotic property that is defined in Subsection **9.1.3**, we postpone the discussion of such an application until the later subsection; see Proposition **9.1.20**.

6.6 Analytic Problems

An error bound theory can be developed based on a deep result of Lojasiewicz pertaining to subanalytic functions. We begin the development with the introduction of functions of this class. Let X be an open set in $\mathbb{R}^n$. A real-valued function f is *analytic* on X if it can be represented locally on X by a convergent infinite power series; in other words, for any $\bar{x} \in X$, there exists an open neighborhood $\mathcal{N}$ of $\bar{x}$ such that

$$f(x) = \sum_{i_1,\ldots,i_n=0}^{\infty} a_{i_1,\cdots,i_n} (x_1 - \bar{x}_1)^{i_1} \cdots (x_n - \bar{x}_n)^{i_n}, \quad \forall x \in \mathcal{N},$$

where each $a_{i_1,\cdots,i_n}$ is a real number. It is clear that an analytic function must be locally Lipschitz continuous. We call a subset $S \subseteq X$ *analytic* if it is finitely representable by a family of inequalities and equations defined by analytic functions on X.

Let us consider sets of the following form:

$$S = \bigcup_{i=1}^{r} \{ x \in \mathbb{R}^n : f_{ij}(x) = 0, j = 1, \ldots q_1; \text{ and } f_{ij}(x) > 0, j = 1, \ldots q_2 \}.$$

If every f_{ij} is a polynomial, S is, by definition, a *semialgebraic set*. We say that a set $S \subseteq \mathbb{R}^n$ is *semianalytic* if for each $\bar{x}$ in $\mathbb{R}^n$ there exist a neighborhood V of $\bar{x}$ and suitable analytic functions f_{ij} for which we can write

$$S \cap V =$$

$$\bigcup_{i=1}^{r} \{ x \in \mathbb{R}^n : f_{ij}(x) = 0, j = 1, \ldots q_1^i; \text{ and } f_{ij}(x) > 0, j = 1, \ldots q_2^i \}.$$

If the graph of a vector function $h : \mathbb{R}^n \to \mathbb{R}^m$ is a semianalytic (semialgebraic) set in $\mathbb{R}^{n+m}$, we say that h is a *semianalytic (semialgebraic) function.*

Clearly, the class of semianalytic sets (functions) strictly contains the class of semialgebraic sets (functions) and semianalytic sets also generalize the class of analytic sets. The local properties of semianalytic sets are similar to the global properties of semialgebraic sets. However, there is a major difference: the image of a semianalytic set under a semianalytic map is not necessarily a semianalytic set (while an analogous property holds for the image of semialgebraic sets under semialgebraic maps). This fact restricts the usefulness of semianalytic sets for our purposes. However, these difficulties disappear if we consider a slightly larger family of sets.

6.6.1 Definition. We say that a subset S of $\mathbb{R}^n$ is *subanalytic* if for each point $\bar{x}$ in S there exist a neighborhood V of $\bar{x}$ and a bounded semianalytic set $A \subset \mathbb{R}^{n+t}$ (for some $t > 0$), such that $S \cap V$ is the projection on $\mathbb{R}^n$ of A. A function $h : \mathcal{D} \subseteq \mathbb{R}^n \to \mathbb{R}^m$ is *subanalytic* if its graph is a subanalytic set in $\mathbb{R}^{n+m}$. □

We stress that the class of subanalytic functions is quite wide; subanalyticity does not even imply continuity. It is conceivable that all interesting variational inequalities encountered in practice are indeed defined by subanalytic functions. In what follows, we say that a VI (K, F) is subanalytic if F is a subanalytic function and K is a subanalytic set. We summarize some basic properties of subanalytic functions and subanalytic sets as follows.

(p1) The class of subanalytic sets is closed under finite union and intersection.

(p2) The Cartesian product of subanalytic sets is subanalytic.

(p3) The distance function to a subanalytic set is subanalytic; i.e., if S is a subanalytic set in $\mathbb{R}^n$, then $\operatorname{dist}(x, S)$ is a real subanalytic function.

(p4) The inverse image of a subanalytic set under a subanalytic map is a subanalytic set; i.e., if $h : \mathbb{R}^n \to \mathbb{R}^m$ is a subanalytic function and $S \subseteq \mathbb{R}^m$ is a subanalytic set, then $h^{-1}(S)$ is a subanalytic set in $\mathbb{R}^n$. In particular, all level sets of subanalytic functions are subanalytic sets.

(p5) If $f : \mathbb{R}^n \to \mathbb{R}^p$ and $g : \mathbb{R}^p \to \mathbb{R}^m$ are subanalytic functions, with f continuous, then $g \circ f$ is subanalytic. In particular, the class of continuous subanalytic functions is closed under algebraic operations.

(p6) Let $S \in \mathbb{R}^n$ be a subanalytic set and suppose that $\bar{x}$ belongs to the closure of S. There exists a continuous curve $g : [0, 1] \to \mathbb{R}^n$ with

$g(0) = \bar{x}$ and $g(s) \in S$ for every $s \in (0, 1]$. That is, every boundary point of a subanalytic set is reachable via a continuous path in the set.

(p7) Piecewise analytic functions with semianalytic pieces are semianalytic (thus subanalytic).

(p8) The pointwise supremum of a finite family of continuous subanalytic functions is subanalytic. In particular, if g is a continuous subanalytic function, then so are g_+ and $|g|$.

(p9) The image of a bounded subanalytic set under a subanalytic map is subanalytic; but this property is not valid if "subanalytic" is replaced by "semianalytic".

The next result describes another important property of subanalyticity.

6.6.2 Lemma. Let S be a closed convex subanalytic set. The Euclidean projector onto S, Π_S, is a subanalytic function.

Proof. Since S is subanalytic, the (Euclidean) distance function $\operatorname{dist}(\cdot, S)$ is subanalytic by (p3). To show that Π_S is subanalytic we have to check that its graph $\mathcal{G}$ is subanalytic. We can explicitly write down the expression of the set $\mathcal{G}$ as

$$\mathcal{G} = (\mathbb{R}^n \times S) \cap T,$$

where

$$T \equiv \{ (x, y) \in \mathbb{R}^{2n} : \| x - y \|^2 - \operatorname{dist}(x, S)^2 = 0 \}.$$

Since $\|x - y\|^2$ is algebraic and therefore subanalytic and $\operatorname{dist}(x, S)^2$ is subanalytic by (p3) and (p5), the function $\|x-y\|^2 - \operatorname{dist}(x, S)^2$ is, again by (p5), subanalytic. Hence, by (p4), T is subanalytic. But then, since $\mathbb{R}^n \times S$ is subanalytic by assumption and (p2), it follows that $\mathcal{G}$ is subanalytic by (p1). □

The following result, known as Lojasiewicz' inequality, is the cornerstone for the derivation of error bounds of subanalytic systems. The proof of this inequality is highly technical; see the Notes and Comments for references.

6.6.3 Theorem. Let $\phi, \psi : X \to \mathbb{R}$ be two continuous subanalytic functions defined on the compact subanalytic set $X \subset \mathbb{R}^n$. If $\phi^{-1}(0) \subseteq \psi^{-1}(0)$, then there exist a scalar $\rho > 0$ and an integer $N^* > 0$ such that for all $x \in X$,

$$\rho \, | \, \psi(x) \, |^{N^*} \leq | \, \phi(x) \, |.$$

Thus, if the reverse inclusion $\psi^{-1}(0) \subseteq \phi^{-1}(0)$ also holds, then the two functions ϕ and ψ are "equivalent" on X in the sense that positive scalars ρ_1 and ρ_2 and positive integers N_1 and N_2 exist such that

$$\rho_1 \,|\, \psi(x) \,|^{N_1} \leq |\, \phi(x) \,| \leq \rho_2 \,|\, \psi(x) \,|^{1/N_2}$$

for all $x \in X$. □

We employ the above theorem to prove the following fundamental error bound for subanalytic systems.

6.6.4 Corollary. Let X be a closed subanalytic set in $\mathbb{R}^n$. Let S denote the set of $x \in X$ satisfying

$$g_i(x) \leq 0, \quad i = 1, \ldots, m, \quad \text{and} \quad h_j(x) = 0, \quad j = 1, \ldots, \ell,$$

where each g_i and h_j are continuous subanalytic functions defined on an open set containing X. If $S \neq \emptyset$, then for every compact subanalytic subset $T \subset X$, there exist positive constants c and γ such that

$$\operatorname{dist}(x, S) \leq c\, r(x)^{\gamma}, \quad \forall\, x \in T,$$

where

$$r(x) \equiv \sum_{i=1}^{m} g_i(x)_+ + \sum_{j=1}^{\ell} |\, h_j(x) \,|$$

is the residual function of the set S.

Proof. The function $r(x)$ is subanalytic by (p5); so is the distance function $\phi(x) \equiv \operatorname{dist}(x, S)$ by (p3) because S is a subanalytic set by (p1) and (p4). Both functions r and ϕ are continuous. Clearly, $(r|_X)^{-1}(0) = S = \phi^{-1}(0)$. The desired error bound follows easily from Theorem **6.6.3**. □

Corollary **6.6.4** is essentially a local error bound result because of the boundedness requirement of the test set T. In general, the exponent γ and the multiplicative constant c depend on the functions g_i and h_j, as well as on the size of the compact set T where the error bound holds. However, the result gives no clue for computing these important parameters in practice.

The next result combines Corollary **6.6.4** with Proposition **6.1.3** to yield a global Hölderian error bound for a bounded feasible set defined by finitely many convex, subanalytic inequalities. The main difference between this result and Proposition **6.1.4** is the absence of the Slater assumption.

6.6.5 Corollary. Let S denote the set of $x \in \mathbb{R}^n$ satisfying

$$g_i(x) \leq 0, \quad i = 1, \ldots, m,$$

where each g_i is convex, subanalytic function defined on $\mathbb{R}^n$. If S is nonempty and bounded, then there exist positive constants c and γ such that

$$\operatorname{dist}(x, S) \leq c\,(\,r(x) + r(x)^{\gamma}\,) \quad \forall\, x \in \mathbb{R}^n,$$

where

$$r(x) \equiv \sum_{i=1}^{m} g_i(x)_+$$

is the residual function for S.

Proof. Since S is compact and each g_i is convex, it follows that for every $\varepsilon > 0$, the level set $r^{-1}(-\infty, \varepsilon]$ is bounded. Hence, by Corollary **6.6.4**, for any $\varepsilon > 0$, there exist positive scalars c and γ such that

$$\operatorname{dist}(x, S) \leq c\,(\,r(x) + r(x)^{\gamma}\,), \quad \forall\, x \text{ satisfying } r(x) \leq \varepsilon.$$

By Proposition **6.1.3**, this can be extended to a global bound with a possibly different multiplicative constant c. □

It is straightforward to apply Corollary **6.6.4** to establish Hölderian error bounds for CPs and KKT systems for VIs defined by subanalytic functions. As an illustration, we present in the next result such an error bound for a CP of the implicit kind. There is no need for a proof.

6.6.6 Proposition. Let $H : \mathbb{R}^{2n+m} \to \mathbb{R}^p$ be a continuous subanalytic function. Suppose that the CP:

$$\begin{gathered} H(x, y, z) = 0 \\ 0 \leq x \perp y \geq 0 \end{gathered}$$

has a solution. With S denoting its solution set, it holds that for every compact set T in $\mathbb{R}^{2n+m}$, positive constants c and γ exist such that

$$\operatorname{dist}((x, y, z), S) \leq c\,r(x, y, z)^{\gamma}, \quad \forall\,(\,x, y, z\,) \in T,$$

where

$$r(x, y, z) \equiv \|\, H(x, y, z)\,\| + \|\, x_-\,\| + \|\, y_-\,\| + |\, x^T y\,|$$

is the residual function of the CP. □

6.7 Identification of Active Constraints

In this section, we discuss a major application of error bounds; namely, the identification of active constraints in a KKT system. The reader can easily apply this approach to CPs of various kinds, such as an NCP or a complementarity problem of the implicit kind.

The identification of active constraints is an important issue throughout mathematical programming. Algorithmically, one of the most challenging aspect of a CP, VI, or NLP is to accurately identify which inequality constraint is binding at a solution that is being computed, without knowing a *priori* that solution. Once all binding constraints are identified, the combinatorial nature of the problem in question disappears and the problem reduces to that of solving a system of smooth equations. For instance, if one can correctly identify the binding constraints of an optimal solution to a linear program, then one can compute that solution exactly by solving a single system of linear equations. In this section, we do not address the algorithmic implications of the results derived herein; instead, we focus on the basic theory behind the identification technique.

The setting is that of Subsection **6.2.2**. Specifically, let $\bar{x}$ be an isolated KKT point of the system (6.2.2) satisfying the MFCQ. We also assume that F is continuously differentiable, and g_i and h_j are twice continuously differentiable. The basic postulate is the existence of an *identification function* $\rho : \mathbb{R}^{n+\ell+m} \to \mathbb{R}_+$ that satisfies the following properties:

(a) ρ is continuous and equal to zero at $(\bar{x}, \mu, \lambda)$ for every $(\mu, \lambda) \in \mathcal{M}(\bar{x})$;

(b) for every $(\bar{\mu}, \bar{\lambda}) \in \mathcal{M}(\bar{x})$,

$$\lim_{\substack{(x,\mu,\lambda)\to(\bar{x},\bar{\mu},\bar{\lambda}) \\ (x,\mu,\lambda)\notin\{\bar{x}\}\times\mathcal{M}(\bar{x})}} \frac{\rho(x,\mu,\lambda)}{\|\, x - \bar{x} \,\| + \operatorname{dist}((\mu,\lambda),\mathcal{M}(\bar{x}))} = \infty.$$

Subsequently, we show how to build, under appropriate error bounds for the set $\{\bar{x}\} \times \mathcal{M}(\bar{x})$, identification functions. In essence, the only key requirement on an identification function is that it approaches zero at a slower rate than the distance function when a triple (x, μ, λ) approaches the set $\{\bar{x}\} \times \mathcal{M}(\bar{x})$ in question.

The theorem below shows that the index set

$$\mathcal{I}_\rho(x,\mu,\lambda) \equiv \{\, i \,:\, g_i(x) \geq -\rho(x,\mu,\lambda) \,\}$$

correctly identifies all the binding constraints at $\bar{x}$, provided that the triple (x, μ, λ) is sufficiently close to $\{\bar{x}\} \times \mathcal{M}(\bar{x})$.

6.7.1 Theorem. Let $\bar{x}$ be an isolated KKT point of the system (6.2.2) satisfying the MFCQ. Suppose that ρ is an identification function satisfying conditions (a) and (b) above. There exists $\varepsilon > 0$ such that

$$\mathcal{I}_\rho(x,\mu,\lambda) = \mathcal{I}(\bar{x})$$

for all triples (x, μ, λ) satisfying

$$\| x - \bar{x} \| + \text{dist}((\mu, \lambda), \mathcal{M}(\bar{x})) \leq \varepsilon.$$

Proof. Since each g_i is continuously differentiable, there exist positive scalars c and $\bar{\varepsilon}$, depending on $\bar{x}$ only, such that for all i,

$$\| x - \bar{x} \| \leq \bar{\varepsilon} \Rightarrow g_i(x) \geq g_i(\bar{x}) - c \| x - \bar{x} \|.$$

Consider a fixed but arbitrary $(\bar{\mu}, \bar{\lambda})$ in $\mathcal{M}(\bar{x})$. By property (b) of ρ, for every constant $c' > c$, there eixsts $\varepsilon_0(\bar{\mu}, \bar{\lambda}) > 0$ such that

$$\rho(x, \mu, \lambda) \geq c' [\| x - \bar{x} \| + \| (\mu, \lambda) - (\bar{\mu}, \bar{\lambda}) \|]$$

for all triples $(x, \mu, \lambda) \notin \{\bar{x}\} \times \mathcal{M}(\bar{x})$ satisfying

$$\| x - \bar{x} \| + \| (\mu, \lambda) - (\bar{\mu}, \bar{\lambda}) \| \leq \varepsilon_0(\bar{\mu}, \bar{\lambda}).$$

Let (x, μ, λ) be a triple satisfying

$$\| x - \bar{x} \| + \| (\mu, \lambda) - (\bar{\mu}, \bar{\lambda}) \| \leq \varepsilon(\bar{\mu}, \bar{\lambda}) \equiv \min(\bar{\varepsilon}, \varepsilon_0(\bar{\mu}, \bar{\lambda})). \tag{6.7.1}$$

Suppose $g_i(\bar{x}) = 0$. If $(x, \mu, \lambda) \in \{\bar{x}\} \times \mathcal{M}(\bar{x})$, then $x = \bar{x}$ and (μ, λ) is in $\mathcal{M}(\bar{x})$. By property (a) of ρ, we have

$$g_i(x) = g_i(\bar{x}) = 0 = \rho(x, \mu, \lambda).$$

Thus $i \in \mathcal{I}(x, \mu, \lambda)$. If $(x, \mu, \lambda) \notin \{\bar{x}\} \times \mathcal{M}(\bar{x}))$, we have

$$\begin{aligned} g_i(x) &\geq -c \| x - \bar{x} \| \\ &\geq -c' [\| x - \bar{x} \| + \| (\mu, \lambda) - (\bar{\mu}, \bar{\lambda}) \|] \\ &\geq -\rho(x, \mu, \lambda). \end{aligned}$$

Consequently, we have shown $\mathcal{I}(\bar{x}) \subseteq \mathcal{I}_\rho(x, \mu, \lambda)$.

Conversely, if $g_i(\bar{x}) < 0$, then $g_i(\bar{x}) < -\rho(\bar{x}, \bar{\mu}, \bar{\lambda})$. By continuity, it follows that by reducing $\bar{\varepsilon}$ if necessary, any triple (x, μ, λ) satisfying (6.7.1) must also satisfy $g_i(x) < -\rho(x, \mu, \lambda)$; thus $i \notin \mathcal{I}_\rho(x, \mu, \lambda)$.

Summarizing the above derivation, we have shown that for every pair $(\bar{\mu}, \bar{\lambda})$ in $\mathcal{M}(\bar{x})$, there exists a scalar $\varepsilon(\bar{\mu}, \bar{\lambda}) > 0$ such that for every triple (x, μ, λ) satisfying (6.7.1), we have $\mathcal{I}_\rho(x, \mu, \lambda) = \mathcal{I}(\bar{x})$. Since $\mathcal{M}(\bar{x})$ is compact, the existence of the desired ε follows readily from a simple covering argument. □

If $\mathcal{M}(\bar{x})$ is a singleton, so that the SMFCQ holds at $(\bar{x}, \bar{\mu}, \bar{\lambda})$, where $\mathcal{M}(\bar{x}) = (\bar{\mu}, \bar{\lambda})$, then it is even possible to identify the set of strongly active constraints, that is the support of the multiplier $\bar{\lambda}$. Define

$$\mathcal{I}_{\rho,+}(x, \mu, \lambda) \equiv \{ i \in \mathcal{I}_\rho(x, \mu, \lambda) \, : \, \lambda_i \geq \rho(x, \mu, \lambda) \}.$$

6.7.2 Theorem. Let $\bar{x}$ be an isolated KKT point of the system (6.2.2). Suppose that $\mathcal{M}(\bar{x}) = \{(\bar{\mu}, \bar{\lambda})\}$. If ρ is an indentification function for the triple $(\bar{x}, \bar{\mu}, \bar{\lambda})$, then there exists $\varepsilon > 0$ such that

$$\mathcal{I}_{\rho,+}(x, \mu, \lambda) = \operatorname{supp}(\bar{\lambda})$$

for all triples (x, μ, λ) satisfying

$$\| x - \bar{x} \| + \| (\mu, \lambda) - (\bar{\mu}, \bar{\lambda}) \| \leq \varepsilon.$$

Proof. By Theorem **6.7.1**, $\mathcal{A}_{\rho,+}(x, \mu, \lambda) \subseteq \mathcal{I}(\bar{x})$ for all (x, μ, λ) sufficiently close to $(\bar{x}, \bar{\mu}, \bar{\lambda})$. If $\bar{\lambda}_i > 0$, then by continuity, it follows that $i \in \mathcal{A}_{\rho,+}(x, \mu, \lambda)$ provided (x, μ, λ) is sufficiently close to $(\bar{x}, \bar{\mu}, \lambda)$. Since $\mathcal{M}(\bar{x}) = \{(\bar{\mu}, \bar{\lambda})\}$, property (b) of ρ becomes

$$\lim_{(\bar{x},\bar{\mu},\bar{\lambda}) \neq (x,\mu,\lambda) \to (\bar{x},\bar{\mu},\bar{\lambda})} \frac{\rho(x, \mu, \lambda)}{\| (x, \mu, \lambda) - (\bar{x}, \bar{\mu}, \bar{\lambda}) \|} = \infty. \tag{6.7.2}$$

If $\bar{\lambda}_i = 0$, then we have for all (x, μ, λ) sufficiently close to $(\bar{x}, \bar{\mu}, \lambda)$,

$$\begin{aligned} \lambda_i &\leq | \lambda_i - \bar{\lambda}_i | \\ &\leq \| (x, \mu, \lambda) - (\bar{x}, \bar{\mu}, \bar{\lambda}) \| \\ &< \rho(x, \mu, \lambda), \end{aligned}$$

where the last inequality follows from the limit (6.7.2). Hence i does not belong to $\mathcal{I}_{\rho,+}(x, \mu, \lambda)$. Consequently, $\mathcal{I}_{\rho,+}(x, \mu, \lambda) = \operatorname{supp}(\bar{\lambda})$ for all (x, μ, λ) sufficiently close to $(\bar{x}, \bar{\mu}, \lambda)$. □

In the next result, we identify two general situations in which an identification function can be constructed from an error bound of the KKT system.

6.7.3 Proposition. Let $\bar{x}$ be an KKT point of the system (6.2.2). In each of the following two cases, the function $\rho(x, \mu, \lambda)$ is an identification function for $\{\bar{x}\} \times \mathcal{M}(\bar{x})$.

(a) If $\bar{x}$ satisfies the assumptions of Proposition **6.2.7**, then $\bar{x}$ is an isolated KKT point, and

$$\rho(x, \mu, \lambda) \equiv \sqrt{r(x, \mu, \lambda)}$$

where

$$r(x, \mu, \lambda) \equiv \| \mathbf{L}(x, \mu, \lambda) \| + \| h(x) \| + \| \min(\lambda, -g(x)) \|$$

is the residual of the KKT system (6.2.2).

(b) If $\bar{x}$ is an isolated KKT point and F, g and h are analytic functions in a neighborhood of $\bar{x}$, then

$$\rho(x,\mu,\lambda) \equiv \begin{cases} 0 & \text{if } r(x,\mu,\lambda) = 0 \\ \dfrac{-1}{\log r(x,\mu,\lambda)} & \text{if } r(x,\mu,\lambda) \in (0, 0.9) \\ \dfrac{-1}{\log 0.9} & \text{if } r(x,\mu,\lambda) \geq 0.9. \end{cases}$$

Proof. (a) By Proposition **6.2.7**, there exists $c > 0$ such that

$$\| x - \bar{x} \| + \operatorname{dist}((\mu,\lambda), \mathcal{M}(\bar{x})) \leq c\, r(x,\mu,\lambda),$$

whenever (x,μ,λ) is sufficiently close to $\{\bar{x}\} \times \mathcal{M}(\bar{x})$. For a fixed but arbitrary $(\bar{\mu},\bar{\lambda})$ in $\mathcal{M}(\bar{x})$, if $\{(x^k,\mu^k,\lambda^k)\}$ is any sequence converging to $(\bar{x},\bar{\mu},\bar{\lambda})$, then (x^k,μ^k,λ^k) gets arbitrarily close to $\{\bar{x}\} \times \mathcal{M}(\bar{x})$ for all k sufficiently large. Therefore, for all such k with $(x^k,\mu^k,\lambda^k) \notin \{\bar{x}\} \times \mathcal{M}(\bar{x})$, we have $r(x^k,\mu^k,\lambda^k) > 0$ and

$$\frac{\rho(x^k,\mu^k,\lambda^k)}{\| x^k - \bar{x} \| + \operatorname{dist}((\mu^k,\lambda^k), \mathcal{M}(\bar{x}))} \geq c^{-1} \frac{1}{\sqrt{r(x^k,\mu^k,\lambda^k)}};$$

the right-hand fraction clearly tends to ∞ as $k \to \infty$.

(b) Let $\varepsilon > 0$ be such that x^* is the only KKT point in $\operatorname{cl} \mathbb{B}(x^*,\varepsilon)$. Consider the KKT system in this neighborhood:

$$\begin{aligned} \mathbf{L}(x,\mu,\lambda) &= 0 \\ h(x) &= 0 \\ 0 \leq \lambda \perp g(x) &\leq 0 \\ (x-\bar{x})^T(x-\bar{x}) &\leq \varepsilon^2. \end{aligned}$$

The solution set of this analytic system is equal to $\{\bar{x}\} \times \mathcal{M}(\bar{x})$. By Corollary **6.6.4**, there exist positive constants c and γ such that for every triple (x,μ,λ) satisfying

$$\| x - \bar{x} \| + \operatorname{dist}((\mu,\lambda), \mathcal{M}(\bar{x})) \leq \varepsilon,$$

we must have

$$\| x - \bar{x} \| + \operatorname{dist}((\mu,\lambda), \mathcal{M}(\bar{x})) \leq c\, r(x,\mu,\lambda)^{\gamma}.$$

To complete the proof, it suffices to note that

$$\lim_{t \downarrow 0} \frac{-1}{t^{\gamma} \log t} = \infty,$$

which is an easy exercise in elementary calculus. □

6.8 Exact Penalization and Some Applications

Another major application of error bounds is in the derivation of exact penalty functions of constrained optimization problems. In this section, we present the basic principle of exact penalization and discuss how it can be used to obtain a necessary and sufficient condition for the existence of a global Lipschitzian error bound for a system of finitely many convex inequalities.

Consider an optimization problem in the form:

$$\begin{array}{ll} \text{minimize} & \theta(x) \\ \text{subject to} & x \in S \equiv W \cap X, \end{array} \tag{6.8.1}$$

where θ is a real-valued function defined on $\mathbb{R}^n$ and W and X are two subsets of $\mathbb{R}^n$. The principle of exact penalization states that if an error bound exists for the set S with residual function $\psi(x)$ on the set X, then the above optimization problem is equivalent to the minimization of $\theta + \eta\psi$ on the set X for all η sufficiently large. The precise statement of this principle is given in the following theorem.

6.8.1 Theorem. Suppose θ is Lipschitz continuous on the closed set X. Let ψ be a residual function of the set S that majorizes the distance function to S on X; i.e., ψ is a nonnegative function on X, vanishes on S, and there is a constant $c > 0$ such that

$$\text{dist}(x, S) \leq c\,\psi(x), \quad \forall\, x \in X.$$

If W is also closed and (6.8.1) has an optimal solution, then there exists $\bar{\eta} > 0$ such that for all $\eta \geq \bar{\eta}$,

$$\text{argmin}\,\{\,\theta(x) \,:\, x \in S\,\} = \text{argmin}\,\{\,\theta(x) + \eta\,\psi(x) \,:\, x \in X\,\}.$$

Proof. Let $L > 0$ be a Lipschitz constant of θ on X; thus

$$|\,\theta(x) - \theta(y)\,| \leq L\,\|\,x - y\,\|, \quad \forall\, x, y \in X.$$

Let $\bar{\eta} > cL$. Let $\bar{x}$ be any minimizer of θ on S. Let $x \in X$ be arbitrary. Let $z \in S$ be such that $\|x - z\| = \text{dist}(x, S)$. We have, for $\eta \geq \bar{\eta}$,

$$\begin{aligned} \theta(x) + \eta\,\psi(x) &\geq \theta(x) + \bar{\eta}\,c^{-1}\,\text{dist}(x, S) \\ &\geq \theta(x) + L\,\|\,x - z\,\| \\ &\geq \theta(x) + \theta(z) - \theta(x) \\ &= \theta(z) \\ &\geq \theta(\bar{x}) = \theta(\bar{x}) + \eta\,\psi(\bar{x}). \end{aligned}$$

Hence $\bar{x}$ is a minimizer of $\theta + \eta\psi$ on X. Conversely, if x^* is a minimizer of $\theta + \eta\psi$ on X and if z is a vector in S closest to x^*, then

$$\begin{aligned} \theta(\bar{x}) &\geq \theta(x^*) + \eta\,\psi(x^*) \\ &\geq \theta(z) + (\,\eta\, c^{-1} - L\,)\,\|\, x^* - z\,\| \\ &\geq \theta(z) \geq \theta(\bar{x}). \end{aligned}$$

Consequently, $x^* = z$, and hence the two argmin sets are equal. □

The reader is asked to prove a local version of the above theorem in Exercise **6.9.14**. In order to state the first application of the theorem, we recall several facts from convex analysis; these can all be deduced from the results in Section **7.1**, which discusses Clarke's nonsmooth calculus. Let S be a closed convex set in $\mathbb{R}^n$. The distance function $\theta(x) \equiv \operatorname{dist}(x, S)$ is convex; its subgradient $\partial\theta(\bar{x})$ at a vector $\bar{x} \in S$ is equal to the intersection of the normal cone $\mathcal{N}(\bar{x}; S)$ with the closed Euclidean unit ball in $\mathbb{R}^n$. If $\{\theta_1, \cdots, \theta_k\}$ is a finite family of convex functions on $\mathbb{R}^n$, then the subgradient $\partial\theta(\bar{x})$ of the maximum function

$$\theta(x) \equiv \max\{\, \theta_i(x) \,:\, i = 1, \ldots, k\,\}$$

is equal to the convex hull of the subgradients $\partial\theta_i(\bar{x})$ of the binding constraints at $\bar{x}$; see Proposition **7.1.9**. We say that a finite collection of sets

$$S_i \equiv \{\, x \in \mathbb{R}^n \,:\, g_i(x) \leq 0\,\}, \quad i = 1, \ldots, k$$

is *linearly metrically regular with respect to their representations* if there exists a constant $c > 0$ such that

$$\operatorname{dist}(x, S) \leq c \max(\, g_1(x)_+, \cdots, g_k(x)_+\,), \quad \forall\, x \in \mathbb{R}^n,$$

where S is the intersection of the sets S_i; in other words, the given k sets are metrically regular with respect to their representations if an error bound holds for the set S with

$$r(x) \equiv \max(\, g_1(x)_+, \cdots, g_k(x)_+\,)$$

as the residual function. If each $g_i(x) \equiv \operatorname{dist}(x, S_i)$, we will simply say that the family $\{S_i\}$ is linearly metrically regular. The following lemma is stated and proved for two sets; its extension to any finite number of sets is straightforward.

6.8.2 Lemma. Let S_i for $i = 1, 2$ be two closed convex subsets of $\mathbb{R}^n$ with $S_i = g_i^{-1}(-\infty, 0]$ for some real-valued convex function g_i. If S_1 and S_2 are linearly metrically regular with respect to their representations, then for any $\bar{x} \in S$ such that $g_1(\bar{x}) = g_2(\bar{x}) = 0$,

$$\mathcal{N}(\bar{x}; S) = \mathrm{I\!R}_+ \, \partial g_1(\bar{x}) + \mathrm{I\!R}_+ \, \partial g_2(\bar{x}).$$

If S_i is linearly metrically regular for $i = 1, 2$, then

$$\mathcal{N}(\bar{x}; S) = \mathcal{N}(\bar{x}; S_1) + \mathcal{N}(\bar{x}; S_2).$$

Proof. Since $g_i(\bar{x}) = 0$, it follows from the convexity of g_i that $\partial g_i(\bar{x})$ is contained in $\mathcal{N}(\bar{x}; S)$. Thus the inclusion

$$\mathcal{N}(\bar{x}; S) \supseteq \mathrm{I\!R}_+ \, \partial g_1(\bar{x}) + \mathrm{I\!R}_+ \, \partial g_2(\bar{x})$$

is obvious. To prove the reverse inclusion, let $v \in \mathcal{N}(\bar{x}; S)$. So $\bar{x}$ is a constrained global minimum of the linear function $-v^T x$ over $x \in S$. By metric regularity and Theorem **6.8.1** applied to $\theta(x) \equiv -v^T x$ and $X \equiv \mathrm{I\!R}^n$, we deduce the existence of a constant $c > 0$ such that $\bar{x}$ is an unconstrained global minimizer of $\theta(x) + c\psi(x)$, where

$$\psi(x) \equiv \max(\, g_1(x)_+, g_2(x)_+ \,) = \max(\, 0, g_1(x), g_2(x) \,).$$

Consequently, $0 \in -v + c\partial\psi(\bar{x})$, or equivalently, $v \in c\partial\psi(\bar{x})$. By the aforementioned facts, since $g_i(\bar{x}) = 0$ for $i = 1, 2$, it follows that $\partial\psi(\bar{x})$ is the convex hull of the origin, $\partial g_1(\bar{x})$, and $\partial g_2(\bar{x})$. This shows that v belongs to the vector sum of $\mathrm{I\!R}_+\partial g_1(\bar{x})$ and $\mathrm{I\!R}_+\partial g_2(\bar{x})$. The last statement of the lemma follows from the aforementioned facts about the subgradient of the distance function. □

Using the above lemma, we present a necessary and sufficient condition for a convex, finitely representable set to possess a global Lipschitzian error bound in terms of a natural residual. Specifically, let

$$S \equiv \{\, x \in \mathrm{I\!R}^n \, : \, g_i(x) \le 0, \; i = 1, \ldots, m \,\}, \tag{6.8.2}$$

where each $g_i : \mathrm{I\!R}^n \to \mathrm{I\!R}$ is a convex function for $i = 1, \ldots, m$. Generalizing the differentiable case, we say that the (nondifferentiable) Abadie CQ holds at $\bar{x} \in S$ if

$$\mathcal{N}(\bar{x}; S) = \sum_{i \in \mathcal{I}(\bar{x})} \mathrm{I\!R}_+ \, \partial g_i(\bar{x}),$$

where as always $\mathcal{I}(\bar{x})$ is the active index set $\bar{x}$.

6.8.3 Theorem. Assume that $S \neq \emptyset$. For any positive scalar $\eta > 0$, the following two statements are equivalent:

(a) for all $x \in \mathbb{R}^n$,

$$\operatorname{dist}(x, S) \le \eta \max_{1 \le i \le m} g_i(x)_+;$$

(b) the (nondifferentiable) Abadie CQ holds at all points of S, and

$$\inf_{\bar{x} \in \partial S} \left\{ \max_{i \in \mathcal{I}(\bar{x})} \left(v^T \sum_{j \in \mathcal{I}(\bar{x})} \lambda_j w^j \, : \, v \in \partial g_i(\bar{x}) \right) : \right.$$
$$\left. \lambda \in \mathbb{R}_+^{|\mathcal{I}(\bar{x})|}, \; \left\| \sum_{j \in \mathcal{I}(\bar{x})} \lambda_j w^j \right\| = 1, \; w^j \in \partial g_j(\bar{x}) \right\} \ge \eta^{-1}. \tag{6.8.3}$$

Proof. (a) $\Rightarrow$ (b). By extending the proof of Lemma **6.8.2** to any finite number of sets, we can show that the (nondifferentiable) Abadie CQ holds at all points of S under (a). To show (6.8.3), write

$$f(x) \equiv \max_{1 \le i \le m} g_i(x)_+,$$

which is a convex function. As such, we have

$$f'(x; d) = \max\{ v^T d \, : \, v \in \partial f(x) \}, \quad \forall x, d \in \mathbb{R}^n.$$

For any $\bar{x} \in \partial S$, we have $f(\bar{x}) = 0$. Hence $\partial f(\bar{x})$ is equal to the convex hull of the origin and the subgradients $\partial g_j(\bar{x})$ for $j \in \mathcal{I}(\bar{x})$. Consequently, it follows that

$$f'(\bar{x}; d) = \max\left(0, \; \max_{j \in \mathcal{I}(\bar{x})} \{ v^T d \, : \, v \in \partial g_j(\bar{x}) \} \right).$$

For any $\lambda \in \mathbb{R}_+^{|\mathcal{I}(\bar{x})|}$ and any $w^j \in \partial g_j(\bar{x})$ for $j \in \mathcal{I}(\bar{x})$, the vector

$$d \equiv \sum_{j \in \mathcal{I}(\bar{x})} \lambda_j w^j$$

belongs to the normal cone $\mathcal{N}(\bar{x}; S)$ by the (nondifferentiable) Abadie CQ. Consequently, for any $\tau > 0$, $\Pi_S(\bar{x} + \tau d) = \bar{x}$. Hence we have

$$\begin{aligned}
\tau \| d \| &= \operatorname{dist}(\bar{x} + \tau d, S) \\
&\le \eta f(\bar{x} + \tau d) \\
&= \eta \, [f(\bar{x}) + \tau f'(\bar{x}; d) + o(\tau)] \\
&= \eta \tau \left(f'(\bar{x}; d) + \frac{o(\tau)}{\tau} \right).
\end{aligned}$$

Therefore, $f'(\bar{x}; d) \ge \eta^{-1} \|d\|$, and (b) holds.

(b) ⇒ (a). Let $x \notin S$ be arbitrary. The projected vector $\Pi_S(x)$ belongs to the boundary of S; moreover, $x - \Pi_S(x)$ belongs to the normal cone $\mathcal{N}(\Pi_S(x); S)$, which is equal to the vector sum of $\mathbb{R}_+ \partial g_j(\Pi_S(x))$ for all $j \in \mathcal{I}(\Pi_S(x))$. Hence there exist nonnegative scalars λ_j and vectors w^j in $\partial g_j(\Pi_S(x))$ such that

$$x - \Pi_S(x) = \sum_{j \in \mathcal{I}(\Pi_S(x))} \lambda_j \, w^j.$$

By (b), there exist an index $i \in \mathcal{I}(\Pi_S(x))$ and a vector $v \in \partial g_i(\Pi_S(x))$ such that

$$\begin{aligned} \operatorname{dist}(x, S) &= \| x - \Pi_S(x) \| \\ &\leq \eta \, v^T \sum_{j \in \mathcal{I}(\Pi_S(x))} \lambda_j \, w^j \\ &= \eta \, v^T (x - \Pi_S(x)) \\ &\leq \eta \, (g_i(x) - g_i(\Pi_S(x))) \leq \eta \, f(x). \end{aligned}$$

Consequently, (a) and (b) are equivalent. □

If each function g_i is in addition quadratic, Theorem **6.8.3** can be sharpened. Before presenting the sharpened result, consider the case where each function g_i is continuously differentiable. In this case, the condition (6.8.3) can be equivalently stated as follows: for every $\bar{x} \in \partial S$ and for all nonnegative scalars λ_j for $j \in \mathcal{I}(\bar{x})$ such that

$$\left\| \sum_{j \in \mathcal{I}(\bar{x})} \lambda_j \, \nabla g_j(\bar{x}) \right\| = 1$$

we have

$$\max_{i \in \mathcal{I}(\bar{x})} \left(\nabla g_i(\bar{x})^T \sum_{j \in \mathcal{I}(\bar{x})} \lambda_j \, \nabla g_j(\bar{x}) \right) \geq \eta^{-1}.$$

in turn, this follows if we can show that for every $\bar{x} \in \partial S$ and for all index subsets $\mathcal{J}$ of $\mathcal{I}(\bar{x})$ for which the gradients

$$\{ \nabla g_i(\bar{x}) \, : \, j \in \mathcal{J} \}$$

are linearly independent, we have,

$$\left\| \sum_{j \in \mathcal{J}} \lambda_j \, \nabla g_j(\bar{x}) \right\| \geq \eta^{-1} \sum_{j \in \mathcal{J}} \lambda_j, \quad \forall \, \lambda \in \mathbb{R}^{|\mathcal{J}|}_+.$$

Recall that for a convex inequality system, the LICQ implies the Slater CQ. The following lemma therefore shows that if each g_i is convex quadratic, a desired constant $\eta > 0$ exists such that the above inequality holds for all vectors $\bar{x}$ and index sets $\mathcal{J}$ as described.

6.8.4 Lemma. Let g_i be a convex quadratic function for $i = 1, \ldots, m$. Suppose that there exists $\hat{x}$ satisfying $g_i(\hat{x}) < 0$ for all i. There exists a positive constant $\xi > 0$ such that

$$\left\| \sum_{i \in \mathcal{I}(x)} \lambda_i \nabla g_i(x) \right\| \geq \xi \sum_{i \in \mathcal{I}(x)} \lambda_i, \quad \forall x \in \mathbb{R}^n, \text{ and } \lambda_i \geq 0,$$

where $\mathcal{I}(x) \equiv \{ i : g_i(x) = 0 \}$. □

The proof of the above lemma is highly technical. We refer the reader to Section **6.10** for references and bibliographical remarks. Summarizing the above discussion, we obtain the following result readily.

6.8.5 Theorem. Let g_i be a convex quadratic function for $i = 1, \ldots, m$. Suppose that the set S in (6.8.2) is nonempty. The following two statements are equivalent.

(a) There exists $\eta > 0$ such that for all $x \in \mathbb{R}^n$,

$$\operatorname{dist}(x, S) \leq \eta \max_{1 \leq i \leq m} g_i(x)_+;$$

(b) The Abadie CQ holds at all points of S. □

6.9 Exercises

6.9.1 Let $f : X \to X$ be a contraction on the closed set $X \subseteq \mathbb{R}^n$; i.e., for some constant $c \in (0, 1)$,

$$\| f(x) - f(y) \| \leq c \| x - y \|, \quad \forall x, y \in X.$$

By Theorem **2.1.21**, f has a unique fixed point, say x^*. Show directly that

$$\| x - x^* \| \leq \frac{1}{1 - c} \| x - f(x) \|, \quad \forall x \in X.$$

Thus $\|x - f(x)\|$ provides a residual for the fixed-point problem of a contractive map.

6.9.2 Let K be a closed rectangle (with possibly infinite bounds). Give an explicit expression for $r_{\mathcal{N}}(x)$ for $x \in K$.

6.9.3 Show that there exists a scalar $\eta > 0$ such that for any two scalars a and b,

$$(ab)_+ + a_- + b_- \leq \eta \sqrt{((ab)_+)^2 + \left(\sqrt{a^2+b^2} - a - b\right)^2}.$$

(Hint: consider the left-hand side as the ℓ_1-norm of the vector

$$\begin{pmatrix} (ab)_+ \\ a_- + b_- \end{pmatrix}$$

and the right-hand side (without the factor η) as the ℓ_2-norm of the vector

$$\begin{pmatrix} (ab)_+ \\ \psi_{\mathrm{FB}}(a,b) \end{pmatrix}.$$

An easy analysis will then yield the desired inequality.) Deduce from this inequality that the error bound (6.3.2) in Proposition **6.3.5** holds with r_{LTKYF} replaced by the residual r_{YYF} defined in (6.3.4) and with a different constant c. Using Lemma **9.1.3**, show that

$$r_{\mathrm{YYF}}(x) \geq \frac{2}{\sqrt{2}+2} \| \min(x, F(x)) \|, \quad \forall x \in \mathbb{R}^n.$$

6.9.4 Let $F : \mathbb{R}^n \to \mathbb{R}^n$ be a continuous mapping and let K be a closed convex subset of $\mathbb{R}^n$. Suppose that SOL(K, F) is nonempty and the VI (K, F) is semistable. Let $r : \mathbb{R}^n \to \mathbb{R}_+$ be such that for every unbounded sequence $\{x^k\}$ satisfying $\mathbf{F}^{\mathrm{nat}}_K(x^k) \neq 0$ for every k,

$$\lim_{k\to\infty} \frac{\| \mathbf{F}^{\mathrm{nat}}_K(x^k) \|}{\| x^k \|} = 0 \Rightarrow \limsup_{k\to\infty} \frac{r(x^k)}{\| x^k \|} > 0.$$

Show that there exists $\eta > 0$ satisfying

$$\mathrm{dist}(x, \mathrm{SOL}(K,F)) \leq \eta \left[\| \mathbf{F}^{\mathrm{nat}}_K(x) \| + r(x) \right], \quad \forall x \in \mathbb{R}^n.$$

6.9.5 The 2-dimensional vector function

$$F(x_1, x_2) \equiv \begin{pmatrix} x_1 + 2x_1x_2 \\ -x_1^2 + x_2 \end{pmatrix}$$

is easily seen to be strongly monotone but not Lipschitz continuous. Show that a global Lipschitzian error bound does not hold for the NCP (F) with the min residual. (Hint: consider the sequence $x^k \equiv (k, k^2)$.)

6.9.6 Let $G : \mathbb{R}^m \to \mathbb{R}^m$ be strongly monotone and Lipschitz continuous on the range of the matrix $A \in \mathbb{R}^{m \times n}$. Let K be a polyhedron in $\mathbb{R}^n$. Show that the VI (K, G, A, b) is stable if and only if it has a nonempty and bounded solution set.

6.9.7 Consider the minimization problem (6.5.2), where S is a closed convex subset of $\mathbb{R}^n$ and $f : \mathbb{R}^n \to \mathbb{R}$ is continuously differentiable on $\mathbb{R}^n$ and bounded below on S; thus the infimum value $f_{\inf}$ of f on S is finite. The stationary point problem of (6.5.2) is the VI $(S, \nabla f)$. Let ∇f_S^{nat} and ∇f_S^{nor} denote the natural map and normal map of this VI, respectively; thus,

$$\nabla f_S^{\text{nat}}(x) \equiv x - \Pi_X(x - \nabla f(x)), \quad x \in \mathcal{D}$$

and

$$\nabla f_S^{\text{nor}}(z) \equiv \nabla f(\Pi_S(z)) + z - \Pi_S(z), \quad z \in \mathbb{R}^n$$

We say that a (possibly unbounded) sequence $\{x^k\}$ is

- *asymptotically feasible* if

$$\lim_{k \to \infty} \operatorname{dist}(x^k, S) = 0;$$

- *naturally stationary* if

$$\lim_{k \to \infty} \nabla f_S^{\text{nat}}(x^k) = 0;$$

- *normally stationary* if $\{x^k\}$ is asymptotically feasible and there exists a sequence $\{z^k\}$ such that $\Pi_S(x^k) = \Pi_S(z^k)$ for all k and

$$\lim_{k \to \infty} \nabla f_S^{\text{nor}}(z^k) = 0;$$

- *asymptotically minimizing* if $\{x^k\}$ is asymptotically feasible and asymptotically optimal, the latter meaning

$$\lim_{k \to \infty} f(x^k) = f_{\inf}.$$

If $S = \mathbb{R}^n$, then natural stationarity is the same as normal stationarity.

Prove the following statements. (Hint: Proposition **1.5.14** is useful in several parts.)

(a) Every naturally stationary sequence must be asymptotically feasible.

(b) If a sequence $\{x^k\}$ is normally stationary, then the projected sequence $\{\Pi_S(x^k)\}$ is naturally stationary. Moreover, if the sequence $\{x^k\}$ is feasible, then $\{x^k\}$ is normally stationary if and only if

$$\lim_{k \to \infty} \operatorname{dist}(-\nabla f(x^k), \mathcal{N}(x^k; S)) = 0.$$

(c) Use Ekeland's Theorem **6.5.1** to show that if f and ∇f are uniformly continuous near a minimizing sequence $\{x^k\}$, then $\{x^k\}$ is naturally stationary. (See the discussion preceding Lemma **9.1.7** for the definition of uniform continuity near a sequence.)

(d) Suppose there exist constants $\gamma \in (0,1]$ and $\delta > 0$ such that for every $x \in S$ with $f(x) > f_{\text{inf}}$,

$$\| x - \Pi_S(x - \nabla f(x)) \| \geq \delta (f(x) - f_{\text{inf}})^{1-\gamma}.$$

Every normally stationary sequence is minimizing. If f and ∇f are uniformly continuous near a naturally stationary sequence $\{x^k\}$, then $\{x^k\}$ is minimizing.

(e) Let $S = \mathbb{R}^n$. In part (c), the uniform continuity of f can be dropped but not that of ∇f. In part (d), the uniform continuity of f and ∇f can both be dropped.

(f) Let $f(x_1, x_2) = \cos^2(x_1 x_2)$ and

$$S \equiv \{ (x_1, x_2) \in \mathbb{R}^2_- : x_1 x_2 \geq \pi/2 \}.$$

We have $f_{\text{inf}} = 0$. Consider the sequence $\{x^k\}$ with $x^k \equiv (1, -k\pi/2)$ for all k. Show that $\{x^k\}$ is not asymptotically feasible and

$$f(x^k) = \begin{cases} 0 & \text{if } k \text{ is odd} \\ 1 & \text{if } k \text{ is even.} \end{cases}$$

Let $\bar{x}^k \equiv \Pi_S(x^k)$. Show that $\{\bar{x}^k\}$ is asymptotically minimizing, naturally and normally stationary.

6.9.8 Let $F : \mathbb{R}^n \to \mathbb{R}^n$ be a continuously differentiable function and $\{x^k\}$ be an arbitrary sequence such that $\theta_{\text{FB}}(x^k) > 0$ for every k. Let $(\theta_{\text{FB}})_{\text{inf}}$ denote the (unconstrained) infimum value of θ_{FB} on $\mathbb{R}^n$. Assume that $\{\theta_{\text{FB}}(x^k)\}$ and $\{JF(x^k)\}$ are bounded and JF is uniformly continuous near $\{x^k\}$. Show that

$$\lim_{k\to\infty} \theta_{\text{FB}}(x^k) = (\theta_{\text{FB}})_{\text{inf}} \Rightarrow \lim_{k\to\infty} \nabla\theta_{\text{FB}}(x^k) = 0;$$

that is, if $\{x^k\}$ is asymptotically minimizing for the function θ_{FB}, then it must be asymptotically stationary.

6.9.9 We say that a convex program

$$\begin{array}{ll} \text{minimize} & \theta(x) \\ \text{subject to} & x \in X, \end{array}$$

where $\theta : X \to \mathbb{R}$ is a convex function on the closed convex set $X \subseteq \mathbb{R}^n$, is *well posed* if the optimal solution set X_{opt} is nonempty and

$$\forall \{x^k\} \subset X, \quad \lim_{k\to\infty} \theta(x^k) = \theta_{\min} \Rightarrow \lim_{k\to\infty} \operatorname{dist}(x^k, X_{\text{opt}}) = 0,$$

where $\theta_{\min}$ is the optimal objective value. Show that the following statements are equivalent.

(a) For any $\delta > 0$, there exists $\eta(\delta) > 0$ such that for all $x \in X$ such that $\operatorname{dist}(x, X_{\text{opt}}) \geq \delta$,

$$\operatorname{dist}(x, X_{\text{opt}}) \leq \eta(\delta)\,(\,\theta(x) - \theta_{\min}\,).$$

(b) The given convex program is well posed.

(c) For any $\varepsilon > 0$, there exists $\zeta(\delta) > 0$ such that for all $x \in X$ such that $\theta(x) - \theta_{\min} \geq \varepsilon$,

$$\operatorname{dist}(x, X_{\text{opt}}) \leq \zeta(\delta)\,(\,\theta(x) - \theta_{\min}\,).$$

6.9.10 This exercise gives counterexamples to Exercise **6.9.8** without the assumptions made therein. Consider the 2-dimensional NCP (F) with

$$F(x_1, x_2) = \begin{pmatrix} \exp(x_1^2 - x_2) \\ 0 \end{pmatrix}, \qquad (\,x_1, x_2\,) \in \mathbb{R}^2.$$

The solution set of the NCP is the nonnegative x_2 axis. Consider the sequence $\{x^k\}$ with

$$x^k = (\,x_1^k, x_2^k\,) \equiv (\,k, k^2 + 0.5\log k\,), \quad \forall\, k.$$

Show that $\theta_{\text{FB}}(x^k) \to 0$ but $\partial\theta_{\text{FB}}(x^k)/\partial x_1 \to 2$ as $k \to \infty$. What is wrong?

6.9.11 Consider the NCP (F) with $F(x_1, x_2) = (x_1 x_2, x_2)$ for (x_1, x_2) in $\mathbb{R}^2$. The function F is semicopositive on $\mathbb{R}^2_+$ (for the definition of such a function; see Exercise **3.7.31**). The solution set of this NCP, denoted S, is the nonnegative x_1 axis. Show by a direct argument that a constant $c > 0$ exists satisfying

$$\operatorname{dist}(x, S) \leq c\,\|\,\mathbf{F}_{\text{FB}}(x)\,\|, \quad \forall\, x \in \mathbb{R}^n.$$

This NCP is not stable because with $q \equiv \varepsilon(-1, 1)$, the perturbed NCP $(F + q)$ has no solution for any $\varepsilon > 0$.

6.9.12 Let $f : \mathbb{R}^n \to \mathbb{R}^n$ be a PC^1 function near a zero x^*. Let

$$\{ f^1, \cdots, f^k \}$$

be the effective C^1 pieces of f near x^*; that is, there exists a neighborhood $\mathcal{N}$ of x^* such that for every x in $\mathcal{N}$, $f(x)$ belongs to the family $\{f^1(x), \cdots, f^k(x)\}$. Suppose that $Jf^i(x^*)$ is nonsingular for all $i = 1, \ldots, k$. Show that there exist positive scalars c and ε such that

$$\| x - x^* \| \leq \varepsilon \Rightarrow \| x - x^* \| \leq c \| f(x) \|.$$

State and and prove a similar pointwise error bound result for a semismooth function f at a zero x^* in terms of the B-subdifferential $\partial_B f(x^*)$. (See Theorem **7.2.10** for a further generalization to a locally Lipschitz function.)

6.9.13 Let $F : K \to \mathbb{R}^n$ be continuous and strongly monotone on the closed convex set $K \subseteq \mathbb{R}^n$. Show that if $\eta > 0$ is a strong monotonicity constant of F, then

$$\| x - x^* \| \leq \sqrt{\eta^{-1}} \sqrt{\theta_{\text{gap}}(x)}, \quad \forall x \in K,$$

where x^* is the unique solution of the VI (K, F). See Exercise **10.5.8** for an improved bound.

6.9.14 Let $W \equiv g^{-1}(-\infty, 0]$, where $g : \mathbb{R}^n \to \mathbb{R}^m$ is a given vector function. Let $x^* \in S \equiv W \cap X$. Suppose there exist positive constants c and γ and a neighborhood V of x^*, within which θ is Lipschitz continuous and such that

$$\text{dist}(x, S) \leq c \| g(x)_+ \|^\gamma, \quad \forall x \in V \cap X.$$

Show that x^* locally minimizes θ on S if and only if x^* locally minimizes on X the function

$$\theta(x) - \frac{1}{\log \| g(x)_+ \|}.$$

The significance of the latter equivalence is that the neither the constant c nor the exponent γ appears in the displayed function.

6.9.15 Let $f : \mathbb{R}^n \to \mathbb{R}$ be convex and let $S \equiv f^{-1}(-\infty, 0]$. Show that the following three statements are equivalent.

(a) There exists a constant $\eta > 0$ such that

$$\text{dist}(x, S) \leq \eta \, f(x)_+, \quad \forall x \in \mathbb{R}^n.$$

(b) For all $\bar{x} \in f^{-1}(0)$, $\mathcal{N}(\bar{x}; S) = \mathbb{R}_+ \partial f(\bar{x})$; furthermore, for all such $\bar{x}$ and all $w \in \partial f(\bar{x})$, there exists $v \in \partial f(\bar{x})$ satisfying

$$v^T w \geq \eta^{-1} \| w \|.$$

(c) For all $\bar{x} \in f^{-1}(0)$,

$$f'(\bar{x}; w) \geq \eta^{-1} \| w \|, \quad \forall w \in \mathcal{N}(\bar{x}; S).$$

Consider further the following four statements.

(d) There exists a vector d satisfying $f_\infty(d) < 0$; equivalently, the recession cone X_∞ has a Slater point.

(e) X has a Slater point and f is well behaved; i.e., the following implication holds:

$$[\lim_{k \to \infty} v^k = 0, \ v^k \in \partial f(x^k) \quad \forall k] \Rightarrow \lim_{k \to \infty} f(x^k) = \inf_{x \in \mathbb{R}^n} f(x).$$

(This implication says that every stationary sequence of f is a minimizing sequence.)

(f) X has a Slater point and there exists a scalar $\delta > 0$ such that for every $\bar{x} \in \partial S$, a vector $\bar{z} \in \operatorname{int} S$ exists satisfying $\|\bar{x} = \bar{z}\| \leq \delta(-g(\bar{z}))$.

(g) The *abstract strong Slater condition* holds; i.e., $0 \notin \operatorname{cl} \partial f(f^{-1}(0))$.

Show that (d) $\Rightarrow$ (g); (e) $\Rightarrow$ (g); (f) $\Rightarrow$ (g); and (g) $\Rightarrow$ (a).

6.10 Notes and Comments

In the mathematical programming literature, Hoffman's 1952 seminal paper [349] is the earliest publication of an error bound for a linear inequality system. Although much of Robinson's work in the 1970s is closely connected to error bounds (such as his 1975 paper [724], whose main result we state as Proposition **6.1.4** in an Euclidean space, and his 1976 paper [726], which lays the foundation of modern theory of metric regularity), the contemporary study of error bounds for mathematical programs was fuelled by Mangasarian's paper [567], which gave a new proof of Hoffman's error bound along with a computable estimate for a multiplicative constant for such a bound. A subsequent paper [568] extends the results to a system of differentiable convex inequalities. As part of its contribution, the latter paper introduced an asymptotic constraint qualification (ACQ) that was generalized by Auslender and Crouzeix [32] to nondifferentiable inequalities and led these authors to consider the class of well-behaved convex functions [33]; see also [30]. The work by Auslender in this direction accumulated in the article [29], which presents an excellent survey of the theory of unboundedness in optimization. The terminology of a (strongly) well-behaved pair to describe the conditions in Corollary **6.5.4** is inspired by the paper [32].

Another major milestone in the theory of error bounds occurred when Mangasarian and Shiau [578] obtained the first error bound for a monotone LCP, using Hoffman's error bound applied to the polyhedral representation of its solution set. Example **6.4.2**, which appeared in this reference, had provided the motivation for much of the subsequent research to seek conditions to remove the square-root term in the error bound (6.4.3) for a monotone LCP. More importantly, this example strongly suggests that global Lipschitzian error bounds are not the norm for nonlinear systems; instead, Hölderian error bounds are more likely to exist unless special conditions prevail.

Several important extensions followed [578]. The paper [570] sharpened the Mangasarian-Shiau error bound in the presence of a nondegenerate solution to the LCP; a global error bound for a monotone AVI was obtained in [571]; see also [239]. Parallel to and independent of Mangasarian's work on the monotone LCP, Pang [662] obtained a global Lipschitzian error bound for a strongly monotone VI with the natural residual (Proposition **6.3.1**) and the gap residual (6.9.13). Subsequently, Mathias and Pang [593] introduced the fundamental constant $\rho(M)$ of a P matrix M (see Exercise **3.7.35**) and detailed the role of this constant in a global Lipschitzian error bound for an LCP of the P type.

In a private communication, Fukushima pointed out to Pang that the Lipschitz continuity of the function F was not explicitly assumed in [662] for the error bound with the natural residual; such continuity turned out to be essential for the validity of the error bound. Example **6.3.4** was constructed by Tseng in May 2000. Prior to that, Yamashita constructed the counterexample in Exercise **6.9.5** in December 1998. Error bounds for the NCP based on the merit function r_{LTKYF} were obtained in [544, 397]; those that are based on the merit function r_{YYF} in (6.3.4) were obtained in [877]; see Exercise **6.9.3**. With all these merit functions as residuals, global Lipschitzian error bounds for the NCP with a uniformly P function hold without the Lipschitz continuity of the defining function.

The interest in local error bounds (see part (a) of Proposition **6.1.2**) stemmed partly from the fundamental result of Robisnon about the upper Lipschitz continuity of polyhedral multifunctions, Theorem **5.5.8**. The fact that every local error bound can be extended to a global error bound by the premultiplication by the factor $(1 + \|x\|)$ in part (b) of the mentioned proposition was first used by Luo, Mangasarian, Ren, and Solodov [531] in connection with a special error bound for the LCP in terms of the implicit Lagrangian function for the NCP [580] (see Section **10.3.1**).

The first paper that studied local error bounds for KKT systems of NLPs under the MFCQ is perhaps [222], where an extension to VIs under the SMFCQ was also mentioned. There, the error bounds provided the cornerstone for the theory of active-set identification. Related references include [322, 873] where local error bounds for degenerate NLPs were derived. Subsection **6.2.2** is a synthesis of these studies; most importantly, the R_0 assumption in Proposition **6.2.7** is the weakest second-order-type condition known for local error bounds of KKT systems to hold.

One of the major applications of error bounds is to establish the rates of convergence of various iterative methods (see Section **12.6**). Luo and Tseng are champions of such an application. Starting with their original work on the linear rate of convergence of matrix splitting methods for the monotone LCP [536], in which they made an ingenious use of the upper Lipschitz continuity of polyhedral multifunctions, Luo and Tseng published a series of papers [530, 537, 539, 540, 541, 543, 847] in which they obtained similar rates of convergence for a host of iterative descent methods for convex optimization problems and monotone VIs. The error bound results in Subsection **6.2.3** are a synthesis of the work of Luo and Tseng, developed in a unified manner herein for the class of linearly constrained, strongly monotone composite VIs.

The global Lipschitzian error bound of an LCP with a P matrix, Exercise **3.7.35**, raises two questions that are answered by the results in Subsection **6.3.2**, which concerns the VI. Originally posed in the context of the LCP, the first question is as follows. What is the class of matrices $M \in \mathbb{R}^{n \times n}$ for which a global Lipschitzian error bound exists for all vectors q in the LCP range of M? Mangasarian and Ren [577] show that such an error bound holds for an LCP with an R_0 matrix. This result provides only a partial answer to the question, because it does not completely characterize the matrix M in terms of the said error bound. Using the theory of recession functions [343, 746], Gowda [301] established a general result of global error bound properties of a PA function via its recession function, from which Theorem **6.3.12** and Corollary **6.3.16** follow readily.

In an effort to generalize the LCP result of Mangasarian and Ren, Chen [117] studied global error bounds for NCPs under various nonlinear R_0-type assumptions. The two-residual result, Proposition **6.1.5**, is inspired by Chen's work; so are Theorem **6.3.8** and Corollary **6.3.9**. Our results are more general than Chen's, which include Remark **6.3.10**.

Luo and Tseng [529] were the first to establish the converse of the Mangasarian and Ren error bound result, i.e., Corollary **6.3.16** with K being the nonnegative orthant. Luo and Tseng also considered the problem of a single multiplicative error bound constant for the LCP corresponding to a fixed matrix M. This problem can be posed more formally as follows. Assume that a global Lipschitzian error bound exists for all vectors q in the LCP range of M. The multiplicative constant of such an error bound is in general dependent on the vector q with M fixed. What is the class of matrices M for which a single multiplicative constant exists that is applicable to all vectors q? As we see from Theorem **6.3.18**, this question is closely related to a Lipschitz property of the solutions with a fixed matrix M. Indeed, inspired by the Lipschitz continuity of the solution function $x(q)$ of the LCP (q, M) with a P matrix M, Pang was curious about the validity of the converse of this property. Namely, suppose that M is a Lipschitzian matrix for which $\text{SOL}(q, M) \neq \emptyset$ for all q; does it follow that M must be a P matrix? Gowda [299] answered this question partially by assuming that M belongs to a special class of matrices; he also studied the connection between the Lipschitzian property of M and the lower semicontinuity of the solutions of the LCP (q, M) as a multifunction of q. Murthy, Parthasarathy, and Sabatini [629] completely settled Pang's question in the affirmative. Stone [815] obtained further properties of a Lipschitzian matrix, relating them to the so-called "INS matrices". These articles motivated Gowda and Sznajder to study the Lipschitzian property of a polyhedral multifunction [311] and the pseudo Lipschitzian property of the inverse of a PA map [312]. In particular, Proposition **6.3.19** is proved by Gowda and Sznajder. See the paper by Penot [691] for the discussion of openness and Lipschitz behavior of general multifunctions.

Extending the notion of a sharp, or strongly unique, minimum [696], Ferris [236] and Burke and Ferris [99] introduced the concept of a weak sharp minimum of a constrained optimization problem with nonunique optimal solutions and studied the existence of such minima for linear and convex quadratic programs and monontone LCPs. The finite termination of iterative methods for convex programs with weak sharp minima was examined in these two references and also in [237]. The minimum principle sufficiency was formally defined by Ferris and Mangasarian [238], who also proved Proposition **6.4.4**. The fact that a solvable linear program always has weak sharp minima, Exercise **3.7.11**, was proved earlier by Meyer and Mangasarian [603].

Much of Section **6.4** is based on the paper by Ferris and Pang [242], which was a synthesis of the work of Mangasarian on monotone LCPs with nondegenerate solutions, of Ferris (and Burke) on weak sharp minima, and of Ferris and Mangasarian on the minimum principle. The origin of Corollary **6.4.7** is the classic Goldman-Tucker lemma in linear programming [294]. The corollary as stated is proved in [242]. A refinement of this corollary was obtained by Sznajder and Gowda [826].

Li [503] studied error bounds for feasible solutions to piecewise convex quadratic programs and obtained results that supplement those in Subsection **6.4.1**. Specifically, a piecewise convex quadratic program is as follows:

$$\begin{array}{ll} \text{minimize} & \theta(x) \\ \text{subject to} & x \in P(A,b), \end{array}$$

where θ is a piecewise convex quadratic function; i.e., there exist finitely many convex quadratic functions $\{\theta_1, \cdots, \theta_k\}$ for some positive integer k such that $\theta(x) \in \{\theta_1(x), \cdots, \theta_k(x)\}$ for all $x \in \mathbb{R}^n$. Li obtained error bounds for feasible solutions to the optimal solution set in terms of their deviations from the optimal objective value; he also specialized the results to the case where θ is a convex, piecewise quadratic function. It is not known whether Li's error bound results can be generalized to infeasible vectors. Li [504] applied his results to establish the linear convergence of descent methods for the unconstrained minimization of convex quadratic splines. In fact, beginning with his joint work with Swetits [508] on convex regression, Li is a champion of using (unconstrained) piecewise quadratic functions for solving (linearly constrained) strictly convex quadratic programs; see the series of papers [505, 509, 510]. For related work, see [548, 549]. Sun [821] studies the structure of convex piecewise quadratic functions; Sun [822] investigates a nonsmooth Newton method for minimizing convex and nonconvex, differentiable and nondifferentiable, piecewise quadratic functions.

Defined originally for optimization problems with unique optimal solutions [495, 831], the theory of well-posed convex optimization problems [196, 527] is closely related to error bounds; see Exercise **6.9.9**, which is based on Deng [178]. Clearly, every convex program with weak sharp minima is well posed; moreover, Li's results in [503] show that every convex, piecewise quadratic program is well posed. As noted in Deng's paper, the well-posedness theory is closely related to the well-behavedness theory of Auslender and Crouzeix [33] and the theory of linear regularity of a finite collection of closed convex sets. In turn, the latter theory has a strong bearing on the convergence of projection methods for solving convex feasibility problems [42, 43] and the (strong) conical hull intersection property [44, 45, 181]. The paper [572] formulates some least-residual problems for the treatment of ill-posed LCPs.

Luo and Luo [528] were the first to study error bounds for convex quadratic inequality systems. They showed that such systems have global Lipschitzian error bounds under the Slater CQ. The significance of this result is that this is an instance of a nonlinear inequality system for which a global Lipschitzian error bound holds without an asymptotic CQ (such as Mangasarian's original ACQ in [568]). Based on the key lemma **6.8.4**, whose proof can be found in [528], Li [506] established the characterization Theorem **6.8.5** for the existence of a global Lipschitzian error bound for a convex quadratic inequality system in terms of the Abadie CQ. Bartelt and Li [40] related Hausdorff strong unicity for best approximations to Abadie's CQ for the associated convex quadratic feasibility problem.

In [528] the authors also studied polynomial systems and obtained a Hölderian error bound for such a system on a compact test set, by applying Hörmander's theory [354]. There are two direct descendants of the work of Luo and Luo. The first is the extension from a polynomial system to a (sub)analytic system [532]; see also [169]. Summarized in Section **6.6**, this extension is made possible by Lojasiewicz' deep result on division by analytic functions [520, 521] (Theorem **6.6.3**). For references on semianalytic and subanalytic sets, the reader can consult [65, 344, 345, 522], where error bounds for these sets can be found. The other direct descendant of the article [528] is a global Hölderian error bound for a convex quadratic system without any CQ [867]. The exponent of the residual needed in the latter Hölderian bound has to do with the degree of singularity of the nonlinear inequalities in the system, which is a quantitative measure of the degree of violation of Slater's CQ. Bartelt and Li [41] further examined this exponent from the point of view of Chebyshev approximations.

In algorithmic design for constrained optimization, the identification of active constraints is an important issue, because a result of this kind will show that eventually an algorithm will behave like an algorithm for unconstrained optimization. Early work [95, 100, 101, 103, 200, 201, 872] in this direction assumed the nondegeneracy concept of Dunn. The article [218] introduced several computational indicators for identifying zero variables in interior-point methods for solving linear programs. Based on error bounds, Facchinei, Fischer, and Kanzow [222] developed the theory of identification functions presented in Section **6.7**; this is by far the most general theory for the constructive identification of active constraints. The use of the latter theory in designing fast convergent algorithms is emerging rapidly; see [223, 225]. Yamashita, Dan, and Fukushima [880] develop an identification scheme for degenerate indices in NCPs based on the proximal point algorithm without assuming nondegeneracy and local solution uniqueness.

Using identification functions and the theory of 2-regular mappings with Lipschitzian derivatives [370], Izmailov and Solodov [371, 372] developed locally superlinearly convergent Newton methods for solving singular equations and KKT systems under weak regularity conditions.

Starting with the pioneering work of Fiacco and McCormick [246], the theory of exact penalization has a long history in constrained optimization. With a focus on how penalization can be used to obtain optimality conditions for constrained optimization problems, the survey [97] gives an elegant, unified approach for deriving many results on exact penalty functions; see also the related paper [96], which relates exact penalization to the concept of "calmness" in nonlinear programming and the prior survey [184]. The principle of exact penalization, Theorem **6.8.1**, was originally proved by Clarke [127], who used the distance function $\operatorname{dist}(x, S)$ instead of a general residual function ψ. The drawback with the distance function is that it is in general not suitable for computational purposes; thus its use is restricted primarily to theoretical analysis. Chapter 2 in [533] presents a comprehensive treatment of the theory of exact penalty functions for MPECs, using the theory of error bounds as the fundamental tool for the derivation of such functions. The paper [168] derives an exact penalty function for optimization problems with subanalytic constraints. Demyanov, Di Pillo, and Facchinei [174] analyze exact penalty functions for nonsmooth optimization problems by using (Dini) Hadamard directional derivative with respect to the constraint set; one of the conditions these authors used is an extended version of the one in Proposition **6.5.3**.

Convex analysis is a useful tool for the study of error bounds for closed convex sets. The paper [501] presents a systematic treatment of this subject in an Euclidean space and obtains a version of Lemma **6.8.2** that forms the basis for the necessary and sufficient condition of the existence of a global Lipschitzian error bound for a convex inequality system. Theorem **6.8.3** appeared in the survey article [670] on error bounds in mathematical programming. Klatte and Li [434] presented a detailed investigation of various ACQs, showing in particular that they are essentially of three types: (i) a "bounded excess condition" as defined in [428, 429], (ii) the Slater CQ together with the ACQ of Auslender and Crouzeix [32], and (iii) a positivity property of the directional derivative (see part (b) in Exercise **6.9.15**). This exercise is drawn from [501]; condition (f) in the exercise was introduced by Mangasarian [573] who called the condition the "strong Slater constraint qualification". Bertsekas [60] employed a simple derivation to obtain an upper bound of the optimal objective value of the dual of a nonconvex program and recovered Mangasarian's bound as a special case.

As is clear from Robinson's early paper [724], the theory of error bounds can be established in abstract spaces. Burke and Tseng [102] used Fenchel duality to give a unified analysis of Hoffman's error bound in a normed linear space. Deng [176, 177] established global error bounds for closed convex sets defined by closed proper convex functions in a Banach space, assuming a Slater condition on the associated recession functions; see part (d) of Exercise **6.9.15**, which is stated in $\mathbb{R}^n$. Wu and Ye [875, 876] studied error bounds for lower semicontinuous, proper, convex functions in Banach spaces and metric spaces, respectively. In particular, Proposition **6.1.3** is inspired by a result of the latter authors [876] that deals with Lipschitzian bounds. The assumption of the Hölderian bound in the proposition is useful for proving Corollary **6.6.5**.

In a series of papers [636, 637, 638, 639, 640, 641], Ng and his Ph.D. students W.H. Yang and X.Y. Zheng study error bounds for convex and nonconvex inequality systems in abstract spaces and obtain sharp results for special systems. In particular, the paper by Ng and Zheng [638] obtains complete error bound results for one quadratic inequality that extend some previous results of Luo and Sturm [534]. Takahashi's condition (see Theorem **6.5.2**) was shown to be equivalent to Ekeland's renowned variational principle [216, 217] by Hamel [323]. Ng and Zheng [641] employ Takahashi's condition to derive error bounds for lower semicontinuous functions in terms of Dini-directional derivatives. Our proof of Ekeland's variational principle in Theorem **6.5.1** follows that in [752, Proposition 1.4.3]. Section **6.5** is based on the work of Ng, Yang, and Zheng.

For a set $X \equiv \{x \in \mathbb{R}^n : g(x) \leq 0\}$ for some real-valued function g, the property of a pointwise, Lipschitzian error bound near a vector $\bar{x} \in X$ with residual $r(x) \equiv g(x)_+$ coincides with the traditional concept of metric regularity of X at $\bar{x}$. Inspired by a result of Li [506] which pertains to the case where g is the pointwise maximum of finitely many differentiable convex functions, Lemma **6.8.2** was proved in [501] where its converse was conjectured. This converse is recently settled by Zheng in his Ph.D. thesis [903], in a form that is rather natural (from a duality point of view) but not quite the same as that originally conjectured. Stimulated by Bauschke and Borwein [42] and Bauschke, Borwein, and Li [44], the joint paper [636] investigates in detail regularity conditions of a finite family of convex sets and their relations to error bounds. The Ph.D. thesis of Zheng and that of Yang [882] contain a wealth of error bound results that extend beyond those reviewed here.

The motivation to study minimizing sequences and stationary sequences of optimization problems is to understand the convergence of iterative methods without assuming a priori the boundedness of the sequences. Auslender [29] provides an excellent survey of this topic. Further references include [28, 336, 361]. With an emphasis toward the role of error bounds, Exercise **6.9.7** is drawn from [123]. Exercises **6.9.8** and **6.9.10** are drawn from the article [273], which studies minimizing and stationary sequences of differentiable merit functions for VIs and CPs; see also [698]. He [335] provided the function $F(x_1, x_2)$ in Exercise **6.9.11** to show that global Lipschitzian error bounds exist for NCPs not of the uniformly P class.

The theory of error bounds continues to flourish. Recent research areas not reviewed here include error bounds for convex conic systems [897], for regularized systems [800, 849], for convex multifunctions [507], for 2-regular maps [370], and for semidefinite programs [179]; see also the above-mentioned Ph.D. theses of Yang and Zheng.

BIBLIOGRAPHY

[1] H.Z. AASHTIANI AND T.L. MAGNANTI. Equilibria on a congested transportation network. *SIAM Journal on Algebraic and Discrete Methods* 2 (1981) 213–226.

[2] H.Z. AASHTIANI AND T.L. MAGNANTI. A linearization and decomposition algorithm for computing urban traffic equilibria. *Proceedings of the 1982 IEEE International Large Scale Systems Symposium* (1982) pp. 8–19.

[3] J. ABADIE. On the Kuhn-Tucker theorem. In J. Abadie, editor, *Nonlinear Programming*, North Holland (Amsterdam 1967) pp. 19–36.

[4] M. AGANAGIC. Variational inequalities and generalized complementarity problems. Technical Report SOL 78-11, Systems Optimization Laboratory, Department of Operations Research, Stanford University (Stanford 1978).

[5] B.H. AHN. *Computation of Market Equilibria for Policy Analysis: The Project Independence Evaluation Study (PIES) Approach*, Garland (New York 1979).

[6] B.H. AHN AND W.W. HOGAN. On convergence of the PIES algorithm for computing equilibria. *Operations Research* 30 (1982) 281–300.

[7] P. ALART AND A. CURNIER. A mixed formulation for frictional contact problems prone to Newton like solution methods. *Computer Methods in Applied Mechanics and Engineering* 92 (1991) 353–375.

[8] A.M. AL-FAHED AND P.D. PANAGIOTOPOULOS. Multifingered frictional robot grippers: a new type of numerical implementation. *Computers & Structures* 42 (1992) 555-562.

[9] A.M. AL-FAHED, G.E. STAVROLAKIS, AND P.D. PANAGIOTOPOULOS. Hard and soft fingered robot grippers. The linear complementarity approach. *Zeitschrift für Angewandte Mathematik und Mechanik* 71 (1991) 257-265.

[10] E. ALLGOWER AND K. GEORG. Simplicial and continuation methods for approximating fixed points and solutions to systems of equations. *SIAM Review* 22 (1980) 28–85.

[11] E. ALLGOWER AND K. GEORG. *Numerical Continuation Methods: An Introduction*, Springer-Verlag (Berlin 1990).

[12] E. ALTMAN, K. AVRACHENKOV, AND C. BARAKAT. TCP network calculus: the case of large delay-bandwidth product. Paper presented at the *IEEE Infocom 2002 Conference*.

[13] E.J. ANDERSON AND S.Y. WU. The continuous complementarity problem. *Optimization* 22 (1991) 419–426.

[14] L.E. ANDERSSON AND A. KLARBRING. Quasi-static frictional contact of discrete mechanical structures, *European Journal of Mechanics A/Solids* 19 (2000) S61–S78.

[15] M. Anitescu, J.F. Cremer, and F. Potra. On the existence of solutions to complementarity formulations of contact problems with friction. In M.C. Ferris and J.S. Pang, editors, *Complementarity and Variational Problems: State of the Art*, SIAM Publications (Philadelphia 1997) pp. 12–21.

[16] M. Anitescu and F. Potra. Formulating dynamic multi-rigid-body contact problems with friction as solvable linear complementarity problems. *ASME Nonlinear Dynamics* 4 (1997) 231–247.

[17] M. Anitescu, F.A. Potra, and D. Stewart. Time-stepping for three-dimensional rigid-body dynamics. *Computer Methods in Applied Mechanics and Engineering* 177 (1999) 183–197.

[18] K.J. Arrow and G. Debreu. Existence of an equilibrium for a competitive economy. *Econometrica* 22 (1954) 265–290.

[19] K.J. Arrow, L. Hurwicz, and H. Uzawa. Constraint qualifications in maximization problems. *Naval Research Logistics Quarterly* 8 (1961) 175–191.

[20] R. Asmuth. Traffic network equilibrium. Technical Report SOL 78-2, Systems Optimization Laboratory, Department of Operations Research, Stanford University (Stanford 1978).

[21] R. Asmuth, B.C. Eaves, and E.L. Peterson. Computing economic equilibria on affine networks with Lemke's algorithm. *Mathematics of Operations Research* 4 (1979) 207–214.

[22] J.P. Aubin. *Mathematical Methods of Game and Economic Theory*, North-Holland (Amsterdam 1979).

[23] J.P. Aubin. Lipschitz behavior of solutions to convex minimization problems. *Mathematics of Operations Research* 9 (1984) 87–111.

[24] J.P. Aubin and I. Ekeland. *Applied Nonlinear Analysis*, John Wiley (New York 1984).

[25] J.P. Aubin and H. Frankowska. *Set-Valued Analysis*, Birkäuser (Boston 1990).

[26] G. Auchmuty. Variational principles for variational inequalities. *Numerical Functional Analysis and Optimization* 10 (1989) 863–874.

[27] A. Auslender. *Optimisation: Méthodes Numériques*, Masson (Paris 1976).

[28] A. Auslender. Convergence of stationary sequences for variational inequalities with maximal monotone operators. *Applied Mathematics and Optimization* 28 (1993) 161–172.

[29] A. Auslender. How to deal with the unbounded in optimization: Theory and algorithms. *Mathematical Programming* 79 (1997) 3–18.

[30] A. Auslender, R. Cominetti, and J.-P. Crouzeix. Convex functions with unbounded level sets and applications to duality theory. *SIAM Journal on Optimization* 3 (1993) 669–687.

[31] A. Auslender and R. Correa. Primal and dual stability results for variational inequalities. *Computational Optimization and Applications* 17 (2000) 117–130.

[32] A. Auslender and J.P. Crouzeix. Global regularity theorems. *Mathematics of Operations Research* 13 (1988) 243–253.

[33] A. Auslender and J.P. Crouzeix. Well behaved asymptotical convex functions. *Analyse Non-linéare* (1989) 101–122.

[34] S.A. Awoniyi and M.J. Todd. An efficient simplicial algorithm for computing a zero of a convex union of smooth functions. *Mathematical Programming* 25 (1983) 83–108.

[35] C. BAIOCCHI AND A. CAPELO. Translated by L. Jayakar. *Variational and Quasi-variational Inequalities: Applications to Free Boundary Problems*, John Wiley (Chichester 1984).

[36] H. BANK, J. GUDDAT, D. KLATTE, B. KUMMER, AND K. TAMMER. *Non-Linear Parametric Optimization*, Birkhäuser Verlag (Basel 1982).

[37] D. BARAFF. Issues in computing contact forces for nonpenetrating rigid bodies. *Algorithmica* 10 (1993) 292–352.

[38] D. BARAFF. Fast contact force computation for nonpenetrating rigid bodies. *Computer Graphics (Proceedings SIGRAPH)* 28 (1994) 23–34.

[39] V. BARBU. *Optimal Control of Variational Inequalities*, Pitman Advanced Publishing Program (Boston 1984).

[40] M. BARTELT AND W. LI. Abadie's constraint qualification, Hoffman's error bounds, and Hausdorff strong unicity. *Journal of Approximation Theory* 97 (1999) 140–157.

[41] M. BARTELT AND W. LI. Exact order of Hoffman's error bounds for elliptic quadratic inequalities derived from vector-valued Chebyshev approximation. *Mathematical Programming, Series B* 88 (2000) 223–253.

[42] H.H. BAUSCHKE AND J.M. BORWEIN. On projection algorithms for solving convex feasibility problems. *SIAM Review* 38 (1996) 367–426.

[43] H.H. BAUSCHKE, J.M. BORWEIN, AND A.S. LEWIS. The method of cyclic projections for closed convex sets in Hilbert space. *Recent developments in optimization theory and nonlinear analysis (Jerusalem 1995) Contemporary Mathematics* 204, American Mathematical Society (Providence 1997) pp. 1–38.

[44] H.H. BAUSCHKE, J.M. BORWEIN, AND W. LI. Strong conical hull intersection property, bounded linear regularity, Jameson's property (G), and error bounds in convex optimization. *Mathematical Programming* 86 (1999) 135–160.

[45] H.H. BAUSCHKE, J.M. BORWEIN, AND P. TSENG. Bounded linear regularity, strong CHIP, and CHIP are distinct properties. *Journal of Convex Analysis* 7 (2000) 395–412.

[46] M.S. BAZARAA, H.D. SHERALI, AND C.M. SHETTY. *Nonlinear Programming: Theory and Algorithms*, second edition, John Wiley (New York 1993).

[47] E.G. BELOUSOV. *Introduction to Convex Analysis and Integer Programming.* (In Russian). Moscow University Publisher (1977).

[48] E.G. BELOUSOV AND D. KLATTE. A Frank-Wolfe type theorem for convex polynomial programs. *Computational Optimization and Applications* 22 (2002) 37–48.

[49] A. BENSOUSSAN. Points de Nash dans le cas de fontionnelles quadratiques et jeux differentiels linéaires a N personnes. *SIAM Journal on Control* 12 (1974) 460–499.

[50] A. BENSOUSSAN. On the theory of option pricing. *Acta Applicandae Mathematicae* 2 (1984) 139-158.

[51] A. BENSOUSSAN, M. GOURSAT, AND J.L. LIONS. Contrôle impulsionnel et inéquations quasi-variationnelles stationnaires. *Comptes Rendus Academie Sciences Paris* 276 (1973) 1279–1284.

[52] A. BENSOUSSAN AND J.L. LIONS. Nouvelle formulation de problèmes de contrôle impulsionnel et applications. *Comptes Rendus Academie Sciences Paris* 276 (1973) 1189–1192.

[53] A. BENSOUSSAN AND J.L. LIONS. Nouvelles méthodes en contrôle impulsionnel. *Applied Mathematics and Optimization* 1 (1974) 289–312.

[54] A. BEN-TAL AND A. NEMIROVSKII. On polyhedral approximations of the second-order cone. *Mathematics of Operations Research* 26 (2001) 193–205.

[55] C. BERGE. *Topological Spaces*, Oliver and Boyd (Edinburgh 1963).

[56] C. BERGTHALLER AND I. SINGER. The distance to a polyhedron. *Linear Algebra and its Applications* 169 (1992) 111–129.

[57] A. BERMAN. *Cones, Matrices, and Mathematical Programming*. Lecture Notes in Economics and Mathematical Systems 79. Springer-Verlay (Berlin 1979).

[58] D. BERNSTEIN AND S.A. GABRIEL. Solving the nonadditive network equilibrium problem. In P. Pardalos, D.W. Hearn, and W.W. Hager, editors, *Network Optimization*, Lecture Notes in Economics and Mathematical Systems 450, Springer-Verlag (Berlin 1997) pp. 72–102.

[59] D.P. BERTSEKAS. *Nonlinear Programming*, second edition, Athena Scientific (Massachusetts 1999).

[60] D.P. BERTSEKAS. A note on error bounds for convex and nonconvex programs. *Computational Optimization and Applications* 12 (1999) 41–52.

[61] D.P. BERTSEKAS AND E. GAFNI. Projection methods for variational inequalities with application to the traffic assignment problem. *Mathematical Programming Study* 12 (1982) 139–159.

[62] D.P. BERTSEKAS AND J.N. TSITSIKLIS. *Parallel and Distributed Computation, Numerical Methods*, Athena Scientific (Massachusetts 1997).

[63] D.N. BESSIS AND F.H. CLARKE. Partial subdifferentials, derivatives and Rademacher's Theorem. *Transactions of the American Mathematical Society* 351 (1999) 2899–2926.

[64] R. BHATIA. *Matrix Analysis*, Springer-Verlag (New York 1997).

[65] E. BIERSTONE AND P.D. MILMAN. Semianalytic and subanalytic sets. *Institut des Hautes Etudes Scientifiques, Publications Mathématiques* 67 (1988) 5–42.

[66] S.C. BILLUPS AND M.C. FERRIS. Solutions to affine generalized equations using proximal mappings. *Mathematics of Operations Research* 24 (1999) 219–236.

[67] S.C. BILLUPS AND K.G. MURTY. Complementarity problems. *Journal of Computational and Applied Mathematics* 124 (2000) 303–318.

[68] G. BJÖRKMAN. The solution of large displacement frictionless contact problems using a sequence of linear complementarity problems. *International Journal for Numerical Methods in Engineering* 31 (1991) 1553–1566.

[69] G. BJÖRKMAN. Path following and critical points for contact problems. *Computational Mechanics* 10 (1992) 231–246.

[70] G. BJÖRKMAN, A. KLARBRING, T. LARSSON, M. RÖNNQVIST, AND B. SJÖDIN. Sequential quadratic programming for non-linear elastic contact problems. In J. Herskovits, editor, *Structural Optimization 93, The World Congress on Optimal Design of Structural Systems* 2 (1993) pp. 301–308.

[71] F. BLACK AND M. SCHOLES. The pricing of options and corporate liabilities. *Journal of Political Economy* 81 (1973) 647–659.

[72] J.F. BONNANS. Local study of Newton type algorithms for constrained problems. In S. Dolecki, editor, Lecture Notes in Mathematics 1405, Springer-Verlag (Berlin 1989) pp. 13–24.

[73] J.F. BONNANS. Rates of convergence of Newton type methods for variational inequalities and nonlinear programming. Manuscript, INRIA (1990).

[74] J.F. BONNANS. Local analysis of Newton-type methods for variational inequalities and nonlinear programming. *Applied Mathematics and Optimization* 29 (1994) 161–186.

[75] J.F. BONNANS, R. COMINETTI, AND A. SHAPIRO. Sensitivity analysis of optimization problems under second order regularity constraints. *Mathematics of Operations Research* 23 (1998) 806–831.

[76] J.F. BONNANS, R. COMINETTI, AND A. SHAPIRO. Second order optimality conditions based on parabolic second order tangent sets. *SIAM Journal on Optimization* 9 (1999) 466–493.

[77] J.F. BONNANS AND A. SHAPIRO. Sensitivity analysis of parametrized programs under cone constraints. *SIAM Journal on Control and Optimization* 30 (1992) 1409–1422.

[78] J.F. BONNANS AND A. SHAPIRO. Optimization problems with perturbations. A guided tour. *SIAM Review* 40 (1998) 202–227.

[79] J.F. BONNANS AND A. SHAPIRO. *Perturbation Analysis of Optimization Problems*, Springer-Verlag (New York 2000).

[80] J.F. BONNANS AND A. SULEM. Pseudopower expansion of solutions of generalized equations and constrained optimization problems. *Mathematical Programming* 70 (1995) 123–148.

[81] K.C. BORDER. *Fixed Point Theorems with Applications to Economics and Game Theory*. Cambridge University Press (Cambridge 1985).

[82] J.M. BORWEIN. Stability and regular points of inequality systems. *Journal of Optimization Theory and Applications* 48 (1986) 9–52.

[83] J.M. BORWEIN AND M.A.H. DEMPSTER. The linear order complementarity problem. *Mathematics of Operations Research* 14 (1989) 534–558.

[84] J.M. BORWEIN AND A.S. LEWIS. *Convex Analysis and Nonlinear Optimization. Theory and Examples*, CMS Books in Mathematics/Ouvrages de Mathmatiques de la SMC, 3. Springer-Verlag (New York 2000).

[85] M.J. BRENNAN AND E.S. SCHWARTZ. Valuation of American put options. *Journal of Finance* 32 (1977) 449–462.

[86] M. BROADIE AND J. DETEMPLE. American option valuation: New bounds, approximations, and a comparison of existing methods. *Review of Financial Studies* 9 (1996) 1211–1250.

[87] M. BROADIE AND J. DETEMPLE. The valuation of American options on multiple assets. *Mathematical Finance* 7 (1997) 241–286.

[88] M. BROADIE AND J. DETEMPLE. Recent advances in numerical methods for pricing derivative securities. In L.C.G. Rogers and D. Talay, editors, *Numerical Methods in Finance*, Cambridge University Press (1997) pp. 43-66.

[89] B. BROGLIATO. *Nonsmooth Impact Mechanics. Models, Dynamics and Control*, Springer-Verlag (London 1999).

[90] B. BROGLIATO, A.A. TEN DAM, L. PAOLI, AND M. ABADIE. Numerical simulation of finite dimensional multibody nonsmooth mechanical systems. Manuscript, Laboratoire d'Automatique de Grenoble (May 2000).

[91] C.G. BROYDEN. On degeneracy in linear complementarity problems. *Linear Algebra and its Applications* 143 (1991) 99–110.

[92] F. BROWDER. The solvability of nonlinear functional equations. *Duke Mathematical Journal* 33 (1963) 557–567.

[93] S. BURER, R.D.C. MONTEIRO, AND Y. ZHANG. Solving a class of semidefinite programs via nonlinear program. *Mathematical Programming* 93 (2002) 97–122.

[94] S. BURER, R.D.C. MONTEIRO, AND Y. ZHANG. Interior-point algorithms for semidefinite programming based on a nonlinear programming formulation. *Computational Optimization and Applications* 22 (2002) 49–79.

[95] J.V. BURKE. On the identification of active constraints II: The nonconvex case. *SIAM Journal on Numerical Analysis* 27 (1990) 1081–1102.

[96] J.V. BURKE. Calmness and exact penalization. *SIAM Journal on Control and Optimization* 29 (1991) 493–497.

[97] J.V. BURKE. An exact penalization viewpoint of constrained optimization. *SIAM Journal on Control and Optimization* 29 (1991) 968–998.

[98] J.V. BURKE AND M.C. FERRIS. Characterization of solution sets of convex programs. *Operations Research Letters* 10 (1991) 57–60.

[99] J.V. BURKE AND M.C. FERRIS. Weak sharp minima in mathematical programming. *SIAM Journal on Control and Optimization* 31 (1993) 1340–1359.

[100] J.V. BURKE AND J.J. MORÉ. On the identification of active constraints. *SIAM Journal on Numerical Analysis* 25 (1988) 1197–1211.

[101] J.V. BURKE AND J.J. MORÉ. Exposing constraints. *SIAM Journal on Optimization* 4 (1994) 573–595.

[102] J.V. BURKE AND P. TSENG. A unified analysis of Hoffman's bound via Fenchel duality. *SIAM Journal on Optimization* 6 (1996) 265–282.

[103] P.H. CALAMAI AND J.J. MORÉ. Projected gradient method for linearly constrained problems. *Mathematical Programming* 39 (1987) 93–116.

[104] M. CAO AND M.C. FERRIS. Lineality removal for copositive-plus normal maps. *Communications on Applied Nonlinear Analysis* 2 (1995) 1–10.

[105] M. CAO AND M.C. FERRIS. A pivotal method for affine variational inequalities. *Mathematics of Operations Research* 21 (1996) 44-64.

[106] M. CAO AND M.C. FERRIS. P_c-matrices and the linear complementarity problem. *Linear Algebra and its Applications* 246 (1996) 231–249.

[107] I. CAPUZZO DOLCETTA. Sistemi di complementarità e disequazioni variazionali. Ph.D. thesis, Department of Mathematics, University of Rome (1972).

[108] I. CAPUZZO DOLCETTA AND U. MOSCO. Implicit complementarity problems and quasi-variational inequalities. In: R.W. Cottle, F. Giannessi, and J.L. Lions, editors, *Variational Inequalities and Complementarity Problems: Theory and Applications*, John Wiley (New York 1980) pp. 75-87.

[109] J. CARDELL, C.C. HITT, AND W.W. HOGAN. Market power and strategic interaction in electricity networks. *Resource and Energy Economics* 19 (1997) 109–137.

[110] M. CAREY. Integrability and mathematical programming models: a survey and parametric approach. *Econometrica* 45 (1977) 1957–1976.

[111] Y. CENSOR, A.N. IUSEM, AND S.A. ZENIOS. An interior point method with Bregman functions for the variational inequality problem with paramonotone operators. *Mathematical Programming* 81 (1998) 373–400.

[112] D. CHAN AND J.S. PANG. The generalized quasi-variational inequality problem. *Mathematics of Operations Research* 7 (1982) 211–222.

[113] D. CHAN AND J.S. PANG. Iterative methods for variational and complementarity problems. *Mathematical Programming* 24 (1982) 284–313.

[114] R. CHANDRASEKARAN, S.N. KABADI, AND K.G. MURTY. Some NP-complete problems in linear programming. *Operations Research Letters* 1 (1982) 101–104.

[115] R.W. CHANEY. Piecewise C^k functions in nonsmooth analysis. *Nonlinear Analysis, Theory, Methods and Applications* 15 (1990) 649–660.

[116] G.S. CHAO AND T.L. FRIESZ. Spatial price equilibrium sensitivity analysis. *Transportation Research* 18B (1984) 423–440.

[117] B. CHEN. Error bounds for R_0-type and monotone nonlinear complementarity problems. *Journal of Optimization Theory and Applications* 108 (2001) 297–316.

[118] B. CHEN, X. CHEN, AND C. KANZOW. A penalized Fischer-Burmeister NCP functions. *Mathematical Programming* 88 (2000) 211–216.

[119] X. CHEN, D. SUN, AND J. SUN. Smoothing Newton's methods and numerical solution to second order cone complementarity problems. Technical report, Department of Decision Sciences, National University of Singapore (2001).

[120] X. CHEN AND P. TSENG. Non-interior continuation methods for solving semidefinite complementarity problems. *Mathematical Programming*, forthcoming.

[121] M.J. CHIEN AND E.S. KUH. Solving piecewise linear equations for resistive networks. *Circuit Theory and Applications* 3 (1976) 3–24.

[122] S.C. CHOI, W.S. DESARBO, AND P.T. HARKER. Product positioning under price competition. *Management Science* 36 (1990) 175–199.

[123] C.C. CHOU, K.F. NG, AND J.S. PANG. Minimizing and stationary sequences of optimization problems. *SIAM Journal on Control and Optimization* 36 (1998) 1908–1936.

[124] P.W. CHRISTENSEN. A semismooth Newton method for elastoplastic contact problems. *International Journal of Solids and Structures* 39 (2002) 2323–2341.

[125] P.W. CHRISTENSEN, A. KLARBRING, J.S. PANG, AND N. STROMBERG. Formulation and comparison of algorithms for frictional contact problems. *International Journal for Numerical Methods in Engineering* 42 (1998) 145–173.

[126] P.W. CHRISTENSEN AND J.S. PANG. Frictional contact algorithms based on semismooth Newton methods. In M. Fukushima and L. Qi, editors, *Reformulation: Nonsmooth, Piecewise Smooth, Semismooth and Smoothing Methods*, Kluwer Academic Publishers (Dordrecht 1999) pp. 81–116.

[127] F.H. CLARKE. *Optimization and Nonsmooth Analysis*, John Wiley (New York 1983).

[128] G. COHEN. Auxiliary problem principle extended to variational inequalities. *Journal of Optimization Theory and Applications* 59 (1988) 325–333.

[129] G. COHEN AND F. CHAPLAIS. *Journal of Optimization Theory and Applications* (1988).

[130] M.Z. COHEN AND G. MAIER, editors. *Engineering Plasticity by Mathematical Programming*. NATO Advanced Study Institute, Pregamon Press (New York 1979).

[131] T.F. COLEMAN, Y. LI, AND A. VERMA. A Newton method for American option pricing. Technical Report CT9907, Cornell Theory Center, Cornell University (New York 1999).

[132] T.F. COLEMAN, Y. LI, AND A. VERMA. Reconstructing the unknown local volatility function. *The Journal of Computational Finance* 2 (1999) 77–102.

[133] C. COMI AND G. MAIER. Extremum theorem and convergence criterion for an iterative solution to the finite-step problem in elastoplasticity with mixed nonlinear hardening. *European Journal Mechanics A/Solids* 9 (1990) 563–585.

[134] A.P. COSTA, I.N. FIGUEIREDO, J.A.C. MARTINS, AND J. JÚDICE. A complementarity eigenproblem in the stability analysis of finite-dimensional elastic systems with frictional contact. In M.C. Ferris, J.S. Pang, and O.L. Mangasarian, editors. *Complementarity: Applications, Algorithms, and Extensions*, Kluwer Academic Publishers (New York 2001) pp. 67–83.

[135] R.W. COTTLE. *Nonlinear Programs with Positively Bounded Jacobians.* Ph.D. thesis, Department of Mathematics, University of California, Berkeley (1964).

[136] R.W. COTTLE. Note on a fundamental theorem in quadratic programming. *Journal of the Society of Industrial and Applied Mathematics.* 12 (1964) 663–665.

[137] R.W. COTTLE. Nonlinear programs with positively bounded Jacobians. *Journal of the Society for Industrial and Applied Mathematics* 14 (1966) 147–158.

[138] R.W. COTTLE. Complementarity and variational problems. *Symposia Mathematica* 19 (1976) 177–208.

[139] R.W. COTTLE AND G.B. DANTZIG. Complementary pivot theory in mathematical programming. In G.B. Dantzig and A.F. Veinott, Jr., editors, *Mathematics of the Decision Sciences, Part 1.*, American Mathematical Society, Providence (1968) pp. 115–136.

[140] R.W. COTTLE AND G.B. DANTZIG. A generalization of the linear complementarity problem. *Journal of Combinatorial Theory* 8 (1970) 79–90.

[141] R.W. COTTLE, F. GIANNESSI, AND J.L. LIONS, editors. *Variational Inequalities and Complementarity Problems: Theory and Applications*, John Wiley (New York 1980).

[142] R.W. COTTLE, J.S. PANG, AND R.E. STONE. *The Linear Complementarity Problem*, Academic Press (Boston 1992).

[143] R.W. COTTLE, J.S. PANG, AND V. VENKATESWARAN. Sufficient matrices and the linear complementarity problem. *Linear Algebra and its Applications* 114/115 (1989) 231–249.

[144] R.W. COTTLE AND A.F. VEINOTT, JR. Polyhedral sets having a least element. *Mathematical Programming* 3 (1972) 238–249.

[145] G.E. COXSON. The P matrix problem is co-NP complete. *Mathematical Programming* 64 (1994) 173–178.

[146] J.P. CROUZEIX. Pseudomonotone variational inequality problems: existence of solutions. *Mathematical Programming* 78 (1997) 305–314.

[147] J.P. CROUZEIX. Characterizations of generalized convexity and generalized monotonicity, a survey. In J.P. Crouzeix, J.E. Martinez-Legaz, and M. Volle, editors, [Proceedings of the 5th International Symposium on Generalized Convexity held in Luminy, June 17–21, 1996], *Nonconvex Optimization and its Applications* 27, Kluwer Academic Publishers (Dordrecht 1998) pp. 237–256.

[148] J.P. CROUZEIX, P. MARCOTTE, AND D. ZHU. Conditions ensuring the applicability of cutting-plane methods for solving variational inequalities. *Mathematical Programming* 88 (2000) 521–539.

[149] A. CURNIER AND P. ALART. A generalized Newton method for contact problems with friction. *Journal de Mcanique Thorique et Applique, Special Issue : Numerical Method in Mechanics of Contact Involving Friction* (1988) 67–82.

[150] A. CURNIER, Q.C. HE, AND A. KLARBRING. Continuum mechanics modelling of large deformation contact with friction. In M. Raous, M. Jean and J. J. Moreau, editors, *Contact Mechanics*, Plenum Press (New York 1995) 145–158.

[151] S.C. DAFERMOS. Traffic equilibrium and variational inequalities. *Transportation Science* 14 (1980) 42–54.

[152] S.C. DAFERMOS. Sensitivity analysis in variational inequalities. *Mathematics of Operations Research* 13 (1988) 421–434.

[153] S.C. DAFERMOS. Exchange price equilibria and variational inequalities. *Mathematical Programming* 46 (1990) 391–402.

[154] S.C. DAFERMOS AND S.C. MCKELVEY. Partitionable variational inequalities with applications to network and economic equilibria. *Journal of Optimization Theory and Applications* 73 (1992) 243–268.

[155] S.C. DAFERMOS AND A. NAGURNEY. General spatial economic equilibrium problem. *Operations Research* 32 (1984) 1069–1086.

[156] S.C. DAFERMOS AND A. NAGURNEY. Sensitivity analysis for the asymmetric network equilibrium problem. *Mathematical Programming* 28 (1984) 174–184.

[157] G.B. DANTZIG AND A.S. MANNE. A complementarity algorithm for an optimal capital path with invariant proportions. *Journal of Economic Theory* 9 (1974) 312-323.

[158] M.H.A. DAVIS AND T. ZARIPHOPOULOU. American options and transaction fees. In M.H.A. Davis, D. Duffie, W.H. Fleming, and S.E. Shreve, editors, *Mathematical Finance*, IMA Volumes in Mathematics and its Applications, # 165, Springer-Verlag (Berlin 1995) pp. 47–61.

[159] O. DAXHELET AND Y. SMEERS. Variational inequality models of restructured electricity systems. In M.C. Ferris, O.L. Mangasarian, and J.S. Pang, editors, *Complementarity: Applications, Algorithms and Extensions*, Kluwer Academic Publishers (Dordrecht 2001) pp. 85–120.

[160] J.C. DE BREMAECKER, M.C. FERRIS, AND D. RALPH. Compressional fractures considered as contact problems and mixed complementarity problems. *Engineering Fracture Mechanics* 66 (2000) 287–303.

[161] M. DE LUCA AND A. MAUGERI. Quasi-variational inequalities and applications to the traffic equilibrium problem–Discussion of a paradox. *Journal of Computational and Applied Mathematics* 28 (1989) 163–171.

[162] T. DE LUCA, F. FACCHINEI, AND C. KANZOW. A semismooth equation approach to the solution of nonlinear complementarity problems. *Mathematical Programming* 75 (1996) 407–439.

[163] B. DE MOOR, L. VANDENBERGHE, AND J. VANDEWALLE. The generalized linear complementarity problem and an algorithm to find all its solutions. *Mathematical Programming* 57 (1992) 415–426.

[164] B. DE SCHUTTER AND B. DE MOOR. The extended linear complementarity problem. *Mathematical Programming* 71 (1995) 289–325.

[165] B. DE SCHUTTER AND B. DE MOOR. Minimal realization in the max algebra is an extended linear complementarity problem. *Systems and Control Letters* 25 (1995) 103–111.

[166] B. DE SCHUTTER AND B. DE MOOR. The linear dynamic complementarity problem is a special case of the extended linear complementarity problem. *Systems and Control Letters* 34 (1998) 63–75.

[167] G. DEBREU. *Theory of Values*, Yale University Press (New Haven 1959).

[168] J.P. DEDIEU. Penalty functions in subanalytic optimization. *Optimization* 26 (1992) 27–32.

[169] J.P. DEDIEU. Approximate solutions of analytic inequality systems. *SIAM Journal on Optimization* 11 (2000) 411–425.

[170] K. DEIMLING. *Nonlinear Functional Analysis*, Springer-Verlag (Berlin 1985).

[171] S. DEMPE. Directional differentiability of optimal solutions under Slater's condition. *Mathematical Programming* 59 (1993) 49-69.

[172] M.A.H. DEMPSTER AND J.P. HUTTON. Fast numerical valuation of American, exotic and complex options. *Applied Mathematical Finance* 4 (1997) 1–20.

[173] M.A.H. DEMPSTER AND J.P. HUTTON. Pricing American stock options by linear programming. *Mathematical Finance* 9 (1999) 229–254.

[174] V.F. DEMYANOV, G. DI PILLO, AND F. FACCHINEI. Exact penalization via Dini and Hadamard conditional derivatives. *Optimization Methods and Software* 9 (1998) pp. 19–36.

[175] M. DENAULT AND J.L. GOFFIN. On a primal-dual analytic center cutting plane method for variational inequalities. *Computational Optimization and Applications* 12 (1999) 127–155.

[176] S. DENG. Computable error bounds for convex inequality systems in reflexive Banach spaces. *SIAM Journal on Optimization* 7 (1997) 274–279.

[177] S. DENG. Global error bounds for convex inequality systems in Banach spaces. *SIAM Journal on Control and Optimization* 36 (1998) 1240–1249.

[178] S. DENG. Well-posed problems and error bounds in optimization. In M. Fukushima and L. Qi, editors, *Reformulation: Nonsmooth, Piecewise Smooth, Semismooth and Smoothing Methods*, Kluwer Academic Publishers (Dordrecht 1999) pp. 117–126.

[179] S. DENG AND H. HU Computable error bounds for semidefinite programming. *Journal of Global Optimization* 14 (1999) 105–115.

[180] K. DERVIS, J. DE MELO, AND S. ROBINSON. *General Equilibrium Models of Development Policy*. Cambridge University Press (Cambridge 1982).

[181] F. DEUTSCH, W. LI, AND J. SWETITS. Fenchel duality and the strong conical hull intersection property. *Journal of Optimization Theory and Applications* 102 (1999) 681–695.

[182] J.N. DEWYNNE, A.E. WHALLEY, AND P. WILMOTT. Path-dependent options and transaction costs. *Philosophical Transactions of the Royal Society of London, A* 347 (1994) 517–529.

[183] J.N. DEWYNNE AND P. WILMOTT. Asian options as linear complementarity problems: analysis and finite-difference solutions. *Advances in Futures and Options Research* 8 (1995) 145–173.

[184] G. DI PILLO AND L. GRIPPO. Exact penalty functions in constrained optimization. *SIAM Journal on Control and Optimization* 27 (1989) 1333–1360.

[185] H. DIETRICH. A smooth dual gap function solution to a class of quasivariational inequalities. *Journal of Mathematical Analysis and Applications* 235 (1999) 380–393.

[186] S.P. DIRKSE AND M.C. FERRIS. MCPLIB A collection of nonlinear mixed complementarity problems. *Optimization Methods and Software* 5 (1995) 319-345.

[187] S.P. DIRKSE AND M.C. FERRIS. The PATH solver: A non-monotone stabilization scheme for mixed complementarity problems. *Optimization Methods and Software* 5 (1995) 123–156.

[188] S.P. DIRKSE AND M.C. FERRIS. A pathsearch damped Newton method for computing general equilibria. *Annals of Operations Research* 68 (1996) 211-232.

[189] A.L. DONTCHEV. Implicit function theorems for generalized equations. *Mathematical Programming* 70 (1995) 91–106.

[190] A.L. DONTCHEV. Characterizations of Lipschitz stability in optimization. In R. Lucchetti and J. Revalski, editors, *Recent Developments in Well-Posed Variational Problems.* Kluwer Academic Publishers (Dordrecht 1995) pp. 95–115.

[191] A.L. DONTCHEV. A proof of the necessity of linear independence condition and strong second-order sufficient optimality condition for Lipschitzian stability in nonlinear programming. *Journal of Optimization Theory and Applications* 98 (1998) 467–473.

[192] A.L. DONTCHEV AND W.W. HAGER. On Robinson's implicit function theorems. *Set-Valued Analysis and Differential Inclusions (Pamporovo, 1990)*, Progress in Systems and Control Theory 16, Birkhuser (Boston 1993) pp. 75–92.

[193] A.L. DONTCHEV AND W.W. HAGER. Implicit functions, Lipschitz maps, and stability in optimization. *Mathematics of Operations Research* 19 (1994) 753–768.

[194] A.L. DONTCHEV AND R.T. ROCKAFELLAR. Characterizations of strong regularity for variational inequalities over polyhedral convex sets. *SIAM Journal on Optimization* 6 (1996) 1087–1105.

[195] A.L. DONTCHEV AND R.T. ROCKAFELLAR. Characterizations of Lipschitzian stability in nonlinear programming. *Mathematical Programming with Data Perturbations.* Lecture Notes in Pure and Applied Mathematics 195, Dekker (New York 1998) pp. 65–82.

[196] A.L. DONTCHEV AND T. ZOLEZZI. *Well-Posed Optimization Problems.* Lecture Notes in Mathematics 1543, Springer-Verlag (Berlin 1993).

[197] J. DUGUNDJI. *Topology*, Allyn and Bacon (Boston 1966).

[198] J.C. DUNN. On recursive averaging processes and Hilbert space extensions of the contraction mapping principle. *Journal of the Franklin Institute* 295 (1973) 117–133.

[199] J.C. DUNN. Convexity, monotonicity, and gradient processes in Hilbert spaces. *Journal of Mathematical Analysis and Applications* 53 (1976) 145–158.

[200] J.C. DUNN. On the convergence of projected gradient processes to a singular critical points. *Journal of Optimization Theory and Applications* 56 (1987) 203–216.

[201] J.C. DUNN. A projected Newton method for minimization problems with nonlinear inequality constraints. *Numerische Mathematik* 53 (1988) 377–410.

[202] B.C. EAVES. On the basic theorem of complementarity. *Mathematical Programming* 1 (1971) 68–75.

[203] B.C. EAVES. Computing Kakutani fixed points. *SIAM Journal of Applied Mathematics* 21 (1971) 236–244.

[204] B.C. EAVES. The linear complementarity problem. *Management Science* 17 (1971) 612–634.

[205] B.C. EAVES. Homotopies for computation of fixed points. *Mathematical Programming* 3 (1972) 1–22.

[206] B.C. EAVES. A short course in solving equations with PL homotopies. In R.W. Cottle and C.E. Lemke, editors, *Nonlinear Programming.* SIAM-AMS Proceedings 9, American Mathematical Society (Providence 1976) pp. 73–143.

[207] B.C. EAVES. Computing stationary points. *Mathematical Programming Study* 7 (1978) 1–14.

[208] B.C. EAVES. Computing stationary points, again. In O.L. Mangasarian, R.R. Meyer, and S.M. Robinson, editors, *Nonlinear Programming 3*, Academic Press (New York 1978) pp. 391-405.

[209] B.C. EAVES. Where solving for stationary points by LCPs is mixing Newton iterates. In B.C. Eaves, F.J. Gould, H.O. Peitgen, and M.J. Todd, editors, *Homotopy Methods and Global Convergence.* Plenum Press (New York 1983) pp. 63–78.

[210] B.C. EAVES. Thoughts on computing market equilibrium with SLCP. *The Computation and Modelling of Economic Equilibria (Tilburg, 1985)*, Contributions to Economics Analysis 167, North-Holland (Amsterdam 1987) pp. 1–17.

[211] B.C. EAVES, F.J. GOULD, H.O. PEITGEN, AND M.J. TODD, editors, *Homotopy Methods and Global Convergence.* Plenum Press (New York 1983).

[212] B.C. EAVES AND U. ROTHBLUM. Relationships of properties of piecewise affine maps over ordered fields. *Linear Algebra and its Applications* 132 (1990) 1–63.

[213] B.C. EAVES AND K. SCHMEDDERS. General equilibrium models and homotopy methods. *Journal of Economic Dynamics and Control* 23 (1999) 1249–1279.

[214] J. ECKSTEIN AND D.P. BERTSEKAS. On the Douglas-Rachford splitting method and the proximal point algorithm for maximal monotone operators. *Mathematical Programming* 55 (1992) 293–318.

[215] S. EILENBERG AND D. MONTGOMERY. Fixed point theorems for multivalued transformations. *American Journal of Mathematics* 68 (1946) 214–222.

[216] I. EKELAND. On the variational principle. *Journal of Mathematical Analysis and Applications* 47 (1974) 324–354.

[217] I. EKELAND AND R. TEMAN. *Analyse Convexe et Probl'emes Variationnels.* Dunod (Pairs 1974). [English edition: North-Holland (Amsterdam 1976).]

[218] A.S. EL BAKRY, R.A. TAPIA, AND Y. ZHANG. A study of indicators for identifying zero variables in interior-point methods. *SIAM Review* 36 (1994) 45–72.

[219] N. EL FAROUQ. Pseudomonotone variational inequalities: Convergence of proximal methods. *Journal of Optimization Theory and Applications* 109 (2001) 311–326.

[220] N. EL FAROUQ AND G. COHEN. Progressive regularization of variational inequalities and decomposition algorithms. *Journal of Optimization Theory and Applications* 97 (1998) 407–433.

[221] F. FACCHINEI. Structural and stability properties of P_0 nonlinear complementarity problems. *Mathematics of Operations Research* 23 (1998) 735–745.

[222] F. FACCHINEI, A. FISCHER, AND C. KANZOW. On the accurate identification of active constraints. *SIAM Journal on Optimization* 9 (1998) 14–32.

[223] F. FACCHINEI, A. FISCHER, AND C. KANZOW. On the identification of zero variables in an interior-point framework. *SIAM Journal on Optimization* 10 (2000) 1058–1078.

[224] F. FACCHINEI AND C. KANZOW. Beyond monotonicity in regularization methods for nonlinear complementarity problems. *SIAM Journal on Control and Optimization* 37 (1999) 1150–1161.

[225] F. FACCHINEI AND S. LUCIDI. Convergence to second order stationary points in inequality constrained optimization. *Mathematics of Operations Research* 23 (1998) 746–766.

[226] F. FACCHINEI AND J.S. PANG. Total stability of variational inequalities. Technical report 09-98, Dipartimento di Informatica e Sistemistica, Università Degli Studi di Roma "La Sapienza" (September 1998).

[227] K. FAN. Fixed-point and minimax theorems in locally convex linear spaces. *Proceedings of the National Academy of Sciences U.S.A.* 38 (1952) 121–126.

[228] S.C. FANG. An iterative method for generalized complementarity problems. *IEEE Transactions on Automatic Control* AC-25 (1980) 1225–1227.

[229] S.C. FANG. Traffic equilibria on multiclass user transportation networks analyzed via variational inequalities. *Tamkang Journal of Mathematics* 13 (1982) 1–9.

[230] S.C. FANG. Fixed point models for the equilibrium problems on transportation networks. *Tamkang Journal of Mathematics* 13 (1982) 181–191.

[231] S.C. FANG. A linearization method for generalized complementarity problems. *IEEE Transactions on Automatic Control* AC29 (1984) 930–933.

[232] S.C. FANG AND E.L. PETERSON. Generalized variational inequalities. *Journal of Optimization Theory and Application* 38 (1982) 363–383.

[233] S.C. FANG AND E.L. PETERSON. General network equilibrium analysis. *International Journal of Systems Sciences* 14 (1983) 1249–1257.

[234] S.C. FANG AND E.L. PETERSON. An economic equilibrium model on a multicommodity network. *International Journal of Systems Sciences* 16 (1985) 479–490.

[235] U. FARAUT AND A. KORÁNYI. *Analysis on Symmetric Cones*, Oxford University Press (New York 1994).

[236] M.C. FERRIS. *Weak sharp minima and penalty functions in nonlinear programming.* Ph.D. thesis, Churchill College, Cambridge University, United Kingdom (Cambridge 1989).

[237] M.C. FERRIS. Finite termination of the proximal point algorithm. *Mathematical Programming* 50 (1991) 359–366.

[238] M.C. FERRIS AND O.L. MANGASARIAN. Minimum principle sufficiency. *Mathematical Programming* 57 (1992) 1–14.

[239] M.C. FERRIS AND O.L. MANGASARIAN. Error bounds and strong upper semicontinuity for monotone affine variational inequalities. *Annals of Operations Research* 47 (1993) 293–305.

[240] M.C. FERRIS, O.L. MANGASARIAN, AND J.S. PANG, editors. *Complementarity: Applications, Algorithms and Extensions*, Kluwer Academic Publishers (Dordrecht 2001).

[241] M.C. FERRIS, A. MEERAUS, AND T.F. RUTHERFORD. Computing Wardropian equilibria in a complementarity framework. *Optimization Methods and Software* 10 (1999) 669–685.

[242] M.C. FERRIS AND J.S. PANG. Nondegenerate solutions and related concepts in affine variational inequalities. *SIAM Journal on Control and Optimization* 34 (1996) 244–263.

[243] M.C. FERRIS AND J.S. PANG. Engineering and economic applications of complementarity problems. *SIAM Review* 39 (1997) 669–713.

[244] M.C. FERRIS AND J.S. PANG, editors. *Complementarity and Variational Problems: State of the Art*, SIAM Publications (Philadelphia 1997).

[245] A.V. FIACCO. *Introduction to Sensitivity and Stability Analysis in Nonlinear Programming.* Academic Press (New York 1983).

[246] A.V. Fiacco and G.P. McCormick. *Nonlinear Programming: Sequential Unconstrained Minimization Techniques.* John Wiley (New York 1968). [Reprinted as SIAM Classics in Applied Mathematics 4 (Philadelphia 1990).]

[247] A. Fischer. A special Newton-type optimization method. *Optimization* 24 (1992) 269–284.

[248] A. Fischer. A Newton-type method for positive-semidefinite linear complementarity problems. *Journal of Optimization Theory and Applications* 86 (1995) 585–608.

[249] A. Fischer. On the superlinear convergence of a Newton-type method for LCP under weak conditions. *Optimization Methods and Software* 6 (1995) 83–107.

[250] A. Fischer, V. Jeyakumar, and D.T. Luc. Solution point characterizations and convergence analysis of a descent algorithm for nonsmooth continuous complementarity problems. *Journal of Optimization Theory and Applications* 110 (2001) 493–514.

[251] M.L. Fischer and F.J. Gould. A simplicial algorithm for the nonlinear complementarity problem. *Mathematical Programming* 6 (1974) 281–300.

[252] M.L. Fischer and J.W. Tolle. The nonlinear complementarity problem: existence and determination of solutions. *SIAM Journal of Control and Optimization* 15 (1977) 612–623.

[253] C.S. Fisk and D. Boyce. Alternative variational inequality formulations of the equilibrium-travel choice problem. *Transportation Science* 17 (1983) 454–463.

[254] C.S. Fisk and S. Nguyen. Solution algorithms for network equilibrium models with asymmetric user costs. *Transportation Science* 16 (1982) 316–381.

[255] R. Fletcher. *Practical Methods of Optimization*, second edition, John Wiley (Chichester 1987).

[256] F. Flores-Bazan. Existence theorems for generalized noncoercive equilibrium problems: The quasi-convex case. *SIAM Journal on Optimization* 11 (2001) 675–690.

[257] M. Florian. Mathematical programming applications in national, regional, and urban planning. In M. Iri and K. Tanabe, editors, *Mathematical Programming: Recent Developments and Applications*, Kluwer Academic Publishers (Tokyo 1989) pp. 57–82.

[258] M. Florian and M. Los. A new look at static spatial price equilibrium models. *Regional Science and Urban Economics* 12 (1982) 579–597.

[259] M. Florian and H. Spiess. The convergence of diagonalization algorithms for asymmetric network equilibrium problems. *Transportation Research, Part B* 16 (1982) 477–483.

[260] I. Fonseca and W. Gangbo. *Degree Theory in Analysis and Applications*, Oxford University Press (Oxford 1995).

[261] P.A. Forsyth and K.R. Vetzal. Quadratic convergence of a penalty method for valuing American options. *SIAM Journal on Scientific Computing* 23 (2002) 2095–2122.

[262] M. Frank and P. Wolfe. An algorithm for quadratic programming. *Naval Research Logistics Quarterly* 3 (1956) 95–110.

[263] V.M. Friedman and V.S. Chernina. An iterative process for the solution of the finite-dimensional contact problems. *U.S.S.R. Computational Mathematics and Mathematical Physics* 7 (1967) 210–214.

[264] T.L. FRIESZ. Network equilibrium, design, and aggregation. *Transportation Research* 19A (1985) 413–427.

[265] T.L. FRIESZ, D. BERNSTEIN, T.E. SMITH, R.L. TOBIN, AND B.W. WIE. A variational inequality formulation of the dynamic network user equilibrium problem. *Operations Research* 41 (1993) 179–191.

[266] T.L. FRIESZ, D. BERNSTEIN, AND R. STOUGH. Dynamic systems, variational inequalities and control theoretic models for predicting urban network flows. *Transportation Science* 30 (1996) 14–31.

[267] T.L. FRIESZ AND P.T. HARKER. Freight network equilibrium: a review of the state of the art. In A. Daughety, editor, *Analytical Studies in Transportation Economics*, Cambridge University Press (1985) pp. 161–206.

[268] T.L. FRIESZ, R.L. TOBIN, T. SMITH, AND P.T. HARKER. A nonlinear complementarity formulation and solution procedure for the general derived demand network equilibrium problem. *Journal of Regional Science* 23 (1983) 337–359.

[269] T. FUJISAWA AND E.S. KUH. Piecewise-linear theory of resistive networks. *SIAM Journal on Applied Mathematics* 22 (1972) 307–328.

[270] O. FUJIWARA, S.P. HAN, AND O.L. MANGASARIAN. Local duality of nonlinear programs. *SIAM Journal on Control and Optimization* 22 (1984) 162–169.

[271] M. FUKUSHIMA. The primal Douglas-Rachford splitting algorithm for a class of monotone mappings with application to the traffic equilibrium problem. *Mathematical Programming* 72 (1996) 1–15.

[272] M. FUKUSHIMA, Z.Q. LUO, AND P. TSENG. Smoothing functions for second-order-cone complementarity problems, *SIAM Journal on Optimization* 12 (2002) 436–460.

[273] M. FUKUSHIMA AND J.S. PANG. Minimizing and stationary sequences of merit functions for complementarity problems and variational inequalities. In M.C. Ferris and J.S. Pang, editors, *Complementarity and Variational Problems: State of the Art*, SIAM Publications (Philadelphia 1997) pp. 91–104.

[274] M. FUKUSHIMA AND J.S. PANG. A penalty approach to generalized Nash equilibria, with application to multi-leader-follower games. Manuscript, Department of Mathematical Sciences, The Johns Hopkins University (October 2002).

[275] M. FUKUSHIMA AND L. QI, editors. *Reformulation: Nonsmooth, Piecewise Smooth, Semismooth and Smoothing Methods*, Kluwer Academic Publishers (Dordrecht 1999).

[276] D. GABAY AND H. MOULIN. On the uniqueness and stability of Nash equilibria in noncooperative games. In A. Bensoussan, P. Kleindorfer and C.S. Tapiero, editors, *Applied Stochastic Control in Econometrics and Management Science*, North-Holland (Amsterdam 1980) pp. 271–292.

[277] S.A. GABRIEL AND D. BERNSTEIN. The traffic equilibrium problem with nonadditive path costs. *Transportation Science* 31 (1997) 337–348.

[278] S.A. GABRIEL, A.S. KYDES, AND P. WHITMAN. The National Energy Modeling System: A large-scale energy-economic equilibrium model. *Operations Research* 49 (2001) 14–25.

[279] S.A. GABRIEL AND J.S. PANG. An inexact NE/SQP method for solving the nonlinear complementarity problem. *Computational Optimization and Applications* 1 (1992) 67–91.

[280] J.W. GADDUM. A theorem on convex cones with applications to linear inequalities. *Proceedings of the American Mathematical Society* 3 (1952) 957–960.

[281] E.M. GAFNI AND D.P. BERTSEKAS. Two-metric projection methods for constrained optimization. *SIAM Journal on Control and Optimization* 22 (1984) 936–964.

[282] A. GALANTAI. The theory of Newton's method. *Journal of Computational and Applied Mathematics* 124 (2000) 25–44.

[283] D. GALE AND H. NIKAIDO. The Jacobian matrix and global univalence of mappings. *Mathematische Annalen* 159 (1965) 81–93.

[284] D.Y. GAO. Bi-complementarity and duality: A framework in nonlinear equilibria with applications to contact problem of elastoplastic beam theory. *Journal of Mathematical Analysis and Applications* 221 (1998) 672–697.

[285] C.B. GARCIA AND W.I. ZANGWILL. *Pathways to Solutions, Fixed Points, and Equilibria.* Prentice–Hall, Inc. (Englewood Cliffs 1981).

[286] M.R. GAREY AND D.S. JOHNSON. *Computers and Intractability: A Guide to the Theory of NP-Completeness.* W.H. Freeman (San Francisco 1979).

[287] J. GAUVIN. A necessary and sufficient regularity condition to have bounded multipliers in nonconvex programming. *Mathematical Programming* 12 (1977) 136–138.

[288] V. GINSBURGH AND J. WAELBROECK. *Activity Analysis and General Equilibrium Modelling*, North–Holland, Amsterdam (1981).

[289] C. GLOCKER. Formulation of spatial contact situations in rigid multibody systems. *Computer Methods in Applied Mechanics and Engineering* 177 (1999) 199–214.

[290] C. GLOCKER. Spatial friction as standard NCP. *Zeitschrift für Angewandte Mathematik und Mechanik* 81 (2001) S665–S666.

[291] R. GLOWINSKI, J.L. LIONS, AND R. TRÉMOLIÈRES. *Analyse Numérique des Inéquations Variationelles*, volumes 1 and 2, Dunod-Bordas (Paris 1976).

[292] D. GOELEVEN. *Noncoercive Variational Problems and Related Problems*, Addison Wesley Longman Inc. (1996).

[293] J.L. GOFFIN, P. MARCOTTE, AND D.L. ZHU. An analytic center cutting plane method for pseudomonotone variational inequalities. *Operations Research Letters* 20 (1997) 1–6.

[294] A.J. GOLDMAN AND A.W. TUCKER. Theory of linear programming. In H.W. Kuhn and A.W. Tucker, editors, *Linear Inequality and Related Systems*, Princeton University Press (Princeton 1956) 53–97.

[295] B. GOLLAN. On the marginal function in nonlinear programming. *Mathematics of Operations Research* 9 (1984) 208–221.

[296] E.G. GOLSHTEIN AND N.V. TRETYAKOV. *Modified Lagrangians and Monotone Maps in Optimization*, John Wiley (New York 1996). [Translation of *Modified Lagrangian Functions: Theory and Related Optimization Techniques*, Nauka (Moscow 1989).]

[297] M.S. GOWDA. Pseudomonotone and copositive-star matrices. *Linear Algebra and its Applications* 113 (1989) 107–110.

[298] M.S. GOWDA. Complementarity problems over locally compact cones. *SIAM Journal on Control and Optimization* 27 (1989) 836–841.

[299] M.S. GOWDA. On the continuity of the solution map in linear complementarity problems. *SIAM Journal on Optimization* 2 (1992) 619–634.

[300] M.S. GOWDA. Applications of degree theory to linear complementarity problems. *Mathematics of Operations Research* 18 (1993) 868–879.

[301] M.S. GOWDA. An analysis of zero set and global error bound properties of a piecewise affine function via its recession function. *SIAM Journal on Matrix Analysis and Applications* 17 (1996) 594–609.

[302] M.S. GOWDA. Inverse and implicit function theorems for H-differentiable and semismooth functions. Manuscript, Department of Mathematics and Statistics, University of Maryland Baltimore County (November 2000).

[303] M.S. GOWDA AND J.S. PANG. On solution stability of the linear complementarity problem. *Mathematics of Operations Research* 17 (1992) 77–83.

[304] M.S. GOWDA AND J.S. PANG. Some existence results for multivalued complementarity problems. *Mathematics of Operations Research* 17 (1992) 657–669.

[305] M.S. GOWDA AND J.S. PANG. Stability analysis of variational inequalities and nonlinear complementarity problems, via the mixed linear complementarity problem and degree theory. *Mathematics of Operations Research* 14 (1994) 831–879.

[306] M.S. GOWDA AND J.S. PANG. On the boundedness and stability of solutions to the affine variational inequality problem. *SIAM Journal on Control and Optimization* 32 (1994) 421–441.

[307] M.S. GOWDA AND T. PARTHASARATHY. Complementarity forms of theorems of Lyapunov and Stein, and related results. *Linear Algebra and its Applications* 320 (2000) 131–144.

[308] M.S. GOWDA AND G. RAVINDRAN. Algebraic univalence theorems for nonsmooth functions. *Journal of Mathematical Analysis and Applications* 252 (2000) 917–935.

[309] M.S. GOWDA AND T.I. SEIDMAN. Generalized linear complementarity problem. *Mathematical Programming* 46 (1990) 329–340.

[310] M.S. GOWDA AND Y. SONG. On semidefinite linear complementarity problems. *Mathematical Programming* 88 (2000) 575–587.

[311] M.S. GOWDA AND R. SZNAJDER. On the Lipschitzian properties of polyhedral multifunctions. *Mathematical Programming* 74 (1996) 267–278.

[312] M.S. GOWDA AND R. SZNAJDER. On the pseudo-Lipschitzian behavior of the inverse of a piecewise affine function. In M.C. Ferris and J.S. Pang, editors, *Complementarity and Variational Problems: State of the Art*, SIAM Publications (Philadelphia 1997) pp. 117–131.

[313] M.S. GOWDA AND R. SZNAJDER. Weak univalence and connectedness of inverse images of continuous functions. *Mathematics of Operations Research* 24 (1999) 255–261.

[314] M.S. GOWDA AND M.A. TAWHID. Existence and limiting behavior of trajectories associated with P_0-equations. *Computational Optimization and Applications* 12 (1999) 229–251.

[315] B. GRUNBAUM. *Convex Polytopes*. John Wiley & Sons, Inc. (New York 1967).

[316] J. GUDDAT AND F.G. VASQUEZ, WITH H.TH. JONGEN. *Parametric Optimization: Singularities, Pathfollowing and Jumps*, John Wiley (Chichester 1990).

[317] O. GÜLER, A.J. HOFFMAN, AND U.G. ROTHBLUM. Approximations to solutions to systems of linear inequalities. *SIAM Journal on Matrix Analysis and Applications* 16 (1995) 688–696.

[318] C.D. HA. Stability of the linear complementarity at a solution point. *Mathematical Programming* 31 (1985) 327–332.

[319] C.D. HA. Application of degree theory in stability of the complementarity problem. *Mathematics of Operations Research* 31 (1987) 327–338.

[320] G.J. HABETLER AND M.M. KOSTREVA. On a direct algorithm for nonlinear complementarity problems. *SIAM Journal on Control and Optimization* 16 (1978) 504–511.

[321] W.W. HAGER. Lipschitz continuity for constrained process. *SIAM Journal on Control and Optimization* 17 (1979) 321–338.

[322] W.W. HAGER AND M.S. GOWDA. Stability in the presence of degeneracy and error estimation. *Mathematical Programming* 85 (1999) 181–192.

[323] A. HAMEL. Remarks to an equivalent formulation of Ekeland's variational principle. *Optimization* 31 (1994) 233–238.

[324] S.P. HAN AND O.L. MANGASARIAN. Exact penalty functions in nonlinear programming. *Mathematical Programming* 17 (1979) 251–269.

[325] T. HANSEN AND T.C. KOOPMANS. On the definition and computation of a capital stock invariant under optimization. *Journal of Economic Theory* 5 (1972) 487–523.

[326] A. HARAUX. How to differentiate the projection on a convex set in Hilbert space. Some applications to variational inequalities. *Journal of the Mathematical Society of Japan* 29 (1977) 615–631.

[327] P.T. HARKER. A variational inequality approach for the determination of oligopolistic market equilibrium. *Mathematical Programming* 30 (1984) 105–111.

[328] P.T. HARKER. Alternative models of spatial competition. *Operations Research* 34 (1986) 410–425.

[329] P.T. HARKER. *Predicting Intercity Freight Flows*, VNU Science Press (Utrecht 1987).

[330] P.T. HARKER. Generalized Nash games and quasivariational inequalities. *European Journal of Operations Research* 54 (1991) 81–94.

[331] P.T. HARKER. Lectures on Computation of Equilibria with Equation-Based Methods: Applications to the Analysis of Service Economics and Operations. CORE Lecture Series, Université Catholique de Louvain (Louvain-la-Neuve 1993).

[332] P.T. HARKER AND J.S. PANG. Finite–dimensional variational inequality and nonlinear complementarity problems: A survey of theory, algorithms and applications. *Mathematical Programming, Series B* 48 (1990) 161–220.

[333] P. HARTMAN AND G. STAMPACCHIA. On some nonlinear elliptic differential functional equations. *Acta Mathematica* 115 (1966) 153–188.

[334] E. HAYNSWORTH AND A.J. HOFFMAN. Two remarks on copositive matrices. *Linear Algebra and its Applications* 2 (1969) 387–392.

[335] Y.R. HE. Error bounds for nonlinear complementarity problems. Manuscript, Department of Mathematics, Chinese University of Hong Kong (December 2000).

[336] Y.R. HE Minimizing and stationary sequences of convex constrained minimization problems. *Journal of Optimization Theory and Applications* 111 (2001) 137–153.

[337] D.W. HEARN. The gap function of a convex program. *Operations Research Letters* 1 (1982) 67–71.

[338] D.W. HEARN, S. LAWPHONGPANICH, AND S. NGUYEN. Convex programming formulations of the asymmetric traffic assignment problem. *Transportation Research, Part B* 18 (1984) 357–365.

[339] W.P.H. HEEMELS. Linear Complementarity Systems: A Study in Hybrid Dynamics. Ph.D. thesis, Department of Electrical Engineering, Eindhoven University of Technology (November 1999).

[340] W.P.M.H. HEEMELS, J.M. SCHUMACHER, AND S. WEILAND. The rational complementarity problem. *Linear Algebra and its Applications* 294 (1999) 93–135.

[341] W.P.M.H. HEEMELS, J.M. SCHUMACHER, AND S. WEILAND. Projected dynamical systems in a complementarity formalism. *Operations Research Letters* 27 (2000) 83–91.

[342] W.P.M.H. HEEMELS, J.M. SCHUMACHER, AND S. WEILAND. Linear complementarity systems. *SIAM Journal on Applied Mathematics* 60 (2000) 1234–1269.

[343] J.B. HIRIART-URRUTY AND C. LEMARÉCHAL. *Convex Analysis and Minimization Algorithms I and II*, Springer-Verlag (New York 1993).

[344] H. HIRONAKA. *Introduction to Real-Analytic Sets and Real-Analytic Maps.* Quaderni dei Gruppi di Ricerca Matematica del Consiglio Nazionale delle Ricerche. *Instituto Matematico "L. Tonelli" dell'Università di Pisa* (Pisa 1973).

[345] H. HIRONAKA. Subanalytic sets. *Number Theory, Algebraic Geometry and Commutative Algebra* (1973) 453–393.

[346] I. HLAVÁČEK, J. HASLINGER, J. NEČAS, AND J. LOVÍŠEK. *Solution of Variational Inequalities in Mechanics*, Applied Mathematical Sciences 66, Springer-Verlag, (New York, 1988).

[347] B.F. HOBBS. Linear complementarity models of Nash-Cournot competition in bilateral and POOLCO power markets. *IEEE Transactions on Power Systems* 16 (2001) 194–202.

[348] B.F. HOBBS, C.B. METZLER, AND J.S. PANG. Strategic gaming analysis for electric power networks: An MPEC approach. *IEEE Transactions on Power Systems*, 15 (2000) 638–645.

[349] A.J. HOFFMAN. On approximate solutions of systems of linear inequalities. *Journal of Research of the National Bureau of Standards* 49 (1952) 263–265.

[350] W.W. HOGAN. Energy policy models for project independence. *Computers and Operations Research* 2 (1975) 251.

[351] W.W. HOGAN. Project independence evaluation system: structure and algorithms. In P.D. Lax, editor, *Mathematical Aspects of Production and Distribution of Energy*, Proceedings of Symposia in Applied Mathematics, American Mathematical Society 21 (Providence 1977) pp. 121–137.

[352] T. HOGGARD, A.E. WHALLEY, AND P. WILMOTT. Hedging option portfolios in the presence of transaction costs. *Advances in Futures and Options Research* 7 (1994) 21–35.

[353] R.H.W. HOPPE AND H.D. MITTELMANN. A multi-grid continuation strategy for parameter-dependent variational inequalities. *Journal of Computational and Applied Mathematics* 2 (1989) 35–46.

[354] L. HÖRMANDER. On the division of distributions by polynomials. *Arkiv for Mathematik* 3 (1958) 555–568.

[355] R.A. HORN AND C. JOHNSON. *Matrix Analysis*, Cambridge University Press, (Cambridge 1985).

[356] R.A. Horn and C. Johnson. *Topics in Matrix Analysis*, Cambridge University Press, (Cambridge 1990).

[357] H. Hu. Perturbation analysis of global error bounds for systems of linear inequalities. *Mathematical Programming, Series B* 88 (2000) 277–284.

[358] Y. Hu, H.S. Cheng, T. Arai, Y. Kobayashi, and S. Aoyama. Numerical simulation of piston ring in mixed lubrication–a nonaxisymmetrical analysis. *Journal of Tribology* 116 (1994) 470–478.

[359] J. Huang and J.S. Pang. Option pricing and linear complementarity. *The Journal of Computational Finance* 2 (1998) 31–60.

[360] J. Huang and J.S. Pang. A mathematical programming with equilibrium constraints approach to inverse pricing of American options: the case of an implied volatility surface. *The Journal of Computational Finance* 4 (2000) 21–56.

[361] L.R. Huang, K.F. Ng, and J.P. Penot. On minimizing and critical sequences in nonsmooth optimization *SIAM Journal on Optimization* 10 (2000) 999–1019.

[362] G. Isac. *Complementarity Problems.* Lecture Notes in Mathematics 1528, Springer-Verlag (New York 1992).

[363] G. Isac. A generalization of Karamardian's condition in complementarity theory. *Nonlinear Analysis Forum* 4 (1999) 49–63.

[364] G. Isac. Exceptional families of elements, feasibility and complementarity. *Journal of Optimization Theory and Applications* 104 (2000) 577–588.

[365] G. Isac, G. V. Bulavski, and V. Kalashnikov. Exceptional families, topological degree and complementarity problems. *Journal of Global Optimization* 10 (1997) 207–225.

[366] G. Isac and A. Carbone. Exceptional families of elements for continuous functions: some applications to complementarity theory. *Journal of Global Optimization* 15 (1999) 181–196.

[367] G. Isac and M.M. Kostreva. The generalized order complementarity problem. *Journal of Optimization Theory and Applications* 19 (1991) 227–232.

[368] G. Isac and M.M. Kostreva. The implicit generalized order complementarity problem and Leontief's input-output model. *Applicationes Mathematicae* 24 (1996) 113–125.

[369] V.I. Istratescu. *Fixed Point Theory*, D. Reidel Publishing Company (Boston 1991).

[370] A.F. Izmailov and M.V. Solodov. Error bounds for 2-regular mappings with Lipschitzian derivatives and their applications. *Mathematical Programming* 89 (2001) 413–435.

[371] A.F. Izmailov and M.V. Solodov. Superlinearly convergent algorithms for solving singular equations and smooth reformulations of complementarity problems. *SIAM Journal on Optimization* 13 (2002) 386–405.

[372] A.F. Izmailov and M.V. Solodov. Karush-Kuhn-Tucker systems: regularity conditions, error bounds, and a class of Newton-type methods. *Mathematical Programming*, forthcoming.

[373] P. Jaillet, D. Lamberton, and B. Lapeyre. Variational inequalities and the pricing of American options. *Acta Applicandae Mathematicae* 21 (1990) 263–289.

[374] R. Janin. Directional derivative of the marginal function in nonlinear programming. *Mathematical Programming Study* 21 (1984) 110–126.

[375] B. JANSEN, C. ROOS, T. TERLAKY, AND A. YOSHISE. Polynomiality of primal-dual affine scaling algorithms for nonlinear complementarity problems. *Mathematical Programming* 78 (1997) 315–345.

[376] H. JIANG. Local properties of solutions of nonsmooth variational inequalities. *Optimization* 33 (1995) 119–132.

[377] H. JIANG, M. FUKUSHIMA, L. QI, AND D. SUN. A trust region method for solving generalized complementarity problems. *SIAM Journal on Optimization* 8 (1998) 140–157.

[378] H. JIANG AND L. QI. Local uniqueness and convergence of iterative methods for nonsmooth variational inequalities. *Journal of Mathematical Analysis and Applications* 196 (1995) 314–331.

[379] K. JITTORNTRUM. Solution point differentiability without strict complementarity in nonlinear programming. *Mathematical Programming Study* 21 (1984) 127–138.

[380] L. JOHANSSON AND A. KLARBRING. The rigid punch problem with friction using variational inequalities and linear complementarity. *Mechanics of Structures and Machines* 20 (1992) 293–319.

[381] L. JOHANSSON AND A. KLARBRING. Study of frictional impact using a nonsmooth equations solver. *Journal of Applied Mechanics* 67 (2000) 267–273.

[382] C. JONES AND M.S. GOWDA. On the connectedness of solution sets in linear complementarity problems. *Linear Algebra and its Applications* 272 (1998) 33–44.

[383] P.C. JONES. Computing an optimal invariant capital stock. *SIAM Journal on Algebraic and Discrete Methods* 3 (1982) 145–150.

[384] P.C. JONES, G. MORRISON, J.C. SWARTS, AND E.S. THEISE. Nonlinear spatial price equilibrium algorithms: a computational comparison. *Microcomputers in Civil Engineering* 3 (1988) 265–271.

[385] P.C. JONES, R. SAIGAL, AND M. SCHNEIDER. Computing nonlinear network equilibria. *Mathematical Programming* 31 (1985) 57–66.

[386] H.TH. JONGEN, D. KLATTE, AND K. TAMMER. Implicit functions and sensitivity of stationary points. *Mathematical Programming* 49 (1990) 123–138.

[387] H.TH. JONGEN, T. MOBERT, J. RÜCKMANN, AND K. TAMMER. On inertia and Schur complement in optimization. *Linear Algebra and its Applications* 95 (1987) 97–109.

[388] D.W. JORGENSON AND P.J. WILCOXEN. Reducing US carbon emmissions: An econometric general equilibrium assessment. *Resource and Energy Economics* 15 (1993) 7–25.

[389] N.H. JOSEPHY. Newton's method for generalized equations. Technical Summary Report 1965, Mathematics Research Center, University of Wisconsin (Madison 1979).

[390] N.H. JOSEPHY. Quasi-Newton methods for generalized equations. Technical Summary Report 1966, Mathematics Research Center, University of Wisconsin (Madison 1979).

[391] N.H. JOSEPHY. A Newton method for the PIES energy model. Technical Summary Report 1977, Mathematics Research Center, University of Wisconsin (Madison 1979).

[392] F. JOURDAN, P. ALART, AND M. JEAN. A Gauss-Seidel like algorithm to solve frictional contact problems. *Computer Methods in Applied Mechanics and Engineering* 155 (1998) 31–47.

[393] R.I. KACHUROVSKII. On monotone operators and convex functionals. (In Russian). *Uspekhi Matematicheskikh Nauk (N.S.)* 15 (1960) 213–215.

[394] I. KANEKO. Piecewise linear elastic-plastic analysis. *International Journal for Numerical Methods in Engineering* 14 (1979) 757–767.

[395] I. KANEKO. Complete solution of a class of elastic-plastic structures. *Computer Methods in Applied Mechanics and Engineering* 21 (1980) 193–209.

[396] C. KANZOW AND M. FUKUSHIMA. Equivalence of the generalized complementarity problem to differentiable unconstrained minimization *Journal of Optimization Theory and Applications* 90 (1996) 581–603.

[397] C. KANZOW, N. YAMASHITA, AND M. FUKUSHIMA. New NCP-functions and their properties. *Journal of Optimization Theory and Applications* 94 (1997) 115–135.

[398] S. KARAMARDIAN. The nonlinear complementarity problem with applications, part 1. *Journal of Optimization Theory and Applications* 4 (1969) 87–98.

[399] S. KARAMARDIAN. Generalized complementarity problems. *Journal of Optimization Theory and Applications* 8 (1971) 161–168.

[400] S. KARAMARDIAN. The complementarity problem. *Mathematical Programming* 2 (1972) 107–129.

[401] S. KARAMARDIAN. Complementarity problems over cones with monotone and pseudomonotone maps. *Journal of Optimization Theory and Applications* 18 (1976) 445–454.

[402] S. KARAMARDIAN. Existence theorem for complementarity problem. *Journal of Optimization Theory and Applications* 19 (1976) 227–232.

[403] S. KARAMARDIAN (in collaboration with C.B. Garcia), editor. *Fixed Points: Algorithms and Applications*, Academic Press (New York 1977).

[404] S. KARAMARDIAN AND S. SCHAIBLE. Seven kinds of monotone maps. *Journal of Optimization Theory and Applications* 66 (1990) 37–46.

[405] S. KARAMARDIAN, S. SCHAIBLE, AND J.P. CROUZEIX. Characterizations of generalized monotone maps. *Journal of Optimization Theory and Applications* 76 (1993) 399–413.

[406] I. KARATZAS. On the pricing of American options. *Applied Mathematics and Optimization* 17 (1988) 37–60.

[407] W. KARUSH. Minima of functions of several variables with inequalities as side conditions. Masters Thesis, Department of Mathematics, University of Chicago (1939).

[408] J. KATZENELSON. An algorithm for solving nonlinear resistive network. *Bell System Technical Journal* 44 (1965) 1605–1620.

[409] Y.H. KIM AND B.M. KWAK. Numerical implementation of dynamic contact analysis by a complementarity formulation and application to a valve-cotter system in motorcycle engines. *Mechanics of Structures and Machines* 28 (2000) 281–301.

[410] D. KINDERLEHRER AND G. STAMPACCHIA. *An Introduction to Variational Inequalities and Their Applications*, Academic Press (New York 1980).

[411] A.J. KING AND R.T. ROCKAFELLAR. Sensitivity analysis for nonsmooth generalized equations. *Mathematical Programming* 55 (1992) 193–212.

[412] A. KLARBRING. *Contact problems with friction - using a finite dimensional description and the theory of linear complementarity*, Linköping Studies in Science and Technology, Thesis No. 20, LIU-TEK-LIC-1984:3.

[413] A. KLARBRING. A mathematical programming approach to three dimensional contact problems with friction. *Computer Methods in Applied Mechanics and Engineering* 58 (1986) 175–200.

[414] A. KLARBRING. On discrete and discretized non-linear elastic structures in unilateral contact (stability, uniqueness and variational principles). *International Journal on Solids and Structures* 24 (1988) 459–479.

[415] A. KLARBRING. Derivation and analysis of rate boundary value problems of frictional contact. *European Journal Mechanics A/Solids* 9 (1990) 53–85.

[416] A. KLARBRING. The rigid punch problem in nonlinear elasticity: formulation, variational principles and linearization. *Journal of Technical Physics* 32 (1991) 45–60.

[417] A. KLARBRING. Mathematical programming and augmented Lagrangian methods for frictional contact problems. In A. Curnier, editor, *Proceedings Contact Mechanics International Symposium*, Presse Polytechniques et Universitaires Romandes (Lausanne 1992) pp. 409–422.

[418] A. KLARBRING. Mathematical programming in contact problems. In M. H. Aliabadi and C. A. Brebbia, editors, *Computational Methods for Contact Problems*, Computational Mechanics Publications (Southampton 1993) pp. 233-264.

[419] A. KLARBRING. Large displacement frictional contact: a continuum framework for finite element discretization. *European Journal of Mechanics A/Solids* 14 (1995) 237–253.

[420] A. KLARBRING. On (non-)existence and (non-)uniqueness of solutions. In F. Pfeiffer and C. Glocker, editors, *Frictional Contact Problems*, IUTAM Symposium on Unilateral Multibody Dynamics, Kluwer Academic Publisher (Dordrecht 1999) pp. 157–168.

[421] A. KLARBRING. Contact, friction, discrete mechanical structures and mathematical programming. In P. Wrigers and P. Panagiotopoulos, editors, *New Developments in Contact Problems (CISM Courses and Lectures No 384)*, Springer (Wien 1999) pp. 55-100.

[422] A. KLARBRING AND G. BJÖRKMAN. A mathematical programming approach to contact problems with friction and varying contact surface. *Computers and Structures* 30 (1988) 1185–1198.

[423] A. KLARBRING AND G. BJÖRKMAN. Solution of large displacement contact problems with friction using Newton's method for generalized equations. *International Journal for Numerical Methods in Engineering* 34 (1992) 249–269.

[424] A. KLARBRING, A. MIKELIĆ, AND M. SHILLOR. Frictional contact problems with normal compliance. *International Journal of Engineering Science* 26 (1988) 811–832.

[425] A. KLARBRING AND J.S. PANG. Existence of solutions to discrete semicoercive frictional contact problems. *SIAM Journal on Optimization* 8 (1998) 414–442.

[426] A. KLARBRING AND J.S. PANG. The discrete steady sliding problem. *Zeitschrift für Angewandte Mathematik und Mechanik* 79 (1999) 75–90.

[427] D. KLATTE. On quantitative stability for $C^{1,1}$ programs. In R. Durier and C. Michelot, editors, *Recent Developments in Optimization (Dijon, 1994)*, [Lecture Notes in Economics and Mathematical Systems 429] Springer Verlag (Berlin 1995), pp. 214–230.

[428] D. KLATTE. Lipschitz stability and Hoffman's error bound for convex inequalities. In J. Guddat, H.Th. Jongen, and F. Nožička, G. Still, and F. Twilt, editors. Proceedings of the Conference Parametric Optimization and Related Topics IV (Enschede 1995), *Approximation and Optimization* 9 (Lang 1997) pp. 201–212.

[429] D. KLATTE. Hoffman's error bound for systems of convex inequalities. In A.V. Fiacco, editor, *Mathematical Programming with Data Perturbations*, Marcel Dekker Publishers (New York 1998) pp. 185-199.

[430] D. KLATTE. Upper Lipschitz behavior of solutions to perturbed $C^{1,1}$ programs. *Mathematical Programming* 88 (2000) 285–312.

[431] D. KLATTE AND B. KUMMER. Strong stability in nonlinear programming revisited. *Journal of Australian Mathematical Society B* 40 (1999) 336–352.

[432] D. KLATTE AND B. KUMMER. Generalized Kojima functions and Lipschitz stability of critical points. *Computational Optimization and Applications* 13 (1999) 61–85.

[433] D. KLATTE AND B. KUMMER. *Nonsmooth Equations in Optimization: Regularity, Calculus, Methods and Applications.* Kluwer Academic Publishers (Dordrecht 2002).

[434] D. KLATTE AND W. LI. Asymptotic constraint qualifications and global error bounds for convex inequalities. *Mathematical Programming* 84 (1999) 137–160.

[435] D. KLATTE AND K. TAMMER. Strong stability of stationary solutions and Karush-Kuhn-Tucker points in nonlinear programming. *Annals of Operations Research* 27 (1990) 285–308.

[436] D. KLATTE AND G. THIERE A note of Lipschitz constants for solutions of linear inequalities and equations. *Linear Algebra and its Applications* 244 (1996) 365–374.

[437] M. KOCVARA AND J.V. OUTRATA. On optimization of systems governed by implicit complementarity problems. *Numerical Functional Analysis and Optimization* 15 (1994) 869–887.

[438] M. KOCVARA AND J.V. OUTRATA. On the solution of optimum design problems with variational inequalities. In D. Du, L. Qi, and R. Womersley, editors, *Recent Advances in Nonsmooth Optimization*, World Scientific Publishers (River Ridge 1995) pp. 171–191.

[439] M. KOCVARA AND J.V. OUTRATA. On a class of quasi-variational inequalities. *Optimization Methods and Software* 5 (1995) 275–295.

[440] M. KOJIMA. Computational methods for solving the nonlinear complementarity problem. *Keio Engineering Reports* 27 (1974) 1–41.

[441] M. KOJIMA. A unification of the existence theorems of the nonlinear complementarity problem. *Mathematical Programming* 9 (1975) 257–277.

[442] M. KOJIMA. Studies on piecewise-linear approximations of pieceiwse-C^1 mappings in fixed points and complementarity theory. *Mathematics of Operations Research* 3 (1978) 17–36.

[443] M. KOJIMA. Strongly stable stationary solutions in nonlinear programs. In S.M. Robinson, editor, *Analysis and Computation of Fixed Points*, Academic Press (New York 1980) pp. 93–138.

[444] M. KOJIMA AND R. HIABAYASHI. Continuous deformation of nonlinear programs. Sensitivity, stability, and parametric analysis. *Mathematical Programming Study* 21 (1984) 150–198.

[445] M. Kojima, N. Megiddo, T. Noma, and A. Yoshise. *A Unified Approach to Interior Point Algorithms for Linear Complementarity Problems*, Lecture Notes in Computer Science 538, Springer-Verlag (Berlin 1991).

[446] M. Kojima, H. Nishino, and T. Sekine. An extension of Lemke's method to the piecewise linear complementarity problem. *SIAM Journal on Applied Mathematics* 31 (1976) 600–613.

[447] M. Kojima, A. Okada, and S. Shindoh. Strongly stable equilibrium points of N-person noncooperative games. *Mathematics of Operations Research* 10 (1985) 650–663.

[448] M. Kojima and R. Saigal. A study of PC^1 homeomorphisms on subdivided polyhedra. *SIAM Journal on Mathematical Analysis* 10 (1979) 1299–1312.

[449] M. Kojima and R. Saigal. On the relationship between conditions that insure a PL mapping is a homeomorphism. *Mathematics of Operations Research* 5 (1980) 101–109.

[450] M. Kojima, M. Shida, and S. Shindoh. Reduction of monotone linear complementarity problems over cones to linear programs over cones. *Acta Mathematica Vietnamica* 22 (1997) 147–157.

[451] M. Kojima, M. Shida, and S. Shindoh. Local convergence of predictor-corrector infeasible-interior-point algorithms for SDPs and SDLCPs. *Mathematical Programming* 80 (1998) 129–160.

[452] M. Kojima, M. Shida, and S. Shindoh. A predictor-corrector interior-point algorithm for the semidefinite linear complementarity problem using the Alizadeh-Haeberly-Overton search direction. *SIAM Journal on Optimization* 9 (1999) 444–465.

[453] M. Kojima, M. Shida, and S. Shindoh. Search directions in the SDP and the monotone SDLCP: generalization and inexact computation. *Mathematical Programming* 85 (1999) 51–80.

[454] M. Kojima and S. Shindo. Extensions of Newton and quasi-Newton methods to systems of PC^1 equations. *Journal of Operations Research Society of Japan* 29 (1986) 352–374.

[455] M. Kojima, S. Shindo, and S. Hara. Interior-point methods for the monotone semidefinite linear complementarity problem in symmetric matrices. *SIAM Journal on Optimization* 7 (1997) 86–125.

[456] C.D. Kolstad and L. Mathiesen. Necessary and sufficient conditions for uniqueness of a Cournot equilibrium. *Review of Economic Study* 54 (1987) 681–690.

[457] C.D. Kolstad and L. Mathiesen. Computing Cournot-Nash equilibria. *Operations Research* 39 (1991) 739–748.

[458] M.M. Kostreva. Block pivot methods for solving the complementarity problem. *Linear Algebra and its Applications* 21 (1978) 207–215.

[459] M.M. Kostreva. Elasto-hydrodynamic lubrication: A non-linear complementarity problem. *International Journal for Numerical Methods in Fluids* 4 (1984) 377–397.

[460] M.A. Krasnoselskii. *Positive Solutions of Operator Equations.* [Translated from the Russian by R.E. Flaherty, edited by L.F. Boron]. Noordhoff (Groningen 1964).

[461] M.G. Krein and M.A. Rutman. Linear operators leaving invariant a cone in a Banach space. (In Russian.) *Uspekhi Matematicheskikh Nauk (N.S.)* 3 (1948)

3–95. [English translation: American Mathematical Society Translation Series I 10 (1962) 199–325.]

[462] J.B. KRUSKAL. Two convex counterexamples: a discontinuous envelope function and a non-differentiable nearest point mapping. *Proceedings of the American Mathematical Society* 23 (1969) 697–703.

[463] D. KUHN AND R. LÖWEN. Piecewise affine bijections of $\mathbb{R}^n$, and the equation $Sx^+ - Tx^- = y$. *Linear Algebra and its Applications* 96 (1987) 109–129.

[464] H.W. KUHN. Nonlinear programming: A historical view. In R.W. Cottle and C.E. Lemke, editors, *Nonlinear Programming* SIAM-AMS Proceedings 9, American Mathematical Society (Providence 1976) pp. 1–26.

[465] H.W. KUHN AND A.W. TUCKER. Nonlinear programming. In J. Neyman, editor, Proceedings of the Second Berkeley Symposium on Mathematical Statistics and Probability, University of California Press (Berkeley 1951) pp. 481-492.

[466] B. KUMMER. Newton's method for non-differentiable functions. In J. Guddat, B. Bank, H. Hollatz, P. Kall, D. Klatte, B. Kummer, K. Lommatzsch, K. Tammer, M. Vlach, and K. Zimmermann, editors, *Advances in Mathematical Optimization*. Akademie-Verlag (Berlin 1988) pp. 114–125.

[467] B. KUMMER. The inverse of a Lipschitz function in $\mathbb{R}^n$: Complete characterization by directional derivatives. IIASA working paper 89-084 (1989).

[468] B. KUMMER. Lipschitzian inverse functions, directional derivatives, and applications in $C^{1,1}$-optimization. *Journal of Optimization Theory and Applications* 70 (1991) 559–580.

[469] B. KUMMER. An implicit-function theorem for $C^{0,1}$-equations and parametric $C^{1,1}$-optimization. *Journal of Mathematical Analysis and Applications* 158 (1991) 35–46.

[470] B. KUMMER. Inverse functions of pseudo regular mappings and regularity conditions. *Mathematical Programming* 88 (2000) 313–339.

[471] L. KUNTZ AND S. SCHOLTES. Structural analysis of nonsmooth mappings, inverse functions, and metric projections. *Journal of Mathematical Analysis and Applications* 188 (1994) 346–386.

[472] L. KUNTZ AND S. SCHOLTES. A nonsmooth variant of the Mangasarian-Fromovitz constraint qualification. *Journal of Optimization Theory and Applications* 82 (1994) 59–75.

[473] L. KUNTZ AND S. SCHOLTES. Qualitative aspects of the local approximation of a piecewise differentiable function. *Nonlinear Analysis* 25 (1995) 197–215.

[474] B.M. KWAK. Complementarity problem formulation of three-dimensional frictional contact. *Journal of Applied Mechanics* 58 (1991) 134–140.

[475] B.M. KWAK. Nonlinear complementarity problem formulation of three-dimensional frictional contact and its numerical implementations. *Contact mechanics, III (Madrid, 1997)*, Computational Mechanics (Southampton 1997) pp. 159–169.

[476] B.M. KWAK AND B.C. LEE. A complementarity problem formulation for two-dimensional frictional contact problem. *Computers and Structures* 18 (1988) 469–480.

[477] J. KYPARISIS. On uniqueness of Kuhn-Tucker multipliers in nonlinear programming. *Mathematical Programming* 32 (1985) 242–246.

[478] J. KYPARISIS. Uniqueness and differentiability of solutions of parametric nonlinear complementarity problems. *Mathematical Programming* 36 (1986) 105–113.

[479] J. KYPARISIS. Sensitivity analysis framework for variational inequalities. *Mathematical Programming* 38 (1987) 203–213.

[480] J. KYPARISIS Sensitivity analysis for variational inequalities and nonlinear complementarity problems. *Annals of Operations Research* 27 (1990) 143–174.

[481] J. KYPARISIS. Solution differentiability for variational inequalities. *Mathematical Programming, Series B* 48 (1990) 285–302.

[482] J. KYPARISIS. Parametric variational inequalities with multivalued solution sets. *Mathematics of Operations Research* 17 (1992) 341–364.

[483] J. KYPARISIS AND C.M. IP. Solution behavior for parametric implicit complementarity problems. *Mathematical Programming* 56 (1992) 71–90.

[484] J. KYPARISIS AND Y.P. QIU Solution differentiability for oligopolisitic network equilibria. *Operations Research Letters* 9 (1990) 395–402.

[485] T. LARSSON AND M. PATRIKSSON. A class of gap functions for variational inequalities. *Mathematical Programming* 64 (1994) 63–80.

[486] S. LAWPHONGPANICH AND D.W. HEARN. Simplicial decomposition of the asymmetric traffic assignment problem. *Transportation Research, Part B* 18 (1984) 123–133.

[487] S.S. LEE. A computational method for frictional contact problem using finite element method. *International Journal for Numerical Methods in Engineering* 37 (1994) 217–228.

[488] H.E. LELAND. Option pricing and replication with transaction costs. *Journal of Finance* 40 (1985) 1283–1301.

[489] B. LEMAIRE. The proximal algorithm. In J.P. Penot, editor, *New Methods in Optimization and their Industrial Uses*, Birkhäuser-Verlag (Basel 1989) pp. 73–88.

[490] C.E. LEMKE. Bimatrix equilibrium points and mathematical programming. *Management Science* 11 (1965) 681–689.

[491] C.E. LEMKE AND J.T. HOWSON, JR. Equilibrium points of bimatrix games. *SIAM Journal of Applied Mathematics* 12 (1964) 413–423.

[492] G. LESAJA. *Interior Point Methods for P_*-Complementarity Problems*. Ph.D. thesis, Department of Mathematics, University of Iowa (1996).

[493] G. LESAJA. Long-step homogenous interior-point algorithm for the P_* nonlinear complementarity problems. *Yugoslav Journal of Operations Research* 12 (2002) 17–48.

[494] A.Y.T. LEUNG, G.Q. CHEN, AND W.J. CHEN. Smoothing Newton method for solving two- and three-dimensional frictional contact problems. *International Journal for Numerical Methods in Engineering* 41 (1998) 1001–1027.

[495] E.S. LEVITIN AND B.T. POLYAK. Convergence of minimizing sequence in conditional extremum problems. *Soviet Mathematics Doklady* 7 (1996) 764–767.

[496] A.B. LEVY. Implicit multifunction theorems for the sensitivity analysis of variational conditions. *Mathematical Programming* 74 (1996) 333–350. [Errata: *Mathematical Programming* 86 (1999) 439–441].

[497] A.B. LEVY. Stability of solutions to parameterized nonlinear complementarity problems. *Mathematical Programming* 85 (1999) 397–406.

[498] A.B. LEVY. Solution sensitivity from general principles. *SIAM Journal on Control and Optimization* 40 (2001) 1–38.

[499] A.B. LEVY, R.A. POLIQUIN, AND R.T. ROCKAFELLAR. Stability of locally optimal solutions. *SIAM Journal of Optimization* 10 (2000) 580–604.

[500] A.B. Levy and R.T. Rockafellar. Sensitivity analysis of solutions to generalized equations. *Transactions of the American Mathematical Society* 345 (1994) 661–671.

[501] A.S. Lewis and J.S. Pang. Error bounds for convex inequality systems. In J.P. Crouzeix, J.-E. Martinez-Legaz and M. Volle, editors, *Generalized Convexity, Generalized Monotonicity: Recent Results*, Proceedings of the Fifth Symposium on Generalized Convexity, Luminy-Marseille, 1996; Kluwer Academic Publishers (Dordrecht 1998) pp. 75–110.

[502] W. Li. The sharp Lipschitz constants for feasible and optimal solutions of a perturbed linear program. *Linear Algebra and its Applications* 187 (1993) 15–40.

[503] W. Li. Error bounds for piecewise convex quadratic programs and applications. *SIAM Journal on Control and Optimization* 33 (1995) 1510–1529.

[504] W. Li. Linearly convergent descent methods for the unconstrained minimization of convex quadratic splines. *Journal Optimization Theory and Applications* 86 (1995) 145–172.

[505] W. Li. A merit function and a Newton-type method for symmetric linear complementarity problems. In M.C. Ferris and J.S. Pang, editors, *Complementarity and Variational Problems: State of the Art*, SIAM Publications (Philadelphia 1997) pp. 181–203.

[506] W. Li. Abadie's constraint qualification, metric regularity, and error bounds for differentiable convex inequalities. *SIAM Journal on Optimization* 7 (1997) 966–978.

[507] W. Li and I. Singer. Global error bounds for convex multifunctions and applications. *Mathematics of Operations Research* 23 (1998) 443–462.

[508] W. Li and J. Swetits. A Newton method for convex regression, data smoothing, and quadratic programming with bounded constraints. *SIAM Journal on Optimization* 3 (1993) 466–488.

[509] W. Li and J. Swetits. A new algorithm for solving strictly convex quadratic programs. *SIAM Journal on Optimization* 7 (1997) 595–619.

[510] W. Li and J. Swetits. Regularized Newton methods for minimization of convex quadratic splines with singular Hessians. In M. Fukushima and L. Qi, editors, *Reformulation: Nonsmooth, Piecewise Smooth, Semismooth and Smoothing Methods*, Kluwer Academic Publishers (Dordrecht 1999) pp. 235–257.

[511] X.W. Li, A.K. Soh, and W.J. Chen. A new non-smooth model for three dimensional frictional contact problems. *Computational Mechanics* 26 (2000) 538–535.

[512] J.L. Lions and G. Stampacchia. Variational inequalities. *Communications on Pure and Applied Mathematics* 20 (1967) 493–519.

[513] J.M. Liu. Sensitivity analysis in nonlinear programs and variational inequalities via continuous selections. *SIAM Journal on Control and Optimization* 33 (1995) 1040–1061.

[514] J.M. Liu. Strong stability in variational inequalities. *SIAM Journal on Control and Optimization* 33 (1995) 725–749.

[515] J.M. Liu. *Perturbation Analysis in Nonlinear Programs and Variational Inequalities*. Ph.D. thesis, Department of Operations Research, The George Washington University (May 1995).

[516] N.G. Lloyd. *Degree Theory*, Cambridge University Press (Cambridge 1978).

[517] G. Lo. *Complementarity Problems in Robotics.* Ph.D. thesis, Department of Mathematical Sciences, The Johns Hopkins University (1996).

[518] H.K. Lo and A. Chen. Reformulating the traffic equilibrium problem via a smooth gap function. *Mathematical and Computer Modelling* 31 (2000) 179–195.

[519] H.K. Lo and A. Chen. Traffic equilibrium problem with route-specific costs: formulation and algorithms. *Transportation Research Part B–Methodological* 34 (2000) 493–513.

[520] M.S. Lojasiewicz. Division d'une distribution par une fonction analytique de variables réelles. *Comptes de Rendus de Séance, Paris* 146 (1958) 683–686.

[521] M.S. Lojasiewicz. Sur la problème de la division. *Studia Mathematica* 18 (1959) 87–136.

[522] M.S. Lojasiewicz. Ensembles semi-analytiques. Institut des Hautes Etudes Scientiques (Bures-sur-Yvette 1964).

[523] F. Lopez de Silanes, J.R. Markusen, and T.F. Rutherford. Complementarity and increasing returns in intermediate inputs. *Journal of Development Economics* 45 (1994) 101–119.

[524] A.V. Lotov. An estimate of solution set perturbations for a system of linear inequalities. *Optimization Methods and Software* 6 (1995) 1–24.

[525] P. Lötstedt. Coulomb friction in two-dimensional rigid body systems. *Zeitschrift Angewandte Mathematik und Mechanik* 61 (1981) 605–615.

[526] P. Lötstedt. Mechanical systems of rigid bodies subject to unilateral constraints. *SIAM Journal on Applied Mathematics* 42 (1982) 281–296.

[527] R. Lucchetti and J. Revalski, editors. *Recent Developments in Well-Posed Variational Problems.* Kluwer Academic Publishers (Dordrecht 1995).

[528] X.D. Luo and Z.Q. Luo. Extension of Hoffman's error bound to polynomial systems. *SIAM Journal on Optimization* 4 (1994) 383–392.

[529] X.D. Luo and P. Tseng. On a global projection-type error bound for the linear complementarity problem. *Linear Algebra and its Applications* 253 (1997) 251–278.

[530] Z.Q. Luo. New error bounds and their applications to convergence analysis of iterative algorithms. *Mathematical Programming, Series B* 88 (2000) 341–355.

[531] Z.Q. Luo, O.L. Mangasarian, J. Ren, and M.V. Solodov. New error bounds for the linear complementarity problem. *Mathematics of Operations Research* 19 (1994) 880–892.

[532] Z.Q. Luo and J.S. Pang. Error bounds for analytic systems and their applications. *Mathematical Programming* 67 (1994) 1–28.

[533] Z.Q. Luo, J.S. Pang, and D. Ralph. *Mathematical Programs with Equilibrium Constraints.* Cambridge University Press (Cambridge 1996).

[534] Z.Q. Luo and J.F. Sturm. Error bounds for quadratic systems. *High performance optimization in Applied Optimization* 33 (Dordrecht 2000) 383–404.

[535] Z.Q. Luo and P. Tseng. A decomposition property for a class of square matrices. *Applied Mathematics Letters* 4 (1991) 67–69.

[536] Z.Q. Luo and P. Tseng. On the convergence of a matrix splitting algorithm for the symmetric monotone linear complementarity problem. *SIAM Journal on Control and Optimization* 29 (1991) 1037–1060.

[537] Z.Q. Luo and P. Tseng. Error bound and convergence analysis of matrix splitting algorithms for the affine variational inequality problem. *SIAM Journal on Optimization* 2 (1992) 43–54.

[538] Z.Q. Luo and P. Tseng. On global error bound for a class of monotone affine variational inequality problems. *Operations Research Letters* 11 (1992) 159–165.

[539] Z.Q. Luo and P. Tseng. On the linear convergence of descent methods for convex essentially smooth minimization. *SIAM Journal on Control and Optimization* 30 (1992) 408–425.

[540] Z.Q. Luo and P. Tseng. On the convergence rate of dual ascent methods for linearly constrained convex minimization. *Mathematics of Operations Research* 18 (1993) 846–867.

[541] Z.Q. Luo and P. Tseng. Error bounds and convergence analysis of feasible descent methods: a general approach. *Annals of Operations Research* 46/47 (1993) 157–178.

[542] Z.Q. Luo and P. Tseng. Perturbation analysis of a condition number for linear systems. *SIAM Journal on Matrix Analysis and Applications* 15 (1994) 636–660.

[543] Z.Q. Luo and P. Tseng. On the rate of convergence of a distributed asynchronous routing algorithm. *IEEE Transactions on Automatic Control* 39 (1994) 1123–1129.

[544] Z.Q. Luo and P. Tseng. A new class of merit functions for the nonlinear complementarity problem. In M.C. Ferris and J.S. Pang, editors, *Complementarity and Variational Problems: State of the Art*, SIAM Publications (Philadelphia 1997) pp. 204–225.

[545] Z.Q. Luo and S. Zhang. On extensions of the Frank-Wolfe Theorems. *Computational Optimization and Applications* 13 (1999) 87–110.

[546] H.J. Lüthi. On the solution of variational inequality by the ellipsoid method. *Mathematics of Operations Research* 10 (1985) 515–522.

[547] H.J. Lüthi and B. Bueler. The analytic center quadratic cut method for strongly monotone variational inequality problems. *SIAM Journal on Optimization* 10 (2000) 415–426.

[548] K. Madsen, H.B. Nielsen, and M.C. Pinar. A finite continuation algorithm for bound constrained quadratic programming. *SIAM Journal on Optimization* 9 (1998) 62–83.

[549] K. Madsen, H.B. Nielsen, and M.C. Pinar. Bound constrained quadratic programming via piecewise quadratic functions. *Mathematical Programming* 85 (1999) 135–156.

[550] T.L. Magnanti. Models and algorithms for predicting urban traffic equilibrium. In M. Florian, editor, *Transportation Planning Models*, North-Holland (Amsterdam 1984) pp. 153–186.

[551] T.L. Magnanti and G. Perakis. A unifying geometric solution framework and complexity analysis for variational inequalities. *Mathematical Programming* 71 (1995) 327–351.

[552] T.L. Magnanti and G. Perakis. Averaging schemes for variational inequalities and systems of equations *Mathematics of Operations Research* 22 (1997) 568–587.

[553] T.L. Magnanti and G. Perakis. The orthogonality theorem and the strong-F-monotonicity condition for variational inequality algorithms. *SIAM Journal on Optimization* 7 (1997) 248–273.

[554] G. MAIER. A quadratic programming approach for certain classes of nonlinear structural problems. *Meccanica* 3 (1968) 121–130.

[555] G. MAIER. A matrix structural theory of piecewise linear elastoplasticity with interacting yield planes. *Meccanica* 5 (1970) 54–66.

[556] G. MAIER. Incremental plastic analysis in the presence of large displacement and physical instabilizing effects. *International Journal of Solids and Structures* 7 (1971) 345–372.

[557] G. MAIER. Inverse problem in engineering plasticity: a quadratic programming approach. *Accademia Nazionale dei Lincei*, Serie VIII, vol. LXX (1981) 203–209.

[558] G. MAIER, G. BOLZON AND F. TIN-LOI. Mathematical programming in engineering mechanics: some current problems. In M.C. Ferris, O.L. Mangasarian, and J.S. Pang, editors, *Complementarity: Applications, Algorithms and Extensions*, Kluwer Academic Publishers (Dordrecht 2001) pp. 201–231.

[559] G. MAIER, F. GIANNESSI, AND A. NAPPI. Indirect identification of yield limits by mathematical programming. *Engineering Structures* 4 (1982) 86–98.

[560] G. MAIER AND G. NOVATI. A shakedown and bounding theory allowing for nonlinear hardening and second order geometric effects with reference to discrete structural models. In *Inelastic Solids and Structures, A. Sawczuk memorial volume*, Pineridge Press (Swansea 1990) pp. 451–471.

[561] O.G. MANCINO AND G. STAMPACCHIA. Convex programming and variational inequalities. *Journal of Optimization Theory and Applications* 9 (1972) 3–23.

[562] O.L. MANGASARIAN. Pseudo-convex functions. *SIAM Journal on Control* 3 (1965) 281–290.

[563] O.L. MANGASARIAN. *Nonlinear Programming*, McGraw-Hill (New York 1969). [Reprinted as SIAM Classics in Applied Mathematics 10 (Philadelphia 1994).]

[564] O.L. MANGASARIAN. Equivalence of the complementarity problem to a system of nonlinear equations. *SIAM Journal of Applied Mathematics* 31 (1976) 89–92.

[565] O.L. MANGASARIAN. Uniqueness of solution in linear programming. *Linear Algebra and its Applications* 25 (1979) 151–162.

[566] O.L. MANGASARIAN. Locally unique solutions of quadratic programs, linear and nonlinear complementarity problems. *Mathematical Programming* 19 (1980) 200–212.

[567] O.L. MANGASARIAN. A condition number for linear inequalities and linear programs. In G. Bamberg and O. Opitz, editors, *Methods of Operations Research*, Proceedings of 6. Symposium über Operations Research, Augsburg, 7-9 September 1981, Verlagsgruppe Athenaum/ Hain/Scriptor/Hanstein (Konigstein 1981) pp. 3–15.

[568] O.L. MANGASARIAN. A condition number for differentiable convex inequalities. *Mathematics of Operations Research* 10 (1985) 175–179.

[569] O.L. MANGASARIAN. A simple characterization of solution sets of convex programs. *Operations Research Letters* 7 (1988) 21–26.

[570] O.L. MANGASARIAN. Error bounds for nondegenerate monotone linear complementarity problems. *Mathematical Programming Series B* 48 (1990) 437–445.

[571] O.L. MANGASARIAN. Global error bounds for monotone affine variational inequality problems. *Linear Algebra and its Applications* 174 (1992) 153–164.

[572] O.L. MANGASARIAN. The ill-posed linear complementarity problem. In M.C. Ferris and J.S. Pang, editors, *Complementarity and Variational Problems: State of the Art*, SIAM Publications (Philadelphia 1997) pp. 226–233.

[573] O.L. MANGASARIAN. Error bounds for nondifferentiable convex inequalities under a strong Slater constraint qualification. *Mathematical Programming* 83 (1998) 187–194.

[574] O.L. MANGASARIAN AND S. FROMOVITZ. The Fritz John necessary optimality conditions in the presence of equality constraints. *Journal of Mathematical Analysis and Applications* 17 (1967) 34–47.

[575] O.L. MANGASARIAN AND L. MCLINDEN. Simple bounds for solutions of monotone complementarity problems and convex programs. *Mathematical Programming* 32 (1985) 32–40.

[576] O.L. MANGASARIAN AND J.S. PANG. The extended linear complementarity problem. *SIAM Journal on Matrix Analysis and Applications* 16 (1995) 359–368.

[577] O.L. MANGASARIAN AND J. REN. New improved bounds for the linear complementarity problem. *Mathematical Programming* 66 (1994) 241–255.

[578] O.L. MANGASARIAN AND T.H. SHIAU. Error bounds for monotone linear complementarity problems. *Mathematical Programming* 36 (1986) 81–89.

[579] O.L. MANGASARIAN AND T.H. SHIAU. Lipschitz continuity of solutions of linear inequalities, programs and complementarity problems. *SIAM Journal on Control and Optimization* 25 (1987) 583–595.

[580] O.L. MANGASARIAN AND M.V. SOLODOV. Nonlinear complementarity as unconstrained and constrained minimization. *Mathematical Programming* 62 (1993) 277–298.

[581] A.S. MANNE. On the formulation and solution of economic equilibrium models. *Mathematical Programming Study* 23 (1985) 1–23.

[582] A.S. MANNE AND T.F. RUTHERFORD. International trade in oil, gas and carbon emission rights: An intertemporal general equilibrium model. *The Energy Journal* 14 (1993) 1–20.

[583] P. MARCOTTE. A new algorithm for solving variational inequalities, with application to the traffic assignment problem. *Mathematical Programming* 33 (1985) 339–351.

[584] P. MARCOTTE. Application of Khobotov's algorithm to variational inequalities and network equilibrium problems. *Information Systems and Operations Research* 29 (1991) 114–122.

[585] P. MARCOTTE AND J.P. DUSSAULT. A note on a globally convergent Newton method for solving monotone variational inequalities. *Operations Research Letters* 6 (1987) 35–42.

[586] P. MARCOTTE AND J.P. DUSSAULT. A sequential linear programming algorithm for solving monotone variational inequalities. *SIAM Journal on Control and Optimization* 27 (1989) 1260–1278.

[587] P. MARCOTTE AND L. WYNTER. A new look at the multiclass network equilibrium problem. *Transportation Science*, forthcoming.

[588] P. MARCOTTE AND D. ZHU. Weak sharp solutions of variational inequalities. *SIAM Journal on Optimization* 9 (1998) 179–189. [Correction: same journal 10 (2000) 942.]

[589] J.A.C. MARTINS AND A. KLARBRING, editors, *Computational Modeling of Contact and Friction. Computer Methods in Applied Mechanics and Engineering* 177, No. 3–4, North-Holland Publishing Company (Amsterdam 1999).

[590] E. MASKIN AND J. TIROLE. A theory of dynamic oligopoly, I: Overview and quantity competition with large fixed costs. *Econometrica* 56 (1988) 549–569.

[591] E. MASKIN AND J. TIROLE. A theory of dynamic oligopoly, I: Price competition, kinked demand curves, and edgeworth cycles. *Econometrica* 56 (1988) 571–579.

[592] M.A. MATAOUI. *Contributions à la Décomposition et à l'Agrégation des Problèmes Variationnels.* Thèse de Doctorat en Mathématiques et Automatique, École des Mines de Paris (Fontainebleau 1990).

[593] R. MATHIAS AND J.S. PANG. Error bounds for the linear complementarity problem with a P-matrix. *Linear Algebra and its Applications* 132 (1990) 123–136.

[594] L. MATHIESEN. Computational experience in solving equilibrium models by a sequence of linear complementarity problem. *Operations Research* 33 (1985) 1225-1250.

[595] L. MATHIESEN. Computation of economic equilibria by a sequence of linear complementarity problem. *Mathematical Programming Study* 23 (1985) 144–162.

[596] L. MATHIESEN. An algorithm based on a sequence of linear complementarity problems applied to a Walrasian equilibrium model: An example. *Mathematical Programming* 37 (1987) 1–18.

[597] G.P. MCCORMICK. Second order conditions for constrained minima. *SIAM Journal on Applied Mathematics* 15 (1967) 641–652.

[598] L. MCLINDEN. Stable monotone variational inequalities. *Mathematical Programming* 48 (1990) 303–338.

[599] N. MEGIDDO. A monotone complementarity problem with feasible solutions but no complementarity solutions. *Mathematical Programming* 12 (1977) 131–132.

[600] N. MEGIDDO On the parametric nonlinear complementarity problem. *Mathematical Programming Study* 7 (1978) 142–150.

[601] N. MEGIDDO AND M. KOJIMA. On the existence and uniqueness of solutions in nonlinear complementarity problems. *Mathematical Programming* 12 (1977) 110–130.

[602] C. METZLER, B. HOBBS, AND J.S. PANG. Nash-Cournot equilibria in power markets on a linearized DC network with arbitrage: Formulations and properties. *Networks and Spatial Economics* (2002), forthcoming.

[603] R.R. MEYER AND O.L. MANGASARIAN. Non-linear perturbation of linear programs. *SIAM Journal on Control and Optimization* 17 (1979) 745–752.

[604] E. MIERSEMANN AND H.D. MITTELMANN. Continuation for parameterized nonlinear variational inequalities. *Journal of Computational and Applied Mathematics* 26 (1989) 23–34.

[605] E. MIERSEMANN AND H.D. MITTELMANN. On the continuation for variational inequalities depending on an eigenvalue parameter. *Mathematical Methods in the Applied Science* 11 (1989) 95–104.

[606] E. MIERSEMANN AND H.D. MITTELMANN. Stability in obstacle problems for the von Krmn plate. *SIAM J. Mathematical Analysis* 23 (1992) 1099–1116.

[607] R. MIFFLIN. Semismooth and semiconvex functions in constrained optimization. *SIAM Journal on Control and Optimization* 15 (1977) 957–972.

[608] G. MINTY. Monotone Networks. *Proceedings of the Royal Society of London (Series A)* 257 (1960) 194–212.

[609] G. MINTY. Monotone (nonlinear) operators in Hilbert Space. *Duke Mathematics Journal* 29 (1962) 341–346.

[610] G. MINTY. Two theorems on nonlinear functional equations in Hilbert space. *Bulletin of the American Mathematical Society* 69 (1963) 691–692.

[611] H.D. MITTELMANN. Nonlinear parameterized equations: new results for variational problems and inequalities. In E.L. Allgower and K. Georg, editors, *Computational Solution of Nonlinear Systems of Equations*, Lectures in Applied Mathematics 26, American Mathematical Society (Providence 1990) pp. 451–466.

[612] F.S. MOKHTARIAN AND J.L. GOFFIN. A path-following cutting plane method for some monotone variational inequalities. *Optimization* 48 (2000) 333–351.

[613] R.D.C. MONTEIRO AND J.S. PANG. On two interior-point mappings for nonlinear semidefinite complementarity problems. *Mathematics of Operations Research* 23 (1998) 39–60.

[614] R.D.C. MONTEIRO AND J.S. PANG. A potential reduction Newton method for constrained equations. *SIAM Journal on Optimization* 9 (1999) 729–754.

[615] R.D.C. MONTEIRO AND S.J. WRIGHT. Local convergence of interior-point algorithms for degenerate monotone LCP. *Computational Optimization and Applications* 3 (1994) 131–156.

[616] R.D.C. MONTEIRO AND P.R. ZANJACOMO. General interior-point maps and existence of weighted paths for nonlinear semidefinite complementarity problems. *Mathematics of Operations Research* 25 (2000) 381–399.

[617] B.S. MORDUKHOVICH. *Approximation Methods in Problems of Optimization and Control.* Nauka (Moscow 1988).

[618] B. MORDUKHOVICH. Complete characterization of openness, metric regularity, and Lipschitzian properties. *Transactions of the American Mathematical Society* 340 (1993) 1–36.

[619] B. MORDUKHOVICH. Lipschitzian stability of constraint systems and generalized equations. *Nonlinear Analysis* 22 (1994) 173–206.

[620] B. MORDUKHOVICH. Stability theory for parametric generalized equations and variational inequalities via nonsmooth analysis. *Transactions of the American Mathematical Society* 343 (1994) 609–657.

[621] B. MORDUKHOVICH. Coderivatives of set-valued mappings: Calculus and applications. *Nonlinear Analysis and Theory* 30 (1997) 3059–3070.

[622] B. MORDUKHOVICH AND J.V. OUTRATA. On second-order subdifferentials and their applications. *SIAM Journal on Optimization* 12 (2001) 139–169.

[623] J.J. MORÉ. Coercivity conditions in nonlinear complementarity problems. *SIAM Review* 16 (1974) 1–16.

[624] J.J. MORÉ. Classes of functions and feasibility conditions in nonlinear complementarity problems. *Mathematical Programming* 6 (1974) 327–338.

[625] J. MORÉ AND W.C. RHEINBOLDT. On P- and S-functions and related class of n-dimensional nonlinear mappings. *Linear Algebra and its Applications* 6 (1973) 45–68.

[626] J.J. MOREAU. Numerical aspects of the sweeping process. Computational modeling of contact and friction. *Computer Methods in Applied Mechanics and Engineering* 177 (1999) 329–349.

[627] U. MOSCO. *Implicit variational problems and quasi-variational inequalities.* In J. Gossez, E.J. Lami Dozo, J. Mawhin and L. Waelbroeck, editors, *Nonlinear Operators and the Calculus of Variations*, Lecture Notes in Mathematics 543, Springer-Verlag (Berlin 1976) pp. 82–156.

[628] F.H. MURPHY, H.D. SHERALI, AND A.L. SOYSTER. A mathematical programming approach for determining oligopolistic market equilibrium. *Mathematical Programming* 24 (1982) 92–106.

[629] G.S.R. MURTHY, T. PARTHASARATHY, AND M. SABATINI. Lipschitzian Q-matrices are P-matrices. *Mathematical Programming* 74 (1996) 55–58.

[630] K.G. MURTY AND S.N. KABADI. Some NP-complete problems in quadratic and nonlinear programming. *Mathematical Programming* 39 (1987) 117–129.

[631] A. NAGURNEY. Competitive equilibrium problems, variational inequalities and regional science. *Journal of Regional Science* 27 (1987) 55–76.

[632] A. NAGURNEY. *Network Economics: a Variational Inequality Approach.* Advances in Computational Economics, 1, Kluwer Academic Publishers (Dordrecht 1993).

[633] A. NAPPI. System identification for yield limits and hardening moduli in discrete elastic-plastic structures by nonlinear programming. *Applied Mathematical Modeling* 6 (1982) 441–448.

[634] J.F. NASH. Equilibrium points in n-person games. *Proceedings of the National Academy of Sciences* 36 (1950) 48–49.

[635] J.F. NASH. Non-cooperative games. *Annals of Mathematics* 54 (1951) 286–295.

[636] K.F. NG AND W.H. YANG. Error bound and its relation to regularity. Manuscript, Department of Mathematics, Chinese University of Hong Kong (December 2001).

[637] K.F. NG AND W.H. YANG. Error bounds for abstract linear inequality system. *SIAM Journal on Optimization* 12 (2002) 1–17.

[638] K.F. NG AND X.Y. ZHENG. Global error bounds with fractional exponents. *Mathematical Programming, Series B* 88 (2000) 357–370.

[639] K.F. NG AND X.Y. ZHENG. Constrained error bounds of quadratic functions and piecewise affine inequality systems. Manuscript, Department of Mathematics, Chinese University of Hong Kong (October 2001).

[640] K.F. NG AND X.Y. ZHENG. The least Hoffman's error bound for linear inequality system. Revised manuscript, Department of Mathematics, Chinese University of Hong Kong (December 2001).

[641] K.F. NG AND X.Y. ZHENG. Global error bounds for lower semicontinuous functions in normed spaces. *SIAM Journal on Optimization* 12 (2001) 1–17.

[642] S. NGUYEN AND C. DUPUIS. An efficient method for computing traffic equilibria in networks with asymmetric transportation costs. *Transportation Science* 18 (1984) 185–202.

[643] J. NOCEDAL AND S.J. WRIGHT. *Numerical optimization.* Springer-Verlag (New York 1999).

[644] K.P. OH. The numerical solution of dynamically loaded elastohydrodynamic contact as a nonlinear complementarity problem. *Journal of Tribology* 106 (1984) 88–95.

[645] K.P. OH. Analysis of a needle bearing. *Journal of Tribology* 106 (1984) 78–87.

[646] K.P. OH. The formulation of the mixed lubrication problem as a generalized nonlinear complementarity problem. *Journal of Tribology* 108 (1986) 598–604.

[647] K.P. OH AND P.K. GOENKA. The elastohydrodynamic solution of journal bearings under dynamic loading. *Journal of Tribology* 107 (1985) 389–395.

[648] K.P. OH, C.H. LI, AND P.K. GOENKA. Elastohydrodynamic lubrication of piston skirts. *Journal of Tribology* 109 (1987) 356–362.

[649] T. OHTSUKI, T. FUJISAWA, AND S. KUMAGAI. Existence theorems and a solution algorithm for piecewise linear resistive networks. *SIAM Journal on Mathematical Analysis* 8 (1977) 69–99.

[650] K. Okuguchi. *Expectations and Stability in Oligopoly Models.* Lecture Notes in Economics and Mathematical Systems 138, Springer-Verlag (Berlin 1976).

[651] Z. Opial. Weak convergence of the sequence of successive approximations for nonexpansive mappings. *Bulletin of the American Mathematical Society* 73 (1967) 591–597.

[652] J.M. Ortega and W.C. Rheinboldt. *Iterative Solution of Nonlinear Equations in Several Variables*, Academic Press (New York 1970).

[653] J. Outrata Optimality conditions for a class of mathematical programs with equilibrium constraints. *Mathematics of Operations Research* 24 (1999) 627–644.

[654] J. Outrata, M. Kocvara, and J. Zowe. *Nonsmooth Approach to Optimization Problems with Equilibrium Constraints: Theory, Applications and Numerical Results*, Kluwer Academic Publishers (Dordrecht 1998).

[655] J.V. Outrata and J. Zowe. A Newton method for a class of quasi-variational inequalities. *Computational Optimization and Applications* 4 (1995) 5–21.

[656] R.S. Palais. Natural operations on differential forms. *Transactions of the American Mathematical Society* 22 (1959) 125–141.

[657] P.D. Panagiotopoulos. *Inequality Problems in Mechanics and Applications*, Birkhäuser Inc. (Boston 1985).

[658] P.D. Panagiotopoulos. *Hemivariational Inequalities*, Springer-Verlag (Berlin 1993).

[659] J.S. Pang. *Least-Element Complementarity Theory.* Ph.D. thesis, Department of Operations Research, Stanford University (1976).

[660] J.S. Pang. The implicit complementarity problem. In O.L. Mangasarian, R.R. Meyer and S.M. Robinson, editors, *Nonlinear Programming 4*, Proceedings of the Conference Nonlinear Programming 4, 1980; Academic Press (New York 1981) pp. 487–518.

[661] J.S. Pang. Asymmetric variational inequality problems over product sets: Applications and iterative methods. *Mathematical Programming* 31 (1985) 206–219

[662] J.S. Pang. A posteriori error bounds for the linearly-constrained variational inequality problem. *Mathematics of Operations Research* 12 (1987) 474–484.

[663] J.S. Pang. Newton's method for B-differentiable equations. *Mathematics of Operations Research* 15 (1990) 311–341.

[664] J.S. Pang. Solution differentiability and continuation of Newton's method for variational inequality problems over polyhedral sets. *Journal of Optimization Theory and Applications* 66 (1990) 121–135.

[665] J.S. Pang. A B-differentiable equation based, globally and locally quadratically convergent algorithm for nonlinear programs, complementarity and variational inequality problems. *Mathematical Programming* 51 (1991) 101–131.

[666] J.S. Pang. Convergence of splitting and Newton methods for complementarity problems: An application of some sensitivity results. *Mathematical Programming* 58 (1993) 149–160.

[667] J.S. Pang. A degree-theoretic approach to parametric nonsmooth equations with multivalued perturbed solution sets. *Mathematical Programming, Series B* 62 (1993) 359–384.

[668] J.S. Pang. Complementarity problems. In R. Horst and P. Pardalos, editors, *Handbook in Global Optimization*, Kluwer Academic Publishers (Boston 1994).

[669] J.S. PANG. Necessary and sufficient conditions for solution stability in parametric nonsmooth equations. In D.Z. Zhu, L. Qi, and R. Womersley, editors, *Recent Advances in Nonsmooth Optimization*, World Scientific Publishers (River Ridge 1995) pp. 260–287.

[670] J.S. PANG. Error bounds in mathematical programming. *Mathematical Programming, Series B* 79 (1997) 299–332.

[671] J.S. PANG AND M. FUKUSHIMA. Complementarity constraint qualifications and simplified B-differentiability for mathematical programs with equilibrium constraints. *Computational Optimization and Applications* 13 (1999) 111–136.

[672] J.S. PANG AND S.A. GABRIEL. NE/SQP: A robust algorithm for the nonlinear complementarity problem. *Mathematical Programming* 60 (1993) 295–338.

[673] J.S. PANG, B.F. HOBBS, AND C. DAY. Properties of oligopolistic market equilibria in linearized DC power networks with arbitrage and supply function conjectures. In E. Sachs, editor, *System Modeling and Optimization XX*, [Proceedings of the IFIP TC7 20th Conference on System Modeling and Optimization, July 23-27, Trier, Germany], Kluwer Academic Publishers (2002) pp.

[674] J.S. PANG AND J. HUANG. Pricing American options with transaction costs by complementarity methods. In M. Avellandeda, editor, *Quantitative Analysis in Financial Markets*, [Collected Papers of the New York University Mathematical Finance Seminar, Volume III], World Scientific Press (New Jersey 2002) pp. 172–198.

[675] J.S. PANG AND L. QI. A globally convergent Newton method for convex SC^1 minimization problems. *Journal of Optimization Theory and Applications* 85 (1995) 633–648.

[676] J.S. PANG AND D. RALPH. Piecewise smoothness, local invertibility, and parametric analysis of normal maps. *Mathematics of Operations Research* 21 (1996) 401–426.

[677] J.S. PANG AND D.E. STEWART. A unified approach to discrete frictional contact problems, *International Journal of Engineering Science* 37 (1999) 1747–1768.

[678] J.S. PANG, D. SUN, AND J. SUN. Semismooth homeomorphisms and strong stability of semidefinite and Lorentz complementarity problems. *Mathematics of Operations Research*, forthcoming.

[679] J.S. PANG AND J.C. TRINKLE. Complementarity formulations and existence of solutions of multi-rigid-body contact problems with Coulomb friction. *Mathematical Programming* 73 (1996) 199–226.

[680] J.S. PANG AND J.C. TRINKLE. Stability characterization of rigid body contact problems with Coulomb friction. *Zeitschrift für Angewandte Mathematik und Mechanik* 80 (2000) 643–663.

[681] J.S. PANG, J.C. TRINKLE, AND G. LO. A complementarity approach to a quasistatic multi-rigid-body contact problem. *Computational Optimization and Applications* 5 (1996) 139–154.

[682] J.S. PANG AND J.C. YAO. On a generalization of a normal map and equation. *SIAM Journal on Control and Optimization* 33 (1995) 168–184.

[683] J.S. PANG AND C.S. YU. Linearized simplicial decomposition methods for computing traffic equilibria on networks. *Networks* 14 (1984) 427–438.

[684] T. PARTHASARATHY. *On global univalence theorems.* Lecture Notes in Mathematics 977, Springer-Verlag (Berlin 1983).

[685] T. PARTHASARATHY, D. SAMPANGI RAMAN, AND B. SRIPARNA. Relationship between strong monotonicity property, P_0-property, and the GUS property in semidefinite linear complementarity problems. *Mathematics of Operations Research* 27 (2002) 326–331.

[686] M. PATRIKSSON. *The Traffic Assignment Problem, Models and Methods*, VSP (Utretch 1994).

[687] M. PATRIKSSON. *Nonlinear Programming and Variational Inequality Problems. A Unified Approach.* Kluwer Academic Publishers (Dordrecht 1999).

[688] M. PATRIKSSON, M. WERME, AND L. WYNTER. Obtaining robust solutions to general VIs. Paper presented at the Third International Conference on Complementarity problems, July 29–August 1, 2002, Cambridge, England.

[689] J.M. PENG, C. ROOS, T. TERLAKY, AND A. YOSHISE. Self-regular proximities and new search directions for nonlinear $P_*(\kappa)$ complementarity problems. Manuscript, Department of Computing and Software, McMaster University (November 2000).

[690] J.M. PENG AND Y. YUAN. Unconstrained methods for generalized complementarity problems. *Journal of Computational Mathematics* 15 (1997) 253–264.

[691] J.P. PENOT. Metric regularity, openness and Lipschitzian behavior of multifunctions. *Nonlinear Analysis, Theory, Methods, and Applications* 13 (1989) 629–643.

[692] A.M. PEREIRA AND J.B. SHOVEN. Survey of dynamic computational general equilibrium models for tax policy evaluation. *Journal of Policy model* 10 (1988) 401–436.

[693] C. PERRONI AND T. RUTHERFORD. International trade in carbon emission rights and basic materials: General equilibrium calculations for 2020. *Scandinavian Journal of Economics* 95 (1993) 257–278.

[694] F. PFEIFFER AND C. GLOCKER. *Multibody Dynamics with Unilateral Contacts*, John Wiley (New York 1996).

[695] E. POLAK. *Optimization: Algorithms and Consistent Approximations.* Springer (New York 1997).

[696] B.T. POLYAK. *Introduction to Optimization.* Optimization Software, Inc. (New York 1987).

[697] P.V. PRECKEL. Alternative algorithms for computing economic equilibria. *Mathematical Programming Study* 23 (1985) 163–172.

[698] H.D. QI. On minimizing and stationary sequences of a new class of merit functions for nonlinear complementarity problems. *Journal of Optimization Theory and Applications* 102 (1999) 411–431.

[699] L. QI. Convergence analysis of some algorithms for solving nonsmooth equations. *Mathematics of Operations Research* 18 (1993) 227–244.

[700] L. QI AND H.Y. JIANG. On the range sets of variational inequalities. *Journal of Optimization Theory and Applications* 83 (1994) 565–586.

[701] L. QI AND J. SUN. A nonsmooth version of Newton's method. *Mathematical Programming* 58 (1993) 353–368.

[702] Y.P. QIU. Solution properties of oligopolistic network equilibria. *Networks* 21 (1991) 565–580.

[703] Y.P. QIU AND T.L. MAGNANTI. Sensitivity analysis for variational inequalities defined on polyhedral sets. *Mathematics of Operations Research* 14 (1989) 410–432.

[704] Y.P. QIU AND T.L. MAGNANTI. Sensitivity analysis for variational inequalities. *Mathematics of Operations Research* 17 (1992) 61–76.

[705] M. QUEIROZ, J. JÚDICE, AND C. HUMES, JR. The symmetric eigenvalue complementarity problem. Manuscript, Department of Computer Science, University of São Paulo, Brazil (February 2002). [Submitted to *Mathematics of Computation.*]

[706] H. RADEMACHER. Über partielle und totale Differenzierbarkeit I. *Mathematical Annals* 89 (1919) 340–359.

[707] V.T. RAJAN, R. BURRIDGE, AND J.T. SCHWARTZ. Dynamics of a rigid body in frictional contact with rigid walls. *IEEE International Conference on Robotics and Automation, Raleigh NC* (March 1987) 671–677.

[708] D. RALPH. A new proof of Robinson's homeomorphism theorem for PL-normal maps. *Linear Algebra and Applications* 178 (1993) 249–260.

[709] D. RALPH. On branching numbers of normal manifolds. *Nonlinear Analysis* 22 (1994) 1041–1050.

[710] D. RALPH. Global convergence of damped Newton's method for nonsmooth equations, via the path search. *Mathematics of Operations Research* 19 (1994) 352–389.

[711] D. RALPH AND S. DEMPE. Directional derivatives of the solution of a parametric nonlinear program. *Mathematical Programming* 70 (1995) 159–172.

[712] D. RALPH AND S. SCHOLTES. Sensitivity analysis of composite piecewise smooth equations. *Mathematical Programming* 76 (1997) 593–612.

[713] G. RAVINDRAN AND M.S. GOWDA. Regularization of P_0-functions in box variational inequality problems. *SIAM Journal on Optimization* 11 (2000) 748–760.

[714] A. REINOZA. The strong positivity conditions. *Mathematics of Operations Research* 10 (1985) 54–62.

[715] W.C. RHEINBOLDT. On M-functions and their application to nonlinear Gauss-Seidel iterations and to network flows. *Journal of Mathematical Analysis and Applications* 32 (1970) 274–307.

[716] W.C. RHEINBOLDT AND J.S. VANDERGRAFT. On piecewise affine mappings in $\mathbb{R}^n$. *SIAM Journal on Applied Mathematics* 29 (1975) 680–689.

[717] V. RICO-RAMIREZ, B.A. ALLAN, AND A.W. WESTERBERG. Conditional modeling. I. Requirements for an equation-based environment. *Industrial and Engineering Chemistry Research* 38 (1999) 519–530.

[718] V. RICO-RAMIREZ AND A.W. WESTERBERG. Conditional modeling. 2. Solving using complementarity and boundary-crossing formulations. *Industrial and Engineering Chemistry Research* 38 (1999) 531–553.

[719] V. RICO-RAMIREZ AND A.W. WESTERBERG. Interior point methods for the solution of conditional models. *Computers and Chemical Engineering* 26 (2002) 375–383.

[720] M. RIVIER, M. VENTOSA, A. RAMOS, F. MARTINEZ-CÓRCOLES, AND A.C. TOSCANO. A generation operation planning model in deregulated electricity markets based on the complementarity problem. In M.C. Ferris, O.L. Mangasarian, and J.S. Pang, editors, *Complementarity: Applications, Algorithms and Extensions*, Kluwer Academic Publishers (Dordrecht 2001) pp. 273–296.

[721] T. ROBERTSON, F.T. WRIGHT, AND R.L. DYKSTRA. *Order restricted Statistical Inference*. John Wiley & Sons, Ltd. (Chichester 1988).

[722] S.M. ROBINSON. Normed convex processes. *Transactions of the American Mathematical Society* 174 (1972) 127–140.

[723] S.M. ROBINSON. Perturbed Kuhn-Tucker points and rates of convergence for a class of nonlinear-programming algorithms. *Mathematical Programming* 7 (1974) 1–16.

[724] S.M. ROBINSON. An application of error bounds for convex programming in a linear space. *SIAM Journal on Control* 13 (1975) 271–273.

[725] S.M. ROBINSON. Stability theory for systems of inequalities. II. Differentiable nonlinear systems. *SIAM Journal on Numerical Analysis* 13 (1976) 497–513.

[726] S.M. ROBINSON. Regularity and stability for convex multivalued functions. *Mathematics of Operations Research* 1 (1976) 130–143.

[727] S.M. ROBINSON. A characterization of stability in linear programming. *Operations Research* 25 (1977) 435–447.

[728] S.M. ROBINSON. Generalized equations and their solutions. I. Basic theory. *Mathematical Programming Study* 10 (1979) 128–141.

[729] S.M. ROBINSON, editor. *Analysis and Computation of Fixed Points*, Academic Press (New York 1980).

[730] S.M. ROBINSON. Strongly regular generalized equations. *Mathematics of Operations Research* 5 (1980) 43–62.

[731] S.M. ROBINSON. Some continuity properties of polyhedral multifunctions. *Mathematical Programming Study* 14 (1981) 206–214.

[732] S.M. ROBINSON. Generalized equations and their solutions. II. Applications to nonlinear programming. *Mathematical Programming Study* 19 (1982) 200–221.

[733] S.M. ROBINSON. Local structure of feasible sets in nonlinear programming. I. Regularity. *Numerical methods (Caracas, 1982)*, Lecture Notes in Mathematics 1005, Springer (Berlin 1983) pp. 240–251.

[734] S.M. ROBINSON. Generalized equations. In A. Bachem, M. Grötschel, and B. Korte, editors, *Mathematical Programming: The State of the Art*, Springer-Verlag (Berlin 1983) pp. 346–367.

[735] S.M. ROBINSON. Local structure of feasible sets in nonlinear programming. II. Nondegeneracy. *Mathematical Programming Study* 22 (1984) 217–230.

[736] S.M. ROBINSON. Implicit B-differentiability in generalized equations. Technical report #2854, Mathematics Research Center, University of Wisconsin, Madison (1985).

[737] S.M. ROBINSON. Local structure of feasible sets in nonlinear programming. III. Stability and sensitivity. *Mathematical Programming Study* 30 (1987) 45–66.

[738] S.M. ROBINSON. Mathematical foundations of nonsmooth embedding methods. *Mathematical Programming, Series B* 48 (1990) 221–229.

[739] S.M. ROBINSON. An implicit-function theorem for a class of nonsmooth functions. *Mathematics of Operations Research* 16 (1991) 292–309.

[740] S.M. ROBINSON. Normal maps induced by linear transformations. *Mathematics of Operations Research* 17 (1992) 691–714.

[741] S.M. ROBINSON. Homeomorphism conditions for normal maps of polyhedra. In A. Ioffe, M. Marcus, and S. Reich, editors, *Optimization and Nonlinear Analysis (Haifa, 1990)*, Pitman Research Notes in Mathematics Series 244, Longman House (Harlow 1992) pp. 240–248.

[742] S.M. ROBINSON. Nonsingularity and symmetry for linear normal maps. *Mathematical Programming* 62 (1993) 415–425.

[743] S.M. ROBINSON. Shadow prices for measures of effectiveness. I. Linear model. *Operations Research* 41 (1993) 518–535. [Erratum: *Operations Research* 48 (2000) 185.]

[744] S.M. ROBINSON. Shadow prices for measures of effectiveness. II. General model. *Operations Research* 41 (1993) 536–548.

[745] S.M. ROBINSON. Differential stability conditions for saddle problems on products of convex polyhedra. In H. Fischer, B. Riedmller and S. Schffler, editors, *Applied Mathematics and Parallel Computing*. Festschrift for Klaus Ritter, Physica Verlag (Heidelberg 1996) pp. 265–274.

[746] R.T. ROCKAFELLAR. *Convex Analysis*, Princeton University Press (Princeton 1970).

[747] R.T. ROCKAFELLAR. Linear-quadratic programming and optimal control. *SIAM Journal on Control and Optimization* 25 (1987) 781–814.

[748] R.T. ROCKAFELLAR. Proto-differentiability of set-valued mappings and its applications in optimization. In H. Attouch, J.P. Aubin, F.H. Clarke, and I. Ekeland, editors, *Analyse Non Linéaire*, Gauthier-Villars (Paris 1989) pp. 449–482.

[749] R.T. ROCKAFELLAR. Computational schemes for large-scale problems in extended linear-quadratic programming. *Mathematical Programming* 48 (1990) 447–474.

[750] R.T. ROCKAFELLAR AND R.J.-B. WETS. A Lagrangian finite generation technique for solving linear-quadratic problems in stochastic programming. *Mathematical Programming Study* 28 (1986) 63–93.

[751] R.T. ROCKAFELLAR AND R.J.-B. WETS. Generalized linear-quadratic problems of deterministic and stochastic optimal control in discrete time. *SIAM Journal on Control and Optimization* 28 (1990) 810–822.

[752] R.T. ROCKAFELLAR AND R.J.-B. WETS. *Variational Analysis*, Springer-Verlag (Berlin 1998).

[753] J.F. RODRIGUES. *Obstacle Problems in Mathematical Physics*. Elsevier Science Publishers B.V. (Amsterdam 1987).

[754] T.F. RUTHERFORD. *Applied General Equilibrium Modeling*. Ph.D. thesis, Department of Operations Research, Stanford University (1987).

[755] T.F. RUTHERFORD. Extensions of GAMS for complementarity problems arising in applied economic analysis. *Journal of Economic Dynamics and Control* 19 (1995) 1299-1324.

[756] V.L. RVACHEV. *Theory of R-Functions and Some Applications*. (In Russian.) Naukova Dumka (1982).

[757] P.A. SAMELSON, R.M. THRALL, AND O. WESLER. A partition theorem for Euclidean n-space. *Proceedings of the American Mathematical Society* 9 (1958) 805–807.

[758] P.A. SAMUELSON. Prices of factors and goods in general equilibrium. *Review of Economic Studies* 21 (1953) 1–20.

[759] H.E. SCARF. The approximation of fixed points of a continuous mapping. *SIAM Journal on Applied Mathematics* 15 (1967) 1328–1343.

[760] H.E. SCARF. (In collaboration with T. HANSEN.) *The Computation of Economic Equilibria*. Yale University Press, (New Haven 1973).

[761] H. SCHEEL AND S. SCHOLTES. Mathematical programs with complementarity constraints: Stationarity, optimality, and sensitivity. *Mathematics of Operations Research* 25 (2000) 1–22.

[762] S. SCHOLTES. Introduction to piecewise differentiable equations. Habilitation thesis, Institut für Statistik und Mathematische Wirtschaftstheorie, Universität Karlsruhe (1994).

[763] S. SCHOLTES. Homeomorphism conditions for coherently oriented piecewise affine mappings. *Mathematics of Operations Research* 21 (1996) 955–978.

[764] S. SCHOLTES. A proof of the branching number bound for normal manifolds. *Linear Algebra and its Applications* 246 (1996) 83–95.

[765] S. SCHOLTES. Convergence properties of a regularization scheme for mathematical programs with complementarity constraints. *SIAM Journal on Optimization* 11 (2001) 918–936.

[766] S. SCHOLTES AND M. STÖHR. Exact penalization of mathematical programs with equilibrium constraints. *SIAM Journal on Control and Optimization* 37 (1999) 617–652.

[767] S. SCHOLTES AND M. STÖHR. How stringent is the linear independence assumption for mathematical programs with complementarity constraints. *Mathematics of Operations Research* 26 (2001) 851–863.

[768] R. SCHRAMM. On piecewise linear functions and piecewise linear equations. *Mathematics of Operations Research* 5 (1980) 510-522.

[769] A. SEEGER. Eigenvalue analysis of equilibrium processes defined by linear complementarity conditions. *Linear Algebra and Its Applications* 299 (1999) 1–14.

[770] H. SELLAMI AND S.M. ROBINSON. Homotopies based on nonsmooth equations for solving nonlinear variational inequalities. In G. Di Pillo and F. Giannessi, editors, *Nonlinear Optimization and Applications*, Plenum Press (New York 1996) pp. 329–343.

[771] H. SELLAMI AND S.M. ROBINSON. Implementation of a continuation method for normal maps. *Mathematical Programming* 76 (1997) 563–578.

[772] W. SEYFFERTH AND F. PFEIFFER. Dynamics of assembly processes with a manipulator. *Proceedings of the 1992 IEEE/RSJ International Conference on Intelligent Robots and Systems* (1992) pp. 1303–1310.

[773] A. SHAPIRO. Sensitivity analysis of nonlinear programs and differentiability properties of metric projections. *SIAM Journal on Control and Optimization* 26 (1988) 628–645.

[774] A. SHAPIRO. On concepts of directional differentiability. *Journal of Optimization Theory and Applications* 66 (1990) 477–487.

[775] A. SHAPIRO. Perturbation analysis of optimization problems in Banach spaces. *Numerical Functional Analysis and Optimization* 13 (1992) 97–116.

[776] A. SHAPIRO. Directionally nondifferentiable metric projection. *Journal of Optimization Theory and Applications* 81 (1994) 203–204.

[777] A. SHAPIRO. Sensitivity analysis of parameterized programs via generalized equations. *SIAM Journal on Control and Optimization* 32 (1994) 523–571.

[778] A. SHAPIRO. On uniqueness of Lagrange multipliers in optimization problems subject to cone constraints. *SIAM Journal on Optimization* 7 (1997) 508–518.

[779] V. SHAPIRO. Theory of R-functions and applications: A primer. Technical Report 91-1219, Computer Science Department, Cornell University (revised June 1991).

[780] V. SHAPIRO. Real functions for representation of rigid solids. *Computer-Aided Geometric Design* 11 (1994) 153–175.

[781] V. SHAPIRO AND I. TSUKANOV. Implicit functions with guaranteed differential properties. *Proceedings of the Fifth ACM Symposium on Solid Modeling and Applications, Ann Arbor, June 9-11, 1999.*

[782] M. SHIDA, S. SHINDOH, AND M. KOJIMA. Centers of monotone generalized complementarity problems. *Mathematics of Operations Research* 22 (1997) 969–976.

[783] J.B. SHOVEN AND A.M. PEREIRA. Economic equilibrium–model formulation and solution. *Journal of Economics–Zeitschrift für Nationalokonomie* 47 (1987) 88–91.

[784] J.B. SHOVEN AND J. WHALLEY. *Applying General Equilibrium.* Cambridge University Press (New York 1992).

[785] A.H. SIDDIQI, P. MANCHANDA, AND M. KOCVARA. An iterative two-step algorithm for American option pricing. *IMA Journal of Mathematics Applied in Business and Industry* 11 (2000) 71–84.

[786] E.M. SIMANTIRAKI. *Interior-Point Methods for Linear and Mixed Complementarity Problems.* Ph.D. thesis, Graduate School of Management, Rutgers University (May 1996).

[787] E.M. SIMANTIRAKI AND D.F. SHANNO. An infeasible-interior-point algorithm for solving mixed complementarity problems. In M.C. Ferris and J.S. Pang, editors, *Complementarity and Variational Problems: State of the Art*, SIAM Publications (Philadelphia 1997) pp. 386–404.

[788] M. SION. On general min-max theorems. *Pacific Journal of Mathematics* 8 (1959) 171–176.

[789] M.L. SLATER. Lagrange multipliers revisited: A contribution to nonlinear programming. Cowles Commission Discussion Paper, Mathematics 403, Yale University (November 1950).

[790] S. SMALE. A convergent process of price adjustment and global Newton methods. *Journal of Mathematical Economics* 3 (1976) 107–120.

[791] Y. SMEERS. Computable equilibrium models and the restructuring of the European electricity and gas markets. *Energy Journal* 18 (1997) 1–31.

[792] Y. SMEERS AND J.Y. WEI. Spatially oligopolistic model with opportunity cost pricing for transmission capacity reservations–A variational inequality approach. CORE Discussion Paper 9717, Université Catholique de Louvain (February 1997).

[793] M.J. SMITH. The existence, uniqueness and stability of traffic equilibria. *Transportation Research* 13B (1979) 295–304.

[794] M.J. SMITH. The existence and calculation of traffic equilibria. *Transportation Research* 17B (1983) 291–303.

[795] M.J. SMITH. An algorithm for solving asymmetric equilibrium problems with a continuous cost-flow function. *Transportation Research* 17B (1983) 365–371.

[796] T.E. SMITH. A solution condition for complementarity problems: with an application to spatial price equilibrium. *Applied Mathematics and Computation* 15 (1984) 61–69.

[797] T.E. SMITH, T.L. FRIESZ, D.H. BERNSTEIN, AND Z.G. SUO. A comparative analysis of two minimum-norm projective dynamics and their relationship to variational inequalities. In M.C. Ferris and J.S. Pang, editors, *Complementarity and Variational Problems: State of the Art*, SIAM Publications (Philadelphia 1997) pp. 405–439.

[798] A.K. SOH, X. LI, AND W. CHEN. Non-smooth method for three-dimensional frictional contact problems. Manuscript, Department of Mechanical Engineering, University of Hong Kong (March 2001).

[799] M.V. SOLODOV. Some optimization formulations of the extended complementarity problem. *Computational Optimization and Applications* 13 (1999) 187–200.

[800] M.V. SOLODOV AND B.F. SVAITER. Error bounds for proximal point subproblems and associated inexact proximal point algorithms. *Mathematical Programming* 88 (2000) 371–389.

[801] Y. SONG. *The P and Globally Uniquely Solvable Properties in Semidefinite Linear Complementarity Problems.* Ph.D. thesis, Department of Mathematics, University of Maryland Baltimore County (May 2002).

[802] Y. SONG, M.S. GOWDA, AND G. RAVINDRAN. On characterizations of P and P_0-properties in nonsmooth functions. *Mathematics of Operations Research* 25 (2000) 400–408.

[803] T. SRINIVASAN AND J. WHALLEY. *General Equilibrium Trade Policy Modelling*, MIT Press (Boston 1986).

[804] G. STAMPACCHIA. Formes bilineares coercives sur les ensembles convexes. *Comptes Rendus Academie Sciences Paris* 258 (1964) 4413–4416.

[805] G. STAMPACCHIA. Variational inequalities. In *Theory and Applications of Monotone Operators (Proc. NATO Advanced Study Inst., Venice, 1968)*, Edizioni "Oderisi" (Gubbio 1969) pp. 101–192.

[806] G.E. STAVROULAKIS, P.D. PANAGIOTOPOULOS, AND A.M. AL-FAHED. On the rigid body displacements and rotations in unilateral contact problems and applications. *Computers and Structures* 40 (1991) 599–614.

[807] D.E. STEWART. A high accuracy method for solving ODEs with discontinuous right-hand side. *Numerische Mathematik* 58 (1990) 299–328.

[808] D.E. STEWART. An index formula for degenerate LCPs. *Linear Algebra and its Applications* 191 (1993) 41–53.

[809] D.E. STEWART. Convergence of a time-stepping scheme for rigid-body dynamics and resolution of Painlevé's problem. *Archive for Rational Mechanics and Analysis* 145 (1998) 215–260.

[810] D.E. STEWART. Rigid-body dynamics with friction and impact. *SIAM Review* 42 (2000) 3–39.

[811] D.E. STEWART AND J.C. TRINKLE. An implicit time-stepping scheme for rigid body dynamics with inelastic collisions and Coulomb friction. *International Journal on Numerical Methods for Engineering* 39 (1996) 2673–2691.

[812] D.E. STEWART AND J.C. TRINKLE. Dynamics, friction, and complementarity problems. In M.C. Ferris and J.S. Pang, editors, *Complementarity and Variational Problems: State of the Art*, SIAM Publications (Philadelphia 1997) pp. 405–424.

[813] J. STOER AND C. WITZGALL. *Convexity and Optimization in Finite Dimensions I*, Springer-Verlag (Berlin 1970).

[814] J.C. STONE. Sequential optimization and complementarity techniques for computing economic equilibria. *Mathematical Programming Study* 23 (1985) 173–191.

[815] R.E. STONE. Lipschitzian matrices are nondegenerate INS-matrices. In M.C. Ferris and J.S. Pang, editors, *Complementarity and Variational Problems: State of the Art*, SIAM Publications (Philadelphia 1997) pp. 440–451.

[816] N. STRÖMBERG. An augmented Lagrangian method for fretting problems. *European Journal of Mechanics, A/Solids* 16 (1997) 573–593.

[817] N. STRÖMBERG, L. JOHANSSON, AND A. KLARBRING. Derivation and analysis of a generalized standard model for contact, friction and wear. *International Journal of Solids and Structures* 33 (1996) 1817–1836.

[818] D. SUN. A further result on an implicit function theorem for locally Lipschitz functions. *Operations Research Letter* 28 (2001) 193–198.

[819] D. SUN AND J. SUN. Semismooth matrix valued functions. *Mathematics of Operations Research* 27 (2002) 150–169.

[820] D.C. SUN. A thermal elastica theory of piston-ring and cylinder-bore contact. *Journal of Applied Mechanics* 58 (1991) 141–153.

[821] J. SUN. On the structure of convex piecewise quadratic functions. *Journal of Optimization Theory and Applications* 72 (1992) 499–510.

[822] J. SUN. On piecewise quadratic Newton and trust region problems. *Mathematical Programming* 76 (1997) 451–467.

[823] J. SUN AND J. ZHU. A predictor-corrector method for extended linear-quadratic programming. *Computers and Operations Research* 23 (1996) 755–767.

[824] S.M. SUN, M.C. NATORI, AND K.C. PARK. A computational procedure for flexible beams with frictional contact constraints. *International Journal for Numerical Methods in Engineering* 36 (1993) 3781–3800.

[825] R. SZNAJDER AND S. GOWDA. The generalized order linear complementarity problem. *SIAM Journal on Matrix Analysis* 15 (1994) 779–795.

[826] R. SZNAJDER AND M.S. GOWDA. Nondegeneracy concepts for zeros of piecewise smooth functions. *Mathematics of Operations Research* 23 (1998) 221–238.

[827] A. TAMIR. Minimality and complementarity properties associated with Z-functions and M-functions. *Mathematical Programming* 7 (1974) 17–31.

[828] M.A. TAWHID On the local uniqueness of solutions of variational inequalities under H-differentiability. *Journal of Optimization Theory and Applications* 113 (2002) 149–164.

[829] M.A. TAWHID AND M.S. GOWDA. On two applications of H-differentiability to optimization and complementarity problems. *Computational Optimization and Applications* 17 (2000) 279–300.

[830] T. TERLAKY. On ℓ_p programming. *European Journal of Operations Research* 22 (1985) 70–100.

[831] A.N. TIKHONOV. On the stability of the functional optimization problem. *Mathematical Physics* 6 (1966) 631–634.

[832] F. TIN-LOI AND M.C. FERRIS. Holonomic analysis of quasibrittle fracture with nonlinear hardening. In B.L. Karihaloo, Y.W. Mai, M.I. Ripley, and R.O. Ritchie, editors. *Advances in Fracture Research*, Pergamon Press (New York 1997) pp. 2183–2190.

[833] F. TIN-LOI AND M.C. FERRIS. A simple mathematical programming method to a structural identification problem. In C.K. Choi, C.B. Yun, and H.G. Kwak, editors. *Proceedings, 7th International Conference on Computing in Civil and Building Engineering (ICCCBE, VII)*, Techno Press (Taejon 1997) pp. 511–518.

[834] F. TIN-LOI AND J.S. MISA. Large displacement elastoplastic analysis of semirigid steel frames. *International Journal for Numerical Methods in Engineering* 39 (1996) 741–762.

[835] F. TIN-LOI AND J.S. PANG. Elastoplastic analysis of structures with nonlinear hardening. *Computer Methods in Applied Mechanics and Engineering* 107 (1993) 299–312.

[836] F. TIN-LOI AND V. VIMONSATIT. First-order analysis of semi-rigid frames: a parametric complementarity approach. *Engineering Structures* 18 (1996) 115–124.

[837] F. TIN-LOI AND S.H. XIA. Holonomic softening: models and analysis. *Mechanics of Structures and Machines* 29 (2001) 65–84.

[838] R.L. TOBIN. Sensitivity analysis for variational inequalities. *Journal of Optimization Theory and Applications* 48 (1986) 191–204.

[839] R.L. TOBIN. A variable dimension solution approach for the general spatial equilibrium problem. *Mathematical Programming* 40 (1988) 33–51.

[840] R.L. TOBIN AND T.L. FRIESZ. Sensitivity analysis for equilibrium network flow. *Transportation Science* 22 (1988) 242–250.

[841] M.J. TODD. *Computation of Fixed Points and Applications* Lecture Notes in Economics and Mathematical Systems 124, Springer-Verlag (Heidelberg 1976).

[842] M.J. TODD. A note on computing equilibria in economics with activity model of production. *Journal of Mathematical Economics* 6 (1979) 135–144.

[843] J.C. TRINKLE AND J.S. PANG. Dynamic multi-rigid-body systems with concurrent distributed contacts. *Proceedings of the 1997 IEEE International Conference on Robotics and Automation, Albuquerque, New Mexico* (1997) pp. 2276-2281.

[844] J.C. TRINKLE, J.S. PANG, S. SUDARSKY, AND G. LO. On dynamic multi-rigid-body contact problems with Coulomb friction. *Zeitschrift für Angewandte Mathematik und Mechanik* 77 (1997) 267–279.

[845] J.C. TRINKLE, J.A. TZITZOURIS, AND J.S. PANG. Dynamic multi-rigid-systems with concurrent distributed contacts. *The Royal Society Philosophical Transactions: Mathematical, Physical and Engineering Sciences* 359 (2001) 2575–2593.

[846] P. TSENG. Further applications of a splitting algorithm to decomposition in variational inequalities and convex programming. *Mathematical Programming* 48 (1990) 249–263.

[847] P. TSENG. On linear convergence of iterative methods for the variational inequality problem. *Journal of Computational and Applied Mathematics* 60 (1995) 237–252.

[848] P. TSENG Merit functions for semidefinite complementarity problems. *Mathematical Programming* 83 (1998) 159–185.

[849] P. TSENG. Error bounds for regularized complementarity problems. In M. Théra and R. Tichatschke, editors, *Ill-Posed Variational Problems and Regularization Techniques.* Lecture Notes in Economics and Mathematical Systems 477, Springer (Berlin 1999) pp. 247–274.

[850] P. TSENG. Error bounds and superlinear convergence analysis of some Newton-type methods in optimization. In G. Di Pillo and F. Giannessi, editors, *Nonlinear Optimization and Related Topics*, Kluwer Academic Publishers (Dordrecht 2000) pp. 445–462.

[851] P. TSENG. Co-NP completeness of some matrix classification problems. *Mathematical Programming* 88 (2000) 183–192.

[852] M. TURKAY AND I.E. GROSSMANN. Logic-based MINLP algorithms for the optimal synthesis of process networks. *Computers and Chemical Engineering* 20 (1996) 959–978.

[853] J.A. TZITZOURIS. *Numerical Resolution of Frictional Multi-Rigid-Body Systems via Fully Implicit Time-Stepping and Nonlinear Complementarity.* Ph.D. thesis, Department of Mathematical Sciences, The Johns Hopkins University (September 2001).

[854] M. ULBRICH. Semismooth Newton methods for operator equations in function spaces. Technical Report 00-11, Department of Computational and Applied Mathematics, Rice University (2000).

[855] M. ULBRICH. Nonmonotone trust-region methods for bound-constrained semismooth equations with a applications to nonlinear mixed complementarity problems. *SIAM Journal on Optimization* 11 (2001) 889–916.

[856] M. ULBRICH. On a nonsmooth Newton method for nonlinear complementarity problems in function space with applications to optimal control. In M.C. Ferris, O.L. Mangasarian, and J.S. Pang, editors, *Complementarity: Applications, Algorithms and Extensions*, Kluwer Academic Publishers (Dordrecht 2001) pp. 341–360.

[857] H. VÄLIAHO. P_* matrices are just sufficient. *Linear Algebra and its Applications* 239 (1996) 103–108.

[858] A.J. VAN DER SCHAFT AND J.M. SCHUMACHER. Complementarity modeling of hybrid systems. *IEEE Transactions on Automatic Control* 43 (1998) 483–490.

[859] L. VANDENBERGHE, B.L. DE MOOR, AND J. VANDEWALLE. The generalized linear complementarity problem applied to the complete analysis of resistive piecewise-linear circuits. *IEEE Transactions on Circuits and Systems* 36 (1989) 1382–1391.

[860] A. VILLAR. *Operator Theorems with Applications to Distributive Problems and Equilibrium Models.* Lecture Notes in Economics and Mathematical Systems 377, Springer-Verlag (New York 1992).

[861] S. VILLENEUVE AND A. ZANETTE. Parabolic A.D.I. methods for pricing American options on two stocks. *Mathematics of Operations Research* 27 (2002) 121–149.

[862] J. VON NEUMANN. Zur Theorie der Gesellschaftspspiele. *Mathematische Annalen* 100 (1928) 295–320.

[863] R.R. WAKEFIELD AND F. TIN-LOI. Large scale nonholonomic elastoplastic analysis using a linear complementarity formulation. *Computer Methods in Applied Mechanics and Engineering* 84 (1990) 229–242.

[864] R.R. WAKEFIELD AND F. TIN-LOI. Large displacement elastoplastic analysis of frames using an iterative LCP approach. *International Journal of Mechanical Science* 33 (1991) 379–391.

[865] D.W. WALKUP AND R.J.B. WETS. A Lipschitzian characterization of convex polyhedra. *Proceedings of the American Mathematical Society* 20 (1969) 167–173.

[866] L. WALRAS. *Eléments d'Economie Politique Pure*, L. Corbaz and Company (Lausanne 1874). [Translated as *Elements of Pure Economics* by W. JAFFE, Allen and Unwin (London 1954).]

[867] T. WANG AND J.S. PANG. Global error bounds for convex quadratic inequality systems. *Optimization* 31 (1994) 1–12.

[868] J.G. WARDROP. Some theoretical aspects of road traffic research. *Proceeding of the Institute of Civil Engineers, Part II* (1952) pp. 325–378.

[869] J.Y. WEI AND Y. SMEERS. Spatial oligopolistic electricity models with Cournot generators and regulated transmission prices. *Operations Research* 47 (1999) 102–112.

[870] P. WILMOTT, J.N. DEWYNNE, AND S. HOWISON. *Option Pricing: Mathematical Models and Computation*, Oxford Financial Press (Oxford 1993).

[871] R.S. WOMERSLEY. Optimality conditions for piecewise smooth functions. *Mathematical Programming Study* 17 (1982) 13–27.

[872] S.J. WRIGHT. Identifiable surfaces in constrained optimization. *SIAM Journal on Control and Optimization* 31 (1993) 1063–1079.

[873] S.J. WRIGHT. Superlinear convergence of a stabilized SQP method to a degenerate solution. *Computational Optimization and Applications* 11 (1998) 253–275.

[874] S.Y. WU, J.C. YAO, AND J.S. PANG. Inexact algorithm for continuous complementarity problems on measure spaces. *Journal of Optimization Theory and Applications* 91 (1996) 141–154.

[875] Z. WU AND J.J. YE. Sufficient conditions for error bounds. *SIAM Journal on Optimization* 12 (2002) 421–435.

[876] Z. WU AND J.J. YE. On error bounds for lower semicontinuous functions. *Mathematical Programming* (2002) 92 301–314.

[877] K. YAMADA, N. YAMASHITA, AND M. FUKUSHIMA. A new derivative-free descent method for the nonlinear complementarity problem. In G. Di Pillo and F. Giannessi, editors, *Nonlinear Optimization and Related Topics* Kluwer Academic Publishers, (Dordrecht 2000) pp. 463-487.

[878] T. YAMAMOTO. Historical developments in convergence analysis for Newton's and Newton-like methods. *Journal of Computational and Applied Mathematics* 124 (2000) 1–23.

[879] Y. YAMAMOTO. Fixed point algorithms for stationary point algorithms. In M. Iri and K. Tanabe, editors, *Mathematical Programming: Recent Developments and Applications*, Kluwer Academic Publishers (Boston 1989) pp. 283–307.

[880] N. YAMASHITA, H. DAN, AND M. FUKUSHIMA. On the identification of degenerate indices in the nonlinear complementarity problem with the proximal point algorithm. *Mathematics of Operations Research* (2002) forthcoming.

[881] N. YAMASHITA AND M. FUKUSHIMA. A new merit function and a descent method for semidefinite complementarity problems. In M. Fukushima and L. Qi, editors, *Reformulation: Nonsmooth, Piecewise Smooth, Semismooth and Smoothing Methods*, Kluwer Academic Publishers (Dordrecht 1998) pp. 405-420.

[882] W.H. YANG. *Error Bounds and Regularity in Mathematical Programming*. Ph.D. thesis. Department of Mathematics, The Chinese University of Hong Kong (December 2002).

[883] Z. YANG. A simplicial algorithm for computing robust stationary points of a continuous function on the unit simplex. *SIAM Journal on Control and Optimization* 34 (1996) 491–506.

[884] J.C. YAO. The generalized quasi-variational inequality problem with applications. *Journal of Mathematical Analysis and Applications* 158 (1991) 139–160.

[885] J.C. YAO. Variational inequalities with generalized monotone operators. *Mathematics of Operations Research* 19 (1994) 691–705.

[886] J.C. YAO. Generalized-quasi-variational inequality problems with discontinuous mappings. *Mathematics of Operations Research* 20 (1995) 465–478.

[887] S.T. YAU AND D.Y. GAO. Obstacle problem for von Karmán equations. *Advances in Applied Mathematics* 13 (1992) 123–141.

[888] J.J. YE. Optimality conditions for optimization problems with complementarity constraints. *SIAM Journal on Optimization* 9 (1999) 374–387.

[889] J.J. YE AND X.Y. YE. Necessary optimality conditions for optimization problems with variational inequality constraints. *Mathematics of Operationa Research* 22 (1997) 977–997.

[890] J.J. YE, D.L. ZHU, AND Q.J. ZHU. Exact penalization and necessary optimality conditions for generalized bilevel programming problems. *SIAM Journal on Optimization* 7 (1997) 481–597.

[891] Y. YE. A fully polynomial-time approximation algorithm for computing a stationary point of the general linear complementarity problem. *Mathematics Operations Research* 18 (1993) 334–345.

[892] J.J. ZAHER. *Conditional Modeling.* Ph.D. thesis, Department of Chemical Engineering, Carnegie-Mellon University (May 1995).

[893] E.H. ZARANTONELLO. Solving functional equations by contractive averaging. Report 160, Mathematics Research Center, University of Wisconsin, Madison (1960).

[894] E.H. ZARANTONELLO. Projections on convex sets in Hilbert space and spectral theory I and II. In E.H. Zarantonello, editor, *Contributions to Nonlinear Functional Analysis*, Academic Press (New York 1971) pp. 237–424.

[895] E. ZEIDLER. *Nonlinear Functional Analysis and its Applications III: Variational Methods and Optimization*, Springer-Verlag (New York 1985).

[896] E. ZEIDLER. *Nonlinear Functional Analysis and its Applications I: Fixed-point Theorems*, Springer-Verlag (New York 1986).

[897] S. ZHANG. Global error bounds for convex conic problems. *SIAM Journal on Optimization* 10 (2000) 836–851.

[898] X.L. ZHANG. Numerical analysis of American option pricing in a jump-diffusion model. *Mathematics of Operations Research* 22 (1997) 668–690.

[899] Y.B. ZHAO AND J.Y. HAN. Exceptional family of elements for a variational inequality problem and its applications. *Journal of Global Optimization* 14 (1999) 313–330.

[900] Y.B. ZHAO, J.Y. HAN, AND H.D. QI. Exceptional families and existence theorems for variational inequality problems. *Journal of Optimization Theory and Applications* 101 (1999) 475–495.

[901] Y.B. ZHAO AND D. LI. Strict feasibility conditions in nonlinear complementarity problems. *Journal of Optimization Theory and Applications* 107 (2000) 641–664.

[902] Y.B. ZHAO AND D. SUN. Alternative theorems for nonlinear projection equations and applications to generalized complementarity problems. *Nonlinear Analysis: Theory, Methods and Applications* 46 (2001) 853-868.

[903] X.Y. ZHENG. *Error Bounds for Set Inclusions and Inequality Systems.* Ph.D. thesis. Department of Mathematics, The Chinese University of Hong Kong (December 2002).

[904] C.Y. ZHU. Modified proximal point algorithm for extended linear-quadratic programming. *Computational Optimization and Applications* 2 (1992) 182–205.

[905] C.Y. ZHU AND R.T. ROCKAFELLAR. Primal-dual projected gradient algorithms for extended linear-quadratic programming. *SIAM Journal on Optimization* 3 (1993) 751–783.

[906] D.L. ZHU AND P. MARCOTTE. New classes of generalized monotonicity. *Journal of Optimization Theory and Applications* 87 (1995) 457–471.

[907] D.L. Zhu and P. Marcotte. Co-coercivity and its role in the convergence of iterative schemes for solving variational inequalities. *SIAM Journal on Optimization* 6 (1996) 714–726.

[908] S.I. Zuhovickii, R.A. Poljak, and M.E. Primak. Two methods of search for equilibrium points of n-person concave games. *Soviet Mathematics Doklady* 10 (1969) 279–282.

[909] S.I. Zuhovickii, R.A. Poljak, and M.E. Primak. Numerical methods of finding equilibrium points of n-person games. In *Proceedings of the First Winter School of Mathematical Programming in Drogobych* (1969) 93–130.

[910] S.I. Zuhovickii, R.A. Poljak, and M.E. Primak. On an n-person concave game and a production model. *Soviet Mathematics Doklady* 11 (1970) 522–526.

[911] S.I. Zuhovickii, R.A. Poljak, and M.E. Primak. Concave equilibrium points: numerical methods. *Matekon* 9 (1973) 10–30.

[912] R. Zvan, P.A. Forsyth, and K.R. Vetzal. Penalty methods for American options with stochastic volatility. *Journal of Computational and Applied Mathematics* 91 (1998) 119–218.

[913] R. Zvan, P.A. Forsyth, and K.R. Vetzal. Robust numerical methods for PDE models of Asian options. *Journal of Computational Finance* 1 (1998) 39–78.

[914] R. Zvan, P.A. Forsyth, and K.R. Vetzal. PDE methods for pricing barrier options. *Journal of Economic Dynamics and Control* 24 (2000) 1563–1590.

Index of Definitions and Results

Numbered Definitions

Main Results in Chapter 1

Proposition **1.1.3.**	VI (K, F) = CP (K, F) when K is a cone.
Proposition **1.2.1.**	KKT system of linear constrained VI.
Theorem **1.3.1.**	(Symmetry principle) F is gradient map $\Leftrightarrow$ JF is symmetric.
Proposition **1.3.4.**	KKT system of VI under Abadie CQ.
Theorem **1.4.1.**	Minimax/maximin problems in saddle problems.
Proposition **1.4.9.**	(Fejer's Theorem) $\mathcal{M}_+^n = (\mathcal{M}_+^n)^*$.
Theorem **1.5.5.**	Basic properties of Euclidean projector.
Proposition **1.5.8.**	Natural equation formulation of VI.
Proposition **1.5.9.**	Normal equation formation for VI.
Proposition **1.5.11.**	GUS of VI = bijection of normal map.
Proposition **1.5.14.**	3 measures of an inexact solution to a VI.

Main Results in Chapter 2

Theorem **2.1.2.**	$\deg(\Phi, \Omega, p) \neq 0 \Rightarrow \Phi^{-1}(p) \cap \Omega \neq \emptyset$.
Proposition **2.1.3.**	Basic properties of degree.
Proposition **2.1.3.**	Degree of a univalent map is ± 1.
Proposition **2.1.6.**	Degree equal sum of indices.
Theorem **2.1.10.**	Palais' global homeomorphism theorem.
Theorem **2.1.11.**	Domain invariance theorem.
Corollary **2.1.13.**	Condition for locally Lipschitz homeomorphism.
Theorem **2.1.18.**	Brouwer's fixed-point theorem.
Theorem **2.1.19.**	Kakutani's fixed-point theorem.
Theorem **2.1.21.**	Banach fixed-point theorem.
Theorem **2.2.1.**	Basic existence theorem for VI via degree
Proposition **2.2.3.**	Broad conditions for existence of a solution to a VI.
Proposition **2.2.7.**	Solution existence and boundedness in a coercive VI.
Proposition **2.2.8.**	Necessary and sufficient condition for $\text{SOL}(K, F) \neq \emptyset$.
Theorem **2.3.3.**	Solution existence and uniqueness in a strongly monotone VI.
Theorem **2.3.4.**	Solvability of a pseudo monotone VI.
Proposition **2.3.12.**	Representation of $\text{SOL}(K, F)$ under F-uniqueness.
Theorem **2.3.16.**	Crouzeix' existence theorem for a pseudo monotone VI.
Proposition **2.4.3.**	Dual of a pointed, closed convex cone is solid, and vice versa.
Theorem **2.4.4.**	For a pseudo monotone CP, strict feas. = sol. exist. + bounded.
Theorem **2.4.7.**	For an affine monotone CP, feasibility = solvability.
Proposition **2.4.10.**	Cross complementarity of solutions of a monotone CP.
Lemma **2.4.14.**	Two invariants of $\text{SOL}(K, q, M)$ with a positive semidefinite M.
Theorem **2.4.15.**	Polyhedral representation of solutions to a monotone AVI.
Proposition **2.5.6.**	CP kernel and solution boundedness.
Theorem **2.5.10.**	Solution existence and boundedness in a copositive CP.
Proposition **2.5.11.**	$\text{int}\, \mathcal{K}(K, M)^* \subseteq \mathcal{R}(K, M)$ with a copositive M.
Theorem **2.5.15.**	Piecewise polyhedrality of solutions to an AVI.
Theorem **2.5.17.**	Conditions for solution convexity in an AVI.
Theorem **2.5.20.**	Solvability in terms of feasibility for a copositive AVI.
Theorem **2.6.1.**	Existence of a solution to a CP.
Proposition **2.6.5.**	R_0 pair and norm-coercivity of $\mathbf{M}_K^{\text{nor}}$ and $\mathbf{M}_K^{\text{nat}}$.
Theorem **2.8.3**	Existence of a solution to a QVI via degree.

Main Results in Chapter 3

Theorem **3.1.1.**	Rademacher's F-differentiability theorem.
Proposition **3.1.6.**	Composition of B-differentiable functions.

Main Results in Chapter 3 (continued)

Proposition **3.2.1.**	Consequences of basic CQs.
Proposition **3.2.2.**	Upper semicontinuity of $\mathcal{M}(x)$ under MFCQ.
Lemma **3.2.3.**	Hoffman's error bound Lemma.
Corollary **3.2.5.**	Lipschitz property of polyhedra
Proposition **3.3.4.**	Isolatedness of a solution to a VI.
Proposition **3.3.7.**	Isolatedness of a solution to a linearly constrained VI.
Corollary **3.3.8.**	Isolatedness of a solution to an AVI.
Corollary **3.3.9.**	Isolatedness of a solution to an NCP.
Theorem **3.3.12.**	Isolatedness in a KKT system.
Proposition **3.3.21.**	Isolatedness of a zero of a B-differentiable function.
Corollary **3.3.22.**	Isolatedness of a solution to a vertical CP.
Corollary **3.3.24.**	Isolatedness in terms of representative matrices.
Proposition **3.4.2.**	Characterizations of a nondegenerate solution to a VI.
Proposition **3.5.1.**	Conditions for solution existence in a partitioned VI.
Proposition **3.5.9.**	Differentiable functions of the P type.
Proposition **3.5.10.**	Sol. existence and uniqueness in a partitioned, uniformly P VI.
Theorem **3.5.11.**	Conditions for solution existence in a partitioned P_0 VI.
Theorem **3.5.12.**	Conditions for sol. exist. + boundedness in a $P_*(\sigma)$ VI.
Theorem **3.5.15.**	Existence of the Tikhonov trajectory in a P_0 VI.
Theorem **3.6.4.**	Basic properties of a weakly univalent function.
Theorem **3.6.6.**	Solution connectedness of a P_0 VI.
Proposition **3.6.9.**	Solution connectedness of a P_0 LCP.

Main Results in Chapter 4

Theorem **4.1.1.**	B-differentiability of polyhedral projection.
Corollary **4.1.2.**	F-differentiability of polyhedral projection.
Proposition **4.1.4**	PL property of polyhedral projection.
Proposition **4.1.5.**	Face properties of the polyhedron $P(A, b)$.
Theorem **4.1.8.**	Properties of the normal manifold.
Proposition **4.2.1.**	Polyhedral subdivision of PA maps.
Proposition **4.2.2.**	Basic properties of PA maps.
Proposition **4.2.5**	Characterization of coherent orientation of PA maps.
Proposition **4.2.6**	Properties of coherently oriented PA maps.
Proposition **4.2.12**	Surjectivity of coherently oriented PA maps.
Proposition **4.2.7.**	Matrix criteria for coherent orientation of $\mathbf{M}_K^{\text{nor}}$ and $\mathbf{M}_K^{\text{nat}}$.
Proposition **4.2.10.**	Coherent orientation of the MLCP map.
Theorem **4.2.11.**	Conditions for Lipschitz homeomorphism of PA maps.
Proposition **4.2.15.**	Openness = coherent orientation, for a PA map.
Theorem **4.3.1.**	Robinson's injectivity theorem for an affine pair (K, M).
Theorem **4.3.2.**	GUS property of an affine pair.
Proposition **4.3.3.**	Inverse of $\mathbf{M}_K^{\text{nor}}$ with K polyhedral and M positive definite.
Theorem **4.4.1.**	B-differentiability of Euclidean projection under SBCQ.
Theorem **4.5.2.**	PC^1 property of Euclidean projection under CRCQ.
Corollary **4.5.5.**	F-differentiability of Euclidean projector under CRCQ.
Lemma **4.6.1.**	Basic properties of PC^1 functions.
Lemma **4.6.3.**	B-subdifferential of PC^1 functions.
Theorem **4.6.5.**	Conditions for locally Lipschitz homeomorphism of PC^1 maps.
Corollary **4.7.3.**	Pointwise continuity of parametric projection under MFCQ.
Theorem **4.7.5.**	[MFCQ + CRCQ] $\Rightarrow$ PC^1 of parametric Euclidean projection.

Main Results in Chapter 5

Proposition **5.1.4**.	Index $\neq 0 \Rightarrow$ local solvability of perturbed VIs.
Theorem **5.2.4**.	For a B-diff. function, nonzero index $\Rightarrow$ stability of zero.
Theorem **5.2.8**.	For H loc. Lip., strong stab. $\Leftrightarrow$ strong reg. $\Leftrightarrow$ loc. Lip. homeo.
Proposition **5.2.9**.	For H str. F-diff., str. stab. $\Leftrightarrow$ stab. $\Leftrightarrow$ reg. $\Leftrightarrow$ $JH(x)$ nonsing.
Theorem **5.2.12**.	For H B-diff., stability under a nonvanishing property.
Theorem **5.2.14**.	For H strongly B-diff., conditions for strong stability.
Proposition **5.2.15**.	Stability in terms of FOA.
Proposition **5.3.3**.	Strong stability $\Leftrightarrow$ strong regularity of $x \in \mathrm{SOL}(K, F)$.
Proposition **5.3.4**.	Stability $\Leftrightarrow$ regularity of a solution to a pseudo monotone VI.
Proposition **5.3.6**.	Strong stab. of $x \in \mathrm{SOL}(K, F)$ $\Leftrightarrow$ strong stab. of $z \equiv x - F(x)$
Theorem **5.3.11**.	Condition for stab. of $x \in \mathrm{SOL}(K, F)$ under R_0 conditions.
Theorem **5.3.14**.	Condition for stability of $x \in \mathrm{SOL}(K, F)$ under copositivity.
Proposition **5.3.15**.	Stability of a sol. to a partitioned VI under semicopositivity.
Theorem **5.3.17**.	Conditions for strong stability of $x \in \mathrm{SOL}(K, F)$.
Corollary **5.3.19**.	Condition for stability of a solution to a P_0 NCP.
Corollary **5.3.20**.	Strong stability $\Leftrightarrow$ strong b-regularity for a solution to an NCP.
Corollary **5.3.22**.	Conditions for strong stability of a KKT triple.
Theorem **5.3.24**.	Strong stability under CRCQ.
Proposition **5.4.3**.	Upper semicontinuity of $\mathcal{M}(x, \mathbf{p})$ under MFCQ.
Theorem **5.4.4**.	Stab. of a sol. to a parametric VI under MFCQ + strict copos.
Theorem **5.4.5**.	Stab. of sol. to parametric VI under MFCQ + CRCQ + R_0.
Theorem **5.4.6**.	Str. stab. $\Rightarrow$ PC^1 parametric sol. under MFCQ + CRCQ.
Theorem **5.4.10**.	Dir. der. of parametric sol. under MFCQ + CRCQ + str. stab.
Theorem **5.4.12**.	PC^1 parametric sol. under MFCQ + CRCQ + SCOC.
Theorem **5.4.13**.	F-diff. of parametric sol. under MFCQ + CRCQ + SCOC.
Corollary **5.4.14**.	F-diff. of parametric sol. with fixed set.
Theorem **5.4.15**.	PC^1 multipliers to parametric VI under LICQ + SCOC.
Lemma **5.4.16**.	Matrix criterion of SCOC under MFCQ.
Proposition **5.5.5**.	Solution set semistability = local error bound for $(\mathbf{F}_K^{\mathrm{nor}})^{-1}(0)$.
Theorem **5.5.6**.	Sol. set semistab. via ε-sol. bounded + pointwise semistab.
Theorem **5.5.8**.	Polyh. multifuncs. are everywhere pointwise upper Lip. cont.
Corollary **5.5.9**.	AVIs are semistable.
Corollary **5.5.11**.	AVIs are stable under nonzero degree.
Corollary **5.5.13**.	[coercivity + semistability] $\Rightarrow$ solution set stability.
Corollary **5.5.14**.	Solution set stability under finite number of solutions.
Theorem **5.5.15**.	For a solvable P_0 VI, perturbed solvability = sols. bounded.
Proposition **5.5.16**.	Continuity of perturbed sols. for a P_0 VI.
Proposition **5.5.17**.	For a P_0 AVI, sol. set stab. = sols. exists and bounded.

Main Results in Chapter 6

Subject Index

www.ingramcontent.com/pod-product-compliance
Ingram Content Group UK Ltd.
Pitfield, Milton Keynes, MK11 3LW, UK
UKHW020652210926
15297UKWH00010B/154
9781475780901